Water Reuse

An International Survey of current practice, issues and needs

Water Reuse

An International Survey of current practice, issues and needs

Edited by Blanca Jiménez and Takashi Asano

Publishing

Published by IWA Publishing, Alliance House, 12 Caxton Street, London SW1H 0QS, UK
Telephone: +44 (0) 20 7654 5500; Fax: +44 (0) 20 7654 5555; Email: publications@iwap.co.uk
Web: www.iwapublishing.com

First published 2008

Cover design by www.designforpublishing.co.uk
Index provided by Indexing Specialists (UK) Ltd

British Library Cataloguing in Publication Data
A CIP catalogue record for this book is available from the British Library

Library of Congress Cataloging- in-Publication Data
A catalog record for this book is available from the Library of Congress

ISBN: 1843390892
ISBN13: 9781843390893

Contents

List of Contributors

Oscar Alfranca
Department of Agribusiness Biotechnology Engineering
Universidad Politécnica de Cataluña
Campus Baix Llobregat - Edif. ESAB
Avda. del canal olimpic
S/N, 08860 Castelldefels, Spain
Email: oscar.alfranca@upc.edu

John Anderson
1 Cumbora Circuit
Berowra,
NSW 2081, Australia
Email: ja@aftonwater.com.au

Nguyen Viet Anh
Water Supply and Sanitation Division
Faculty of Environmental Engineering
Hanoi University of Civil Engineering (HUCE)
DHXD, 55 Giai Phong Road
Hanoi, 10000, Vietnam
Email: vietanhctn@yahoo.com

Takashi Asano
1125 Dartmouth Place
Davis, CA 95616-2311, USA
Email: tasano@ucdavis.edu

Sue Baggett
School of Water Sciences
Cranfield University
Cranfield, Bedfordshire
MK43 0AL, UK
Email: s.b.baggett@cranfield.ac.uk

Akiça Bahri
IWMI Regional Office
International Water Management Institute (IWMI)
PMB CT 112, Cantonments,
Accra, Ghana
Email: a.bahri@cgiar.org

Dr.-Ing. Maike Beier
Leibniz University of Hannover
Welfengarten 1
D-30167 Hannover, Germany

Jim Bradley
MWH New Zealand Ltd
PO Box 12-941, Penrose
Auckland, New Zealand
Email: jim.w.bradley@mwhglobal.com

Dipl. -Ing. Jörg Brinkmeyer
Leibniz University of Hannover
Welfengarten 1
D-30167 Hannover, Germany
Email: brinkmeyer@deencon.de

François Brissaud
Hydrosciences, MSE
Université Montpellier II
34095 Montpellier Cedex 05, France
Email: brissaud@msem.univ-montp2.fr

Luis Cabrera
Técnica y Proyectos S.A. (TYPSA)
Mallorca 272 - 276, 3er.
08037, Barcelona, Spain
Email: lcabrera@typsa.es

Guan-Hao Chen
Department of Civil Engineering
The Hong Kong University of Science & Technology
Clear water Bay, Kowloon,
Hong Kong, China
Email: ceghchen@ust.hk

Peter Cornel
Institut WAR, Technische Universität Darmstadt
Petersenstrasse 13
D-64287 Darmstadt, Germany
Email: p.cornel@iwar.tu-darmstadt.de

Cheryl K. Davis
San Francisco Public Utilities Commission
City and County of San Francisco
Water Supply and Treatment Division
1000 El Camino Real, P.O. Box 730
Millbrae, California 94030, USA
Email: ckd@sfwater.org

Peter Dillon
CSIRO Land and Water
PMB2 Glen Osmond
SA 5064, Australia
Email: peter.dillon@csiro.au

Dominique Endamana
International Institute of Tropical Agriculture, Cameroun
BP 2008 Messa
Yaoundé, Cameroun
Email: d.endamana@cgiar.org

Jeroen Ensink
Department of Infectious and Tropical Diseases
London School of Hygiene and Tropical Medicine
Keppel Street, London WC1E 7HT, UK
Email: Jeroen.ensink@lshtm.ac.uk

Kirsten Exall
National Water Research Institute
Water Science and Technology Directorate
Environment Canada
867 Lakeshore Rd
Burlington, ON
Canada L7R 4A6
Email: Kirsten.Exall@ec.gc.ca

Naoyuki Funamizu
Department of Environmental Engineering
Hokkaido University
Kita 13, Nishi 8, Kita-ku, 060-8628,
Sapporo, Japan
Email: funamizu@eng.hokudai.ac.jp

Alan Godfree
Public Health Team
United Utilities Water
Lingley Mere Business Park, Great Sankey
Warrington WA5 3LP, UK
Email: alan.godfree@uuplc.co.uk

Samuel Godfrey
Project Officer - Water and Environmental Sanitation UNICEF Bhopal Field Office E-7/650 Arera Colony, Bhopal 462016
Madhya Pradesh, India
Email: sgodfrey@unicef.org
Steffen Gruenheid
Taubenstr. 7
38106 Braunschweig, Germany
Email: sgruenheid@oewa.de

Shigeki Hatori
Sanki Engineering Co. Ltd
1742-7,Shimotsuruma, Yamato-Shi
Kanagawa 242-0001, Japan
Email: shigeki_hatori@eng.sanki.co.jp

Ivanildo Hespanhol
CIRRA/IRCWR, University of São Paulo
Av. Prof. Lúcio Martins Rodrigues 120
Cidade Universitária
05508-000, São Paulo, Brazil
Email: ivanhes@usp.br

Wim van der Hoek
International Water Management Institute
127, Sunil Mawatha
Pelawatte
Battaramulla, Sri Lanka
Email: w.vanderhoek@compaqnet.nl

Xia Huang
Dept. of Environmental Science and Engineering
Tsinghua University
Beijing 100084, China
Email: xhuang@tsinghua.edu.cn

Frans Huibers
Associate Professor Health and Environment
Irrigation and Water Engineering Group
Wageningen University
P.O. Box 47
6700 AA Wageningen, The Netherlands
Email: Frans.Huibers@wur.nl

Raquel Iglesis
I+D de la División Tratamiento y Depuración
Centro de Estudios Hidrográficos del CEDEX
C/ Paseo Bajo Virgen del Puerto,3
28005 Madrid, Spain
Email: raquel.iglesias@cedex.es

Namjung Jang
Environmental Eng. Dept.
Dohwa Consulting Engineers Co. Ltd
#736-6 Yeoksam-dong, Gangnam-gu
SEOUL, 135-080, South Korea
Email: skawjd2@hanmail.net

Paul Jeffrey
School of Water Sciences
Cranfield University
Cranfield, Bedfordshire,
MK43 0AL, UK
Email: p.j.jeffrey@cranfield.ac.uk

Prof. Dr.-Ing. Martin Jekel
Dept. of Water Quality Control
Technical University Berlin
Strasse des 17. Juni 135
10623 Berlin, Germany
Email: martin.jekel@TU-Berlin.DE

Hu Jiangyong
Center for Water Research, Civil Engineering Department
National University of Singapore
10 Kent Ridge Crescent, Singapore 119260
Email: cvehujy@nus.edu.sg

Blanca Jiménez
Apdo Postal 70472
Ciudad Universitaria
04510 Coyoacan DF, Mexico
Email: bJiménezc@iingen.unam.mx

Marcelo Juanico
Juanicó - Environmental Consultants Ltd
Aliah 2 St., 18392 Afula, Israel
Email: juanico@juanico.co.il

Ives Magloire Kengne
Wastewater Research Unit
Faculty of Science
University of Yaounde I
P.O. Box: 8404
Yaounde, Cameroon
Email: ives_Kengne@yahoo.fr

In S. Kim
Director of Center for Seawater Desalination
Prof. of Dept. Of Env. Sci. And Eng.
Gwangju Institute of Science and Technology
1 Oryong-dong, Buk-gu,
Gwangju 500-712, South Korea
Email: iskim@gist.ac.kr

Erich H. Konig
Windhoek Goreangab Operating Company
P.O.Box 3149
Windhoek, Namibia
Email: ekonig@wingoc.com.na

Valentina Lazarova
Suez Environnement – CIRSEE
38 rue du president Wilson
78230 Le Pecq, France
Email: valentina.lazarova@suez-env.com

Darla Hatton MacDonald
Policy & Economic Research Unit
CSIRO Land & Water
PMB 2 Glen Osmond
SA 5064, Australia
Email: darla.hattonmacdonald@csiro.au

June Marks
School of Social Sciences
Flinders University, Adelaide
GPO Box 2100
Adelaide, SA 5001, Australia
Email: june.marks@flinders.edu.au

Jiri Marsalek
National Water Research Institute
Water Science and Technology Directorate
Environment Canada
867 Lakeshore Rd
Burlington, ON, Canada L7R 4A6
Email: Jiri.Marsalek@ec.gc.ca

Linda McPherson
CH@M HILL
825 NE Multnomah, Suite 1300
Portland, OR 97232-2146, USA
Email: lmacpher@ch2m.com

Alessandro Meda
Institut WAR, Technische Universität Darmstadt
Petersenstrasse 13
D-64287 Darmstadt, Germany
Email: a.meda@iwar.tu-darmstadt.de

Juergen Menge
Department of Infrastructure, Water and Waste Management
PO Box 59
Windhoek, Namibia
Email: jgm@windhoekcc.org.na

Takuya Onitsuka
5-48-16 Sakuragaoka Setagaya-ku
Tokyo, Japan
Email: t-onitsuka@suiki.co.jp

Enrique Ortega
Jefe de la División Tratamiento y Depuración
Centro de Estudios Hidrográficos del CEDEX
C/ Paseo Bajo Virgen del Puerto,3
28005 Madrid, Spain
Email: enrique.ortega@cedex.es

Piet du Pisani
PO Box 59
Windhoek, Namibia
Email: pdp@windhoekcc.org.na

Wendy Proctor
Policy & Economic Research Unit
CSIRO Land & Water
Clunies Ross Street
Black Mountain
Australian Capital Territory, Australia 2601
Email: wendy.proctor@csiro.au

John Radcliffe
CSIRO
PMB 2, Glen Osmond
South Australia 5064, Australia
Email: john.radcliffe@csiro.au

Liqa Raschid-Sally
International Water Management Institute
PMB CT 112, Cantonments
Accra, Ghana
Email: l.raschid@cgiar.org

Mark Redwood
Senior Program Officer, Urban Poverty and Environment
International Development Research Centre (IDRC)
250 Albert Street
Ottawa, On, K1G 3H9, Canada
Email: mredwood@idrc.ca

Eric Rosenblum, PE, BCEE
Division Manager, South Bay Water Recycling
City of San Jose Environmental Services Department
3025 Tuers Road San Jose,
California 95121, USA
Email: eric.rosenblum@sanjoseca.gov

Prof. Dr.-Ing. Karl-Heinz Rosenwinkel
Leibniz University of Hannover
Welfengarten 1
D-30167 Hannover, Germany
Email: rosenwinkel@isah.uni-hannover.de

Lluís Sala
Consorci de la Costa Brava
Plaça Josep Pla 4, 3er 1a
E-17001 Girona, Spain
Email: lsala@ccbgi.org

Miquel Salgot
Institut de Recerca de l'Aigua de la Universitat de Barcelona
Facultat de Farmàcia – Edafologia
Joan XXIII, s/n.
08028 Barcelona, Spain
Email: salgot@ub.edu

Karl Schaefer
National Water Research Institute
Water Science and Technology Directorate
Environment Canada
867 Lakeshore Rd
Burlington, ON
Canada L7R 4A6
Email: Karl.Schaefer@ec.gc.ca

Frederik Schutte
Department of Chemical Engineering
University of Pretoria, South Africa
Email: frik.schutte@up.ac.za

Luis Segui
Tecnología del Medio Ambiente S.A. (TECNOMA)
Mallorca 272 - 276, 3er.
08037, Barcelona, Spain
Email: lsegui@tecnoma.es

Robert W. Simmons
International Water Management Institute
c/o ICRISAT
Patancheru 502324
Andhra Pradesh, India
Email: r.simmons@cgiar.org

Ben van der Merwe
Environmental Engineering Services
P.O. Box 6373
Auspannplatz
Windhoek, Namibia
Email: benvdm@iafrica.com.na

Saravanamuthu Vigneswaran
Faculty of Engineering, University of Technology,
Sydney (UTS),
550 room, Building 2, City campus,
PO Box 123 Broadway
NSW 2007, Australia
Email: vigid@eng.uts.edu.au

Chettiyappan Visvanathan
Environmental Engineering and Management Program,
Asian Institute of Technology, PO Box 4,
Klong Luang, 12120 Pathumthani, Thailand
Email: visu@ait.ac.th

Introduction

As world demand for water grows, water reclamation and reuse becomes increasingly important and an indispensable component of integral water resources management to enhance water supply reliability. In industrialized countries, water reuse is practiced to preserve freshwater sources, to protect the environment and to economically use treated water at increasingly higher water quality standards. This way, planned water reuse projects are multiplying, mostly in water stressed areas, for municipal and industrial purposes. In contrast, in developing countries, the need for increasing water supplies in arid or semi-arid areas and the use of polluted water supplies due to the lack of sanitation is forcing unplanned water reuse by necessity. In this context, interest in promoting planned water reuse projects is driven by the need to address the negative observed effects on health and the environment together with the need to keep providing a supply of water. Whatever the motivation, it is certain that water reuse will increase worldwide in the future with a view to countering problems already known, but also to counter a new threat arising from climate change according to the fourth assessment of the International Panel on Climate Change.

To obtain maximum benefits from water reuse it is important to understand the different forms, advantages, drawbacks and even controversial opinions of practices around the world. Water reclamation and reuse is being practiced worldwide, influenced by historical, cultural and economic realities producing a wide spectrum of challenges and opportunities. The Water Reuse Specialist Group of the International Water Association has, on several occasions, pointed out these differences which have encouraged interesting discussions among specialists about planning methodologies, appropriate technologies, applicable standards, and even acceptable definitions of water reuse. The goal of this book is to share such information. By presenting different situations around the world, the needs and limitations of the diverse reuse options, the emerging and controversial issues in the reuse field and contrasting case studies, the book provides a well-documented landscape of the

©2008 IWA Publishing. *Water Reuse - An International Survey of current practice, issues and needs* by B. Jiménez *et al.* ISBN: 9781843390893. Published by IWA Publishing, London, UK.

realities of water reclamation and reuse in different regions. Highlighting the known practices and less conventional ones will no doubt renew the reader's perception of water reuse.

The book contains 32 chapters divided into four sections: (1) a world overview; (2) stakeholders' points of view; (3) emerging topics; and (4) a selection of contrasting case studies from different countries. Throughout the book, care has been taken to maintain a balance between developed and developing countries' needs and practices so as to offer a view on water reuse from different and even opposite perspectives.

In Section One, Chapter 1: Water Reuse in the world, *Blanca Jiménez* and *Takashi Asano* summarize the main findings from this international survey, describing different motivations and reuses around the world. The chapter particularly highlights the relative importance of unplanned versus planned reuse in an attempt to draw attention to the need to implement safer projects as policy. Chapter 2 presented by *Akiça Bahri* describes the situation in the region of the world suffering the most acute water shortage problems: The Middle East and North Africa. Chapter 3, written by *Marcelo Juanicó* and *Miquel Salgot*, details the situation in the north of the Mediterranean region, a water shortage area with high demand during the tourist season. Due to the variety of countries forming this region, reuse strategies go from simple planning programs to very aggressive ones with a high degree of implementation, as is the case in Israel. *Kirsten Exall, Blanca Jiménez, Jiri Marsalek and Karl Schaefer* present the situation in the United States and Canada in Chapter 4 where the largest number of water reuse projects exists, particularly in the US states of California, Arizona, Florida, and Texas. Water reuse in Australia and New Zealand is described in Chapter 5 by *John Anderson, Jim Bradley*, and *John Radcliffe*. Although both countries are not water stressed in most basins, water reuse projects have been implemented as part of a policy to achieve sustainable use of water resources. In particular, Australia considers water reuse as a key part of the country's integrated water resources management policy and for that reason has set specific national targets for reuse with the participation of the government, water utilities, academics, and NGOs. *Peter Cornel* and *Alessandro Meda* prepared Chapter 6 to illustrate the situation in Central Europe. In this region, agriculture is rain fed and water reuse is practiced for municipal, industrial and ecological purposes to fulfil water demand in a high population density and a highly industrialized area. Chapter 7 written by *Naoyuki Funamizu, Xia Huang, Guan-Hao Chen, Hu Jiangyong* and *Chettiyappan Visvanatha* covers Asia, a wide region with highly variable water availability as a result of the variety of climates and an uneven population distribution. In Asia, several countries are reusing water due to significant water shortages and/or increasing water demands due to rapid population growth. Also, water reuse is performed as a political strategy as is the case of Singapore, which relies on Malaysia for more than 50% of its supply. The Central and South African situation is described by *Frederick Schutte* in Chapter 8. The region with 10 million square kilometres and 280 million inhabitants has a noticeable lack of information for many of the countries. Nevertheless, by describing water reuse practices in Kenya, Zimbabwe, Namibia and South Africa, the main needs of the area can be discovered. Chapter 9, written by *Blanca Jiménez*, covers the Latin America and Caribbean region, a water rich area with the exception of some countries and islands where reuse is very important for agricultural purposes and environmental protection as a means of promoting tourist activities.

In Section Two, information on water reuse from the "user's point of view" or "applications perspective" is presented. In Chapter 10, *Akiça Bahri* and *Valentina Lazarova* present water reuse practices for agriculture, different irrigation applications, water quality needs to protect crops and soil, irrigation methods and treatment options. The special case of agricultural reuse concerning urban agriculture is analyzed by *Mark Redwood* and *Frans*

Huibers in Chapter 11. Although urban agriculture is frequently considered as marginal in an urban economy, it turns out that it is an important source of livelihood in several cities in developing countries. Along with this, practical recommendations to prevent negative effects on health arising from this practice are also described. Municipal water reuse is analyzed in Chapter 12 by *John Anderson*. Different ways of reusing water for public and municipal purposes are presented in order to render urban water use more efficient. *In Kim*, *Sarvanamuthu Vigneswaran* and *Namjung Jang* describe in Chapter 13 the current practices of water reuse in industries. Industrial water use is forecast to rise by more than 60% in 30 years and water reuse will be in many cases the only way to cope with this demand. In Chapter 14, *Peter Dillon* and *Blanca Jiménez* analyze the indirect reuse of water through aquifer recharge confronting intentional and non intentional practices. The chapter contains information of a variety of settings, purposes, methods and outcomes in different situations.

Section Three, containing topics currently at debate in the water reuse field, begins with Chapter 15: Ethical Dilemmas in Water Recycling. *Cheryl Davis* makes a profound and inquisitive review of different considerations concerning water reuse. She analyses the role that water professionals should play in providing a good water reuse service, taking into consideration technical points of view but also ethical ones. Chapter 16, written by *Darla Hatton, Mac Donald* and *Wendy Proctor*, elaborates on the economic dilemmas in water management and reuse describing why economics makes water resource management an unusual and complex problem in the water sector, using Australia as an example. In Chapter 17, *John Radcliffe* makes important considerations on the role that government needs to play to plan and implement water reuse projects. The chapter entitled *Public Policies and Institutional Capacity Building for Innovation in Recycling* highlights the need of making innovative changes in public policies to reuse wastewater. Public acceptance is a major aspect for a project to succeed. For this reason community consultation and engagement in developing and managing water reuse projects is needed. Public acceptance of water reuse is analyzed by *John Anderson*, *Sue Baggett*, *Paul Jeffrey*, *Linda McPherson*, *June Marks* and *Eric Rosenblum* in Chapter 18. To end Section Three, Chapter 19 entitled *Water Reuse Criteria: Environmental and Health Risk based Standards and Guidelines* written by *Alan Godfree* and *Samuel Godfrey* offers a complete review of the international criteria and representative national standards on water reuse.

Section Four of the book contains case studies illustrating different and contrasting situations, uses and topics. Chapter 20 written by *Naoyuki Funamizu*, *Takuya Onitsuka* and *Shigeki Hatori* describes water reuse in Japan; the chapter offers new, flexible and even fancy ideas to reuse treated wastewater in cities. In contrast, *Jeroen Ensink*, W*im van der Hoek*, and *Robert Simmons* present in Chapter 21 the case study: *Livelihoods from Wastewater in Faisalabad, Pakistan*. The text presents findings of a study made to evaluate the risks and benefits of non treated wastewater use in Faisalabad, where farmers pay an annual 7,500 USD fee to use it because "it seems to be a sustainable and profitable affair". Chapter 22: *Indirect Reuse for Human Consumption in Germany* prepared by *Martin Jekel* and *Steffen Grunheid* shows how in the City of Berlin, 56% of the drinking water is derived from bank filtration of a treated and highly monitored effluent. An opposite example is presented in Chapter 23: *Unplanned Reuse of Wastewater for Human Consumption in the Tula Valley, Mexico,* by *Blanca Jiménez*. This case exemplifies the non planned reuse for human consumption of raw wastewater infiltrated to an aquifer to supply 400,000 people. To continue with the subject of reuse for human consumption, Chapter 24 reviews the water reuse in Windhoek, Namibia: 37 years on and still the only case of water reuse for human consumption. This chapter prepared by *Ben van der Merwe*, *Piet Du Pisani*, *Juergen Menge*, and *Erich Koning* describe the evolution of the unique project to directly reclaim water for human consumption. The chapter concludes that based on the success of the project "water should be judged by its quality, not its history". In Chapter 25: *Industrial Water Resource Management and Recycling Experiences from Germany, Case Studies from the Food and Beverage Industry*, *Karl-Heinz Rosenwinkel* and *Jorg Brinkmeyer* present the idea that industrial water recycling, besides saving water resources is a valuable economic alternative to end-of-pipe-technology. To illustrate industrial reuse in developing countries Chapter 26 written by *Ivanildo Espanhol* describes the Sao Paolo, Brazil situation. This industrialized city is no longer able to cope with growing water demands and therefore, to promote industrial recycling and liberate water for municipal uses, the price of water for industries has been considerably increased. In Chapter 27 *Marcelo Juanico* presents the case study of Israel. The text summarizes the developments of wastewater reuse practices, the experience gained and the issues that remain controversial in this country that reuses an important amount of the wastewater produced. Chapter 28: *Economic Analysis of Wastewater Reuse Projects: A Methodology for Private Reuse and Public Reuse Cases* presents two practical cases for applying economic methodologies for determining the cost/benefit ratio considering external costs. It has been prepared by *Luis Segui*, *Luis Cabrera* and *Oscar Alfranca*. Chapter 29: *Wastewater reuse and Reclamation in Spain* written by *Lluis Sala*, *Raquel Iglesias* and *Enrique Ortega* describes the rapid evolution of reuse practices in that country. The difficulties of setting common guidelines within a region are described by *Francois Brissaud* and *Akiça Bahri* in Chapter 30: *Trying to set a Common Framework to Rule Water Reuse in the Mediterranean Region*. In Chapter 31: *Wastewater Use in High Rainfall Riverine Cities: Comparison from Cameroon, Nepal and Vietnam*, *Liqa Raschid-Sally*, *Ives Magloire Kenge*, *Nguyen Viet Anh* and *Dominique Endamana* show that water reuse is important even in water rich countries. Finally, in Chapter 32: *Case Studies in Middle Eastern and North African countries*, *Akiça Bahri* analyses the different approaches taken by oil-rich and low-income countries, presenting case studies from seventeen of the territories.

By offering the variety of subjects contained in this book contributed by many experts, the editors aim to highlight the diverse nature of water reclamation and reuse practices occurring in the world. It is our belief that while water reclamation and reuse options may be quite diverse and whether reuse is practiced intentionally or unintentionally, directly or indirectly it should be acknowledged and paid equal attention so that local needs, capabilities and limitations are recognized, in order to effectively move to planned and controlled water reuse practices and better protection of public health and the environment. Planned and controlled water reuse should in the future be part of a truly sustainable water resources management approach worldwide.

Blanca Jiménez, Mexico City, Mexico
Takahi Asano, Davis, California, USA.

SECTION ONE

A WORLD OVERVIEW

1

Water reclamation and reuse around the world

Blanca Jiménez and Takashi Asano

1.1 INTRODUCTION

This chapter describes water reuse at a global level based on the findings contained in this book. It summarizes and compares the information considered in Chapters 2 to 9, which describe the situation in the Middle East and North Africa, the Northern Mediterranean, United States and Canada, Australia and New Zealand, Central Europe, Asia, Central and Southern Africa and Latin America and the Caribbean, with the aim of establishing general trends and main differences in the world. A preliminary survey was performed, from which the content and the structure of the book were defined. Regions were defined considering the need to assemble together countries which in principle would have similar reuse patterns produced by the social response to adapt to the local water availability. The regions with low water availability are considered first. From the beginning of the assessment, the need to have common definitions quickly became evident and these are therefore presented as the first part of this chapter. The comparison among the regions covers: (a) water availability, (b) the state of reuse around the world including an overview of the general situation per region; (c) the main drivers to reuse wastewater; (d) the list of the main countries reusing wastewater; and (e) the main objectives to reuse wastewater for agriculture, municipal and industrial uses.

1.2 DEFINITIONS

Technical terms used to describe reuse practices may have different meanings from country to country simply because they have been developed in parallel with water reclamation and reuse

©2008 IWA Publishing. *Water Reuse - An International Survey of current practice, issues and needs* by B. Jiménez *et al.* ISBN: 9781843390893. Published by IWA Publishing, London, UK.

activities. The standardization of reuse terms is not an easy task because, in many cases, they have been incorporated into national legislations making it difficult to adopt new definitions. Thus for the purpose of this survey, the definitions used are presented in Table 1.1.

Table 1.1. Definitions used for the international survey with partial reference to AQUASTAT in Earth Trends, 2001 and Earth Trends, 2005.

Term	Definition
Dependency ratio	Part of the total renewable water resources originating outside the country, expressed as a percentage.
Direct reuse	Reuse of treated or untreated wastewater by directly transferring it from the site where it is produced to the conveyance facilities for its use.
Direct reuse for agriculture	Reuse of treated or untreated wastewater for irrigation without prior dilution with another water source.
Direct reuse for human consumption or direct potable reuse	Reuse of treated wastewater that has been conveyed directly from a water reclamation plant to the water supply network.
First water use	Water coming from a source receiving non-identifiable wastewater discharges.
Indirect reuse	Reuse of treated or untreated wastewater after it has been discharged into a natural surface water or groundwater body, from which further water is taken.
Indirect reuse for agriculture	Reuse of treated or untreated wastewater for irrigation after its discharge in a water body from which it is taken to be used once again.
Indirect reuse for human consumption	Reuse of treated or untreated wastewater to increase a water source that is used as a municipal supply. It can be performed intentionally or unintentionally.
Intentional or planned reuse	Use of wastewater as part of a planned project. It is always performed intentionally and using reclaimed water.
Internal renewable water resources (IRWR)	These are the sum of the internal groundwater recharge (total volume of water entering aquifers within a country's borders from endogenous precipitation and surface water flow) plus the internally-produced surface water (average annual flow of rivers generated from endogenous precipitation and base flow generated by aquifers) minus the overlap between them (the contribution of aquifers to surface flow and the recharge of aquifers by surface run-off).
Natural renewable water resources (NRWR)	Is the sum of the IRWR and the natural flow that on average would enter into a country without human intervention.
Reclaimed water	Wastewater that has been treated to the point that it is suitable for safe reuse.
Recycling	Utilization of treated or untreated wastewater for the same purpose that generated it, i.e. it does not involve a change of user. For instance, recycling the effluents of the pulp and paper industry.
Reuse	Utilization of treated or untreated wastewater for a purpose other than the one that generated it, i.e. it involves a change of user. For instance, the reuse of municipal wastewater for agricultural irrigation.
Unplanned or incidental indirect potable reuse	Subsequent use of treated or untreated wastewater after it has been discharged into surface water or groundwater from which water is taken for drinking purposes or another use. Initially, it always occurs as a subconscious activity; with time it might occur consciously but not as part of a planned project in which wastewater is properly treated and water quality monitored for this specific purpose.
Water availability per capita	Is the ratio of renewable water resources within a region to the number of people living in that region.
Water Intensity Use Index (Water Stress Index)	Is the ratio of the mean total annual water withdrawal to the total renewable freshwater resources within a region on an annual basis.
Water reclamation	The treatment of wastewater to make it reusable for one or more applications. The process produces reclaimed water.
Water withdrawals	The total water removed from natural resources for drinking, municipal, industrial and agricultural irrigation uses on a single year basis, not counting evaporative losses from storage basins. They include abstractions from groundwater and surface water sources as well as from inflows from other countries and fresh water produced by desalination plants.

1.3 THRESHOLDS ON WATER AVAILABILITY

For a long time researchers have looked for an easy way of defining whether a region presents water scarcity problems or not. By analyzing the experience of moderately developed countries located in arid zones, different authors have identified different parameters for doing so, as presented in Table 1.2. These parameters do not represent absolute conditions but are indicative of different water stress situations where reuse can be or is an important activity.

Table 1.2. Threshold values used to characterize water stress within a region (with information from Falkenmark and Widstrand, 1992; Earth Trends, 2001; Lazarova *et al.*, 2001; Lehner *et al.*, 2001 and Bixio *et al.*, 2006).

Characteristic	Threshold	Situation	Influence on water reuse
Based on the per capita water availability of renewable freshwater, m^3/capita·yr			
Water stress	<1,700	The region begins to experience water stress and the economy or human health may be harmed.	Under these circumstances, developing a water reuse program is recommended
Chronic water scarcity	<1,000	The region experiences frequent water supply problems, both short and long-term	Reuse activities need to be put in place
Absolute water stress	<500	The region completes its water supply by desalting seawater, over-exploiting aquifers[1] or performing unplanned water reuse	Urgent planned water reuse measures need to be implemented.
Minimum survival level	<100	Water supply for domestic and commercial uses is compromised, since the total availability is not enough to fulfill demand for all uses (municipal, agricultural and industrial)	Under such circumstances the current economic development model is unsustainable
Based on the Water Intensity Use Index (WIUI) or Water Stress Index (WSI)			
Water stress	>20%	The region is experiencing severe water supply problems that are addressed by reusing wastewater (planned or not), overexploiting aquifers (by 2-30 times), or desalinating seawater	Integral water management programs including planned wastewater reuse and recycling are vital to the economy

[1] although groundwater overexploitation has been observed even in countries with water availability over 4000 m^3/*capita*·yr

The water availability per capita index only represents the potential usable water per person. Actually, water is used differently in each country depending on their economic activities, available water infrastructure and social patterns. Therefore, a complementary way to measure the extent of use of water within a region is the Water Intensity Use Index (WIUI), which expresses in percentage terms the relation between the amount of water used (extracted from the environment) and the total renewable water available for a region. It is considered that a value above 20% indicates water stress problems.

1.4 WORLD SITUATION

1.4.1 Water availability

At a global level, water availability for 2006 was 8,462 m^3/capita·yr but at a regional level it went from as little as 1,380 m^3/capita·yr in the Middle East and North Africa to almost

53,300 m^3/capita·yr in Oceania (see Table 1.3). These figures do not reflect the situation of individual countries (Annex 1 contains data at country level) or within a country. A list of the countries that are water scarce according to the water availability per capita index and the WIUI index are presented on Tables 1.4 and 1.5, respectively. Note that the list of countries change according to the index used. No country from Oceania, North America or South America is contained on this list although it is well known that for some of them, at a local level, water problems do exist, for example in Australia, United States and Mexico. In some cases WIUI values for some countries are even well above the threshold value of 20%, as is the case for Kuwait and the United Arab Emirates with a WIUI index above 2,000%. Even in regions with good water availability water problems may exist, as is the case, for instance, in Europe with 10,680 m^3/capita·yr where 11 countries out of 38 are under water stress conditions.

Water availability per capita in developed countries is 58% higher than that of developing ones, but water intensity use is similar in both groups. This means that if more water is used in the future in developing countries, for instance to produce more food to feed an increasing number of people, the Water Intensity Use index in some developing countries will be higher than in the developed countries.

Table 1.3. Water Availability and Water Intensity Use indices per region (with information from Earth Trends, 2007).

Region	Water Availability Index In 2006 m^3/capita·yr	Water Intensity Use Index In 2000 %
Middle East and North Africa	1,383	62.8
Asia (excluding Middle East)	3,990	19.3
Mexico, Central America & the Caribbean	6,740	8.5
United States and Canada	19,649	9.3
Europe	10,680	6.4
Sub-Saharan Africa	7,209	3.1
Oceania	53,290	1.6
South America	45,400	1.3
Developed Countries	11,392	9.0
Developing Countries	7,693	8.9
High Income Countries	10,554	10.1
Middle Income Countries	10,171	6.9
Low Income Countries	5,894	12.1
World	8,462	8.9

At the present time it is estimated that around 700 million people (*i.e.* almost 11% of the total world population) in 43 countries live with less than 1000 m^3/capita·yr. By the year 2025, 38% of people (more than 3 billion) will live in such conditions, increasing the number to nearly half of the population and 149 countries by the year 2050 (UN, 2006). By that time, the Middle Eastern countries will be reaching the minimum water survival level of 100 m^3/capita·yr. According to Figure 1.1, the regions suffering the most will be North and South Africa, western Asia, the North China Plain, western and southern India, Pakistan, central and southern Mexico, the western coast of the United States, the Mediterranean region, and a large part of Australia. According to the IPCC (2007), climate change will be an additional factor that modifies water availability and could favour drought in regions already water scarce.

Figure 1.1. Projected annual renewable water supply per person by river basin for the year 2025. From: Water Resources eAtlas, 2007

Table 1.4. List of water stressed countries according to the water availability per capita index (prepared with information from Earth trends, 2006)

Minimum survival <100	Absolute water stress 100-500	Chronic water stress 500-1000	Water stress 1000-1700
Middle East and North Africa			
Kuwait	Libyan	Egypt	Lebanon
Gaza Strip	Jordan	Morocco	Syria
United Arab Emirates	Bahrain	Cyprus	
Qatar	Yemen		
Saudi Arabia	Israel		
	Algeria		
	Oman		
	Tunisia		
Asia (excluding Middle East)			
Maldives	Singapore	-	Pakistan
			Korea, Republic
			India
Central America and Caribbean			
Bahamas	Barbados	S. Kitts & Nevis	Haiti
		Antigua & Barbuda	
Europe			
-	Malta	-	Denmark
			Czech Rep.
			Poland
Subsaharan Africa			
-	Djibouti	Cape Verde	Rwanda
		Kenya	South Africa
		Burkina Faso	Malawi
			Eritrea
			Comoros
			Zimbabwe
			Ethiopia
			Burundi
			Lesotho

Table 1.5. List of water stressed countries according to the Water Intensity Use Index

1000-3000%	500-1000%	100-500%	50-100%	20-50%
Middle East and North Africa				
United Arab Emirates	Saudi Arabia Libyan Qatar	Yemen Oman Israel Jordan Iraq Syrian	Tunisia Algeria Iran	Morocco Afghanistan Lebanon Cyprus
Asia (excluding Middle East)				
Turkmenistan	-	Uzbekistan Azerbaijan Pakistan	Bangladesh India Japan	Kazakhstan Armenia Korea, Rep. Sri Lanka China Thailand Singapore
Central America and Caribbean				
-	-	Barbados	-	Cuba
Europe				
-	-	Malta Hungary Moldova, Rep.	Belgium Netherlands Romania Ukraine	Bulgaria Spain Germany Poland Italy Denmark France Portugal
Sub-Saharan Africa				
-	-	Mauritania Sudan	Niger Somalia	Mauritius South Africa Swaziland Zimbabwe Eritrea

1.4.2 Water reuse

As result of water scarcity several countries are reusing wastewater. Figure 1.2 presents the countries reporting any type of reuse. At a regional level differences on reuse practices can be observed and are highlighted in Boxes 1.1 to 1.5 (below). Concerning reuse options, agriculture is by far the most important in terms of volume, simply because it is the activity that demands most water around the world. This reuse is expected to increase because the potential to reuse wastewater is still high (even agricultural reuse only represents <1% in volume of the total demand of water by the sector).

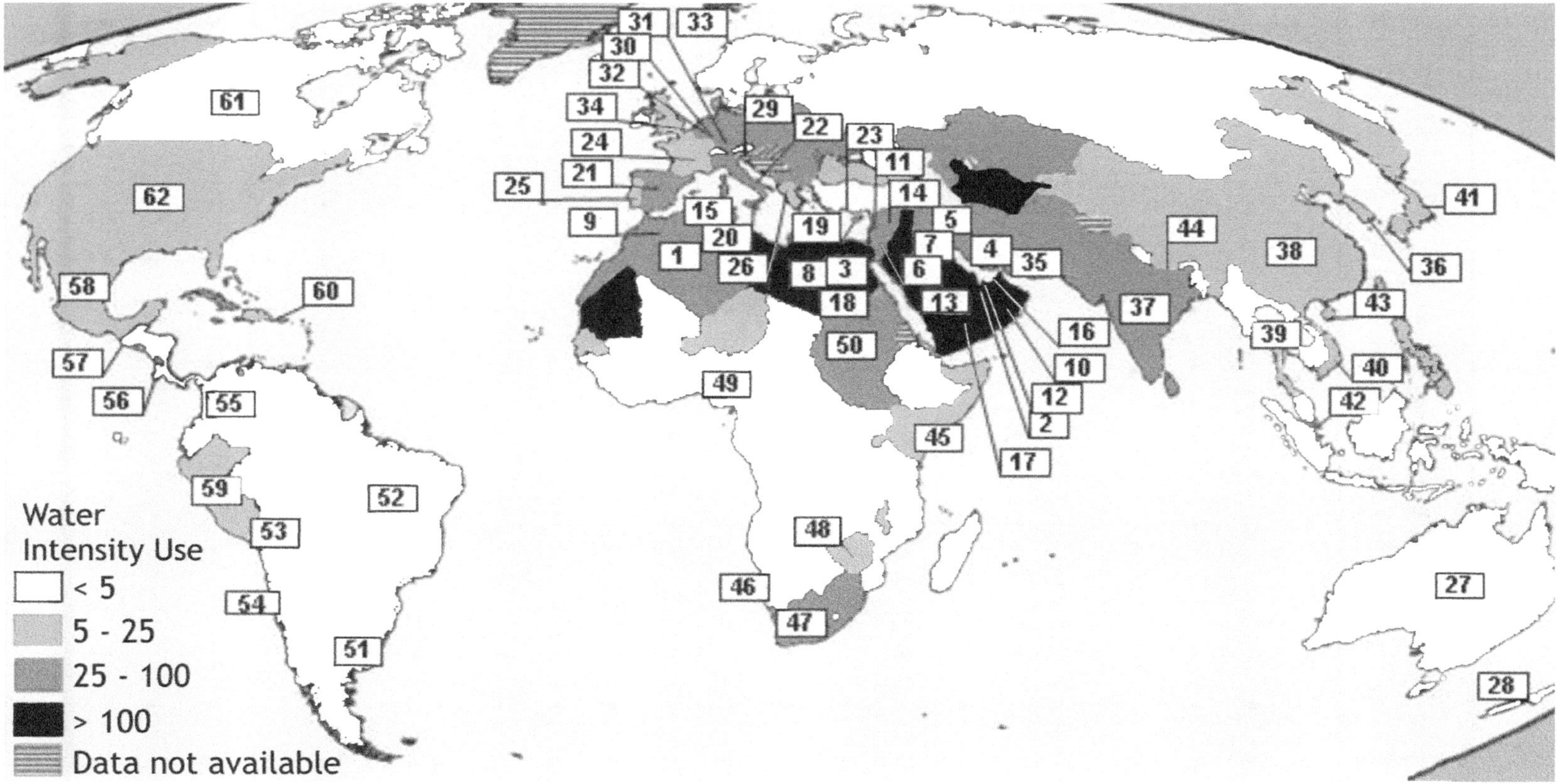

Figure 1.2 Types of wastewater reuse reported in different countries (see table on following pages for full details).

Country	Number	Agriculture	Municipal	Aquaculture	Potable unplanned indirect reuse	Groundwater recharge	Industrial	Environ-mental	Planned	Non-planned	Future plans	Desalination	Other type
Algeria	1										•		
Argentina	51	•									•		
Australia	27	•	•		•	•	•	•	•		•		•
Austria	29						•		•				
Bahrain	2	•	•						•		•	•	
Belgium	30		•		•	•					•		
Bolivia	53	•							•				
Brazil	52	•	•						•	•	•		
Cameroon	49	•								•			
Canada	61	•	•					•					
Chile	54	•					•		•		•		
China	38	•	•		•	•	•	•		•	•		•
Colombia	55	•								•	•		
Costa Rica	56	•								•	•		
Cyprus	19	•	•					•			•	•	
Dominican Rep.	60	•											
Egypt	3	•				•			•	•	•		
France	24	•	•		•				•	•	•		•
Germany	31		•		•	•	•	•	•		•	•	
Greece	26	•					•		•		•		
Guatemala	57	•							•				
Hong Kong	43		•						•				
India	37	•				•				•	•	•	

Country	Number	Agriculture	Municipal	Aquaculture	Potable unplanned indirect reuse	Groundwater recharge	Industrial	Environ-mental	Planned	Non-planned	Future plans	Desalination	Other type
Iran	4	•				•		•	•	•	•	•	
Iraq	5											•	
Israel	18	•	•			•		•	•		•		
Italy	22	•	•			•	•		•	•	•	•	
Japan	41	•	•				•	•	•		•	•	•
Jordan	6	•				•		•	•		•		
Kenya	45	•					•		•	•	•		
Korea, Rep.	36		•				•		•		•	•	
Kuwait	7	•	•			•	•	•	•		•		
Libya	8	•									•	•	
Malta	20	•					•		•				
Mexico	58	•	•		•	•	•	•	•	•	•		
Morocco	9	•	•						•	•	•		
Namibia	46		•			•			•		•		
Nepal	44	•								•			
Netherlands	32		•				•	•	•		•	•	
New Zealand	28		•					•	•				
Oman	10	•	•			•	•	•			•		
Pakistan	35	•								•			
Palestine	11	•				•			•	•	•		
Peru	59	•				•			•	•	•		
Portugal	25	•	•				•	•	•	•	•		
Qatar	12	•	•					•	•				
South Africa	47	•	•		•	•	•	•			•		

Country	Number	Agriculture	Municipal	Aquaculture	Potable unplanned indirect reuse	Groundwater recharge	Industrial	Environ-mental	Planned	Non-planned	Future plans	Desalination	Other type
Saudi Arabia	13	•	•			•	•	•	•		•	•	
Singapore	42	•	•				•		•		•	•	
Spain	21	•	•			•		•	•	•	•	•	
Sudan	50	•								•			
Sweden	33	•											
Syria	14	•				•			•	•			
Thailand	39	•		•		•	•		•	•	•		
Tunisia	15	•	•			•		•	•				
Turkey	23	•	•					•	•	•			
United Arab Emirates	16	•	•					•	•		•	•	
United Kingdom	34		•		•	•	•	•	•	•			
United States	62	•	•			•	•	•	•		•	•	
Vietnam	40	•				•				•			
Yemen	17	•	•			•	•	•	•	•	•		
Zimbabwe	48	•			•			•	•	•	•		

The tendency to reuse wastewater is most common among those activities demanding the most water (Figure 1.3). The following examples illustrate this:

(1) Pakistan and Tunisia, with a water use for agricultural purposes of 96% and 86% respectively, reuse a very large amount of their wastewater to irrigate (although in Pakistan it is non-treated wastewater while in Tunisia it is treated).

(2) Namibia and Singapore, which use 29% and 45% of total water extracted for municipal purposes respectively, are the two countries with the most important water for human consumption reclamation projects.

(3) Finally, the USA, Singapore and Germany, where 45%, 51% and 69% of water is used for industrial purposes respectively, have a large number of recycling and reuse projects across industries.

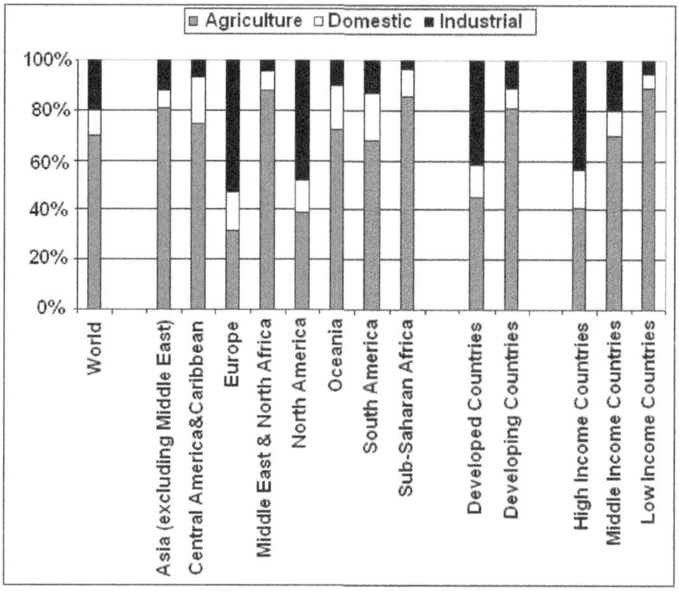

Figure 1.3. Water withdrawals by sector for 2006 with information from Earth Trends, 2007.

Box 1.1 Highlights for Africa

MIDDLE EAST AND NORTH AFRICA

- This region has the largest number of water stressed countries.
- Mean Water Availability in the region is 1,383 m^3/capita·yr with a WIUI of 62.8% but Water Availability varies from 7 m^3/capita·yr (in Kuwait) to 2,880 m^3/capita·yr (in Turkey), while the WIUI ranges from 17% (in Turkey) to 7,500% (in Bahrain).
- The main driver for reuse is water scarcity.
- There are notable differences between high, and middle and low income countries in the region.

High income countries:

- Use desalinated seawater for water supply.
- Have water reuse schemes for agricultural and landscape irrigation.
- Use reclaimed water for uses with a high water demand.
- Use standards inspired by the California Title 22.
- *Fatwas* (religious statements) have been made to promote planned water reuse projects.

Middle-low to low-income countries:

- Perform reuse with partially treated or untreated wastewater.
- Agricultural irrigation is the main reuse activity.
- Basically follow WHO guidelines.

CENTRAL AND SOUTH AFRICA

- Little available information on reuse practices.
- Mean Water Availability is 7,209 m^3/*capita*·yr and WIUI is 3.1% but Water Availability ranges from 372 m^3/*capita*·yr for Djibouti to 221,035 m^3/*capita*·yr in Congo while the WIUI goes from near 0% in Congo, to 125% in Sudan and 425% in Mauritania.
- Water reuse is driven by water scarcity and a lack of sanitation.
- Wastewater is appreciated as a reliable water source and for its nutrient content.
- In this region, the world's single technological project involving wastewater reuse for human consumption is located in Namibia.

Box 1.2 Highlights for Europe

- Water scarcity occurs in some countries as result of a high population density.
- Mean Water Availability is 10,680 m^3/capita·yr with an WIUI of 6.4% but Water Availability varies from 125 m^3/capita·yr in Malta, to 82,274 in Norway and 572,390 in Iceland; while, the WIUI index goes from values near 0% for Iceland, to 127% in Hungary and 231% in the Republic of Moldova.
- Water reuse is driven by: (a) water demand in high density population areas (competition between industrial and municipal water demand in the northern part and tourist and agricultural demand vs. municipal demand in the southern part); and (b) stringent effluent discharge regulations.
- ***In northern countries*** water reuse is performed for environmental protection (50%) and industries (33%) and rarely for agriculture, since rainfall suffices for this purpose.
- ***In southern countries*** (the Mediterranean Region), reuse is performed for agricultural irrigation (44%) and environmental applications (37%).
- 2% (21 m^3/s) of the total treated effluents are reused, and reuse is increasing at a rate of 25% per year.
- There are more than 200 water reuse projects in operation and a lot more at various planning stages.
- European countries follow either WHO guidelines or California's Title 22 standards.

Box 1.3 Highlights for the United States and Canada

UNITED STATES AND CANADA

- Reuse is only practiced in some states/provinces of both countries due to chronic and temporary water shortage, fast growing water demand in urbanized areas, more stringent standards for wastewater discharge, the increased cost of mobilizing new water resources and environmental constraints.
- Mean Water Availability is 19,649 m^3/*capita*·yr and WIUI is 9.3% but Water Availability is 10,135 m^3/*capita*·yr in the USA and 89,111 m^3/*capita*·yr in Canada while the WIUI is 1.6% in Canada and 17.1% in the USA.

In the USA:

- At least 88 m^3/s of treated wastewater are reused, mainly in Florida, Texas, New Mexico, Arizona, California, Colorado, Nevada and Hawaii.
- Agricultural irrigation is the most common form of reuse with several projects ranging from 0.012 m^3/s to 2.2 m^3/s.
- Landscape and golf course irrigation using reclaimed water is growing rapidly.
- The first standards for water reuse in the world were established in the state of California in 1918. This legislation evolved into the Title 22 standards, which are stringent because of the high level of public health protection required in the state.
- 22 out of the 50 states comprising the USA have water reuse standards. Some follow the Title 22 standards' style but others do not.
- In 2005, USEPA released new water reuse criteria.

CENTRAL AND SOUTH AMERICA AND THE CARIBBEAN ISLANDS

- Latin America is a water rich area because of the abundance of water in South America, while the Caribbean islands, Mexico and Central America are in general water scarce areas.
- Mean Water Availability for Mexico, Central America and the Caribbean islands is 6,740 m^3/*capita*·yr with a WIUI of 8.8%, but the Water Availability ranges from more than 67,000 m^3/*capita*·yr in Belize to only 61 m^3/*capita*·yr in Bahamas, while the WIUI goes from 0.6% in Panama to 112% in Barbados.
- Mean Water Availability for South America is of 45,400 with a WIUI of 1.3%. The maximum WIUI in the region is 10.6% for Argentina with the minimum water availability in the region being 20,800 m^3/capita·yr.
- Water reuse is driven by the interest in recycling nutrients contained in wastewater in poor soil areas, the lack of sanitation that makes raw sewage available for irrigation, and water scarcity in the Caribbean islands, Mexico and Peru.

- Wastewater is frequently used untreated and to irrigate crops directly or indirectly, and it is appreciated by farmers because it is a reliable water source, because of its nutrient content and since it has a low or zero cost.
- Public policies tend to control unplanned reuse rather than promote planned use.
- Most of the countries follow WHO guidelines but have problems implementing them.
- Before 1970, reuse technologies were predominantly stabilization ponds but are being replaced by activated sludge, advanced primary treatment, anaerobic filters and UASB processes.
- Non-intentional reuse of wastewater for human consumption as a result of the incidental recharge of aquifers due to agricultural irrigation has been documented in Mexico and Peru.
- Other types of planned reuse (industrial, environmental enhancement and municipal) are also carried out but with a much smaller volume (4 m^3/s).

Box 1.4 Highlights for Asia

- Around the region there is a significant variation in water availability and water shortage areas coincide with high urban population density.
- Mean Water Availability is 3,990 m^3/capita·yr with a WIUI of 19.3%; but water availability varies from 89 m^3/capita.yr in the Maldives to 55,049 m^3/capita·yr in Laos while the WIUI goes from < 1% in Bhutan to over 323% in Pakistan or 1,812% in Uzbekistan.
- Water reuse is driven by water scarcity, lack of sanitation, demand in high population density areas and, in one case (Singapore) by international political pressure on water resources.
- There are two groups of countries: those performing reuse for agriculture and aquaculture, and those that also reuse water for municipal and industrial purposes (like Japan and Korea).
- Municipal reuse is for activities requiring low quality water (like toilet flushing) but also for human consumption (only in Korea).
- Environmental reuse with high quality water is often practiced in Japan.
- In Singapore, the NEWater project indirectly reuses water for human consumption. The decision was made to secure water supply in a country where 50% of the water resources come from another country.

Box 1.5 Highlights for Oceania

- Within the region water scarcity exists only where population density is high.
- Mean Water Availability is 53,290 m^3/*capita*·yr and WIUI is 1.6%; water availability ranges from 33,341m^3/*capita*·yr in Fiji to 133,478 m^3/*capita*·yr in Papua and New Guinea while the WIUI goes from near 0% (in Papua New Guinea) to 4.9% (in Australia).
- Water reuse is driven by regional water scarcity and stringent effluent discharge conditions to protect ocean, coastal and surface water ecosystems.
- Australia is undertaking important water reuse programs. It has developed a new water policy based on mandatory measures and incentives for promoting water reuse.
- In 2000, less than 11% of the wastewater and 3% of the urban storm water were recycled and reused for non-potable uses in Australia. Reuse is now increasing at a rate of 10-17% per year.
- Reclaimed water is used for mining (32%), agricultural irrigation (28%), recreation (24%), electricity and gas industries (5%), the metal industry (3%), water supply (3%), forestry (2%), and others (3%).
- Australia has developed *ad hoc* regulation using a combination of natural treatment with advanced technology.
- Advances in reuse are quickly achieved thanks to a national coordinated water reuse program, in which NGOs, professional associations, universities, water utilities and research centres are involved.
- Water reuse schemes have been developed with subsidies, where the recycled water cost has been set at 30% of the cost of potable water.

1.4.3 Main wastewater reuse drivers

Different factors (physical, economic, social and political) were detected during this survey as triggering mechanisms for water reuse, albeit with variations between developed and developing countries. The main factors are summarized in Table 1.6. The most important in terms of volume stem from the lack of water, a high level of local water demand and the need for reliable sources of water. The origin of these drivers may be physical (defined by the characteristics of the particular environment), social (arising from the public), economic, political or as a consequence of water management policies.

Table 1.6. Main observed wastewater reuse drivers.

DRIVER	Developed countries	Developing countries
Due to the physical situation		
Lack of water	+++	+++
Drought management and to ensure a reliable source of water	+++	++
Generation of wastewater near agricultural fields demanding water	+++	+++
Lack of sanitation inducing unintentional reuse of wastewater		+++
As result of water management policies		
For environmental protection to control negative fall-out from the disposal of treated effluents (mainly in coastal and tourist areas or in highly sensitive aquatic ecosystems) water reuse is being practiced	++	
To improve ecological conditions in environments with low water quality by reusing properly treated wastewater	++	
Convenience of reusing wastewater in order to use first use water for drinking water supply	++	
Growing recognition among water and wastewater managers of the economic and environmental benefits of using recycled water	+	
For health protection in regions where unintentional reuse is being performed with low water quality.		+
Greater recognition of the environmental and economic costs of water storage facilities such as dams and reservoirs	++	
The growing numbers of successful water recycling projects in the world	+	
As result of social pressure		
Preference for supporting water reuse programs rather than to increasing water prices to transport water from other sources or to cover advanced wastewater treatment costs	++	
Increasing awareness of the environmental impact associated with the overuse of water supplies	+	
To reclaim compounds in reused water (such as nitrogen and phosphorus) at no cost	+	+++
Community enthusiasm for the concept of water reuse	+	
Economic reasons		
As an option to partially recover treatment costs to meet stringent standards	++	
Economic reasons and water management policies		
To employ reuse as a lower-cost disposal option	++	+
Environmental protection in tourist areas	++	++
Physical, social and economic reasons		
High water demand in local areas, mainly for urban and industrial use	+++	+++

+++: High ++: Medium +: Low importance

1.4.4 Main countries reusing wastewater

It is difficult to establish which are the main countries reusing wastewater for two reasons: (a) because reuse is measured differently in different countries (as the total volume reused, reuse per capita, as a percentage of total water use, or as a percentage of treated wastewater); and (b) total country values may hide the importance of reuse at local level. For instance, it is well known that California and Florida are important wastewater reusers, but total figures for the United States do not reflect this fact. Table 1.7 contains a classification of the countries reporting figures on wastewater (treated or not), reused based on different criteria. The numbers are uncertain especially where non-treated wastewater is involved, because censuses are rarely performed on un-planned practices, or figures are simply hidden for political and economic reasons. In total volume, China, Mexico and the United States are the countries with the largest quantity of wastewater reuse, but in the first two cases non-treated wastewater is involved, and for China the value reported is certainly underestimated considering the total number of hectares irrigated using non-treated wastewater reported in other references and shown in Figure 1.4. Also, the figure concerning the total reuse of wastewater somehow also reflects the size of the country. If the reuse per inhabitants is considered, Qatar, Israel and Kuwait are the countries highest ranked, while when reuse is considered as the percentage of the total water used, Kuwait, Israel and Singapore become the most important. Whichever way, these countrywide figures in most cases represent necessarily a mixture of the reuse situation in water rich and water poor areas within the same country.

Table 1.7. Ranking of countries reusing wastewater, using three different criteria.

Rank	Country	Total reuse m^3/d	Country	Reuse, m^3/d per million capita	Country	Reuse/ Extraction, %
1	China	14,817,000	Qatar	170,323	Kuwait	35.2
2	Mexico	14,400,000	Israel	166,230	Israel	18.1
3	USA	7,600,000	Kuwait	163,330	Singapore	14.4
4	Egypt	1,920,000	Mexico	136,235	Qatar	13.3
5	Saudi Arabia	1,847,000	UAE	126,713	Cyprus	10.4
6	Syria	1,014,000	Cyprus	88,952	Jordan	8.1
7	Israel	1,014,000	Saudi Arabia	75,081	UAE	8.0
8	Chile	840,600	Bahrain	56,301	Malta	7.8
9	Spain	821,920	Syria	55,109	Tunisia	7.1
10	Japan	573,800	Chile	52,211	Mexico	6.7
11	Tunisia	512,328	Tunisia	51,233	Saudi Arabia	5.5
12	UAE	506,850	Jordan	40,179	Namibia	4.3
13	Peru	505,100	Malta	27,400	Bahrain	4.2
14	Australia	456,100	Oman	27,385	Chile	2.4
15	Iran	455,700	Egypt	26,301	Oman	1.9
16	Korea, Rep	430,000	US	25,486	Syria	1.9
17	Kuwait	424,657	Australia	22,805	Bolivia	1.1
18	Iran	422,000	Spain	20,436	Egypt	1.0
19	Jordan	225,000	Namibia	19,733	Libya	0.9
20	Turkey	136,986	Libya	18,966	Peru	0.9
21	Argentina[(1)]	129,600	Peru	18,327	China	0.9
22	Colombia	129,600	Singapore	17,442	Korea, Rep	0.8

Rank	Country	Total reuse m^3/d	Country	Reuse, m^3/d per million capita	Country	Reuse/ Extraction, %
23	Italy	123,288	China	11,268	Spain	0.8
24	Libya	110,000	Korea, Rep	9,024	Australia	0.6
25	Qatar	105,600	Iran	6,510	US	0.6
26	Morocco	104,110	Dominican R.	6,502	Colombia	0.4
27	Germany(1)	86,400	Iran	6,072	Yemen	0.3
28	South Africa(1)	82,195	Greece	5,136	Morocco	0.3
29	Singapore	75,000	Bolivia	4,800	Greece	0.3
30	Oman	71,200	Japan	4,479	Dominican R.	0,3
31	Cyprus	68,493	Argentina(1)	3,375	South Africa(1)	0.2
32	Yemen	63,480	Morocco	3,358	Japan	0.2
33	Dominican R.	57,024	Yemen	3,067	Iran	0.2
34	Greece	54,800	Colombia	2,885	Argentina(1)	0.2
35	Bolivia	43,200	Palestine	2,734	UK(1)	0.2
36	Pakistan(1)	42,200	Italy	2,163	Lebanon	0.1
37	Bahrain	41,100	Turkey	2,011	Turkey	0.1
38	UK(1)	40,000	South Africa(1)	1,771	Italy	0.1
39	Namibia	35,520	Lebanon	1,528	Germany(1)	0.1
40	Brazil(1)	23,330	Germany(1)	1,048	Belgium(1)	<0.1
41	France	19,178	Belgium(1)	671	Guatemala	<0.1
42	Malta	10,960	UK(1)	662	France	<0.1
43	Palestine	8,750	France	320	Brazil(1)	<0.1
44	Belgium(1)	6,950	Pakistan(1)	267	Pakistan(1)	<0.1
45	Lebanon	5,500	Brazil(1)	127	Poland	<0.1
46	Poland	2,740	Guatemala	119		
47	Guatemala	1,460	Poland	71		

UAE: United Arab Emirates

(1) Total reuse figures may be greater in practice; data presented here come from projects reported in this book.

Table 1.8 contains only data on reuse of treated wastewater. In total volume, United States, Saudi Arabia and Egypt are the most important countries, but when the amount of treated wastewater reuse per capita is considered, Qatar, Israel, and Kuwait are in the first three places; when reuse is expressed as a percentage of the total use Kuwait, Israel, and Singapore become the most important ones.

Table 1.8. Rankings of countries reusing treated wastewater, using three different criteria.

Rank	Country	Treated waste-water reused m^3/d	Country	Treated wastewater reused m^3/million capita	Country	Treated wastewater reuse as % of the total water extraction
1	USA	7,600,000	Qatar	170,323	Kuwait	35.2
2	S. Arabia	1,847,000	Israel	166,230	Israel	18.1
3	Egypt	1,780,821	Kuwait	163,330	Singapore	14.4
4	Syria	1,014,000	UAE	126,713	Qatar	13.3
5	Israel	1,014,000	Cyprus	88,952	Cyprus	10.4
6	Spain	821,920	S. Arabia	75,081	Jordan	8.1
7	Mexico	767,280	Bahrain	56,301	UAE	8.0
8	China	670,000	Syria	55,109	Malta	7.8
9	Japan	573,800	Tunisia	51,233	Tunisia	7.1
10	Tunisia	512,328	Jordan	40,179	S. Arabia	5.5
11	UAE	506,850	Malta	27,400	Namibia	4.3
12	Australia	456,100	Oman	27,385	Bahrain	4.2
13	Korea, Rep	430,000	USA	25,486	Oman	1.9
14	Kuwait	424,657	Egypt	24,395	Syria	1.9
15	Iran	420,000	Australia	22,805	Bolivia	1.1
16	Chile	320,000	Spain	20,436	Egypt	1.0
17	Peru	280,100	Chile	19,876	Libya	0.9
18	Jordan	225,000	Namibia	19,733	Chile	0.9
19	Turkey	136,986	Libya	18,966	Korea, Rep	0.8
20	Argentina[(1)]	129,600	Singapore	17,442	Spain	0.8
21	Italy	123,288	Peru	10,163	Australia	0.6
22	Libya	110,000	Korea, Rep	9,024	USA	0.6
23	Qatar	105,600	Mexico	7,259	Peru	0.5
24	Yemen	92,000	Iran	6,000	Yemen	0.5
25	Germany[(1)]	86,400	Bolivia	4,800	Mexico	0.4
26	SA[(1)]	82,195	Japan	4,479	SA[(1)]	0.2
27	Singapore	75,000	Yemen	4,444	Japan	0.2
28	Oman	71,200	Argentina[(1)]	3,375	Iran	0.2
29	Cyprus	68,493	Palestine	2,734	Argentina[(1)]	0.2
30	Bolivia	43,200	Greece	2,624	UK[(1)]	0.2
31	Bahrain	41,100	Italy	2,163	Lebanon	0.1
32	UK[(1)]	40,000	Turkey	2,011	Turkey	0.1
33	Namibia	35,520	SA[(1)]	1,771	Greece	0.1
34	Greece	28,000	Lebanon	1,528	Italy	0.1
35	Brazil[(1)]	23,330	Germany[(1)]	1,048	Germany[(1)]	0.1
36	France	19,178	Belgium[(1)]	671	China	0.0
37	Malta	10,960	UK[(1)]	662	Belgium[(1)]	0.0
38	Palestine	8,750	China	510	Guatemala	0.0
39	Belgium[(1)]	6,950	France	320	Morocco	0.0
40	Morocco	6,600	Morocco	213	France	0.0
41	Lebanon	5,500	Brazil[(1)]	127	Brazil[(1)]	0.0
42	Poland	2,740	Guatemala	119	Poland	0.0
43	Guatemala	1,460	Poland	71		

UAE: United Arab Emirates SA: South Africa S. Arabia: Saudi Arabia

[(1)] Total reuse figures may be greater in practice; data presented are from projects reported in this book..

1.4.5 Agricultural reuse

Food production is a water intensive activity (76% of the total use given to water). Under prevailing land and water management practices a balanced diet represents a depleting water use of 1,300 m^3/capita·yr, which is 70 times more than the 50 L/p day used for basic household water needs (SIWI-IMWI, 2006). Agriculture represents an important demand on water and, as a consequence, is the biggest reuser of wastewater in volume among all the different uses of water. Figure 1.4 contains the 20 countries reusing the greatest absolute quantities for agriculture and Figure 1.5 the 20 countries with the largest wastewater reuse per one million inhabitants; in both cases data for treated and non-treated wastewater are included. Even though the information used to construct Figures 1.4 and 1.5 is heterogeneous and incomplete it does give some appreciation of the relative importance of the reuse of non-treated wastewater. In several countries a dependency on the use of this kind of water has been created for many reasons (not only lack of water), and this needs to be understood to proper set measures to control negative effects whilst also preserving the positive. Annex 2 presents in detail the total data gathered during this survey for this type of reuse.

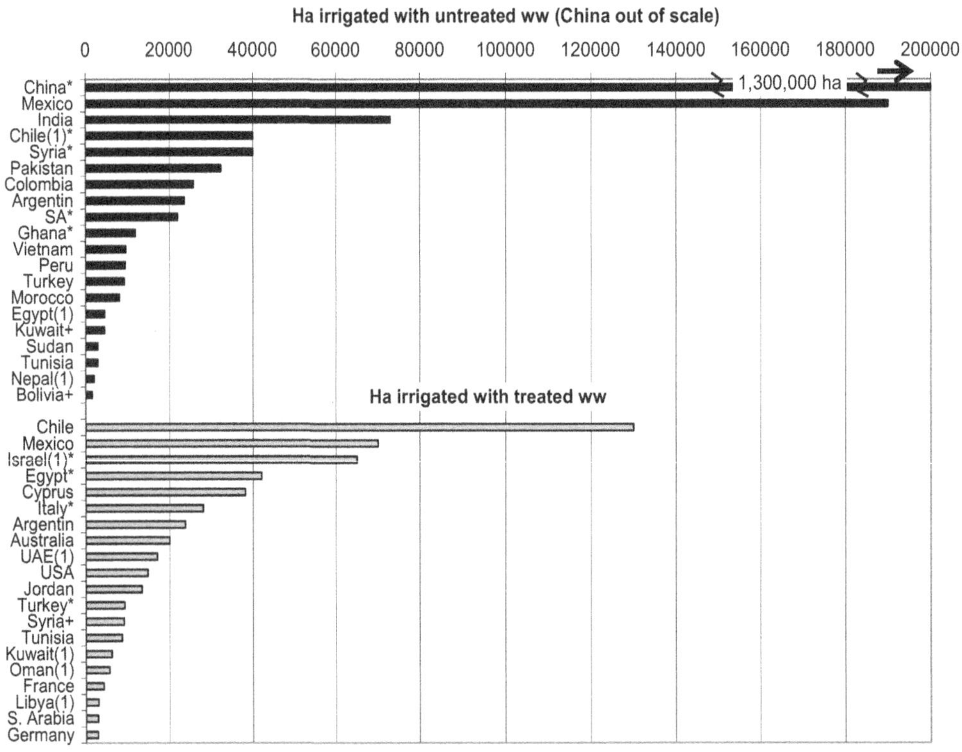

Note: Information may vary from source to source. Some countries report agricultural wastewater use without mentioning the amount of hectares involved.

*Data are confusing.

+No data are available, although the practice is reported.

$^{(1)}$Surface might be greater.

Figure 1.4. The 20 countries reporting the largest surface irrigated with treated and untreated wastewater (with information from Jiménez, 2006).

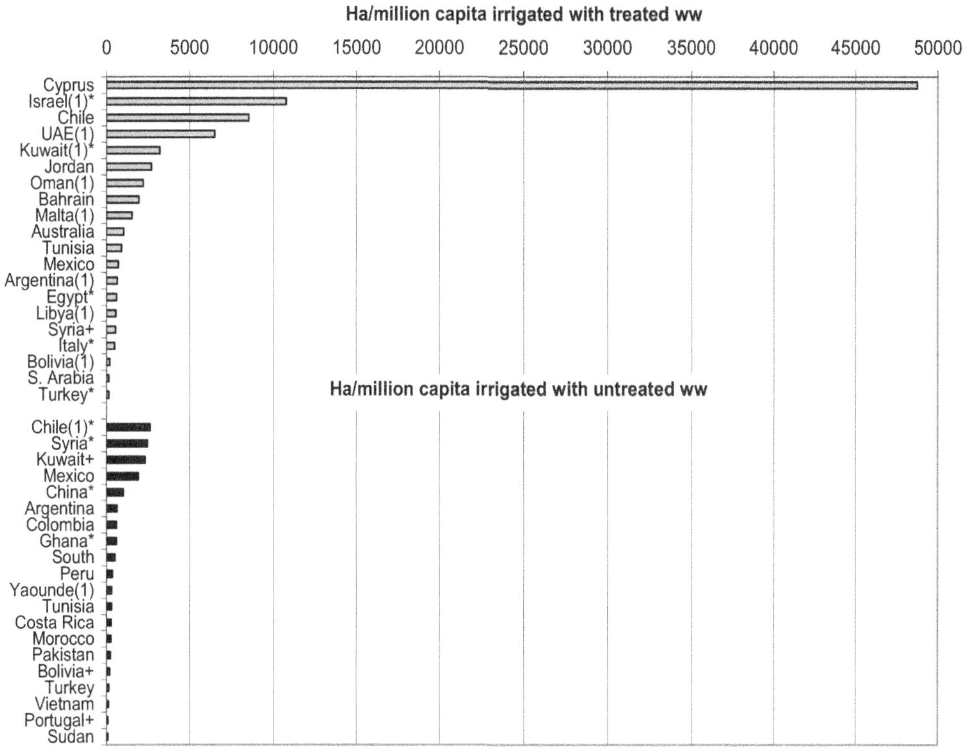

Note: Information may vary from source to source. Some countries report agricultural wastewater use without mentioning the amount of hectares involved.

*Data are confusing. +No data are available, although the practice is reported. [(1)]Surface might be greater.

Figure 1.5. The 20 countries reporting the highest Ha/million capita irrigated with untreated and treated wastewater (with information from Jiménez, 2006).

Irrigation is a key input for agriculture in developing countries[1]. Due to lack of sanitation, there are frequent examples of irrigation performed using wastewater in developing countries and not only in dry but in humid areas also. Although in most cases it occurs as an unintentional activity, with time it is becoming a recognized and increasing activity. The possibility of using wastewater is attractive to farmers for several reasons, such as its reliability, its low or zero cost and its nutrient content which increases crop production without adding artificial fertilizers (IMWI, 2003 and Jiménez and Garduño, 2001). Figures on the reuse of wastewater for irrigation are difficult to obtain, mostly due to economic penalties a country might suffer in trading its produce. However, it is estimated that at least 20,000,000 ha in 50 countries are irrigated with polluted water (United Nations, 2003), either directly or indirectly, and that 10% of the world's population consumes crops produced with wastewater (Smit and Nasr, 1992). The relative importance of this practice varies by country; in Hanoi, Vietnam, for instance, up to 80% of the vegetables consumed are produced with wastewater (Ensink *et al.*, 2004b).

Data from some countries are presented in Figures 1.4 and 1.5 to provide an idea of the relative situation of the reuse of treated and non-treated wastewater in irrigation around the world.

[1]Actually three quarters of the total irrigated area is located in developing countries (UN, 2003).

Finally, Table 1.9 lists the reuse of wastewater in volume for different countries, considering here only treated wastewater.

Table 1.9. Countries reporting the largest volumes and volumes/million capita of treated wastewater reuse for irrigation (with information from Jiménez 2006).

Rank	Country	Treated wastewater used for irrigation, m^3/d	Country	Treated wastewater used for irrigation, $m^3/d\cdot$million capita
1	Mexico	4,492,800	Kuwait(1)*	225,455
2	Egypt	1,917,808	Qatar(1)*	141,593
3	China(1)*	1,238,860	Israel	127,007
4	Syria*	1,182,000	Cyprus	87,364
5	Spain*	931,507	UAE(1)*	76,746
6	USA(2)	911,000	Syria*	73,013
7	Israel	767,123	Malta*	66,667
8	Italy*	741,262	Jordan*	45,727
9	S. Arabia(1)*	594,521	Mexico	45,441
10	Kuwait(1)*	431,520	Bahrain(1)*	42,188
11	Iran	421,918	S. Arabia(1)*	29,221
12	Chile(1)	380,000	Egypt	28,251
13	Jordan*	224,658	Oman	26,399
14	UAE(1)*	200,000	Chile(1)	24,982
15	Turkey	136,986	Spain*	23,340
16	Argentina(1)	129,600	Libya(1)	20,794
17	Tunisia*	117,802	Italy*	12,885
18	Libya(1)	110,000	Tunisia*	12,454
19	Qatar(1)*	80,000	Iran	5,999
20	Cyprus	68,493	Bolivia(1)	5,187
21	Oman	67,000	Argentina(1)	3,500
22	Yemen*	50,000	USA(2)	3,216
23	Bolivia(1)	43,200	Yemen*	2,725
24	Japan	35,300	Turkey	2,055
25	Bahrain(1)*	27,000	Greece(1)*	1,888
26	Malta*	26,000	Lebanon	1,567
27	Brazil(1)	23,328	China(1)*	966
28	Greece(1)*	20,030	Ecuador	376
29	France(1)*	19,178	France(1)*	324
30	Morocco(1)*	6,600	Japan	278
31	Lebanon	5,479	Morocco(1)*	221
32	Ecuador	4,752	Guatemala	144
33	Guatemala	1,642	Brazil(1)	137
34	Kyrgyzstan	384	Kyrgyzstan	78

UAE: United Arab Emirates S. Arabia: Saudi Arabia

*Data are confusing. +No data are available, although the practice is reported.

(1)Value might be greater. (2)California and Florida

The term "urban agriculture" has evolved to describe wastewater reuse in urban and peri-urban areas from developing countries in small sized parcels (0.5-2 ha) to produce fruit trees, fodder, flowers, and vegetables. Occasionally, it also involves a small amount of fish production. Urban agriculture is performed in arid and wet countries and stems from the combination of wastewater availability, demand for fresh produce by city dwellers and the presence of marginalized people (Cockram and Feldman, 1996). Generally it contributes to food security, improved nutrition and health, which in turn improves living standards (van der Hoek *et al.*, 2001). In spite of improved nutrition and with its resistance to diseases, urban agriculture also causes diarrheic diseases. In order to control such diseases, low technology and affordable methods have been developed at local level. These methods preserve the beneficial properties of the wastewater, including its low or zero cost and the possibility of obtaining fertilizers "for free" while removing an important amount of pathogens (Jiménez, 2006). Unfortunately, these methods are poorly understood internationally, as well as having weak scientific recognition making their implementation difficult in different parts of the world (Ensink *et al.*, 2004a; Jiménez and Garduño, 2001). Currently, there are several million farmers in developing countries practicing urban agriculture. It is estimated that 10-70 percent of people living in different cities depend on it and that up to 50 percent of the market vegetables in cities from Asia to Africa are being produced this way (Cornish and Lawrence, 2001 and IMWI, 2003). By the year 2025, it is estimated that there will be 292 cities in the world with more than 1 million people, most of them located in low-income countries (UN, 2006), and if sanitation and poor conditions are not improved, urban agriculture will increase, and with it the negative effects on public health. In short, the current risky unplanned approach to urban agriculture should and could be transformed into a planned approach to preserve its economic and social benefits while controlling the associated health issues.

1.4.6 Municipal reuse

In 20 years, 60 percent of the world's population will be living in cities (UN, 2006), thus: (a) more water will be needed for municipal use, and consequently (b) additional municipal wastewater will be produced, all in the same place and within a limited area. This situation represents both a risk and an opportunity to better use water, for example by increasing and diversifying municipal wastewater reuse. Opportunities to reuse wastewater in cities are classified into two groups: (a) those demanding relatively low quality water because they involve low health risks, and (b) those demanding high quality water because of their association with health risks. In the first group, there are several types of use, such as the filling of recreational lakes, car, truck or street washing and park irrigation. Among options demanding high water quality, reuse for drinking supply is included. There are successful experiences around the world of both types of reuse, the first group being the most common while the second, although less common, are no less important (particularly at a local level where the lack of water represents a serious barrier to the development of cities). Water reuse projects implemented for drinking supply have been thoroughly studied, are highly monitored and have wide social acceptance in countries such as Namibia and Singapore. But there are also some examples of planned projects that have never been implemented due to public rejection, as has occurred in specific areas of California (Seah, 2002) and in Australia (AASTE, 2004). It would be beneficial to spread information on successful cases in order to increase public acceptance of reused water; in fact, this should be done for any kind of reuse project.

Unplanned reuse schemes for municipal purposes are obviously not commonly reported in the literature. Nevertheless, there are some examples involving indirect reuse that have been reported. These cases are of two kinds: (a) those where effluents of municipal wastewater treatment plants after being discharged into and mixed with fresh water sources are used once again, and (b) those where non-treated wastewater is discharged into the environment and is reused after being mixed with first use water; commonly the water body acting as a receiver is an aquifer. As can be imagined, the latter are only reported in areas where sanitation has a low coverage. The implications of both types of reuse activities are varied, although not as bad as might be expected for the latter. Three different examples of water reuse for drinking purposes in very different situations are presented in Chapters 23, 24 and 25. Although their detailed analysis is beyond the scope of this chapter, what needs to be mentioned here is that the number of regions where water availability is reaching critical values is increasing, and even though there is a tendency to leave planned direct wastewater reuse for drinking purposes as a last option, it is highly probable that it will be a common practice in the foreseeable future, so research and pilot projects on direct or indirect potable reuse should be intensified in order to be adequately prepared to cope with the scientific, technical, economic, social and legal dimensions of this issue.

1.4.7 Industrial reuse

Industrial reuse and recycling differs from other types of reuse (*e.g.* municipal and agricultural), as it involves the private sector which has its own rules and well-defined needs. Industrial reuse happens in industrialized areas mainly located in developed countries, where the amount of water used by industry is substantial. Industrial recycling and reuse is driven by economic forces. Internal recycling is always performed first (before reuse) because the savings are immediate. Reuse is the second option because it normally involves greater investment needs, implies the fulfilment of treatment standards supervised by the government and the need for negotiation between different parties. To promote industrial reuse, the best government strategies have proven to be the use of incentives rather than compulsory regulations. That way, each industry defines its own needs and possibilities in terms of water quantity and quality according to its situation, as well as economic criteria and availability of financial support.

Of the different industrial reuse options, cooling is the most common because of its high water demand, the relatively low water quality required, the ease with which the practice may be applied to different types of industries and simplicity of implementation. As a consequence, power plants are the most involved in water reuse schemes. The technology used is based on secondary treatment plants, sometimes complemented with filtration or softening processes to avoid cooling tower clogging or scaling.

Considering how vital water is to many key economic activities, the risks of water scarcity may take on strategic importance for businesses and their financial backers around the world (UNEP, 2004). In most countries, water is national property and the state grants water use licenses or allocates water resources. In times of drought or water crisis, governments designate domestic use as a top priority for allocation, often followed by agriculture, for reasons related to food security. Industry is often the last priority, which increases the business risk for companies operating in water scarce and water stressed areas (UNEP, 2004). Under such circumstances, preparing and implementing water reuse programs is a strategic issue for economic development. Particularly in China, India, Russia, Poland, Mexico, and Brazil, which are countries listed in the 10 global contenders for business growth and industrial development, water stress is a problem for industrialization (UNEP, 2004). Therefore, UNEP has called on financial institutions to become aware of the need to prepare programs to finance projects to address water scarcity, for example water reuse projects.

1.4.8 Legislation

Water reuse, in contrast to potable water, has no universal standards. There are three reasons for this: (a) it covers different uses; (b) it is a new human practice; and (c) it has been developed locally in different ways to address specific needs which are difficult to extrapolate to other conditions.

Agricultural irrigation, being the first water reuse option recognized, was also the first to be considered by WHO guidelines. In 1989 international guidelines for reuse of water for agricultural irrigation were published but they did not influence as many countries as the drinking water guidelines. In fact, standards set by one state of the United States (California) were more influential in many countries through the legislation known as California Title 22. WHO guidelines were recently reviewed (in 2006) with the aim of harmonizing different approaches so far developed. The idea was to combine the two guidelines in a uniform framework flexible enough to be applied in different conditions. Its application and therefore its impact are still in progress.

Even though there is a wide variety of projects concerning municipal and industrial reuse around the world there is not an international proposal to regulate, although some countries have already established norms such as United States, Australia, Japan, Israel and some European countries.

There is no doubt that in order to increase public acceptance of reuse, there will first need to be worldwide accepted standards, and much work still needs to be done by professionals to produce those standards.

1.5 CONCLUDING REMARKS

Water reuse is practiced in developed and developing countries in very different ways. In the former, it is mostly a planned activity while in the latter it frequently occurs unplanned. Nevertheless, both are wastewater reuse practices and should be acknowledged as such (Jiménez and Asano, 2004), and in all areas, the reuse of wastewater is considered economically interesting. In developed countries reuse is considered viable for reasons including the application of stringent standards and the use of expensive technology and economic incentives. By contrast, in developing countries the use of untreated wastewater to produce goods and beneficially recycle nutrients also makes reuse attractive. The reason why these different approaches converge is because wastewater is being considered as a resource. Nevertheless, this is not always officially recognized in national legislations or policies making the implementation of reuse projects difficult in practice. Whether water reuse will reach its full potential depends on a number of variable factors in each country, such as economic considerations, potential uses for reclaimed water, the stringency of wastewater discharge requirements and public policies for conservation and protection. Local strategies need to be flexible and promote wastewater reuse in the best and safest possible conditions.

1.6 REFERENCES

AASTE, Australian Academy of Technological Sciences and Engineering (2004) Water recycling in Australia: A review undertaken by the Australian Academy of Technological Sciences and Engineering Water Recycling in Australia.

Bixio, D., De Koning, J., Savic, D., Wintgens, T., Melin, T. and Thoeye, C. (2006) Wastewater reuse in Europe. Desalination. 187, 89-101

Cockram, M. and Feldman, S. (1996) The beautiful city: gardens in Third World. *African Urban Quarterly* **11**(2-3), 202-208.

Cornish, G. A. and Lawrence, P. (2001) Informal Irrigation in peri-urban areas: A summary of findings and recommendations, Report OD 144 HR Wallingford/DFID.

Earth Trends: Earth Trends, the Environmental Information Portal (2005, 2006 and 2007)
http://earthtrends.wri.org/pdf_library/data_tables/fre1_2003.PDF,
http://earthtrends.wri.org/pdf_library/country_profiles/,
http://earthtrends.wri.org/searchable_db/index.php?theme=2 and

http://earthtrends.wri.org/searchable_db/index.php?theme=2&variable_ID=694&action=select_countries

Earth Trends: The Environmental Information Portal (2001) Water Resources and Freshwater Ecosystems, 1999-2000. World Resources Institute, http://earthtrends.wri.org/datatables/

Ensink, J., Mahmood, T., van der Hoek, W., Raschid-Sally, L. and Amerasinghe, F. (2004b) A nationwide assessment of wastewater use in Pakistan: an obscure activity or a vitally important one? *Wat. Policy* **6**, 197–206.

Ensink, J., Simmons, J. and van der Hoek, W. (2004a) Wastewater use in Pakistan: The cases of Haroonabad and Faisalabad. Chapter 8 in: Scott, C.A., Faruqui, N.I. and Raschid-Sally, L. (eds) Wastewater use in irrigated agriculture. CAB International, 91-102.

Falkenmark, M. and Widstrand, C. (1992) Population and water resources: A delicate balance. *Population Bulletin*, *Population Reference Bureau*, **47**(3), 1-36.

IMWI, International Management Water Institute (2003) Putting research knowledge into action. *Wa. Policy Briefing*, (9), IMWI Ed., Sri Lanka.

IPCC, International Pannel of Climate Change (2007). Climate change 2007: The Physical Science Basis; Summary for Policy Makers. Cambridge. From: http://www.ipcc.ch

Jiménez, B. (2005) Treatment Technology and Standards for Agricultural Wastewater Reuse: A Case Study in Mexico. *Irrigation and Drainage* **54**(1), 23-35.

Jiménez, B. (2006) Irrigation in developing countries using wastewater. *International Review for Environmental Strategies (IRES)* **6**(2), 229-250.

Jiménez, B. and Asano, T. (2004) Acknowledge all approaches: The global outlook on reuse. *Wat. 21* (December), 32-27.

Jiménez, B. and Garduño, H. (2001) Social, political and scientific dilemmas for massive wastewater reuse in the world in navigating through waters: Ethical issues in the water industry. Davis and McGin editors. Edited by AWWA.

Lazarova, V., Levine, B. Sack, J., Cirelli, G., Jeffrey, P., Muntau, H., Salgot, M. and Brissaud, F. (2001) Role of water reuse for enhancing integrated water management in Europe and Mediterranean countries. *Wat. Sci. Technol.* **43**(10), 23-33.

Lehner, B., Henrichs, T., Döll, P. and Alcamo, J. (2001) EuroWasser – Model-based assessment of European water resources and hydrology in the face of global change. Kassel World Water Series 5, Center for Environmental Systems Research, University of Kassel, Germany.

Seah, C. (2002) Media blitz on the yuck factor. The Star. Retrieved June 13, 2003 from http://www.singapore-window.org/sw02/020721st.htm

SIWI-IMWI (2006) Water – more nutrition per drop. Towards sustainable food production and consumption patterns in a rapidly changing World Stockholm International Water Institute (SIWI) and the International, Stockholm, Sweden, 36 pp.

Smit, J. and Nasr, J. (1992) Urban agriculture for sustainable cities: Using wastes and idle land and water bodies as resources. *Environ. Urbanization* **4**(2), 141–152.

United Nations (2003) Water for people, water for life. The United Nations world water development report. Barcelona: UNESCO.

United Nations (2006) http://esa.un.org/unup/p2k0data.asp

UNEP, United Nations Environmental Program (2004) Challenges of water scarcity: A business case for financial institutions. Financial Initiative for sustainability. 36 pp.

van der Hoek, W., Sakthivadivel, R., Renshaw, M., Silver, J.B., Birley, M.H. and Konradsen, F. (2001) Alternate wet/dry irrigation in rice cultivation: A practical way to save water and control malaria and Japanese encephalitis? Research Report 47. International Water Management Institute, Colombo, Sri Lanka, 30 pp.

Water Resources eAtlas (2007) Watersheds of the World: global maps. Annual renewable water supply per person by basin for 1995 and projections for 2025. Chapter 15. http://www.iucn.org/themes/wani/eatlas/pdf/gm/gm15.pdf

WHO (2006) Guidelines for the Safe use of Wastewater, Excreta and Greywater in Agriculture and Aquaculture. Vol.1, 2, 3 and 4. World Health Organization, Ed. Paris, France.

2

Water reuse in Middle Eastern and North African countries

Akiça Bahri

2.1 INTRODUCTION

Water reuse in Middle Eastern and North African countries is an old and common practice, which has gone through different development stages with time, knowledge of the processes, treatment technology, and regulation evolution. It has been practiced for centuries, from the Phoenician and Roman times to more recently by the Mediterranean and Middle Eastern civilizations. The city of Tunis built her first wastewater treatment plant in 1929 and irrigation of plantations with secondary effluent begun in Kuwait in 1956 before further development in the region.

In the Middle Eastern and North African region, the volume of wastewater is increasing. Consequently, there is a major potential use of reclaimed water in the region. Large areas could be supplied with reclaimed water which could also be used for different other purposes depending on the demand, the water characteristics and its suitability, etc. It is, however, essential that the development of water reuse in agriculture and other sectors be based on scientific evidence of its effects on the environment and public health.

Water reuse is an important component of many local economies with many large economic benefits from the irrigated areas, and reduction in wastewater discharge and pollution of the water bodies and the environment in general. However, application of raw or partially treated wastewater is not without causing serious public health consequences and negative environmental impacts in many locations all over the world. Therefore, wastewater must be treated and used in such a way as to ensure only acceptable risks for users, workers, consumers, and the environment. This chapter gives an overview of the wastewater

 Water Reuse - An International Survey of current practice, issues and needs by B. Jiménez *et al.* ISBN: 9781843390893. Published by IWA Publishing, London, UK.

reclamation and reuse situation in the region and of the development and management issues facing it.

2.2 THE MIDDLE EASTERN AND NORTH AFRICAN REGION

The Middle Eastern and North African region considered here includes 19 countries or territories: Afghanistan, Algeria, Bahrain, Egypt, Iran, Iraq, Jordan, Kuwait, Lebanon, Libya, Morocco, Oman, Palestine, Qatar, Saudi Arabia, Syria, Tunisia, United Arab Emirates, and Yemen. Three sub-regions may be distinguished:

- North Africa: Algeria, Libya, Morocco, Tunisia;
- Middle East: Egypt, Iran, Iraq, Jordan, Lebanon, Palestinian territories of Gaza and the West Bank, Syria, Yemen; and
- Gulf countries: Bahrain, Kuwait, Oman, Qatar, Saudi Arabia and United Arab Emirates.

Most of the countries of the Middle Eastern and North African Region are arid to semi-arid and are facing increasingly more serious water shortage problems. They are characterized by limited rainfall and recurring droughts. The average annual rainfall varies considerably across the region, ranging from less than 30 mm in the Arab area, Jordan and the Syrian Desert, to 1,000 mm in the northern part of Lebanon.

Total population of the region is actually around 358 million inhabitants (Table 2.1). The population growth rate is moderate to high. The demographic evolution of population is aggravated by very intensive urbanization. The internal renewable water resources range from a negligible amount in Kuwait to 128,500 Mm^3/yr in Iran. Due to rapid population growth, the average annual per capita renewable water has rapidly decreased since 1950. It varies across a wide range – from 7 m^3/yr in Kuwait to 2,917 m^3/yr in Iraq (Table 2.1). Except Afghanistan, Iraq, Iran, Syria, and Lebanon, the water availability in all the other countries is below the level of "chronic water scarcity" (< 1000 m^3/*capita*.yr), due mainly to very high population growth. Twelve countries among the 19 under consideration are already well below the "absolute water stress" with a per capita water availability of renewable freshwater resources of less than 500 m^3/yr (Figure 2.1) and which will reach, by 2050, the minimum survival level of 100 m^3/*capita*.yr. In such places, conventional water resources will be insufficient to even meet the domestic water demand.

Available water resources are becoming increasingly scarce, vulnerable and threatened by over-exploitation and different pollution sources. In some countries, actual water consumption already exceeds the renewable conventional water resources (Figure 2.2) and it is estimated that no more known water resources can be developed. Some countries of the Middle East and Gulf Region have few naturally available fresh water resources and rely mainly on groundwater and desalinated seawater (Table 2.2). Desalination of brackish and seawater is already under implementation or planned in some other countries, despite its cost. Surface waters are already in most cases utilized to their maximum capacity. Groundwater aquifers are often over-drafted and sea and brackish water intrusion in coastal areas has reached threshold limits in some locations. Non-renewable deep or fossil aquifers are being tapped to varying degrees. Exploitation of non-renewable fossil aquifers is intensive in Libya, Egypt, Tunisia, Algeria, Saudi Arabia, and United Arab Emirates (UAE). Exploitation indexes are over 50% and reaching 100%, and current pressures on water resources are increasing in most countries (Margat and Vallée, 2000). According to FAO (2000), groundwater mining as a percentage of total water withdrawal is around 15 in Qatar, 17 in Jordan, 40 in Bahrain, 46 in Kuwait, 71 in the UAE, 80 in Saudi Arabia and is up to 90% in

Libya. As a consequence, several problems appear such as water and soil salinization, desertification, increasing water pollution, and unsustainable land and water use.

The Middle Eastern and North African region is nowadays depending for its economic and social development on agriculture and tourism for some of the countries and, secondarily, on industry and other economic activities. Agriculture remains an important contributor to the gross domestic product of the North African countries and to Egypt, Syria and Palestine. It also represents a large share of the employment and of the export earnings for some of the countries of the region. It is the principal user of water in all the considered countries. Its share of the water withdrawals ranges from less than 70% in Algeria, Bahrain, Kuwait, Lebanon, Palestine and the United Arab Emirates to over 90% in Afghanistan, Iran, Iraq, Oman, Syria, and Yemen (Table 2.1). The domestic share of water varies from over 20% in Algeria, Bahrain, Jordan, Kuwait, Lebanon, Palestine, Qatar, and the United Arab Emirates to 5% or less in Afghanistan, Iraq, Syria, and Yemen. The industrial share of water is generally less than 5% except in Algeria, Egypt, Iraq, and the United Arab Emirates (FAO, 2000).

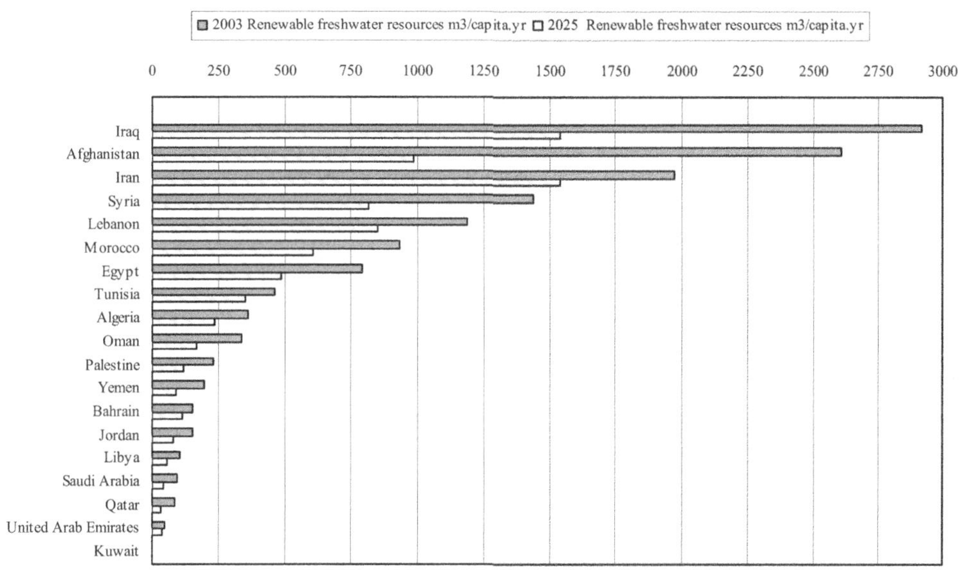

Figure 2.1. Renewable freshwater resources available per capita for the years 2003 and 2025.

Table 2.1 Renewable freshwater resources and water withdrawals in the Middle Eastern and North African Region (Aquastat database, FAO, 2000).

Country	Area [km^2]	Population [thousands]	Renewable freshwater resources [Mm^3/yr]	Renewable freshwater resources available per capita [$m^3/capita.yr$]	Dependency ratio* [%]	Total water withdrawal [Mm^3/yr]	Total water withdrawal per capita [$m^3/capita.yr$]	Total water withdrawal/ Renewable water resources [%]	Total water withdrawal by sector: Agriculture [%]	Domestic [%]	Industry [%]
Afghanistan	652,209	24,926	65,000	2,608	15.4	23,260	933	36	98	2	0
Algeria	2,381,740	32,339	11,670	361	3.6	6,070	188	52	65	22	13
Bahrain	698	739	116	157	96.6	300	406	259	57	40	3
Egypt	1,001,450	73,390	58,300	794	96.9	68,300	931	117	86	8	6
Iran	1,648,200	69,788	137,500	1,970	6.6	72,880	1,044	53	91	7	2
Iraq	438,320	25,856	75,420	2,917	53.3	42,700	1,651	57	92	3	5
Jordan	88,780	5,614	880	157	22.7	1,010	180	115	75	21	4
Kuwait	17,820	2,595	20	7	100	440	170	2200	52	46	2
Lebanon	10,400	3,708	4,407	1189	0.8	1,380	372	31	67	33	1
Libya	1,759,540	5,659	600	106	0	4,268	754	711	83	14	3
Morocco	446,550	31,064	29,000	934	0	12,600	406	43	87	10	3
Oman	309,500	2,935	985	336	0	1,360	463	138	90	7	2
Palestine (Gaza Strip & West Bank)	7,477	3,200	739	231	-	293	92	40	60	40	-
Qatar	11,000	619	53	86	3.8	290	468	547	72	24	3
Saudi Arabia	2,149,690	24,919	2,400	96	0	17,320	695	722	89	10	1
Syria	185,180	18,223	26,260	1441	80.3	19,950	1,095	76	95	3	2
Tunisia	163,610	9,937	4,595	462	8.7	2,640	266	57	82	14	4
United Arab Emirates (UAE)	83,600	3,051	150	49	0	2,300	754	1533	68	23	9
Yemen	527,970	20,733	4,100	198	0	6,630	320	162	95	4	1
		357,595	422,195	738		283,991	590		80	16	4

*Dependency ratio: Internal renewable water resources / Renewable freshwater resources.

Table 2.2 Sources of water in the Middle Eastern and North African Region (adapted from FAO/RNE and WHO/EMRO, 2003).

Country	Desalinated water		Reused treated wastewater		Use of desalinated water and treated wastewater		Reclaimed water reused, as a percentage of total water withdrawal
	Year	Mm^3/yr	Year	Mm^3/yr	Mm^3/yr	As a percentage of total withdrawal	
Afghanistan		0	-	-	-	-	-
Algeria	1998	17	-	-	17	0.3	-
Bahrain	1998	58	1991	15	73	24	5.0
Egypt	2002	100	1993	700	800	1.2	1.0
Iran	1991	3	1999	154	157	0.2	0.2
Iraq	-	-	-	-	-	-	-
Jordan	1993	2	1997	82	84	8.3	8.1
Kuwait	1993	231	1997	80	311	71	18.2
Lebanon	-	-	1997	2	2	0.1	0.1
Libya	1998	18	1999	40	58	1.4	0.9
Morocco	1998	7	1994	38	45	0.4	0.3
Oman	2001	51	1991	26	77	5.7	1.9
Palestine (Gaza Strip & West Bank)	-	-	2002	4	4	1.4	1.4
Qatar	1993	99	1994	34	133	46	11.7
Saudi Arabia	1998	949	1992	217	1166	6.7	1.3
Syria	1993	0	2000	370	370	1.9	1.9
Tunisia	1998	13	2003	43	56	2.1	1.6
United Arab Emirates	1993	385	1999	185	570	25	8.0
Yemen	1988	10	2000	6	16	0.2	0.1

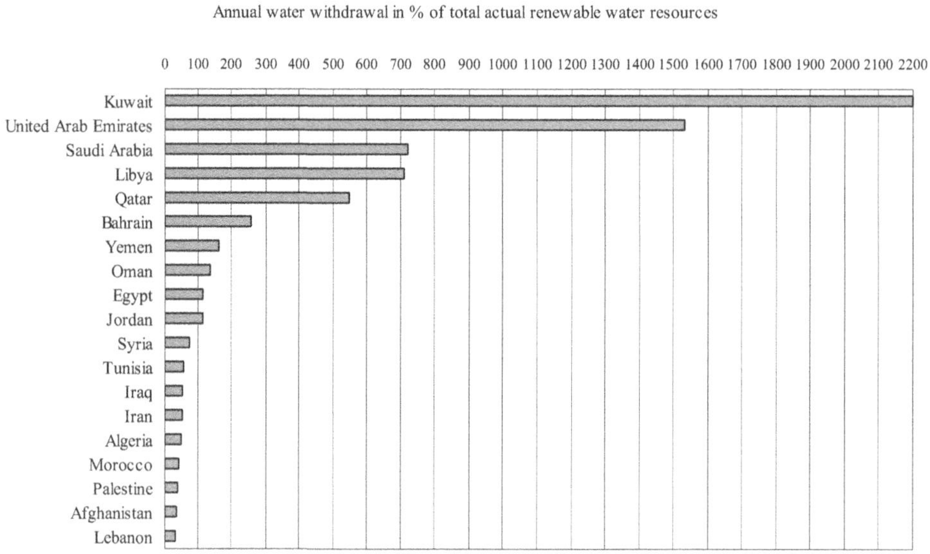

Figure 2.2. Water stress index.

With population growth and accelerated urbanization, improvement of living standards and socio-economic conditions, and development of irrigation, industry and tourism, the problems of water scarcity will intensify. Continuing increase in demand by the urban sector has led to increased utilization of fresh water, for domestic, industrial and tourism purposes on the one hand, and generation of large volumes of wastewater on the other. The irrigated sector, in competition with other sectors, will face increasing problems of water quantity and quality considering increasingly limited conventional water resources and growing future requirements and a decrease in the volume of fresh water available for agriculture. Around the cities of the region, competition with other sectors often makes water the main factor that limits agricultural development.

The agricultural sector is also influenced by the agreements of agricultural exchange and the future free trade area. Expansion of the irrigated area will continue with increasing demand for food and from the development of agricultural production for export markets. On the other hand, the irrigated sector will have to face major challenges with the future scenario of agricultural trade liberalization; a part of the water resources may be reallocated to high added-value export products instead of basic production or to industrial activities, tourism, and domestic water supply. Providing water quantities and qualities in compliance with the needs is one of the challenges facing the region.

Policy makers have then been compelled to develop additional water resources as well as to preserve the existing ones. Water conservation, reclamation and recycling are therefore designed to encourage integrated and efficient management and use of water resources. Water reuse is becoming a substantial component of the national resources policy, not from the point of view of water supply (its contribution is at most up to 18% as in the case of Kuwait (Table 2.2.)) but from the strategic point of view. It may indeed substitute good quality water for less quality demanding uses, and allow a reduction of local groundwater over-drafting and seawater intrusion in coastal areas.

2.3 THE REUSE CONDITIONS

A large range of approaches to water reuse policy may be found depending on the socio-economic, institutional, and technological conditions. Differences between countries occur in environmental and public health policies. They also occur in existing wastewater collection, treatment and disposal facilities, in human resources, in equipment and material resources, and in financial resources (USEPA, 1992). Many cities in the Middle Eastern and North African region are still unsewered; when sewers are available, they often discharge untreated effluents to the environment. There is a potential for reuse when water constraints reach critical levels and when pollution becomes a threat to surface and underground water resources, wetlands and the marine ecosystems on the one hand, and to public health on the other hand. The need to develop additional water resources then becomes important, and collection systems and wastewater treatment plants have therefore been built. However, in several cases, these plants are often not functioning or are overloaded and thus discharge effluents not suitable for reuse applications. In some other situations where conditions for reuse are met, reuse policies have been set up and treated effluents are being reused for different purposes.

2.3.1 Driving forces, benefits and concerns of water reuse

The driving forces for water reuse development in the Middle Eastern and North African region are related to different issues such as water scarcity (limited renewable freshwater resources threatened by pollution, droughts), availability of raw or treated wastewater, economical issues (cost-effectiveness of reclaimed water), regulatory policies, and environmental concerns (gradually more stringent water quality discharge regulations).

The benefits, potential health risks and environmental impacts resulting from water reuse and the management measures aimed at using wastewater within acceptable levels of risk for the public health and the environment are acknowledged in several documents (Shuval *et al.*, 1986; Mara and Cairncross, 1989; Asano, 1998; Crites and Tchobanoglous, 1998; Angelakis *et al.*, 1999; Blumenthal *et al.*, 2000; Angelakis and Bontoux, 2001; Lazarova and Bahri, 2005).

Water reuse is meant to help close the water cycle and therefore enable sustainable reuse of available water resources. When integrated to water resources management, water reuse may be considered as an important component of the national resources policy and as an integral part of environmental pollution control and water management strategy. It may present benefits to public health, the environment and economic development. Reclaimed water may provide significant additional renewable, reliable amounts of water and contribute to the conservation of fresh water resources. It may be considered as a valuable source of water and nutrients in agricultural schemes and therefore contributes to reducing use of chemical fertilizers and to increasing agricultural productivity. Reuse of reclaimed water, if properly managed, may alleviate pollution of water resources and sensitive receiving bodies. It may also contribute to desertification control and desert reclamation. Saline water intrusion may be controlled in coastal aquifers through groundwater recharge operations. Other social and economic benefits may result from such schemes, such as employment and products for export markets.

It is, however, essential that the development of reuse prevents negative effects on the environment and public health. Application of raw or partially treated wastewater has generated the existence of endemic, and quite epidemic diseases, and constitutes a threat to public acceptance of reuse. Indeed, water reuse for irrigation raises issues both as a potential resource of nutrients and as a source of pollution. Wastewater content in organic matter, nitrogen, phosphorus, and potassium may improve soil fertility and enhance plant development. However, contents in mineral and organic trace substances and pathogens represent a risk for human health and the environment. Wastewater biological composition imposes crop restrictions and constraints for the users. Content in microorganisms is of essential concern for residential, urban and recreational uses, as far as humans may be in contact, directly or indirectly, with wastewater. Ingestion or inhalation of pathogens may result in infection and disease. Organic and inorganic trace elements may present an environmental concern because of their potential harmful effects on biota. They may accumulate in the surface soil layers, be transported to underlying groundwater systems or be removed through plant uptake. They may then induce metabolism problems in plants and animals and consequently contaminate food chains. These elements may be transferred to animals or humans through different pathways and cause human health effects depending on their concentration. The lack of health protection measures for farmers, workers, and consumers in the region where raw wastewater is being applied is a reason for concern.

2.3.2 The current state and significance of water reuse in the region

Approximately 20-30% of treated effluent is being reused in Tunisia, 78% in Kuwait, and 85% in Jordan. The significance of water reuse may be evaluated through the comparison of water reuse potential with total water use or agricultural water use. It is generally small compared with total water withdrawal, ranging from 0.1 to 18.2% (Figure 2.3) but is expected to increase significantly. In Tunisia, the volume of treated effluent accounted for 4.3% of available water resources in the year 1996, and may reach 11% in the year 2030. Compared to irrigation water resources, it is actually about 7% in Tunisia, 8% in Jordan, and 32% in Kuwait. In Jordan, treated effluent may represent 40% of agriculture water share in 2020.

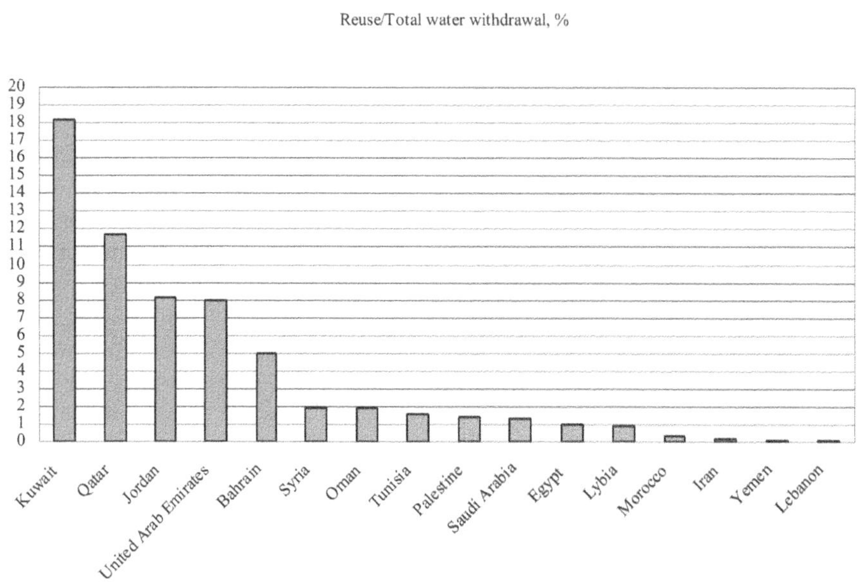

Figure 2.3. Reclaimed water reused as a percentage of total water withdrawal.

In all countries of the region, wastewater is recognized as a potential and reliable water source, which can contribute significantly to available water supplies in the future. In order to meet water demands, reclaimed water volumes must increase more than 4 times by the year 2010 in Jordan, and by more than 10 times by the year 2025 in Egypt (USEPA, 2004). Some countries in the Middle East are planning to reuse 50 to 70% of their total wastewater volume.

In the Gulf countries, treated wastewater provides an additional source, which allows limited fresh water resources to be reserved for potable supply and other priority uses. The total capacity of the wastewater treatment facilities is more than 1,100 Mm^3/yr and most of the treatment facilities have advanced tertiary treatment. The total treated effluent is about 918 Mm^3/yr, out of which about 400 Mm^3/yr (44% of the treated effluent) is used mainly for irrigating non-edible and fodder crops and urban landscaping (gardens and road highway). Treated effluent is also used for some industrial purposes. About 60% of the secondary treated effluent is discharged to the sea or into wadis (UNEP, 1999). The amount of recycled water (44% of the total treated wastewater), contributes 1.8% of the total water supply. According to Al-Zubari (1997), it is estimated that if only 50% of domestic water supplies are treated and recycled for agricultural purposes, recycled waters will have the potential to meet more than 11% of the total water demand, and satisfy more than 14% of the agricultural sector demand. This could reduce fossil groundwater withdrawal by more than 15% by the year 2020.

The volumes of wastewater generated, treated and reused in the different countries are reported in Table 2.4.

Table 2.4 Wastewater generated, treated and reused in the different countries (adapted from USEPA, 2004).

Country	Year	Wastewater generated Mm^3/yr	Wastewater treated Mm^3/yr	Wastewater reused Mm^3/yr
Afghanistan	-	-	-	-
Algeria	2004	600-700	66	-
Bahrain	1996	84-103	56	9-15
Egypt	2000	1955	940	700
Iran	1999	1859	219	154
Iraq	1999	-	425	-
Jordan	2002	180	82	70
Kuwait	1997	179	103	80
Lebanon	1997	165	4	2
Libya	1999	546	40	40
Morocco	2000	550	40	38
Oman	2001	78	37	26
Palestine (Gaza Strip and West Bank)	2002	50	-	4
Qatar	2001	110	43	34
Saudi Arabia	2000	845	674	217
Syria	2000	785	370	370
Tunisia	2003	240	187	43
United Arab Emirates	1999	500	193	185
Yemen	2000	74	34	6

2.3.3 Technologies for the treatment, storage and application of wastewater

The whole range of treatment levels and technologies can be found in the region depending on the size of the community, the quality of the wastewater, the location of the treatment plant and the availability and cost of the land, the expected quality of the effluent, the reuse opportunities, the existing regulations, and the economic conditions of the country.

In most of the cases and more particularly for large cities with high populations, mechanical wastewater treatment processes have been adopted for treating wastewater. This has occurred independently of the type of reuse, the general approach adopted up to now being based on producing an effluent in compliance with water quality discharge requirements. In the Gulf countries, advanced treatment technologies have been adopted for food crop irrigation, landscape irrigation or groundwater recharge. Chlorine is mainly used for disinfection in the Gulf countries, but UV radiation is developing in countries such as the UAE, Kuwait, and Tunisia. Disinfection using maturation ponds is practiced in a few cases in Morocco and Tunisia. The use of membranes for dissolved solids and pathogen removal through ultrafiltration and reverse osmosis has recently been put into operation in the Sulaibiyah treatment plant in Kuwait for unrestricted non-potable uses. Ozonation is also used as disinfection in Saudi Arabia and in Abu Dhabi, UAE.

Small communities or hotels in tourist areas along the beaches in Jordan, Egypt, Tunisia, Saudi Arabia, and other Gulf countries use small-scale technologies, especially highly compact facilities. These units are generally run by private owners. Oman has more than 250 small treatment plants with capacity ranging from 8-15,000 m^3/d (UNEP, 2001). Most of them are mechanical treatment plants with sludge recycling. Countries such as Jordan, Morocco and Palestine have implemented different small-scale units such as small-scale trickling filters, duckweed-based ponds, infiltration percolation systems, etc.

Waste stabilization ponds and aerated lagoons are widespread in countries such as Jordan and Tunisia. There is however a clear trend toward intensive treatment technologies when more attention should be given to properly designed treatment processes, as well as to decentralized treatment technologies, intended for reuse.

Wastewater treatment plants in countries with trained manpower and with mechanisms for cost recovery are producing effluents complying with their regulations; in other situations, deficiencies in operation and maintenance and financial constraints result in low quality effluent or non-operational plants. Operation, maintenance and delivery services in order to provide the users with a quality effluent appropriate for reuse need to be improved as they limit the development of water reuse operations.

Storage of reclaimed water is also an important treatment step in the provision of reclaimed water. It is important for peak-equalization and the possibility to bridge between periods, as demand for irrigation water is mainly during the dry season. Storing reclaimed water in reservoirs or aquifers would lead to more reliable supplies, water quality improvement, an increase in the rate of reuse and prevention of coastal waters contamination. The absence or insufficiency of storage is one of the limiting factors for water reuse development in Tunisia. Mixing reclaimed water with natural runoff in the King Talal Reservoir in Jordan is a common practice which contributes to the improvement of the irrigation water. Closed reservoirs have been built in the UAE and Kuwait with chlorine disinfection provided at the inlet and outlet of the storage step.

Different irrigation systems are used for the application of wastewater. Some countries however regulate or recommend the application system to be used depending on the crop to be irrigated. Sprinklers, for example, are not recommended for fruit trees and for vegetables in Kuwait.

2.3.4 Water reuse in the Middle Eastern and North African region

In most of the countries of the Middle Eastern and North African region, wastewater is widely reused to different extents within planned or unplanned systems. In many cases, raw wastewater is applied. In other cases, wastewater treatment plants are often not functioning or overloaded and thus they discharge insufficiently treated effluents not suitable for reuse applications. This leads to the existence of health risks and environment impacts and to the prevalence of water-related diseases. In some other situations where conditions for reuse are met, wastewater is then submitted to adequate reclamation systems and treated effluents are being reused for different purposes without presenting risks for human health. In these cases, reclaimed water is an important alternative resource for sustainable development and food production.

A wide variety of approaches to water reuse policy may therefore be found because of the difference in the capacity to implement such policies and depending on the socio-economic, institutional, and technological conditions. Differences between countries occur in their environmental and public health policies. They also occur in existing wastewater collection, treatment and disposal facilities, in human capacities, and in equipment, material, and financial resources (USEPA, 1992). Thus, countries such as Jordan, Tunisia and the Gulf states, have included water reuse in their water resources planning and have official policies calling for water reuse. Some have also issued water reuse strategies and started implementing them.

The main reuse operations in the region are for agricultural irrigation, landscape irrigation, and groundwater recharge (Table 2.5). Urban and landscape irrigation is given the first priority in the Gulf countries. Reuse of treated effluent for environmental purposes, such as artificial and constructed wetlands, and for industrial operations is very seldom practiced. There are several examples in the region of dual distribution system for landscape irrigation and other non-potable purposes and some for toilet flushing such as in Palestine, at Bir Zeit University where reclaimed water is substituted to potable water, or in Saudi Arabia where ablution water is recycled for toilet flushing at the two holy mosques in Mecca and Medina.

A large range of crops are irrigated with raw or treated effluent in the region. In Kuwait for example, the following crops are irrigated with tertiary treated effluent:

- fodder plants - alfalfa, Sudan grass, corn, vetch, barley, etc.;
- field crops - corn, barley, wheat and oats;
- fruit trees - date palms, olive, zyziphus and early salt-tolerant vines;
- vegetables - potatoes, dry onions, garlic, beet and turnip are irrigated by any method;
- vegetables to be cooked before consumption, such as egg plant, squash, pumpkin, cabbage, cauliflower, sweet corn, broad beans, Jews mallow, Swiss chard, etc.
- there are some limitations regarding strawberries and salad crops.

Most of the countries have regulations prohibiting the irrigation of vegetables with raw or treated effluent but they are often not enforced. It should also be noticed that several research and pilot studies have been conducted in the region. The information gained from such studies has allowed the development of treatment and reuse for the specific conditions of the region. The implementation of large-scale reuse schemes has resulted in significant technical and operational experience in reclaimed water reuse. But it remains a big challenge because of the specificities of the reclaimed water market and of the different issues that have to be taken into account. Water reuse market assessments are still lacking in the region when they should be carried out at the planning stage to identify cost-effective reuse opportunities and subsequent treatment processes. Planning and management of agricultural reuse operations imply institutional, organizational, legal, regulatory, socio-economic, financial, environmental, and technical constraints. Reuse projects are also complex and expensive operations with economic and financial uncertainties, which may be critical: some costs and benefits are sill difficult to quantify (impact on public health protection, economic development, etc.). A large diversity in planning approaches can therefore be found among the countries of the region and different management structures are set up depending on the type of agency: water-, wastewater-agency, coordination body, or other types. In any case and for all new wastewater treatment projects, a reuse component should be included whenever feasible, such as in Tunisia or Jordan.

Table 2.5 Application of wastewater use practices in the Middle Eastern and North African Region.

	Use of treated effluent for							
Country Practices	Un-restricted agricultural irrigation	Restricted agricultural irrigation	Land-scape irrigation	Ground-water recharge	Industrial reuse	Parks and roads	Toilet flushing	Use of untreated effluent
Afghanistan								✔
Algeria		✔						✔
Bahrain		✔	✔					
Egypt		✔	✔	✔				✔
Iran								✔
Iraq								✔
Jordan		✔						
Kuwait	✔		✔	✔				
Lebanon								✔
Libya		✔						
Morocco		✔						✔
Oman	✔		✔	✔	✔	✔		
Palestine		✔		✔			✔	✔
Qatar	✔		✔			✔		
Saudi Arabia	✔		✔		✔	✔	✔	
Syria		✔						✔
Tunisia		✔	✔	✔				
UAE	✔		✔					
Yemen		✔						✔

2.3.5 Acceptance of water reuse and public awareness

Use of wastewater is a widespread practice in the region. Public acceptance and the willingness to implement such projects are highly connected to the degree of water scarcity in the country and the degree of contact with reclaimed water. Farmers have switched to wastewater wherever conventional water sources became inaccessible or more costly to access, for example due to groundwater overdraft. They may, however, be reluctant to use reclaimed water because of crop restrictions preventing them from growing vegetables. They are not aware, in general, of the content of the wastewater they use for irrigation and of the eventual health and environmental impacts. They ask for more reliability of the water supply and for higher flow rates of water. Consumers are usually not aware that agricultural products are irrigated with untreated wastewater or with reclaimed water. Awareness campaigns and programs should be carried out to educate and orient farmers on the precautions of water reuse, and to inform consumers about the safety of agricultural products irrigated with well-managed reclaimed water.

There is therefore a need for more transparency, information sharing and involvement of the farmers and local communities in the decision-making process in water reuse projects. Water quality data must be widely available and freely shared with customers of water and the general public. Extension services should have particular training concerning water reuse. Public outreach and education programs to generate greater acceptance of reuse projects are an essential component of the planning of reclaimed water service. Awareness and educational training programs should also be extended to include schools curricula.

In order to address the religious concerns, *Fatwas* (i.e., legal statements issued by religious scholars on specific issues to settle questions where Islamic jurisprudence is unclear) have been issued in Saudi Arabia and in the UAE concerning the use of wastewater, in agreement with Islamic law and in order to achieve effective water demand management in the interests of the community and its natural resources. These *Fatwas* stated that wastewater may be used for irrigation provided that the impurities present in raw wastewater are removed.

2.3.6 The institutional framework

A large number of ministries and agencies, either governmental or private or both and at national, regional or local levels (ministries of agriculture, environment, water resources, public works, health, ministries of finance and economic planning, national water and sanitation utility, municipalities, and water users associations) are involved in water reuse operations. The institutional framework is therefore generally fragmented in particular between sanitation utilities and ministries of agriculture, with responsibilities overlapping, lack of well-defined responsibilities and insufficient coordination. There is an absence of national water reuse policies in most countries. There are a few examples of inter-ministerial committees at national and regional levels such as in Tunisia where the liaison and respective responsibilities of the concerned institutions are defined. Thus, there is a strong need for improving cooperation among stakeholders through inter-institutional coordination structures and various other mechanisms such as local level partnerships between stakeholders (for example between farmers and treatment plant operators), or private-public partnerships (O&M, BOT, etc.). In Tunisia, private companies are operating and maintaining wastewater treatment plants and providing effluent complying with the existing regulations. The Build Operate and Transfer system is under consideration in Tunisia and Jordan to produce appropriate quality effluent for reuse and to sell it to farmers or to other users such as golf courses. In Kuwait, a 30-year concession contract was signed by the government in May 2002 to a consortium to design, built, own, operate, and maintain the Sulaibiyah facility. In Oman, the government is encouraging the participation of the private sector in the construction and operation of wastewater collection and treatment facilities. A water reuse agency or an organization such as the Office of Water Recycling in California are among the institutional settings that may help the development, planning and management of water reuse operations.

2.3.7 The existing water reuse regulations

Several water reuse guidelines and manuals have been published (Chang *et al.*, 1995, 1998; Pettygrove and Asano, 1985; UNEP, 1991; USEPA, 1992 and 2004; WHO, 1989 and 2005). These guidelines have been supporting many countries of the Middle Eastern and North African region to implement or upgrade environmentally sound and safe wastewater reclamation and reuse systems adapted to their own technical, socio-economic and cultural conditions. Some countries have also implemented water reuse strategies and issued standards for pathogens, and organic and inorganic pollutants.

Most of the countries of the region have regulations or guidelines related to water reuse (Table 2.6). The first ones were issued in Kuwait in 1976. The standards vary from country to country, as do their derivation. The main differences are in their enforcement. Wastewater effluent discharge standards have generally been set up but some could not be met as they were too stringent and therefore inappropriate. Countries where reuse is developing on a rational basis, within an organized institutional setting, have elaborated and implemented their own regulations and precise standards. In other countries, reference is just made to health standards. Some of the significant discrepancies in the standards are, in part, due to differences in approaches to public health and environmental protection. For example, some countries have taken the approach of minimizing any risk, and have elaborated regulations close to California's Title 22 water recycling criteria (1978 and 2000), whereas the approach of other countries is essentially a reasonable anticipation of adverse effects resulting in the adoption of a set of water quality criteria based on the WHO guidelines (1989 and 2005). Table 2.7 shows significant different physical-chemical and microbial standards for reclaimed water reused for irrigation of food crops for human consumption. Some regulations include both treatment and water quality requirements. Except for some of the Gulf countries such as Kuwait, Saudi Arabia and the UAE, which refer to total coliforms, most of the countries with water reuse regulations have set limits for fecal coliforms or *Escherichia coli*. None include limits for viruses. Vegetables eaten uncooked are excluded from irrigation with treated effluent in most countries. However, the practice of using raw wastewater is widespread, as vegetables constitute cash crops for farmers. Regulations also differ in the general requirements, management practices, operational standards, frequency of monitoring requirements, etc. A variety of approaches have been taken by different agencies to regulate water quality for water reuse systems. These differences pertain mostly to the existing irrigation practices, local soil conditions, desire to protect public health, choice of irrigation or wastewater treatment technologies and needs to keep costs down.

Countries can, then, be classified into four categories according to the origin of the standards they apply, especially when dealing with reuse for irrigation purposes:

1. Those with no guidelines or standards specific to water reuse: Afghanistan, Iraq, Lebanon, and Libya.
2. Those with no specific regulations dealing with water reuse but with comprehensive guidelines for managing wastewater: Algeria, Egypt, and Syria.
3. Those which are referring to or have adopted a set of public health water quality criteria based on the WHO guidelines (1989): Iran, Morocco, Palestine, Tunisia, and Yemen.
4. Those which elaborated regulations close or equivalent to California's Title 22 water recycling criteria or the USEPA guidelines: Bahrain, Kuwait, Jordan, Oman, Qatar, Saudi Arabia and the United Arab Emirates.

Table 2.6. Regulations or guidelines related to water reuse in the Middle Eastern and North African region.

Country	Legislation / Guidelines
Afghanistan	There are no guidelines or standards related to water reuse.
Algeria	The Water Law in Algeria prohibits the use of raw and treated wastewater for the irrigation of vegetables to be eaten raw; but allows it for the production of fodder crops, pasture and trees. Authorization is required for the reuse of non-conventional waters. There are no reuse standards.
Bahrain	For unrestricted agricultural irrigation and landscape irrigation: tertiary treatment plus disinfection is required. The microbiological criteria are the following: <2.2 coliforms/100 mL, ≤1 helminth egg/L and periodical quality control tests are conducted.
Egypt	There are not yet specific regulations dealing with water reuse but comprehensive guidelines for managing reclaimed water, focusing on restricted irrigation, are under development. Water pollution from municipal and industrial effluents is mainly regulated by three laws: Law No. 48/1982 for the protection of the Nile River, Law No. 12/1984 on irrigation and drainage, Law No. 4/1994 on environmental protection. Law No. 48 (1982) requires secondary treatment of effluent prior to discharge into the Nile river and irrigation canals, and specifies limits for maximum allowable fecal coliform of 5,000/100 mL and BOD of 60 mg/L. The 1984 martial law regulation prohibits the use of treated effluent for irrigating vegetables and the use of effluent for crops unless treated to the required standards for agricultural drainage water. There are no reuse standards.
Iran	The "Effluent Discharge Standard" developed by the Department of the Environment in 1994 is the only existing wastewater code that sets effluent discharge limits to surface waters, cesspits, and agricultural irrigation. This standard does not define any criteria to use treated effluent for industrial, fisheries and recreational activities. International guidelines developed by the WHO and by the USEPA are currently used to regulate water reuse.
Iraq	Neither guidelines nor a code of practice for the use of wastewater in irrigation have been adopted so far.
Jordan	The previous Jordanian quality standard on reuse, JS893/1995, prohibited irrigation of vegetables eaten raw. It listed 47 specific constituents and prescribes quality standards on discharge to streams and catchment areas and on water reuse for each of the six following uses: irrigation of vegetables eaten cooked, irrigation of fruit trees, forests, industrial crops, and grains, artificial recharge of groundwater, use in aquaculture (fish hatcheries), irrigation of public parks, and irrigation of fodder. Fecal coliforms for irrigation of cooked vegetables had to be <1000 MPN/100 mL, BOD_5 <150 mg/L, and helminth egg ≤1/L. Necessary presence of residual chlorine in treated effluent was required. New standards have recently been approved and were enacted in 2003 under the title JS893/2003, in which the irrigation of vegetables eaten raw with reclaimed water remains prohibited whatever treatment is applied. *E. coli* should not exceed 100 count/100 mL for cooked vegetables. Groundwater recharge is not allowed where aquifers are used for drinking purposes. The helminth egg criterion has been maintained for all uses.
Kuwait	Effluent quality standards for water reuse require advanced level treatment with effluent's total coliforms count not exceeding 100/100 mL, and BOD and TSS <10 mg/L.
Lebanon	There are currently no guidelines or regulations for water reuse. However, Lebanese regulations prohibit the reuse of treated effluents for irrigation of fruits and vegetables. National environmental standards for discharge of treated effluents into surface water and sea have been established.
Libya	Under preparation.

Morocco	There are not yet any specific water reuse regulations. Reference is usually made to the WHO recommendations. Reuse standards are under preparation. Relevant regulations are Decree No 2-97-875/ 1998 and Water Law N° 10-95. Quality standards for water to be used for irrigation have been issued: No 1276-01 of October 17, 2002.
Oman	There is a General Water Resources Law - Royal Degree No. 83/88. Treatment is required and two main rules regulate water reuse: (1) water reuse, discharge and sludge disposal that include physical-chemical parameters, and (2) wastewater standards related to biological characteristics. Two categories of water reuse are distinguished: For Category A (which includes irrigation of vegetables and fruit to be eaten raw, landscape areas with public access, controlled aquifer recharge, and spray irrigation), effluent quality should be such that fecal coliforms are less than 200/100 mL, and helminth eggs less than 1/L. For Category B (which includes irrigation of cooked vegetables, fodder, cereals, and areas with no public access), the standard is <1000 fecal coliforms/100 mL, and <1 helminth egg/L.
Palestine	The draft Water Law and reuse standards are based on the WHO guidelines.
Qatar	Advanced treatment is required as well as strict standards.
Saudi Arabia	The Royal Order No. M/34 issued in 1980 forms the legislative framework for the protection of water resources including wastewater. National wastewater quality standards have since been approved. Advanced wastewater treatment is requested for unrestricted irrigation. Concentrations in the effluent should not exceed 2.2 count/100 mL for total coliforms, 10 mg/L for BOD and TSS, and 1 NTU for turbidity. Irrigation of any kind of vegetable is forbidden.
Syria	There are currently no guidelines or regulations for water reuse. Resolution No. 2823, dated 29/8/1990, issued by the Ministry of Agriculture excludes vegetable farming from irrigation with polluted water sources of all origins. Use of wastewater is restricted to fodder and industrial crops and fruit trees on small areas.
Tunisia	Water reuse is regulated by (1) the 1975 Water Law, (2) the 1989 decree (Decree no. 89-1047), and other standards: • Reclaimed water quality standards for reuse (Tunisian standard 106.03, 1989); • Wastewater disposal standards (Tunisian standard 106.002, 1989); • Code for crop selection and agricultural practices. Reclaimed water quality criteria for agricultural reuse were developed using the guidelines of FAO (1985) and WHO (1989) for restricted irrigation (less than 1 nematode egg/l), and other Tunisian standards related to irrigation or water supply.
United Arab Emirates	Advanced treatment is required as well as strict effluent standards, which are as follows: BOD_5 <10 mg/L, TSS <10 mg/L, Total coliforms <100/100 mL. Complete nitrification is requested. In Sharjah: total coliforms <23/100 mL and 1 mg/L residual chlorine as Cl_2.
Yemen	There are draft national water reuse standards proposing a uniform standard for all crops.

Table 2.7 Overview of some national standards for selected chemical and microbial parameters for reclaimed water reused for irrigation of food crops for human consumption.

Parameter	Unit	Kuwait	Jordan**	Oman		Saudi Arabia	Tunisia	UAE	California	USEPA	WHO
			2003	A*	B*	1989					
pH		-	6–9	6–9	6–9	6.0–8.1	6.5–8.5	-	-	6–9	-
BOD_5	mg/L	10	30	15	20	10	30	10	-	10	-
COD	mg/L	40	100	-	-	-	90	-	-	-	-
TSS	mg/L	10	50	15	30	10	30	10	-	-	-
Turbidity	NTU	-	-	-	-	<1	-	-	-	-	-
Chlorine residual	mg/L	1	0.5-1.0	-	-	-	-	-	Present	1	-
Total coliforms	MPN/100mL	<100	-	-	-	<2.2	-	<100	<2.2	-	-
Fecal coliforms	MPN/100mL	-	<100	<200	<1000	-	-	-	-	<200	<1000
Nematodes	eggs/L	-	-	-	-	-	< 1	-	-	-	≤ 1
Helminths	eggs/L	-	≤ 1	< 1	< 1	-	-	-	-	-	-

**: cooked vegetables.

A*: irrigation of vegetables and fruit to be eaten raw, landscape areas with public access, controlled aquifer recharge, and spray irrigation.

B*: irrigation of cooked vegetables, fodder, cereals, and areas with no public access.

2.3.8 Economics of water reuse

The economic value of reusing water, the public health and safety contribution, and the environmental benefits of planned utilization of reclaimed water, compared to discharge in the environment, are among the most important factors to be taken into consideration in determining the potential for water reuse.

Economic analysis, together with environmental assessments and sound technical studies which can contribute to the assessment of various options for wastewater treatment and reuse, are very seldom carried out in the region. Choices appear to be made almost exclusively on the basis of technical considerations and, to some degree, environmental concerns. This may have led in some cases to the adoption of high-cost alternatives. The full potential for water reuse may not have been realized in all cases, thus depriving farmers or other users from potential benefits. Therefore, economic analysis assessing the cost effectiveness and the financial feasibility of reuse operations should become better integrated at an early stage in the design of reuse policies and reuse projects. Cost-benefit analyses of reuse operations, which would assess the positive and negative aspects of water reuse, and which would take into account the different existing alternatives and indirect gains, are required. The different reuse opportunities should be equally assessed and the most cost-effective one selected. Assessment of cost differentials between treatment for reuse, water transfer, and treatment and discharge through sea-outfalls, as well as the cost of providing additional quantities of fresh water, should be made as they emphasize the economic viability of water reuse operations.

The marginal cost of providing additional good quality water of the same volume as that of the wastewater produced is generally higher than that of the wastewater. The cost of construction and operation of sewerage systems and treatment plants is high in the region. It varies from US\$1/m^3 for ordinary concrete pipelines and waste stabilization ponds to about US\$4-6/m^3 for good quality sewers and advanced treatment processes. Conventional gravity sewers constitute the major part (80-90%) of the total cost of wastewater facilities. The wastewater treatment cost varies from US\$0.10 to US\$0.50/m^3 depending on the size of the plant, the treatment and the extent of the process, but remains less than that of desalination. For example, in Bahrain, cost of tertiary treated effluent is about US\$0.32/m^3 while the cost of desalinated water is about US\$0.79/m^3.

2.3.9 Financing wastewater treatment and reuse

Wastewater reclamation and reuse projects require significant levels of funding for both capital and operation and maintenance costs. Therefore, at least O&M costs should be recovered through two types of tariffs. The O&M costs of sewerage and treatment up to standards for discharge into the environment should be recovered through sanitation tariffs, and the O&M costs of possible additional treatment, costs of storage and conveyance and distribution of reclaimed water to farms should be covered through irrigation tariffs. However, in most countries of the region, wastewater reclamation tariffs are far from covering O&M costs, resulting in poor performance of treatment plants. In addition to this, farmers generally do not pay for the use of reclaimed water or wastewater except in Jordan and Tunisia (World Bank, 2001). In Palestine, farmers have expressed their willingness to pay a small fee (Abdo, 2001). Golf courses or industries are expected to be willing to pay fees for treated effluent if the water quality is adequate and no alternative lower-cost sources of water are available. It is also generally recognized in the region that a tariff for irrigation with wastewater should be lower than the tariff for freshwater. A pricing policy is nevertheless needed. Payment of tariffs supposes a relationship of trust established between users, entities in charge of distribution of reclaimed water and plant operators, reliability of water services (quality and quantity), an appropriate tariff for alternative freshwater sources, no crop restrictions, and a functioning extension service.

Moreover and according to the World Bank, the nutrient value in treated effluents (N, P, and K) is about 3 US cents/m^3 which can save the farmer about US$120/ha.yr in fertilizer costs. But, farmers often do not take advantage of the nutrient content of the water and of potential cost savings in the application of mineral fertilizers; they often apply fertilizers in addition to the nutrients present in the effluent.

2.4 CONCLUSION

The amount of wastewater is rapidly growing with population growth, urbanization, and social and economic development of many of the countries of the region. Countries are unequally developed, several being already equipped with wastewater treatment plants while others have virtually no equipment. This, combined with the absence of the basic infrastructure in some cities or current poor facilities, represents a major challenge for the region. Disposal of untreated sewage from main cities to water courses poses serious health risks because it affects downstream uses: e.g. drinking water supplies and irrigation. Countries need to upgrade existing unplanned or semi-planned reuse schemes, and to make water reuse an integrated component of new sanitation projects. They need to improve sewage and treatment facilities, develop wastewater master plans for cities and urban areas, which will establish targets for providing wastewater collection systems, treatment facilities, and reuse operations. Wastewater treatment and reuse will provide additional water resources, also ensuring health protection and enhanced environmental conditions.

Several challenges have therefore still to be overcome in terms of wastewater treatment and reuse. Technological, scientific, institutional, or legal aspects of collection, treatment, and reuse of urban wastewater are partially mastered and consequently require particular attention. Where uncontrolled direct or indirect reuse of raw wastewater is currently occurring, water quality should be improved so as to prevent health hazards and environmental problems. Intermediate steps should be undertaken to mitigate the negative impacts, such as introducing crop restrictions and standards for effluent reused for irrigation and other uses, applying source control of contaminants, use of appropriate irrigation, agricultural, harvest and public health practices that limit risks, improving extension and outreach activities to all stakeholders, and upgrading the effluent quality from treatment plants. The medium-term goal should be prohibition of all irrigation use of untreated wastewater. Where crop restriction is applied, upgrading wastewater treatment plants should be promoted for more economical pollution abatements, for unrestricted use with wider crop rotation, and higher valorization and payback of the investments made in the agricultural sector. Local arrangements with farmers' associations specifying mutual rights and responsibilities should be negotiated. Source control of contaminants needs to be enforced. Reclaimed water uses should be expanded. Tariffs for reclaimed water have to be introduced, and public awareness campaigns directed at the general public should be conducted.

Most of the countries of the region have regulations or guidelines related to pollution control and water reuse and there is a wide variety of regulations and standards. The main differences are in their enforcement. Experiences gained in different countries will help develop a common approach to wastewater treatment and reuse based on the documentation and evaluation of existing practices, and establish risk-based legislation governing the reuse of treated effluent in agriculture and for other uses.

Further development of water reuse calls for development of technologies and innovative water reclamation systems well adapted to the socio-economic conditions. In order to protect water resources and ensure safer reuse operations, the links between water resources, drinking water, wastewater and reclaimed water have to be better investigated. Involvement and participation of all stakeholders from the early stages of reclamation and reuse projects design, up to the implementation stage, are key factors for any water reuse project's success and for gaining public acceptance. A survey recently conducted by the Global Water Research Coalition (Crook *et al.*, 2005) identified six key overarching factors to the success of future development of water reuse: public trust; pricing and economics; public health and environmental protection; guidelines and regulations; planning, management, and applications; and improved technologies and methods.

Institutional strengthening, with the establishment of national committees, is required for a better planning and coordination among the different bodies concerned. Future reuse projects will depend on a better planning and management of reuse operations based on real water demand. This means a better institutional, regulatory, and organizational setting, as well as a better assessment of economic and financial feasibility of reclaimed water reuse applications. Technical aspects also need further study, along with applied research for specific applications. Education, information, and training of farmers and extension services also play an important role in promoting these practices aiming to achieve higher agricultural production without adverse impacts on the environment.

The implementation of an integrated water resources management strategy based on the elaboration and enforcement of water quality standards and policy guidelines for wastewater reclamation and reuse, the selection and application of appropriate wastewater treatment technology, human resources training, sensitizing and public awareness creation, and regional cooperation strengthening may solve the problems of water scarcity and wastewater management in the region.

2.5 REFERENCES

Abdo, K.M. (2001) Water Reuse in Palestine. Ministry of Agriculture, General Directorate of Soil & Irrigation, Ramallah, Palestine, 22 p., Paper presented at the Regional Workshop on Water Reuse in the Middle East and North Africa, organized by the World Bank's MENA Regional Water Initiative in Cairo, Egypt, July 2-5, 2001.

Al-Zubari, W. (1997) Towards the establishment of a total water cycle management and re-use program in the GCC countries. Paper presented at the Regional Workshop on the Technologies on Wastewater Treatment and Reuse, Bahrain, 2-4 June.

Angelakis, A.N., Marecos do Monte, M.H., Bontoux, L. and Asano, T. (1999) The status of wastewater reuse practice in the Mediterranean Basin. *Wat. Res.*, **33**(10), 2201-2217.

Angelakis, A.N. and L. Bontoux (2001). Wastewater Reclamation and Reuse in Eureau Countries. Water Policy 3: 47-59.

Asano, T. (1998). Wastewater Reclamation and Reuse. Water Quality Management Library Vol. 10. Technomic Publishing Inc., Lancaster, Pennsylvania, USA.

Blumenthal, U.J., Mara, D.D., Peasey, A., Ruiz-Palacios, G. and Scott, R. (2000) Guidelines for the microbiological quality of treated wastewater used in agriculture: recommendations for revising WHO guidelines. *Bulletin of the WHO* **78** (9), 1104-1116.

Chang A.C., Page A.L. and Asano T. (1995) Developing human health-related chemical guidelines for reclaimed water and sewage sludge applications in agriculture, World Health Organization, Geneva, Switzerland.

Chang, C., Page, A.L. and Asano, T. (1998) Evaluating methods of establishing human health-related chemical guidelines or cropland application of municipal wastewater. In Asano T. ed. *Wastewater Reclamation and Reuse*. Lancaster, PA, Technomic Publishing, 581-626.

Crook, J. Mosher, J. and Casteline, J.M. (2005) Status and Role of Water Reuse: An International View, Report prepared for the Global Water Research Coalition, March 2005, 84 p. + Appendixes.

Crites, R. and Tchobanoglous, G. (1998) Small and Decentralized Wastewater Management Systems; WCB/McGraw-Hill: New York, 1998.

FAO (2000) Aquastat-FAO's information system on water and agriculture, http://www.fao.org/ag/agl/aglw/aquastat/dbase/index.stm

FAO/RNE and WHO/EMRO (2003) Proceedings of the expert consultation for launching the regional network on wastewater re-use in the Near East. TC/D/Y5186E/12.03/100. Food and Agriculture Organization of the United Nations. Regional Office for the Near East World Health Organization. Regional Office for the Eastern Mediterranean. 138 p. + appendixes. Cairo, 2003.

Lazarova, V. and Bahri, A. Ed. (2005) Water reuse for irrigation: Agriculture, Landscapes, and Turf Grass, Catalog no. 1649, ISBN: I-56670-649-I, CRC PRESS, 456 p.

Mara, D. and Cairncross, S. (1989). Guidelines for the safe use of wastewater and excreta in agriculture and aquaculture, Measures for public health protection, WHO, Geneva, Switzerland.

Margat, J. and Vallée, D. (2000) Mediterranean vision on water, population and the environment for the 21st century. Blue Plan. Mediterranean Action Plan/UNEP, GWP/MEDTAC, Montpellier, France.

Pettygrove, G.S., and Asano, T., (ed.). (1985) Irrigation with reclaimed municipal wastewater - A guidance manual. Lewis Publishers, Inc., Chelsea, Mississipi.

Shuval, H., Adin, A., Fattal, B., Rawitz, E. and Yekutiel P. (1986). Wastewater irrigation in developing countries: Health effects and technical solutions. World Bank Technical paper 51, The World Bank, Washington, D.C.

State of California. (1978). Wastewater Reclamation Criteria, An Excerpt from the California Code of Regulations, Title 22, Division 4, Environmental Health, Dept. of Health Services, Sacramento, California.

State of California (2000). Title 22, Code of regulations, 24 p. November 2000.

United Nations Environment Program (1991) Environmental guidelines for municipal water reuse in the Mediterranean Region. Mediterranean Action Plan, Priority Actions Program, United Nations Environment Program in cooperation with Food and Agriculture Organization of the United Nations, PAP-3/G.1.

United Nations Environment Program (1999) Overview on land-based sources and activities affecting the marine environment in the ROPME Sea Area. United Nations Environment Program, Regional Seas Reports and Studies No. 168.

United Nations Environment Program (2001) Overview of the socio-economic aspects related to the management of municipal wastewater in West Asia (including all countries bordering the Red Sea and Gulf of Aden). UNEP/PERSGA/ROPME Workshop on Municipal Wastewater Management in West Asia Bahrain, 10-12 November 2001. UNEP/ROWA-GPA SEWAGE.RW.1/5 2, October 2001. 59 pp.

U.S. Environmental Protection Agency (1992) Guidelines for water reuse. EPA/625/R-92/004, U. S. Environmental Protection Agency, Washington D.C. 247 pp.

U.S. Environmental Protection Agency (2004) Guidelines for water reuse. EPA/625/R-04/108, U. S. Environmental Protection Agency, Washington D.C. 286 pp. + Appendixes.

World Bank (2001) Regional Water Initiative: Water reuse in the Middle East and North Africa, Proceedings, Regional Workshop on Water Reuse in the Middle East and North Africa, Sponsored by the World Bank and the Swiss Agency for Development and Cooperation, July 2-5, 2001, Cairo, Egypt, The World Bank, Washington, D. C.

World Health Organization (1989) Health guidelines for the use of wastewater in agriculture and aquaculture, Report of a WHO Scientific Group. WHO Technical report series 778. World Health Organization, Geneva, Switzerland.

World Health Organization (2005) WHO guidelines for the safe use of wastewater in agriculture, Draft Report of the World Health Organization, Geneva, Switzerland.

3

Water reuse in the Northern Mediterranean Region

Marcelo Juanicó and Miquel Salgot

3.1 INTRODUCTION

Northern Mediterranean countries fulfil all the conditions to perform massive water reuse:

- water scarcity;
- increasing water demand;
- seasonal water demand;
- large quantities of sewage treated to relatively high standards;
- a policy for protection of the environment requiring decreasing discharge of effluents to rivers and sea;
- the technological, institutional and financial resources to develop massive water reuse under controlled conditions.

Thus, the potential for reuse of treated wastewater in the region is huge. Most countries are in different stages of developing official policies on the issue, some of them with specific plans for the next decade. Spain and Italy are expected to have the most conspicuous developments in the short-term. The adoption of long-awaited regulations for reuse in the urban sector will boost urban reuse; thus, agriculture is not expected to be the sole consumer of treated effluents in the future. In-factory water recycling will probably continue to increase in the industrial sector. Groundwater recharge could also become important, especially against seawater intrusion.

©2008 IWA Publishing. *Water Reuse - An International Survey of current practice, issues and needs* by B. Jiménez *et al.* ISBN: 9781843390893. Published by IWA Publishing, London, UK.

3.2 GENERAL SITUATION

3.2.1 Climate

Temperate-rainy winters and hot-dry summers characterize the Northern Mediterranean basin; agriculture, parks, gardens and artificial landscapes require summer irrigation. Precipitations are above 800-1000 mm/y in the Northern areas of European countries, dropping to below 300-500 in the Southern areas of Portugal, Spain, Italy, Greece and Israel. Precipitations have a high annual variation with frequent droughts in the driest areas.

3.2.2 Water resources

Water scarcity is a constraint shared by most countries in the region. Some countries have water resources below the chronic water scarcity level estimated at 1500 m^3/*capita*.yr. Others (*e.g.* Greece, Turkey, France, Italy and Portugal) have enough water resources at a national level, but long distances and/or topography create serious regional or local problems of water scarcity. Finally, some countries (*e.g.* Cyprus) seem to have enough water only on theoretical terms while actual exploitation capacity is much lower. Water availability is also threatened by the influx of more than 150 million tourists to the Mediterranean coast each summer. The situation will worsen even more in the future due to population growth, further tourism development and increasing living standards. Spain, Malta, Cyprus and Israel have lately launched massive seawater desalination programs and other countries will probably implement similar plans.

One approach to evaluate water scarcity is the WSI – Water Stress Index: the ratio between the annual water withdrawals and the available renewable water resources. It is supposed that when WSI exceeds 20% water management starts to become stressed, complicated and expensive. According to Figure 3.1, half of the Northern Mediterranean countries are suffering water stress at national level (and most of them have specific areas with high WSI).

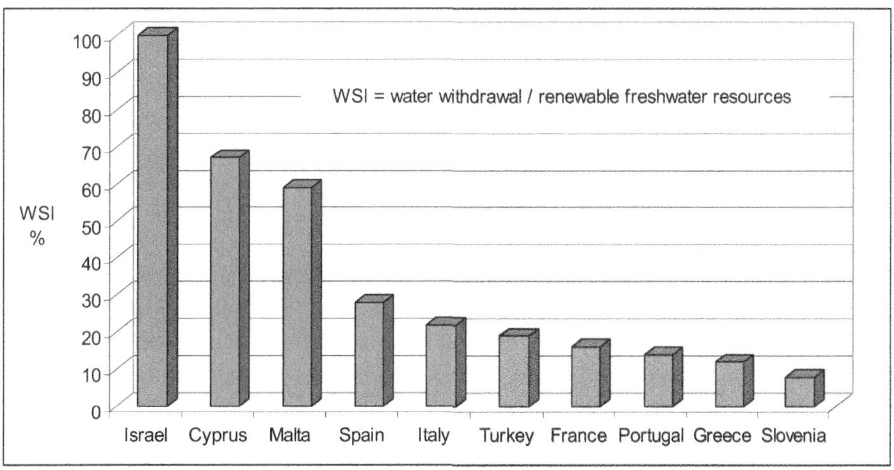

Figure 3.1. Water stress index at national level in Northern Mediterranean countries (differences within a country may lead to local problems of water stress). A WSI above 20% points out the need for comprehensive water management efforts. (Source: Modified from AQUAREC (2005) plus updates and additions).

Another approach to evaluate water scarcity is to divide renewable water resources by the population. It is supposed that a value of 1500 m^3/*capita*.yr indicates a level of chronic water scarcity. Table 3.1 indicates that, according to this ratio, at least three countries (Cyprus, Israel and Malta) suffer from chronic water scarcity. Again, the calculations at the national level hide water scarcity problems at specific locations within the countries.

3.2.3 Regulations

Cyprus, Israel and Italy have regulations following the "California school", while Andalusia, Catalonia and Balearic Islands have regulations following the "WHO school". Albania, ex-Yugoslavia countries, Greece, Malta, Portugal, Spain and Turkey have no regulations yet. In Cyprus, Portugal and Spain, regulations are being developed or reviewed. New guidelines for the whole Mediterranean basin have been drafted recently (see Chapter 30) with the aim of overcoming the discrepancies between regulations in the region; the draft is presently being discussed. So far, there are no supra-national regulations on water reuse in Europe.

3.2.4 Wastewater Reuse practice: today

In most Northern Mediterranean countries, 70-80% of water demand goes to agricultural and landscape irrigation. Thus, wastewater agricultural irrigation became a common practice even without appropriate treatment. With the exception of Albania and Ex-Yugoslavia countries, wastewater is reused to different extents all over the region. However, only Cyprus and Israel have developed an official policy defining treated wastewater as an integral part of their water resources. Much effort has been made in reporting large projects above 5 Mm^3/yr, but thousands of unreported small projects exist below 0.1 Mm^3/y (many of them artisanal) generating an important social, economic and environmental impact. Urban and golf courses irrigation is also an increasing practice. Industrial in-factory water recycling is growing due to increasing costs for supplied water and wastewater treatment; some industrial parks have reduced their water consumption by a half during the last decade.

3.2.5 Wastewater Reuse practice: future perspectives

The potential for reuse of treated wastewater in the region is huge. Most countries are in different stages of developing official policies on the issue, some of them with specific plans for the next decade. Spain and Italy are expected to have the most conspicuous developments at short-term. The adoption of long-waited regulations for urban reuse will boost the reuse of treated wastewater in this sector; thus, agriculture is not expected to be the almost sole consumer of treated effluents in the future.

However, wastewater reuse implementation is facing obstacles and delays mainly due to lack of clear policies defining treated wastewater as part of the water resources of the countries, legal and institutional frameworks for the implementation of reuse projects, proper training to farmers and information to public, and multi-national standards. A problem observed in some countries (*e.g.*, France, Italy and Spain) is that while water scarcity and potential for water reuse are much more conspicuous in the Southern areas of the countries (*e.g.*, Provence, Sicilia, Andalusia) the water reuse policy and regulations are developed, approved and controlled by a central administration located far away (Paris, Rome, Madrid) and generally disconnected from the scenarios where the policy and regulations will apply.

Table 3.1 Water resources and reuse in Northern Mediterranean countries.

Country	Area *** [km^2]	Renewable freshwater resources [Mm^3 /yr *]	Freshwater availability				Wastewater Reuse	
			2003		2025 (1994 UN forecasting)		2002 ± 3 [Mm^3 /yr]	Amounts planned to reach during next decade [Mm^3/yr]
			Population [millions]	Availability [m^3/*capita*.yr] (sorting column)	Population [millions]	Availability [m^3/*capita*.yr]		
Malta	320	30	0.40	75	0.42	70	4	9
Israel	20,300	2,000	6.10	330	7.80	260	280	380
Cyprus	9,250	(300) 900	0.77	(390) 1,170	0.93	(320?) 890	2.5?	30
Italy	300,000	85,000	57.00	1,500	52.00	1,600	45	250
Level of chronic water scarcity				1,500				
Spain	500,000	111,000	40.22	2,760	37.57	3,500	(?) 300	1200
Turkey	771,000	203,000 110,000 feasible	68.11	2,980 1740 feasible	90.90	1,910	50	no data found
Albania	27,400	21,000	3.580	5,870	4.67	3,990	negligible	no plans set
France	544,000	150,000	44.80	3,300	45.00	3,300	6-7	no plans set
Greece	131,000	69,000	10.67	6,470	9.97	6,970	>10	>15
Portugal	92,000	66,000	10.10	6,530	9.68	7,220	not clear	>20
Ex-Yugoslavia**	255,000	265,000	22.5	11,800	24.60	10,840	negligible	no plans set
Total	**2,320,400**	**884,000**	**199.950**		**222.39**		**>700**	**~ 2000**

* Mm^3: million cubic meters. It does not include sea water desalination or water imports.

** No specific data available yet on the new countries formed from Ex-Yugoslavia.

*** Land area, does not include territorial sea waters.

Water stress in the Northern Mediterranean imposes a whole family of innovations in the water sector. Water reuse is only one of the elements of this family:

- maximum exploitation of conventional water resources;
- strict water saving policy and practice
- in-factory water recycling;
- exploitation of non-conventional water resources: e.g. brackish and fossil waters;
- full integration of treated effluents to the water resources;
- desalination of sea and brackish water;
- allocation of water resources for the recovery and preservation of natural aquatic habitats: rivers, lakes, wetlands, etc.

The lack of coordination between the different agencies that traditionally regulate the water sector is a common problem, both vertically (different stages of the water cycle) and horizontally (health, agriculture, industry, municipalities, environment, etc.). It seems that the old structure of the water sector is not made to match the challenges of the sophisticated management required by the above quoted family of innovations.

3.3 COUNTRY INFORMATION

3.3.1 Italy

Water reuse in Italy has been recently reviewed by Barbagallo *et al.* (in press). The summary presented here is based on that work.

Water resources

Although water resources are large in theory, the usable amount of water is estimated to be limited to 85,000 Mm^3/yr while existing infrastructure allows the use of only 55,000 Mm^3/yr (Barbagallo *et al.*, *op. cit.*). This figure locates Italy as a whole in the limit of chronic water scarcity: 1500 m^3/*capita*.yr. Precipitations are evenly distributed in space and according to population (Table 3.2) but Southern Italy and main islands suffer much serious water scarcity due to their warmer and drier climate, unevenly distribution of the precipitations during the year (short rainy winters) and frequent droughts.

Table 3.2. Distribution of area, population, precipitations and water resources in Italy (data from Barbagallo *et al. op. cit.*).

	Area	Population	Precipitations	Usable water resources
North	40%	45%	40%	53%
Central	24%	22%	24%	40%
South	19%	22%	24%	
Main islands	17%	11%	12%	7%

Management of water resources

Water resources are public property and regulated by the state that owns most water distribution systems. Water for agricultural irrigation is usually not metered and farmers pay water according to the irrigated surface (*e.g.* 100€/ha.year in Consorzio Villoresi – Northern Italy). It is planned that treated effluents for irrigation will be metered and paid by m^3.

The water sector was fragmented into thousands of different poorly coordinated agencies causing inadequacy, inefficiency and high costs. The "Galli" law of 1994 "erased" the municipal limits and unified the fragmented water sector into *Optimum Territorial Areas*, but this reform is still being implemented.

Sewage treatment

Most existing sewage treatment plants are based on activated sludge and some new ones are based on submerged aerobic filters. About 10-15% of the existing plants perform "tertiary" treatment, *i.e.* nutrient removal, according to EU urban wastewater Directive. About 80% of the treatment plants discharge to rivers and 3-4% to sea and lakes. The remaining wastewater is discharged to undefined places (that may include reuse).

Regulations

A new set of standards for water reuse in irrigation was promulgated on June 12th, 2003 (Ministry Decree, D.M. n. 185/03) (Table 3.3). Barbagallo *et al.* (*op. cit.*) criticize the standard noting that it addresses 54 parameters, 11 of them with the same values required by the drinking water standard, 20 of them not addressed by the drinking water standard, "*and some other parameters (for instance biocides and pesticides) are difficult to be explained in an agricultural environment*".

Table 3.3 Quality requirements for wastewater irrigation in Italy, approved in 2003.

Parameter	standard	Parameter	standard
pH	6.0-9.5	Vanadium [mg V/L]	0.1
SAR - Sodium Ads. Ratio	10.0	Zinc [mg Zn/L]	0.5
Coarse solids	absent	Cyanides [mg CN/L]	0.05
TSS [mg/L]	10.0	Sulphides [mg H_2S/L]	0.5
BOD_5 [mg/L]	20.0	Sulphites [mg SO_3/L]	0.5
COD [mg/L]	100.0	Sulphates [mg SO_4/L]	500
Phosphorus [mg P/L] (total)	2.0	Chlorine residual [mg/L]	0.2
Total Nitrogen [mg N/L]	15.0	Chlorides [mg Cl/L]	250
Ammonia [mg NH_4/L]	2.0	Fluorides [mg F/L]	1.5
EC_W [□S/cm]	3,000	Oils & fats [mg/L]	10.0
Aluminium [mg Al/L]	1.0	Mineral oils [mg/L]	0.05
Arsenic [mg As/L]	0.02	Phenols [mg/L] (total)	0,1
Barium [mg Ba/L]	10.0	Pentachlorophenol [mg/L]	0.003
Boron [mg B/L]	1.0	Aldehydes(total) [mg/L]	0.5
Cadmium [mg Cd/L]	0.005	Tetra/tricloro-ethylene [mg/L]	0.01
Cobalt [mg Co/L]	0.05	Chlorinated solvents [mg/L]	0.04
Chromium [mg Cr/L] (total)	0.1	Total THM [mg/L]	0.03
Chrom. hexavalent [mg Cr/L)	0.005	Aromatic solvents [mg/L]	0.001
Iron [mg Fe/L]	2.0	Benzene [mg/L]	0.01
Manganese [mg Mn/L]	0.2	Benzo(a)pyrene [mg/L]	0.00001
Mercury [mg Hg/L]	0.001	Organic N solvents [mg/L]	0.01
Nickel [mg Ni/L]	0.2	Surfactants [mg/L] (total)	0.5
Lead [mg Pb/L]	0.1	Chlorinated biocides [mg/l]	0.0001
Copper [mg Cu/L]	1.0	Any phosphorated pesticide [mg/l]	0.00001
Selenium [mg Se/L]	0.01	Other pesticides [mg/L] (total)	0.05
Tin [mg Sn/l]	3.0	E. coli [CFU/100mL] (80% of samples)	10
		Constructed wetlands	50
		Stabilisation ponds	100
Thallium [mg Tl/L]	0.001	Salmonellae [CFU/100mL]	absent

Reuse practice

Table 3.4. Main reuse projects in Italy (data from Barbagallo *et al., op. cit.*).

Place	Project
Valle d'Aosta	In construction. Landscape irrigation and fire protection. 150,000 hab. equivalent
Alto Adige	Planned. 2,500 hab. equivalent
Friuli-Venezia	In construction. Industrial reuse.
Lombardia	In construction. Industrial reuse.
Veneto	Planned. Industrial reuse: 50,000 m^3/d Irrigation: 6000 m^3/d
Emilia-Romagna	Running. Orchads irrigation (400 ha)
Toscana, Prato	Running. Industrial reuse: 15,000 m^3/d
Toscana, Piombino	Running. Industrial reuse: 10,000 m^3/d
Toscana, Pistoia	Pilot. Irrigation.
Campania, Sarno	Planned. Irrigation.
Puglia	Starting by 2004-2005 Irrigation: 250,000 m^3/d
Sicilia, Grammichele	Running. Citrus irrigation: 1,550 m^3/d.
Sicilia, San Michele	Running. Olive trees irrigation: 1000 m^3/d
Sicilia, Palermo and Gela	In construction. Irrigation: 30,000 m^3/d
Siciclia, Catania	Feasibility. Irrigation: 110,000 m^3/d
Sardegna, Is Arenas	Running. Irrigation: 95,000 m^3/d
Sardegna, Villasimius	In construction. Irrigation: 3000 m^3/d

Table 3.5 Representative reuse projects for irrigation in southern Italy (1998 survey, data taken from Barbagallo et al. op. cit.). C: citrus trees; Ch: cherry trees; Fr: fruit orchards; F: fodder; G: greenhouses; O: olive trees; P: pasture; V: vineyards; Vg: vegetables.

Region	Site	Wastewater [Mm^3/y]	Irrigable area (Ha)	Main crops	Construction cost in 10^3 €
SICILIA	Caltagirone	2.5	500	C	1,945
	Catania	47.9	9,600	C	3,113
	Enna	1.5	290	O V	0,840
	Mineo	0.5	100	C O	1,421
	S.M.di Ganzaria	0.3	150	O	930
	Vittoria	6.1	1,500	C G	5,188
SARDEGNA	Dorgali	0.57	162	V O	1,348
	Rio Posada	0.66	270	F V	4,791
	Sassari	29.1	7,000	Various	16,006
	Tortoli	9.5	2,000	Various	2,971
PUGLIA	Conversano	2	477	O Fr	13,489
	Lecce	11.1	2,450	W P	20,753
	Martina Franca	5.8	2,086	O Ch F V	10,579
	Putignano	3.9	1,000	Vg	8,892
	Tricase	2.2	300	O Ch F Fr P	4,929

3.3.2 France

Water resources

Waste resources are estimated as 3,300 m^3/*capita*.yr (World Resources 2000-2001), well above the stress level. Nevertheless, during the last ten years there has been a 20% increase in water demand (mainly for agriculture and resort areas) coupled with a series of drought years. In more than one-third of the country, water tables are falling as the autumn and winter rains are no longer making up for the amounts drawn off in spring and summer, a situation that forced authorities to impose restrictions to water use (Angelakis *et al.*, 2000; USEPA, 2004).

Sewage treatment

France treats about 4000 Mm^3/yr of sewage per year. There are more than 10 MBR-Membrane Biological reactor projects where industrial wastewater is treated for reuse (Angelakis *et al.*, 2005). Disinfection is accomplished either by chlorine or UV, with a current trend towards UV. Several research projects and demonstration studies are underway addressing treatment technologies, water quality, and integrated water management. These studies include research on pharmaceuticals and personal care products in reclaimed water used for potable purposes, cost-effective technologies for agricultural reuse, MBR treatment, sustainable water management, and desalination of groundwater (Crook *et al.*, 2005).

Regulations

The country's regulatory framework (Circular no. 51 of July 22, 1991, of the Ministry of Health) is based on the WHO guidelines (1989), but it is more stringent having additional requirements concerning irrigation management, timing, distance and other measures for preventing health risks related to human exposure and negative environmental impacts. For example, in the case of spray irrigation, a 100 m distance must be respected beyond the reach of the spraying system for residences, sport and leisure areas and motorway toll gates. In addition to the above microbiological standards, the document requires: (a) guarantees that the aquifer will not be contaminated; (b) knowledge of the treated wastewater effluent quality and fertilising capacity; and (c) trained operation and control personnel (Lazarova *et al.*, 2000, USEPA, 2004).

A government decree of 1994 provides the basis for water reuse rules in France. First, it clearly states that treated effluents can be used for agricultural purposes only if this is conducted without any risk for the environment or the public. Second, wastewater treatment requirements, irrigation modalities, and monitoring programs must be defined according to recommendations from the Conseil Supérieur d'Hygiene Publique de France (CSHPF), and the Inter-ministerial Water Mission through ordinance of the Ministry of Health, the Ministry of Environment, and the Ministry of Agriculture. Third, the approval of a wastewater reuse project still depends on the approval of the local representative of the Ministry of Health (Lazarova *et al.*, 2000).

These rules and regulations should allow wastewater reuse projects to become part of France water resource management policy. Yet, local French sanitary authorities impose very strict controls on wastewater reuse projects. The water quality required for these projects is often more stringent than those required by the regulations approved by the CSHPF, and even more stringent than California Title 22 in the case of some urban applications (grey water recycling, for example). In addition, the heavy administrative procedures required for the approval of any water reuse project has slowed down the emergence of water reuse in France (Lazarova *et al.*, 2000).

New water reuse guidelines are under preparation. They may introduce some new parameters for unrestricted irrigation (i.e. *Salmonella*, *Taenia* eggs and viruses).

Reuse practice

France has irrigated crops with wastewater for more than a century, in particular around Paris. Until 1940, it was the only method of treating and disposing wastewater of the Greater Paris conurbation. This practice is still going on in the Achères region, where some of the wastewater is used after screening and settling, but is likely to be discontinued soon.

Interest in water reuse rose again in the early 1990s for several reasons:

- the development of intensive irrigated farming (such as maize), in particular in south-western France and the Paris region;
- the fall of water tables after several recent severe droughts which have paradoxically affected the regions traditionally considered to be the wettest (western and north-western France);
- pollution control in recreational and shellfish farming areas along the Atlantic coast; and
- industrial in-factory reuse by MBR technology allow for less water consumption and effluent discharge (more than 10 projects in France).

Potable unplanned indirect reuse is common in France, where surface water diluted with wastewater is used for potable supply. An example is Aubergenville, in the Paris region, where the Seine River, which is 25% wastewater effluent, is treated and used to recharge the drinking water aquifer.

Presently, there are at least 30 water reuse projects in France, half of which use reclaimed water for agricultural irrigation covering more than 3000 ha; the other 15 projects use reclaimed water for golf courses and urban area irrigation (Angelakis *et al.*, 2003, 2005, USEPA, 2004).

The Clermont-Ferrand recycling scheme implemented in 1999, where 10,000 m^3/d of effluents treated by activated sludge followed by maturation ponds are used for irrigation of over 700 ha of maize, is today considered to be one of the largest projects in Europe.

One of the first examples in France of integrated water management with water reuse is on Noirmoutier Island. The lack of water resources, the 10-fold increase in tourist population during the summer and intensive agricultural activities required water reuse. Wastewater treatment on the island is achieved through two treatment plants with a total capacity of 6,100 m^3/d. The plants have activated sludge systems followed by maturation ponds for storage and disinfection. Thirty percent of the treated wastewater (0.33 Mm^3/yr) is used for the irrigation of 500 ha of vegetable crops. There are plans to reuse 100% of the wastewater flow in the near future (USEPA, 2004).

3.3.3 Spain

Spain can be divided into humid and arid areas. The humid Atlantic basins do not have water scarcity problems while the Mediterranean basins, mainly arid or semi-arid, are suffering structural or occasional water scarcity. The main archipelagos (Balearic and Canary islands) also suffer from water scarcity.

In the Mediterranean basins, periodic droughts or excess demand are unbalancing the water distribution among agriculture, industry, ecology, recreational activities and urban needs. Because the development of the country has been linked to an important extent to tourism activities, including different aspects such as mass and golf tourism, water demand has increased accordingly. Peak demands for agriculture, tourism and leisure coincide with the dry season, the summertime, when a strong pressure is exerted on existing resources. The problem has been managed until now by overexploiting groundwater resources or fully diverting surface waters. Several water transport infrastructures have been created, increasing water holding capacity through dams or reallocating resources diverting rivers into neighbouring basins.

Nevertheless, such solutions reached a limit during the 1990s, because of the lack of resources to be diverted, increasing demands (resident population increase, tourism, agriculture, and industry), heavy droughts and bad distribution systems. Several islands needed to develop new resources, desalting sea and brackish water using reverse osmosis or electrodialysis. Numerous attempts at developing a National Hydrologic Plan did not succeed and at present, some solutions based on river flows diversion, which include national rivers and the extension of the Rhône river water carrier from Montpellier (France) to Barcelona, have been discarded while an increase of seawater desalination, reclaimed water reuse and other minor issues are under consideration and planning.

Planned wastewater recycling and reuse is now being publicized as being part of the possible solution, although translation into common practice has still not been applied. Unplanned reuse used to be a classical solution for arid and semi-arid areas all around the Spanish Mediterranean coastline. There are several causes for this illegal practice. Perhaps, one of the most important is the lack of reuse regulation at state level. In Spain, the government issued 14 years ago one Law and one Decree where wastewater reuse was indicated as a possibility and a minimal statement appeared indicating the need for an administrative concession and a compulsory report of the Health Authorities. An indication was made that further legal developments would be needed.

Raw sewage has been used for agricultural irrigation, but the efforts to develop wastewater treatment according to EU rules are reducing the amount of available raw sewage and increasing the amount of treated wastewater to be disposed of. Many rivers on the Mediterranean coastline have no running water during the summer (due to the lack of rain during these months) except for treated or untreated wastewater. If wastewater is reclaimed and reused, some of the rivers will become dry. This is leading "ecological" (green) movements to ask for wastewater not to be diverted for irrigation, a request that will lead to problems when trying to reuse wastewater in the inner agricultural areas of the country. Conversely, the majority opinion is that wastewater reuse all along the coastlines is a good solution and needs to be promoted (Salgot, 2002).

Spain is presently reusing about 300 Mm^3/y of wastewater. The main approved uses are agriculture and golf courses irrigation, and secondarily groundwater recharge and industrial reuse. There is also strong pressure for discharging treated wastewater into rivers, but after advanced treatment in order to improve running water quality. Recycling systems are today relying on tertiary classical technologies (coagulation-filtration plus disinfection) and extensive natural technologies (wetlands, lagooning and infiltration-percolation). The most used disinfection technologies are UV and chlorination. The areas where recycling and reuse are most operative at present are Balearic Islands (golf courses, urban parks and groundwater recharge), Canary Islands (golf courses and agriculture) and the entire Mediterranean coastline (agriculture, golf courses and leisure activities other than golf), and Vitoria in the Basque country where municipal wastewater is reclaimed and reused for agricultural purposes.

The water suppliers become increasingly concerned with the business opportunities that the reclamation and reuse of wastewater are opening. Several companies and administrations support research and development (R&D) activities in collaboration with the universities, and a number of experimental recycling facilities are already operative. AEAS (the Spanish chapter of EUREAU, the association of water and wastewater treatment companies) is also supporting innovation on theoretical and control tools to improve reclaimed water quality. University research groups on wastewater reuse are spread all around the Mediterranean basin and are mainly working on wastewater tertiary treatments (advanced treatments in Catalonia, natural treatments in Andalusia, Catalonia, Murcia, and Valencia) and rules and regulations in Catalonia and Andalusia.

There is a bright future for wastewater reuse in Spain, but at present it is compromised owing to the fact that while new projects are proposed, a lot of difficulties arise due to the lack of a more complete legal definition. In Spain there is a strong tendency to decentralize administration and give more power to the "Autonomous Governments" (regional governments). The decisions and permissions for wastewater reuse are given now in a case per case basis, depending on the Regional Administrations.

Since it is difficult to get such approvals without having definite legal health regulations, several Regional Health Authorities have decided to develop their own guidelines for wastewater reuse for irrigation. By 2005, three guidelines (Balearic Islands, Catalonia, and Andalusia) were operative. Draft guidelines for the Spanish national regulation were proposed in 1996, taking an approach more similar to the California standards than to the WHO guidelines. However, this draft was never approved and the present trend is not to adapt the Californian criteria (Salgot, 2002).

A new "White Book" on water was prepared and published in 1998, incorporating wastewater reuse into the recognized available water resources. On that basis, a group of experts jointed by the Ministry of Environment elaborated a proposal of minimal criteria (physical-chemical and microbiological) for wastewater reuse, criteria that follow the WHO guidelines (Angelakis *et al.*, 2001). It was submitted to the government for approval. By 2005 these minimal criteria had not been approved and a revision was undertaken to adapt the draft to new findings in wastewater reclamation and reuse. The main idea at present is to define five water qualities for different uses.

3.3.4 Israel

Israel is included as a case study in Chapter 25, thus the following lines are only an abstract.

Israel has performed massive reuse of effluents for agricultural irrigation since the early 1970s and is presently reusing almost 75% of all the sewage produced in the country. Sewage is defined by law as an integral part of the water resources of the country. The numerous reuse projects in the country are not similar, with differences in size, technology, irrigated crops, etc. This diversity is positive. Part of the success of the wastewater reuse practice in Israel is due to the capacity of the well-organized and informed farmers to adapt quickly to the switch from water to wastewater. Early regulations adopted an approach of "different qualities for different crops" with a single poor level of environmental protection. Recently approved (2005) new regulations switched to a single high standard for "unrestricted irrigation" and address sustainable environmental protection. Salination is one of the sustainability related problems of long-term massive water reuse. The country has implemented several measures to reduce the concentration of salts and boron in municipal sewage.

3.3.5 Cyprus

Cyprus has developed an official policy defining treated wastewater as an integral part of their water resources.

Water resources

The average annual precipitation, including snowfall, amounts to approximately 500 mm, but during the past 30 years (1973-2003) this amount was reduced to 480 mm (Cyprus Water Development Program, 2005).

Precipitation increases from the south-western windward slopes to the top of the central massif, from less than 450 mm to nearly 1100 mm. On the leeward slopes precipitations decrease to 300-350 mm. The potential annual water resources have been estimated at 900 Mm^3/yr, of which 600 Mm^3/yr are surface water and 300 Mm^3/yr are groundwater. However, the water amounts that are presently available are less than 300 Mm^3/yr (Medaware, 2004; Cyprus Water Development Program, 2005).

Table 3.6. Cyprus - Annual water demand by sector for the year 2000 (Source: Medaware, 2004).

Demand Sector	**Mm^3/y**	**%**
Agriculture	182.4	69
Domestic	67.5	25
Industry	3.5	1
Environment	12.5	5
Total	265.9	100

Cyprus has no rivers with perennial flow while rainfall is highly variable and droughts occur frequently. Since groundwater is reliable, clean and most importantly cheap, water resources development in Cyprus initially focused on groundwater, and until 1970 groundwater was the main source of water for both drinking and irrigation purposes. As a result almost all aquifers were seriously depleted because of overpumping and seawater intrusion was observed in most of the coastal aquifers. At the same time large quantities of surface water were lost to the sea.

In order to face the situation, desalination units have been constructed since 1997 aiming at rendering the water supply of the major residential and tourist centres independent of rainfall. Desalination units at present contribute up to 33.5 Mm^3 per year. The restrictions on water consumption were lifted by 2001. (Medaware 2004; Cyprus Water Development Program, 2005).

Reuse of treated sewage effluent amounts to only about 3 Mm^3 per year, from which 2 Mm^3/y is used for agriculture and the rest for landscape irrigation (Medaware 2004).

Management of water resources

All water resources including wastewater belong to the state, but the laws give the right to private individuals to construct wells for ground water abstraction after getting a permit from the District Officer. The laws also give the right to individuals to form Irrigation Divisions or Associations to construct irrigation works, and to villages and towns to form their own Commissions for constructing their own waterworks for domestic supply, and their own Sewage and Drainage Boards for the collection, treatment and disposal of sewage effluents. Presently, about 60% of the annual amount of water for irrigation purposes is provided from Government Irrigation Schemes and the sources of irrigation water are surface water, groundwater and reclaimed water. The water demand in the non-government schemes is satisfied mainly by groundwater.

The laws, except in the cases of Irrigation Divisions, Association, Commission and Sewage Boards, do not mention the administrative authority which keeps the water resources inventory, or evaluates, or allocates and controls the use of water. This seems to be a basic setback of the water legislation (Medaware, 2004).

There are three main institutions involved in the production, treatment and reuse of wastewater. The Water Development Department, which is responsible for the implementation of water policy, is responsible for tertiary treatment as well as allocation and distribution to the farm level. The Department of Agriculture is responsible for the training of farmers in all matters related to agricultural production with the use of treated wastewater. It has also the mandate of monitoring the code of practice at farm level. The sewerage boards have the responsibility of operation and maintenance of the main sewer systems (pipes, pumping stations and treatment plants) (Bazza, 2003).

Sewage treatment

The number of Sewage Treatment Plants (STPs) currently in operation is around 30 with a total capacity of about 20 Mm^3/year (Table 3.7). These STPs cover the four major greater urban areas and some large tourist centers (Lefkosia, Lemesos, Larnaka, Pafos and Agia Napa–Paralimni) and part of the rural areas. Centralized sewage networks now serve 12% of the rural population. There are also a large number of small STPs in the hotels. In addition there exists a programme for building such units in 28 large rural centers (having a population of more than 2000) and in sensitive mountain villages by the year 2012. This is in accord with the instructions/principles of the Urban Wastewater Directive 91/271/EEC of the European Union (EU). It is worth noting that all plants have provisions for tertiary treatment (Medaware, 2004).

Table 3.7. Cyprus - Sewage Effluent Treatment Plants (Source: Medaware, 2004).

Name	WW treated m^3/Year	Treatment	Use
Lefkosia Sewage Board	3,650,000	Secondary	Diverted to Pedieos River
Anthoupolis-Lefcosia	127,750	Secondary	Stored in open Reservoir for evaporation
Larnaca Sewage Board	912500	Tertiary	Landscape Irrigation
Agia Napa – Paralimini	2,500,000	Tertiary	Landscape-Forest
Lemesos Sewage Board	3,000,000	Tertiary	Agriculture-Landscape of Hotels
Pafos Sewage Board	4,895,000	Tertiary	Agriculture
Bathia Gonia	803,000	Tertiary	Agriculture
Dhali-Nisou	182,500	Tertiary	Agriculture
Platres	73,000	Tertiary	Not operating-Agriculture
Carlsberg	146,000	Tertiary	Agriculture
Lefkosia New Hospital	182,500	Tertiary	Not operating-Landscape
Lemesos Hospital	47,450	Tertiary	Landscape
Alassa (new village site)	18,250	Tertiary	Agriculture
Palechori	73,000	Tertiary	Diverted to the River
Apostolos Loucas	25,550	Secondary	Used by The Agr. Res. Instit.
Kofinou	65,700	Secondary	Agriculture
Zenon-Kamares II	109,500	Secondary	Landscape irrigation
Agglisides	365,000	Secondary	Agriculture
Kornos	25,550	Tertiary	Landscape Irrigation
Stavrovouni	25,550	Tertiary	Landscape Irrigation
Agios Ioannis	17,900	Tertiary	Landscape Irrigation
Malounda	7,300	Tertiary	Landscape Irrigation
Klirou	26,300	Tertiary	Landscape Irrigation
Kyperounda	109,500	Tertiary	Agriculture
Troodos	8,800		Landscape
TOTAL Maximum	**19,829,850**		

Regulations

Cyprus regulations regarding wastewater quality for irrigation follow the California school although with some small differences. *De facto*, most STPs are using intensive processes with tertiary treatment.

Reuse practice

Recycled domestic water is presently used for the watering of football fields, parks, hotel gardens, landscape, road islands and forestation (1.5 Mm^3/yr) and for agricultural irrigation, mainly permanent crops (Citrus Olives Vines) and fodder (3.5 Mm^3/yr). It is estimated that by the year 2012 an amount of approximately 30 Mm^3 of treated sewage effluent will be available for agriculture and landscape irrigation (Medaware, 2004).

Cyprus is facing two major obstacles in its continued development: 1) a growing scarcity of water resources in the semi-arid regions of the country; and 2) degradation of water at its beaches. The government has recognized that a water reuse program would address both problems. In addition, it is expected that reclaimed water will provide a reliable alternative resource for irrigation, which draws about 80% of the total water demand. The government has recently launched a program of implementation of new sewerage, wastewater treatment and reuse of treated wastewater in two major tourist areas; Limassol in the southern coast and Larnaca and Ayia Napa-Paralimni on the south-eastern coast.

It is expected that with the completion of these and other central collection and treatment plants in the cities and villages by the year 2012, the volume of treated wastewater will reach 25-30 Mm^3 per year. The reclaimed water will be collected and used for irrigation after tertiary treatment. Since transmission costs will be high, most of the reclaimed water (about 55-60%) will most likely be used for urban and tourist resorts irrigation. A reclaimed water supply of about 10 Mm^3/year is conservatively estimated to be available for agricultural irrigation (Bazza, 2003; USEPA, 2004).

Treatment plants in villages are considered as a basic instrument for agricultural, social and environmental policy of the country. The government covers 75-100% of the cost of construction and operation of tertiary treatment plants, as well as the cost of treated wastewater distribution to farmers (Bazza, 2003).

Saline soils in Cyprus are limited. However secondary salinization due to the use of semi-saline water for irrigation is common particularly under greenhouse production. To overcome the problem, leaching is practiced during the growing season and more substantial leaching before planting (Medaware, 2004).

3.3.6 Turkey

All the information on Turkey has been summarized from Medaware (2004).

Water resources and demand

Although Turkey is situated in a Mediterranean geographical location where climatic conditions are quite temperate, the diverse nature of landscape and in particular the existence of the mountains that run parallel to the coasts result in drastic regional differences in climatic conditions. While the coastal areas bear relatively mild climates, the inland Anatolian plateau has a continental climate with hot, dry summers and long lasting, cold winters with limited rainfall.

The average annual precipitation is 643 mm. Annual precipitation in the Aegean and Mediterranean coasts varies from 580 to 1300 mm, depending on location. The Black Sea coast receives the highest amount of rainfall (up to 2200 mm) and its eastern part is the only region of Turkey that receives rainfall throughout the whole year. The central dry area has precipitations of less than 400 mm per year.

Nominal renewable water resources have been estimated in 200,000–230,000 Mm^3 per year, but the technically and economically usable surface and ground water potential would be limited to 110,000 Mm^3.

Present population is 78 million inhabitants and total annual water withdrawal is 42,000 Mm^3 for the whole country (2000 data) (see Table 3.8). Turkey has enough water at the national level, but the large territory and irregular distribution of rains geographically and during the year create numerous problems of local water scarcity.

Table 3.8. Water demand in Turkey by sectors.

Year	Demand (Mm^3)	Sectoral Consumption		
		Domestic (Mm^3)	Irrigation (Mm^3)	Industrial (Mm^3)
1990	30,600	5,141	22,016	3,443
1992	31,600	5,195	22,939	3,466
1995	33,500	5,300	24,700	3,500
2000	42,000	6,400	31,500	4,100
2002	38,900	5,700	29,200	4,000

The SHW (General Directorate of State Hydraulic Works) has constructed a total area of 2,296,350 ha for irrigation, of which 10% is operated by SHW itself and 70% by other end-user organizations (irrigation associations, village authorities, municipalities, etc.). The remaining 20% is not actually operational due to different problems.

Sewage treatment

Until the late 1990s, the National Bank of Provinces has been in charge of funding wastewater treatment plants within the framework of annual investment programmes according to instructions of the related municipalities. Those constructed plants were then transferred to the municipalities for operation. However, today many associations (such as the Greater Metropolitan Municipalities, Water and Sewerage Administrations, The Ministry of Tourism, Southeast Anatolian Project (SAP) Administration, The General Directorate of Special Protection Areas, and General Directorate of Massive Housing) deal with the investment of wastewater treatment plants. The Greater Metropolitan Municipalities, especially those with high urban populations, prefer to solve their wastewater treatment problems by utilizing foreign credits and by managing the investment period.

Between the years 1970-1980, 11 wastewater treatments plants were installed and operated by the Bank of Provinces (BoP). This number increased in the following years: 60 plants in 1980-1990, 82 plants by the end of 1998, and 118 plants by the end of 2001.

By 2001, the 118 wastewater treatment plants in operation had a total annual capacity of 2550 Mm^3/year, but the actual amount of treated effluents was only 1245 Mm^3/year, just less than 50% of the total treatment capacity; 38% (468 Mm^3/year) of the treated effluents receive physical treatment, whereas 50% (618 Mm^3/year) receive biological treatment and 12% (159 Mm^3/year) advanced treatment.

The most common wastewater treatment type is the extended aeration of activated sludge system. Package units are popular in villages with a population < 1000. Secondary effluents are usually disinfected by the addition of chlorine.

Regulations

Turkey has not yet approved regulations on water reuse.

Reuse practice

Irrigation with wastewater has been estimated at 50 Mm^3/y in 2000. But the country does not have a reuse policy and agricultural irrigation with wastewater is not conducted offically.

Wastewater agricultural irrigation is widespread in South East Anatolia, mainly for the irrigation of vegetables. For example, in Siverek, located in South East Anatolia, domestic wastewater discharged into streams is being reused for agricultural applications on cotton, wheat and various vegetables including eggplants, peppers, tomatos, cabbage, carrots and spinach. The total area irrigated with wastewater is 165 ha and the consumption of irrigation water was 1.9 Mm^3 in 2001.

Tourist villages and resorts, especially those along the Aegean and the Mediterranean coasts of the country, are building their own treatment facilities. Shortage of water resources and increased water demand by tourism in summer months forced these tourism centres to reuse their effluents. The major form of reuse is irrigation of gardens and parks.

Because of insufficient sewerage facilities and lack of satisfactory treatment, an enormous amount of domestic wastewater has been discharged into rivers. This discharged wastewater is indirectly used for irrigation. For example, in Trakya, located in North-West Anatolia, 1,560,620 m^3 per year of domestic wastewater are discharged into the river Evros and 9,000 ha of agricultural area are irrigated with water extracted from the polluted river.

3.3.7 Greece

Water resources

Greece has a population of about 11 million inhabitants and renewable freshwater resources are evaluated at about 70,000 Mm^3 per year. Thus, water resources are abundant at the national level and the water stress index is very low. However, Greek geography is "complicated" with hundreds of islands, isolated peninsulas and valleys separated by a mountainous topography. Water scarcity at the local level is a common problem in numerous places within the country.

Water demand has conspicuously increased during the last 50 years due to development, increase in living standards and massive tourism.

Sewage treatment

About 65% of the population is connected to 350 centralised WWTP with a total capacity of over 1.45 Mm^3 per day (~ 530 Mm^3 per year) (Kamizoulis, 2003).

Regulations

No guidelines or criteria for wastewater reclamation and reuse have yet been adopted. A preliminary study is underway on the need for criteria for the reuse of treated wastewater. In this study, six basic categories of reuse (non-potable urban, agriculture, aquaculture, industrial, environmental, and groundwater recharge) are considered (Angelakis *et al.*, 2000; Kamizoulis, 2003).

Reuse practice

More than 80% of the treated effluents in Greece are produced in regions with deficient water balance. Thus, wastewater reuse in these areas would satisfy a real water demand. The distribution of treated domestic wastewater effluents in deficient water balance regions, as a function of the average distance from the agricultural land which is available for irrigation, have been analyzed: about 90% of the treated effluents are discharged from WWTPs, which are located at a distance of less than 5 km from the available farmland. Therefore, the additional cost for transport to the irrigation fields is low (Angelakis *et al.*, 2000).

Currently, only few small reuse projects are actually running, most of them at a pilot level, but over 15 wastewater treatment plants are planning to reuse their effluents for agricultural irrigation. The major ones are listed in Table 3.9 (Kamizoulis *et al.*, 2003).

Industrial reuse is negligible.

Table 3.9. WWTPs planning reuse in Greece (from Kamizoulis et al., 2003).

Plant Name	m3/day	Uses
Levadia	3,500	Irrigation of cotton
Amfisa	400	Olive tree irrigation
Palecastro	280	Storage, olive tree Irrigation
Chalkida	13,000	Landscape and Forestry irrigation
Karistos	1,450	Landscape and Forestry irrigation
Ierisos	1,200	Landscape and Forestry irrigation
Agios Konstantinos	200	Landscape and Forestry irrigation

3.3.8 Malta

Water resources

Malta has suffered from acute water stress for decades. Scarcity of water was already highlighted in a report by Knights of St. John in 1524 "...except for few springs in the middle of the Island, there was no running water nor even wells, the want of which the inhabitants supplied by cisterns". A 16th century Knights decree required that "...all houses were to have wells for collection of rainwater from roof-tops," (Pizzuto, 2003).

The islands have a total area of almost 320 km^2 and a population of almost 400,000 inhabitants. Additionally, Malta receives more than 1 million tourists per year. The renewable water resources (30 Mm^3 per year of groundwater) hardly cover half of the water demand. Desalination of sea water by reverse osmosis closes the negative water balance.

Sewage treatment

Due to low water consumption per inhabitant, the raw sewage in Malta is strong (BOD_5=530 mg/L and SS=445 mg/L) and has a high salinity (sodium and chloride) due to high levels of these ions in the domestic water supply.

The main sewage treatment plant (Sant Antnin) had a current capacity of 13,000 m^3/d of effluent, but it was increased to 26,000 m^3/d in the mid 1990s. The plant uses an activated sludge process followed by rapid sand filters (9 m3/m^2·h). The effluent is then disinfected with gaseous chlorine (20 mg/L and contact time 30 mins) and pumped into irrigation reservoirs with a free chlorine residual of 2 mg/L (Kamizoulis, 2003; Pizzuto, 2003).

Another three smaller sewage treatment plants are planned by the government (Pizzuto, 2003).

Regulations

There are not yet any regulations for water reuse.

Reuse practice

Because agriculture is the main source of income and lack of water the main constraint for agriculture, wastewater irrigation has been practiced since the late 17th Century (Angelakis *et al.*, 1999).

Since 1983, the effluent of the Sant Antnin sewage treatment plant has been used for irrigation. The effluent is used to irrigate 600 ha of crops by furrow and spray irrigation. The effluent quality is suitable for unrestricted irrigation and is used to produce potatoes, tomatoes, broad and runner beans, green pepper, cabbages, cauliflower, lettuce, strawberries, clover, etc.

Despite the high salinity, there are no problems with crops. This is probably associated with high permeability of the calcareous soil. Soil monitoring has shown a salt accumulation in the top soil during the irrigation season followed by leaching to the groundwater with the winter rains (Kamizoulis, 2003).

Industrial reuse is limited to an industrial laundry.

3.3.9 Portugal

Water resources

Portugal has an almost Mediterranean climate with more than 70% of the annual rainfall falling during half of the year.

Renewable water resources are about 66,000 Mm^3 per year, while water demand is only 12,000 Mm^3 per year. With a population of about 10 million inhabitants, 80% of water demand is for agricultural irrigation (FAO-AQUASTAT, 2003).

Sewage treatment

According to a survey performed in 2000, sewerage coverage is only 55%, with only 36% of sewage receiving treatment (USEPA, 2004), but this situation is presently changing quickly in order to comply with the requirements of the EU directives.

Regulations

Draft regulations are presently being discussed for approval (Marecos do Monte, personal communication, 2004).

Reuse practice

A survey carried out in 1998 assessed the actual practice of water reuse for irrigation in Portugal and showed (Marecos de Monte, personal communication, 2004) that:

- 35 water reuse systems were identified over 27 out of 305 municipalities;
- most of the cases were of municipal wastewater reuse. Industrial wastewater was reused for irrigation only in 5 cases;
- 5 cases out of 35 reused untreated wastewater;
- spray irrigation was the most used irrigation method, followed by drip irrigation, furrow irrigation and truck tanks;
- landscape irrigation was the main application;
- other applications were forage crops, vegetables, fruit trees and sport fields (golf courses and football grounds).

A new large WWTP (460,000 inhabitants) near Lisbon plans to irrigate 1000 ha with tertiary treated wastewater. But at least 3 large projects in the area of Lisbon (Beirolas, Frielas, Costa do Estoril) are suffering from delays in their implementation (Marecos do Monte, personal communication, 2004).

Treated wastewater is a potential source of water for irrigation and should soon reach 580 Mm^3/yr, which is approximately twice as much as today. Even without storage, this amount could be enough to cover about 10% of the water needs for irrigation in a dry year. The use of treated wastewater for irrigation could significantly contribute to the agricultural development in the driest Portuguese provinces (Beja, Evora, Setubal, Lisboa and Santarem). Roughly, between 35,000 and 100,000 ha, depending on storage capacity, could be irrigated with recycled water. Interest is also growing for the irrigation of golf courses (Angelakis, 2003), which is now implemented in the south of the country on a significant scale.

An important research project on waste water reuse for irrigation was carried out by the Laboratorio Nacional de Engenharia Civil (LNEC) and the Laboratorio Quimico Agricola Rebelo da Silva (LQARS). The main objectives of the project were to assess and compare the effects of irrigation of various types of treated urban waste water versus the same crops irrigated with potable water and given commercial fertilizers, in order to provide experimental data to support the production of Portuguese guidelines for waste water reuse for irrigation.

The main conclusions of the study were the following: primary and secondary effluents were found to be suitable for well drained soils and salinity-tolerant crops but unsuitable for sensitive crops; the nitrogen content of both effluents seem to enable the avoidance of the use of commercial fertilizers, as identical yields were obtained for the three treatments (primary and secondary effluents and control potable water); the facultative pond effluent appears to be of higher fertilizing capacity than primary and secondary effluents, since increased yields were obtained comparatively to the water irrigated crops (Angelakis *et al.*, 2000).

Acknowledgements

The preparation of this chapter was possible thanks to the valuable collaboration of IWA colleagues around the Mediterranean Basin: Maria H. Marecos do Monte, Portugal; Giuseppe Cirelli and Salvatore Barbagallo, Italy; Lluis Sala, Spain; C. A. Kambanellas and Despo Fata, Cyprus; and Anton Pizzuto, Malta.

3.4 REFERENCES

Angelakis A., L. Bontoux, and V. Lazarova (2003) Challenges and prospectives for water recycling and reuse in EU countries. *Water Science and Technology: Water Supply* **3**(4): 59–68.

Angelakis, A., B. Durham, M.H.F. Marecos do Monte, M. Salgot, T. Wintgens, C. Thoeye, and T. Peitchev. 2005. Water Recycling and reuse in Eureau Countries: With Emphasis on Criteria Used. *EU1/2-05-WR-26(1),* draft report dated 16 February 2005.

Angelakis A., M. Marecos do Monte, and T. Asano (2000) Wastewater Reuse Practice in the Mediterranean Region. *In: Advanced Wastewater Treatment and Reuse. Design, Operation and Maintenance. Training material.* EW Leonardo da Vinci Programme, Chapter 9: pp.1-37.

Barbagallo S., Cirelli G. and Nurizzo C. (in press) Wastewater Reclamation and Reuse in Italy.

Bazza, M. (2003) Wastewater recycling and reuse in the Near East Region: experience and issues. *Water Science and Technology: Water Supply* 3(4): 33–50.

Crook J.; Mosher J and Casteine J (2005) – Status and Role of Water Reuse: an International View. *Report to the Global Water Research Coalition*, pre-publication version 3/2005, 133 pp.

Cyprus Water Development Program (2005) Official Website at http://www.moa.gov.cy/wdd/

Fatta D. and I. Skoula (2004) Evaluation of existing situation related to the operation of urban wastewater treatment plants and the effluent disposal practices with emphasis on the reuse in the agricultural production. *In: Medaware program: Development of Tools and Guidelines for the*

Promotion of the Sustainable Urban Wastewater Treatment and Reuse in the Agricultural Production in the Mediterranean Countries. Task 2, report March 2004, 126 pp.

Feigin, A., Ravina, I. and Shalhevet, J. (1991) Irrigation with Treated Sewage Effluent. Adv. Ser. Agr. Sci. 17, Springer-Verlag, Berlin.

Friedler, E. and Juanicó, M. (1996) Treatment and storage of wastewater for agricultural irrigation. *Water Irrig. Review* **16**(4), 26-30.

Ganoulis J (2003) Evaluating alternative strategies for wastewater recycling and reuse in the Mediterranean area. *Water Science and Technology: Water Supply* 3(4):11–19

Juanicó, M. and Dor, I. (Eds.) (1999) Reservoirs for Wastewater Storage and Reuse: Ecology, Performance and Engineering Design. Springer-Verlag, Environmental Science Series, Berlin, 394 pp.

Kamizoulis, G.; Bahri A., Brissaud F and Angelakis A. (2003) Wastewater Recycling and Reuse Practices in the Mediterranean Region: recommended guidelines. *MED-REUNET, Case Studies. www.med-reunet.com*

Medaware (2004-Drafts) Development of Tools and Guidelines for the Promotion of the Sustainable Urban Wastewater Treatment and Reuse in the Agricultural Production in the Mediterranean Countries. EU Programe Medaware, Reports Task 1 A to F, and Task 2. Drafts kindly provided by Dr. Despo Fatta from the University of Cyprus.

Nurizzo C. (2003) Reclaimed water reuse in the Mediterranean region: some considerations on water resources, standards and bacterial re-growth phenomena. *Water Sci. Technol.* 3(4):317-324.

Pizzuto A. (2003) Case Study – Sewage Treatment Plant (Malta). Presentation to the Worshop on Integrated management of water resources, Izmir, Turkey, Sep. 2003.

Sala, L. and X. Millet (1997) Aspectos básicos de la reutilización de las aguas residuales regeneradas para el riego de campos de golf. *Consorci Costa Brava Publishers*, 126 pp.

Tsagarakis, K.P., P. Tsoumanis, D. Mara and A.N. Angelakis (2000) Wastewater Treatment and Reuse in Greece: Related Problems and Prospectives. *IWA, Biennial International Conference,* Paris, France.

USEPA (2004) Guidelines for water reuse. *Report EPA/625/R-04/108*, 450 pp.

Weber, B. and Juanicó, M. (2004) Salt reduction in municipal sewage allocated for reuse: the outcome of a new policy in Israel. *Wat. Sci. Technol.* **50**(2):17–22.

4

Water reuse in the United States and Canada

Kirsten Exall, Blanca Jiménez, Jiri Marsalek and Karl Schaefer

4.1 UNITED STATES

According to the World Resources Institute (WRI), in 2007 water availability in the United States was 9,215 meters cubed (m^3) per capita per year. Water availability can also be measured using an index of the degree to which the resource has already been developed. The water intensity use index, as it is known, is derived by dividing the average regional consumptive use by the renewable water supply, which is the sum of precipitation and imports of water, less the water lost through natural evapotranspiration and exports. The water intensity use index for the United States in 1995 was 17.1%. This varied from approximately 1% in New England to as much as 100% in Lower Colorado (Figure 4.1).

The total amount of water used, without considering thermoelectric power withdrawals, was 1 billion m^3/d or 213 billion gallons per day. Agriculture used nearly two-thirds of this total (64%), followed by public supply (20%) and self-supplied industry i.e. industry not taking its water from the municipal network (9%). The remaining 6% was used by self supply domestic users, aquaculture, livestock and mining. Of the total amount of water used, approximately 15% came from saline water sources. California, Texas and Florida accounted for one quarter of the total national water withdrawals. Combined withdrawals from California, Idaho, Colorado and Nebraska represented half of the total water used for irrigation. California, Texas, New York, Florida and Illinois accounted for 40% of the total used in public supply (US Geological Survey, 2007).

 Water Reuse - An International Survey of current practice, issues and needs by B. Jiménez *et al.* ISBN: 9781843390893. Published by IWA Publishing, London, UK.

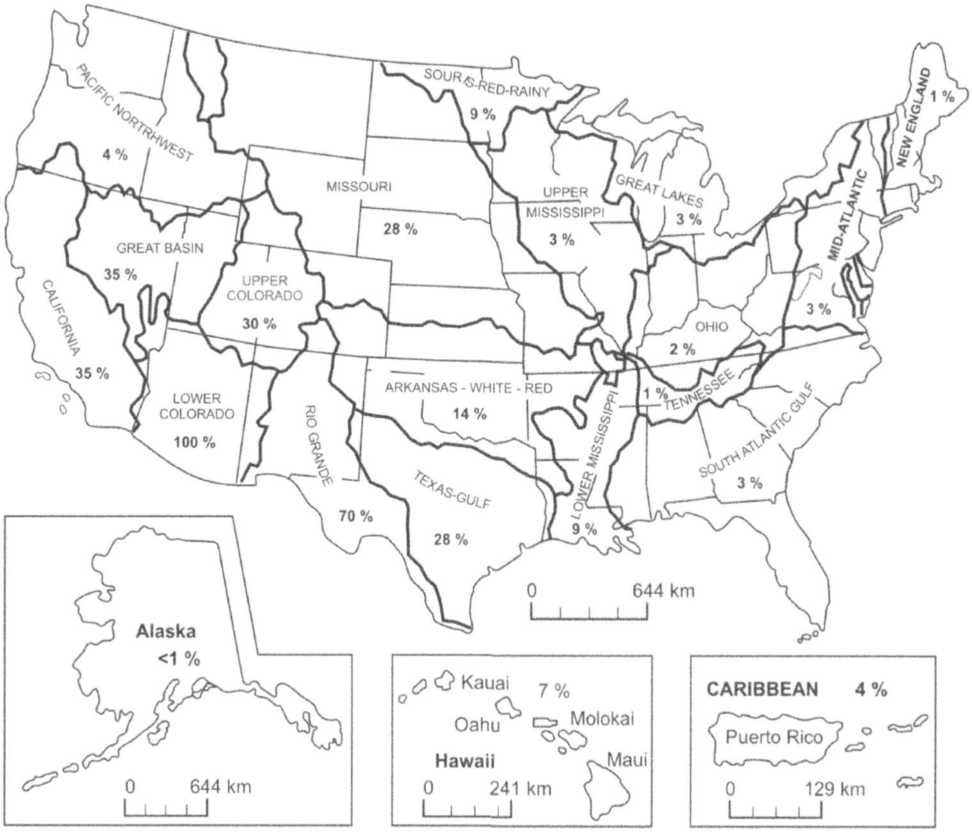

Figure 4.1 Water intensity use in the US basins, with data from Asano *et al.*, 2007.

4.1.1 History and evolution of water reuse

The earliest reference concerning water reuse in the United States appears in a California State bulletin from 1906. The bulletin discusses the water quality requirements for a septic tank effluent to be used for irrigation, in order to take advantage of its fertilizing properties (Asano *et al.*, 2007). One of the first projects to be reported concerning industrial reuse was the use of a chlorinated wastewater effluent for the steel processing industry in Baltimore, Maryland in 1942. Urban water reuse began in the early 1960s in California, Colorado and Florida. California is considered to be the pioneer state in promoting water reclamation and reuse, being the first to regulate it for agricultural irrigation in 1918. Table 4.1 lists several projects that opened up the reuse field in the US.

Table 4.1 Selected historical examples and milestones of water reuse in the US, with information from Asano *et al.*, 2007; USEPA, 2004 and Florida DEP, 2007.

Year	Location	Type of Project
1906	California	Reuse of water for irrigation (first reference in literature).
1912	Golden Gate Park, San Francisco, California	Watering of lawns and ornamental lake supply.
1926	Grand Canyon National Park, Arizona	Toilet flushing, lawn sprinkling, cooling water and boiler feed water.
1929	Pomona, California	Irrigation of lawns and gardens.
1942	Baltimore, Maryland	Reuse of a chlorinated effluent in a metal cooling and steel processing plant at the Bethlehem Steel Company.
1960	Sacramento, California	California legislation encourages water reclamation and reuse in the State Water Code.
1961	Irvine Ranch Water District, California	A project to reclaim water for irrigation, industrial and domestic uses begins. Later the project includes toilet flushing in high-rise buildings, distributing reclaimed water using a dual-distribution system.
1962	County Sanitation Districts of Los Angeles, California	The effluent of the Whittier Narrows plant is used to recharge an aquifer through spreading basins at the Montebello Forebay.
1965	San Diego County, California	The Santee recreation lakes, supplied with reclaimed water, are opened for swimming and catch-and-release fishing.
1976	Orange County Water District, California	Water Factory 21 is the first project to study the feasibility of making sewage potable.
1977	St. Petersburg, Florida	One of the oldest municipal dual-distribution systems and the largest urban water reclamation system in the United States. Reclaimed water is provided for residential lawns, commercial developments, industrial parks, a resource recovery power plant, golf courses, a baseball stadium and schools. At 2007, highly treated recycled water is used for irrigation at 9,992 residential lawns, 61 schools, 111 parks and 6 golf courses.
1976	Monterey, California	The first large-scale study designed to investigate the risks and effects of irrigating with reclaimed water on food crops, including raw-eaten vegetables. The project started in 1976 and results were published in 1987.

4.1.2 Motivations

Water scarcity was the original driver behind the development of water reclamation projects in the US. With time, however, other motivators have appeared. Asano *et al.*, 2007 and the USEPA, 2004, have suggested the following additional drivers:

- Compliance – Stringent water and wastewater standards at a federal and local level have encouraged greater water reuse in the US. At the federal level, the Water Pollution Control Act, more commonly referred to as the Clean Water Act (CWA), has effectively increased the attractiveness of water reuse. The CWA determines the degree and type of wastewater treatment necessary to meet prescribed effluent standards – whether that effluent is to be reclaimed and reused, or discharged to a receiving water body. The CWA promotes zero discharges of pollutants into navigable, fishable and waters where people swim, and therefore indirectly promotes water reuse by necessitating extensive water treatment that results in increasing the value and usability of effluents. Due to the stringent requirements, some municipalities prefer to use highly treated effluents rather than to discharge them into water bodies. Additionally, the implementation of the CWA has prompted the centralization of urban wastewater treatment plants; producing treated effluents near sites with high water demand.

At a local level, several reuse programs have been implemented in response to standards that require extensive treatment at a significant cost, such as those needed to remove nitrogen and phosphorus. One example can be found in the state of Georgia, where in order to protect aquatic organisms the discharge of treated effluents into surface water bodies during the warm season is prohibited. As a consequence, a water reuse program has been implemented.

- Viable source substitution – In many cases, the development of cities, agriculture and industries is limited by the lack of water. Recycled water represents an alternative source that is readily available and may, in many instances, provide the most economical and or feasible substitute source of water for uses such as urban irrigation, air conditioning and toilet flushing. This may mean that it is possible to delay or eliminate the need to expand potable water supply and treatment facilities.
- Localized water demand increases – In the arid west of the US, rapid population growth experienced since the 1960s has resulted in increased demand for water. Where growth has outstripped the availability of traditional water sources, reuse is increasingly viewed as a new water source alternative. The same is true for subtropical climatic regions, such as in the state of Florida, where continuous urban development has put considerable pressure on water resources despite it having the highest average precipitation of all US states.
- Societal pressures – A growing public awareness of the true value of water, as well as greater confidence in the safety and reliability of reuse projects has resulted in the development of legal frameworks and institutional arrangements that permit the implementation of new water reuse projects.

4.1.3 Current status of water reuse in the United States

According to USEPA, in 2002, when 6.4 x 10^6 m^3/d of water was reused, Florida contributed to this amount over one-third with 34% (2.2 x 10^6 m^3), California contributed 31% (2.0 x 10^6 m^3), Texas contributed 13.6% (8.7 x 10^5 m^3) and Arizona's contribution was 12% (7.6 x 10^5 m^3). While these four states accounted for the majority of the water reused in the US, at least 27 other states were also reusing water (USEPA, 2004).

From a geographical point of view, most of the reuse projects are located in the arid and semi-arid western and south-western states where water supplies are limited. An increasing number of projects are being implemented in the humid regions due to rapid growth and urbanization.

4.1.3.1 Uses of reclaimed water

Agricultural irrigation – Agricultural irrigation is the biggest water reuse activity in the US, and the state of California uses the largest portion of reclaimed water for this purpose: 46% of the total volume of reclaimed water produced in the state (California State Water Resources Control Board, 2002). There are several examples of water reuse projects for agriculture around the country. Table 4.2 lists some of them.

Industrial Reuse – Industrial reuse has increased substantially in the US since the early 1990s. California, Arizona, Texas, Florida, and Nevada have major industrial facilities using reclaimed water for cooling water and process/boiler-feed requirements (USEPA, 2004). Some examples are listed in Table 4.3.

Table 4.2 Selected examples of water reuse for agricultural irrigation in the US (Sheikh *et al.*, 1999 and Asano *et al.*, 2007).

Location	Description
Irvine Ranch Water District, California	Irrigation, industrial and domestic uses, later including toilet flushing in high-rise buildings. Water reuse began in 1961.
Monterey County Water Recycling Project, California	One of two projects that emerged from the Monterey Wastewater Reclamation Study for Agriculture (MWRSA). Located in the Salinas Valley, it involves a regional wastewater treatment plant, pumping station, storage facilities, pipelines and other distribution systems, and environmental mitigation for the water reclamation system. Around 25 x 10^6 m^3/year (68,500 m^3/d) of disinfected tertiary treated water is used to irrigate 4,700 hectares (ha) of food crops.
Tallahassee, Florida	The Tallahassee agricultural reuse system, in operation since 1966, is a cooperative between the city, which owns and maintains the irrigation system and a commercial enterprise, which is contracted to operate the farming service. The total area irrigated is 850 ha. The permitted application rate of the site is 8 cm per week, for a total capacity of 111,400 m^3/d. Major crops produced include corn, soybeans, coastal Bermuda grass and rye.
Conserv II, City of Orlando and the Orange County, Florida	The first project in Florida to use reclaimed water to irrigate food crops, it originated in response to a 1979 court order that the City of Orlando and Orange County cease discharging effluents into the Shingle Creek by March 1988 in order to protect the aquatic life of the Tohopekaliga Lake. Conserv II began operations in December 1986 and is a combination of agricultural irrigation and rapid infiltration basins (RIB). The average flow produced by two wastewater treatment plants (one for Orlando and the other for Orange County) is 189,200 m^3/d. Reclaimed water is distributed to 76 agricultural and commercial customers (60% of the flow) and to the RIB sites (40%). The major crop produced is citrus.

Municipal reuse – Landscape irrigation currently accounts for the largest portion of urban use of reclaimed water in the United States. Other municipal reuses include residential lawns irrigation and toilet flushing in buildings (USEPA, 2004).

Recreational, Aesthetic and Environmental reuse – Uses of reclaimed water for recreational purposes range from landscape impoundments and water hazards on golf courses to full-scale development of water-based recreational impoundments, such as those for fishing and boating and those for swimming and wading. Environmental uses comprise wetland recharge and stream augmentation. Some examples are presented in Table 4.4.

Groundwater Recharge – Many examples of groundwater recharge can be found in the United States. Two methods are used for this purpose: vadose zone injection; and direct injection. Vadose zone injection wells are the older method of the two, developed in the 1990s in different areas of Phoenix, Arizona. Direct injection is used mainly to create a barrier to saline intrusion along the US coast. Some examples of groundwater recharge projects are presented in Table 4.5.

Table 4.3 Examples of industrial reuse in US, from USEPA, 2004.

Place	Description
Phoenix, Arizona	The Arizona Public Service 1,270-MW Palo Verde nuclear power plant, located 88 km from the City of Phoenix, uses the majority of the City of Phoenix and area cities' reclaimed water at an average rate of 248,757 m^3/d for cooling purposes.
Bethlehem Steel Company in Baltimore	378,432 m^3/d of reclaimed water produced in the City of Baltimore is used in once-through cooling systems.
Rawhide Energy Station utility power plant in Fort Collins, Colorado	The energy station reclaims 929,000 m^3/d of water in a once-through cooling system for a 250-MW power plant.
Xcel Energy Cherokee Station, Denver Colorado	The Metro Treatment Plant, producing 170,000 m^3/d of secondary treated effluent, distributes water to be reused for the irrigation of parks, golf courses, schools and a local zoo, and for the Xcel power plant. The treatment process includes biological treatment with nitrification, coagulation-flocculation, sedimentation, granular medium filtration and chlorine disinfection. The energy plant began using reclaimed water in 2004 and, due to lower phosphorus content in the reused water compared to the original water used, a scale problem in cooling towers was controlled.
Burbank, California	Since 1967, 18,922 m^3/d of municipal secondary treated effluent has been successfully utilized for cooling water in the city's power generating plant.
Irvine, California	In 1997, a local carpet manufacturer in Irvine, California, retrofitted carpet-dyeing facilities to use 1,901 m^3/d of reclaimed water. The new process is as effective as earlier methods and is saving up to 2,273 m^3/d of freshwater.
Tuftex Carpets Industries in Santa Fe Springs, California	Instead of using water from the Colorado River, the State Water Project or groundwater, reclaimed water is now being used for carpet and textile dyeing, metal finishing, concrete mixing and cooling tower supply.
West Basin Municipal Water District, California	This reclamation project has the peculiarity of producing six different qualities of reclaimed water for different users, the majority of which are industrial. The district purchases secondary treated wastewater from the Hyperion Wastewater Treatment Plant, one of the largest in the country. Reclamation began in 1995, distributing water for parks, sport fields, manufacturing processes and oil refineries. At the present time, reclaimed water is used for landscape irrigation of parks, playgrounds and commercial areas, including landscaping at the Home Depot National Training Center, Toyota and Goodyear. Water is also sent to Exxon-Mobil, Chevron and Carson where it is further treated for specific industrial purposes. Due to the success in reusing water, the project is in its fourth expansion phase.
Curtis Stanton Energy Facility, Orlando, Florida	36,368 m^3/d of reclaimed water from Orange County is used for cooling in the power plant.
City of Las Vegas, Nevada	The City of Las Vegas and Clark County Sanitation District use 340,416 m^3/d of a secondary treated effluent to supply 35% of the water demand in power generating stations operated by the Nevada Power Company. The power company provides additional treatment consisting of a two-stage lime softening, filtration and chlorination prior to use in the cooling tower. A reclaimed water reservoir provides backup for the water supply.
Seattle, Washington	In a partnership between the King County Department of Metropolitan Services (Seattle, Washington), The Boeing Company and Puget Sound Power and Light Company, a new 55,740 m^2 Customer Service Training Center is cooled using a chlorinated secondary treated effluent.

Table 4.4 Examples of water reuse for environmental and aesthetic use, with information from Asano *et al.*, 2007; USEPA, 2004; Water Pollution Control Federation, 1989; and Iron Bridge Water Reclamation Facility, 2006.

Place	Description
Tucson, Arizona (recreational lake)	Lakeside Lake is a 6 ha urban impoundment built in the 1970s in the Atterbury Wash to provide for fishing, boating and other recreational activities. Groundwater and surface runoff initially sustained the lake, but in 1990 it began receiving reclaimed water from the Roger Road Treatment Plant (up to 205 m^3/d). The installation of a mechanical diffuser in 1992 improved dissolved oxygen concentrations.
Lubbock, Texas (recreational lake)	Reclaimed water was used to restore the one time dump, Yellow House Canyon. Four lakes and man-made waterfalls were built for recreational purposes. The lakes are used for fishing, boating and waterskiing, however swimming is restricted. The Canyon Lakes Park system uses approximately 18,200 m^3/d of reclaimed water.
Arcata, California (wetland recharge)	The Arcata Marsh and Wildlife Sanctuary is one of the most well-known examples of water reuse in wetlands and wildlife habitats. The City of Arcata was treating its wastewater to a secondary level and discharging it into the Humboldt Bay, which is one of the major oyster farming sites in California. To protect the bay, two wetland systems were built. The first system is a treatment wetland, while the second is a reclaimed water habitat marsh. In the first system, three parallel water surfaces are able to produce secondary treated quality effluent, which is then chlorinated. The three units cover an area of 12.5 ha and have a mean depth of 0.6 m. The marsh attracts more than 200 species of bird and provides a fish hatchery for salmon. The habitat marsh is also used for recreational purposes.
Petaluma, California (wetlands park)	The City of Petaluma is reusing part of its treated wastewater to irrigate 320 ha of agricultural lands and a golf course, and to feed 12 ha of wetlands. A citizen's organization called the Petaluma Wetlands Park Alliance has promoted the construction of another 54 ha of polishing wetlands as an environmental restoration project, which will begin operations in 2009. The project includes the expansion of the riparian zone, the planting of native vegetation, restoration/enhancement of the tidal marsh, and cultivation of crops for butterfly and bird foraging. Landscaping on the wetlands site will be irrigated with reclaimed water.
Orlando, Florida (wetland recharge)	The City of Orlando's Easterly Wetlands project was established in the late 1980s to reduce nitrogen and phosphorus concentrations from tertiary treated wastewater from the Iron Bridge Regional Water Pollution Control Facility bound for the St. Johns River. The wetlands are located 20 km from the wastewater treatment plant over an area of approximately 486 ha of improved pasture. The original design consisted of three major vegetative areas. The first area, comprising approximately 166 ha of deep marsh and consisting primarily of cattails and bulrush, has a primary function of nutrient removal. The second area is 154 ha of mixed marsh, comprising over 60 submergent and emergent herbaceous species used for nutrient removal and wildlife habitat. Submergent and emergent plants are those growing entirely beneath the water surface and those standing above the surface, respectively. The final 162 ha area was planted as a hardwood swamp with a herbaceous understory, however the trees did not establish as planned and the area is now managed as a deep marsh habitat. The wetlands had an average discharge rate of 66,583 m3/d in 2006, which is just over half the capacity permitted by the Floridian Department of Environmental Protection. The wetland began receiving flow from the Iron Bridge facility in 1987 and reclaimed water flows through the wetlands and is discharged into the St. Johns River.

Place	Description
Ironbridge STP, Orlando, Florida (wetland recharge)	This nature reserve used for hiking, jogging and biking includes a 502 ha wetland using about 23,185 m^3/d of reclaimed water.
Cochrane Memorial Park, Washington (wetland recharge)	The Yelm, Washington project, is a 3 ha city park featuring constructed surface and submerged wetlands designed to polish the reclaimed water prior to groundwater recharge. This project includes a rainbow trout pond.
San Antonio Water System, Texas (stream augmentation and aesthetics)	Reclaimed water replaces groundwater to sustain the San Antonio river. The river flows through a city park, zoo and downtown river walk. A second stream augmentation flows to Salado Creek, where reclaimed water replaces the flow from an abandoned artesian well. A decorative fountain at the City Convention Center also uses reclaimed water.
San Luis Obispo, California, (stream augmentation)	The City of San Luis Obispo reclaims water for multiple purposes including the enhancement of in-stream habitats. The potable supply of the city consists of imported water from outside the San Luis Obispo water shed. The San Luis Obispo River is a 24-km river that runs through the city and has a flow that varies with the season of the year. During the dry period the flow is almost entirely treated water. An ambitious reuse program for agricultural and municipal uses was initiated due to the lack of water; however the discovery of endangered and rare aquatic species shifted the focus from water reuse for conventional purposes to in-stream environmental enhancement.

Table 4.5 Examples of groundwater recharge projects, with information from USEPA, 2004 and El Paso Water Utilities, 2007.

Location	Description
Mesa Northwest, Arizona	The City of Mesa has two water reclamation plants and reuses water on golf courses, for crop irrigation, industrial uses and freeway landscape watering, as well as for groundwater recharge. The northwest reclamation plant produces 68,000 m^3/d of tertiary effluent and discharges into two recharge sites and into the Salt River, which also recharges the aquifer. Recharge is an integral part of the City of Mesa's 100-year water supply plan.
West Basin Municipal Water District, California	Since the 1950s the district has been injecting treated water into the coastal South Bay aquifers to mitigate salt water intrusion. A secondary effluent is treated with reverse osmosis and disinfection prior groundwater recharge.
Orange County, California	Orange County Water District's advanced water treatment facility, known as Water Factory 21, has been injecting highly treated water into coastal aquifers to act as a saltwater intrusion barrier since 1976.
Gainesville, Florida	32,277 m^3/d of recycled water is used to recharge the Floridian Aquifer through deep wells and a further 1,400 m^3/d is used to irrigate the Botanical Gardens.
Fred Hervey Water Reclamation Plant, El Paso, Texas	Facilities are located in Northeast El Paso. Effluent, which is treated to drinking water quality, is used to recharge the Hueco Bolson aquifer using a series of injection wells. Water is not used for drinking purposes. In 2004, 2.6 million m^3/d of reclaimed water was returned to the Hueco Bolson. The Fred Hervey plant also supplies the El Paso Electric Company with approximately 4 million m^3/d of reclaimed water for their cooling towers on an annual basis.

Augmentation of Potable Supplies – A limited number of reuse projects have the specific purpose of augmenting potable supplies. This is performed by recharging surface or groundwater sources with highly treated effluent. Indirect potable reuse is defined in the US (USEPA, 2004) as the augmentation of a community's raw water supply with treated wastewater followed by an environmental buffer. The treated wastewater is mixed with surface and or groundwater, which is typically subjected to additional treatment before entering the water distribution system. Table 4.6 contains some examples of projects of this type.

Table 4.6 Examples of indirect potable water reuse in the US, with information from McEwen, 1998; Choon Nam et al., 2002; Law, 2003 and Department of Water Resources of Georgia, 2007.

Project	Type of project
Upper Occoquan Sewage Authority, Virginia	Since 1978, Virginia's Upper Occoquan Sewage Authority (UOSA) has been reclaiming water for discharge into the (40 million m^3) Occoquan Reservoir. This reservoir supplies up to one million people in Northern Virginia and, during droughts, receives up to 90% of its water from the reclaimed supply. At all other times, reclaimed water typically represents 15% of the reservoir volume. The current technology involves secondary treatment followed by lime treatment, clarification, recarbonation, sand filtration, granulated activated carbon (GAC), ion exchange and chlorination. Ongoing quality monitoring is carried out by an independent panel of review.
Los Angeles County, California	The Montebello Forebay Groundwater Recharge Project, located in southeastern Los Angeles County, is the primary source of replenishment for the Central Basin, the main body of water underlying the greater Los Angeles metropolitan area. Since 1962 the county has been surface-spreading reclaimed secondary treated and chlorinated water into the Whittier Narrows groundwater basin at Montebello Forebay. Reclaimed water is dispersed through a system of rivers and channels to the two shallow, 3.1 m (10 ft) deep spreading basins. Potable water is subsequently withdrawn. It has been estimated that up to 20-30% of the potable water is indirectly recycled water (McEwen, 1998, Khan and Roser, 2007).

Direct potable reuse is defined as the introduction of treated wastewater directly into a water distribution system without intervening storage or passing through an environmental buffer (pipe-to-pipe) (USEPA, 2004). There are no examples of direct potable reuse in the US at this time.

The above examples are of projects that intentionally produce water to indirectly reuse it for water supply augmentation, but many communities in the US are currently using surface water sources subject to a significant number of upstream discharges. For example, more than two dozen major water utilities in the country use water from rivers that receive wastewater discharges. At times of low flow, up to 50% or more of the stream flow may be recycled water (Swayne *et al.*, 1980). This practice is sometimes called unplanned reuse or *de facto* potable reuse (Asano *et al*, 2007).

4.1.4 Water reuse case studies

4.1.4.1 California

California is the third largest state by area in the US and currently ranks as the most populous (Wikipedia, 2007). As a result of the high population and a lack of water resources, there are several water reuse projects in operation. Agricultural holdings have benefited from irrigation sourced from reclaimed water since 1890 (Asano *et al.*, 2007). At least 35 communities were reclaiming water for farm irrigation in 1910 and in 1912 the first example of landscape irrigation began operations at the Golden Gate Park in San Francisco (Asano *et al.*, 2007). By 1952, there were 107 communities in California using reclaimed water for agricultural and or landscape irrigation (Asano *et al.*, 2007). At present, there are several hundred reuse projects involving effluents of a very high quality.

Current status of water reuse – At the end of 2001, reclaimed water use in California exceeded 1.7 million m^3/d (State of California, 2002). By 2002, water reuse was practiced in around 4,800 locations using effluents from more than 200 wastewater treatment plants (Asano *et al.*, 2007).

The Central Valley and the South Coastal Regions of California represented in volume nearly 80% of the reclaimed water used, while the Santa Barbara County and the desert and eastern Sierra Nevada regions accounted for the remaining 20% (Asano *et al.*, 2007). The Central Valley of California, being an agricultural area, reused water mostly for irrigation, while urban uses were dominantly in the South Coastal Region, where about half of the state's population resides. An exception to this trend is the City of San Diego, where only limited water reclamation and reuse projects have been implemented (Asano *et al.*, 2007).

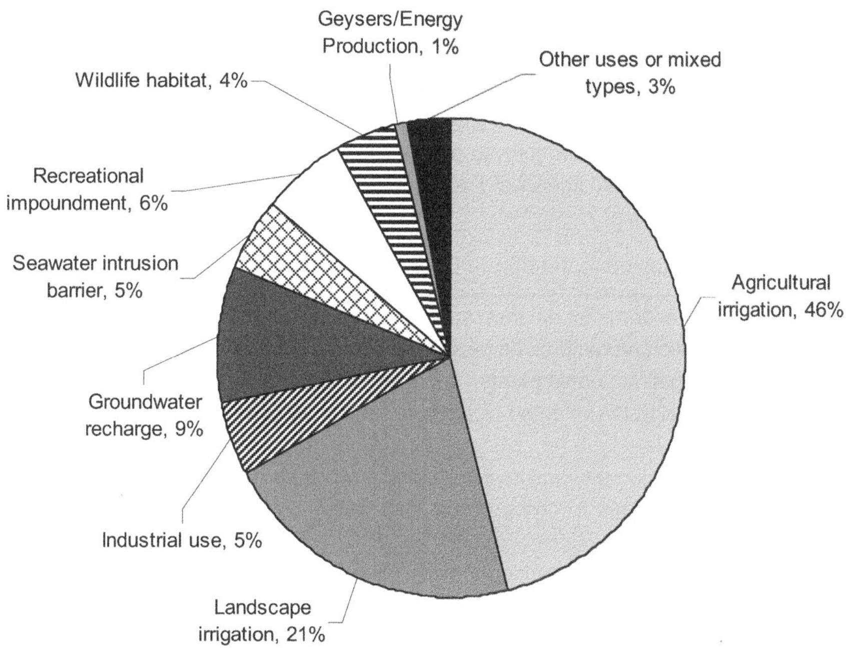

Figure 4.2 Reuses of water in California, with information from: Asano *et al.*, 2007.

Reclaimed water is predominantly used in agricultural applications in California (Figure 4.2). At least 20 varieties of food crops are grown using reclaimed water, including vegetables eaten raw such as lettuce, celery and strawberries. Eleven non-food crops, predominantly pasture and feed for animals, as well as nursery products, are also watered using reclaimed water. Landscape irrigation is the second most common use for reclaimed water. It is used primarily on turf, including over 125 golf courses and many parks, schoolyards and for freeway landscaping. Industrial and commercial uses include cooling towers in power stations, boiler feed water in oil refineries, carpet dying and recycled newspaper processing. Reclaimed water is also used in office and commercial buildings for toilet and urinal flushing (California State Water Resources Control Board, 2003; State of California, 2003b; Crook, 2004; Levine and Asano, 2004; and Asano *et al.*, 2007).

Another common reuse practice in California is groundwater recharge which is performed to balance groundwater extraction and/or to prevent seawater intrusion in coastal aquifers. The most notable example for this purpose is the Montebello Forebay Groundwater Recharge Project. Located near Whittier in Los Angeles County, it has been in operation since 1962. Additionally, reclaimed water is used within the state for recreational and environmental reuse, accounting for 10% of total water reuse (California State Water Resources Control Board, 2002).

Water reuse policies and regulations – The State Board of Public Health introduced in California regulation governing use of sewage for irrigation purposes in 1918 (Committee on the Use of Treated Municipal Wastewater Effluents and Sludge in the Production of Crops for Human Consumption, National Research Council, 1996). This is considered as the first regulation promoting planned reuse in the United States. In 1978, a set of public health laws were developed to regulate the use of recycled water: the Health and Safety Code; the Water Code; and Titles 22 and 17 of the California Code of Regulations (State of California, 1978). Details of the health laws relating to recycled water are published in a book commonly referred to as the Purple Book (State of California, 2001). Box 4.1 contains a summary of the material available in the Purple Book.

Besides regulations, the State of California also establishes incentives to promote reuse. Major incentives include (California State Water Resources Control Board, 2003):

- Grants for local public agencies to study the feasibility of water recycling projects;
- Water Recycling Construction Program loans at low interest rates; and
- Grants to local public agencies for the design and construction of water recycling facilities.

Box 4.1 Information contained in the Purple Book (Department of Health Services, 2001).

- Water quality standards for different approved reuses.
- Public consultation procedures necessary to gain approval for any recycling proposal.
- Emergency situation requirements, such as facilities reserved for the purpose of storing or disposing of untreated or partially treated wastewater for at least a 24-hour period.
- Operational procedures to ensure a high degree of reliability of the recycling systems.
- Pipeline considerations to convey reclaimed water, and standards to prevent cross connections.
- Water rights as a consideration to promote water reuse.
- Considerations for setting recycled water prices.

Public opinion – In California, water reuse is a relatively well accepted practice. According to one California survey there is, in general, a positive opinion about reusing water for fire fighting, irrigation of golf courses and parks, street cleaning, toilet flushing, and drought protection (Filice, 1992). Similar results have been observed in four separate surveys performed in Clark County (Las Vegas, Nevada), where a majority of respondents were very positive concerning the overall benefits obtained by using reclaimed water for golf course irrigation, park irrigation, industrial cooling, and in decorative water features (Alpha Communications Inc., 2001). Water reuse acceptance considerably diminishes, however, when direct and indirect potable reuse are considered – the former to a notably greater extent (USEPA, 2004). In a study performed by Hall and Rubin in 2002 in which 50 wastewater operators and managers were surveyed, 70% of the participants believed that reuse would be an important part of their operation in five years.

Future use of reclaimed water – It is expected that by 2030, use of reclaimed water will increase almost four-fold relative to the amount used in 2002 (Asano *et al.*, 2007). Of this reuse, it is anticipated that around 82% will be used for non-potable purposes, while 18% will be used for potable purposes (Asano *et al.*, 2007).

Selected examples of water reuse – Selected examples of water reuse projects in California are presented in Table 4.7.

Table 4.7 Selected examples of water reuse projects in California, with information from: California Energy Commission, 2002; USEPA, 2004; Crook, 2004; and Asano *et al.*, 2007.

Site	Description
City of Pomona	Began distributing reclaimed water to the California Polytechnic University in 1973. The project has since expanded its delivery of recycled water to two paper mills, roadway landscaping, a regional park and a landfill with an energy recovery facility.
El Dorado Hills	The rapidly expanding suburb of El Dorado Hills, located east of the City of Sacramento introduced a reclaim project for residential irrigation. Wastewater of mostly domestic origin is treated to a tertiary level through two treatment plants, producing 20,700 m^3/d of reclaimed water. The water is used for irrigating two golf courses, public parks, schools, extensive greenbelts and residential front yards. The price of reclaimed water is fixed to 80% of the potable water rate.
Irvine Ranch Water District	Based on an economic study, the county enacted an ordinance requiring all new buildings over 17 meters high to install a dual-distribution system for toilet flushing in areas where reclaimed water is available. The ordinance was put in place in 1991. The cost of dual plumbing systems has been estimated to increase plumbing capital costs in buildings over seven stories by only 9%.
Orange County Water District	The Orange County Water District was formed by an act of the California Legislature in 1933 to protect and manage the depleted groundwater basin that underlies the northwest half of the county and supplies about 75% of the District's total water demand. In the mid 1960s the District established a pilot-scale reclamation project that developed into Water Factory. The first blended reclaimed water from Water Factory 21 was injected into the coastal barrier in October 1976. The recycled product water from the plant meets drinking water standards after treatment using advanced processes. The blended injection water not only protects the basin from saltwater intrusion, but also replenishes aquifers from which 50% of the county's water is drawn.
Monterrey County	Overexploitation of groundwater caused partly by agricultural irrigation had depleted the water source by the mid-1970s. The Monterey Regional Water Pollution Control Agency sponsored an 11-year, $7 million pilot and demonstration project to grow several crops using reclaimed water. Results showed that produce grew well and was free of viruses. No side effects were observed in soils or on plant tissues. As a result, in 1998, 4,855 ha of artichokes, lettuce, cauliflower, broccoli, celery, and strawberries were irrigated using reclaimed water. Currently, 95% of the project irrigation area is voluntarily using reclaimed water.
Santa Rosa Geysers Recharge Project	The City of Santa Rosa treats to a tertiary level the effluents of five different cities in the Sonoma County. More than half of the water produced (50 million m^3/d) is used to irrigate approximately 2,310 ha of farmlands, including pastures, hay crops, vineyards, and row crops. The reclaimed water is also used to irrigate golf courses, parks, school grounds and both public and private urban landscaping. To beneficially use the water produced during the winter months when there is no demand for irrigation, the effluent is used to recharge an aquifer and produce electricity using the energy from geysers. Energy production using reclaimed water began in 2003, using 50,006 m^3/d of water. One of the major benefits of the Geysers Recharge Project is the flexibility afforded by year-round water reuse.

4.1.4.2 Florida

Unlike California, Florida is a relatively water abundant state. Its motivation to reuse water was initially associated with reducing the environmental impact on the sensitive natural surface water sources of the state caused by treated wastewater discharges. Later, increased water demand as result of rapid population growth (a 23.5% increase between 1990 and 2000) also became a driver for reuse. Reuse in the state grew by 53% between 1986 and 2003.

Current status of water reuse – In 2003, approximately 834 million m^3 (2.3 million m^3/d) of water was reclaimed, representing about 54% of the total permitted domestic wastewater treatment capacity (State of Florida, 2004). Water reclamation is practiced throughout the state, with the largest reuse sites located in central Florida (Orlando-Lakeland area), the Tampa Bay area, southwestern Florida, and in some of the Atlantic coast counties (Asano *et al.*, 2007). Sixty-three of its 67 counties reclaim water. Compared with California, Florida reclaims water more intensely for landscape irrigation (45% of the total) and less for agricultural use (16%). A further 6% is used for wetland enhancement and restoration (Florida DEP, 2002a). In 2003, 154,234 residences, 427 golf courses, 4,986 parks and 213 schools were beneficially reclaiming water (State of Florida, 2004). Figure 4.3 shows Florida's reclaimed water usage.

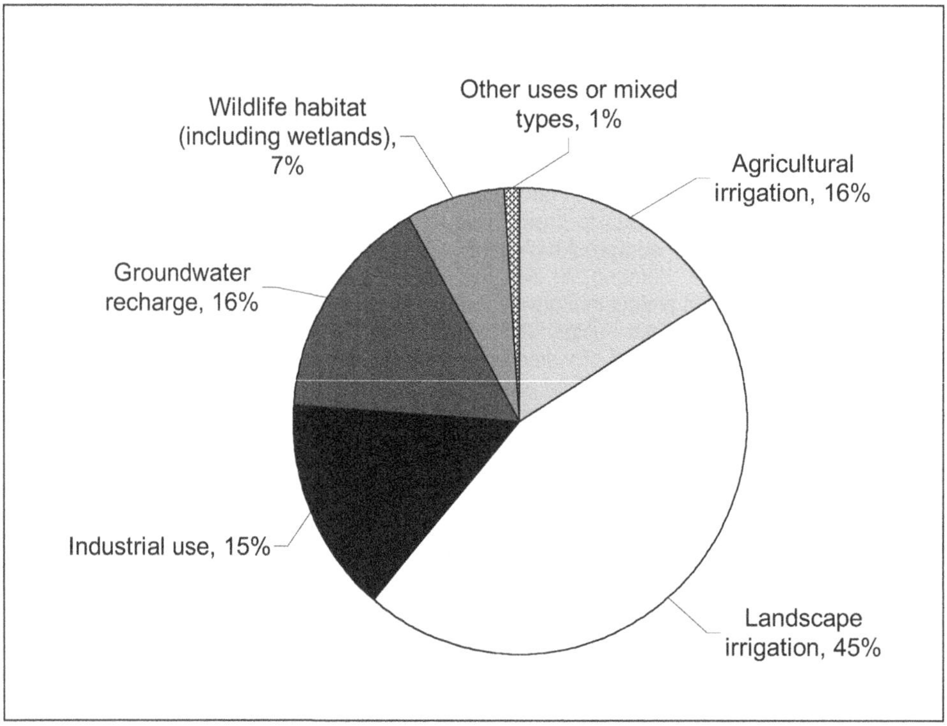

Figure 4.3 Uses of reclaimed water in Florida, State of Florida, 2004.

Water reuse policies and regulations – Florida has established in legislature "the encouragement and promotion of reclaimed water reuse and water conservation" as a formal state objective. A program to promote reuse began in 1987 and in 1988, reuse was made mandatory in Water Resource Caution Areas (WRCA) that were defined as areas having or projected to have critical water supply problems within the next 20 years. In WRCAs, reuse is mandatory unless it is proven to be not economically, environmentally or technically feasible (York and Wadsworth, 1998). To control reuse, Chapter 62610 of the Florida Administrative Code establishes rules for landscape irrigation, agricultural irrigation, industrial uses, groundwater recharge, indirect potable reuse and a wide range of urban reuse activities (Florida DEP, 1999). These rules also address the reclamation of water for aquifer storage and recovery (ASR), the blending of demineralization concentrates with reclaimed water, and the use of supplemental water supplies.

The State of Florida, in recognizing reclaimed water as a valuable water resource, has sought to ensure consistency with statutory requirements related to the linkage between reuse requirements in consumptive use permits and Department of Environmental Protection permits in order to encourage the practice. In June 2003, the Reuse Coordinating Committee produced a report that set out 16 strategies for increasing the efficiency and effectiveness of the use of reclaimed water in Florida, as directed by the 2001 Florida Water Plan and as part of Phase II of the Water Conservation Initiative. These strategies are summarized in Box 4.2.

Box 4.2 Strategies to promote water reuse in Florida (State of Florida, 2004).

1. Encourage water metering and volume-based rate structures.
2. Implement viable funding programs.
3. Facilitate seasonal reclaimed water storage, including aquifer storage and recovery (ASR).
4. Encourage use of reclaimed water in lieu of other water sources in agricultural irrigation, landscape irrigation, industrial/commercial/institutional, and indoor water use sectors.
5. Link water reuse to regional water supply planning (including integrated water resources planning).
6. Develop integrated water resources education programs.
7. Encourage groundwater recharge and indirect potable reuse.
8. Discourage effluent disposal to emphasize that large quantities of water are being wasted.
9. Provide water use permitting incentives for utilities that implement water reuse programs.
10. Encourage reuse in Southeast Florida (where significant amounts of treated wastewater is wasted).
11. Encourage use of supplemental water supplies from all sources including treated stormwater.
12. Encourage efficient irrigation practices.
13. Encourage interconnection of reuse systems to provide greater flexibility and reliability.
14. Enable the redirecting of existing reuse systems to more desirable reuse options as a means of motivating utilities.
15. Use reclaimed water at government facilities.
16. Ensure the continued safety of water reuse.

Public opinion – In Florida, reuse is perceived as a tool to environmentally dispose of wastewater, reducing the negative impacts produced by the conventional means of disposing of treated effluents, and a way to recharge and augment existing available water supplies.

Future use of reclaimed water – To meet water demand for the growing population in Florida, the vision is that:

water reuse will be employed by all domestic wastewater treatment facilities having capacities of 380 m^3/d and larger;

around 65% of all domestic wastewater will be reclaimed and used for beneficial purposes throughout the state;

effluent disposal using ocean outfalls, other surface discharges and deep injection wells will be limited to facilities that serve as backups for water reuse facilities;

groundwater recharge and indirect potable reuse projects will become commonplace;

sewer mining will become a common practice, particularly in larger urban areas, as a means of enabling effective use of reclaimed water; and reclaimed water will be used widely to flush toilets in commercial facilities including hotels and motels, industrial facilities, and multiple-family residential units. Florida aims to be reclaiming 4.5 million m^3/d by 2010.

Selected examples of water reuse – Table 4.8 presents selected examples of water reuse projects in Florida not mentioned previously in this chapter.

Table 4.8 Selected examples of water reuse in Florida, with information from Altamonte Springs, 1997, USEPA, 2004; Mantovani *et al.*, 2001; and Asano *et al.*, 2007.

Site	Description
Water Reuse at Reedy Creek (Walt Disney World Resort Complex and additional services)	Water is reclaimed for five golf courses, highway medians, a water park and a 45-ha tree farm growing horticultural materials for use within the Walt Disney Resort Complex. The project also includes the washdown of impervious surfaces, construction (such as concrete mixing and cleanup), cooling tower make-up and fire fighting (suppression and protection). Customers are metered individually at the point of service. Rates are typically set at 75-80% of the potable water rate based on volumetric consumption to discourage wasteful practices and encourage connection and use. The total size of the water reuse system for potable and non-potable uses ranges from 81,830 to 113,650 m^3/d. Reclaimed water utilization represent 25-30% of the total water demand for the resort complex.
Apricot Project in Altamonte Springs City	Project Apricot, an acronym for 'A Prototype Realistic Innovative Community of Today', provides recycled tertiary treated effluent through a dual-distribution system for household irrigation and car washing in Altamonte Springs. The water is also used for the fire mains, ornamental fountains and ponds and for toilet flushing in commercial buildings. Above ground outdoor taps are prohibited to prevent misuse of the water. All established commercial and multiple-family dwellings and all new single-family dwellings were required to connect to the system within 90 days of its establishment in 1989. Commercial users and condominiums were charged 0.18 USD/m^3, while single dwellings paid a flat fee of 10 USD per month for recycled water. The scheme has led to a 30% reduction in potable demand, but with the reclaimed water priced at 40% of potable rates, shortages occur in hot weather. Sewage is imported from other utilities for treatment to meet demand.
The City of West Palm Beach, Florida Wetlands-Based Water Reclamation	The water supply to the City of West Palm Beach was reduced as part of the Everglades restoration program. As result, a program to use highly treated wastewater for beneficial reuse including augmentation of their drinking water supply was developed. The project combines advanced wastewater treatment with habitat restoration. Effluent from the city's East Central Regional Wastewater Treatment Plant, approximately 159,110 m^3/d, is injected into the groundwater (boulder zone) using six deep wells at a depth of 914 meters. To protect and preserve its surface water supply system and to develop this reuse system to augment the water supply, the city purchased a 610-ha wetland reuse site. This site consists of a combination of wetlands and uplands dominated by Melaleuca trees. A portion of this property was used for the construction of a standby wellfield and up to 53,040 m^3/d of highly treated reclaimed water is sprinkled onto the marshy expanse.

4.1.4.3 Other States

Water reuse is practiced in several other states of the United States also. Table 4.9 presents examples of reuse projects in other US states.

Table 4.9 Description of reuse projects in different US states.

State	Reuse projects examples
Texas	In the state of Texas there is a program to progressively exchange potable water for reclaimed water developed out of need, due to declining reserves of fresh groundwater and an unreliable supply of surface water. The program comprises the irrigation of golf courses, municipal parks and school grounds (Ornelas and Brosman, 2002). Some of the projects have implemented special management procedures to handle the high saline effluents produced (TDS up to 1200 ppm) to avoid foliar damage in plants (Miyamoto and White, 2002). Reclamation is also performed for aquifer recharge to augment potable sources, such as at the Fred Hervey Water Reclamation Plant, located in Northeast El Paso.
Arizona	Reclamation has become an important feature of water planning for the Phoenix Water Services Department to meet increasing demand from the rapidly developing area of North Phoenix (USEPA, 2004). For example, 30,240 m^3/d of a tertiary treated effluent is used to irrigate golf courses, parks, schools and cemeteries. The reclaimed water is sold to customers at 80% of the potable water rate.
Massachusetts	In the tourist town of Yarmouth, Massachusetts, additional drinking water supplies were needed to handle peak summer demands. Consequently, when an irrigation source was needed for a 9-hole expansion of the Bayberry Hills Golf Course, which was part of a landfill closure/reuse plan, the community investigated the use of reclaimed water. In the spring of 1996, the town entered into discussions with the Department of Environmental Protection about utilizing the effluent from the adjacent Yarmouth-Dennis Septage Treatment Plant as the source of irrigation water. Since water reclamation had not previously been practiced in the Commonwealth of Massachusetts, the state needed to develop its own guidelines and improve treatment facilities in order to produce the quality of effluent desired.
Nevada	The population of Las Vegas has doubled over the past 15 years and there are over 40 million visitors annually. The Las Vegas Valley Water District, Nevada, operates close to the limit of its entitlement of 360 million m^3 of water per year from Lake Mead (Hoover Dam), but has secured increased water by return-flow credits through discharges from its three sewage treatment plants. "Return-flow credits" allow wastewater from the city to be treated, then returned to Lake Mead and used again. Southern Nevada's water supply has increased by one-third as a result of these credits. The system of giving credit for return flows has acted to reduce the incentive for moving to direct recycling, however reclaimed water is used for irrigation at parks, cemeteries, golf courses and highway landscaped areas. In fact, Las Vegas legislation states that any new golf courses must use reclaimed water for turf irrigation when it becomes available and the water must be provided at or for less than the cost of potable water.

State	Reuse projects examples
New York	The Oneida Indian Nation-owned Turning Stone Casino & Resort, located in Central New York, required 2540 m^3/d of water to irrigate a planned golf course complex. The casino is located in an area with limited water resources and so reclaimed water was identified as a viable irrigation source. The City of Oneida, 34 miles east of Syracuse, operates an advanced wastewater treatment facility that serves a population of approximately 10,000. Daily flows to the plant average 2.5 million gallons per day. No reclaimed water quality or treatment standards for either restricted or unrestricted urban reuse exist in the state of New York and reuse is uncommon. Implementing water reclamation required inter-governmental cooperation between the Nation and the City of Oneida.
Washington	One of Washington's fastest growing cities, Yelm originally used septic tanks as its sanitation system. Rather than building a tertiary treatment plant to discharge the city's effluent to the Nisqually River where different species of salmon and trout live, a reclamation project was promoted. The facility began operations in 1999, producing approximately 1,137 m^3/d and with a capacity to produce up to 4,546 m^3/d (Cupps and Morris, 2005; and USEPA 2004). Central to the project is 3-ha memorial park and fishing pond. At the park, a constructed wetlands system de-chlorinates, re-oxygenates, and further cleans, screens and moves the water through a series of ponds. Reclaimed water is primarily used for seasonal urban landscape irrigation at local schools and churches, city parks and for treatment plant equipment washdown and process water, fire fighting, street cleaning, and dust control. During the winter rains, demand from irrigation decreases. The excess water is utilized in power generation at the Centralia Power Canal (Washington State Department of Ecology, 2003; and City of Yelm, 2003). The project has won awards from the American Public Works Association, the Association of Washington Cities and in 2002 the Department of Ecology presented the City with an Environmental Excellence Award.
Denver	In 1983, the water supply agency of Metropolitan Denver established an experimental plant to examine technologies for the production of recycled water suitable for direct potable reuse. It built an advanced water reclamation plant and conducted a comprehensive health effects study of the effluent produced. The results of this $30 million program, published in 1993, unequivocally verified the ability of advanced water treatment processes to reliably remove a broad spectrum of pollutants and render a product that satisfied every then-known measure of drinking water safety (McEwen, 1998). A full-scale project was not implemented.
Colorado Springs, Colorado	To reduce dependence on water from the western slopes of the mountains, the City of Colorado Springs implemented a dual-distribution system in which reclaimed water is used for irrigation purposes. This is one of the oldest operating systems in the United States (since 1955) in which reclaimed water is used for urban landscape irrigation (USEPA, 1985). Reclaimed water is used for irrigating parks, golf courses, cemeteries and commercial properties, and for the 280-MW Martin Drake Power Plant.

4.1.5 Legal framework

Many states within the United States have their own water reuse legislation. The legal framework may include regulations, guidelines and incentives. The extensiveness of such frameworks varies considerably from state to state. Several states consider reclaimed water to be a viable water source alternative and have developed regulations specifying water quality requirements and or treatment processes for a full spectrum of reuse applications (USEPA, 2004). In contrast, other states have developed water reuse regulations with the primary intent of providing a disposal alternative to surface water discharge. In the few states with no specific regulations or guidelines on water reclamation and reuse, programs may still be permitted with approval on a case-by-case basis. The USEPA Guidelines for Water Reuse were developed for application in states with no regulations for all or some type of reuse. Table 4.10 shows the number of states having regulations or guidelines for specific reuse purposes.

Table 4.10 Number of states with regulations or guidelines for each type of reuse application (from USEPA, 2004).

Type of Reuse	Number of States
Unrestricted urban	28
Irrigation	28
Toilet flushing	10
Fire protection	9
Construction	9
Landscape impoundment	11
Street cleaning	6
Restricted urban	34
Agricultural (food crops)	21
Agricultural (non-food crops)	40
Unrestricted recreational	7
Restricted recreational	9
Environmental (wetlands)	3
Industrial	9
Groundwater recharge (non-potable aquifer)	5
Indirect potable reuse	5

Water quality criteria and treatment requirements for different reuse applications, as specified in the USEPA Guidelines for Water Reuse (2004), are contained in Table 4.11.

Box 4.3 details the rationale behind the use of certain parameters and values contained in the guidelines. The USEPA guidelines were developed based on the assumption that the sources of reclaimed water are limited to the effluents from municipal or other domestic wastewater treatment facilities, having a limited input of industrial waste.

Table 4.11 Treatment needs and water quality requirements defined by the USEPA Guidelines for Water Reuse (2004).

Types of Reuse	Treatment	Reclaimed Water Quality[2]	Setback Distances[3]
Urban Reuse (landscape irrigation, vehicle washing, toilet flushing, fire protection systems, commercial air conditioners and uses with similar access or exposure to the water)	-Secondary[4] -Filtration[5] -Disinfection[6]	- pH = 6-9 - ≤ 10 mg/L BOD[7] - ≤ 2 NTU[8] - No detectable fecal coli/100 mL[9,10] - 1 mg/L Cl_2 residual (minimum)[11]	15 m to potable water supply wells
Restricted Access Area Irrigation (sod farms, silviculture sites, and other areas where public access is prohibited, restricted or infrequent)	-Secondary[4] -Disinfection[6]	-pH = 6-9 -≤ 30 mg/L BOD[7] -≤ 30 mg/L TSS -≤ 200 fecal coli/100 mL[9,13,14] - 1 mg/L Cl_2 residual (minimum)[11]	-90 m to potable water supply wells -30 m to areas accessible to the public (if spray irrigation)
Agricultural Reuse – food crops not commercially processed[15] (surface or spray irrigation of any food crop, including crops eaten raw)	-Secondary[4] -Filtration[5] -Disinfection[6]	-pH = 6-9 -≤ 10 mg/L BOD[7] -≤ 2 NTU[8] -No detectable fecal coli/100 mL[9,10] -1 mg/L Cl_2 residual (minimum)[11]	- 15 m to potable water supply wells
Agricultural Reuse – food crops commercially processed[15] (surface irrigation of orchards and vineyards)	-Secondary[4] -Disinfection[6]	-pH = 6-9 -≤ 30 mg/L BOD[7] -≤ 30 mg/L TSS -≤ 200 fecal coli/100 mL[9,13,14] -1 mg/L Cl_2 residual (minimum)[11]	-90 m to potable water supply wells -30 m to areas accessible to the public (if spray irrigation)
Agricultural Reuse – nonfood crops (pasture for milking animals; fodder, fiber, and seed crops)	-Secondary[4] -Disinfection[6]	-pH = 6-9 -≤ 30 mg/L BOD[7] -≤ 30 mg/L TSS -≤ 200 fecal coli/100 mL[9,13,14] -1 mg/L Cl_2 residual (minimum)[11]	-90 m to potable water supply wells -30 m to areas accessible to the public (if spray irrigation)
Recreational Impoundments (incidental contact (e.g., fishing and boating) and full body contact with reclaimed water allowed)	-Secondary[4] -Filtration[5] -Disinfection[6]	-pH = 6-9 -≤ 10 mg/L BOD[7] -≤ 2 NTU[8] -No detectable fecal coli/100 mL[9,10] -1 mg/L Cl_2 residual (minimum)[11]	-150 m to potable water supply wells (minimum) if bottom not sealed

Types of Reuse	Treatment	Reclaimed Water Quality[2]	Setback Distances[3]
Landscape Impoundments (aesthetic impoundment where public contact with reclaimed water is not allowed)	-Secondary[4] -Disinfection[6]	-≤ 30 mg/L BOD[7] -≤ 30 mg/L TSS -≤ 200 fecal coli/100 mL[9,13,14] -1 mg/L Cl_2 residual (minimum)[11]	-150 m to potable water supply wells (minimum) if bottom not sealed
Construction Use (soil compaction, dust control, washing aggregate, making concrete)	-Secondary[4] -Disinfection[6]	-≤ 30 mg/L BOD[7] -≤ 30 mg/L TSS -≤ 200 fecal coli/100 mL[9,13,14] -1 mg/L Cl_2 residual (minimum)[11]	
Industrial Reuse (once-through cooling)	-Secondary[4] -Disinfection[6]	-pH = 6-9 -≤ 30 mg/L BOD[7] -≤ 30 mg/L TSS -≤ 200 fecal coli/100 mL[9,13,14] -1 mg/L Cl_2 residual (minimum)[11]	-90 m to areas accessible to the public
Recirculating cooling towers	-Secondary[4] -Disinfection[6] (chemical coagulation and filtration[5] may be needed)	-Variable depends on recirculation ratio - pH = 6-9 -≤ 30 mg/L BOD[7] -≤ 30 mg/L TSS -≤ 200 fecal coli/100 mL[9,13,14] -1 mg/L Cl_2 residual (minimum)[11]	-90 m to areas accessible to the public. May be reduced or eliminated if a high level of disinfection is provided.
Other Industrial Uses	**Depends on site specific uses**		
Environmental Reuse (wetlands, marshes, wildlife habitat, stream augmentation)	Variable -Secondary[4] and Disinfection[6] (minimum)	Variable but not exceed: -≤ 30 mg/L BOD[7] -≤ 30 mg/L TSS -≤ 200 fecal coli/100 mL[9,13,14]	
Groundwater Recharge (by spreading or injection into aquifers not used for public water supply)	-Site-specific and use dependent -Primary (minimum) for spreading -Secondary[4] (minimum) for injection	-Site-specific and use dependent	-Site-specific
Indirect Potable Reuse (groundwater recharge by spreading into potable aquifers)	-Secondary[4] -Disinfection[6] -May also need filtration[5] and/or advanced wastewater treatment[16]	-Secondary[4] -Disinfection[6] -Meet drinking water standards after percolation through vadose zone	-150 m to extraction wells. May vary depending on treatment provided and site-specific conditions.

Types of Reuse	Treatment	Reclaimed Water Quality[2]	Setback Distances[3]
Indirect Potable Reuse (groundwater recharge by injection into potable aquifers)	-Secondary[4] -Filtration[5] -Disinfection[6] -Advanced wastewater treatment[16]	Includes, but not limited to, the following: -pH = 6.5-8.5 -≤ 2 NTU[8] -No detectable total coli/100 mL[9,10] -1 mg/L Cl_2 residual (minimum)[11] -≤ 3 mg/L TOC -≤ 0.2 mg/L TOX -Meet drinking water standards	-600 m to extraction wells. May vary depending on site-specific conditions.
Indirect Potable Reuse (augmentation of surface supplies)	-Secondary[4] -Filtration[5] -Disinfection[6] -Advanced wastewater treatment[16]	Includes, but not limited to, the following: -pH = 6.5-8.5 -≤ 2 NTU[8] -No detectable total coli/100 mL[9,10] -1 mg/L Cl_2 residual (minimum)[11] -≤ 3 mg/L TOC -Meet drinking water standards	-Site-specific

1. These guidelines are based on water reclamation and reuse practices in the US, and they are especially directed at states that have not developed their own regulations or guidelines. While the guidelines should be useful in many areas outside the US, local conditions may limit the applicability of the guidelines in some countries. It is explicitly stated that the direct application of these suggested guidelines will not be used by USAID as strict criteria for funding.
2. Unless otherwise noted, recommended quality limits apply to the reclaimed water at the point of discharge from the treatment facility.
3. Setback distances are recommended to protect potable water supply sources from contamination and to protect humans from unreasonable health risks due to exposure to reclaimed water.
4. Secondary treatment processes include activated sludge processes, trickling filters, rotating biological contractors, and may include stabilization pond systems. Secondary treatment should produce effluent in which both the BOD and TSS do not exceed 30 mg/L.
5. Filtration means the passing of wastewater through natural undisturbed soils or filter media such as sand and/or anthracite, filter cloth, or the passing of wastewater through microfilters or other membrane processes.
6. Disinfection means the destruction, inactivation, or removal of pathogenic microorganisms by chemical, physical, or biological means. Disinfection may be accomplished by chlorination, UV radiation, ozonation, other chemical disinfectants, membrane processes, or other processes. The use of chlorine as defining the level of disinfection does not preclude the use of other disinfection processes as an acceptable means of providing disinfection for reclaimed water.

7. As determined from the 5-day BOD test.
8. The recommended turbidity limit should be met prior to disinfection. The average turbidity should be based on a 24-hour time period. The turbidity should not exceed 5 NTU at any time. If TSS is used in lieu of turbidity, the TSS should not exceed 5 mg/L.
9. Unless otherwise noted, recommended coliform limits are median values determined from the bacteriological results of the last 7 days for which analyses have been completed. Either the membrane filter or fermentation tube technique may be used.
10. The number of fecal coliform organisms should not exceed 14/100 mL in any sample.
11. Total chlorine residual should be met after a minimum contact time of 30 minutes.
12. It is advisable to fully characterize the microbiological quality of the reclaimed water prior to implementation of a reuse program.
13. The number of fecal coliform organisms should not exceed 800/100 mL in any sample.
14. Some stabilization pond systems may be able to meet this coliform limit without disinfection.
15. Commercially processed food crops are those that, prior to sale to the public or others, have undergone chemical or physical processing sufficient to destroy pathogens.
16. Advanced wastewater treatment processes include chemical clarification, carbon adsorption, reverse osmosis and other membrane processes, air stripping, ultrafiltration, and ion exchange.
17. Monitoring should include inorganic and organic compounds or classes of compounds, that are known or suspected to be toxic, carcinogenic, teratogenic, or mutagenic and are not included in the drinking water standards.

Box 4.3 Rationale used to select some parameters and values in the USEPA 2004 guidelines.

Fecal and Total coliform

Fecal coliform concentrations not exceeding 200 MPN/100 mL were considered as a limit for reuse applications with no direct public or worker contact because:

- under such conditions, most bacterial pathogens are destroyed or reduced to low or insignificant levels in the water and the concentration of viable viruses will be somewhat reduced;
- disinfection of secondary effluent to this coliform level is readily achievable at a minimal cost;
- significant health-related benefits associated with disinfection to lower, but not pathogen-free, levels are not obvious.

For uses where direct or indirect contact is likely or expected, and for dual-distribution systems where there is a potential for cross-connections with potable water lines, a non-detectable level of fecal coliforms is recommended. This limit is considered applicable in conjunction with tertiary treatment and other water quality limits, such as a turbidity less than or equal to 2 NTU in order to produce reclaimed water that is essentially free of measurable levels of bacterial and viral pathogens.

For planned indirect potable reuse, total coliform is recommended, in preference to measuring fecal coliform, to be consistent with the National Primary Drinking Water Regulations (NPDWR) that regulate drinking water standards for producing potable drinking water. The NPDWR were produced pursuant to section 1412 of the Public Health Service Act, as amended by the Safe Drinking Water Act (Pub. L. 93–523). A non-detectable level of total coliform is recommended to reclaim water in order to increase the total volume of water available for potable use.

Viruses

The USEPA (2004) guidelines do not suggest limits for parasites or viruses because:

- there is no evidence to indicate that parasites are a problem at water reuse facilities in the US operating at the treatment and quality limits recommended in these guidelines, although there has been considerable interest regarding the occurrence and significance of Giardia and Cryptosporidium in reclaimed water in recent years;
- there is substantial evidence available to indicate that viruses are reduced or inactivated to low or immeasurable levels by way of appropriate wastewater treatment, including filtration and disinfection;
- the task of identification and cataloguing viruses found in wastewater is made difficult by relatively low virus recovery rates, the complexity and high cost of laboratory procedures, and the limited number of facilities with the requisite personnel and equipment to perform the analyses;
- it takes approximately 14 days for the laboratory culturing procedure to determine the presence or absence of viruses in a water sample and an additional 14 days are required to identify the viruses;
- while recombinant DNA technology provides new tools to rapidly detect viruses in water, current methods are neither able to quantify viruses nor differentiate between infective and non-infective virus particles;
- There is a lack of consensus among virus experts regarding the relevance to human health of low levels of viruses in reclaimed water; and
- never has there been a documented case of viral disease resulting from the reuse of wastewater at any of the water reuse operations in the US.

Particulate matter

The suspended matter is related to the presence of microorganisms and to the reduction of the disinfection processes' efficiency by serving as a shield to bacteria and viruses. Organic suspended matter consumes chlorine, reducing the amount available for the purposes of disinfection. As a consequence, reducing particulate matter to 2 NTU or 5 mg/L TSS is recommended, prior to disinfection, to ensure reliable destruction of pathogenic microorganisms. Typically, suspended solids measurements are performed on a composite sample and, therefore, reflect only an average value. Due to the significance of the suspended matter content in treatment reliability, continuously monitored turbidity is deemed to be a superior measurement.

Organic Matter

The type of reuse activity will determine the need to remove organic matter. Adverse effects associated with organic substances include the potential for them to be aesthetically displeasing (malodorous or colored), to provide food for microorganisms, adversely affect disinfection processes, and consume oxygen. The recommended BOD limit is intended to indicate that the organic matter has been stabilized, is non-putrescible, and has been lowered to levels commensurate with anticipated types of reuse.

Besides water quality criteria and treatment need specifications, the planning of reuse projects needs to consider water rights laws, and water and land use regulations. Water rights are especially important because the rights allocated by the states can either promote reuse measures, or they can be an obstacle. For example, in water-constrained areas, where one might expect water reuse to be most attractive, water rights laws might prohibit the use of potable water for non-potable purposes, while at the same time restricting uses of reclaimed water that would result in it not being returned to the same water body from which it came (USEPA, 2004).

In the US, state laws allocate water based on two types of rights: the appropriative doctrine and the riparian doctrine. Appropriative rights operate in most western states and particularly in those areas that are water-constrained. The system governs how water is assigned or delegated to the consumer and essentially functions on a first-in-time, first-in-right basis. Senior users have a continued right to the water. New users may generally submit an application to the state for a water allocation, which is granted with a permit or license according to availability of water after all previously approved withdrawals have been made. An appropriator may not divert more water than can be used and any appropriated water not used is lost (USEPA, 2004).

In contrast, riparian water rights are most commonly found in the East and in areas of water abundance. Rights are allocated according to the customer's proximity to the water source and may be acquired through the purchase of land. Thus, riparian users have an equal claim to water where available. Under riparian rights, a user may not make withdrawals that would cause substantial depletion to the stream flow or significant degradation to the quality of the stream, and use can only be for a legal and beneficial purpose. Water allocated under a riparian right may only be used on the riparian land and cannot be extended to another property but the right to the unused water can be held indefinitely and without forfeiture. This limits the ability of the water authority to quantify the amount of water earmarked and may therefore lead to over-allocations.

In the dry western US, reclaimed water often constitutes a more reliable supply than the allocated surface water or groundwater; particularly where a user with low-priority rights finds them curtailed or denied in times of shortage (USEPA, 2004). Consequently, reclaimed water is not only an attractive alternative water source, but also the largest block of unappropriated water in the West. Who retains control of the reclaimed water among the discharger, water supplier, other appropriators and environmental interests is also a factor. For example, the City of Walla Walla in Washington State was taken to court by a local irrigation district to continue to discharge its wastewater effluent into Mill Creek for irrigation use. On two occasions the court decreed that the City must discharge all of its wastewater effluent - at all seasons of the year - into the creek (USEPA, 2004).

Water supply and use legislation determines how an agency or entity with water rights may decide to distribute that supply to various parties. It is now common for governments at all levels to set standards for the way in which water is used as a condition of supplying water to its customers, including the extent to which it must be conserved or reused. Frequently, standards serve to promote reuse by requiring water users to reduce their total or per capita water use in observance of an established baseline. Some uses of potable water (i.e., irrigation, power plant cooling) are considered "unreasonable" and as a result prohibited, unless it is shown that other non-potable sources are "environmentally undesirable or economically unsound" (USEPA, 2004). There are three main types of water supply and use rules:

- water supply reductions - often imposed during periods of drought;
- water efficiency goals - that can be either mandatory or voluntary; and
- water use restrictions - that may either prohibit the use of potable water for certain purposes, or require the use of reclaimed water in place of potable water.

Land use policies, usually made at a local level but with regard to federal environment laws and state planning regulations; determine the development and use of property which might be served by reclaimed water systems. Communities are required to adopt water management plans and identify additional supplies to support new developments in several western states. Such rules further encourage the implementation of reuse projects that reduce demand for limited resources.

4.1.6 Treatment

Treatment standards or guidelines for one or more types of water reuse application exist in the majority of states. Some require specific treatment processes, others impose effluent quality criteria, and some prescribe both. Many states also include requirements for failsafe procedures to ensure treatment reliability and prevent the distribution of any reclaimed water not adequately treated due to an interrupted process, power outage, or equipment failure (USEPA, 2004). Dual-distribution systems (i.e., reclaimed water distribution systems that parallel a potable water system) must also incorporate safeguards to prevent cross-connections of reclaimed water and potable water lines and the misuse of reclaimed water.

In the US, reclamation plants typically require conventional water and wastewater treatment technologies that are already widely practiced and readily available. In general, the minimum treatment level required for any type of reuse is secondary treatment with disinfection. Where unrestricted public exposure is likely in the reuse application, wastewater must be treated to a high degree prior to its application.

Together with the water quality criteria, the USEPA Guidelines for Water Reuse (2004) suggest a list of treatment processes (Table 4.9). The selection was made based on:

- water reuse experience in the US and elsewhere;
- research and pilot plant or demonstration study data;
- technical material from the literature;

- various states' reuse regulations, policies, or guidelines;
- attainability; and
- sound engineering practice.

The USEPA guidelines advocate a combination of treatment and quality requirements known to produce reclaimed water of acceptable standard, thereby rendering as unnecessary the monitoring of the finished product for certain constituents (e.g., health-significant chemical constituents or pathogenic microorganisms). Pilot tests are recommended to fully characterize the reclaimed water to be produced and to compare its quality to other water sources in the area. Pilot testing is also considered a useful tool with which to demonstrate the ability of the selected unit processes to meet project objectives and to refine the design of sophisticated treatment trains. Additionally, a pilot plant enables information regarding the content of emerging pollutants (microbiological or chemical) to be obtained.

The reliability of a treatment method to reclaim water is just as important as its efficiency. The "Federal Guidelines: Design, Operation and Maintenance of Waste Water Treatment Facilities," developed in 1970 by the then Federal Water Quality Administration (now Environmental Protection Agency), recognized early on the importance of treatment reliability (USEPA, 2004). The guidelines included the following principles regarding treatment plant reliability:

- all water pollution control facilities should be planned and designed to provide for maximum reliability at all times;
- each facility should be capable of operating satisfactorily during power failures, flooding, peak loads, equipment failure, and maintenance shutdowns;
- such reliability can be obtained through the use of various design techniques that will result in a facility that is virtually "fail-safe" (USEPA, 2004).

Reclamation plants, like any other treatment facility, need to fulfil these requirements, which in practice provide for redundant or oversized unit processes. In addition, reuse facilities need to assure reliability based on public health considerations. Due to the significant cost burden of such requirements, implementation of small reuse projects is often limited. Box 4.4 presents a list of requirements.

Box 4.4 A summary of equipment and additional reliability requirements for water reclamation plants, from USEPA, 1974 and 2004.

Equipment requirements (Class I reliability requirements for wastewater treatment plants)

- For mechanically cleaned units, a back-up bar screen shall be provided.
- A back-up pump shall be provided for each set of pumps that perform the same function. Design flow will be maintained with any 1 pump out of service.
- If comminution is provided, an overflow bypass with bar screen shall be provided.
- In primary sedimentation basins there shall be sufficient capacity such that a design flow capacity of 50% of the total capacity will be maintained with the largest unit out of service.
- Filters shall be in a sufficient number of units of a size such that a design capacity of at least 75% of the total flow will be maintained with 1 unit out of service.
- At least 2 aeration basins of equal volume will be provided.
- At least 2 mechanical aerators shall be provided. Design oxygen transfer will be maintained with 1 unit out of service.
- At least 2 chemical flash mixer basins or a back-up means of mixing chemicals separate from the basins shall be provided.
- There shall be a sufficient number of final sedimentation tanks of a size such that 75% of the design capacity will be maintained with the largest unit out of service.
- At least 2 flocculation basins shall be provided.
- There shall be sufficient number of disinfectant contact basins of a size such that the capacity of 50% of the total design flow may be treated with the largest unit out of service.

Additional requirements for reclamation plants

- Piping and pumping flexibility - Flexibility plans should permit the necessary degree of treatment to be obtained under varying conditions. No pipes or pumps should be installed that would circumvent critical treatment processes and possibly allow inadequately treated effluent to enter the reclaimed water distribution system. The facility should be capable of operating during power failures, peak loads, equipment failures, treatment plant upsets and maintenance shutdowns.
- Emergency storage or disposal - Besides normal operational or seasonal storage, there is a need to provide for the containment of treated effluent whenever the quality is not suitable for use.
- Alarms - Minimum instrumentation should consist of alarms at critical treatment units to alert an operator of a malfunction. This concept requires that the plant either be constantly attended, or that an operator be on call (with a remote sounding device) whenever the reclamation plant is in operation. In addition to the alarm system, it is critical to have a means available to take corrective action for each situation, which has caused the alarm to be activated.
- Instrumentation and Control - A reclamation plant must be provided with the necessary equipment to analyze the appropriate parameters; maintain, calibrate and verify accuracy of on-line instruments; monitor and control the treatment process performance; monitor and control the reclaimed water distribution; survey the reliability control system; and consider operator interface and system maintenance. System controls may be manual, automated, or a combination of manual and automated systems.

4.1.7 Implementation of projects

Due to there being guidelines and regulations at multiple levels of government, the legal framework for water reuse in the US is complex. Furthermore, in addition to the requirements for building and operating a reuse project, other rules and regulations exist that are common to any type of infrastructure project. A thorough understanding of which organizations and institutions are concerned with what aspects of a proposed reuse project is necessary. This undertaking should include the piecing together of an inventory of required permits and agency review requirements prior to initiating construction and operation of the reuse system. An evaluation of the institutional structure based on local conditions and expertise is also required to effectively operate the project (Table 4.12).

Table 4.12 Some common institutional patterns (from USEPA, 2004).

Type of institutional arrangement	Production	Wholesale distribution	Retail distribution
Separate authorities	Wastewater treatment agency	Wholesale water agency	Retail water company
Wholesaler/Retailer system	Wastewater treatment agency	Wastewater treatment agency	Retail water company
Joint powers authority (for production and distribution only)	Joint powers authority	Joint powers authority	Retail water company
Integrated production and distribution	Water/Wastewater authority	Water/Wastewater authority	Water/Wastewater authority

4.1.8 Funding

The first source of funding considered in the US comes from those receiving the reclaimed water service. Since fees are set to promote water reuse, however, it is generally difficult to create a totally self-supporting reuse program. To meet initial expenditure requirements, the majority of the construction and related capital costs are financed through long-term water and wastewater revenue bonds, which spread the cost over multiple decades. Supplemental funds may also be provided by grants or developer contributions to mitigate or offset the annual revenue requirement (USEPA, 2004). The range of externally generated capital funding source alternatives includes:

- Local Government Tax-Exempt Bonds – The total capital cost of construction activities for a reuse project could be financed from the sale of long-term (20-30 year) bonds;
- Grants and State Revolving Fund (SRF) Programs – Capital needs could be funded partially through state or local grants programs or through SRF loans, particularly those programs designed specifically to support reuse;
- Capital Contribution – At times, special agreements are reached with developers or industrial users requiring the contribution of either assets or money to offset the costs of a particular project (USEPA, 2004).

Additionally, the Clean Water Act of 1977, as amended, supports water reuse projects through the following provisions:

- Section 201 of PL 92-500 was amended to ensure that municipalities are eligible for "201" funding only if they have "fully studied and evaluated" techniques for "reclaiming and reuse of water". A 201 facility plan study must be completed to qualify for state revolving loan funds.
- Section 214 stipulates that the EPA administrator "shall develop and operate a continuing program of public information and education on water reclamation and reuse of wastewater".
- Section 313, which describes pollution control activities at federal facilities, was amended to ensure that wastewater treatment facilities will utilize "recycle and reuse techniques: if estimated life-cycle costs for such techniques are within 15% of the most cost-effective alternative" (USEPA, 2004).

4.1.9 Conclusion

Water reuse has been practiced in the United States in some form or another since the beginning of the 20th Century. The numbers of reuse projects continues to grow rapidily, aided by legislative developments, and pressured by bourgeoning populations and unreliable weather conditions. A number of drivers for reuse have been identified and presented above.

The end uses for reclaimed water are also diverse. The states of California and Florida lead the country in the uptake of water reuse and illustrate the various motivations for uptake. Geographical particulars make individual projects unique from one another in some instances but complexities in applications can also stem from the layers of governance or a lack thereof. The experiences of these states, among others, will prove valuable for a growing number of communities whose water resources are stretched by population increases and drought.

Gaining public support for water recycling will prove vital to the scope of its uptake. A rigorous testing regime will also be an important component of any reuse project. However, the potential for water reuse is great in the US and it is likely that the practice will become widespread in the future.

4.2 MOTIVATION FOR REUSE IN CANADA

Many municipalities must provide water supply to a growing population in competition with other sectors of the economy, while relying on finite supplies and controlling wastewater discharge into receiving waters. Water reuse is increasingly seen as an opportunity to ease the current stresses on water supply caused by increasing water demands, depletion and degradation of high-quality water sources, reduced supply reliability caused by climate change, ageing infrastructure and limited funding for its expansion. Water reuse simultaneously promotes environmental sustainability through conservation of water resources and reduces wastewater discharges to sensitive receiving waters. A brief review of reuse in Canada starts with an overview of the natural background and water resources and their utilization, followed by a summary of reuse practices, and description of the institutional framework.

4.2.1 Natural background

Canada is the second largest country in the world with respect to her territory reported as 9,985,000 km^2 (including 891,200 km^2 of freshwater surface) (Natural Resources Canada, 2001). This large territory is inhabited by a relatively small population of 33 million (July 2007). Eleven climatic regions with widely varying climatic conditions are recognized in Canada.

With respect to precipitation, both the Atlantic and Pacific coastal regions are characterized by abundant precipitation of up to 3,500 and 1,500 mm, respectively, but the eastern slopes of Rocky Mountains and the central plateau region are extremely arid (less than 300 mm of precipitation/year). The Arctic Tundra Region receives even less precipitation (about 100 mm/year). On average, Canada's annual precipitation is estimated at 535 mm/year (Statistics Canada, 2000), including 193 mm of snowfall, and this precipitation amount falls below the world average of 690 mm/year (Statistics Canada, 2000). Thus, the total volume of annual precipitation over the Canadian territory is about 5.5 x 10^{12} m^3 of water, which produces the total annual streamflow of 105,000 m^3/s representing about 9% of the world's renewable water supply (Pearse *et al.*, 1984).

4.2.2 Water resources and their use

Water resources and their utilization by individual countries are typically quantified at the national level by various indices reported in the literature. Generally, these indices describe the country's natural water resources and their utilization for various purposes, and the most recent metrics also rate the country's ability to protect and manage water resources and implement environmental sustainability. Several such indices are introduced and enumerated below. Whenever feasible, the older data used in calculating the indices were adjusted for the growing Canadian population.

One of the most commonly used water resources/water use indices is the Water Stress Index (WSI), which is also called the Relative Water Stress Index (RWSI; UN, 2006), and is defined as the ratio between the total annual water withdrawals and the available renewable water resources. According to the UN classification, when WSI exceeds 0.4, water resources are considered as stressed, advanced forms of water management are needed, and the provision of water services becomes costly. For WSI < 0.4, the conditions are termed as "relatively unstressed".

To calculate the WSI, one can use the readily available data, which are often somewhat dated. However, minor errors introduced by using older Canadian data are insignificant with respect to the overall conclusion – a great abundance of water resources in Canada. Taking the 1996 total annual withdrawals as 44.75 x 10^6 km^3 (Environment Canada, 2002), calculating the corresponding annual withdrawal per capita as 1,510 m^3 (the 1996 population was estimated at 29.6 million by Statistics Canada), assuming that this withdrawal remained constant between 1996 and 2007 (such constancy was reported e.g. between 1991 and 1996), and adjusting the Natural Renewable Water Resources in Canada per capita from 92,810 m^3 (2002 data) to the current magnitude 87,185 m^3 (based on the 2007 population of 33.0 million), the 2007 Canadian WSI can be estimated as 1,510/87,185 = 0.0173, or less than 2%. Thus, the Canadian WSI is extremely low and indicates, on a national basis, abundance of water resources.

Another measure of water availability is the annual volume of renewable water per capita. A value of 1,500 m^3/capita·yr indicates a threshold level of chronic water scarcity with serious consequences for country development. As indicated above, the Canadian water resources per capita exceed this threshold almost 60 times and indicate plentiful water resources.

While the water availability indices (such as WSI) are useful for a quick assessment of water availability on the national scale, more complex indicators are needed to reflect the country's ability to deal with water management issues. One such an indicator is the International Water Poverty Index (Lawrence *et al.*, 2002), which represents an interdisciplinary measure linking household welfare with water availability and indicating the degree to which water scarcity impacts on human population. It facilitates the ranking of countries (and communities within countries) by accounting for both physical and socio-economic factors associated with water scarcity. In this international ranking, Canada received a score of 77.7 out of 100 (water resources = 15.5; access = 20.0; capacity to manage = 18.7; use = 6.9 and environment = 16.5), which was the second best after Finland (WPI = 78.0; resources = 12.2; access = 20.0; capacity = 18.0; use = 10.6 and environment = 17.1). This high score indicates not only a great availability of water in Canada, but also good access to water resources, capacity to deal with water management issues, a relatively low level of water resources exploitation, and very good overall environmental conditions.

In summary, on a national scale, Canada is blessed with large renewable water resources and possesses great capacity to deal with water management issues related to the use of such resources and the protection of their environmental sustainability. Regional distribution of Canadian water resources often does not match that of water demands, and these results in regional disparities in the availability of water of suitable quality, particularly during periods of droughts. In the future, such conditions may be further exacerbated by two major factors: (a) population growth accompanied by increasing water demands, particularly in urban areas, and (b) climate change. Climate change and variability lead to increasing air temperatures, resulting in further increases of aridity in the low precipitation regions and growing demands on irrigation water, and changes in precipitation depths and patterns are also expected. The resulting conditions pose challenges for water management in some Canadian regions and call for a greater use of such innovative measures as soft path for water (including more rational use of water of appropriate quality and water conservation) and water reuse.

4.2.3 Water reuse in Canada

The interest in water reuse in Canada emerged at least 25 years ago, when the Canada Mortgage and Housing Corporation (CMHC) sponsored one of the first large-scale Canadian projects on this subject (Canviro Consultants and MacLaren Engineers, 1984) and concluded that water reuse for practically all purposes (including potable water supply) was technologically feasible. Since that time, new chemicals of concern have been identified (e.g., endocrine disruptors, pharmaceuticals and personal care products) and adoption of water reuse is still limited. While water reuse is mainly conducted on an experimental basis, examples exist of decentralized and centralized reuse applications.

Decentralized wastewater reclamation and water reuse is practised for individual homes and clusters of homes, or isolated industries, service operations and institutional facilities. In such cases, the most common types of decentralized water reuse are landscape irrigation and toilet flushing. A number of research projects regarding residential greywater reuse applications have been reported (e.g., Waller *et al.*, 1998). Examples of residential recycling and reuse in Canada include the Toronto Healthy House system, a four storey duplex built in 1996, and the CMHC Conservation Co-op in Ottawa, an eight-unit apartment complex. Both sites reported initial difficulties in maintaining water quality and required process adjustments after start-up. Greywater reuse systems have also been considered for isolated service operations and areas in northern Canada to reduce dependence on trucked water supply and sewage disposal services (Waller *et al.*, 1998).

The main applications of centralized wastewater reclamation and distributed water reuse in Canada are agricultural land, golf course and urban landscape irrigation. The practice of reclaimed water irrigation is quite well established in Western Canada, and experimental effluent irrigation projects have been conducted in Canada for over thirty years (Exall, 2004). Since 1977, the City of Vernon, British Columbia has operated a water reuse system for effluent disposal, irrigating agricultural, silvicultural and recreational lands during the irrigation season of April to October. Originally implemented as a means of protecting a sensitive lake from phosphorus inputs, 100% of annual flows are collected in a storage reservoir located 7 km from the plant. Daily flow of secondary effluent at the wastewater reclamation plant has averaged 13,000 m^3. In 2005, biological nutrient removal was added to protect sensitive receiving waters in case of emergency discharge, although the benefits of nutrients in the reclaimed water are lost to the irrigated crops.

Industrial water recycling is fairly common in Canada. Canadian industry accounts for over 80% of the total water intake, and of this total intake, approximately 40% is recycled.

Many industries employ recirculation of their own process waters for use in such areas as cooling tower make-up water. The recirculation rate, defined as the volume of water recirculated as a percentage of total water intake, varies considerably by the manufacturing sector, from a low of 22% in the wood products group, to a high of 292% in plastic products (Scharf *et al.*, 2002).

Industrial use of municipal reclaimed water is less common, although the City of Edmonton, Alberta has recently begun to supply a local industrial partner with reclaimed municipal wastewater for use in hydrogen and steam production. After the Gold Bar Wastewater Treatment Plant was upgraded to include microfiltration and a 5.5 km pipeline was built to the partner's site across the North Saskatchewan River (see Figure 4.4), the partner began using 5,000 m^3 of reclaimed water in 2006. Industrial reuse is expected to increase to 15,000 m^3 by 2008, and additional prospective users for the reclaimed water include a local ski hill and golf course.

Figure 4.4 (a) Installation of a microfiltration cassette at Edmonton's Gold Bar WWTP and (b) reclaimed water pipeline crossing the river. Photos: City of Edmonton Drainage Services.

Rainwater and roof runoff harvesting is quite commonly practised on a household scale for garden watering and a number of municipalities have encouraged this practice by providing subsidies on rain barrels. For balancing water needs over the irrigation period, larger storage volumes (700 L in a particular study) are needed (Derry, 2006). Less frequently, municipal stormwater is reused for golf course irrigation. Stormwater runoff is a more contaminated water source, with washoff of such roadway pollutants as heavy metals and PAHs. Heavy road salt use during the winter can lead to its buildup in stormwater treatment facilities, and the influence of road salts on the growth of turf grass and other plants is of particular concern (Marsalek *et al.*, 2002).

4.2.4 Institutional framework

Unlike the US, Australia and many countries in the Middle East and Europe, there are currently no national guidelines for water reuse in Canada. At the provincial level, the provinces of British Columbia, Alberta and Saskatchewan, and the Atlantic Canada region have some form of regulatory guidance on reuse.

Although British Columbia is viewed as having vast quantities of fresh water resources, it has also had considerable experience with water reuse including range land irrigation, silviculture applications, stream augmentation and toilet flushing. The Municipal Sewage Regulation, enacted in 1999 (BC MELP, 1999), is a performance-based approach that sets standards for the treatment of municipal wastewater, water reuse, and disposal of treated effluent. Authorized uses are grouped into two categories: unrestricted public access and restricted public access, the former of which necessitates significantly higher water quality standards. The Regulation stipulates design, operation and monitoring requirements. In May 2001, the Province published a Code of Practice for the Use of Reclaimed Water (BC MELP, 2001) which serves as a guidance document for the use of reclaimed water in British Columbia and is designed to support the regulatory requirements prescribed in the Municipal Sewage Regulation.

In Alberta, municipal wastewater reuse for irrigation is a regulated activity governed under Alberta's current Environmental Protection and Enhancement Act (EPEA), and all applications for this purpose within Alberta require formal approval (Alberta Environment, 2000). Under Alberta's EPEA, uses for municipal wastewater irrigation include: golf courses; municipal parkland and boulevards; forested woodlots under special approval consideration; and agriculture lands where used for pasture, forage, coarse grains, turf, and oil seeds. Any other crops to be considered must be first supported by scientific studies that ensure no risk to human health or the environment.

Table 4.13 Comparison of general characteristics of various water reuse guidelines and regulations (from Exall *et al.*, 2006)

Jurisdiction	Applications considered	Minimum treatment recommended	Unrestricted irrigation coliform limit (per 100 mL)[a]
CANADA			
Alberta	Agricultural, landscape irrigation	Best practicable to achieve required quality	≤200 FC (geo. mean), ≤1000 TC (geo. mean)
Atlantic Canada (4 provinces)	Agricultural, landscape irrigation	Secondary with six months storage	≤200 FC, ≤1000 TC
Saskatchewan	Agricultural, landscape irrigation	Secondary	≤200 FC (median), ≤400 FC (2 consecutive samples)
British Columbia	Agricultural, landscape, urban, industrial, environmental	Secondary with disinfection	≤2.2 FC (median), ≤14 FC (single sample)
USA			
California legislation	Agric., landscape, urban, industrial, environmental, groundwater recharge	Secondary	≤2.2 TC (MPN), ≤23 TC (single sample)
USEPA guidelines	Agric., landscape, urban, industrial, environmental, groundwater recharge, indirect potable	Secondary [b]	No detectable FC (median), ≤14 FC (single sample)
INTERNATIONAL			
WHO guidelines	Agriculture, aquaculture	Secondary+ [c] or waste stabilization ponds	<200 FC

[a] FC = fecal coliform; TC = total coliform; MPN = most probable number

[b] Primary treatment for surface spreading for groundwater recharge only.

[c] Secondary treatment followed by filtration and disinfection or by polishing ponds.

Both Saskatchewan's "Treated Municipal Wastewater Irrigation Guidelines" (2004) and the chapter on reclaimed water use in the Atlantic Canada Standards and Guidelines Manual (CBCL, 1996) are intended to assist the owners of wastewater treatment works and consultants considering or practicing irrigation as a method of treated wastewater reuse. Permitted irrigation uses include food and non-food crops, and golf courses.

Water quality guidelines in Canadian provinces engaged in water reuse typically include pathogens and chemical and physical constituents such as organics, nutrients, metals, suspended solids and residual chlorine. Applications, treatment and quality criteria in the provinces are compared with those of other jurisdictions in Table 4.13.

More recent institutional developments include a Canadian Standards Association (CSA) standard (Bill 128), for the "Design and Installation of Non-Potable Water Systems/Maintenance and Field Testing of Non-Potable Water Systems". For the first time, Canada has a standard for dual plumbing for non potable water systems, though provinces will need to adopt (or not adopt) these individually into their respective codes of practice. Also, the Canadian Council of Ministers of the Environment has led a draft "Canadian Guidelines for Household Reclaimed Water for Use in Toilet and Urinal Flushing" to support the above Standard. Consultation on this draft ended in November 2007 and new national guidance on this issue is expected soon.

4.3 Summary

As a whole, Canada enjoys abundant water resources, but regional disparities exist and water shortages do occur. In order to meet increasing water demands while protecting sensitive receiving waters from wastewater effluents, water reuse is slowly being adopted. Examples of both decentralized and centralized water reuse projects have been discussed, with applications ranging from toilet flushing and industrial water use to agricultural land and landscape irrigation. A number of provinces have developed guidance documents for water reuse and while no general national water reuse guidelines have been produced to date, standards and guidelines for specific reuse applications are being developed.

4.4 Acknowledgement

The authors of this chapter would like to thank sincerely the excellent review and editing of the manuscript by the WateReuse Association. Their contributions are extensive and, without their help, this chapter could not have been published.

4.5 References

Alberta Environment (2000) Guidelines for municipal wastewater irrigation. Edmonton, AB. April 2000.

Alpha Communications, Inc. (2001) Water Reclamation Public Opinion Surveys. Researched for: Clark County Sanitation District. Las Vegas, Nevada.

Altamonte Springs (1997) Fact Sheet - Project Apricot 10/97 and others, City of Altamonte Springs, Florida.

Asano, T., Burton, F., Leverenz, H., Tsuchihashi, R. and Tchobanoglous, G. (2007) Water Reuse: Issues, Technologies and Applications. McGraw Hill, New York, 1570 pp.

BC MELP (British Columbia Ministry of Environment, Lands and Parks) (1999) Waste Management Act - Municipal Sewage Regulation. Victoria, B.C. July 1999.

BC MELP (British Columbia Ministry of Environment, Lands and Parks). (2001) Code of practice for the use of reclaimed water: A companion document to the Municipal Sewage Regulation. Victoria, B.C. May 2001.

California Energy Commission (2002) Santa Rosa Geysers Recharge Project: GEO-98-001. Final Report. 54 pp. http://www.energy.ca.gov/reports/2003-03-01_500-02-078V1.PDF

California State Water Resources Control Board (2000) California Municipal Wastewater Reclamation Survey. http://www.swrcb.ca.gov/recycling/recyfund/munirec/ index.html

California State Water Resources Control Board (2002) California Municipal Wastewater Reclamation Survey. http://www.waterboards.ca.gov/recycling/docs/wrreclaim_attb.pdf

California State Water Resources Control Board (2003) http://www.swrcb.ca.gov/recycling/recyfund/

Canviro Consultants Ltd. and MacLaren Engineers Inc. (1984) CANWEL, The Canadian Water Energy Loop. Summary Report to Canada Mortgage and Housing Corporation, Ottawa, ON.

CBCL Limited (1996) Atlantic Canada Standards and Guidelines Manual for the Collection, Treatment, and Disposal of Sanitary Sewage.

Choon Nam, O., Rose, J., Lauer, W.B., Jern, N.W., Suok Kai, C., Tam, J.P., Singh, M., Bloodworth, B.C. and Hian Kee, L. (2002) Singapore Water Reclamation Study – Expert Panel Review and Findings. June 2002. http://www.pub.gov.sg/NEWater_files/download/review.PDF

City of Yelm (2003) 2003 Ground Water Monitoring and Use Information. Provided by Badger, S. (City Administrator) and Yanasek, J. (Operator-in-Charge). Yelm, Washington.

Committee on the Use of Treated Municipal Wastewater Effluents and Sludge in the Production of Crops for Human Consumption, National Research Council (1996). Use of Reclaimed Water and Sludge in Food Crop Production. http://www.nap.edu/catalog.php?record_id=5175

Crook, J. (2004) Innovative Applications in Water Reuse: Ten Cases Studies, WateReuse Association, Alexandria, V.A.

Cupps, K. and Morris, E. (2005) Case Studies in Reclaimed Water Use. Creating new water supplies across Washington State. Publication Number: 05-10-013. Washington State Department of Ecology. http://www.ecy.wa.gov/pubs/0510013.pdf

Department of Water Resources of Georgia (2007) http://www.mde.state.md.us/assets/document/WatRues07-7%20Gelot.pdf

Derry, A.E. (2006) Effects of roof type on stormwater quality and runoff in the City of Kingston, Ontario. M.Sc. Thesis, Queen's University, Kingston, Ontario.

Department of Health Services, DHS (2001) California Health Laws Related to Recycled Water. State of California Department of Health Services, Sacramento CA, June 2001 edition.

El Paso Water Utilities (2007) http://www.epwu.org/wastewater/fred_hervey_reclaimation.html, consulted October 2007.

Environment Canada (2002) Urban water indicators: municipal water use and wastewater treatment. Available at: http://www.ec.gc.ca/soer-ree/English/Indicators/ Issues/ Urb H2O/

Exall, K. (2004) A review of water reuse and recycling, with reference to Canadian practice and potential: 2. Applications. *Wat. Qual. Res. J. Canada.* **39**(1), 13-28.

Exall, K., Marsalek, J. and Schaefer, K. (2006) Water reuse and recycling in Canada: opportunities and challenges. In: Hlavinek, P., Kukharchyk, T., Marsalek, J. and Mahrikova, I. (Eds) Integrated Urban Water Resources Management. NATO Security through Science Series – C: Environmental Security. Springer, pp. 253-262.

Filice, F.V. (1996) Using Public Opinion Surveys to Measure Public Acceptance of a Recycled Water Program – San Francisco, CA. Water Reuse Conference Proceedings. AWWA. Denver, Colorado.

Florida DEP (Florida Department of Environmental Protection) (2007) http://www.dep.state.fl.us/water/reuse/project.htm

Florida DEP (Florida Department of Environmental Protection) (1999) Reuse of Reclaimed Water and Land Application. Chapter 62-610, F.A.C. Tallahassee, Florida. http://www.dep.state.fl.us/legal/rules/wastewater/62-610.pdf

Florida DEP (Florida Department of Environmental Protection) (2002a) 2001 Reuse Inventory. Tallahassee, Florida.

Florida DEP (Florida Department of Environmental Protection) (2002b) Florida Water Conservation Initiative. Water Conservation: Preventing and reducing wasteful, uneconomical, impractical, or unreasonable use of water resources, Section 62-40.412(1), F.A.C. Tallahassee, Florida. http://www.dep.state.fl.us/water/waterpolicy/docs/WCI_2002_Final_Report.pdf

Hall, W.L. and Rubin, A.R. (2002) Reclaimed Water: A Public Perception. WEFTEC 2002, Proceedings of the thAnnual Conference and Exposition, Chicago, Illinois.

Khan, S and Roser, D. (2007) Risk assessment and health effect studies of indirect potable reuse schemes. Centre for Water and Waste Technology School of Civil and Environmental Engineering University of New South Wales. Australia.

Law, I.B. (2003) Advanced Reuse – From Windhoek to Singapore and beyond. *Wat.* **30**(5), 31-36.

Lawrence, P., Meigh, J. and Sullivan, C. (2002) The Water Poverty Index: an International Comparison. Keele Economic Research Papers 2002/19, ISSN 1352-8955, Keele University, UK. Available at: www.keele.ac.uk/depts/ec/kerp

Levine, A. and Asano, T. (2004) Recovering sustainable water from wastewater. *Environ. Sci. Tech.* **38**(11), 201A-208A.

Mantovani, P., Asano, T., Chang, A and Okun, D.A. (2001) Management practices for nonpotable water reuse. Project 97-IRM-6, Water Environment Research Foundation, Alexandria VA.

Marsalek, J., Schaefer, K., Exall, K., Brannen, L. and Aidun, B. (2002) Water reuse and recycling. Canadian Council of Ministers of the Environment, Winnipeg, Manitoba. CCME Linking Water Science to Policy Workshop Series Report No. 3, 39 pp.

McEwen, B. (1998) Indirect potable reuse of Reclaimed Water. In: Asano, T. (Ed.) Wastewater Reclamation and Reuse. Chapter 27. Water Quality management Library, Vol. 10, Technomic Publishing, Lancaster, PA.

Miyamoto, S. and White, J. (2002) Foliar Salt Damage of Landscape Plants Induced by Sprinkler Irrigation. Texas Water Resources Inst. TR 1202.

National Research Council. (1996) http://www.nap.edu/catalog.php?record_id=5175

Natural Resources Canada. (2001). Canada Centre for Remote Sensing, GeoAccess Division, Ottawa. Available at: http://www.nrcan.ca

Ornelas, D. and Brosman, D. (2002) Distribution System Startup Challenges for El Paso Water Utilities. 2002 Reuse Symposium. Orlando, Florida.

Pearse, P.H., Bertrand, F. and MacLaren, J.W. (1984) Currents of Change: Final Report. Inquiry on Federal Water Policy, Environment Canada, Ottawa.

Saskatchewan Environment (2004) Treated municipal wastewater irrigation guidelines. Saskatchewan. June 2004.

Scharf, D., Burke, D.W., Villeneuve, M. and Leigh, L. (2002) Industrial water use, 1996. Environmental Economics Branch, Environment Canada, Ottawa, Ontario.

Sheikh, B., Cooper, C. and Israel, K. (1999) Hygienic evaluation of reclaimed water used to irrigate food crops- A case Study. *Wat. Sci. Tech.* **40**(4-5), 261-267.

State of California (1978) Wastewater Reclamation Criteria. Title 22, Division 4, California Code of Regulations. State of California, Department of Health Services. Sanitary Engineering Section. Berkeley, California.

State of California (2001) http://www.dhs.ca.gov/ps/ddwem/publications/waterrecycling/purplebookupdate6-01.PDF

State of California (2002) Statewide Recycled Water Survey, Office of Water Recycling, State Water Resources Control Board, Sacramento, Ca. htpp://www.waterboards.ca.gov/recycling/munirec.html

State of California (2003b) California Codes – Water code Section 13550, subdivision (a).http://www.leginfo.ca.gov/

State of Florida (2004) 2003 Reuse Inventory Department of Environmental Protection, Division of Water Resources Management, Tallahassee, Florida.

Statistics Canada (2000) Human activity and the environment 2000. Statistics Canada, Ottawa, Ontario.

Swayne, M., Boone, G., Bauer, D. and Lee, J. (1980) Wastewater in receiving Waters at water Supply Abstraction Points, EPA-60012-80-044, US Environmental Protection Agency, Washington, D.C.

US Environmental Protection Agency. (1974) Design Criteria for Mechanical, Electric, and Fluid Systems and Component Reliability. EPA-430-99-74-01. EPA Office of Water Program Operations, Municipal Construction Division. Washington, D.C.

UN (United Nations) (2006) Water – A Shared Responsibility. The United Nations World Water Development Report 2. UNESCO, Paris. Available at: http://www.unesco.org/water/wwap /wwdr2/

USEPA (1985) Water Reuse Via Dual Distribution Systems POTW 0196, Environment Reosurces management, Inc.

US Geological Survey (2007) http://water.usgs.gov/watuse/misc/consuse-renewable.html

USEPA (United States - Environmental Protection Agency) (2004) Guidelines for Water Reuse EPA 625/R-04/108. Washington, D.C.

Waller, D.H., Mooers, J.D., Samostie, A. and Sahely, B. (1998) Innovative residential water and wastewater management. Report to Canada Mortgage and Housing Corporation, Ottawa, ON.

Washington State Department of Ecology (2003) Facility Information. Provided by Pieritz, G. (Regional Engineer) and Cupps, K. (Water Reuse Lead). Olympia, Washington.

Water Pollution Control Federation (1989) Water Reuse (Second Edition). Manual of Practice SM-3. Water Pollution Control Federation. Alexandria, Virginia.

Wikipedia (2007) The Free Encyclopedia, California. http://en.wikipedia.org/wiki/California

World Resources Institute (2007) http://earthtrends.wri.org/searchable_db/index.php?step=countries&ccID%5B%5D=5&cID%5B%5D=190&theme=2&variable_ID=689&action=select_years Earth Trends The Environmental Information

York, D. and Wadsworth, L. (1998) Reuse in Florida: Moving towards the 21st Century, *Florida Wat. Resources J.* **11**, 31-33.

5

Water reuse in Australia and New Zealand

John Anderson, Jim Bradley and John Radcliffe

5.1 INTRODUCTION

5.1.1 Australia's Water Resources

Australia lies between the South Pacific and Indian Oceans, between latitudes 10S and 44S and longitudes 113E and 154E. Australia has an area of 7,682,000 km^2, about five-sixths that of the United States of America, but is sparsely populated, with only 20 million people. A high proportion of the population lives in urban centres located in high rainfall areas on the temperate southern margins of the continent between latitudes 25°S and 43°S.

Australia is among the driest countries on Earth. Australia has a variety of climatic areas ranging from sub-tropical, through temperate, to arid. The northern parts of the continent receive predominantly summer rainfall from tropical rainfall systems while the southern parts receive predominantly winter rainfall from the southern ocean rainfall systems. Weather systems are strongly influenced by the El Niño/La Niña ocean circulation and sea surface temperature phenomena in the Pacific Ocean, and also by Indian Ocean and Southern Ocean weather systems. The annual rainfall is extremely variable, ranging from over 3,000mm in the tropics to less than 200mm in central Australia. About 63% of the continent has less than 400mm rainfall. The average rainfall is 455mm. Because of high evaporation rates, surface runoff and groundwater recharge equate to only 12% of average rainfall. As well as the geographic variations, the temporal variability of the rainfall is greater in Australia than in any other continent. Consequently, runoff is also highly variable and much of the country is prone to droughts.

 ISBN: 9781843390893. Published by IWA Publishing, London, UK.

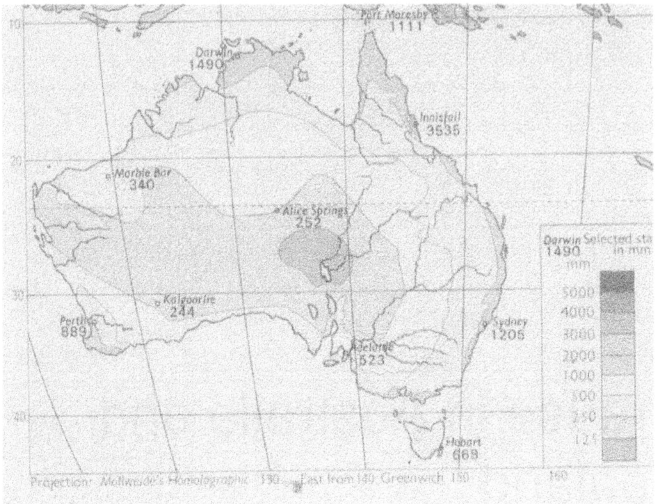

Figure 5.1. Australian Rainfall.

Although Australia has 5% of the global land area, it has only 1% of the global river runoff. Mean annual runoff is about 50 mm, a total of 400,000 Mm^3/yr. About 70% of this runoff occurs as floods. About half of the 120,000 Mm^3/yr of harvestable water occurs north of the Tropic of Capricorn and Tasmania, remote from the main centres of population. Of the remainder, about 26,000 Mm^3/yr is harvested for agricultural, industrial and urban use. The water stress index for Australia is about 7%.

Because of its sparse population, Australia is nominally a water-rich country with a mean *per capita* water availability of about 6,000 m^3/yr (harvestable water). However this broad statistic disguises the fact that in the river basins which are home to most of Australia's population and agriculture, the available water resources are almost fully committed. Most water use in Australia has been on a "once-through" basis before return to the natural water cycle by discharge or evaporation. About 67% is used for agricultural irrigation, 5% for mining and manufacturing, 17% for the service industries, 9% for household supply and 2% for water supplied to the environment (ABS, 2004). Australia *per capita* annual water use of about 1,300 m^3/yr is high because of the high use of water for irrigation of agricultural exports.

In the past, Australia has been extremely liberal in its approach to water rights and water use and this has led to some over-allocation. A number of the key river basins are showing signs of environmental stress, manifested through declining water quality. The rivers of the Murray-Darling Basin, which supply more than 50% of current water use in Australia, have deteriorated markedly since European settlement late in the 18th Century. The deterioration has occurred principally because of agricultural land use practices and diversion of water for irrigation, but also because of nutrients in urban wastewater and stormwater discharges. Similar deterioration has been observed in the Hawkesbury-Nepean Basin which drains western Sydney, caused principally by urban water diversions and urban runoff. These two cases highlight the impacts which occur as water use approaches, or exceed, the limits of sustainability in individual catchments. This highlights the constraints imposed on the use of water resources by the combined impact of water diversions and urban runoff.

Australia is a federation of states. Historically, the management of water resources has been the responsibility of the state governments. Concern over declining river water quality has led to new public policy measures to achieve sustainable management of Australia's water resources. In 1994, the Council of Australian Governments (CoAG) comprising the Prime Minister, State

Premiers, Territory Chief Ministers and the President of the Australian Local Government Association agreed to implement a Water Reform Agenda. In 2004-5, the federal and all state governments except Western Australia adopted a new National Water Initiative (CoAG, 2004). Provisions of the National Water Initiative include:

- water access entitlements and a planning framework;
- water markets and trading;
- best practice water pricing and institutional arrangements;
- integrated management of environmental water;
- water resource accounting;
- urban water reform, with greater use of recycled water;
- community partnerships and adjustment;
- knowledge and capacity building.

The state governments have introduced complementary water reform measures and further changes will be introduced to meet the requirements of the National Water Initiative by 2011. Measures included in the state water reform packages include:

- consolidation of existing water legislation into a new water management acts;
- the development of a water management plan for each catchment, incorporating environmental water quality objectives and river flow objectives which share water between users and the environment. These water sharing plans define licensed water access rights under volumetric allocations, which may be reduced in dry times;
- the development of integrated water planning for urban areas incorporating water conservation and recycling. Complementary pollution control measures impose load-based licence fees for discharges with rebates for water recycling;
- introduction of water efficiency in new buildings. The New South Wales Building Sustainability Index (BASIX) requires a 40% reduction in water use in new houses compared to the current Sydney average as a condition of development consent. (DIPNR, 2004a).

Several state governments – the governments of Victoria, South Australia, Western Australia and the Australian Capital Territory – have set water reuse targets. For example, the Western Australia State Water Strategy has set a target to reuse 20% of Perth's treated wastewater by 2012 (Gallop, 2003).

5.1.2 New Zealand's Water Resources

New Zealand lies in the South-West Pacific to the east of Australia between latitudes 35°S and 47°S and longitudes 167°E and 179°E. Unlike Australia, most of New Zealand does not have the extreme restrictions of fresh water availability to serve urban areas and agriculture. Generally rainfall is moderate and surface and groundwater are reasonably available. Accordingly there is not the need to conserve water across the country to the extent that in Australia leads to a significant amount of water reuse. Notwithstanding this, there are some areas of New Zealand that do suffer from dry periods. These include parts of the Canterbury Plains which are fed by ground water aquifers, parts of Marlborough/Nelson and Hawke's Bay.

While there is often some soil conditioning, fertilisation and irrigation value realised from a number of the wastewater land disposal schemes in New Zealand, this is not strictly speaking water reuse as is generally understood in Australia and other parts of the world where water scarcity is often the prime driver for reuse.

New Zealand does, however, have an environmental sustainability ethic, which together with the special cultural standards of the native Maori leads to a preference for land disposal and land treatment of treated wastewater. The position of Maori is that the direct discharge of (treated) human sewage to natural water is spiritually and culturally offensive. As a result, disposal of treated municipal and industrial wastewater to land is generally the favoured disposal option where it is feasible. In many cases, however, the natural rainfall is so high and/or the soils and topography are not appropriate, so that it is not feasible to irrigate treated wastewater.

There are also a number of primary processing industries that dispose of their treated industrial wastewater onto land and accordingly achieve some reuse value from the nutrients and irrigation effect. These industries include dairy factories, meat processing plants, wool scours and wineries.

5.2 WATER REUSE

5.2.1 Water Reuse in Australia

A national survey of water recycling in Australia was undertaken in 2003-2004 by the Australian Academy of Technological Sciences and Engineering (ATSE, 2004). Water reuse in 2001-02 was less than 10% of treated wastewater flows – see Table 5.1 (Radcliffe, 2007). The quantity of reuse has approximately doubled in the last 10 years with over 500 Waste Water Treatment Plants in Australia found in the Academy review to be recycling at least a proportion of their throughput, mostly for amenity horticulture and agriculture.

Water reuse is more common in the non-metropolitan rural areas away from the coast because of lower rainfall and increased opportunities for agricultural reuse. The proportion of water reused in the state capital cities (Table 5.2) was generally lower than the respective state figures in 2001-2 which incorporate water reuse in the country towns and agricultural districts.

Table 5.1 Annual water reuse from water utility STPs in Australia (Radcliffe, 2007).

Region	1996-97			2001-02			2005-06		
	Effluent GL/yr	Reuse GL/yr	%	Effluent GL/yr	Reuse GL/yr	%	Effluent GL/yr	Reuse GL/yr	%
QLD	328[2]	38[2]	11.6[2]	339	38	11.2	360	47.5	13.2
NSW	548[1]	40.1[1]	7.3[1]	694	61.5	8.9	679[5]	47.6[5]	7.0[5]
ACT	31[2]	0.25[2]	0.8[2]	30	1.7	5.6	31	2.1	6.7
VIC	367	16.9	4.6	448	30.1	6.7	439	81	18.4
TAS	43	1	2.3	65	6.2	4.5[4]	68	3.6	5.3
SA	91[2]	9[2]	9.9[2]	101	15.2	15.1	96	18.9	19.6
WA	109	5.5	6.1	123[3]	12.0[3]	9.7[3]	139	16	11.5
NT	21[2]	1[2]	4.8[2]	21	1.1	5.2	16	2.0	12.3
Aust.	1538	112.9	7.3	1824	166.5	9.1	1828*	218*	12.0*

[1]1996; [2]1998; [3]Minor correction since original publication; [4]Corrected for 5% approved, not operational; [5]2004-5; *assuming NSW 2004-5 data for 2005-6. (Source, see Radcliffe, 2007: 1996-9: [6], 2001-2: [2], 2005-6: ACT [7], NSW [57], Victoria [58], Queensland - B Davis, SA - L Sickerdick, Tas - S. Gallagher, NT - P Sherman.)

Table 5.2 Annual water reuse in Australian capital city STPs in 2001-02 and 2004 (Radcliffe, 2005).

State Capital	2001-02 (%)	2004 (%)
Sydney	2.3	2.6
Melbourne	2.0	14.0
Brisbane	6.0	3.5
Adelaide	11.1	19.2
Perth	2.3	4.1
Hobart	0.1	negligible

5.2.2 Water Reuse in New Zealand

Based on the commonly used definition of water reuse, there is very little true water reuse in New Zealand as such. There are two golf courses that reuse treated wastewater, one in Tauranga using low pressure spray irrigation and one in Omaha Beach using sub-surface irrigation. There is one very small community of approximately 40 houses at Golden Valley, Kuaotuna in the Coromandel using recycled wastewater for controlled non-potable use. The Tauranga City Council proposes making wastewater available for industrial reuse as part of their holistic and sustainable approach to wastewater management.

In terms of land disposal systems for domestic wastewater, there are presently about 60 Community Schemes in New Zealand. The largest three of these Schemes are Rotorua, Levin and Taupo, which in total cover approximately 450 ha of the 800 ha approximately covered by the 60 Schemes. Rotorua disposes of secondary and nutrient removed treated wastewater into pinus radiata forests growing on porous volcanic soils. Levin is similarly pinus radiata forests located on coastal sandy country, and Taupo is a cut and carry hay pasture operation.

5.3 REGULATIONS AND GUIDELINES

5.3.1 Australia

A set Australian national water reuse guidelines were published as part of the National Water Quality Management Strategy in 2000 (ARMCANZ/ANZECC, 2000a). These guidelines had four classes of recycled water corresponding to classes A to D in Table 5.3.

At this stage Australian authorities are operating under various state guidelines with up to five classes of reclaimed water (DEC 2004, EPA Victoria 2003, QEPA 2004, SADHS and EPA 1999). The Queensland example is shown in Table 5.3. The states are expected to move to bring state guidelines in line with the proposed new national guidelines, although water reuse local planning and approval requirements differ to match local administrative arrangements.

Table 5.3: Typical recycled water classes in Australian water reuse guidelines (QEPA, 2004).

Class of Water (High to Low)	Criteria for Class of Water					Highest Acceptable Uses for Class of Water
	E. coli (median) CFU/ 100mL	Physical criteria				
		Particulates		Chlorine residual, mg/L 100 percentile	pH	
		Turbidity NTU 100 percentile	TSS mg/L median			
A+	<10	2 (5)	—	1	6-8.5	High human contact. Medium risk of ingestion. Requires treatment that will achieve 5 log removal of viruses and protozoa in primary settled wastewater.
A	<10	2 (5)	—	1	6-8.5	High human contact. Low risk of ingestion.
B	<100	—	30	(1)	6-8.5	Medium human contact. Minimal risk of ingestion.
C	<1000	—	30	(1)	6-8.5	Low human contact. Minimal risk of ingestion.
D	<10,000	—	—	—	6-8.5	Non-food chain. No human contact.

New national water recycling guidelines for Australia were circulated for discussion in November 2005 (Cunliffe, Bursill and Hooy, 2005; EPHC, 2005) and adopted by Federal and State Ministerial Councils in November 2006 (NRMMC, EPHC, AHMC, 2006). Phase 2 of the guidelines – covering stormwater reuse, managed aquifer recharge and recycled water to supplement drinking water sources – were released as a discussion draft in June 2007 (EPHC, 2007).

It is expected that all states in the Australian federation will adopt the new national guidelines. A feature of the new approach to recycled water guidelines is the use of the Hazard Analysis and Critical Control Point (HACCP) approach to risk management to ensure the reliable delivery of safe recycled water.

Also of interest in the evolution of water reuse in Australia are the NSW Guidelines for Urban and Residential Use of Reclaimed Water (NSWRWCC, 1993). These guidelines introduced the A+ class of recycled water and made possible the landmark residential water reuse projects at Rouse Hill and Sydney Olympic Park in Sydney.

It has been practice in most states to require environmental assessment and irrigation studies to assess the long-term sustainability of water reuse projects. Factors considered include salt and heavy metal contents of the recycled water, depths to water tables and the accumulation of salts, nutrients and heavy metals in the environment.

5.3.2 New Zealand

The regulations covering the reuse of treated wastewater in New Zealand depend on the purpose for which the wastewater is to be reused. Reuse in domestic dwellings and industry is covered by the Health Act 1956 and the Building Act 2004. For discharge into the natural environment *i.e.* reuse for agricultural irrigation by land disposal and land treatment, the appropriate legislation is the Resource Management Act 1991 which requires resource consents to be issued for the discharge of treated wastewater (which contains contaminants) onto, or into, land.

Guidelines for the reuse of treated wastewater are set out in the New Zealand Ministry of Health's Guidelines dated 1991, titled Public Health Guidelines for the Safe Use of Sewage Effluent and Sewage Sludge on Land 1992. These Guidelines have been reprinted and are an Appendix to the New Zealand Guidelines for the Utilization of Sewage Effluent on Land 2000, which was prepared by the New Zealand Land Treatment Collective and the Forest Research Institute (NZLCFR, 2000; NZMFTE, 2003).

The Public Health Guidelines referred to above take precedence over the guidelines for sewerage systems reclaimed water that were prepared in 2000 by the Agriculture and Resource Management Council of Australia and New Zealand, Australian and New Zealand Environment and Conservation Council and National Health and Medical Research Council (ARMCANZ/ANZECC, 2000).

5.4 CASE STUDIES

5.4.1 Urban Reuse

Sydney Olympic Park, NSW - A water recycling scheme has been installed at Sydney Olympic Park where the Sydney Olympic Games were staged and has since been extended to the new suburb of Newington. Up to 7,000m^3/d of recycled water from stormwater and treated wastewater sources is used for toilet flushing in sporting venues, irrigation of open space areas, and is supplied to 2000 residential houses for garden watering and toilet flushing.

Figure 5.2. Dual membrane water reclamation plant, Sydney Olympic Park.

Microfiltration and reverse osmosis treatment processes are used to achieve the required water quality. The system effectively controls surface water from storm events and maximises its beneficial use. The scheme reduces demands on Sydney's freshwater supplies by about 2300 m^3/d. The scheme commenced operation in April 2001 (Listowski, 2000 and 2002; Rathjen *et al.*, 2003; Chapman, 2005).

Rouse Hill, NSW - Water is recycled for residential uses at Rouse Hill, a new housing development area in north-western Sydney. The area is ultimately planned to have 300,000 people with the first stage of the development catering for 100,000 people in 35,000 houses. A second reticulation system has been installed to supply recycled water for toilet flushing and garden watering in residential dwellings. The first section of the scheme commenced operation in August 2001 and currently supplies 15,000 houses. The first stage of the Rouse Hill water reclamation plant treats 4,400m^3/d for reuse with ozonation, microfiltration and superchlorination to meet the NSW Recycled Water Coordination Committee's guidelines (1993). The recycled water has met the guideline values for all microbiological parameters. The major risk/issue associated with the introduction of the recycling scheme has been the quality of household plumbing work. Pre-commissioning inspections have detected numerous errors. Recycled water (all uses) accounts for approximately 35% of total water consumed at Rouse Hill (de Rooy and Engelbrecht, 2003).

Mawson Lakes, SA - The Mawson Lakes housing development in Adelaide will house 10,000 people in 3700 houses and also serves a university and a commercial and industrial estate. Treated wastewater from the nearby Bolivar Wastewater Treatment Plant is treated and recycled for toilet flushing, garden watering and landscape irrigation. Stormwater from the site is collected, treated and recycled to provide lakes and water features within the estate and supplementary water for landscape irrigation. Aquifer storage and recovery is used to store surplus winter flows of stormwater to supply peak irrigation needs during summer.

New Haven Village, SA – New Haven Village is a development of 65 medium density dwellings on 2 ha at Le Fevre Peninsula, some 20 km northwest of central Adelaide, implemented in 1995. Its water and wastewater management system includes on-site treatment and reuse of household sewage (black and grey water) and a stormwater system which collects the first 50 m^3 of a given rain event, the remainder being diverted to a sports field acting as a retention basin. Recycled water from the treatment plant is used for house gardens, toilet flushing and an adjacent oval with irrigation systems that are sub-surface, although the recycled water meets the South Australian Health Commission's 'Class A' quality for unrestricted irrigation reuse. Some occasional difficulties have been experienced with the reclaimed water quality, colour and odour (Thomas *et al.*, 2000). In a recent survey of residents (Marks *et al.*, 2002), although 95% of respondents initially said they have no concerns, all described problems relating to water quality at some point in the interview.

Australian Capital Territory - Water reclaimed by secondary treatment at the Fyshwick plant has been used for over ten years on the playing fields at the Duntroon Military College as part of the North Canberra Effluent Reuse Scheme (NCERS). Microfiltration is used to provide exceptionally high quality reclaimed water. The plant provides 1600m^3/d of recycled water. The plant supplies irrigation water for 40 ha of public parks and sporting fields in O'Connor, Ainslie, Braddon, Reid, Campbell and Australian National University as well as 20 ha at the Australian Defence Force Academy. Capacity can be expanded to 3200m^3/d, potentially providing water for all of inner North and South Canberra.

Southwell Park, ACT - ACTEW has operated a 300m^3/d sewer mining ("watermining®") unit at Southwell Park in Canberra since 1995. The watermining unit has screening, lime settling, microfiltration and chlorination, the output being used on 11ha of local playing fields (ActewAGL, 2003b).

Urban Reuse, New Zealand - Apart from the Golden Valley Sub-division, Kuaotuna in the Coromandel, where there is one very small sub-division of approximately 40 houses using recycled wastewater for controlled non-potable use, there is no urban reuse of treated wastewater presently in New Zealand. A number of studies including the Tauranga City Council's Papamoa East Urban Water Sustainability Project have investigated urban reuse for non-potable purposes and decided against such a technique being adopted for new sub-divisions, at least at this stage. The health authorities have been involved in these investigations and cautioned against even non-potable re-use where there is an adequate local authority water supply available.

5.4.2 Industrial Reuse

Eraring Power Station, NSW - Recycled water from the Dora Creek sewage treatment works is pumped to the Eraring Power Station at Lake Macquarie, about 100 km north of Sydney. There it is further treated by microfiltration and reverse osmosis to produce a water of potable grade which is then further treated in the existing demineralisation plant to produce purified water, which is used as boiler feed to provide steam for the power station turbines. This recycled water replaces 2,740 m^3/d of potable water previously supplied from the urban water supply system (Cole and Deans, 1994).

Port Kembla, NSW - Recycled water is supplied to the Port Kembla steelworks for a variety of process water uses including cooling water, quenching of the blast furnace slag and quenching at the coke ovens. It is currently planned to expand the water recycling system at the Port Kembla steelworks in Australia to at least 35,000m^3/d. The first 20,000m^3/d stage was commissioned in 2006 (Hird, 2005).

Luggage Point, Brisbane - Brisbane City Council has a commercial contract with BP to provide between 10,600 and 14,000m^3/d of recycled water for 20 years. The recycled water, which is produced through an advanced microfiltration and reverse osmosis process, is used at the refinery as boiler feedwater (Hopkins and Barr, 2002). The Luggage Point installation was an attractive option for supplying the adjacent refinery when compared with the cost of the alternative of bringing in a large new main for the carriage of potable water, apart from also representing a significant saving in potable water.

Kwinana, Perth - The Kwinana Water Recycling Project was commissioned in 2004. It produces about 13,700 m^3/d of water below 50mg/L TDS by microfiltration and reverse osmosis at the Woodman Point STP for industrial use. Mining, power generation, chemical fertiliser and petroleum companies have already committed to use some of the output. Currently, the top 10 Kwinana industries use 20,000 m^3/d mains drinking water/annum and have entitlements to 64,000 m^3/d of groundwater per year (Turner, Davies and Ryan, 2003). New industries in the Kwinana Industrial Strip will be obliged to access the recycled water for industrial use as they will be unable to access groundwater, and will be discouraged or mandated from using potable supplies.

Toowoomba, Queensland - one of the largest cities in the Murray Darling Basin with a population of around 100,000 people supplies about 2,740 m^3/d from its Wetalla treatment plant through an 80 km pipeline to the Millmerran power station for use as cooling water.

Western Corridor Scheme, Queensland - The Queensland Government has recently adopted a major water recycling scheme to address water shortages caused by the current drought. Recycled water from Brisbane and Ipswich will receive dual membrane treatment and be pumped through a new Western Corridor recycled water grid to supply water to the Swanbank and Tarong power stations. In 2008 a pipeline extension will be constructed to Brisbane's Wivenoe Dam to enable natural inflows to be supplemented if necessary during severe droughts.

5.4.3 Agricultural Reuse

There is widespread use of treated wastewaters from urban, industrial and agricultural sources for beneficial reuse in agriculture in Australia and New Zealand.

Virginia, South Australia - A major scheme has been constructed to supply up to 65,000 m^3/d of recycled water, from the Bolivar sewage treatment plant in Adelaide, to the Virginia area north of Adelaide for irrigation of horticultural crops in an area where traditional groundwater supplies for irrigation were under stress. Early grower contracts were for up to 57,000 m^3/d in total, with initial usage of about 41,000 m^3/d. The scheme includes a 120,000m^3/d water reclamation plant that incorporates dissolved air flotation and filtration processes (Marks *et al.*, 1998). Recent investigations have been undertaken into the potential to develop an aquifer storage and recovery system which may increase the amount of water available for reuse annually to about 110,000 m^3/d (Martin *et al.*, 2002).

Shoalhaven Recycled Water Management Scheme - Shoalhaven City Council is developing a long term scheme to recycle water for irrigation of agricultural land on the Shoalhaven river floodplain near Nowra, 160 km south of Sydney. A public consultation process supported a proposal that additional rates be charged to support community objectives to eliminate discharges to the Jervis Bay estuary, and minimise ocean discharges. The scheme will reuse about 80% of the recycled water from local sewage treatment plants to irrigate dairy pastures, with the development staged in two parts over about ten years. The first stage will irrigate about 700 ha with possible later expansion to about 1,800 ha. The initial section to irrigate 350 hectares was commissioned in April 2002 and supplied 5,500 m^3/d in 2003. The recycled water is provided through purple pipes, without charge to local dairy farmers. They have converted from dryland to irrigated dairying, leading to a major boost in production and efficiency. This has allowed the farmers to adjust to the dairy industry deregulation reforms (Gould *et al.*, 2003).

Hervey Bay, Queensland - Wide Bay Water at Hervey Bay has implemented two recycled water irrigation schemes to reduce ocean discharges to sensitive waters. The schemes supply recycled water to irrigate 400 ha sugar cane, 60 ha of native pastures, 60ha of native woodland, a turf farm, golf course, sports field and the airport, currently achieving 75% reuse, but aiming for 100% by 2007 (Just, 2001). The recycled water is prepared by active sludge secondary treatment, is chlorinated and held in a lagoon for 30 days. Initially, it was used experimentally on a cane farm with traditionally low production, but which with the use of recycled water, became the highest producing property in the district. This change generating a considerable interest in and demand for recycled water by other growers. Average annual flow is about 7,900 m^3/d. Wide Bay Water has also instituted an innovative stormwater reuse program by harvesting stormwater from flood retardation basins at night and channelling the water through its sewers when there is very little sewage flow, for treatment at its STPs. This project has a number of advantages, including minimising odour production and corrosion within the sewerage network, reduced urban flooding, increased water for agricultural irrigation and increased farm profitability (Heron and Lever, 2001).

Maclaren Vale, SA - The Willunga Basin Water Company, which was formed by a group of grape growers seeking additional water for their vineyards, negotiated access without charge for 10 years to chlorinated class B/C effluent from the activated-sludge Christies Beach STP. A 13 km pipeline and delivery system has been built and supplies about 10.900 m^3/d. Recycled water is delivered at 65% of the current SA Water mains supply price. Salinity varies, but at 750-900 mg/L, is less than that of the now fully allocated groundwater resources of the Willunga Basin (ATSE, 2004).

Wagga Wagga, NSW - Wagga Wagga City Council, working with the CSIRO (Divisions of Water, Forestry and Soils) conducted a major woodlot irrigation trial, on a site that became known as *Flushing Meadows* (Myers *et al.*, 1992). In 1991, on a 7 ha site, 4,544 trees were planted in four blocks: a eucalypt rates trial, a pine rates trial, a pine clone trial and a species trial. The species trial evaluated a total of 60 species/provenances, also to identify an optimum choice for effluent irrigation. The trials demonstrated that *Eucalyptus grandis* grew faster and used more water than the pines. Water use plateaued once the canopy closed, and nutrient application rates, at the higher irrigation rates, exceeded the ability of any of the trees to take them up. Overall, the study found that there are several *Eucalyptus* species suitable for effluent woodlot applications, but that selection had to be based upon site-specific conditions. The outcome of the trials were used to prepare an Australian guideline for effluent irrigated tree plantations (Myers *et al.*, 1999)

Dubbo NSW - The City of Dubbo in the NSW Central West operates a water recycling scheme at its Troy Junction STP. Depending on rainfall for the year, varying amounts of effluent are used to grow fodder crops (lucerne, sorghum, maize and pastures) under centre-pivot irrigation. The Troy Junction reclaimed water is also combined with that from Fletcher International Exports' abattoir/woolscour and the blended water is applied to fodder crops. The overall scheme has been in operation since 1986, but the joint venture was implemented in 1995.

Armidale NSW - Armidale City Council has been applying effluent to pastures for more than 30 years. The reuse scheme was using 50% of total effluent flow by 2000 and the intention was to achieve 100% (6,000 m^3/d) ultimately. Flood irrigation is used to apply the water and operations are by a combination of a lessee and nearby farmers. Use of effluent has resulted in improved pasture growth rates and a careful evaluation of monitoring results has enabled a gradual reduction in the frequency and cost of monitoring

Narrabri NSW - Narrabri Shire Council applies effluent to 130 ha of cotton, by flood irrigation. SAR conditions were not ideal, so gypsum was applied to the site at the rate of 5 tonnes/ha to start. The annual water reuse by the scheme is 2,100 m^3/d. The scheme is operated in association with the Narrabri High School as an agricultural education project. Proceeds from the farm operation support educational activities at the school.

Barwon Water, Victoria - Barwon Water supplies Victorian Class C effluent for the irrigation of golf courses, grape vines, flowers, potatoes, tomatoes and turf production from its Black Rock STP. Reverse osmosis of part of the stream from this plant is being considered to lower the salinity of effluent being used for flower growing, turf and potato production (Barwon Water, 2003).

Gerringong-Gerroa, NSW - The Gerringong-Gerroa sewerage scheme has been developed for a previously unsewered area 100km south of Sydney. The STP has secondary and tertiary treatment using a biological reactor, clarification and sand filtration, followed by advanced tertiary treatment involving ozonation, biologically activated carbon, microfiltration and UV disinfection. An obligatory minimum of 80% of effluent is used for pasture production on an adjacent dairy farm that currently irrigates 70 ha, but has a capacity to irrigate 120 ha. A 50,000m^3 holding dam is used to manage the flow. Surplus wet weather flows are discharged to an infiltration pond in nearby sand dunes adjacent to the ocean (Boake, 2005).

Dairy Farms and Piggeries, New Zealand - Most dairy farms and many piggeries in New Zealand irrigate their wastewater onto the surrounding farm land. Whilst this is primarily done as a means of environmentally acceptable disposal rather then re-use, farmers usually appreciate the value of the nutrients in the wastewater as well as the water in boosting pasture growth. In 2000, the New Zealand Dairy Industry under the New Zealand Dairy Board developed and adopted the Dairy Industry Environment and Annual Welfare Policies. One of these policies banned the spreading of human waste to pastures that was grazed by, or harvested for feed of dairy animals. Following growing demand from a wide range of authorities to utilize land application of treated human sewage the policy has been reviewed. A level of treatment established for which it is acceptable to apply treated sewage or sewage derived material to pasture destined for consumption by diary animals has now been defined. The level of treatment is equivalent to the Title 22 of the California Health Law. This requires a high level of secondary treatment and disinfection along with a management plan being developed for the pasture, or product in question.

Fonterra Dairy Factory, New Zealand - The Fonterra dairy factory at Edendale in the South Island of New Zealand is the second largest milk processing plant in the country. For much of the year all of the processing wastewater is irrigated onto land owned by Fonterra around the processing plant or onto another farm specially purchased for the purpose see Figure 5.3). However, the plant also has a wastewater treatment facility that can provide a high standard of treatment of the wastewater if it has to be discharged to the nearby Mataura River at times when the land is too wet to sustainably accept the treated wastewater.

Figure 5.3. Irrigation of dairy factory wastewater on land near the Fonterra Edendale factory.

Other Industrial Process Wastewaters, New Zealand – At the Oringi meat processing plant, near Dannevirke in the North Island of New Zealand, the processing wastewater is treated in an anaerobic lagoon and the effluent then used to irrigate a plantation of eucalypts and poplar. The trees are grown in a short rotation cycle and harvested by a chipper to provide fuel for the plant boiler.

The Rayonnier medium density fibreboard (MDF) mill near Mataura in the south of New Zealand also irrigates the plant wastewater onto land, as does the Canterbury Meat Packers near Ashburton. The Feltex Kakariki woolscour in the North Island of New Zealand irrigates treated flowdown liquor onto land and is proposing to also discharge rinsewater onto land.

5.4.4 Environmental Reuse

Taronga Zoo, Sydney, NSW - A small water recycling scheme (250m^3/d) has been installed at the Taronga Zoological Gardens in Sydney. Wastewater from the zoo is treated in a conventional activated sludge process followed by microfiltration and disinfection. The recycled water is used for landscape irrigation, moats and the washing down of animal enclosures.

Hawkesbury Environmental Flows, NSW - The NSW Government is evaluating proposals to use up to 110,000 m^3/d of water from the western Sydney treatment plants to restore environmental flows in the Hawkesbury and Nepean Rivers. The project would be part of a wider project which would also supply up to 110,000 m^3/d for urban and agricultural uses in the north-west and south-west sectors of Sydney over the next 25 years (DIPNR, 2004b).

Wetland Infiltration, New Zealand - Treated wastewater effluent from Otaki, located in the North Island south of Levin, is discharged to land via an infiltration wetland, which is essentially a natural form of wetland where the effluent can seep away through the base of the wetland. Native flax grows vigorously in the constructed wetland with the high nutrient content of the effluent. A similar infiltration wetland covering 30ha is used to dispose of treated wastewater from the town of Kaiapoi in the South Island of New Zealand.

Mixed Land Infiltration and River Flows, New Zealand - There are a small number of wastewater schemes, both municipal and industrial that use a "mix and match" approach to treated wastewater disposal and discharge. These schemes are based on the concept using the seasonal assimilative capacity of the soil and waters in a way to maximize the environmental sustainability, while still disposing/discharging treated wastewater. In dry, normally summer, periods disposal is to land when the land is drier and can accept the wastewater and when the streams and rivers are lower in flow; in wetter, normally winter, periods discharge is to the stream or river which has greater assimilative capacity and the land being wetter does not. In some cases return of treated wastewater to a stream or river is used as a means to supplement low stream flows and thereby maintain minimum environmental flows in that water body.

5.4.5 Decentralised Water Harvesting and Water Recycling

Australia is also seeing the development of decentralised water harvesting and water recycling systems as part of new approaches to water sensitive urban development and sustainable water systems (Mitchell, 2002).

Figtree Place, Newcastle, NSW - An urban redevelopment project at Figtree Place in the suburb of Hamilton incorporates rainwater tanks and cleansed stormwater into an integrated water-management system. The project, located on the site of a former tram depot has 27 residential dwellings with an effective housing density of 45 units per ha. The planning provided for the scheme to supply 50% of the in-house needs for hot water and toilet flushing, all of the domestic irrigation needs of the site, and surplus water to supply all the bus-washing needs of the adjacent Hamilton bus depot. Rainwater collection on the site has reduced mains water consumption on the site by 54%, while stormwater run-off from the site has been eliminated (Coombes, Argue and Kuczera, 2000).

Carlton, Vic - The refurbished building at 60 Leicester Street, Carlton, occupied by the Australian Conservation Foundation, incorporates waterless urinals and an in-house biological STP, with water from it being used to flush toilets and irrigate the internal and rooftop gardens. The design brief included collected rainwater to replace 100% of normal mains water unless health considerations dictated otherwise, 100% on-site treatment of 'grey water' and 'black water', and to establish a new benchmark for low water consumption in a commercial building (Krockenberger, 2003).

St Kilda, Vic - On the site of the former St Kilda Municipal Depot at Inkerman Street, St Kilda, a 236 unit housing complex has been developed. The complex incorporates recycling of domestic greywater (bathroom basins, baths and showers), from about half the units in four buildings using an activated-sludge (aeration) tank, with secondary filtration in a 400 m^2 native wetland and sand filtration on the site. The combined grey/stormwater is recycled for sub-surface garden irrigation and toilet flushing across the entire development (Melbourne Water, undated).

Salisbury, South Australia - The City of Salisbury in Adelaide has been very innovative in stormwater management, having established 36 wetlands, and subsequently adopted Aquifer Storage and Recovery (ASR) for wetland-treated water, initially for its own parks and gardens. Recently, it has established a joint venture to store and treat stormwater on Parafield Airport to provide over 2700 m^3/d to G. H. Michell & Sons Australia Pty Ltd, Australia's largest wool processor. (Hains, 2003).

5.5 DISCUSSION

5.5.1 Australia

Australia has made great advances in developing water reuse over the last 15 years. In a series of national workshops in 2003, however, water recycling practitioners identified a number of barriers to achieving greater water reuse. As a result, the Australian Water Association, an affiliate of the International Water Association, made a submission to the Commonwealth National Water Initiative recommending ten national water recycling innovation and development goals to be implemented as part of the Commonwealth's National Water Initiative (see Figure 5.4).

5.5.2 New Zealand

In the future it can be expected that there will be more reuse of treated wastewater in New Zealand, although unlike Australia and other countries there is unlikely to be the same water shortage driver except in some drier geographic areas. In many of the land disposal schemes there are benefits achieved in the crop growth by the irrigation (watering) and uptake of nutrients and organic material. There are, however, concerns with some schemes about their long term sustainability from the perspective of groundwater contamination (particularly nitrogen level build up and pathogenic micro-organisms), soil deterioration, health and safety issues of crops and the personnel associated with the activities.

More sustainable approach to the integration of the three urban waters (water supply, wastewater, stormwater) is expected to result in more reuse for industrial, irrigation and other non-potable uses.

In 2003 the New Zealand Government adopted a programme of action for Sustainable Development. That programme has as one of its four key issues the quality and allocation of freshwater. Activities in the programme on this key issue included a wide-ranging assessment of how to better manage and protect fresh water in New Zealand. This assessment includes wastewater assessment in terms of reuse for water resource efficiency, use of ecosystem services in wastewater treatment and treated wastewater ecosystem re-entry.

Table 5.4: AWA'S Ten-Point Plan for More Water Reuse in Australia.

Integrated Water Cycle Management	
1.	To establish integrated water cycle management plans in all Australian jurisdictions, linked to catchment planning, with water saving targets and incorporating water sensitive urban design, water conservation and water reuse initiatives.
2.	To develop new project evaluation methodologies which quantify the external benefits of water conservation and water recycling.
3.	To identify alternative urban water pricing frameworks which: address externalities, provide effective price signals and provide incentives for consumers and developers to invest in integrated water cycle management, water conservation and water recycling initiatives.
Guidelines and Approvals	
4.	To develop uniform water laws throughout Australia including clear enunciation of laws on rights and access to recycled water and stormwater, and water accounting systems which recognise return flows as a resource and allow trading or allocation credits consistent with downstream needs and entitlements.
5.	To develop revised uniform national water reuse guidelines, covering recycled water and stormwater, as part of the National Water Quality Management Strategy.
6.	To develop a uniform water reuse project approval process throughout Australia to simplify project approvals, provide greater project certainty and allow greater investment in project implementation rather than regulation and compliance.
Best Practice Projects and Technologies	
7.	To develop an Australian best practice water reuse project casebook, linked to an international casebook developed in partnership with the International Water Association, regional groups and other national associations.
8.	To develop best practice guidelines for selected water reuse applications and technologies.
Education	
9.	To develop a national water education program incorporating: consistent positive water terminology; vocational, school and community programs; a National Water Education Centre; and technical tours of leading Australian water facilities and international water facilities such as Singapore's NEWater water recycling visitor centre.
10.	To develop a knowledge network for water reuse and sustainable water management to deliver water reuse conferences, technical tours, newsletters and information to all stakeholders, in partnership with the International Water Association and other groups.

5.6 CONCLUSIONS

Although Australia and New Zealand have low levels of water stress in most catchments, there is strong interest in developing water reuse projects as a means of achieving more sustainable use of national water resources. There has been a large increase in the amount of water reuse in Australia over the last 15 years.

Australia is developing new national water recycling guidelines, which it expected the states will adopt. Several Australian states have set water recycling targets. It is expected that there will be continuing strong growth in water reuse over the next two decades.

5.7 REFERENCES

ABS (2004) Water Account Australia 2000-01. Australian Bureau of Statistics Cat. No. 4610.0, Canberra

ActewAGL (2003) Watermining. http://www.actewagl.com.au/projects/watermining.cfm.

ARMCANZ/ANZECC (2000). National Water Quality Management Strategy – No. 14. Australian Guidelines for Sewerage Systems – Reclaimed Water. Agriculture and Resource Management Council of Australia and New Zealand, and Australian and New Zealand Environment and Conservation Council, Canberra ACT.

ATSE (2004) Water Recycling in Australia, J.C. Radcliffe, author, Australian Academy of Technological Sciences and Engineering: Melbourne 233 pp ISBN 1875618 80 5.

AWA (2004). AWA Ten-point plan for more water recycling in Australia – Submission to intergovernmental deliberations on the National Water Initiative. Australian Water Association, Artarmon, NSW, 2004

Barwon Water (2003). Barwon Water's use of Recycled Water. Barwon Water, Geelong, February 2003, 12 pp

Bencke, W (2001). Domestic Wastewater Recycling trial – implementing on-site wastewater recycling in urban areas – lessons learned. ActewAGL, Canberra, ACT. June 2001

Boake, M.J. (2005) Recycled Water Case Study: Gerringong-Gerroa, in Khan, S.J., Muston, M.H. & Schafer, A.I. (eds), *Integrated Concepts in Water Recycling*, Wollongong, February 2005.

Bovill I & Simpson J (1998), Water Education Project, Proc RWWC, *6th NSW Recycled Water Seminar*, Sydney November 1998, pp126-131

Chapman, H. (2005), WRAMS, Sustainable Water Recycling, in Khan, S.J., Muston, M.H. & Schafer, A.I. (eds), *Integrated Concepts in Water Recycling*, Wollongong, February 2005.

CoAG (2004) Intergovernmental Agreement on the National Water Initiative http://www.nwc.gov.au/NWI/docs/iga_national_water_initiative.pdf.

Cole, C. & Deans, G. (1994), From Effluent to Power: A Major Recycling Initiative by Hunter Water and Pacific Power, *Proc RWCC 4th NSW Recycled Water Seminar,* Newcastle May 1994, 28-35.

Coombes, P.J., Argue, J.R. and Kuczera, G. (1999) Figtree Place: a case study in water sensitive urban development (WSUD), *Urban* Water, Vol 1 (1999) 335-343.

Cunliffe, D.A., Bursill, D. & Hooy, T. (2005), Developing National Guidelines on Water Recycling, in Khan, S.J., Muston, M.H. & Schafer, A.I. (eds), *Integrated Concepts in Water Recycling*, Wollongong, February 2005.

De Rooy, E. and Engelbrecht, E. (2003) Experience with Residential Water Recycling at Rouse Hill, in T. Gardner and D. McGarry (eds) Water Recycling Australia 2nd National Water Recycling Conference, AWA, Brisbane, 1-3 September 2002.

DEC (2004), Environmental Guideline: Use of Effluent by Irrigation, NSW Depeartment of Environment and Conservation, October 2004, ISBN 174137 076 0

DIPNR (2004a). BASIX – Building Sustainability Index. NSW Department of Infrastructure, Planning and Natural Resources, May 2004, www.basix.nsw.gov.au.

DIPNR (2004b). Meeting the challenges – Securing Sydney's water future: The Metropolitan Water Plan 2004. NSW Department of Infrastructure, Planning and Natural Resources, October 2004, www.dipnr.nsw.gov.au.

EPA Victoria (2003). Guidelines for environmental management – Use of reclaimed water. Publication 464.2. EPA Victoria, Southbank, Victoria, September 2002.

EPHC (2005) National Guidelines for Water Recycling – Managing Health and Environmental Risks. Draft for Public Consultation. Natural Resource Management Ministerial Council, Environment Protection and Heritage Council, October 2005 http://www.ephc.gov.au/ephc/water_recycling.html

EPHC (2007), Australian Guidelines for Water Recycling: Managing Health and Environmental Risks (Phase 2): Augmentation of Drinking Water Supplies, Commonwealth of Australia, June 2007.

Gallop, G. (2003). Securing our water future – A state water strategy for Western Australia. Government of Western Australia, Perth, February 2003.

Gould, J., Lee, P., Ryl, J. and Mulligan, B. (2002). Shoalhaven Reclaimed Water Management Scheme – Clever Planning Delivers Bigger Environmental Benefits. IWA 3rd World Water Congress, Melbourne 7-12 April 2002, paper e21703.

Hains, S. (2003) Environmental diversity and economic opportunities – the Salisbury wetland experience. Planning Institute of Australia Congress, Adelaide, March 31 2003

Heron, D. and Lever, M. (2001). Urban Rural Environmental Partnership. In: Water Recycling Australia, P.J. Dillon, Ed, CSIRO and AWA, Canberra ACT

Hird, W. (2005), Recycled Water Case Study: BlueScope Steel Port Kembla Steelworks, in Khan, S.J., Muston, M.H. & Schafer, A.I. (eds), Integrated Concepts in Water Recycling, Wollongong, February 2005.

Hopkins, L., and Barr, K. (2002). Operating a water reclamation plant to convert sewage effluent to high quality water for industrial use. 3rd World Water Congress of the IWA, Melbourne 7-12 April 2002, paper e21595.

Just, G. (2001). Is your sewage effluent the sweetener in my coffee – one lump or two? CD ROM, Australian Water Association 19th Convention, Canberra, paper 061, Australian Water Association, Sydney.

Krockenberger M. (2003). Building our hopes and dreams. *Habitat* 31: (1) 11-13 February 2003.

Listowski, A., Phillips, B.C. and Steele, P. (2000) Reusing Stormwater at Homebush Bay, In Hydro2000 3rd International Hydrology and Water Resources Symposium, IEAust. Perth, 20-23 November 2000.

Listowski, A. (2002) Recycled Water System for Future Urban Development, IWA World Water Congress. Melbourne, April 2002.

Marks, J., Cromar, N., Howard, F., Oemcke, D. and Zadoroznyj, M. (2002) Community Experience and Perceptions of Water Reuse, In Environ 2002 IWA World Water Congress. Melbourne, April 2002.

Marks, R., Wright, C., Kracman, B. and Thomas, R. (1998). Development of Australia's largest high quality effluent release scheme – Bolivar, SA. 11th IWSA-ASPAC Conference, Sydney Nov 1998, pp 564-570.

Martin, R., Clarke, D., Dennis, K., Graham, J., Dillon, P., Pavelic, P. and Barry, K. (2002). Bolivar Water Reuse Project. - NHT Project 974517. Report 2002/02, Department of Water, land and Biodiversity conservation, Adelaide, July 2002.

Melbourne Water (undated) Site 1 – Inkerman Oasis, St Kilda, accessed 310/2003, www.melbournewater.com.au/content/library/wsud/case_studies/inkerman_oasis.pdf

Mitchell, C. and White, S. (2003) Sustainable Urban Water Service Systems, Innovations in Water: Ozwater Convention & Exhibition, Perth, 6-10 April 2003.

Mitchell, V.G. (2002) Developing and Assessing New Approaches to Urban Water Services Provision, In 2nd National Conference on Water Sensitive Urban Design, 2-4 September 2002, Brisbane.

Myers, B.J., Falkiner, R.A., and Crane, W.J.B. (1992), The Wagga effluent plantation project – a CSIRO study of sustainable water and nutrient management in effluent-irrigated plantations, Proc. Recycled Water Seminar, NSW Recycled Water Co-ordination Committee, Wagga Wagga, May 1992

Myers, B.J., Bond, W.J., Benyon, R.G., Falkiner, R.A., Polglase, P.J., Smith, C.J., Snow, V.O. and Theiveyanathan, S. (1999). Sustainable Effluent-Irrigated Plantations: An Australian Guideline. CSIRO Forestry and Forest Products, Canberra, ACT., 293pp.

NRMMC, EPHC, AHMC (2006), Australian Guidelines for Water Recycling: Managing Health and Environmental Risks (Phase 1), Commonwealth of Australia, November 2006

NSWRWCC, (1993), NSW Guidelines for Urban and Residential Use of Reclaimed Water, NSW Recycled Water Coordination Committee, 1st Edition, May 1993.

NZLCFR (2000), New Zealand Guidelines for Utilisation of Sewage Effluent on Land, New Zealand Land Collective and Forest Research 2000.

NZMFTE (2003), Sustainable Wastewater Management, Ministry for the Environment, New Zealand Government, June 2003.

QEPA (2004), Queensland Guidelines for the safe use of recycled water - Public consultation draft,
Envirопment Protection Agency, Brisbane, www.epa.qld.gov.au/water_recycling_guidelines

Radcliffe, J.C. (2003). "An overview of water recycling in Australia – Results from the recent ATSE study". In Australian Water Association – *Water Recycling Australia 2nd National Conference*, Brisbane, 1-2 September 2003.

Radcliffe, J.C. (2005), The Future Directions for Water Recycling in Australia, in Khan, S.J., Muston, M.H. & Schafer, A.I. (eds), Integrated Concepts in Water Recycling, Wollongong, February 2005.

Radcliffe, J. C. (2007). Advances in Water Recycling in Australia 2003-2007. In: Water Reuse and Recycling 2007 SJ Khan, RM Stuetz and JM Anderson (Eds), University of NSW, Sydney. Pp 387-406.

Rathjen, D., Cullen, P., Ashbolt, N., Cunliffe, D., Langford, J., Listowski, A., McKay, J., Priestley, A. and Radcliffe, J. (2003), *Recycling Water for Our Cities*. Report to the Prime Minister's Science Engineering and Innovation Council, 2003.

SADHS & EPA (1999). South Australian Reclaimed Water Guidelines. SA Department of Human Services & Environmental Protection Agency, Government of South Australia, April 1999. http://www.environment.sa.gov.au/epa/pdfs/reclaimed.pdf

Tanner, C. (2002) Water Cycle Management and Integration Case Study of Manly West ESDR, 2nd National Conference on Water Sensitive Urban Design, 2-4 September, Brisbane.

Turner, N., Davies, G. and Ryan, C. (2003).Reclaimed Water - Towards a sustainable city for Perth. CD ROM, Australian Water Association 20th Convention, Perth, April 2003, paper 184, Australian Water Association, Sydney.

6

Water reuse in Central Europe

Peter Cornel and Alessandro Meda

6.1 INTRODUCTION

Europe is one of the most densely populated areas in the world. The climate in Europe varies from sub-arctic in the far north to Mediterranean (with hot dry summers) in the south, a predominantly maritime climate in the west to a more continental climate in Eastern Europe. Thus among the European countries, most of the northern countries have abundant water resources whereas in southern European countries water resources are limited.

For this reason, discussion of water reuse in Europe has been divided up in this book: the southern European countries such as Portugal and all the countries bordering the Mediterranean Sea (Spain, France, Italy, Malta, Slovenia, Croatia, Bosnia and Herzegovina, Serbia and Montenegro, Albania, Greece, Cyprus and Turkey) have been addressed in Chapter 3 (Water Reuse in the Northern Mediterranean Region); this chapter will cover Central Europe, especially the countries of the European communities from Ireland in the west, Norway and Sweden in the north to Bulgaria in the south-east. It will cover islands in the Atlantic Ocean such as Iceland, Ireland and the United Kingdom, as well as Alpine countries like Switzerland and Austria, and the water-rich Scandinavian countries.

The density of population is low in Norway, with roughly 15 *capita* /km^2 and is highest in The Netherlands with around 485 *capita*/km^2. And last but not least, the availability of statistical data concerning e.g. water resources, water consumption, water use and reuse, differ considerably among the surveyed countries.

©2008 IWA Publishing. *Water Reuse - An International Survey of current practice, issues and needs* by B. Jiménez *et al.* ISBN: 9781843390893. Published by IWA Publishing, London, UK.

6.2 OVERVIEW

Table 6.1 lists the surveyed countries in alphabetical order together with some basic data such as land area, population and population density, arable land as well as irrigated land and average annual precipitation.

Table 6.1. Surveyed countries with some characteristic data (The World Factbook, 2005).

	Land area	Population	Pop. density	Arable land	Irrigated land		Average precipitation[1)]
Country	1,000 km^2	10^6 *capita*	*capita*/km^2	1,000 km^2	1,000 km^2	%[2)]	mm/year
Austria	82.4	8.19	99	13.93	0.457	3.28	1,110
Belgium	30.3	10.36	342	7.05	0.040[3)]	0.57[3)]	847
Bulgaria	110.6	7.45	67	44.24	8.000	18.08	608
Czech Republic	77.3	10.24	133	30.76	0.240	0.78	677
Denmark	42.4	5.43	128	22.90	4.760	20.78	703
Estonia	43.2	1.33	31	6.93	0.040	0.58	626
Finland	304.5	5.22	17	21.89	0.640	2.92	537
Germany	349.2	82.43	236	118.21	4.850	4.10	700
Hungary	92.3	10.01	108	46.25	2.100	4.54	589
Iceland	100.3	0.30	3	0.07	n.d.	n.d.	978
Ireland	68.9	4.02	58	10.47	n.d.	n.d.	1,118
Latvia	63.6	2.29	36	18.87	0.200	1.06	641
Lithuania	65.2	3.60	55	29.48	0.090	0.31	656
Luxembourg	2.6	0.47	181	0.60	0.040[4)]	6.64[4)]	934
The Netherlands	33.9	16.41	484	9.05	5.650	62.43	778
Norway	307.9	4.59	15	8.84	1.270	14.37	1,120
Poland	304.5	38.64	127	139.78	1.000	0.72	600
Romania	230.3	22.33	97	94.02	28.800	30.63	637
Slovakia	48.8	5.43	111	14.72	1.740	11.82	824
Sweden	410.9	9.00	22	26.88	1.150	4.28	624
Switzerland	39.8	7.49	188	4.14	0.250	6.03	1,537
UK	241.6	60.44	250	56.68	1.080	1.91	1,220
Sum	3,050.4	315.66	-	725.77	62.357	-	-

[1)] Average precipitation from (FAO 2005)
[2)] Percentage irrigated land to arable land
[3)] Including also Luxemburg
[4)] Including also Belgium
n.d.: no data

Table 6.1 shows the extreme variety in size and population of the countries and demonstrates that agricultural irrigation is a minor issue in most of the countries, as usually wet summers deliver sufficient rainfall for the growth of agricultural products. More than 60% of irrigated land in The Netherlands emphasizes the specific situation of this country, with its highly mechanized agricultural sector, in greenhouses, which provides large surpluses for the food-processing industry and for exports.

The economic data of the surveyed countries show a large span in Gross Domestic Product (GDP) per capita ranging from 62,700 US$ in Luxembourg to 8,300 US$ in Romania. Only three countries, namely Bulgaria, Iceland and Romania, generate more than 10% of the GDP from the agricultural sector, none more than 15% (The World Factbook 2005).

6.3 CURRENT WATER SITUATION IN EUROPE

Water consumption patterns are closely related to the level of income, to industrial activities but also to climatic conditions, which directly affect water consumption. Strong seasonal peaks in municipal water demand might occur in regions where tourism plays a major role or where e.g. golf courses and sport fields have to be irrigated in summer or swimming pools are operated, which all consume especially high amounts of water causing serious stress on resources, especially in Mediterranean countries.

Industrial activity is relatively high in Europe. This might stress water resources in some regions. The degree of dependence of industrial production on sufficient water became obvious in the hot and dry summer of 2003, when power plants had to shut down temporarily, mainly in southern Europe, as cooling water fell short.

The largest water consumer world-wide is agriculture for food production, but in Central Europe irrigation plays no dominant role since the climate is mainly temperate with some rain in summer throughout the period of plant growth. Nevertheless, there is some irrigation with a growing demand caused by more intensive agricultural methods such as, for example, in greenhouses.

6.3.1 Water resources (water availability)

Table 6.2 provides an overview of the renewable water resources for the surveyed European countries. The data presented in this table are taken from FAO (2005).

The natural renewable water resources (NRWR) for a country consider not only the water resources originating from the precipitations occurring within a country's borders, but also the rivers flowing into the country, which originate in other countries.

It is interesting to compare the average annual amount of precipitation with the natural renewable water resources referred to the surface area of the country. In Figure 6.1, the surveyed European countries are ranked on the x-axis with decreasing natural renewable water resources expressed in millimetres (litres per square meter). It can be noted that, in some countries, the amount of natural renewable water resources is higher than the average precipitation. This is, in particular, the case for The Netherlands and Hungary, where the water resources brought by large rivers (Rhine, Meuse and Schelde in The Netherlands, and Danube in Hungary) are predominant.

Considering a more detailed spatial grid, such as a river basin, it can be noted that there is a large spatial variability in water availability among river basins in Europe. Figure 6.2(a) shows the average annual water availability in European river basins as calculated using a global water assessment model called WaterGAP. The WaterGAP consists of two main components, a Global Water Use Model and a Global Hydrology Model, which are applied to compute water use and availability on the river basin level (Lehner *et al.*, 2001).

Figure 6.2(a) "shows that the annual water availability ranges between well above 1,000 mm/year (West Norway, Britain's West Coast, South Iceland) to below 100 mm/year (parts of Spain, Sicily, large parts of the Ukraine, South Russia, large parts of Turkey). In most parts of Europe this reflects current patterns of precipitation – while in other parts river discharge is carried through streams into more arid regions (Hungary, for example, receives most of its water from outside the country borders through the Danube river)" (Lehner *et al.*, 2001).

Table 6.2. Water resources among the surveyed European countries (FAO, 2005).

	Internal renewable water resources (IRWR)		Natural renewable water resources (NRWR) (including flows from other countries)		Dependency ratio
	total	*per capita*	total	*per capita*	
Country	Mm^3/yr	m^3/yr	Mm^3/yr	m^3/yr	%
Austria	55,000	6,720	77,700	9,616	29
Belgium	12,000	1,158	18,300	1,786	34
Bulgaria	21,000	2,819	21,300	2,680	1
Czech Republic	13,150	1,284	13,150	1,280	0
Denmark	6,000	1,105	6,000	1,128	0
Estonia	12,710	9,536	12,810	9,195	1
Finland	107,000	20,486	110,000	21,268	3
Germany	107,000	1,298	154,000	1,878	31
Hungary	6,000	600	104,000	10,433	94
Iceland	170,000	572,898	170,000	609,319	0
Ireland	49,000	12,201	52,000	13,673	6
Latvia	16,740	7,309	35,450	14,642	53
Lithuania	15,560	4,326	24,900	6,737	38
Luxembourg	1,000	2,132	3,100	7,094	68
The Netherlands	11,000	670	91,000	5,736	88
Norway	382,000	83,170	382,000	85,478	0
Poland	53,600	1,387	61,600	1,596	13
Romania	42,300	1,894	211,930	9,445	80
Slovakia	12,600	2,320	50,100	9,279	75
Sweden	171,000	18,996	174,000	19,679	2
Switzerland	40,400	5,395	53,500	7,462	24
UK	145,000	2,399	147,000	2,465	1
Sum	1,450,060	-	1,973,840	-	-

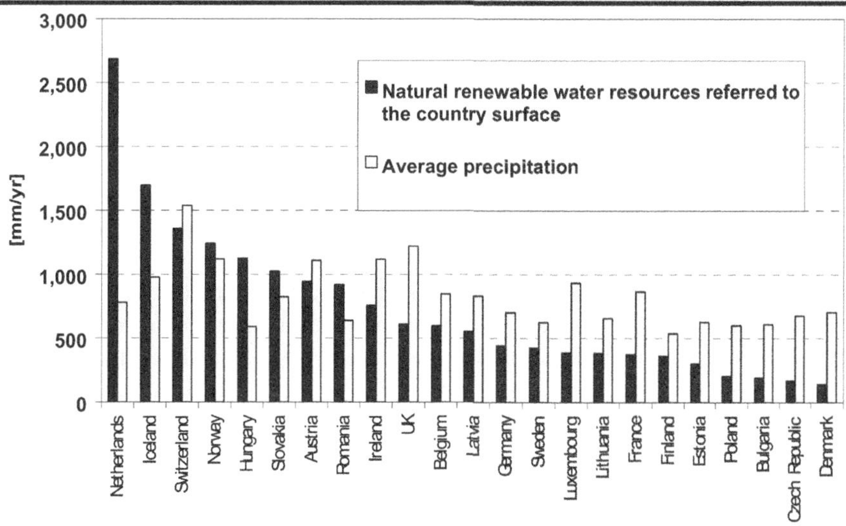

Figure 6.1. Comparison between natural renewable water resources and average annual precipitation for the surveyed European countries.

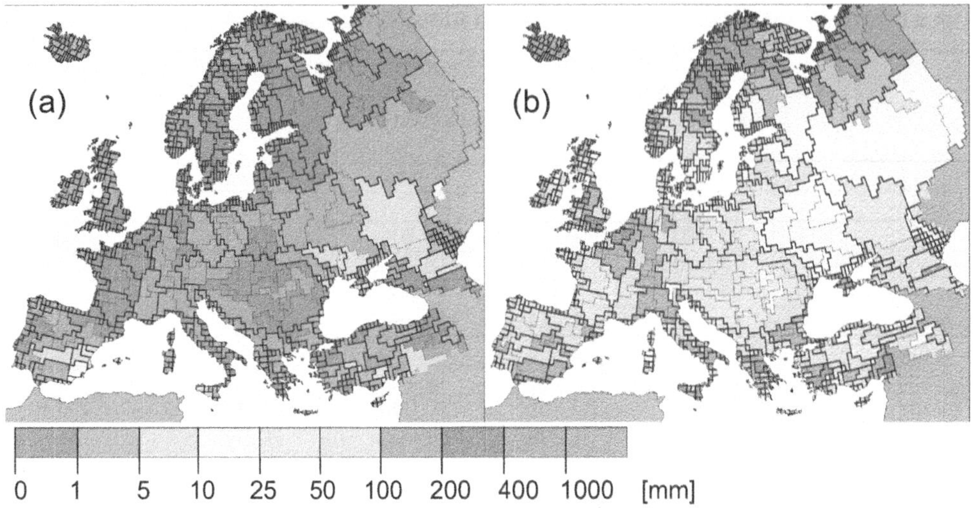

Figure 6.2. (a) Average annual water availability in European river basins and (b) average annual water withdrawals from European river basins in 1995 (Lehner *et al.* 2001).

6.3.2 Water withdrawals

According to the data displayed in Table 6.3, the total water withdrawal of the countries considered in this chapter amounts to about 149,000 Mm^3 per year (data are estimates by FAO for the year 2000). From this withdrawn amount, about 22% are used in agriculture, 65% in industry and 13% for domestic needs. These percentages are calculated by dividing the total sectoral withdrawn amount over total withdrawn amount. Among the countries there are great differences in the amount of water withdrawal and in the use of the withdrawn water. The need for water in industry dominates in most European countries, ranging from 15 to 85%, with especially high percentages in the north-western industrialized countries such as Belgium, Finland, Ireland and the United Kingdom, but also in eastern countries such as Poland and Bulgaria. The water share for agricultural irrigation ranges from 0 to 57%, whereby it should be noted that, for 14 of a total of 22 of the considered countries (corresponding to two thirds), it is equal to or lower than 10%. The water share for agriculture is high in some south-eastern European countries such as Romania, Hungary and Bulgaria and in some northern European countries with well-developed intensive agriculture, such as Denmark, The Netherlands and parts of Germany.

Considering a more detailed spatial grid, it can be noted in Figure 6.2(b) that, in Europe, there is a large spatial variability of water withdrawal, even within the same country: it ranges from nearly zero (in the thinly populated areas of sub-polar Scandinavia and Russia) to well above 400 mm/year (in the most densely populated urban regions) (Lehner *et al.*, 2001).

Table 6.3. Water withdrawals among the surveyed European countries (FAO, 2005).

	Water withdrawals				
	total	*per capita*	withdrawals by sector (as a percentage of total)		
Country	Mm^3/yr	m^3/yr	agriculture (%)	industry (%)	domestic (%)
Austria	2,110	261	1	64	35
Belgium, inc Luxembourg	8,980	840	1	85	13
Bulgaria	10,500	1,321	19	78	3
Czech Republic	2,570	250	2	57	41
Denmark	1,270	239	42	26	32
Estonia	160	115	5	39	56
Finland	2,480	480	3	84	14
Germany	47,050	574	20	68	12
Hungary	7,640	766	32	59	9
Iceland	150	538	0	66	34
Ireland	1,130	297	0	77	23
Latvia	290	120	12	33	55
Lithuania	270	73	7	15	78
Luxembourg (see Belgium)					
The Netherlands	7,940	501	34	60	6
Norway	2,190	490	10	67	23
Poland	16,200	420	8	79	13
Romania	23,180	1,033	57	34	9
Slovakia	-	-	-	-	-
Sweden	2,970	336	9	54	37
Switzerland	2,570	358	2	74	24
UK	9,540	160	3	75	22
Total	149,190	-	21.9[1)]	64.5[1)]	13.6[1)]

n.d.: no data
[1)] Average withdrawal by sector

6.3.3 Water stress

The water stress index (WSI) referred to a whole country can vary considerably, depending on the water resources and water withdrawal data used to calculate it. In Figure 6.3 (from Bixio *et al.*, 2005), the European countries are ranked according to their water stress index. It can be recognized that some countries, e.g. Cyprus and Malta, have a very high WSI (60% and higher), as could be expected because of the climatic conditions and, consequently, the limited water resources. In these countries, where tourism plays a very important economic role, strong seasonal peaks in municipal water demand have to be expected, due to the water consumption of the tourists and also for other special uses such as for golf courses and sport fields irrigation and operation of swimming pools. In these countries, the water need for agricultural irrigation is also very high, due to the arid or semi-arid climatic conditions. It can be noted that some central European countries are also ranked with a high water stress index, for example, Belgium and Germany. In these cases, the high water stress is due to high water extraction: in these very densely populated regions, water withdrawals for municipal and industrial purposes are particularly high.

"Uneven distribution and seasonal variations of water resources make the semi-arid coastal areas and the highly urbanised areas particularly affected by water stress. Changing global weather patterns will make the situation worse, in particular for the southern European countries, susceptible to drought conditions that can be a major environmental, social and economic problem" (Bixio *et al.*, 2005). Common problems in water-stressed countries are groundwater over-extraction with consequent water table depletion and salt-water intrusion in coastal aquifers (Belgium, Greece, and Spain). Over-extraction from surface water may also endanger wetlands (e.g. in Spain and the Czech Republic) (Bodo, 2004).

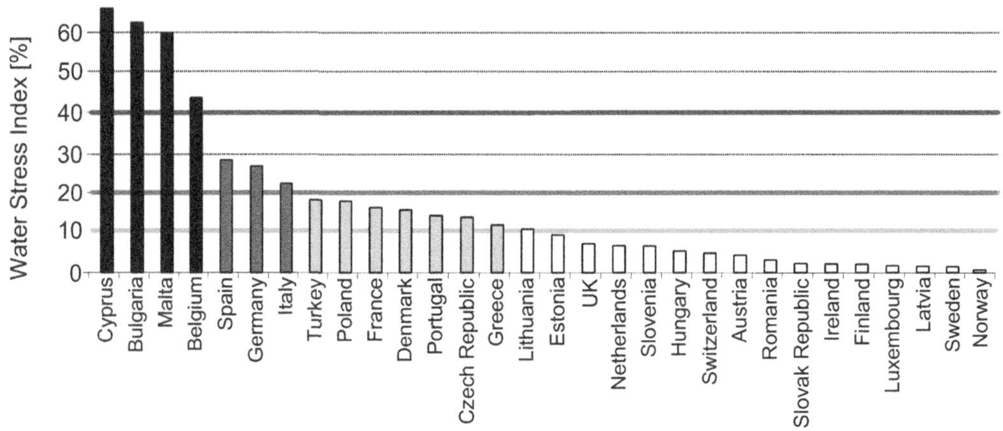

Figure 6.3. Water stress index (WSI) for European countries (Bixio *et al.*, 2005).

In a study carried out by the University of Kassel, Germany (Lehner *et al.* 2001), the water stress was calculated on a river basin basis. The data used for calculating the WSI were the average annual water availability in the respective European river basins based on the 30-year climate time series 1961-90 and the average annual water withdrawals in 1995. The results of this calculation are shown in Figure 6.4. This more detailed spatial consideration of the water stress permits the recognition of regional variations, which can not be detected just on a country base alone.

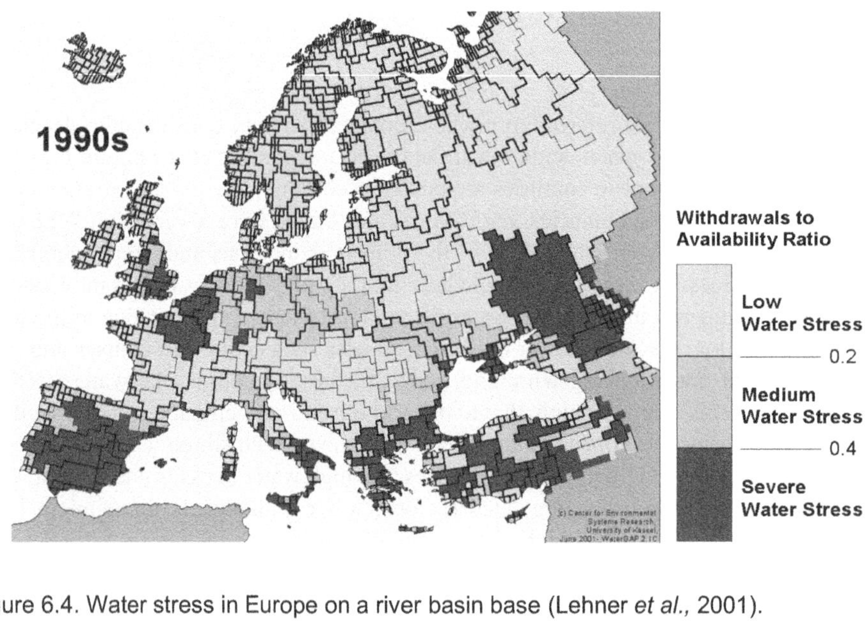

Figure 6.4. Water stress in Europe on a river basin base (Lehner *et al.*, 2001).

"River basins identified to be experiencing severe water stress are - among others - the Don, the Seine, the Meuse, the Thames, as well as most river basins in South Italy, Spain, Greece, and Turkey. All in all, about one fifth of European river basin shows a water stress index higher than 40%, so that these areas can be classified as being under severe water stress. However, river basins may be in the severe water stress category for very different reasons" (Lehner *et al.*, 2001). Whereas in south European countries, the irrigation water needs, combined with dry weather periods, can cause severe water stress, the numbers indicate, for some water-rich densely populated catchment areas, that the high demand for industrial use causes severe water stress.

However, statistics have to be interpreted carefully and it does not seem appropriate to compare water stress index figures of Belgium (43%) and Germany (28%) with those of Spain (30%) and Italy (24%), as can be read from Figure 6.3 (see Box 6.1).

As already mentioned, the water stress index varies considerably depending on the water uses and the wastewater treatment. In many countries, e.g. Belgium and Germany, the largest amount of water is withdrawn for cooling purposes in power plants. The only alteration in the quality of the discharged water is a slightly higher temperature, which make it possible to use the discharged water for other purposes. In other countries, most of the water is extracted for 'consumptive uses' (especially irrigation), so that it is not available any longer for other uses, and as a consequence there is a higher pressure on water resources (See Box 6.1).

Box 6.1 Can statistics tell lies?

Does Germany have a water stress similar to that in Spain, for example, as the water stress indexes in Figure 6.3 indicate? Everybody who knows these countries is surprised by this result. On the one hand, Germany with a green landscape and forests throughout the whole summer, and enough rain to almost abandon irrigation; on the other hand, the dry landscape of Spain, where millions of Germans spend their holidays each year because of the nice, rain-free weather and where intensive agriculture is not thinkable without irrigation. Why does this subjective difference not match the statistical data?

One key might be that the withdrawal data do not distinguish between 'consumptive' and 'non-consumptive' uses. Water extracted for consumptive uses - especially for irrigation, where it is evaporated by plants - is no longer available for other uses, and as a consequence puts higher pressure on water resources than non-consumptive uses like cooling in power plants, where most of the water is returned to the water body with almost no quality deterioration and can be used again. The same is true for most municipal and industrial waters, which do not disappear by use but are returned to the rivers as (treated) wastewater and might be used again downstream.

In addition to the fact that water, after non-consumptive use, is still available and, as a result, puts less pressure on water resources, it has to be considered that in the published statistical data this amount of water is counted as 'leaving the countries by rivers' and, as a consequence, lowers the calculated water resources and increases the water stress index. As this example shows, the method for calculating water stress indexes is questionable and needs at least careful interpretation.

The numbers for Germany clearly illustrate this fact. Using the data of the FAO (the data of other sources e.g. Statistisches Bundesamt (2001) are up to 15% lower) the water stress index for Germany can be calculated as 31% by dividing the withdrawal of 47,050 Mm^3/yr (Table 6.3) by the natural renewable water resources of 154,000 Mm^3/yr (Table 6.2). The statistical yearbook of Germany states that about 26,000 Mm^3/yr of the withdrawn water is cooling water (Statistisches Bundesamt, 2001) which is mainly returned immediately to the same water body from which it was taken. Subtracting the amount of cooling water, the water stress index would drop form 31% to (47-26)/154 = 14%.

Table 6.4. Comparison between water resources and water withdrawal data of Germany and Spain.

		Unit	Germany	Spain	(Source)
Average precipitation (1961-90) IPCC		mm/year	700	636	1
Natural renewable water resources (NRWR)	total	Mm^3/yr	154,000	111,500	1
	per capita	$m^3/year$	1,878	2,794	1
Population (available data in 2004)		10^6 cap	82.43	40.34	2
Water withdrawals	total	Mm^3/yr	47,050	35,630	1
	per capita	m^3/yr	574	883	1
	withdrawals by sector (as a percentage of total)	agriculture (%)	**20**	**68**	1
		industry (%)	68	19	1
		domestic (%)	12	13	1
Computed water stress index		%	31	32	2
Land	land area	1,000 km^2	349.2	499.5	2
	arable land	%	34	26	2
	arable land	1,000 km^2	118.2	130.2	2
	permanent crops	%	0.59	9.87	2
	irrigated area	year of the data	est. 1998	est. 1998	2
		km^2	4,850	36,400	2
		irrigated to arable land ratio (%)	**4**	**28**	1

Sources: 1: (FAO, 2005); 2: (The World Factbook, 2005)

In addition, 98% of the roughly 18,000 Mm^3/yr of all industrial and municipal wastewaters are treated adequately and returned to the surface water. Together with the cooling water, these used waters amount to 44,000 Mm^3/yr out of 47,000 Mm^3/yr which are still available after they have been used. Statistically they might be counted twice, once as withdrawal and a second time as 'leaving the country by rivers'.

Keeping that in mind, the water stress index of Germany would drop far below 5% and be in accordance with the feeling of the population and the experts, that Germany as a whole is a country with sufficient water. Nevertheless, regional differences and future developments mean it is necessary always to be alert as far as water resources are concerned.

In Table 6.4, water resources and water withdrawal data for Germany and Spain are compared. As mentioned above, the computed water stress index for Spain (32%) is just slightly higher then for Germany (31%). But considering the irrigated land area and water use, a very large difference can be noted. Whereas in Germany 20% of the withdrawn water is used in agriculture (according to the data of FAO), in Spain the amount of water used for irrigation rises to 68%. This means that about two thirds of the withdrawn water are destined for 'consumptive' uses, so that they are not available any more and implying a high pressure on water resources.

Conclusion: statistics do not tell lies, but need to be carefully interpreted.

Figure 6.5 compares different water indexes such as:

- water stress index as computed on the basis of the data of the FAO (Figure 6.5A);
- water stress index as referred to in Bixio *et al.* (2005) (Figure 6.5B);
- water exploitation index WEI as referred to in EEA (2005) (Figure 6.5C);
- water exploitation index not considering water withdrawal for energy production (WEI - energy) (EEA 2005) (Figure 6.5D);
- water consumption index WCI (EEA 2005) (Figure 6.5E).

In Figure 6.5, the countries are ranked on the x-axis with decreasing water stress index as computed on the basis of the data of the FAO.

In the first three (A to C) diagrams of Figure 6.5, the y-axes are scaled to the same maximal value of 70%, whereas in the last two diagrams (D and E), the y-axes represent 20% respectively to facilitate the comparison between the different indexes.

The water consumption index is the total water consumption divided by the long term freshwater resources of a country. It considers only the 'consumptive uses', also only those uses through which the water is really consumed, so that it is removed from the (liquid) water cycle, for example by evaporation in irrigated fields. This index highlights those regions where higher consumptive uses such agriculture are predominant. "For the purpose of this assessment it has been assumed that 80% of total water abstracted for agriculture, 20% for urban use, 20% for industry and 5% for energy production is consumed and not returned to the water bodies from where it was abstracted. These figures have been widely accepted, though they may vary by about 5 to 10% depending on the sectors and other factors. For example, actual consumption in agriculture, the largest water-consuming sector, depends on climatic conditions, crop composition and irrigation techniques. Energy is the least consuming sector, returning 95-97% of the abstracted water" (EEA, 2005).

It can be noted from the first three diagrams (A to C), that there are some relevant discrepancies in the water stress index according to different sources. This reflects the difficulty of obtaining trustworthy data about water resources and water withdrawal for the different countries. Another difficulty is the estimation of the water resources of bordering countries. "Inflows from boundary watersheds can add significant percentage up to the freshwater resources in a country, either as surface flow or as groundwater flow. In most cases, the availability of these external resources is regulated by treaties between the water-sharing countries. The correct allocation of the flow along borders is decisive in the water balance, since it is the main source of discrepancies when comparing data of water balances in neighbouring countries, particularly for those along the Rhine, Danube and Oder rivers" (EEA, 2005).

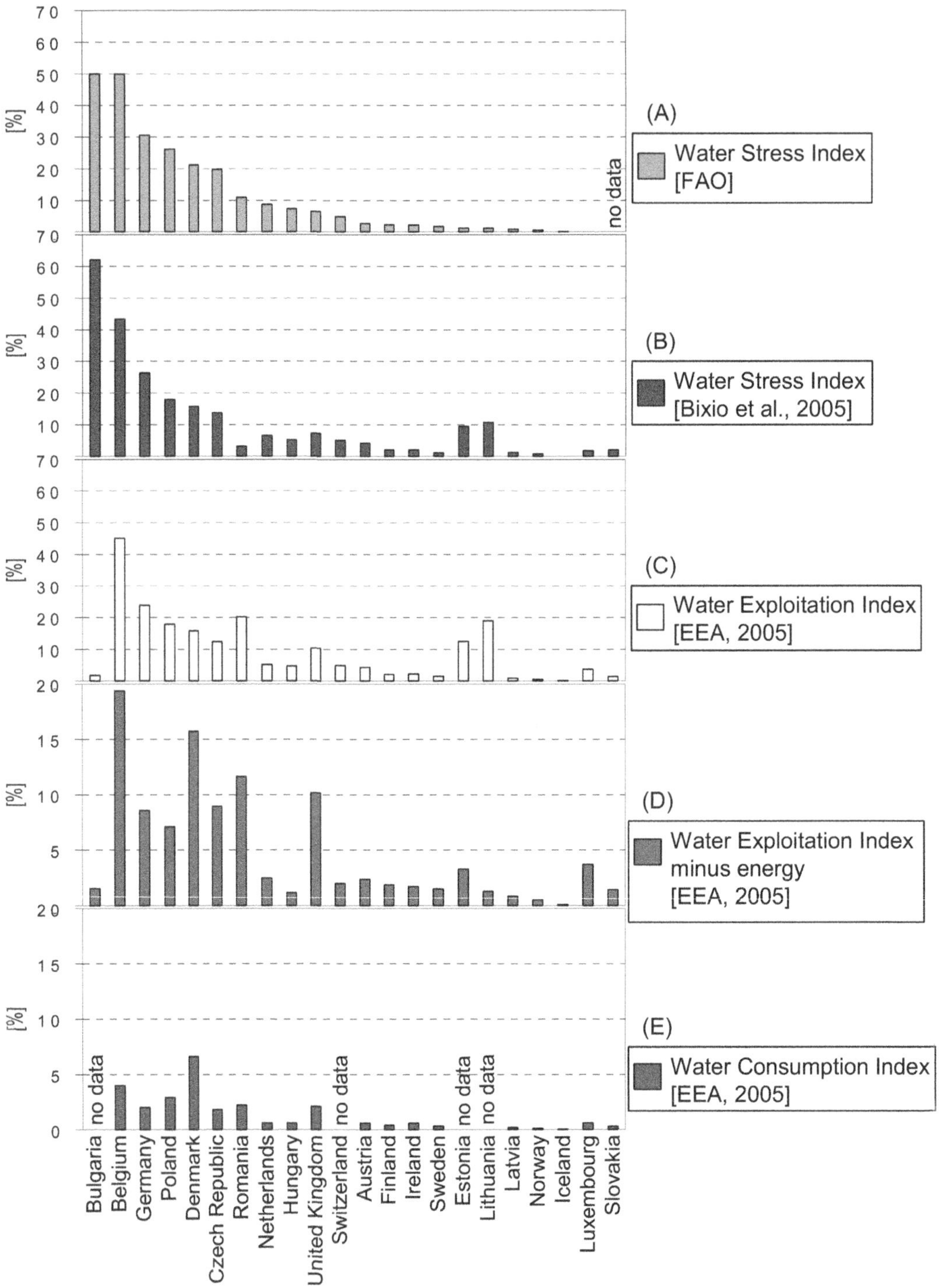

Figure 6.5. Comparison between differently defined water indexes.

6.3.4 Wastewater treatment

6.3.4.1 Connection rate and wastewater treatment

Over the past 15 years, significant progress was made in Europe regarding the proportion of the population connected to wastewater treatment and in wastewater treatment technology. At the end of the 1990s about 80% of the population of the European Union were connected to public sewers and 77% to a wastewater treatment plant. In contrast to this, in the accession countries, only 55% of the population were attached to a public sewage system and 45% to a wastewater treatment plant. In the last 15 years, the focus of attention in the EU shifted from the mechanical (first step) to the biological treatment (second step) and/or to further treatment systems (third step), while in the accession countries, mechanical and biological treatment processes are still predominant (Pau Vall, 2001).

The Urban Wastewater Directive (Directive 91/271/EEC) establishes that the wastewater of all settlements with more than 2,000 PE (PE: Population Equivalent, 1 PE 0,06 kg BOD_5/d) must undergo at least a second step treatment, and that discharges in areas that were classified by the member countries as particularly sensitive must undergo a further treatment (second and third step treatment)

Table 6.5 indicates the status of sewer connection and treatment for various countries as it occurred in the late 1990s (Eurostat, cited in Pau Vall, 2001). The data might be partially out-dated as they do not show the improvement of recent years in connection rates and the construction of new wastewater treatment plants as, for example, in Brussels, Antwerp, Milan, etc.

Table 6.5. Connection rate to a sewer system and to a wastewater treatment in some European countries. (Source: Eurostat, cited in Pau Vall, 2001).

		Connection rate to a public sewer system					Connection rate to a non public sewer system	There from autonomous treatment
		treated						
Country	year	mechanical treatment	biologic treatment	further treatment	total	un-treated		
Austria	1998	0.5	17.2	67.3	81.4	0.1	18.5	18.5
Belgium	1998	-	22.0	16.1	38.1	44.4	17.3	n.d.
Bulgaria	1998	0.9	35.0	-	36.7	30.0	33.3	n.d.
Czech Republic	1999	n.d.	n.d.	n.d.	64.8	9.8	25.4	n.d.
Denmark	1998	1.6	3.4	84.0	89.0	0.0	10.9	10.9
Finland	1999	-	-	80.0	80.0	-	20.0	-
Germany	1995	4.1	12.2	72.3	91.5	0.6	7.9	n.d.
Hungary	1998	3.0	20.0	3.0	26.0	22.0	52.0	17.0
Ireland	1995	24.0	31.8	1.8	57.6	n.d.	32.0	n.d.
Luxembourg	n.d.	n.d.	n.d.	n.d.	93.0	-	7.0	7.0
The Netherlands	1998	-	19.6	78.1	97.7	-	2.3	n.d.
Norway	1999	21.0	1.0	51.0	73.0	7.0	20.0	20.0
Poland	1999	4.3	32.0	15.7	52.0	6.0	42.0	n.d.
Romania	n.d.	n.d.	n.d.	n.d.	n.d.	n.d.	n.d.	n.d.
Slovakia	1998	n.d.	n.d.	n.d.	48.8	5.2	46.0	n.d.
Sweden	1998	-	6.0	87.0	93.0	-	7.0	n.d.
Switzerland	1999	-	22.0	73.8	95.8	-	4.2	n.d.
UK	1997	12.0	52.0	20.0	84.0	10.0	6.0	n.d.

n.d.: no data
-: negligible

The available data show that independent (private) wastewater treatment has some relevance in Norway, Austria, Hungary, Denmark and Luxembourg, with respectively 20%, 18.5%, 17%, 10.9%, and 7% of population connected (last column in Table 6.5). In this context, independent wastewater treatment means all private wastewater treatment plants (e.g. septic tanks) in areas where there is not a public sewer system. Independent decentralized wastewater treatment can be effective in rural areas and in scattered settlements. Nevertheless, no data about the employed treatment processes are available.

Figure 6.6 depicts the connection rate to sewer systems. A distinction has been made between private and public sewer systems. These are further divided into sewer systems with and without subsequent wastewater treatment. The countries are ranked on the x-axis with decreasing connection to a public sewer system with subsequent wastewater treatment.

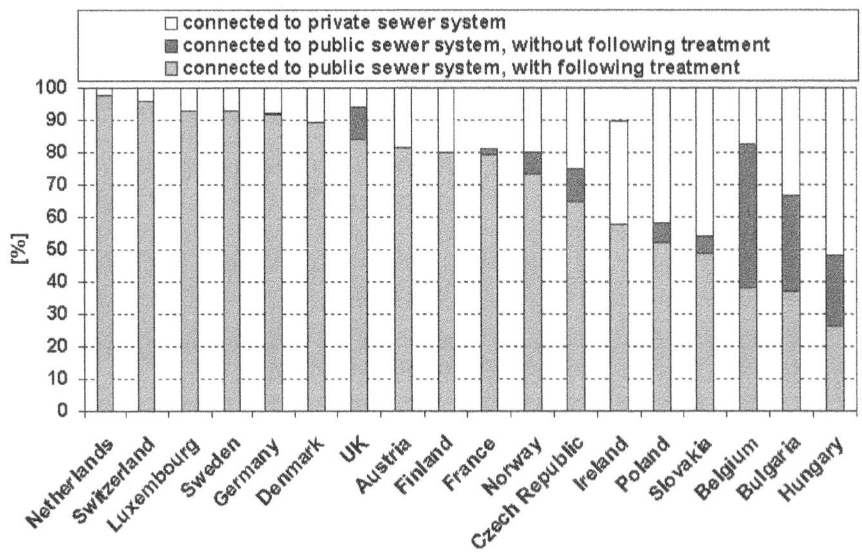

Figure 6.6. Percentage of people connected to a sewer system in some European countries. (Source: Eurostat, cited in Pau Vall, 2001)

6.3.4.2 Wastewater amounts

Wastewaters from point sources come mainly from private households (households and small businesses) and from industry. They are discharged into the environment either directly or after treatment.

Table 6.6 contains data of some countries which describe the wastewater quantities of private households and of the manufacturing industry on a *per capita* basis. The data show a large amount of wastewater from manufacturing industry in Sweden, Finland and Norway, while among the indicated countries, The Netherlands, Austria, and Slovakia show higher wastewater amounts from private households.

Table 6.7 shows wastewater quantities from different manufacturing businesses from selected countries. The manufacturing businesses need a considerable amount of water for cooling purposes. Process water and cooling water are included in the data. The water is partially discharged into the public sewer system, but mostly treated directly in the industry.

Table 6.6. Wastewater quantities in m^3/*capita*·yr (Source: Eurostat, cited in Pau Vall, 2001).

Country	year	Manufacturing industry	Private households
Austria	1998	n.d.	77
Belgium	1998	88	n.d.
Bulgaria	1998	43	33
Czech Republic	1999	n.d.	59
Finland	1998	128	n.d.
Germany	1995	76	45
Luxemburg	1999	16	20
The Netherlands	1990	30	81
Norway	1999	122	n.d.
Poland	1999	14	28
Romania	1999	8	n.d.
Slovakia	1997	60	94
Sweden	1995	233	n.d.

n.d.: no data

Table 6.7. Wastewater quantities in Mm^3/yr in different branches of the manufacturing industry for some selected countries (Source: Eurostat, cited in Pau Vall, 2001).

Country	year	Manufacturing industry in total	Food industry	Metal production	Vehicle construction	Textile industry	Paper industry	Chemical industry, mineral oil production
Belgium	1998	896.6	60.8	477.9	n.d.	15.9	52.0	226.9
Bulgaria	1998	355.3	37.7	95.0	3.1	13.0	30.5	126.0
Finland	1998	658.1	4.0	53.3	-	0.4	521.6	78.8
Germany	1995	6224.0	394.9	897.9	13.9	198.8	670.9	3537.9
The Netherlands	1990	447.5	n.d.	8.0	n.d.	16.4	16.4	89.4
Norway	1999	540.2	3.1	152.2	-	0.9	1.4	301.9
Poland	1999	560.3	39.1	198.6	5.4	43.8	89.8	113.2
Romania	1999	496.8	22.6	155.9	6.5	11.6	n.d.	300.2
Slovakia	1998	321.0	8.9	28.4	n.d.	1.6	39.0	179.7
Sweden	1995	2054.6	70.1	n.d.	n.d.	11.3	882.9	513.6

n.d.: no data

The wastewater quantity depends strongly on the structure of the manufacturing industry. For the member countries from which data are available, most wastewater originates from the branch 'Chemical industry, mineral oil production': in Germany 57% and in The Netherlands 20%. The paper industry produces the most wastewater in Finland (79%) and in Sweden (43%). Also in the accession countries, the greater amount of wastewater stems from the branch 'Chemical industry, mineral oil production' (Pau Vall, 2001).

However, industrial water demand and wastewater quantities might differ largely as internal water recycling and water reuse within the manufacturing industry is often not included in the available data (see Box 6.2).

BOX 6.2. What are the right numbers for industrial water use?

Various statistics provide data for water quantities in industry. Many different terms are used, e.g. provided fresh water, utilized water, wastewater, etc. Each of them may or may not include cooling water. Further, the definition of industry may differ as power plants and the mining industry are considered or not.

The example of Germany might illustrate these facts. Table 6.8 lists quantities in Mm^3/yr for the year 1998 (Statistisches Bundesamt, 2001).

Table 6.8. Industrial water quantities in Mm^3/yr.

	Wastewater		Water supply			
Industry	total	Of which cooling water	provided	utilized	Of which for cooling	use factor
Manufacturing industry	6,008	4,243	6,207	30,226	22,486	4.9
Food	363	162	416	1728	834	4.2
Metal	822	667	873	6,018	4,925	6.9
Vehicle	86	45	93	1,989	1,092	21.5
Textile	175	131	183	242	172	1.3
Paper	547	264	610	3485	816	5.7
Chemical	3,455	2,639	3,422	11,836	10,594	3.6
Power plants (public)	25,984	25,842	26,559	67,734	57,457	2.6
Municipal wastewater	9,695	-	-	-	-	-

Table 6.8 indicates the large quantities of cooling water used in power plants. The use factor for water including cooling water is defined as the quotient of utilized to provided water, and it varies between 1.3 in the textile industry up to 21.5 in the vehicle industry. Use factors indicate the extent of water reuse among different industry branches.

The values are increasing, e.g. in the food industry from 3.5 in 1980 to 4.2 in 1998. They might vary in different countries and even in different regions as the driving force for water reuse in industry is quite often economics. Thus, reuse depends on water price and wastewater fees on the one hand, and on water treatment costs for adequate standards for internal reuse purposes on the other hand.

6.4 WATER REUSE IN CENTRAL EUROPE

6.4.1 Status of reuse of municipal wastewater in Europe

"More than 200 water reuse projects exist in Europe and many others are in an advanced planning phase. This is a particularly large figure considering that in the early 90s municipal water reuse was limited to few cases, mostly incidental, *i.e.* related to the proximity of the wastewater treatment plant to the point of use" (Bixio *et al.*, 2005).

Figure 6.7 shows the geographic distribution of the identifiable water reuse projects, including their size and intended use. The areas of application are split into four categories: 1) agriculture; 2) industry; 3) urban, recreational and environmental uses, including aquifer recharge and 4) combinations of the above (mixed uses). The scale of the projects is also split into four classes: very small (< 0.1 Mm^3/yr), small (0.1-0.5 Mm^3/yr), medium (0.5-5 Mm^3/yr) and large (> 5 Mm^3/yr) (Bixio *et al.*, 2005). As can be seen, most of the reuse takes place in southern Europe, including France, whereas in Central Europe the reuse schemes are mainly small scale and for urban or environmental applications (51% of the projects) or industrial uses (33% of the projects) (Bixio *et al.*, 2005). Although the size of many projects in Germany, The Netherlands and the United Kingdom is not stated, one can assume that the majority of the projects mentioned are small, some being more testing or research facilities than large scale applications.

"Only one water reuse project has been identified for potable water production. The project was set up to reduce the extraction of natural groundwater for potable water production and to hold back the saline intrusion at the Flemish coast of Belgium (...) In Europe there is an escalating interest for artificial groundwater recharge with reclaimed wastewater to hold back saline intrusion in coastal aquifers. This can be seen by the involvement of the WHO regional office for Europe in addressing the specific health risks of this practice. Two large-scale projects are available in the Barcelona area and in the Northern part of London; several other medium sized projects are also available" (Bixio *et al.*, 2005).

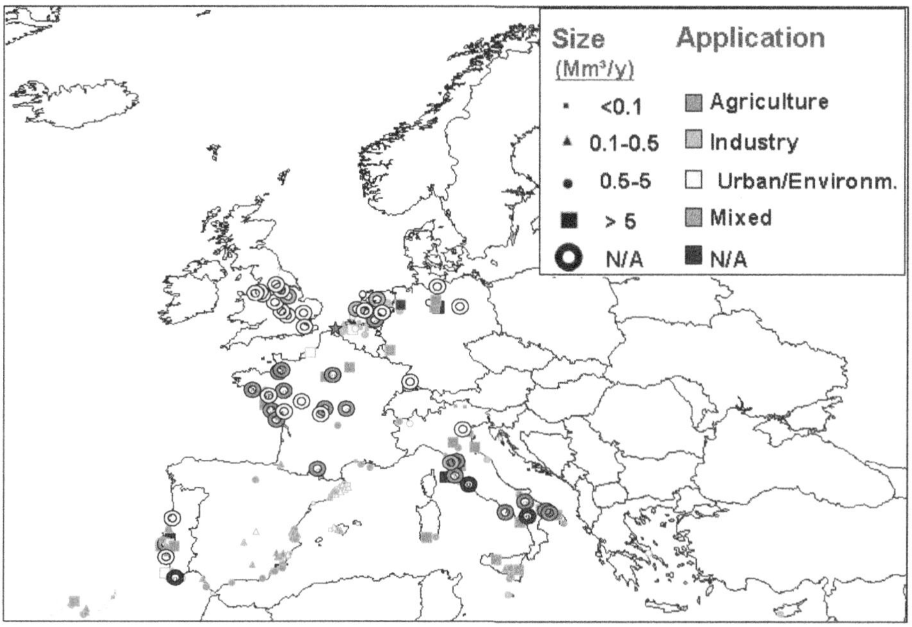

Figure 6.7. Identifiable water reuse projects in Europe, including their size and intended use (Bixio *et al.*, 2005).

Therefore, one can summarize that the intentional reuse of treated municipal wastewater is not widespread in central European countries. In most of the countries of this survey, no reuse projects have been identified.

As mentioned above, water reuse in industry is far more developed. As shown in Box 6.2 for Germany – a country with quite high reuse rates in industry – the recycled water in the manufacturing industry amounts to roughly 24,000 Mm³/yr (30,226 - 6,207 = 24,019, see Table 6.8), a number which exceeds the total amount of municipal wastewater of Germany by a factor 2.4.

In the densely populated countries of Central Europe, indirect reuse has to be addressed. It is obvious that using the large European rivers as receiving water bodies for treated wastewater on the one hand and as a source to produce water for human consumption on the other, indirect reuse occurs. Depending on population density and the size of the river, one can calculate a small ratio of treated wastewater, for the Rhine it is approximately 1 to 3%, depending on how far downstream. And it is known that in some densely populated areas, measurable amounts of non-degradable wastewater constituents are found in surface waters (Putschew *et al.*,2003; Ziegler *et al.*, 2001a; Ziegler *et al.*, 2001b).

6.4.2 Water reclamation technologies

Since municipal wastewater in Central Europe is reused only in a few cases, there is no commonly accepted technology. Usually filtration/membrane filtration and disinfection steps are added to the conventional advanced wastewater treatment. Multi-barrier concepts are being discussed to produce water qualities comparable to drinking water.

The treatment of industrial wastewater for reuse depends on the branch of industry concerned. The treatment might include a variety of different biological treatment steps, precipitation, flocculation, flotation or sedimentation, filtration and membrane filtration. Sometimes ultrafiltration/microfiltration followed by nanofiltration or reverse osmosis are common techniques to produce high quality service water.

6.5 REGIONAL POLICIES AND PRACTICES

6.5.1 Legislation

It is essential that the development of water reuse in agriculture and other sectors be based on scientific evidence of its effects on the environment and public health. Although several studies have been conducted on wastewater quality and for different purposes, at this time there are no guidelines, best practices or regulations of water reuse at an EU level other than the Urban Wastewater Directive (Directive 91/271/EEC) which states that 'treated wastewater should be reused whenever appropriate'. Further work is needed to develop suitable guidelines and definition of 'whenever appropriate' (Muston, 2004).

Despite the fact that no guidelines or regulation exist yet at the level of the European Union, several countries or autonomous regions have published their own standards or regulation, such as, for example Cyprus, France, Italy and Spain, which are indeed not included in this chapter. Other countries rely on external guidelines or standards such as the WHO (currently being revised together with new guidelines in draft form for aquifer recharge) or California (Title 22) standards (Muston, 2004).

Since agricultural reuse plays a secondary role in Central Europe, there are fewer efforts for reuse standards. In Germany, for example, there is a norm on hygienic concerns of irrigation water (DIN 19650 1999) which has no legally binding character. The aim of this norm is to avoid any health hazards for human beings and animals which could accumulate from agricultural irrigation. Thus the focus is not specifically on the use of treated wastewater as irrigation water, but rather any water resource for irrigation and its possible sanitary risks are considered. This norm contains a table with maximum permissible values for different hygienic-microbiological parameters (such as faecal Streptococcus, *E. coli* and *Salmonella*). The thresholds are different for different destinations of the water, e.g. for the irrigation of crops to be eaten raw or for the irrigation of crops which are not eaten directly but are used for the preparation of foodstuffs or for animal feed. This norm does not contain any indication on how to accomplish the maximum permissible values.

On the other hand, some countries have standards for service water or water for non potable use in households. In Germany, for example, there is a norm on the use of prepared service water in buildings (Sen Bau Wohnen, 1995). The aim of this norm is to ensure the sanitary harmlessness of the water and the well-being of the users. This norm contains a list of maximum permissible values for different hygienic-microbiological parameters and for some physico-chemical parameters such as e.g. pH-value, BOD_5, colouration and odour. This norm, too, has no legally binding character and contains no indication on how to accomplish the requirements. Normally the manufacturer of service water reuse plants should guarantee that the quality requirements are reached.

The situation for water recycling and reuse in industry is quite different. Whereas for the reuse of water for irrigation in agriculture or as service water in households and hotels, the focus is on the sanitary harmlessness of the water, for water recycling in industry, the quality requirements for the water depend on the process in which the water is used. Hence the quality requirements differ basically from those for water for irrigation, and are specific for the specific production processes. For example, for recycling of water in cooling towers, the content of suspended solids and salts is the deciding parameter; in the paper industry, a very important parameter could be the colour of the water.

It becomes evident that it is not possible to set up general quality requirements for the recycling of water in industry, due to the plurality of uses and processes in which water could be reused.

6.6 INFLUENCING FACTORS

6.6.1 Ownership and Finance

Ownership and financing of water supply, sewers and wastewater treatment plants, as well as water for agricultural use have developed quite differently in different European countries. An inhibitor for agricultural reuse is the differing interests between urban authorities, who are normally responsible for urban wastewater treatment and disposal, and the potential (normally agricultural) users of reclaimed water (Muston, 2004).

There is considerable discussion on how water reuse projects should be managed, in particular who should take the leadership and how the responsibilities/liabilities should be divided. However, as long as the need for water reuse is not widely accepted and the cost benefits are low, all questions of suitable ownership structure, local circumstances, political will, legislation, institutional structure and regulation are not the real issue. Reuse in industry works because responsibility and gains are clear.

6.6.2 Public acceptance

"The use of treated and recycled wastewater in agricultural, municipal, or domestic applications is quite properly a source of concern for a variety of consumer groups." (Jeffrey and Jefferson, 2004).

Generally, one can assume that consumers usually do not know if the food they are consuming was irrigated with (properly treated) recycled wastewater. This is true within a country and even more likely for internationally traded food, as no one is really aware how foods are grown in far-away countries. Nevertheless, major food companies are now addressing this topic and working on water quality standards for water used in food production (ILSI Europe, 2005).

A totally different story can be told about the reuse of treated wastewater within households, e.g. for toilet flushing or gardening. Even if the authorities would favour water reuse and all technical measures are fulfilled properly, no incentives will work without a general acceptance by the stakeholders, *i.e.* the water and sanitation companies, the community and consumers alike. Public education combined with transparent, clear, and reliable information is the key aspect to the development of public confidence.

To explore some of the attitudinal determinants of public acceptance of water recycling in the United Kingdom, a study was carried out by Paul Jeffrey and Bruce Jefferson (School of Water Sciences, Cranfield University) on a sample of over 300 responders living in England and Wales.

"The most significant finding from the study is that a large majority of respondents were supportive of water recycling as a concept. For example, 89% of respondents agreed with the statement '*I have no objections to water recycling as long as safety is guaranteed.*'" (Jeffrey and Jefferson, 2004). Another part of the questionnaire tested the willingness of the respondents to flush their toilet with recycled water from different sources. Most of the respondents (about 92%) were willing to flush their toilet with recycled water from their own bath and shower. But if the recycled water were to be collected also from the bath and shower of a neighbour, the willingness decreased to about 58%. Using water from the entire street/neighbourhood produced a further small decrease in willingness, to about 50% of respondents (Table 6.9).

Table 6.9. Support for using recycled water from different sources for toilet flushing (Jeffrey and Jefferson, 2004).

System details	Non-metered respondents (%)
Water sourced from own bath/shower	91.7
Water sourced from next door	58.2
Water sourced from whole street/neighbourhood	50.2

The survey also tested awareness of the use of water and the sensitivity to issues related to the saving of water. It was found that the willingness to use recycled water from municipal sources was higher among metered households than among non-metered households.

The survey also tested the willingness to use recycled water depending on the aesthetic water quality and on the potential of contact or ingestion. "Poor aesthetic water quality (as measured by increases in turbidity, colour and suspended solids) has only a minor detrimental effect on the frequency of acceptance. Turbid water is consistently considered less acceptable than either water which is coloured or has high suspended solids content. Increasing potential for human contact and ingestion has a negative impact on acceptability levels (although it is only when asked if they would drink the water that acceptance levels fall below 50%)." (Jeffrey and Jefferson, 2004).

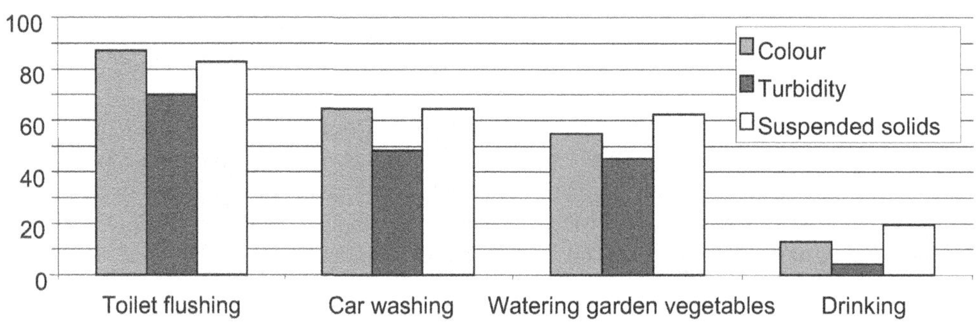

Figure 6.8. Willingness to use recycled water of poor aesthetic quality (Jeffrey and Jefferson, 2004).

6.7 RESEARCH AND DEVELOPMENT

Water authorities in Europe are represented by EUREAU, an organisation established in 1975 and now representing water authorities in 27 countries, including some of the new accession countries. One of the working groups within EUREAU is concerned with water reuse (Muston, 2004). The Commission of the European Communities supports, through its General Directorate Research, a set of research projects.

6.8 ACKNOWLEDGEMENTS

The authors would like to thank Bruce Durham for his collaboration and in particular for the data and articles he provided about water reuse in Europe.

6.9 REFERENCES

Bixio, D, De Koning, J., Savic, D., Wintgens, T., Melin, T., Thoeye, C. (2005) Wastewater reuse in Europe. *Desalination,* in press.

Bodo, B. (2004) AQUAREC - General maps on water supply and demand, April 2004 http://www.ivt.rwth-aachen.de/Eng/Forschung/wpackages.html

DIN 19650 (1999) Bewässerung – Hygienische Belange von Bewässerungswasser, Normenausschuss Wasserwesen (NAW) im DIN Deutsches Institut für Normung e.V.

EEA (2005) European Environmental Agency, Indicator Fact Sheet, (WQ01c) Water exploitation index http://themes.eea.eu.int/Specific_media/water/indicators/WQ01c%2C2003.1001/WEI_101003v2.pdf

FAO (2002) Food and Agriculture Organization of the United Nations (FAO): Water Resources, Development and Management Service. 2002. AQUASTAT Information System on Water in Agriculture: Review of Water Resource Statistics by Country. Available on-line at: http://www.fao.org/waicent/faoinfo/agricult/agl/aglw/aquastat/water_res/index.htm.

FAO (2005) AQUASTAT FAO's Information System on Water and Agriculture http://www.fao.org/ag/agl/aglw/aquastat/dbase/index2.jsp

ILSI Europe (2005) International Life Science Institute, Brussels, Assessing the suitability of water for intended use, unpublished.

Jeffrey, P., Jefferson, B. (2004) Public receptivity regarding 'in-house' water recycling: results from a UK survey, paper presented at the Enviro 2002 Convention and Exhibition and IWA 3rd World Water Congress, April 07th -12th 2002, Melbourne, Australia

Lehner, B., Henrichs, T., Döll, P., Alcamo, J. (2001) EuroWasser – Model-based assessment of European water resources and hydrology in the face of global change. Kassel World Water Series 5, Center for Environmental Systems Research, University of Kassel, Germany

Muston, M. (2004) Aquarec, Outcome report of the International Workshop on Implementation of Municipal wastewater Reuse Plants, 11th – 12th March 2004, Thessaloniki, Greek

Pau Vall, M. (2001) Statistik kurz gefasst, Umwelt und Energie, Thema 8 – 14/2001, Abwasser in den europäischen Staaten, European Community, ISSN 1562-3092

Putschew, A., Mania, M., Jekel, M. (2003) Occurrence and source of brominated organic compounds in surface waters. In *Chemosphere* 52 (2003), p. 399-407

Sen Bau Wohnen (1995) Merkblatt „Betriebswassernutzung in Gebäuden", Senatsverwaltung für Bau- und Wohnungswesen, Berlin, 1995

Statistisches Bundesamt (2001) Statistisches Jahrbuch 2001 für die Bundesrepublik Deutschland, Wiesbaden, September 2001 ISBN: 3-8246-0640-2

The World Factbook (2005) http://www.odci.gov/cia/publications/factbook/index.html (Feb. 2006)

UN (2002) Population Division of the Department of Economic and Social Affairs of the United Nations Secretariat, 2002. World Population Prospects: The 2000 Revision, New York, United Nations

UN (2003) UN World Water Development Report, Barcelona 2003

Ziegler, D., Harti, C., Wischnack, S., Jekel, M. (2001 a) Behaviour of dissolved organic compounds and pharmaceuticals during lake bank filtration in Berlin. In *IAWR-Proceedings of the Intern. Riverbank Filtration Conference Rheinthemen* (2001), Nr. 4, p. 151-160

Ziegler, D., Hartig, C., Wischnack, S., Jekel, M. (2001 b) Organic substances in partly closed water cycles. In *Journal Env. Eng.* (2001).

7

Water reuse in Asia

Naoyuki Funamizu, Xia Huang, Guan-Hao Chen, Hu Jiangyong and Chettiyapan Visvanathan

7.1 INTRODUCTION

7.1.1 Climate and water resources in Asia

The climates in Asian region has every wide variety from cold and humid, mild and humid regions in north-east Asia; hot and humid with or without a dry season in south-east and south Asia; to semi-arid and arid in west Asia. This wide diversity of climates along with uneven distribution of population gives the wide range of values in renewable water resources per capita as shown in Figure 7.1.

The value of renewable water resource ranges from less than 2000m^3/capita/yr (India, Republic of Korea, Uzbekistan) to more than 20,000 m^3/capita/yr (Bhutan, Cambodia, Lao People's Democratic Republic, Malaysia, Myanmar). The water stress index value (percentage of the annual withdrawal to renewable water resource) in Figure 7.1 also ranges from less than 1% to more than 100% at country level. Since the water scarcity depends on local or regional conditions, detailed spatial information is desirable. Figure 7.2 shows the distribution of the annual withdrawal to water availability ratio in 0.5 degree x 0.5 degree grid cell scale. The north and west part of China, India and Pakistan appear darkest on the map (red on the original). Actually, 400 of 668 cities in China are short of water resources and 136 cities in particular are in a severe situation. It should be noted that the "red" spots are scattered across the map. This is mainly because of high population density in urban areas. The water demands in large cities have stressed the reliability of the water supply and claimed the development of new water resources. In Singapore, half of the water supply comes from a river located in a neighboring country. Hong Kong has been using seawater, a readily-available water resource, for toilet flushing since late 1950s and seawater consumption amounts to 0.64×10^6 m^3/d, which is about 20% of the total water demand.

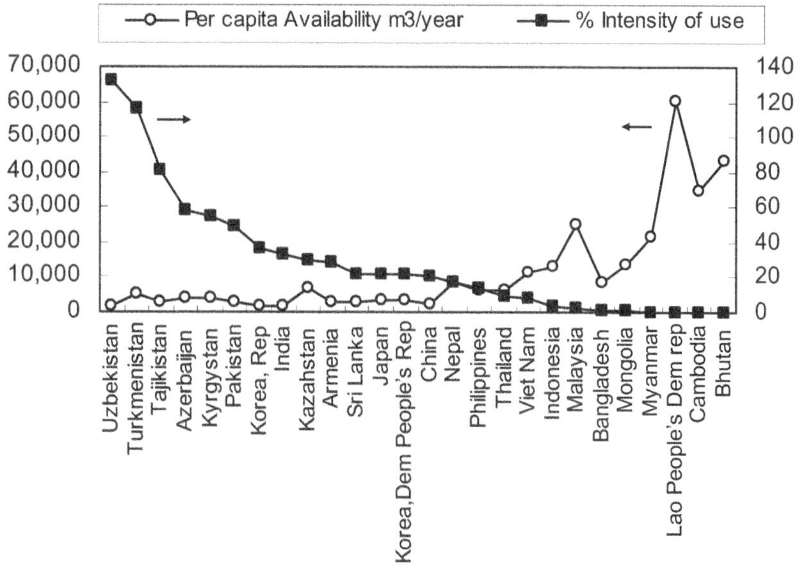

Figure 7.1 Fresh water resources and water stress index of Asian countries. (Source: World Resources 2002-2004, Data Table 11 Freshwater Resources).

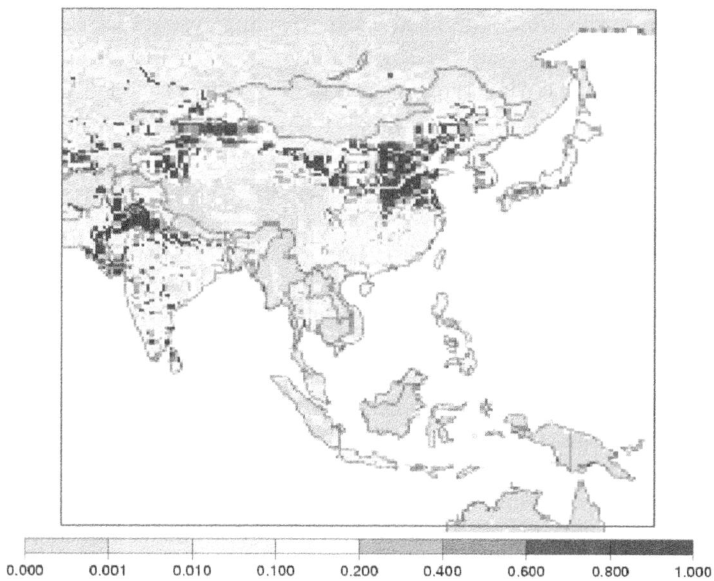

Figure 7.2 Annual withdrawal to Availability Ratio (Source: Oki, 2006).

Figure 7.3 shows water use by three sectors in each country. Most of the countries in Asia use water for agricultural purposes. The countries with high percentages for industry and domestic use are Korea Rep., Armenia, Mongolia, Bhutan and Singapore.

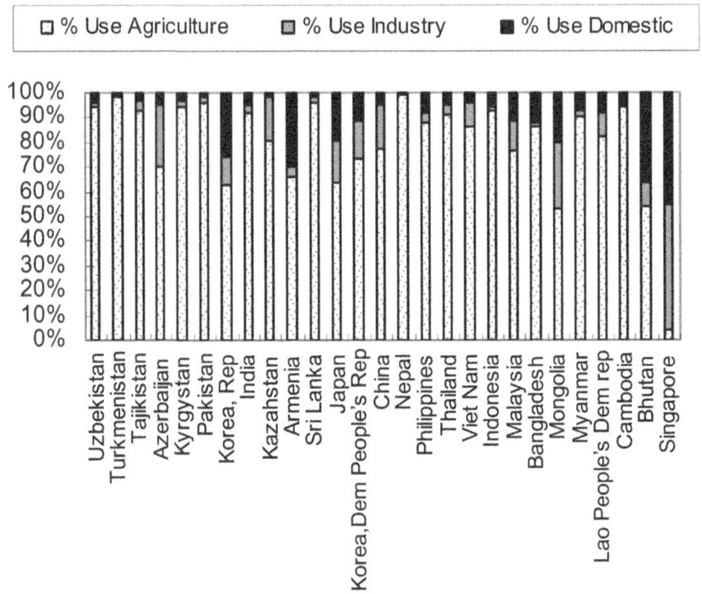

Figure 7.3 Water use by three sectors: agriculture, industry and urban.

7.1.1 Current water reuse practice in Asian countries – an overview

The type of reuse in Asia includes municipal, industrial and agricultural applications. The schemes for water reuse ranges from individual water recycling systems to watershed scale systems.

Industrial reuse is mainly for cooling and washing purposes and process water. A typical example of industrial water reuse in Korea is the individual facility water recycling system in a steel company in Pohang for cooling water. A sugar industry in Thailand has an individual water recycling system with three stage membrane filtration comprising of MF, NF and RO membranes for process water. Average water recycling percentage in Japanese industry is about 75% by individual recycle systems. Approximately 23 x 10^6 m^3 of reclaimed water from municipal wastewater treatment plants are used for industries annually and this volume occupies about 9% of the total volume of reclaimed water in Japan.

Non-potable and indirect potable use of reclaimed wastewater is practiced in Asian countries. Tianjin city in China has a wastewater reclamation plant which treats 50,000 m^3/d of secondary effluent to serve about 160,000 residents. Korean practice is characterized by onsite water recycling systems for toilet flushing, cleaning and cooling. Approximately 430,000 m^3/d of reclaimed water is used and this is about 4% of total water supply in Korea (Noh *et al.*, 2003). In 2003, 367 wastewater treatment plants in Japan supplied approximately 486,200 m^3/d (180 Mm^3/yr) of reclaimed water. In addition, 1,058 on-site individual building and block-wide wastewater recycling systems generated toilet flush water in commercial buildings and apartment complexes. An interesting reuse example is NEWater in Singapore for indirect portable reuse. Since 2003, secondary effluent is further treated at an advanced water reclamation plant (*i.e.* multi-tiered membrane system (MF and RO) and UV disinfection), the high quality water fulfilling all requirements for drinking water based on USEPA and WHO drinking water standards is channeled back to a reservoir. The total capacity of the three NEWATER plants is 75,000m^3/d.

Unplanned water reuse for agricultural irrigation occurs in China when rivers downstream from cities are used for irrigation. Irrigation by raw wastewater is also practiced. In India, over 73,000 hectares of land were irrigated with wastewater from at least 200 sewage farms in 1985. And there has been a dramatic increase in wastewater volumes discharged and used for agricultural irrigation in India. There are studies in Iran examining the use of treated effluent for irrigation water in suburban farms, mainly for fodder crops. The use of untreated wastewater for irrigation is also common in Pakistan (USEPA, 2004).

This chapter, in Sections 7.2 to 7.5, illustrates practices in Asian countries by describing projects and water reuse potential in China, Hong Kong, Singapore and Thailand. Water reuse in Japan is described in Chapter 25. Since data on water reuse in other Asian counties are not well-reported, the very brief descriptions of India, Iran and Pakistan reported by USEPA (2004) are summarized below as an overview of Asian practices in water reuse.

India (USEPA, 2004)

As a result of the fast-growing urban population, service infrastructure is insufficient and about 50% is not serviced by sanitary sewers. There have been some attempts at rectifying this situation in large cities such as Delhi, Mumbai and Calcutta. The Ganges River program is to include treatment facilities for 6 cities in Uttar Pradesh that will incorporate reuse for agriculture and forestry.

Direct reuse of untreated wastewater is very common in India. The law prohibits the irrigation of salad vegetables with wastewater, yet the prohibited practice is widespread and government agencies reportedly do not actively enforce regulations governing reuse.

Iran (USEPA, 2004)

In 1994, the volume of municipal wastewater generated in Iran was estimated to be 3100 x $10^6 m^3$/yr, and projected to increase to 5900 x $10^6 m^3$/yr by 2021. These wastewaters are now largely disposed without treatment. Planned water reuse projects currently produce 154 x $10^6 m^3$/yr.

Recently, the government of Iran approved a recommendation to establish and implement programs for comprehensive reclamation and use of non-conventional water resources. The public also accepts water reclamation and reuse as a sensible way to maximize the use of a limited resource. In the past, effluent was used primarily to fertilize the soil, but now wastewater effluent is increasingly used for improving water use efficiency, surface and groundwater pollution prevention, and to compensate for a shortage of irrigation water. Iranian farmers generally consider wastewater an acceptable water resource for irrigation.

Despite governmental edicts prohibiting the use of untreated wastewater in irrigation and agriculture, there are still some places in Iran where farmers use raw wastewater, due to a shortage of fresh water. Unplanned use of wastewater is observed in cities with no sewage systems and no wastewater treatment plants.

Pakistan (USEPA, 2004)

The use of untreated wastewater for agricultural irrigation is common in Pakistan. It is practiced in 80% of all towns and cities. Reasons why untreated wastewater is used for irrigation include the high reliability of wastewater, the nutrient value of the wastewater, lack of access to other water sources and the profits made by selling crops at the local market.

In Faisalabad, the third largest city in Pakistan, over 2000 hectares of agricultural land are irrigated with untreated wastewater. Farmers consider that the untreated wastewater is more nutrient rich and less saline than treated wastewater. The local water and sanitation agency sells the wastewater to groups, or to a community of farmers.

7.2 WATER REUSE IN CHINA

7.2.1 The availability of water resources in China

In terms of total water resources, China should not be a country with water shortage problems. The average annual total water resource in China is about 2,800 x 10^9 m^3/yr, which is fourth in the world. However, with the largest population in the world, China has a very limited per capita water capability, which is only 2,200 m^3/yr, about one fourth of the world average. Moreover, rainfall has an uneven geographical and seasonal distribution. In general, rainfall precipitation is rich in the south-eastern areas (more than 1,000 mm/yr) but is scarce in the north-western and north-eastern areas (even less than 200 mm/yr). The concentration of rainfall in summer seasons also creates more challenges in water resource management in the country.

On the other hand, the total discharge amount of wastewater reached 46 x 10^9 m^3/yr in China (SEPA, 2003). A lot of wastewater was not properly treated and directly flowed into water bodies, with many rivers and lakes thus polluted, to different extents. According to monitored results in 2003, of the 407 key sections of main rivers prescribed by the government, 29.7% of them were lower than the lowest class of national water standard (only usable for agricultural irrigation). Water pollution is critically damaging the water resource.

7.2.2 Main types of water reuse

Apart from long distance water transfer and rare water resources exploration, wastewater reuse can in many ways be considered as a reliable, practical and economic alternative water resource. To alleviate water shortage problems, the Chinese government has made various effort to promote wastewater treatment and reuse since the 1980s. With rapid economic development, shortage of water resources caused by increased water demand attracted more attention to wastewater reclamation and reuse. Up to 2002, the discharged amount of municipal wastewater was 33.8 x 10^9 m^3/yr, and the treatment ratio was around 40% with different degrees of treatment. Of treated municipal wastewater, about 2.12 x 10^9 m^3/yr of water was reused, accounting for 6.3% of the discharged wastewater.

The types of reclaimed water reuse include: (1) industrial applications for cooling and washing purposes and process water; (2) municipal applications, such as toilet flushing, car washing, landscape irrigation, etc.; (3) agricultural irrigation; and (4) supply to environmental waters. The water reuse for agricultural irrigation occurs when rivers downstream from cities are used for irrigation. The schemes for water reuse range from onsite wastewater recycling systems to large area wastewater reuse systems. Water reuse has been actively promoted particularly in water-short cities or regions. Table 7.2 lists several wastewater reclamation and reuse practices in some water-short cities in China. It can be seen that reclaimed water in these instances is mainly used for municipal use and industrial cooling water.

Table 7.2 Wastewater reclamation and reuse practice in some water-short cities of China.

Wastewater treatment plant	Capacity (m^3/d)	Reclaimed water uses	Operation date
Taiyuan Beijiao WWTP	10,000	cooling water	1992
Taiyuan Chemical plant	24,000	cooling water	1992
Dalian Chunliuhe WWTP	10,000	cooling water, supply to boiler water	1992
Dalian Malanhe WWTP	40,000	municipal uses	2001
Shandong Laizhou WWTP	20,000	municipal uses and cooling water	1996
Qingdao Haibohe WWTP	40,000	municipal uses	2003
Beijing Gaobeidian WWTP	300,000	municipal uses and cooling water	2003
Tianjin Jizhuangzi WWTP	50,000	municipal uses and cooling water	2003
Tianjin TEDA WWTP	25,000	municipal uses, cooling and process water, supply to boiler water	2002
Xi'an Beishiqiao WWTP	50,000	municipal uses and cooling water	2003
Hefei Wangxiao WWTP	100,000	municipal uses and cooling water	2005

7.2.3 Some water reuse examples in China

Water reuse examples in Tianjin (Chen Qisi and Hu Wenli (2004) and Li Jian et al. (2003))

Tianjin is an extremely water-short city with an average per capita water resource of 160m^3/yr, about one sixteenth of the whole country average. Wastewater reclamation and reuse is thus paid particularly high attention. The wastewater reclamation project of Tianjin Jizhuangzi municipal wastewater treatment plant is the first large-scale facility for municipal and industrial use in China. It is composed of (1) the construction of a wastewater reclamation plant having the treatment capacity of 50,000 m^3/d and (2) the construction of the pipe system, 52 kilometers in length. The reclaimed water serves 158,000 residents in domestic blocks for non-potable uses, plus other industrial applications such as cooling water.

The water source for reclamation comes from the secondary effluent of Jizhuangzi municipal wastewater treatment plant. Of the reclamation capacity of 50,000 m^3/d, 20,000 m^3/d of water is treated by a combined system of continuous micro-filtration (CMF) and ozonation units, while the other 30,000 m^3/d is treated by the conventional process of coagulation, sedimentation and filtration. The reclaimed water by the combined system has a high quality, with turbidity <5 NTU, TSS<5 mg/L, which is reused as toilet flushing, park greening, street spraying and scenic environment water. The quality of the reclaimed water by the conventional process is not as good as that from the combined system; however, it is mainly reused for the production process of a paper mill and as cooling water for an electric power plant.

The diagram of the combined system is shown in Figure 7.4.

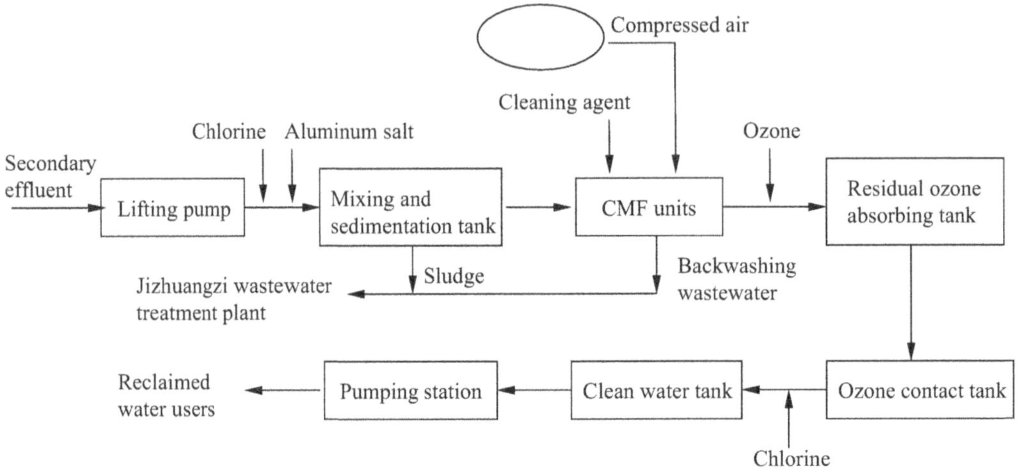

Figure 7.4 Diagram of the combined system of CMF and ozonation in Tianjin Jizhuangzi wastewater reclamation plant.

Another water reuse example in Tianjin is the plant established in Tianjin Economy Developing Area (TEDA). TEDA is located on the coast of the Bohai Sea. Therefore, its wastewater has a higher concentration of salt. The concentrations of Cl^- and total salt in wastewater are 1220 mg/L and 2290 mg/L respectively in dry seasons, and 2550 mg/L and 5075 mg/L respectively in rainy seasons. To solve this problem, a dual membrane (CMF and RO) system was adopted for wastewater reclamation (see Figure 7.4). The capacity of the CMF unit is 25,000 m^3/d; 15,000 m^3/d of the effluent is reused for municipal miscellaneous uses, and the other 10,000 m^3/d is supplied to industrial cooling and process water after desalination by RO.

The reclaimed water cost is 0.3 US\$/m^3 lower than the current price of tap water in Tianjin (0.36 US\$/m^3).

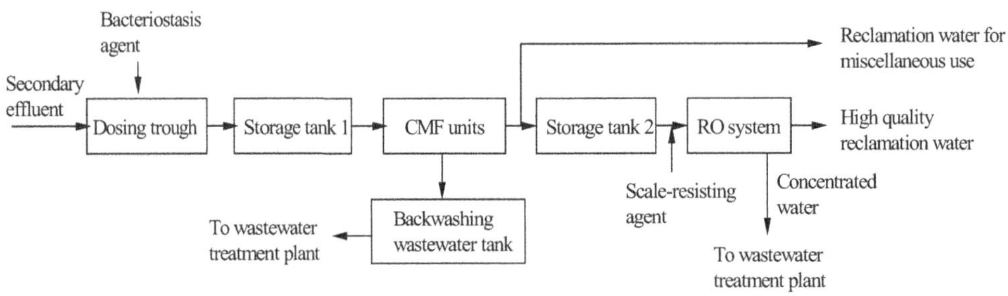

Figure 7.4 Diagram of the dual membrane system for wastewater reclamation in TEDA.

Water reuse examples in Beijing (Gan Yiping, 2004)

Beijing, the capital of China, is also a city short of water. The average per capita water resource is only 300 m^3/yr, about one eighth of the whole country's average.

Gaobeidian WWTP reclamation plant is the largest in Beijing, up to now. Its current capacity is 300,000 m^3/d. Part of the secondary effluent is supplied to the Gaobeidian lake, serving as cooling water for a thermoelectric power plant, and some of the water (after advanced treatment) is supplied to municipal miscellaneous uses, such as park greening, street spraying, etc. A pumping station and 34.8 kilometres of pipelines were constructed to deliver the reclaimed water. For the Beijing Olympiad in 2008, 15 wastewater reclamation plants will be constructed and the total reclamation capacity will reach over 600,000 m^3/d.

Onsite water recycling is another characteristic in the city. Beijing Municipality promulgated a regulation for wastewater reclamation. It is regulated that wastewater reclamation facilities should be constructed in buildings with a construction area larger than 50,000 m^2, or in residential blocks with a recoverable water amount larger than 150 m^3/d. There are over 120 onsite wastewater reclamation plants in operation with total treatment capacity of about 45,000 m^3/d and more than 100 plants in construction. Reclaimed water is used for toilet flushing, car washing, and landscape irrigation.

7.2.4 Reclaimed water quality criteria and water reuse legislation in China

There are a series of national standards for reclaimed water in China: (1) The reuse of urban recycling water — classified standard, GB/T 18919-2002; (2) The reuse of urban recycling water — water quality standard for urban miscellaneous water consumption, GB/T 18920-2002; (3) The reuse of urban recycling water — water quality standard for scenic environment use, GB/T 18921-2002; (4) The reuse of urban recycling water — water quality standard for supply to water source; and (5) The reuse of urban recycling water — water quality standard for industrial water.

In GB/T 18919-2002, reuse of urban reclaimed water has been classified into five categories: (i) water consumption in agriculture, forestry, farm and fishery; (ii) urban water consumption; (iii) industrial water consumption; (iv) water for environment use; (v) supply to the water source. At present, the water quality standards for urban miscellaneous water consumption (GB/T 18920-2002) and for scenic environment use (GB/T 18920-2002) have been issued.

In China, there is a law specially drawn up to manage the water resources of the country. It is "The water law of the people's republic of China", which was revised in 2002. In the 52nd item of this law, wastewater reclamation and reuse are encouraged.

Additionally, much local legislation encourages water reuse.

7.3 WATER REUSE IN HONG KONG

Hong Kong imports about 2 Mm^3/d fresh water from the Dongjiang River of Guangdong Province every day, accounting for 73% of its total fresh water supply. Seawater is supplied to around 80% of the 6.8 million populations for toilet flushing, which amounts to some 0.7 Mm^3/d.

Reuse of sewage treatment works effluent has just started in Hong Kong. The first project is to build a tertiary treatment plant with a capacity of 3,000 m^3/d at Ngong Ping of Lantau Island, which was completed in mid 2005. The reclaimed water, with an *E. coli* level below 100 counts/100 mL, will be used for flushing public toilets and restricted irrigation. The second project is the Shek Wu Hui Effluent Reuse Demonstration Scheme to be carried out from 2005 to 2008, during which a pilot reclaimed water production system with capacity of 150 m^3/d will be operated to demonstrate the public acceptability of treated effluent reuse for toilet flushing and unrestricted irrigation purposes. The effluent quality of this system will follow the USEPA guideline.

In Hong Kong, effluent reuse is controlled under the Environmental Impact Assessment

Ordinance. Any new effluent reuse project has to obtain an Environmental Permit from the Environmental Protection Department for construction and operation. Standards for general reuse of sewage treatment works effluent in Hong Kong will be established in the near future.

7.4. WATER REUSE IN SINGAPORE

7.4.1 General situation on availability of water in Singapore

The water sources in Singapore come from local reservoirs and Singapore's neighboring country, Malaysia. Being located in a tropical region, Singapore has a high annual rainfall. With an approximate area of 680 Km^2, Singapore receives 254 cm of rain per year. Today, about half of Singapore is being utilized as catchments for rainwater collection. 14 impounded reservoirs have so far been constructed. However the amount of water that could be stored in the reservoirs only contributes to 50% of raw water needed. The other half of the water comes from a river located in the neighboring country. The land constraint as well as increasing water demand attributable to economic development and increased population poses considerable water management challenges, as well as opportunities (WaterNet, 2004).

Measures to be undertaken for mining additional water resources include the construction of additional reservoirs and new rainwater storage facilities, introduction of the latest technology for seawater desalination, and recycling of used water. With technological advances, Singapore is able to produce high quality NEWater, launched in 2003, to boost her water supply (ENV News Release, FEB 2003a, and MEWR website). The commissioning of 3 full-scale NEWater Production Plants (NPP) provides a combined capacity of 92,000 m^3/d. With the impending 4th NPP, the combined capacity will be further increased to about 208,000 m^3/d. To expand the existing local sources, desalinated water is supplied, starting from 2005. A seawater desalination plant is currently being constructed that has a capacity of over 136,000 m^3/d, which will supply about 10% of Singapore's water demand from seawater. To augment Singapore's water supply, the Public Utilities Board (PUB) has planned to further collect storm water from residential new town developments as well as capture surface runoffs from highly urbanized catchments (PUB website (a), and Ong *et al.*, 2004). One example is the unique Marina reservoir scheme, which is scheduled for completion by 2009. With the completion of the Marina barrage across Marina channel, a freshwater reservoir, Marina reservoir, will be formed behind the barrage. This reservoir will collect run-off that drains into the Marina basin (PUB annual report, 2003). Surface water from one-sixth of Singapore's land will drain into the Marina reservoir. The new Marina reservoir will augment local sources and increase Singapore's land utilized for water catchment from half to two-thirds of the country (WaterNet, 2004).

Singapore has also entered into a water agreement with Indonesia to further develop the water supply to Singapore (PUB website b).

The result is that instead of depending on just two water sources - local catchments and river water from its neighboring country- four water sources have been available in Singapore since 2005. This makes Singapore's water supply more robust and resilient (ENV News Release, 2003b).

7.4.2 Main types of reuse involved

Water reuse has been practiced over the years at different levels. The type of reuse includes municipal, agricultural and industrial applications. Industrial reuse is mainly for cooling, dying in textile industries, general washing purposes and as raw water for production of high grade water for non-potable use. Figure 7.5 shows the breakdown of industrial water usage. Treated effluent from domestic sewage treatment works has undergone tertiary treatment and reused as industrial water since 1966. Industrial reclaimed water treatment involves conventional sand filtration and chlorination before it is pumped to a service reservoir for distribution to the industries. Currently, there are over 130 km of underground industrial water pumping and distribution mains, as well as 7 service reservoirs, supplying industrial water to industries in the western part of Singapore (PUB website f). The current demand for industrial water is about 90,000 m^3/d (PUB website b).

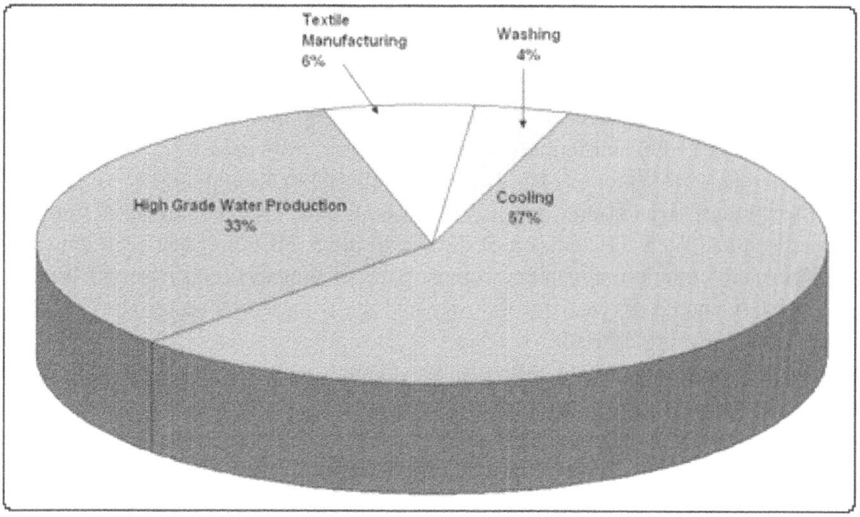

Figure 7. 5. Breakdown of industrial water usage in year 2003 (PUB website e).

Industries are also encouraged to reuse their process water through various means such as cascading systems, counter flow systems, reverse osmosis treatment, etc. As an incentive to encourage industries to reuse and recycle their process water, a water conservation fund has been set up as part of the PUB's efforts to encourage and assist companies, such as small and medium enterprises, to invest in technologies to reduce the use of potable water for industrial processes. Industries are also being actively encouraged to substitute their potable water requirement with alternative sources such as sea water and NEWater wherever possible (ENV News Release, b).

Water reuse has also been practiced for agro-industry due to the typically high requirement. Using proper water recycling technologies, water quality at the fish and prawn ponds was maintained at acceptable levels, even with a complete recycling of the pond water. Treatment technologies such as biofiltration, sedimentation, UV disinfection were successfully employed (Chin *et al.* 1993; Chin and Ong, 1994).

To meet its water demand, Singapore has been exploiting the use of advanced water treatment technologies to derive high quality product water from non-conventional sources to augment its currently available natural water resources. The non-conventional water sources being exploited include used water (via water reclamation). In 2003 treated secondary effluent from domestic sewage treatment works was further treated with advanced water reclamation processes (i.e. multi-tiered membrane systems and UV disinfection), and reused for both indirect potable purpose as well as high-end industrial reuse applications. Reclaimed water has also been used as raw water for high grade water production which is used as process water (such as ultrapure water). Demand from wafer fabrication plants has now reached about 13,000 m^3/d, and is still growing. PUB has been supplying NEWater for air-con cooling to industrial and commercial buildings and centers. Indirect potable usage contributes less than 1% of the amount of water used daily at this juncture (PUB website f).

7.4.3 Interesting reuse examples in Singapore

The most exciting reuse example is NEWater. The treated secondary effluent from domestic sewage treatment works undergoes advanced water reclamation processes which include MF, RO, and UV processes. The finished water is much cleaner than tap water and fulfills all quality requirements stipulated for drinking water based on USEPA and WHO drinking water standards. Singapore has conducted a comprehensive pilot-scale study to confirm the feasibility and viability of NEWater. Following the success of the pilot-scale study, three full-scale NEWater production plants have been in operation since 2003; the fourth has been operated since 2007 and provides NEWater to meet the strong demand from industry.

For industrial reuse, NEWater has been used for wafer fabrication plants as well as air-con cooling for their processes and other non-potable uses. It will serve the industrial hub as well as commercial centres. With the addition of the fourth plant, NEWater has been delivered to industrial estates and commercial centers in most parts of Singapore. In time, NEWater will become the primary source of water for industry and the commercial sector for various non-potable uses (ENV News Release b).

When using NEWater for domestic purposes, an indirect approach was adopted for NEWater as a source of drinking water. For indirect potable reuse, 7,500 m^3/d (about 0.5% of total water consumption) of very high quality reclaimed water (known as NEWater) was channeled back to a reservoir (as raw water for potable use) before going through further water purification processes in February 2003. The amount of NEWater to be introduced for indirect potable reuse will be gradually increased to 2.5% of the total water consumption by 2011(Ong *et al.*, 2004).

PUB encourages the use of NEWater for air conditioning cooling towers in buildings and for suitable process water use in industrial premises. A separate NEWater pipeline network is being built which will eventually reach these buildings and industrial premises. In 2003, 41 km of pipeline was completed. A NEWater service reservoir was also commissioned (PUB annual report, 2003). All new commercial and industrial development proposals are required to provide a dedicated NEWater pipe system to take in NEWater when the supply becomes available. For locations where the NEWater pipeline network will only be available after 2011, space provisions need to be made for the new development to facilitate NEWater pipe system installation in future (PUB website c).

7.4.4 Reuse practices, philosophies, or standards in Singapore

PUB has put in great effort to ensure that the quality surpass all the requirements stipulated by the USEPA and WHO drinking water standards. In addition, PUB has commissioned two Audit Panels (local and international) to audit the operations of NEWater production plants and to ensure that all required standards are fully complied with.

The success of the NEWater Initiative is also attributed to the comprehensive scientific studies funded mainly by PUB and the meticulous efforts made in public education, spearheaded by the government. The NEWater Visitor Center (as seen in Figure 7.6), which highlights how Singapore leverages on advances in technology to reclaim water, is the focus of public education on NEWater (PUB annual report, 2003). Public support towards NEWater has been overwhelming. Through a series of public education programmes, the majority of public support the idea of adopting NEWater for indirect potable use.

Figure 7.6 NEWater visitor center in Singapore (PUB annual report, 2003)

7.4.5 Reuse legislation in Singapore

In 2002, the "Singapore Green Plan 2012" was formulated setting forth targets for the next ten years. The Plan presents a blueprint for environmental policy with 2012 as the target year. To meet 25% of water demand by desalination and recycling of used water was included as one of the targets. In order to achieve these targets, various programs and incentive measures are currently being implemented (Overseas Environmental Measure of Japanese Companies Singapore, 2003).

7.5 WASTEWATER REUSE POTENTIALS IN THAILAND

7.5.1 National water resources

Thailand receives an average annual rainfall of 1630 mm that translates into 836 x 10^9 m^3 of water. The precipitation is not uniform being confined to the monsoon season (85%) between May and October. It is concentrated mostly in the central and the southern regions causing floods, while the northern regions experience drought. This uneven distribution of water is not in accordance with the distribution of water demand throughout the country.

The average per capita water availability is the lowest in Asia, being only 1854 m^3. The surface water volume per person is also the lowest in Asia, indicating that the country is under stress in water resources (TEM, 2001). The total annually available renewable reserve of water is estimated at 199 x 10^9 m^3 of surface and groundwater. Of this, 80-90% is generated during the rainy season (Visvanathan and Cippe, 2002). The major provider of surface water is the basin of the Chao Phraya River, which begins in the north and flows into the Gulf of Thailand passing through the major industrial belt of the country, the Bangkok Metropolitan Region (BMR) and its suburban areas. Groundwater abstraction is estimated to be 8.99 x 10^9 m^3, mostly confined to the BMR and it suburbs.

7.5.2 Wastewater Treatment

There are 83 wastewater treatment plants (WWTPs) throughout Thailand with a total capacity of treating 2.84 x 10^6 m^3/d of wastewater . However, currently only half of them are in good operational condition (ONEP, 2001). The number of WWTPs may reach 250 within the next 20 years. BMR has six central WWTPs in operation to treat almost a million m^3 of domestic wastewater and serve an area of 191.7 km^2 with a population about 3 million. Table 7.3 shows the municipal WWTPs of Bangkok with their capacity, service area and population served. There are 3 new proposed projects that would treat an additional 1.12 x 10^6 m^3 of wastewater and provide service to 177 km^2 of BMR. Four more projects are being included in the 6th Bangkok Metropolitan Development Plan. Bangkok Metropolitan Administration (BMA) also operates 12 community WWTPs with a total capacity of 25,700 m^3/d. A portion of the treated effluent is used for street washing and watering plants along the street as a demonstration pilot project, while the major portion of the treated effluent is being discharged either directly into the Chao Phraya River or into canals to help improve their water quality (UNEP, 2001).

Onsite systems are widely practiced for domestic wastewater treatment in Thailand, mainly for large buildings. Most of the residential units in a municipality use septic tanks and soak-pits to treat toilet wastewater while grey water is directly discharged into storm drains.

Table 7.3. Wastewater Treatment Plants of Bangkok Municipal Region (UNEP, 2001).

Plant	Capacity (m^3/d)	Area served (km^2)	Population
Si Phraya	30,000	2.7	120,000
Rattanakosin	40,000	4.1	70,000
Dindaeng (BMA-1)	350,000	37.0	1,080,000
Chongnonsi	200,000	28.5	580,000
Nongkhaem-Phasicharoen	157,000	44.0	520,000
Ratburana	65,000	42.0	177,000
Bangkok-4	150,000	33.4	432,500
Total	992,000	191.7	2,979,500

There are 30 industrial estates in Thailand and each is equipped with a Central Effluent Treatment Plant (CETP). The total effluent treatment capacity is more than 300,000 m^3/d in addition to individual treatment units for some of the industries. A portion of the treated effluent is used on an ad-hoc basis within the industrial estate and the excess is discharged into water bodies.

The largest CETP in Thailand is under construction in Samut Prakarn province, Thailand's primary industrial region. It was envisaged to manage 450,000 m^3/d of discharge from the industrial, commercial and residential sectors with a BOD loading of 150 tons per day (as in 1995). The Samut Prakarn Wastewater Management Project (SWMP) would be able to collect industrial effluents and treat 525,000 m^3/d (Asian Development Bank, 2003).

The primary problem in the non-functional treatment plants is lack of manpower, budget constraints and lack of revenue raising methods. Though many of the CEPTs in industrial estates have been able to organize user charges, most of the municipal WWTPs do not have such a mechanism and are considering the option of levying user charges.

7.5.3 Wastewater reuse

The tariff of treated water has increased from US$ 0.095–0.313 (THB 4.00–13.00) in recent years (TEM, 2001). A further increase is anticipated and consumers will have to find options of reducing net water consumption by adopting reclamation of wastewater and its reuse. In many industrial sectors, this change in water tariff was found to be the driving force behind promoting internal wastewater reuse. The present scenario of treated wastewater being discharged into water bodies (river and canals) provides an insight into the quantity available for reuse for which necessary guidelines and initiatives are required.

Industrial effluent reuse

The industrial effluent from the 30 CETPs is being used within the industrial estates for secondary purposes. Source removal of industrial pollutants where technically feasible would facilitate recycling and reuse of the pollutant constituents as well as the reuse of the wastewater. This method of wastewater management would shift from the conventional end-of-pipe approach to a source treatment approach by creating a closed loop (recycle) water metabolism. There are some examples of wastewater reuse in pulp and paper and the sugar industry which have used the concept of cleaner production and closed loop for reclamation and reuse of effluent.

Agricultural reuse

Most often, the peri-urban areas use reclaimed municipal wastewater and nutrients for agricultural purposes for which a large proportion of water supplied to cities can be collected and redirected to the surrounding environment. This source of water is reliable with a more or less constant flow rate. Treatment and reuse of wastewater solves the urban waste problem, maximises water use efficiency and sustains or promotes agriculture. Wastewater can be used as a source of irrigation water as well as a source of plant nutrients, allowing farmers to reduce or even eliminate the purchase of chemical fertilizers. Many of the WWTPs in Thailand are located in agricultural areas at the outskirts of the urban areas with a high potential of effluent reuse in agriculture. Reuse of treated municipal effluent would help reduce fertilizer demands (presence of nutrients in the effluent) and act as a reservoir to alleviate shortage of water during the dry seasons. Wastewater is also a potential source of fertilising elements (UNEP, 2001). The treated effluents from the CETP are being informally used for irrigation of the industrial estates and parks.

Another example of wastewater reuse is in fish farming that assists in the treatment of wastewater by reducing the levels of suspended solids and algal growth in wastewater. This improves the quality of the final effluent that can be used for crop irrigation and other uses. Wastewater treatment using fishponds is a natural process that degrades and stabilizes organic wastes. The organic wastes fertilize a fishpond and stimulate growth of microorganisms to serve as fish food.

In Thailand, there is a tremendous potential for agri-aquaculture reuse, and it has been practiced for generations with a more informal/ad-hoc approach. Most of the peri-urban agri-aquaculture activities depend to a large extent on partially treated domestic and industrial wastewater. Nevertheless, promotion of reuse by the government with appropriate technical and legislative assistance would pave the way for increased reuses, enabling added benefits to the farming community.

Groundwater recharge

Ground water supply serves one-fifth of Thailand's 200 towns and cities, and half of 700 municipalities with more than 200,000 wells of 7.55×10^6 m^3/d capacity. This is estimated to be 75% of domestic supply to about 35 million people (Gupta and Babel, 2002). The withdrawal can be replenished with recharge of the aquifers by the effluent from treatment plants, with infiltration from aerated lagoons and stabilization ponds (Kijjanapanich and

Karnchanawong, 2002). Other methods of groundwater recharge can be using percolation tanks, trenching, well injection etc. In developing countries, discharge into the environment of untreated or inadequately treated wastewater can result in unplanned and uncontrolled infiltration to aquifers in arid climates. This incidental infiltration becomes a significant reuse of urban wastewater. Though this is a natural treatment of wastewater, it can also pollute the aquifers used for potable water supply unless the aquifer recharge is monitored.

The amount of wastewater treated and discharged in BMR and Samut Prakarn Wastewater project would be able to provide about 3 million m^3 of treated effluent (when in full operation) which potentially could be used for agricultural purposes both for crops and ornamental plantations, aquaculture (freshwater fisheries), and groundwater recharge in the province. The coastal zone and BMR is facing land subsidence due to excessive withdrawal of ground water by industry. BMR has been making restrictions on groundwater withdrawal by raising the price of groundwater from US$ 0.075–0.20 (THB 3-8) per m^3 over a period of 4 years and has identified areas for recharging the aquifers using reclaimed water (Seemakachorn, 2002). This raise in price will discourage the industries that depend on groundwater and divert them to public surface water supplies.

Other reuses

In Thailand, the use of treated or untreated effluent is not formalized but is widely practiced primarily during the dry season, as water availability decreases for the peri-urban areas and unplanned segments of the urban habitats. These peri-urban irrigated areas are devoted to the production of different crops, vegetables and to the practice of aquaculture. To use reclaimed water for the irrigation of fodder crops, field crops (cereals, industrial crops) or forest trees, a secondary treated effluent should be of sufficient quality. Unless the right combination of wastewater treatment is used, protection of public health cannot be achieved for its use in agriculture. Wastewater and sewage sludge can be used in the peri-urban areas with wastewater reuse as an integral part of overall environmental pollution control and a country's overall water resources balance. Other reuses of wastewater that are currently practiced in a more ad-hoc manner are:

- irrigation of urban environments: parks, gardens, golf courses, clubs, residential areas, cemeteries, green belts, lawns and flowers beds;
- environmental and recreational uses: lakes and reservoirs, wetlands and increase of runoff; and
- urban non-potable uses: fire fighting and toilet flushing.

7.5.4 Administrative and Legal Aspects

Thailand is experiencing an increased water demand in its three sectors (agriculture, industry and domestic) and the requirement exceeds the readily available supply. There is a need for research and development (R&D) to project the future demand for water and potentials for reuse. Setting up water rights and allocation plans would be able to tackle and minimize the conflicts in water distribution and reuse. The administration needs to organize means for legislating relevant regulations for water use and administering them vis-à-vis R&D in the sector.

Administrative aspects

Thailand has established 30 industrial estates and parks to promote industrial growth and help pollution monitoring and control. Common effluent treatments plants (CETP) have been established to meet the stringent effluent standards set by the Pollution Control Department (PCD). Taking into consideration the water demand and wastewater disposal issues, various water reuse potentials need to be studied with a view to develop sustainable industrial water reclamation and reuse. For effective planning of the reuse of wastewater, be it from industries, agriculture or domestic sources, the issues that have to be addressed to optimize the reuse potentials are:

- project planning and adoption of reuse technology along with its economic feasibility and advantage over the use of fresh water;
- establishment of market for reclaimed water; and
- R&D of economically attractive technology.

Further, the Royal Thai Cabinet has initiated privatization of water resource management by forming a corporate agency, East Water Company, under the Provincial Water Authority (PEA) for water management in the eastern region and the coastal areas. Similarly, in the BMR, the Metropolitan Waterworks Authority manages the water supply. Privatizing or corporatizing of water supply schemes would be a positive approach for water management in the country, which would fix user charges and promote reuse presently practiced informally by the East Water Company. Such a method would spearhead development of further reuse of wastewater.

Legal aspects

There is no well-planned legal framework in Thailand for water resource inventory, planning, use, reuse, quality control and protection (Gupta and Babel, 2002). Watercourses are freely available for withdrawal and use. Legislative provisions do not enforce reuse of wastewater. What is practiced on an ad-hoc basis needs to be formalized under set legislative and administrative provisions. Thailand has, over the past 15 years, focused on the management of water pollution with infrastructure for domestic and industrial wastewater treatment systems and the setting up of effluent discharge standards. It is imperative for the environmental agencies to develop appropriate water reuse guidelines and set effluent standards for various reuses to prevent the associated potential environmental and health risks. The increasing demand for water and the limited availability of surface water and over-used ground water has brought about water stress in the country.

The use of wastewater would affect the handlers and consumers of produce (agriculture and aquaculture products) due to crop-contamination and disease linked to water-based pathogens and bioaccumulation of the toxic metals and substances through plants that absorb them from the soil or from wastewater. Therefore, legislation should make stringent effluent standards and guidelines to enable reuse and to protect human and environmental health, to:

- remove or detoxify the contaminants in wastewater and disinfect it;
- modify agri–aquacultural practices (to prohibit irrigation of leafy vegetables with untreated sewage and ensure the use of holding ponds for aquaculture);
- regulate human consumption of certain products; and
- educate handlers and consumers for protective practices.

At the same time, legislation should take into consideration the reduction of fresh water usage and promotion of reclaiming and recycling activities within the framework of industry, by adopting cleaner production strategies with internal and external reuses. PCD has prioritized high water consuming industries for R&D in water use and options for its reduction by encouraging in-plant waste segregation, recycling and reuse. An initial step for waste minimization and reduction of water consumption by as much as 50% is targeted to be achieved.

Legislation and administrative aspects should consider the role of stakeholders in wastewater reuse, basically the following three groups: potential end users, relevant government agencies that monitor the water usages, and the community. A common consensus to identify water reuse possibilities would pave the way for effective management and control of water demand as well as decreasing the pollution load, thus enabling a better control over water quality.

7.5.5 Potentials for Wastewater Reuse

Treated effluent from industries and municipalities can be targeted for several potential uses. That would significantly decrease or stabilize the freshwater consumption that increases with the population while the available resources are limited. These reuses are:

1. Reuse can be within an industry, by a closed loop water metabolism or recycling by which the industrial units categorize the use of water based on the water quality and the industrial processes concerned. Secondary purposes could be for cooling, boiler feed, surface cleaning, dust suppression, toilet flushing and gardening.
2. Reuse of the effluent from a CETP in an industrial estate/park within it or by another industry within a close range depends upon the quality of water requirement. This internal use has greater potentials. All industrial estates or parks are required by law to make provisions for greenery within its boundary, which are being irrigated to a certain extent by treated effluents. Cooling water for power plants, metal smelting industries, cutting and fabrication units, brick kilns etc. could use treated wastewater.
3. Treated municipal and industrial effluents can be used to recharge the aquifers as excessive withdrawal of ground water has caused land subsidence (up to 10cm/year in BMR) and seawater intrusion (some wells in the coastal zones report up to 3000 mg/L of dissolved solids). Plans are being made in BMR and Samut Prakarn province.
4. In the BMR, effluents are being discharged into canals to replenish the water level that decreases during the dry season and to avoid silting, which also helps recharge aquifers.
5. Municipal effluents could be safely used for street cleaning, watering plants and parks within a city or municipality with adequate transport arrangements rather using canal water for the purpose.
6. Thailand has numerous golf courses which have an average water demand of $1000m^3/d$ and could be the ideal end users of treated effluents.

The primary constraints and reservations for reuse of municipal effluent in parks, golf courses, watering of plants within street premises and street cleaning would be the odor problem, microbial quality and economical transport of reclaimed water.

The use of CETP effluents for secondary purposes within the industrial estates has similar reservations with regard to odor problems and toxic constituents. In addition to the regulation of standards, the cost involved for advanced treatment processes, transport costs (conveyance by pipelines or tankers) need to be attractive and less than the abstraction costs from the ground or centralized supply. Recharge of aquifers would require stringent effluent standards to prevent contamination of good aquifers by toxic substances and dissolved salts for which standard guidance is required.

Technologies can be developed for the reclamation and reuse of building wastewater as in Japan. The treated wastewater can be utilized to flush toilets, irrigating gardens etc. Using biological treatment and membrane separation, sludge generation can be reduced with methane gas recovery in a compact facility. This reuse is primarily suitable for large commercial complexes, shopping malls, hotels and even institutions.

Desalination

Salinization has become a major concern in the northeast and south of Thailand along the coast caused by saline water intrusion into the aquifers due to excessive groundwater withdrawals, saline water irrigation and expansion of brackish water aquaculture. Areas with scarcity of fresh water would require adopting desalination of the brackish water and seawater in the coastal zones using modern methods like reverse osmosis, which in particular is reliable, consumes less energy, and is easy to operate and maintain. Industries too could adopt the desalination process for obtaining fresh water. Thailand has a desalination capacity of 24,075 m^3/d. One such industrial unit is the salination plant operated by Esso Sriracha Refinery, Chonburi.

REFERENCES

Asian Development Bank (2003). Samut Prakarn Wastewater Treatment Project will be Benchmark for Tackling Pollution in Thailand. News Release of Independent Review Panel, issued Bangkok, Thailand, 15 June 2001.

Chen Qisi and Hu Wenli: The current status and development of reclaimed water uses in Tianjin, Tianjin Jianshe Keji, No.1, 39-41, 2004. [In Chinese]

Chin, K.K. , Ong, S. L. and Foo, S. C. (1993) A water treatment and recycling system for intensive fish farming. *Wat. Sci. Tech.* **27**(1): 141-148.

Chin, K.K. and Ong, S.L. (1994) Treatment and reuse of water for prawn cultivation. *Wat. Sci. Tech.* **30**(9):255-258.

ENV News Release No, 005/2003, Feb 2003 a), http://app.mewr.gov.sg/view.asp?id=SAS196

ENV News Release No, 004/2003, Feb 2003 b), http://app.mewr.gov.sg/view.asp?id=SAS197

Gan, Y. (2004) The current status of wastewater reclamation and reuses in Beijing , *Urban Management & Technology*, **5**(4): 160-161 [In Chinese]

Gupta, A.D. and Babel, M. S. (2002) Institutional Reform for Effective Water Resources Management: Thailand Perspective. Water Engineering and Management Program, School of Civil Engineering, Asian Institute of Technology, Thailand..

Kijjanapanich, V. and Karnchanawong, S. (2002) *Water quality of infiltrate from laboratory-scale plots irrigated by effluent from domestic wastewater treatment plants,* A Report. Department of Environmental Engineering, Chiang Mai University, Chiang Mai 50200, Thailand.

Li, J. Li, F. , Guan, D. (2003) Wastewater reclamation project with a dual membrane system in Tianjin Economy Developing Area, *China Water & Wastewater*, **19**(11): 96-97 [In Chinese].

MEWR website, http://app.mewr.gov.sg/home.asp?cid=167&nid=167&id=SAS763

Noh, S., Kwon ,I., Yang, H.M. and Kim, H (2003) Current Status of Water Reuse Systems in Korea, full paper in CD-ROM, 4th International Symposium on Wastewater Reclamation and Reuse organized by the International Water Association, November 12-15, 2003, Mexico City.

Oki, T., Agata, Y., Kanae, S., Saruhashi, T., Yang, D. and Musiake, K. (2001) Global Assessment of Current Water Resources using Total Runoff Integrating Pathways, *Hydrol. Sci. J.*, **46**:983-996.

ONEP: Executive Summary: *Thailand Environment Report 2001.* Office of Natural Resources and Environment Policy and Planning, Ministry of Natural Resources and Environment, Bangkok, Thailand, 2001. Website: http://www.onep.go.th/eng/soe2001_3.asp

Ong, S.L, Lee, L.Y., Hu, J.Y., Song, L.F., and Ng, H.Y. A holistic approach toward overcoming limited water resources constraint. Presented at the 1st International Workshop on Asian Approach Towards Sustainable Urban Regeneration, Tokyo, Japan, 4-7 September, 2004.

Overseas Environmental Measure of Japanese Companies (Singapore), 2003, http://www.env.go.jp/earth/coop/oemjc/singa/e/contents.html

PUB annual report, 2003

PUB website a): http://www.pub.gov.sg/our_services/WaterWSHistoryandFuture.php?l1=2&l2=8&l3=4-2&l4=4

PUB website b): http://www.pub.gov.sg/home/Faq.php
PUB website c), http://www.pub.gov.sg/conservation/ConservationMeasuresIntroduction.php
PUB website d): http://www.pub.gov.sg/info_center/IcFfWaterReclamationFigures.php?l1=4&l2=22&l3=28
PUB website e): http://www.pub.gov.sg/our_services/IndustrialWaterProductionProcess.php?l1=2&l2=9&l3=9
Seemakachorn W. : *Possible Reuse of Effluent from Wastewater Treatment System in Samut Prakarn Area.* AIT Thesis RSPR No. EV 02-10, Asian Institute of Technology, Thailand, 2002.
SEPA. State of the environment, China 2003.
TEM : *Thailand Environment Monitor 2001.* The World Bank Group, Washington D.C., USA.,2001.
US-EPA (2004) Guideline for water reuse, EPA/625/R-04/108 September, Municipal Support Division Office of Wastewater Management Office of Water Washington, D.C.
UNEP (2001) Bangkok State of the Environment 2001, United Nations Environmental Programme. ISBN: 92-807-2143-1.
Visvanathan, C. and Cippe, (2002) A.: *Strategies for Development of Industrial Wastewater Reuse in Thailand*, A Report. School of Environment, Resources and Development, Asian Institute of Technology, Thailand.
WaterNet: PUB, Sep. 2004
WaterNet: PUB, Dec 2004
World Resources 2002-2004, Data Table 11 Freshwater Resources.

8

Water reuse in Central and Southern Regions of Africa

Frederik Schutte

8.1 INTRODUCTION

The central and southern regions of Africa cover a large land area (about 10 million km^2) with a total population of about 280 million people. The climate and water availability vary over a wide spectrum from countries with very high rainfall and large surface water resources in the central parts, to arid and semi-arid areas in the south-western parts of South Africa and Namibia. The economies of the countries in the area also vary over a wide spectrum. South Africa is highly industrialised with very large industrial and mining sectors. The Democratic Republic of the Congo (DRC) is a large oil producer, while Angola also has large oil resources. Most of the other countries have agriculture-based economies while mining also contributes to the economies of a number of countries.

The countries that form part of this region are members of the Southern African Development Community (SADC) and the East African Community (EAC). This chapter presents a general view of water reuse in the region with some examples from different countries.

The countries are listed in Table 8.1 which gives information on the water resources and water abstraction of each country together with land area and estimates of population (Gleick, 1998).

Table 8.1: Water sources, use, population, area of SADC and EAC Countries.

	Estimated freshwater sources: km^3/yr	Total freshwater withdrawal: km^3/yr	Population: million	Estimated withdrawal: m^3/capita/yr	Land area: km^2 X10^6
Angola	184	0.48	14	38	1.25
Botswana	14.7	0.11	1.7	70	0.58
Democratic Republic Congo	1019	0.36	54.9	7	2.44
Kenya	30.2	2.05	34.3	68	0.58
Lesotho	5.2	0.05	2.2	22	0.03
Madagascar	337	16.30	16.4	937	0.59
Malawi	18.7	0.94	11.5	85	0.12
Mauritius	2.2	0.36	1.2	305	0.01
Mozambique	216	0.61	18.1	31	0.80
Namibia	45.5	0.25	1.8	144	0.84
South Africa	50	13.31	46.4	288	1.22
Swaziland	4.5	0.66	1.1	667	0.02
Tanzania	89	1.17	33.6	35	0.95
Uganda	66	0.2	24.4	9	0.24
Zambia	116	1.71	10.7	187	0.75
Zimbabwe	20	1.22	11.6	98	0.39
Total	2218		283.9		10.81

8.2 GENERAL SITUATION

Most of the countries in the southern part of the region have a relatively low rainfall with limited water resources. The economies and infrastructure of many of these countries are also not very well developed. The result is that in many parts water supply and sanitation services are poorly developed. Water supply to cities and urban areas is mostly of a reasonable standard, while most rural areas rely for their water supply on wells and water abstracted from streams with limited or no treatment at all. The situation is similar with respect to sanitation, where wastewater from urban centres is normally treated to some extent before being discharged, while on-site sanitation is mostly the rule in rural areas.

The result of this general situation is that treated or partially treated wastewater from urban areas is discharged to receiving rivers or streams and this blend of treated wastewater and runoff is generally abstracted by communities living downstream. Indirect reuse is therefore taking place in practically all countries in the region where abstracted water that is partially derived from treated return flows is used for domestic purposes, gardening and irrigation. Abstracted water is mostly treated for domestic use in larger communities but is often used untreated in smaller settlements and rural areas.

In countries with high rainfall and natural surface runoff, a large degree of dilution takes place which provides some protection to down stream users. However, in the semi-arid countries of the region, the runoff in most streams and rivers during the dry season consists to a large degree of return treated or partially treated wastewater and this constitutes a health risk to downstream consumers.

In view of the fact that the reuse situation is similar in many countries only a few examples are discussed further on. The rest of the chapter focuses on the reuse situation in South Africa and a number of reuse examples are discussed in more detail. The reasons are, firstly, that water reuse (especially indirect reuse) plays a vital role in balancing water supply and demand in South Africa. Secondly, extensive research on different aspects of water reclamation and reuse has been conducted over many years in South Africa and a large amount of literature is available on these aspects. Thirdly, a relatively large number of direct reuse applications in industry exists in South Africa, as well as examples of reuse for landscape irrigation.

8.3 KENYA

Kenya has a population of more than 30 million, most of whom (80%) live in rural areas. Farming and pastoralism are therefore the predominant means of livelihood in the country (Tole, 2006). Industries are mostly for processing agricultural goods. Other types of manufacturing industries are just beginning to be developed.

Kenya is also a water scarce country, with a per capita availability of water at around 600 m^3 per year. Various methods of water conservation have therefore been adopted out of necessity, including wastewater reuse.

In formal agricultural and industrial systems, there is some amount of water reuse practiced, in order to:

- meet the water needs amidst a situation of scarcity;
- save on costs of fresh water;
- meet environmental regulations for waste water discharge.

An example is the Oserian Development Company in Naivasha, Kenya which uses vast quantities of water for cultivation of fresh cut flowers for export to Europe. The water is drawn from Lake Naivasha, a fresh water resource of limited capacity. The hydroponic system for watering and feeding the flowers allows for water drained from the soils to be recycled, thereby conserving both water and nutrients through recycling of both. This lowers the costs of provision of water, and also the costs of fertilizer inputs.

In industry, where water is used for washing raw materials before processing, water reuse is also practiced. For example at Magadi Soda Company in the Rift Valley, which mines trona and processes it to soda ash, washing removes soluble and insoluble impurities, and this water is recycled several times before being discarded. Similarly in the energy industry, water from cooling towers is recycled to condense steam in the condensers.

An interesting wastewater reuse is the use of wastewaters in towns and cities for cultivation of food crops. Waste water in drains is used for irrigating crops such as kales, tomatoes, and arrow roots, which are then sold in the markets in the cities. One therefore finds many small cultivated fields running parallel to open drainage systems, and even along the highly polluted Nairobi River, which is a receptacle for waste waters from factories along its course. Similarly, adjacent to sewage treatment works, there are fields of cultivated crops that rely on sewage waste waters for irrigating crops that are sold in markets. Some studies have identified high levels of heavy metals in food crops cultivated using such waste waters in Kenya.

The Splash-Carnivore constructed wetland was designed to treat waste waters from the Carnivore Restaurant and the Splash Water World entertainment facilities in Nairobi. The wetland removes most pollutants to levels of better than 90%. It is intended that the treated wastewater be used to irrigate gardens in the facilities after chlorination to remove bacteriological contamination.

8.4 ZIMBABWE

The central and southern parts of Zimbabwe have limited water resources in contrast to the northern parts with abundant water from the Zambezi river and Lake Kariba. Indirect water reuse takes place generally in Zimbabwe whereby treated wastewater discharged to rivers and streams is used by down stream communities. There are also a large number of specific reuse applications in Zimbabwe. The following sections give detailed information on reuse of water in some towns (Madyiwa, 2007).

8.4.1Harare

The city of Harare has a number of wastewater treatment works and the treated effluent from these works is used as discussed below.

Firle Treatment Works

Effluent from the Firle treatment works is used for the irrigation of pastures on which about 3000 head of cattle is raised for meat production. Flood disposal and irrigation of pastures at Firle farm has been practiced for over 30 years as a form of tertiary treatment of sewage whereby grass takes up nutrients for metabolic purposes. Firle farm employs 32 farm workers and supports 3000 beef cattle, all of them born and bred on the farm. The cattle are sent to abattoirs for slaughter and sale of beef to the population at large. The pastures at Firle farm comprise star grass and kikuyu grass and both grass species are perennial. While star grass reportedly grows faster than kikuyu, the former is considered to be more resilient to droughts. As a result, the two grasses have been grown together, in order to reduce the risk of failure of pasture. There is no data available on the application rates of sewage sludge and effluent on land at Firle farm, but based on annual sewage output and area irrigated, the application rate is about 48 megalitres/ha/year or 126-167 t solids/ha/year.

Firle Wastewater Treatment Plant processes both industrial and domestic effluent. The plant services the southern residential areas and many industries within the Firle catchment area. Firle Wastewater Treatment Plant comprises of six units with a capacity to process 144 megalitres of effluent per day. Two of the units use the conventional wastewater biological trickling filtration system that produces low quality water (ca. 48 megalitres) used for irrigation of pastures. The other four units use the Biological Nutrient Removal System (BNRS) that produces high quality water (ca. 96 megalitres), which is discharged into the natural river system. The low quality water produced from the biological filtration system is mixed with digested sewage sludge from all the processing units to produce a slurry (3-4% solids) used to irrigating 860 hectares of pasture.

Other Treatment works

- Effluent from Crowborough is used for irrigation of pastures.
- Effluent from the Donnybrook Treatment works is used to irrigate a variety of crops.
- Effluent from the Malborough Treatment works is used to irrigate gum trees.
- Wastewater from Chitungwiza is used to irrigate a variety of crops including maize and wheat.

8.4.1 Mutare

Liquid sludge is used to irrigate gum plantations. The city has also applied to utilise its wastewater for agricultural production.

8.4.2 Kwekwe

The city had plans as far back as 2001 to utilise wastewater for agriculture but was not sure of the implications to the environment.

8.4.3 Irrigation reuse

Water used to irrigate conventional irrigation schemes normally comes from sources such as rivers, dams and boreholes. Unlike wastewater, this water meets quality standards for irrigation. In some smallholder irrigation, reuse of drainage water at the lower end of the irrigation scheme has been introduced on a trial basis. Two reuse systems are being tried currently. These are:

Irrigation of fruit trees using drainage water. Fruit trees have been grown in drainage areas of smallholder irrigation schemes to utilise drainage water and in the process also improve on the nutrition of farmers.

Use of drainage water for breeding fish in fish ponds. Several fish ponds (typically small ponds with dimensions of about 4m by 2m) have been constructed in drainage areas and fish have been grown in the ponds. The idea was to reuse water for improving nutrition for smallholder farmers. This has been done on several smallholder irrigation schemes.

8.5 BOTSWANA

Botswana is a country with very limited water resources. Treated wastewater presents a resource that is available and that can be used with limited additional treatment. In the light of erratic rainfall aggravated by long drought episodes (five years is not uncommon), it is prudent to maximize wastewater reuse. Since independence, Botswana has been experiencing frequent droughts and as demand for water increases the government and the people will be forced to optimise the use and reuse of all water sources and minimising losses and wastage in the systems and process. The view was expressed by the Department of Water Affairs (at a conference 'Towards an efficient economy in a drought prone environment' in Gaberone in 2005) that there is need for a complete turn-around on the misconception that some water sources are not culturally acceptable. There is adequate and consistent effluent water that could be treated for reuse in irrigation, industrial application and drinking.

Treated effluent will constitute a major waste resource in major centres such as Gaborone, Francistown, Selebi-Phikwe, Lobatse, and Orapa, and in other major villages. In Gaborone and Francistown, by 2030 the available new water will have increased to 28000 megalitre per year and 11427 megalitre per year respectively, representing approximately 42% of the demand in those centres. The Department also expressed the view that wastewater reuse should assume equity with other water sources as a means of meeting appropriate components of the growing demand.

Established and possible uses of wastewater in Botswana include (UNEP newsletter July 2006):

- peri-urban farming around Gaborone to meet high demand by farmers;
- demand by present and future industries;
- demand for landscaping by Gaborone City Council;
- meeting water quality requirements for Water Utilities Corporation (WUC) for blending and treatment for potable use;
- provision of water demanded by the Ministry of Agriculture (MOA) for extensive agricultural research projects at Seble College.

8.6 NAMIBIA

Namibia is the only country where water reclamation for direct municipal reuse has been practised successfully for almost forty years. Chapter 25 of this book is devoted to the Windhoek reuse project and for this reason it will not be considered any further in this chapter.

8.7 SOUTH AFRICA

8.7.1 Climate

South Africa is located in a predominantly semi-arid part of the world. The climate varies from semi-desert in the west to very high rainfall along the eastern coastal area. The average rainfall for the country of about 450 mm per year is well below the world average of about 860 mm/a, while evaporation is comparatively high. As a result, South Africa's water resources are relatively limited. The country has no large or navigable rivers. The combined flow of all the rivers in the country amounts to approximately 50 000 million m^3 per year, less than half of that of the Zambezi River.

Due to the poor spatial distribution of rainfall the natural availability of water across the country is very uneven. This situation is compounded by the strong seasonality of rainfall. As a result, surface water runoff is highly variable with low stream flow in most South African rivers most of the time. Sporadic high flows occur from time to time, but this limits the proportion of stream flow that can be impounded and be relied upon. To aggravate the water supply situation, most urban and industrial settlements have developed remote from large watercourses dictated by the occurrence of mineral resources and associated industrial development. The result is that in several river basins the water demand exceeds availability and large-scale transfer of water across catchments have already been implemented in past decades (Basson, 1997; Department of Water Affairs and Forestry, 2004).

To meet the country's growing water requirements, water resources are highly developed and fully utilised in many catchments of the country. As a result of the many flow control structures, the abstraction of water and return flows to rivers, as well as the impacts of land use, the flow regime in many rivers has been significantly altered. In some instances this has resulted in severe degradation of water quality. The economic and industrial growth of the economy and rapid urbanisation of the population is expected to result in further deterioration of the country's water resources unless effective corrective measures are taken at national and catchment levels.

In the interior of the country most of the water used in a non-consumptive manner is returned to the rivers after treatment and therefore available for reuse. Many industries also practise direct internal water reuse for a variety of purposes such as for cooling, process water, material conveyance, etc. In urban and industrial areas such as Pretoria and Johannesburg approximately 50% of the total water requirements becomes available as return flow and is re-used. Similar return flows are available in coastal cities such as Cape Town and Durban, but only about 5 and 15% respectively is re-used in these cities.

8.7.2 Indirect water reuse

The South African National Water Act stipulates that water that has been abstracted must be purified to certain standards after use and be returned to the stream from which it was abstracted. This implies that indirect reuse of water is compulsory in terms of legislation. Indirect reuse therefore takes place in all areas of South Africa where downstream users abstract water which has been returned after use and purification. Very limited exemption is allowed in terms of the National Water Act on this aspect.

As a result of the increasing return flows in the country's surface waters, quality deterioration is occurring at an alarming rate. A high priority is therefore given by the relevant authorities to managing and controlling the quality of return flows. In the context of planned indirect reuse, South Africa has adopted a strategy for water quality management which is much more complicated than in countries with abundant water supplies. In such countries, it is possible to adopt water pollution control policies which focus exclusively on the prevention of pollution, without taking quantitative considerations into account (e.g. effluent may not be allowed to be discharged into water courses at all). South Africa has been forced to adopt both a Receiving Water Quality Objectives (RWQO) approach and a Pollution Prevention approach. The latter approach involves the reduction of pollution at source, water recycling, and control over the manner in which wastes are handled and disposed of.

The RWQO approach entails that the water quality requirements of different user categories must be documented and taken into account, when decisions have to be made regarding the quality of the resource as well as the standards that must be imposed upon effluent dischargers. To underpin decision making processes, the Department of Water Affairs and Forestry published a set of water quality guidelines for the recognized water uses: domestic, industrial, agricultural, recreational and aquatic ecosystems. The target water quality range stated within these guidelines are considered the ideal situation and are accepted as the objective for water quality management purposes.

Central to the RWQO approach is the concept of waste load allocation and assimilative capacity of natural waters. In principle, waste load allocation involves assignments of allowable discharges in such a way that the water quality objectives of the designated water uses can be met. It requires determination of:

- water quality objectives for desirable water uses;
- understanding of the relationship between pollutant loads and water quality;
- economic impacts;
- socio-political constraints; and
- cost/benefit relationships.

Assimilative capacity is the concept that water bodies can tolerate the input of some wastes without the deterioration of water quality to the point where water uses are adversely affected. A water body provides many mechanisms to modify, move, or otherwise transform material discharged into it, but this capacity should not be strained.

Assimilative capacity for a constituent differs in fundamental ways, depending on whether the constituent can be considered conservative or non-conservative:

- Conservative constituents are not lost due to chemical reactions or biochemical degradations. Such constituents may include, for example, total dissolved solids and chlorides. Conservative constituents accumulate along the length of a water body in the direction of motion, so that amounts added at the most upstream point are still present at the most downstream point. Concentrations of conservative constituents can be reduced only by dilution with water with a lower concentration.
- Non-conservative constituents, on the other hand, decay with time due to such mechanisms as chemical reactions, bacterial degradation, radioactive decay, or settling of particulates out of the water column. Many constituents exhibit non-conservative behaviour, including oxidisable organic matter, nutrients, volatile chemicals and bacteria. The amount of a non-conservative constituent decrease with time and/or distance from the point of input.

In line with its precautionary approach, the Department has adopted an hierarchical decision-making procedure in considering applications for effluent discharge to a receiving water body. The steps are the following:

- Options for preventing and minimizing waste through source reduction, recycling, detoxification and neutralizing of wastes must be thoroughly investigated. Caution should be taken that, in this process, one is truly avoiding or minimising waste and not simply shifting it from one environmental medium to another, for example, from water to land, or from water to air.
- If, after all the practical options to prevent and/or mimimise waste have been exhausted, there is still waste or effluent, such effluent will be required to meet whichever is the strictest: the current general effluent standards (or special effluent standards, where applicable); or standards based on receiving water quality objectives, and taking into account the projected impact of the effluent discharged on the quality of the receiving water.
- Exemptions from minimum effluent standards or receiving water quality based standards will be considered under special circumstances, and as a last resort, but will require sufficient justification on technological, economic and socio-political grounds. Such exemptions may not always be granted, will in most cases be temporary, and will almost certainly be withheld if the effluent discharge investigation shows that the receiving water's fitness for the maintenance of the natural aquatic environment, and for intended water uses, will be significantly reduced.

Should it compare favourably with other options, greater reuse could therefore be a substantial source of water, especially for coastal cities. Return flows normally have a significant impact on the quality of the receiving waters and this aspect needs to be specially managed. Where return flows are re-used directly, sophisticated treatment processes and proper management may be required, depending on the quality of the return flow and its intended applications.

Quality problems resulting from indirect reuse

The main water quality problems arising from indirect reuse are eutrophication caused by nutrients in treated effluents and salination due to inorganic salts in treated wastewater and industrial effluents discharged to surface water. These problems became apparent in South Africa in the 1960s and 70s due to increases in return flows and gave rise to extensive research efforts to ameliorate these problems:

- The eutrophication problem was addressed by research into and development of biological nutrient removal (BNR) processes to reduce nutrients (nitrogen and phosphorus) in the effluent of wastewater treatment plants. Early work by Barnard and co-workers at the CSIR in Pretoria on the Bardenpho and similar processes paved the way for researchers at the University of Cape Town (UCT) and later also by the Johannesburg City Council to make major contributions in this field. Most of this research was sponsored by the Water Research Commission (WRC) in Pretoria and the UCT group became a world leader in this field. The research by the UCT group and the Johannesburg City Council group resulted in the development of advanced biological processes which removed N and P from treated wastewater to low levels in order to limit emissions of nitrogen and phosphorus to receiving water courses. Although this research led to the development of advanced technology for the removal of nutrients, the overall effect on the trophic status of most of the inland water sources into which treated effluent is discharged is rather limited. Algal blooms still occur, resulting in taste and odour problems in treated water (unless activated carbon is used) and also giving rise to the presence of algal toxins in the water.

- The second aspect of salination is mainly a result of industrial and mining activities. Underground mine water, blow-down from cooling systems and some industrial discharges are the main sources of inorganic salts that lead to increasing TDS levels in receiving water courses. Research to develop and apply technology also started in the 1970s to develop local technology and to adopt and apply available desalination technology to solve this problem. This research focussed on desalination of effluents containing high TDS levels so that the desalinated effluent could be reused and the brine containing the bulk of the dissolved solids could be disposed of in an environmentally sustainable manner. Some of these projects are discussed in more detail below.

A further water quality problem relevant to indirect water reuse for municipal applications is the recycling of so-called environmental pollutants that occur at very low concentrations, the effects of which are not known at this stage. Examples of these substances include endocrine disruptors, residual pharmaceuticals, halogenated compounds, etc. Although the long-term health effects of these substances are not known, worldwide concern about their presence in drinking water supplies is growing.

8.7.3 Direct water reuse

As far as direct reuse of reclaimed water is concerned, South Africa has been in the forefront of developments for many years. The research that led to the erection of the Windhoek water reclamation plant was conducted in South Africa in the late 1960s and continued for many years. This research was conducted by the National Institute for Water Research of the CSIR and later expanded to the Pretoria, Cape Town and Port Elizabeth City Councils where relatively large scale plants where erected and operated as demonstration plants. This research was funded mainly by the Water Research Commission in Pretoria. The objectives of the research were to further develop the technology, to serve as demonstration plants for reclamation technology, and to promote public acceptance of direct reuse. As a result of these efforts major advances in reclamation technology were made and it was demonstrated that through the approach of a multiple-barrier system, reclaimed water of the highest quality could be produced consistently. However, as long as alternative sources of water were available, direct reuse of reclaimed water for domestic use would not be accepted mainly due to prejudice and perceptions about the 'history' of the reclaimed water and not due to quality considerations. It is ironic that most of these 'alternative sources' contain large proportions of treated effluents resulting in indirect reuse of the treated effluents.

The only direct reuse of treated municipal wastewater currently practised in South Africa is reuse by industry and reuse for landscape irrigation. There are a large number of industrial reuse examples in South Africa, and a few of these are discussed below.

8.7.4 Municipal reuse

There is only one example of direct reuse of treated industrial/mining effluent for municipal use in South Africa and that is the treatment of acid mine drainage in the Witbank area to produce water of drinking quality (20000 m^3/d) that is sold to the local authority to supplement the municipal supply (Water Wheel, 2006).

The Witbank area is a major coal producing area in South Africa. Coal mining exposes sulphur-bearing pyrite which, upon oxidation results in the formation of acidic water with a high sulphate content, so-called acid mine drainage (AMD). Contamination of the water in the area by AMD results in low pH water (around pH 3) and TDS of around 4500 mg/*l*, the bulk of which consists of sulphate (around 3500 mg/*l*).

The process that was developed and that was demonstrated on a pilot scale (120 m^3/d) consists of a neutralisation process using the CSIR's lime/limestone treatment process to increase pH and allow precipitation of metals such as iron, aluminium and manganese. This is followed by clarification and ultrafiltration to remove any remaining metals and micro-organisms. This is followed by reverse osmosis using spiral membranes to reduce the TDS to low levels. The final steps are disinfection (for residual protection in the distribution system) and chemical stabilisation to produce water that complies with the highest quality requirements for drinking water. The final water has a TDS of around 135 mg/*l* and sulphate is reduced to about 80 mg/*l*. Two identical process trains have been established, each with a capacity of 14000 m^3/d.

Another interesting aspect of the project is the fact that it is a joint venture between two competing mining companies to develop a joint solution to their water management problems in a manner that will not only benefit the environment but communities in the area where water shortages are experienced from time to time.

A similar project is now in the pipeline for the Middelburg area, where the same types of mine water problems exist. It is anticipated that a regional water supply company may be established to manage the projects and the distribution of the reclaimed water.

8.7.5 Industrial reuse

There are two systems of industrial water reuse being practised in South Africa. The first is where secondary treated municipal wastewater is upgraded by tertiary treatment and then distributed as 'industrial water' to industrial users in an area. The second is internal reuse of its own effluent (with or without treatment) by an industry in order to reduce intake of fresh water and/or to eliminate problems related to discharging of effluent that does not comply with discharge standards, or to comply with zero liquid effluent discharge requirements.

An example of the first type of reuse is the Durban South reuse scheme where secondary effluent is upgraded and distributed to industrial users in the adjacent industrial area. A further example is the upgrading of treated effluent from the Potsdam treatment works near Cape Town for reuse by the Chevron refinery.

Durban Water Recycling

The Durban Water Recycling Works is located to the south of Durban in the grounds of the city's Southern Wastewater Treatment Works (SWTW) where primary treated wastewater is discharged to the Indian Ocean via a marine outfall pipeline. In the early 1990s, the staff of the municipal utility, eThekwini Water Services (formerly Durban Metro Water Services) carried out planning for the future augmentation of the capacity of the southern marina outfall pipeline. The estimated capital costs for duplicating the pipeline were prohibitive, and alternatives had to be investigated to prevent large increases in the costs of wastewater disposal in the greater Durban area (Gisclon *et al.*, 2002).

At that time Mondi Paper approached eThekwini Water Services (EWS) to investigate the possibility of increasing Mondi's off take of 8000 m^3/d of low grade reclaimed water to a substantially larger volume of high quality reclaimed water. In 1993, EWS staff began work on a technical pre-feasibility study into the production of high quality reclaimed water at

SWTW. A reclamation process was developed and was tested at laboratory and pilot scales in 1994 and 1995. The early process development work resulted in a tertiary treatment process consisting of chemical phosphate removal, sand filtration, ozonation, granular activated carbon adsorption and chlorination. The pilot trials were successful in proving that high quality reclaimed water, meeting Mondi Paper's exacting standards, could be reliably produced from the wastewater entering SWTW.

Primarily to alleviate the capacity limitations of the marine outfall pipeline and with a view to a possible water reclamation project in the future, EWS constructed a 50000 m^3/d primary and secondary treatment plant at SWTW in 1996. The secondary plant consisted of an activated sludge reactor with three clarifiers and produced an effluent that complies with the General Standard.

Despite the proven technical feasibility of the reclamation project, the economic feasibility remained in doubt until 1997, when EWS tasked a merchant Bank to access the project. The economic assessment was positive. The costs, technical complexity and the operational risks associated with the proposed tertiary water treatment plant were considered to be beyond the normal functions of EWS. It was therefore decided that a project for the tertiary water treatment works would best be implemented as a Public Private Partnership (PPP).

In 1998, EWS invited proposals worldwide to pre-qualify prospective tenderers for a 20-year concession contract to finance, design, construct and operate the tertiary water treatment plant. This was followed by a formal bidding phase and the contract was awarded in December 1999.

Funding for the project was provided entirely by the private sector. DWR shareholder equity and loans from commercial banks and the Development Bank of South Africa provided the necessary capital. The investment funds provided for the construction of the tertiary plant at a cost of R74 million (around US$10 m) as well as the purchase of the EWS assets used for the project.

Construction on the tertiary plant started in 2000. The construction phase included upgrading the activated sludge process from 50 Ml/d to 77 Ml/d, the construction of the 47.5 Ml/d tertiary plant, tying in with pre-existing and decommissioned municipal assets, refurbishment of the SWTW high level storage tank and the installation of the reclaimed water reticulation system. Construction was finished and the plant began commissioning in April 2001. Full water production began in May 2001.

The wastewater treatment plant consists of primary and secondary treatment. The activated sludge process is of conventional design and serves to remove 95% of the incoming COD and 98% of the incoming ammonia loads. The process is operated in the nitrogen removal mode with typical effluent nitrate and nitrite concentrations of 5 mg/l and 0.02 mg/l respectively.

Tertiary treatment consists of ferric chloride, lime and polymer dosing to enhance the flocculation process. A dual-media filtration step is the last solids barrier in the process. These rapid gravity filters have a bed depth of 3 m, the top meter of the media being anthracite and the rest silica sand.

The ozonation step is used to oxidise non-biodegradable organic compounds, including colour causing compounds. The ozone contacting time is 15 minutes. This is followed by granular activated carbon (GAC) adsorption as a polishing step and for the removal of ozone degraded organics. To ensure no biological re-growth in the reticulation system a free chlorine residual of 0.8 mg/l is maintained at the point of reclaimed water use.

Chlorinated reclaimed water is stored in the 21 Ml high level storage tank before distribution to the industrial users, i.e. Mondi and Sapref. The reticulation pipe work length is approximately 2 km. Mondi Paper uses the reclaimed water directly for the production of fine paper and is extremely sensitive to process water quality and its impact on paper brightness. Mondi Paper's reclaimed water specification includes 23 parameters that are measured in the South African potable water standard (SANS 241:2005).

For Mondi Paper and Sapref the project has the following significant economic implications:

- 44% reduction in water tariff (2001) representing a significant cost saving in Mondi's paper production;
- the likelihood that the price of recycled water will escalate at a lower rate than potable water;
- significantly enhanced drought water supply security.

Recycling of Potsdam effluent by Chevron Refinery

The Potsdam wastewater treatment works of the City of Cape Town has a capacity of 40 Ml/day. About 17 Ml/d is available for reuse to replace municipal water while a further 20 Ml/d will be available for irrigation purposes (City of Cape Town 2006). The Chevron refinery in the vicinity uses about 7.5 Ml/d of municipal water and started a project about 4 years ago to replace the municipal water with reclaimed wastewater (Water Sewage and Effluent 2006). The project has been completed with the commissioning in May 2006 of the treatment plant to upgrade treated effluent from Potsdam.

The plant consists of a series of processes to treat the effluent from Potsdam to the quality required for use in the refinery. The key processes include dissolved air flotation to reduce suspended solids, followed by ultrafiltration which reduces the suspended solids to very low levels and also remove pathogens from the water. The final process is reverse osmosis that reduces the total dissolved solids in the effluent from around 1000 mg/*l* to about 150 mg/*l*.

Internal water recycling by industries

Examples of internal water recycling by some of the largest industries in South Africa are discussed in this section. Table 8.2 gives a list of industries where direct water recycling with minimal treatment is practised.

Sasol

Sasol is a very large petroleum and chemical manufacturing industry that uses vast amounts of fresh water. Sasol has a very complex water system that employs a high degree of reuse and recycling. In most cases advanced treatment technology is used to treat the water to a suitable quality for reuse and recycling. Examples include the use of reverse osmosis (both tubular RO and spiral RO) as well as electrodialysis reversal (EDR) to desalinate mine water as well as some high total dissolved solids (TDS) industrial streams. The low TDS permeate is reused while the brine is further concentrated by means of evaporative processes before final disposal. One of the interesting examples of reuse is the use of so-called stripped gas liquor as cooling water.

The Sasol One factory produces synthesis gas from coal, together with oxygen and steam as main feedstock. As a by-product of the gas production, a liquid is formed. This gas liquor is treated to separate the hydrocarbon phase (oils and tars) and the solid phase from the aqueous phase. The aqueous phase is further treated by means of liquid-liquid extraction process to remove the bulk of the phenolic compounds from the water, as well as distillation and stripping to remove the solvent and to strip the bulk of the ammonia from the water. Stripped gas liquor (SGL) is produced and then transported to the water treatment section where the SGL is used.

In the original Sasol design SGL was transported directly to the Sasol One sewage treatment works. This sewage treatment works uses trickling filter technology to treat a variety of industrial waste water streams together with domestic sewage from the Sasolburg town. The final effluent from the sewage treatment works is used to transport ash in the Sasol One ash transport system (Riedel *et. al.*, 2004).

Table 8.2: Examples of industrial use of secondary effluent with no or little tertiary treatment (Odendaal 1998).

Industry	Purpose	Treatment	Quantity
AECI (chemicals manufacturer), Modderfontein	Gas washing in ammonia from coal process. Thereafter used on a cascade basis and finally for ash slurry	- Orbal activated sludge plant - Chlorination - Sand filtration	1200 m^3/d
Sasol (oil from gas), Sasolburg	Ash transport	- Biofiltration	12 000 m^3/d
General Hide (tanning), Harrismith	Hide processing and dilution of final, high TDS effluent for irrigation	- Activated sludge nutrient removal	200 m^3/d
Galvanizing Techniques, Port Elizabeth	Cooling and rinsing of steel fabrications	- Activated sludge - Sand filtration - Chlorination	15 m^3/d
Beatrix Goldmine, Welkom	Gold reduction works and slurrying to slimes dams	- Activated sludge - Chlorination	3 000 m^3/d
Kelvin power station, Johannesburg	Cooling	- Biofiltration and nutrient removal activated sludge	35 000 m^3/d
Cape Fellmongering, Port Elizabeth	Process water (completely replaced potable water for this purpose)	- Activated sludge - Sand filtration - Chlorination	360 m^3/d
Fishwater Flats wastewater treatment plant, Port Elizabeth	Compressor cooling, bearing cooling, bearing lubrication, cleaning and irrigation	- Activated sludge - Sand filtration - Chlorination	4 000 m^3/d
Nampak (tissue paper manufacturer), Bellville	Process (replaced 85% of potable water)	- Orbal activated sludge - Sand filtration	800 m^3/d

During 1998, a project was implemented to utilise SGL as make-up in a cooling system, rather than direct treatment in the sewage treatment works. Two process cooling towers and a section of the cooling water reticulation system were converted to run on SGL instead of the normal cooling water blend. In the cooling system, phenol and ammonia was biologically reduced. The cooling system concentrated the SGL before the blow-down was treated in the sewage treatment works. The load (hydraulic and chemical) to the sewage treatment works was significantly reduced by this action. This system proved to have a positive impact on the water discharged from Sasol One to the Vaal River through the reduction of the COD, ammonia and phenol concentrations and the volume of final discharge. A further positive impact was a reduction in the amount of water abstracted from the Vaal River as cooling water make-up. Raw water abstraction was reduced by approximately 9 Ml/d while the discharge was reduced by approximately 6 Ml/d. The project also had a number of negative impacts, mainly related to increased levels of fouling in heat exchangers.

Eskom

Eskom is the national body responsible for generation of electrical power in South Africa and uses very large quantities of fresh water mainly for evaporative cooling purposes at its inland power stations. These recycling systems necessitate blow-down of some of the recycled cooling water to maintain the concentration of chemical species in the re-circulating water within predetermined limits to prevent fouling of condenser systems. The blow down containing relatively high TDS levels cannot be discharged to the water environment.

An acceptable way of disposal is to use the blow down to convey ash to the ash dumps. In a 'wet ashing' system, the ash is removed from the boiler in the form of a wet slurry and conveyed to the ash dam. The water that is used to convey the ash percolates through the ash in the ash dam and collects in ash water recovery dams. The water is then pumped back to the power station to repeat the cycle of conveyance. Wet ashing is a partly consumptive use of blow down since about 60% of the water used is trapped in the ash and 40% becomes available for reuse.

Where dry ashing systems are used, other means of disposal of blow down are required. On a dry ash system, water is used to condition the ash to form hydrated minerals. Thus, although consumptive, less water is used than in a wet ashing system. The excess blow down therefore has to be managed in another manner. The only feasible alternative for Eskom to treat the blow down is by means of desalination technology. Eskom had already initiated research into desalination of cooling water blow down in the early 1980s and installed the first full-scale desalination plant in the mid 1980s at the Lethabo power station. This tubular reverse osmosis system had a capacity of 9000 m^3 per day and produced high quality permeate that was reused in the power station. The tubular RO plant has since been replaced by a spiral RO system. Soon after the initial desalination plant was installed at the Lethabo power station, an electrodialysis reversal (EDR) plant was installed at the Tutaka power station for the same purpose producing about 6 Ml/d of high quality desalinated water for reuse.

Mondi Paper Mill

In 1994 Mondi Kraft, a leading international supplier of pulp, paper and board products, installed a system for the treatment of "black liquor" effluent streams generated in the production of pulp (Odendaal, 1998). The system was designed for the integrated pulp and liner-board facility at Piet Retief. The decision was motivated by a need to conserve water, to reduce operating costs and minimise the impact of the mill's effluent stream on the environment. The effluent treatment process was the result of more than 3 years of on-site pilot plant and laboratory studies.

The treatment plant currently treats 1 700 m^3/d, with facilities to upgrade to 2 400 m^3/d. The treatment process consists of three stages:

- tubular ultrafiltration for the removal of suspended solids and organic compounds with high molecular weight;
- ion exchange for the removal of low molecular weight compounds;
- reverse osmosis, whereby the majority of remaining organic materials and dissolved salts are removed;

The effluent stream is treated until the water is sufficiently clean for reuse in the factory. The final concentrated waste, in the form of salt cake, is produced in saleable form.

Columbus Stainless Steel Plant

The Columbus Stainless Steel factory at Middelburg produces various effluent streams during the manufacturing and beneficiation process, and has taken proactive steps to comply with the discharge requirements laid down by the Department of Water Affairs and Forestry.

The effluent streams have been categorised into two broad streams, namely strong and weak effluents. After pilot trials, it was decided to treat the weak effluents stream with the aid of reverse osmosis. This effluent stream derives mainly from contaminated site run-off water and ion exchange regeneration effluents, and has a flow rate of 3 300 m^3/d. The first reverse osmosis plant with a capacity of 1 500 m^3/d was commissioned during July 1994, and a second with a capacity of 1 800 m^3/d in May 1995 (Odendaal, 1998).

Pretreatment consists of screening, coagulation, sand filtration, pH adjustment with sulphuric acid and cartridge filtration. The permeate is reused in the factory as process water and the brine is further concentrated in an evaporator-crystallizer.

8.7.6 Irrigation reuse

Policy relating to the reuse of treated or partially treated wastewater for irrigation in South Africa differs from that of many other countries. In general the disposal of purified effluent by means of irrigation is not allowed. The requirement by law is that the treated effluent must be returned to the source from which the water was abstracted to allow use by down stream users. Exceptions include the irrigation of public parks and sport fields that would otherwise require abstraction of water from available freshwater supplies and in coastal areas where wastewater would otherwise be lost to sea. Temporary exceptions are given to irrigate (dispose of) wastewater where it is technically and economically impossible to meet the effluent standards for discharging to water courses.

The South African Department of Health formulated guidelines with respect to the public health effects of irrigation with treated municipal wastewater. In general, no irrigation with treated sewage effluent is allowed on crops that may be eaten raw.

The city of Cape Town has embarked on an ambitious project to determine the potential to increase the reuse of treated wastewater from its municipal wastewater works which would otherwise be lost to sea. A comprehensive study was undertaken to establish the potential for using treated wastewater to replace municipal water for irrigation of golf courses and other public areas such as parks. The results from the study indicate that a potential exists to replace 170 Ml/d of municipal water by treated effluent for irrigation purposes (City of Cape Town, 2006).

Cape Town has a growing population of around 3.2 million and suffered from a severe drought and extensive water restrictions in 2004/5. This gave rise to investigations to improve water management, to investigate desalination of seawater to supplement water sources and to investigate the increased use of treated effluents for applications where fresh water could be replaced.

More than 60% of the water consumed in Cape Town ends up in 20 treatment works across the city from where the treated effluent is discharged back to the environment and eventually to the sea. A fraction of this water is already reused for irrigation of golf courses and by a limited number of industries. The total existing daily summer reuse is about 30 Ml/d, which is about 7% of the total effluent available.

The city recently undertook a project to investigate requirements for the refurbishment and expansion of wastewater treatment works and infrastructure for distribution. This resulted in a number of works being upgraded, such as Bellville, Parow and Kraaifontein. The largest project being undertaken at present is at the Potsdam treatment works which will increase the treated effluent available for reuse to 24 Ml/d.

Some of the existing users of treated effluent from the city's treatment works include:

- Milnerton golf course;
- Sappi Paper;
- Sports fields in Milnerton and Table View;
- Table View beachfront dune project.

New users on the reuse scheme include:

- Chevron refinery;
- Kynoch Chemical;
- Agricultural use.

8.8 CONCLUSIONS

Indirect water reuse is taking place in practically all countries in the region where abstracted water that is partially derived from treated return flows is used for domestic purposes, gardening and irrigation. In the semi-arid countries of the region, the runoff in most streams and rivers during the dry season consists to a large degree of return treated or partially treated wastewater.

Most of the countries in the southern part where water resources are limited rely on indirect reuse for sufficient water to meet their requirements. South Africa specifically is highly dependent on return flows to meet its water needs.

South Africa has a large number of direct reuse schemes, mainly in industries that experience difficulties to treat their effluents to the required standards. Many industries therefore recycle effluents for internal use and often employ advanced treatment that produces water of such high quality that it could be reused for high quality applications.

8.9 REFERENCES

Basson, M.S. (1998) Overview of water resources availability and utilisation in South Africa. DWAF Report P RSA/00/0197. CTP Book Printers, Cape Town.

City of Cape Town (2006) The City of Cape Town brings treated effluent to the forefront. Press release.

Department of Water Affairs and Forestry (2004) National Water Resources Strategy. DWAF Pretoria.

Gisclon, A. , McCarley, S. and McNally, K. (2002) The Durban water recycling project – the vision becomes reality. Proceedings Biennial Conference of the Water Institute of Southern Africa, Durban.

Madyiwa, S. (2006) Peronal communication. Department of Agricultural Services, PO Box CY 639, Causeway, Harare, Zimbabwe.

Madyiwa S (2007) Department of Natural Resources, Harare, Zimbabwe. Personal communication.

Odendaal, P.E., Van der Westhuizen, J.L.J. and Grobler, G.J. (1998) In: Wastewater reclamation and reuse. Edtor T.K. Asano. pp1163 - 1191 Technomic Publishing Co. Inc., Lancaster, USA.

Gleick, P.H. (1998) The World's Water 1998-1999. Island Press, Washington, DC.

Riedel, K.J., Gubuza, D.M., Booysen, M.G., Edwards, W. (2004) Use of industrial effluent as process cooling water. Proceedings WISA conference, Durban.

Tole, P. M. (2006) Personal communication. Kenyatta University, PO Box 83844, Nairobi, Kenya.

Water Sewage and Effluent (2006) Closing the loop? Brooke Patrick Publications. Johannesburg.

Water Wheel (2006) Mine water becomes a commodity. Water Research Commission, Pretoria.

9

Water reuse in Latin America and the Caribbean

Blanca Jiménez

9.1 WATER AVAILABILITY

Latin America is considered a water-rich area with a mean water availability of 46,894 m^3/yr per capita (WRI, 2007) for its approximately 550 million people. Water resources are mainly located in South America, a region that possesses 47% of the world's water resources, notably in Brazil. In contrast, the Caribbean islands and countries such as Mexico, parts of Chile and the Northwest of Brazil are water scarce. Mean water availability does not reflect the situation in individual countries where availability ranges from only 300 m^3/yr per capita in Barbados to around 320,000 m^3/yr per capita in Guyana (Figure 9.1). The water intensity use index is also very variable for the countries in the region, ranging from 0.3% for Bolivia and Paraguay to 45% for the Dominican Republic and as much as 101% for Barbados.

Agricultural use demands the most water in Latin America at around 76% of the total (Figure 9.2), followed by domestic (15%) and industrial uses. Agricultural use drops to 66% in the Caribbean region where water is used mostly for municipal supply, including hotels in this tourist area.

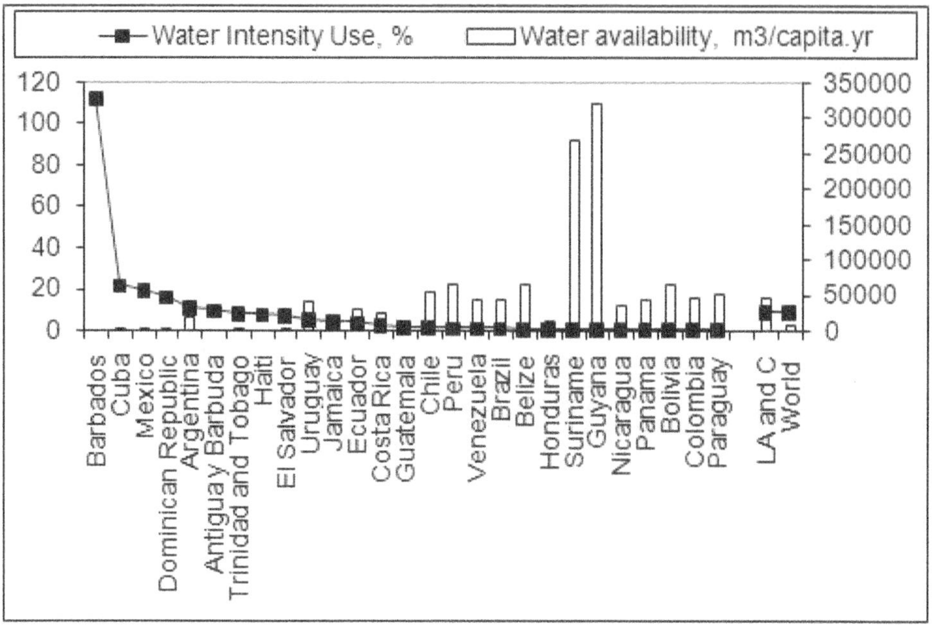

Figure 9.1 Intensity of Use Index and water availability in Latin American and the Caribbean countries, with data from 2005 using information from the World Resources Institute (2007).

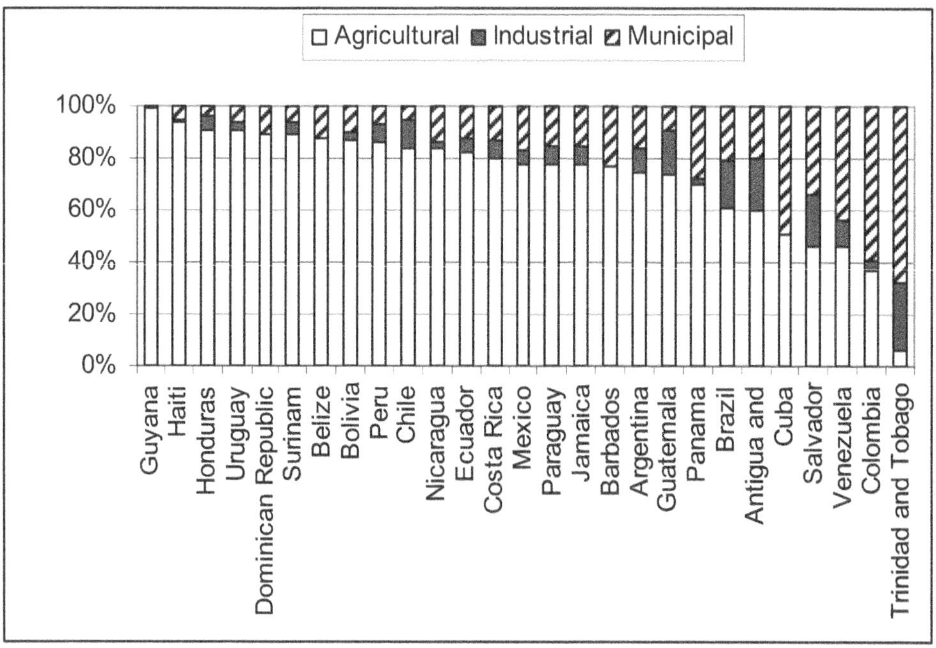

Figure 9.2 Water uses in Latin American and the Caribbean countries (WRI, 2007 using data from Earth Trends, 2005).

Water quality is a common problem in Latin America and the Caribbean that has been addressed differently by each country. Using a normalized version of the water quality index (on a 0-10 scale instead of the original varying from -2.25 to 1.85 as minimum and maximum values respectively) developed by Esty and Cornelius (2002) it is possible to have a general idea of the water quality in the region. Esty and Cornelius have used this index to rank water quality in 122 countries around the world. Latin America has on average a water quality score of 5.3 using the normalized values. Surprisingly there is no correlation between the economic situation in a country and its water quality (Figure 9.3). For instance, Trinidad and Tobago and Mexico with relatively high Gross Domestic Products per capita do not score particularly well in terms of water quality.

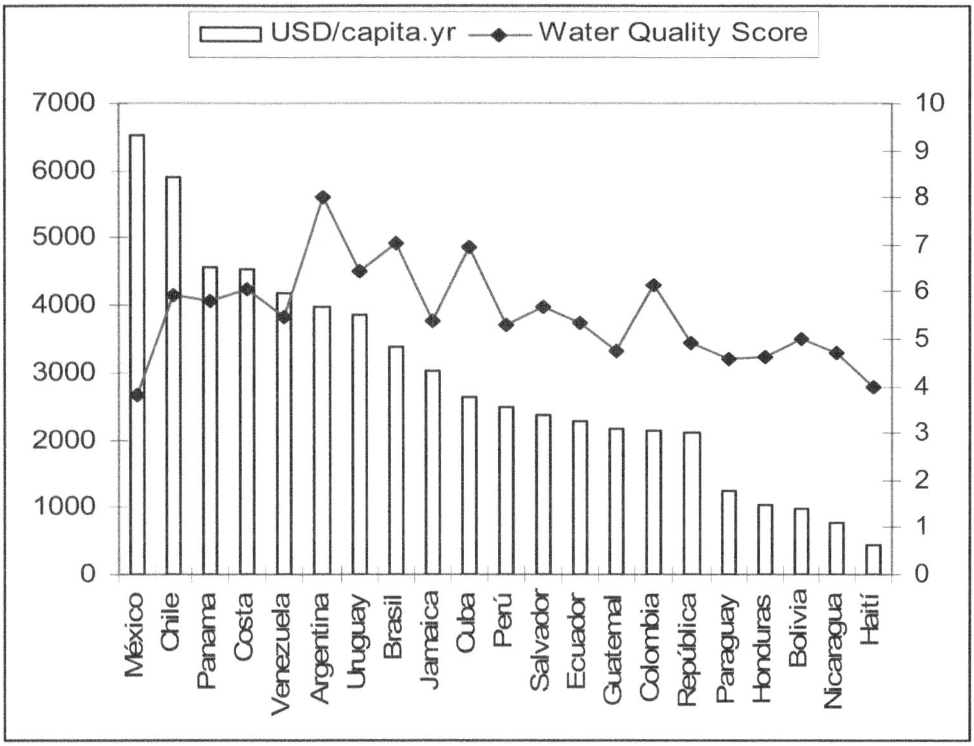

Figure 9.3 Gross Domestic Product per capita and Water quality index normalized on a scale of 0-10 in selected Latin American countries.

9.2 CURRENT REUSE PRACTICES AND MOTIVATIONS

Wastewater reuse is practiced in water stressed countries frequently involving low water quality, but it also exists as an unplanned activity in water rich countries. Both situations arise due to the low level of sanitation in Latin America, only 22% according to WWF (2006). Most of the wastewater is reused, intentionally or unintentionally, for agricultural irrigation (Table 9.1) due to an increasingly high urban population (currently at 81%) that, on one hand produces a large amount of untreated wastewater and, on the other demands food.

Table 9.1 Non-treated wastewater-irrigated areas in Latin American Countries.

Country	Description	Reference
Argentina	3,700 ha in direct irrigation[(1)] 20,000 ha with treated and diluted water	Bartone and Arlosoroff, 1987 Kotlik, 1998
Brazil	The practice is acknowledged but there is no precise data on the total area irrigated with waste or polluted water	Rodrigues-Pimentel *et al.*, 2001
Chile	16,000 ha of direct irrigation	Bartone and Arlosoroff, 1987
Colombia	26,000 ha of indirect irrigation	Moscoso-Cavallini and Egocheaga-Young, 2000
Mexico	260,000 ha of direct[(1)] and indirect[(2)] irrigation with 13.8 Mm^3/d (160 m^3/s) of wastewater, mainly without treatment	CNA, 2004
Peru	6,800 ha in Lima and 2,546 ha other sites	Bartone and Arlosoroff, 1987 and van der Hoek, 2004

[(1)] Direct wastewater irrigation: Use of wastewater to irrigate

[(2)] Indirect wastewater irrigation: Use of wastewater to irrigate after dilution with surface water

Irrigation enables 13% of the arable land to produce 30% of total sector income and is therefore a key factor for most of the economies within the region. This is why the total irrigated area in Latin America continues to increase. In the 30 years to 1995 it grew 96% to reach a value of 9.1 million hectares (World Bank, 2002). Given its water-intensive nature, agriculture very easily incorporates wastewater produced by cities for irrigation either directly (without dilution) or indirectly (after discharge into rivers or lakes that are used for irrigation). Based on the patchy information available it is estimated that more than 510,000 ha are directly irrigated with wastewater and 2,500,000 ha are irrigated using polluted water (Bartone, 1990 and WHO, 1989) and overall one third of the total irrigated area uses polluted water.

Besides the lack of "clean" water, reasons for using wastewater for agricultural irrigation are the following:

- seasonal variations in "first use" water availability;
- its fertilizing properties;
- its low or zero cost;
- as part of the historical process of water use, i.e. the consistent use of water from the same source which, due to the lack of sanitation, becomes polluted with time.

Frequently, farmers not only accept the use of wastewater but request it to boost agricultural productivity and hence their income, although frequently they are unaware of the health risks associated. The majority of the irrigated area in Latin America is concentrated in Mexico, Brazil, Peru, Chile and Argentina (79% of the total) and it is these countries which report the greatest area irrigated with wastewater or polluted water. According to Moscoso-Cavallini and Egocheaga-Young (2004) irrigation with wastewater is performed on non-food produce or crops consumed cooked or after industrialization (81%), fruit trees (11%) and vegetables (7%). Mexico is the country with the largest area irrigated with wastewater (see Box 9.1) and, unlike most of the countries where public acceptance of the use of wastewater to irrigate is low, Mexico's farmers appreciate it. The government has openly recognized this practice as a first step towards controlling the problem. Additionally, it is introducing several programs to progressively control reuse. Following the cholera pandemic of 1989, the government prohibited the use of wastewater to irrigate crops consumed raw and also limited the area that could be irrigated using wastewater, with poor results. Then in 1996 it developed a regulation to control the quality of the wastewater for irrigation at low cost, combining different methods to control the spread of diseases with the use of feasible and affordable technology. This action reduced the area under irrigation with wastewater by 20% and increased wastewater treatment for agricultural reuse to 1,728,000 m^3/d (20 m^3/s according to Jimenez, 2005).

Even in areas with high water availability where large urban areas dispose of their wastewater into lakes, reuse is becoming an interesting option because of environmental concerns. Secondary treatment plants that were originally designed to control pollution in surface water remove only organic matter, not nutrients, the result being that their effluents are causing eutrophication in lakes. Consequently, agricultural reuse is promoted to remove nutrients from the treated effluents prior to its disposal in surface reservoirs.

Extensive use of wastewater in agriculture has encouraged a lot of research in several Latin American countries. It has been shown that different crops are not polluted in the same way. Some crops, such as garlic and onion, have compounds that limit, at least partially, bacterial growth. Others, such as rice, have been found to possess important wastewater treatment capabilities through soil filtration while still others (such as zucchini, coriander, parsley) are reported to concentrate pathogens and bacteria.

Box 9.1 Wastewater reuse in Mexico: The Mezquital Valley case

In Mexico a total of 17.5 Mm^3/d of municipal wastewater is collected of which 30% is treated in 1,360 wastewater treatment plants (40% in activated sludge systems, 20% in stabilization ponds, 20% using advanced primary treatment, and the rest with aerated ponds, primary settlers, oxidation ditches, biofilters and septic or Imhoff tanks). The volume of treated wastewater has been increasing at a rate of 7% per year since 1996, and by the year 2008 it is expected that 60% of the municipal wastewater produced will be treated. It is estimated that of the total amount of municipal wastewater produced, 16.4 Mm^3/d is reused, either directly (3.2 M^3/d) or indirectly (13.2 m^3/d), and with (4.5 Mm^3/d) or without (11.9 Mm^3/d) treatment, mainly to irrigate 260,000 ha (CNA, 2004). The majority of the water is reused for agricultural irrigation, while only a small fraction is reused in industries (1.4%) or for urban purposes (0.004%).

The biggest continuous irrigated area using untreated wastewater in the world is located in Mexico, where the practice began in 1896 as a method for disposing of Mexico City's wastewater. This area is colloquially known as the Mezquital Valley, but its formal name is the Tula Valley. The Mezquital Valley is a semi-arid area, with only 550 mm/yr of rainfall and 1,750 mm/yr of evapotranspiration. The region used to be very poor and agriculture could not be developed due to the lack of water. With the arrival of Mexico City's wastewater the situation soon changed and economic conditions improved (Jimenez, 2005). Consequently, in 1912, irrigation using wastewater was officially acknowledged. Currently 73,632 farmers irrigate 76,119 ha (Jimenez, 2005). The main crops are alfalfa and maize (60%), but oats, barley, wheat, beans and some vegetables (chilli, Italian squash and tomatoes) are also grown. In order to irrigate, a complex hydraulic system has gradually been constructed by the government. It consists of three perennial rivers, nine dams (three of fresh water and six of wastewater) and more than 800 km of unlined channels to distribute around 4.5 Mm^3/d of untreated wastewater (Jimenez, 2004). Wastewater is greatly appreciated by Mezquital farmers. They know that due to its fertilizing properties, land leasing prices increase (1 ha can be rented at 455 US$/yr where wastewater is available instead of 183 US$/yr for rain-fed agricultural land) and that it is possible to grow two or three crops per year instead of just one (Jimenez, 2005). Wastewater contributes some 2,400 kg of organic matter per hectare, 195 kg/ha of nitrogen and 81 kg/ha of phosphorus to the soil each year, increasing maize productivity by 150%, barley by 100% and tomatoes by 94%.

The disadvantage is the effect on health: a 16-fold increase in morbidity by helminths in children in comparison to unexposed nearby areas has been reported. To date, the WHO has obtained most of its information used to establish limits on helminth eggs content for agricultural reuse of wastewater from the Mezquital area (Cifuentes *et al.*, 1992; WHO, 1989; Blumenthal *et al.*, 1996 and 2000).

Based on the situation in the Mezquital Valley, it has been recommended that in order to protect farmers', crops' and the health of consumers while maintaining short and long-term positive impacts on soil productivity, wastewater must be treated and managed wisely. This management must include an appropriate treatment of Mexico City's wastewater to remove pathogens while partly preserving nutrients and organic matter (Jimenez and Chavez, 1997). In particular, it is recommended not to reduce the contribution of organic matter to the soil in order to avoid metal mobilization (Siebe and Fisher, 1996; Downs *et al.*, 2000). Also pre-treatment or segregation of industrial discharges to Mexico City's sewers is recommended because wastewater used to irrigate is infiltrated to an aquifer serving as a water supply in the Mezquital Valley. To meet these demands, treatment methods at one third of the cost of conventional options have been developed (Jimenez and Chavez, 1997).

Examples of industrial reuse in Latin American are still isolated and limited to a few countries. Mexico, with a significant lack of water and a relatively high level of industrialization, reuses almost 240,000 m^3/d of wastewater for industrial purposes, with some projects operating since 1956 (in Monterrey and Mexico City). Wastewater is reused mainly for power plant cooling; a single plant reclaims 129,600 m^3/d of secondary treated effluent from Mexico City for this purpose (Jiménez-Cisneros, 2003). Sao Paulo in Brazil is a highly industrialized city with an ambitious program to reuse water. The main driving force is economic, as the cost of fresh water is about 17 times the cost of reclaimed water (Rodrigues-Pimentel *et al.*, 2001). In Antofagasta, Chile, 5% of the municipal treated wastewater is reused, mainly for mining. For Latin America as a whole wastewater reuse in industries and in municipalities represents a much smaller volume, estimated at around 345,600 m^3/d (4 m^3/s), than that used for irrigation. In most countries, industrial reuse potential is considered limited by the subsidized low cost of first use water and a low enforcement capacity in relation to deterring illegal groundwater abstraction.

Municipal reuse data is very scarce for Latin America. Mexico reports a total reclaimed water use for the whole country of 78,000 m^3/d and Brazil claims that several municipalities are making extensive use of wastewater without reporting figures (Espanhol, 2006). In both countries wastewater is used for lawn irrigation, public parks, recreational lakes and car washing. Several Brazilian municipalities are distributing treated wastewater by tanker trucks for washing streets and road construction dust control. Reclaimed water is sold at a cost of 0.12 US$/$m^3$ (Espanhol, 2004).

An additional type of reuse is the so-called "non intentional reuse" due to the incidental recharge of aquifers that are being used for municipal supply by the infiltration of wastewater transported in unlined channels and irrigation drains, or due to the infiltration of wastewater used to irrigate. This phenomenon has been documented in Mexico and Peru (Foster *et al.*, 2004). Fortunately, during infiltration, passage of the wastewater through the soil significantly improves its quality (Jimenez and Chavez, 2004); however periodical surveys must be performed to detect potential problems and take required measures. Chapter 23 describes in detail the unintentional use of infiltrated wastewater for drinking purposes.

9.3 REUSE ACTIVITIES AND PROJECTS IN LATIN AMERICA

Argentina (Barbeito Anzorena, 2001). Rainfall in Argentina varies from only 50 mm/yr (La Rioja and San Juan provinces) to around 2,000 mm/yr (in the Patagonic Andes, Selva Tucumano-Oranense and Selva Misionera). The pluvial precipitation in the Patagonic plateau and the Puna region is commonly < 200 mm/yr. Due to the lack of water, unintentional and intentional wastewater reuse exists in agriculture (with both untreated and treated wastewater), the latter expanding and the former close to extinction. Planned reuse is practiced in Mendoza (for agriculture), Chubut (for forest irrigation), Villanueva (for horticulture, floriculture and forest irrigation) and La Rioja (for irrigation) to reclaim both water and nutrients. Untreated wastewater is used to irrigate indirectly once it has been discharged into rivers from where it is taken once again after dilution takes place with fresh water.

Campo Espejo, Gran Mendoza, Argentina (Barbeito Anzorena, 2001). Prior to 1976, untreated wastewater was used directly for agricultural irrigation. Since then, irrigation has taken place using wastewater treated in a stabilization pond system. Reclaimed water is used in well-defined areas (Special Restricted Crops Area, or ACRE by their acronym in Spanish) where only certain types of crop can be grown. The soil is slightly saline and therefore reclaimed water is also used to wash salts from it. In total 129,600 m^3/d of treated water is reused to irrigate 1,900 ha.

Caribbean islands (Meganck *et al.*, 2006). Extreme lack of water combined with significant tourist activity has led to most of the Caribbean islands running water reuse projects, especially in hotels. Wastewater is reclaimed for lawn irrigation after secondary treatment. No data on the total amount of water reused is available.

Brazil (Rodrigues-Pimentel *et al.*, 2001). Although Brazil is the country with the largest average water availability in Latin America, there exists unplanned, and sometimes unintentional agricultural reuse due to the lack of sanitation and water availability at a local level (some regions have <200 m^3/capita.yr). When reuse is planned, stabilization pond effluents are used for agricultural irrigation and aquaculture.

Fortaleza, Brazil (Rodrigues-Pimentel *et al.*, 2001). Fortaleza has an uneven rainwater distribution and groundwater resources are scarce. Since 1992, a stabilization pond effluent (23,328 m^3/d) has been used to irrigate bananas, sugar cane, maracuja, cocoa, alfalfa, coriander, onions and fodder as well as for aquaculture (Tilapia fish production).

Sao Paulo, Brazil (Rodrigues-Pimentel *et al.*, 2001). Wastewater is reused to clean public roads and monuments of the city (605 m^3/d). In Sao Paulo, drinking water costs 2.7 US$/$m^3$ while treated water costs 0.17 US$/$m^3$ (2.3 and 0.36 Reals respectively). Reuse is growing rapidly due to this price difference.

Bolivia (Coronado-Rocha *et al.*, 2001). This country is mostly rural. Wastewater (treated or untreated) is used for irrigation and appreciated by farmers because of its reliability and availability. Wastewater reuse for direct irrigation occurs in the La Mayca region but there is no data on the size of the irrigated area.

Cochabamba, Bolivia (Coronado-Rocha *et al.*, 2001). Treated wastewater comes from facultative ponds, which, owing to population growth are overloaded and produce an effluent with a high faecal coliform ($2x10^6$ MPN/100 mL) content. A total volume of 43,200 m^3/d of treated wastewater is used to irrigate 1,560 ha; distribution takes place through farmer associations. There are also reports on the use of untreated wastewater to irrigate but no figures are available.

Chile (Mena-Patri, 2001). Untreated wastewater was used directly for agricultural irrigation until 1992 when, as a result of the cholera outbreak, it was forbidden. Irrigation with treated wastewater is currently accepted but only for certain types of crops. Untreated wastewater discharged into rivers is used to irrigate all kind of crops indirectly, once diluted with fresh water due to the lack of sanitation (around 30% of the wastewater is treated) and the surface water concession to farmers. A total of 130,000 ha are irrigated with untreated wastewater in Chile, mainly in the Maipo and Maipocho areas, usually once diluted with first use water; a small part (3% of the area) applies wastewater directly to fields.

Antofagasta, Chile (Lam-Esquenazo *et al.*, 2001 and Meganck *et al.*, 2006). This is one of the most arid regions in the country, where it is very hard to promote economic development due to the lack of water. Just 3% of the collected wastewater is treated and the rest is disposed of into the ocean. Five percent of the treated wastewater (19,000 m^3/d) is reused to irrigate 65 ha of maize, carrots, garlic, lettuce and vegetables. The remaining treated wastewater is reused for mining and industry which are activities that can afford the cost of the treated wastewater. Additionally, in Antofagasta there are 180 ha irrigated with untreated wastewater. Although wastewater reuse could be extended in the region (to irrigate plants that are currently consuming fresh water such as eucalyptus and plants for landscaping) this is not done. Besides, produce quality is decreasing due to the content of boron, chlorides and sodium in water. Therefore, wastewater reuse is expected to increase only for industrial and commercial activities.

Santiago, Chile (Mena-Patri, 2001). Santiago is a very industrialized area with significant agricultural activity. Unfortunately, precipitation is low (312 mm/yr) and occurs in only five months of the year. Wastewater reuse for irrigation is a consequence of the combination of the increased wastewater production with the increased water demand. Prior to the 1980s, irrigation of any kind of crop was performed with wastewater, but since then the irrigation of ground level crops that are consumed raw has been prohibited. A long-term program to build three treatment plants was begun in order to continue reclaiming wastewater. The first, consisting of an activated sludge system with a capacity of 302,400 m^3/d, began to produce treated wastewater in 2002. The second, also an activated sludge system of 760 320 m^3/d, started operation in 2003. The third, scheduled for 2009, will treat 475,200 m^3/d of wastewater through oxidation ponds. In nearly half of the total irrigated area (110,000 ha) vegetables are grown in small farms, and in another area representing almost one half, fruit trees and vineyards are cultivated. Reclaimed water is applied via flooding and furrows. The remains - around 4,000 ha - are irrigated with wastewater for growth of a limited range of crops. In addition to wastewater reuse, the reuse of biosolids in agriculture is encouraged.

Ibagué, Colombia (Vanegas-Gálvez, 2001). Annual precipitation in Ibagué is 1,490 mm/yr. Its wastewater is discharged without treatment into rivers, which have been used as a source of water for irrigation for more than 40 years. This provides a permanent water supply throughout the year and has high organic and nutrient content that reduce the need for fertilizers. The polluted water is collected from the river in storage reservoirs to regulate the flow prior to its application to fields. These reservoirs serve as a primary treatment with effects similar to facultative ponds. Rice is the main crop grown, as well as sorghum, tobacco, subsistence crops (corn, yucca and banana), fruit trees (papaya, plum, tangerine, and mango) and fodder. Pollution by heavy metals and pathogens has not been reported on produce. Before 2000, a total amount of 129,600 m^3/d of non treated wastewater was used to irrigate 8,000 ha. A programme to irrigate a total area of 9,000 ha using treated wastewater has involved the construction of a wastewater treatment plant (with a total flow of 13,133 m^3/d) and the current construction of a second plant (3,888 m^3/d), according to Fundación Sustentable (2007).

Costa Rica (Abarca-Garbanzo, 2001). Given the high water availability in Costa Rica, reuse is not considered an important objective. However, in the Central Valley, where 55% of the population lives, untreated wastewater is discharged into a river and used to irrigate a variety of crops over an area of around 1,200 ha.

Porto Viejo, Ecuador (Castro-Merizalde, 2001). Water availability is limited by the low precipitation (491 mm/yr), its marked seasonality and the high evapotranspiration rate (1,575 mm/yr). The main economic activity in Porto Viejo is commerce followed by agriculture. Eighty-five percent of the city sewage is treated in a stabilization pond. Part of its effluent (4,752 m^3/d) is discharged into a river to be used later for irrigation and since 1963 the rest has been used directly to irrigate 80 ha. Effluent quality is not high and the river water has an *E. coli* content of >10^4 MPN/100 mL. Irrigated crops are corn, tomatoes and peppers.

Guatemala (Sánchez de León, 2001). At the national level, the use of treated wastewater for irrigation is incipient. It is only performed in isolated cases for grazing pasture.

Sololá, Guatemala (Sánchez de León, 2001). Sololá is located on the Guatemalan plateau and has an annual precipitation of 1,827 mm. Sololá's wastewater is treated with an UASB (upflow anaerobic sludge blanket reactor) and an anaerobic filter system to remove organic matter, but not nutrients and pathogens. The reuse of the effluent is promoted to avoid the eutrophication of Lake Atitlán, caused by the disposal of the effluent. Even though legislation forbids the reuse of effluents to irrigate crops consumed raw, some plots with beans and tomatoes are being watered. In total 1,642 m^3/d of partially treated effluent is used to directly irrigate 5 ha.

Mexico (CNA, 2004). Eighty-five percent of the total municipal wastewater produced in the country is reused directly or indirectly for agricultural irrigation with treatment (35%) or without treatment (65%). The lack of water combined with the lack of sanitation has promoted unintentional wastewater reuse for agricultural irrigation in 30 medium and big cities. In total, 13,900,000 m^3/d of treated and untreated wastewater is reused mostly for the irrigation of 200,000 ha of agricultural land.

Ciudad Juarez, Mexico (Garza-Almanza, 2001). Ciudad Juarez is located on the USA border in the state of Chihuahua. It has around 1.3 million inhabitants and a rainfall of only 300 mm/yr. The maquila[1] industry, commerce and agriculture are the main local activities, but the latter is declining due to a boom of the maquila industry and the emigration of farmers to the USA. For more than 60 years, untreated wastewater was used to irrigate the Juárez Valley where soil is saline and has a low nutrient content. Since 2000, two advanced primary treatment plants have been processing 302,400 m^3/d of wastewater. The effluent along with 95,040 m^3/d (1.1 m^3/s) of untreated water continue to be used to irrigate 26,000 ha of cotton, alfalfa (for fodder), wheat, sorghum, pasture, walnut, nuts and apple trees. Some plots, on which are grown corn, oats, chillies and zucchini, are also watered. Wastewater (treated or untreated) is free for farmers who pay only for the maintenance of the irrigation infrastructure. Irrigation takes place through furrows or flooding at a rate of 1,300 mm/yr. In the future, industrial reuse is expected to increase, as long as fresh water continues to become more expensive and treated wastewater is available. The water tariff for 2006 was 1.8 USD/m^3, among the highest tariffs in Mexico.

Mezquital Valley, Mexico (Jiménez, 2005). The Mezquital Valley is located 100 km north of Mexico City. In this area more than 76,000 ha are being irrigated with Mexico City's wastewater. This use of wastewater began more than 110 years ago. The total area irrigated reached a maximum of 90,000 ha in the 1990s, and since then it has been decreasing. Wastewater is highly appreciated by farmers due to the lack of fresh water resources and its organic and nutrient content that fertilize the local soil. Water is applied by furrows and flooding. Thanks to the use of wastewater, the region has improved its economic situation. Nevertheless, at the same time the helminthiasis rates among children under 16 years old have been found to be 16-times higher than in unexposed children. This area is considered to be the largest compact irrigation field in the world using 4.5 Mm^3/d of untreated wastewater. More details are presented in Box 9.1.

Texcoco Ex-Lake, Mexico, (Muciño, 2001). Located to the east of Mexico City, in the ex-Texcoco lake, this semi-arid region has a high evaporation rate and four months of sporadic

[1] Industry located in the US border receiving parts of products to assembled them in order to send the final product to the US to commercialize it.

rainfall. Since 1971, almost all of the effluent of an activated sludge plant (86,400m^3/d) and a facultative ponds system (43,200 m^3/d) have been used to fill a lake of 36 Mm^3 used for environmental and recreational purposes. The lake was originally planned as a maturation pond to allow the later use of the effluent for agricultural irrigation. However, the high evaporation rate and soil salinity of the bottom part of the lake render the effluent too saline for use by farmers. Therefore only a small part of the lake effluent is used to for agricultural irrigation during the dry season. Additionally, a small fraction of the secondary effluents (4,320 m^3/d) is transported to a tertiary physico-chemical treatment plant to recharge the aquifer by injection into the subsoil.

Luque, Paraguay (Coronel, 2002). Paraguay is a rural country with almost no wastewater treatment. Most of the wastewater is discharged into streams that flow to the Ypacaraí lake or to the Paraguay river. Both water bodies are used for irrigation. The main crops irrigated are sugar cane, strawberries, corn, manioc, beans and fodder. Detailed information on reuse practices is scarce.

Peru (del Carpio, 2000). In Peru, a total of 87,350 m^3/d of wastewater has been reported for irrigation of 1,515 ha of different kinds of crop.

Miraflores, Peru (Acosta *et al.*, 2001). In Miraflores, rainfall is only 25 mm/yr. Wastewater from the Lima Metropolitan area began to be treated in 1959 using stabilization ponds. There are 21 stabilization ponds over a 21 ha area, some of them having aeration systems since 1984. Between 172,800 and 259,200 m^3/d of the treated wastewater is used to irrigate 632.5 ha in the peri-urban area as well as for aquaculture (Tilapia fish production). Additionally, there are another 522 ha irrigated with untreated wastewater. Irrigated crops include maize, chala, alfalfa, king grass, yellow corn and fruit trees (apples and avocado). Irrigation is also performed for forestry (eucalyptus and grevillea) and landscaping. Agricultural reuse is promoted due to the low cost of treated and non treated wastewater, the need to use large amounts of water to wash the saline soil, and to avoid the use of fertilizers. Irrigation takes place through furrows and flooding. Although it is forbidden to apply untreated wastewater to vegetables, the practice happens due to the difficulty associated with surveying lots of farmers working on very small plots (around 1 ha each). Due to the unsanitary conditions, the helminthiasis rate is 144.5/1,000 persons, a third of which are Ascariasis. Infiltration of the treated and untreated wastewater to the groundwater has been reported.

San Agustín, Peru (Manrique-Arce *et al.*, 2001). San Agustín produces 224,640 m^3/d of untreated wastewater. Part of it is discharged to the Rímac River and then used to irrigate produce destined for the city of Lima such as celery, garlic, pore, onion, cabbage, coriander and Italian squash. It is estimated that the ratio of wastewater to river water is about 2.6 to 1. Additionally, part of the untreated effluent is used to directly irrigate 155 ha of the Oquendo farmland. According to Castro de Esparza and Flórez-Muñóz (1990), 49% of the crops contain faecal coliforms, 91% helminths eggs and 66% *Entamoeba coli*. Irrigation is performed through furrows while flooding is only occasionally used because it causes mosquito proliferation. Waste and polluted water are used because of the lack of fresh water, rather than for reclamation of its nutrient content as the soil is of good quality. It has been reported that wastewater used to irrigate has filtered into the aquifer. The aquifer is at least 4 m deep, and recharge has been beneficial because it controls seawater intrusion.

Tacna, Peru (Del Carpio, 2000). Tacna is a farming city with enough water (annual rainfall is 1,900 mm/yr) but there are conflicts between agricultural and municipal users. Prior to 1978 untreated wastewater was used to irrigate 738 ha, but since this date aerated and facultative ponds with a capacity of 19,872 m^3/d have been installed. These ponds remain in operation but are overloaded, producing a low quality effluent that is still used for irrigation. There are now an additional 129 ha using 4,925 m^3/d of untreated wastewater. Irrigated crops are olives trees, grapes, bell peppers, paprika, onion, pear trees, tomatoes, squash and fodder (alfalfa, chala and corn). Wastewater is also reused in forestry (cultivated and natural). The soil is relatively saline and has low organic matter content. As in other cases, there are many farmers involved in irrigation practices, owning between 3.5 and 12 ha, making any control difficult. Irrigation is performed through flooding.

La Vega, Dominican Republic (Mercedes-Matos, 2001). Since 1973, 90% of the sewage (57,024 m^3/d) has been transported 80 km from Santo Domingo to La Vega via the River Poton. In La Vega, diluted wastewater is used to irrigate rice, corn, tomatoes and fodder over an area of around 250 ha. During the dry season the flow in the Poton River is so low that irrigation is performed almost exclusively with wastewater. Farmers reuse wastewater because it is free and, although soil is of good quality, it is recognized that wastewater increases productivity. There is no data on the impacts on health. Farmers own small plots (< 2 ha) and perform irrigation through flooding.

Maracaibo, Venezuela (Rosillo, 2001). Wastewater reuse is not a current practice. The metropolitan area of Maracaibo has 1.7 million inhabitants. In 2000, the city began to treat 293,760 m^3/d of wastewater with stabilization ponds that do not eliminate nitrogen and phosphorous. The effluent is discharged into the Maracaibo Lake and is causing eutrophication. As there is a large agricultural area near Maracaibo, agricultural reuse is being proposed to avoid pollution of the lake.

9.4 REGULATIONS

In general, few Latin American countries consider reuse as part of the country's water resources. Nevertheless, in most of them unplanned reuse is recognized, albeit indirectly, by including regulations to control it, notably for agricultural irrigation (Table 9.2). Therefore water resource management strategy is oriented to readdress current reuse practices rather than to promote new planned reuse projects.

Table 9.2 Reuse Regulations in some Latin American countries.

General Legislation	Water reuse regulations
Argentina	
Reuse is considered as a general goal	In 1996, the Mendoza province enacted the 778/96 resolution setting parameters to reuse wastewater for agriculture and industry. This regulation defines the ACRES (restricted cultivation areas, in Spanish) as areas where only certain types of crops can be grown when reclaimed water is used. Reclaimed water standards follow the WHO criteria. The implementation of the regulation depends on the General Department of Irrigation.
Brazil	
Since 1977, the National Policy of Hydraulic Resources considers water reuse as high priority and has recognized reclaimed water as a water resource.	Since 1997 there have been regulations (NBR 13.969/97) for agricultural, municipal and industrial reuse. Wastewater agricultural reuse standards are based on the WHO guidelines. Reclaimed water reuse is forbidden on crops consumed uncooked, and on vegetables growing at ground level. The types of crop that can be irrigated using treated wastewater have been well defined since 1978. Limits for metals and salt content are also considered. A new resolution is being developed to define legal and institutional aspects and to establish new limits for agricultural reuse as well as for other types of reuse (municipal, industrial, aquaculture, artificial underground recharge and environmental protection).

General Legislation	Water reuse regulations
Bolivia	
Discharging wastewater into water bodies with more than $1x10^3$ faecal coliforms is prohibited (Environmental regulations, 1995)	There are none.
Chile	
Resolution 207 of the General Water Department establishes conditions for the use of surface and groundwater as well as for water reuse.	Rule NCh 1.333 prohibits irrigation of produce to be consumed uncooked with wastewater. To reclaim treated wastewater it sets a limit for faecal coliforms of 1,000 MPN/100 mL for the irrigation of vegetables and fruits that grow at ground level and are usually eaten raw. There are no limits for the helminth eggs content.
Colombia	
The law 373 from 1997 establishes a program to efficiently use and save water.	The legislation on "Water and Residual Discharge Uses" stipulates that water used for the irrigation of unpeeled fruits for human consumption as well as short-stemmed vegetables should not contain more than 5,000 total coliforms/100 mL and 1,000 faecal coliforms/100mL. There are no other specific regulations for water reuse.
Costa Rica	
There is a whole set of laws and institutions related directly or indirectly to wastewater reuse, making further water regulation implementation difficult.	In chapter VI article 32 of the Water Discharge regulation, conditions for eight types of reuse including agriculture are established. Wastewater reuse for agricultural irrigation is permitted as long as the quality of soil and that of the groundwater is not affected. Crops that can be watered with reclaimed water are limited to those that undergo any type of industrial process as well as produce that is not for human consumption, such as grazing pasture, fodders, fiber and seed crops. Faecal coliforms are the only parameter considered for agricultural reuse with a limit of < 100 MNP/100 mL. Municipal reuse is permitted in greenbelts, car washing, fire fighting, forestry, recreational activities and for construction.
Ecuador	Article 20, Chapter III of the Regulation to Prevent and Control Water Pollution, published in 1989, establishes the quality criteria for the reuse of wastewater destined for agriculture. This takes into account metals, faecal coliforms < 1,000/100 mL, the absence of helminth eggs, pH and grease and oil limits. It does not include SAR (sodium absorption ratio) or organic matter (BOD). The regulation was completed in 1992 and it defines water quality in terms of the wastewater treatment process to be used: a. For the irrigation of products for raw consumption (faecal coliforms < 1,000/100 mL and 0 helminth eggs/L) stabilization ponds are required. b. For the irrigation of cereals, crops that are used in industrial processes, animal feed, grazing land and trees (0 helminth eggs /L and no limits for faecal coliforms) stabilization ponds with a minimum retention time of 10 days are required. c. For localized irrigation (such as drip irrigation) of the above-mentioned products with no exposure to workers or the public to the wastewater there are no biological limits or treatment methods. Furthermore, farmers are required to undertake studies to determine the effects of water reuse on soil (in particular on salinity, infiltration, drainage, etc.), water quality (toxicity problems and crop tolerance), and on crops. They also need to define the irrigation methods, storage needs, infrastructure, costs and profitability.

General Legislation	Water reuse regulations
Guatemala	
The use of raw wastewater for agricultural irrigation is prohibited.	Governmental agreement No 236 dating from 2006 considers three type of reuse: agricultural irrigation, aquaculture and recreational activities. Water quality standards are set for each case. For agricultural irrigation, the emphasis is on reclamation of nutrients. Biological limits only consider the faecal coliform content.
Mexico	
Since 1982, water reuse has been considered part of the national policy and has been regulated as part of the wastewater discharging criteria.	Wastewater reuse is controlled through two regulations. In the first, it is considered part of the wastewater discharge to national surface and groundwater bodies' regulations. Agricultural reuse is considered as a special case of soil disposal. For this reuse, there are no limits for the BOD or the TSS content, but there are for metal and biological pollutants. For crops that are used after industrialization or crops that are eaten cooked, up to 5 helminth eggs/L and 1,000 faecal coliforms/100 mL are allowed. The higher value of helminth eggs compared to that recommended by WHO guidelines was adopted considering that conventional technology was not able to produce treated water with less than 1 egg/L from the high helminth eggs content normally found in wastewater. For crops consumed raw the regulation establishes that treated wastewater must contain <1 helminth egg/L and <1,000 faecal coliforms/100 mL. A value than can be achieved with any secondary biological or physicochemical process (in this latter case APT is also included) combined with a filtration step (Jiménez and Chavez, 1997). The organic matter and the solids content are not considered in the irrigation criteria, as the former is considered beneficial to soil, while the latter must be removed anyway in order to meet the stipulated helminth eggs values. The second reuse regulation controls wastewater reuse in public works; the main parameters are biological ones, leaving definition of other parameters to agreements established between users.
Nicaragua	
There are no reuse regulations; only a wastewater discharge regulation.	There are none.
Peru	
Standards to control wastewater discharges and reuse are considered in different regulations and laws.	Law 17,752 dating from 1969 and modified in 1983, establishes that wastewater to be used for agricultural irrigation must have <1 nematode egg/L. Furthermore, faecal coliforms must be <1,000 CFU/100 mL and < 5,000 total coliforms CFU/100 mL when it is used to irrigate vegetables consumed uncooked. The regulation also includes limits for nitrates, metals, organic matter, pH, and some toxic compounds, requiring a tertiary level treatment.
Dominican Republic	
NORDOM 634 regulates the quality of the wastewater discharged into receiving bodies and considers reuse as an option.	In the case of vegetables, wastewater with >5000 faecal coliforms MPN/100 mL must not be used; helminth eggs content is not limited, BOD_5 must be <5 mg/L, TSS <1,000 mg/L and floatable solids must be absent.
Venezuela	Irrigation with untreated wastewater is prohibited. The reuse of treated wastewater is allowed on certain types of crops. A value of <100 faecal coliforms/100 mL and 1,000 total coliforms are set to irrigate vegetables for human consumption. A monthly average value of 5,000 total coliforms and 1,000 faecal coliforms per 100 mL are allowed on other types of crops and for crops used for animal feed.

Wastewater discharge regulations control reuse for agricultural irrigation because of the frequency with which non-treated wastewater is used to irrigate. Based on the observed health effects, norms almost always follow WHO guidelines. Although helminth eggs content is considered an important parameter, it is sometimes excluded from regulations due to the lack of skilled personnel to perform the required analysis. Of those countries that do set a limit for helminth eggs, this is often not followed for the same reason. Therefore, to enforce the law, serious training efforts ought to be made in this respect. Data concerning faecal or total coliforms content are widely considered and used to enforce regulations, even though it is recognized that they are not useful indicators of parasites. Additionally, given the unaffordable treatment costs for water intended for agricultural purposes, almost all legislation attempts to limit the types of crop that may be irrigated, although in practice this has proven to be difficult to enforce due to the large number of farmers responsible for small plots of land where waste or polluted water is used. In practice, the overall regulatory effectiveness varies from country to country.

9.5 REUSE TECHNOLOGIES

Diverse types of technologies are used to treat wastewater for reuse in Latin America. While between 1960 and 1970 there was a generalized trend towards the use of stabilization ponds to reclaim water for agriculture, nowadays other technologies are preferred in some countries such as the Caribbean islands, Chile, Costa Rica, and Mexico. Stabilization ponds have been replaced or not selected, particularly in arid and semi-arid regions, due to:

- the need to increase the treatment capacity in areas where little land is available, for instance, near cities or in agricultural zones with high land cost;
- important water losses in areas with high evaporation rates;
- mosquito proliferation and foul odors when ponds are overloaded, as is frequently the case in cities with fast population growth;
- deficiencies in removing pathogens, in particular bacteria, due to poor design or insufficient maintenance. Frequently, the accumulation of sediments in ponds promotes by-passes, reducing the hydraulic contact time and hence the disinfection capacity; and
- soil with significant water infiltration or high saline or sulfate content (notably in Chile) that degrades the effluent quality.

Where stabilization ponds are used, a hydraulic retention time of 21 days is preferred. Sometimes, as happens in Bolivia and Brazil, stabilization ponds are placed after UASB systems to produce better effluent quality. Stabilization pond costs in Argentina, Brazil, Nicaragua and Peru are reported to be around 0.03 to 0.17 US\$/m^3 (Barbeito Anzorena, 2001; Rodrigues-Pimentel *et al.*, 2001, Gámez, 2001, Del Carpio, 2000 and Manrique–Arce *et al.*, 2001).

Processes that are replacing stabilization ponds are activated sludge, advanced primary treatment (or chemically enhanced primary treatment), anaerobic filters and UASB (up-flow anaerobic sludge blanket). The last two have proven inefficient at removing helminth eggs and bacteria (von Sperling *et al.*, 2002; Jimenez, 2007). Activated sludge costs vary from 0.14 to 0.2 US\$/m^3 in Chile and Mexico, while the APT cost, reported only for Mexico City, is 0.05 to 0.09 US\$/m^3. Furthermore, because of the high costs of conventional wastewater treatment methods (according to Colling, 2000, 145 US\$/per capita) efficient and affordable technology is still needed. New approaches to control pathogens in wastewater and sludge reused for agriculture have been developed in Brazil, Chile and Mexico.

For municipal reuse, activated sludge is the treatment method most frequently used, sometimes complemented by tertiary filtration. For industrial reuse a mixture of processes are applied, ranging from physicochemical processes to membrane ones.

9.6 FUTURE REUSE OF WASTEWATER

Countries such as Argentina, Brazil, Chile, Colombia, Costa Rica, Mexico and Peru have planned or are planning to increase water reuse not only to face water scarcity either at country or a regional level, but also as a strategy to profitably enhance wastewater treatment (Table 9.3). Additionally, in the Caribbean region reuse is promoted in conjunction with sanitation to face up to water scarcity and to protect natural resources that are an important part of tourist activities.

Table 9.3 Countries with plans to reuse water in the future.

Country	Future reuse interests
Argentina	For agriculture and industry.
Brazil	For crops with high economic yields or for fodder irrigation, for aquifer recharge and for industrial purposes as the price of fresh water rises.
Chile	For agriculture and industry.
Colombia	Interest in reuse in agriculture in some areas of the country.
Costa Rica	Interest in agricultural reuse for crops that are not eaten raw.
Mexico	Reuse is very important for agriculture, municipal and industrial purposes, particularly in the central and northern areas of the country.
Peru	Agriculture reuse for high yield products that can be exported.

In periurban areas of Latin American cities, the interest in promoting reuse for urban agriculture has been stated in the Declaration of Quito 2000 (PNUD, 2001), signed by some Latin America cities, the Food and Agriculture Organization (FAO) and the United Nations Development Programme (UNDP). This Declaration states that the reuse of urban water for agriculture is a reliable way of ensuring food provision and improving the populations' standard of living, as well as protecting surface water bodies from pollution. This declaration is similar to the Hyderabad Declaration.

Several studies undertaken by CEPIS/OPS demonstrated that in Latin America, the benefit/cost ratio of agricultural reuse projects (including the cost of wastewater treatment) varies from 1.1 to 2, depending on the type of treatment, the profitability of the crop, and the type of irrigation system. Where flooding and furrow irrigation are used, wastewater treatment processes that remove suspended solids are not considered necessary; where crops are used after industrialization or where flowers are grown, water quality criteria are less stringent. Furthermore, governments are increasingly supporting the use of modern irrigation methods and better wastewater treatment systems to produce cash crops like vegetables to increase farmers' income in poor regions.

In addition to interest in wastewater reuse in agriculture, there is also interest to reuse of wastewater for reforestation, aquifer recharge, recreation and aquaculture, but the volumes involved are much smaller.

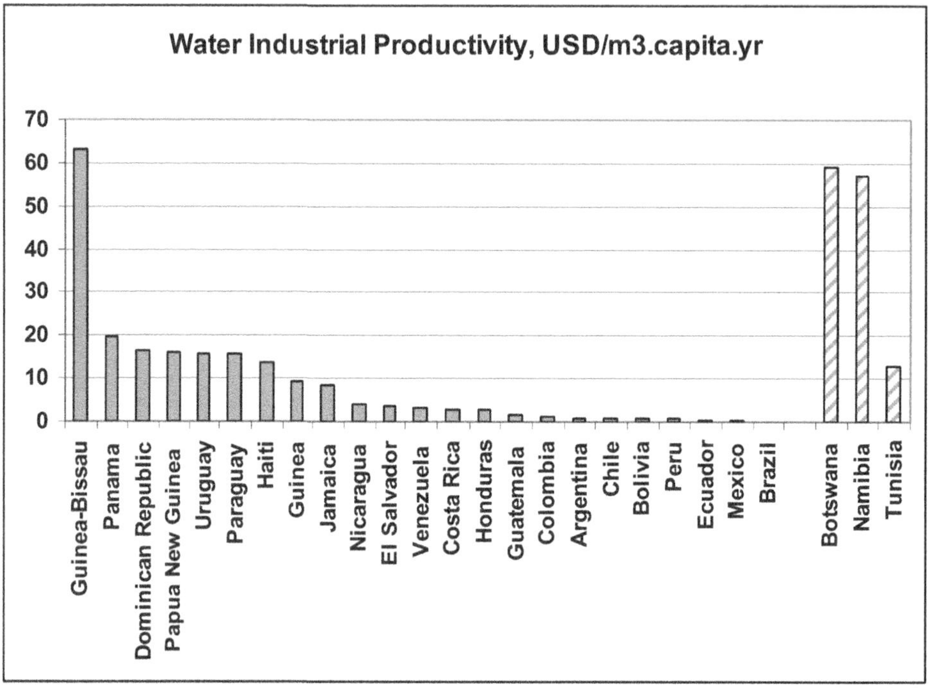

Figure 9.4 Intensity water productivity index and water intensity use per country (with data from UN, 2003).

In future, reuse may be important for industries at a local level. However, at the country level it is important to make more beneficial use of water in economic terms, particularly in countries where water is already scarce or there are conflicts between water users. Figure 9.4 shows the industrial water productivity index (calculated as the ratio of the added value generated by the industry per cubic meter per capita; World Bank, 2001) for selected countries. It can be appreciated that water industrial productivity is highly variable and that some water scarce countries use a larger amount of water to produce a lower income. Water availability in Latin America is considered an asset for industrial development. At the world level it is estimated that by the year 2025 industrial water use will increase 60% to account for 24% of total global water extraction (Shiklomanov, 1999). According to the UN (2003), as a consequence of globalization and market liberalization, this increase will occur mainly in developing countries. This might represent a threat to countries with poor legal control over the quality and quantity of their water resources. In South America, it is predicted that by the year 2025 water extraction will increase to 70%, mainly for industrial and municipal uses (Davis, 1996).

9.7 References

Abarca-Garbanzo, A. (2001) Case Study: Puntarenas, Costa Rica. Regional Project on Treatment and Integral Use of Wastewater in Latin America: Actual situation and potential. Agreement: IDRC – OPS/HEP/CEPIS 2000 – 2002 (In Spanish).

Acosta, L., Alvaro, H., Jiménez, H., Manrique, R. and Torralba, R. (2001) Case Study: Villa del Salvador, Lima, Peru. Regional Project on Treatment and Integral Use of Wastewater in Latin America: Actual situation and potential. Agreement: IDRC – OPS/HEP/CEPIS 2000 – 2002 (In Spanish).

Acuerdo Gubernamental (2006) from Guatemala http://www.bvsde.paho.org/bvsacd/cd38Guatemala/A236-06.pdf, consulted on 7 September 2007

Barbeito Anzorena, E. (2001) Case Study: Campo Espejo del aglomerado Gran Mendoza, Argentine Republic. Regional Project on Treatment and Integral Use of Wastewater in Latin America: Actual situation and potential. Agreement: IDRC – OPS/HEP/CEPIS 2000 – 2002 (In Spanish).

Bartone, C. (1990) International perspective on water management and wastewater reuse-appropriate technologies. IAWPRC Biennial International Conference and Water Reuse Seminar, Kyoto, 29 July-3 August.

Bartone, C. and Arlosoroff, S. (1987) Reuse of pond effluents in developing countries. *Wat. Sci. Tech.* **19**(12):289-297.

Blumenthal, U., Mara, D., Ayres, R., Cifuentes, E., Peasey, A., Stott, R., Lee, D. and Ruiz-Palacios, G (1996) Evaluation of the WHO nematode egg guidelines for restricted and unrestricted irrigation. *Wat. Sci. Tech.* **33**(10-11):277-283.

Blumenthal, U., Peasey, A., Ruiz-Palacios, G. and Mara, D. (2000) Guidelines for wastewater reuse in agriculture and aquaculture: recommended revisions based on new research evidence. WELL Study, Task No.: 68, Part 1. Water and Environmental Health at London and Loughborough, London, UK. http:/www.lboro.ac.uk/well/resources/well-studies/well-studies.htm.

Castro de Esparza, M.L. and Flórez-Muñóz, A. (1990) Health Risks evaluation of wastewater use in agriculture. Rojas-Vargas, R. (Reviewer). Volume 1, CEPIS-OPS. Lima-Perú. 60 p. (In Spanish)

Castro-Merizalde, B. (2001) Case Study: Portoviejo, Ecuador. Regional Project on Treatment and Integral Use of Wastewater in Latin America: Actual situation and potential. Agreement: IDRC – OPS/HEP/CEPIS 2000 – 2002 (In Spanish).

Cifuentes, E., Blumenthal, J.; Ruiz-Palacios, G. and Beneth, S. (1992) Health Impact Evaluation of Wastewater in Mexico. *Public Health Revue.* **19**:243-250.

CNA, Comisión Nacional del Agua. (2004) Statistics of water in Mexico. Water National Commission (in Spanish).

Colling, D. (2000). Water Management and wastewater management in large to medium-sized urban centers, experiences from the developing world. Client. Sida Stockholm, August.

Coronado-Rocha, O., Moscoso-Agreda, O. and Ruiz-Hurtado, R. (2001) Case Study: Cochabamba, Bolivia. Regional Project on Treatment and Integral Use of Wastewater in Latin America: Actual situation and potential. Agreement: IDRC – OPS/HEP/CEPIS 2000 – 2002 (In Spanish).

Coronel, D. (2002) San Lorenzo y Luque, Cuenca del lago Ipacaraí, Paraguay Regional Project on Treatment and Integral Use of Wastewater in Latin America: Actual situation and potential. Agreement: IDRC – OPS/HEP/CEPIS 2000 – 2002 (In Spanish).

Davis, D. (1996) Water Resources Assessment - The Tool for a Sustainable Future. In: Water Resources Assessment and Management Strategies in Latin America and the Caribbean. Proceedings of the WMO/IDB Conference, San José, Costa Rica.

Del Carpio, O. (2000) Case Study: Tacna, Peru. Regional Project on Treatment and Integral Use of Wastewater in Latin America: Actual situation and potential. Agreement: IDRC –OPS/HEP/CEPIS 2000 – 2002 (In Spanish).

Downs, T., Cifuentes, E., Ruth, E. and Suffet, I. (2000) Effectiveness of natural treatment in a wastewater irrigation District of the Mexico City region: a synoptic field survey. *Wat. Environ. Res.* **72**(1):4-21.

Earth Trends (2005) http://earthtrends.wri.org/country_profiles/index.php?theme=2

Espanhol, I. (2004) Wastewater as a resource for beneficial uses in Brasil. Conferência De Chefes De Estado Da América Do Sul E Dos Países Árabes Seminário Semi-Árido/Recursos Hídricos Fortaleza, CE/Petrolina, PE, 29.09-01.10.2004, http://www2.mre.gov.br/aspa/semiarido/data/ivanildo_hespanhol.htm

Espanhol, I. (2006) A New Paradigm for Urban Water Management and how Industry is Coping With it. Paper presented in the workshop Closing the urban water cycle, organized by UNESCO Guanajuato, Mexico, 21-25 November, 2005.

Esty, D.C. and Cornelius, P.K. (eds) (2002) Environmental performance measurement Global report 2001-2002, New York, Oxford, University Press.

Foster, S., Garduño, H., Tuinhof, A., Kemper, K. and Nanni, M. (2004) Briefing Note 12. Urban Wastewater as Groundwater Recharge evaluating and managing the risks and benefits. World Bank Ed.

Fundación Sustentable (2007) http://www.fundacionsustentable.org/article1167-Ibagu%C3%A9-ser%C3%A1-proyecto-piloto-en-aguas-residuales.html, consulted on 6 September 2007

Gámez, S. (2001) Case Study: Jinotepe, Carazo, Nicaragua. Regional Project on Treatment and Integral Use of Wastewater in Latin America: Actual situation and potential. Agreement: IDRC – OPS/HEP/CEPIS 2000 – 2002 (In Spanish).

Garza-Almanza, V. (2001) Case Study: valle de Juárez, México. Ciudad Juarez, Chihuahua. Regional Project on Treatment and Integral Use of Wastewater in Latin America: Actual situation and potential. Agreement: IDRC – OPS/HEP/CEPIS 2000 – 2002, (In Spanish)

Jiménez, B. (2004) El Mezquital, Mexico: The biggest irrigation district that uses wastewater, in Water Reuse for irrigation: Agriculture, Landscape and turf grass. Lazarova and Bahri editors. Editorial CRC Press

Jimenez, B. (2005) Treatment, Technology and Standards for Agricultural Wastewater Reuse: A Case Study in Mexico. *Irrigation and Drainage* **54**(1):23-35.

Jimenez, B. (2007) Helminth Ova Removal from Wastewater for Agriculture and Aquaculture Reuse *Wat. Sci. Tech*. **55**(1–2):485–493

Jimenez, B. and Chávez, A. (1997) Treatment of Mexico City Wastewater for Irrigation Purposes. *Environ. Tech*. **18**:721-730.

Jimenez, B. and Chávez, A. (2004) Quality assessment of an aquifer recharged with wastewater for its potential use as drinking source: "El Mezquital Valley" case. *Wat. Sci. Tech*. **50**(2):269-273.

Jiménez-Cisneros, B. (2003) The use of Water in the Valley of Mexico. *Páramo del Campo y de la Ciudad, Poverty and Marginality Study Center*, **1**(1):26-33 (In Spanish).

Kotlik, L. (1998) Water reuse in Argentina. *Wat. Supply* **16**(1/2):293-294.

Lam-Esquenazo, A., Rocco de la Fuente, F. and Romero, L. (2001) Case Study: Antofagasta, Chile. Regional Project on Treatment and Integral Use of Wastewater in Latin America: Actual situation and potential. Agreement: IDRC – OPS/HEP/CEPIS 2000 – 2002 (In Spanish)

Manrique-Arce, R., Acosta, L., Alvaro, H., Jiménez, H. and Torralba, R. (2001) Case Study: San Agustín, Perú. Regional Project on Treatment and Integral Use of Wastewater in Latin America: Actual situation and potential. Agreement: IDRC – OPS/HEP/CEPIS 2000 – 2002 (In Spanish).

Meganck, R., Rast, W. and Rodgers, K.P. (consulted on 2006) Source Book of Alternative Technologies for Freshwater Augmentation in Latin America and the Caribbean United Nations Environment Programme http://www.oas.org/dsd/publications/Unit/oea59e/ch01.htm#TopOfPage

Mena-Patri, M. (2001) Case Study: Santiago, Chile. Regional Project on Treatment and Integral Use of Wastewater in Latin America: Actual situation and potential. Agreement: IDRC –OPS/HEP/CEPIS 2000 – 2002 (In Spanish).

Mercedes-Matos, L. (2001) Case Study: Concepción de la Vega, Dominican Republic. Regional Project on Treatment and Integral Use of Wastewater in Latin America: Actual situation and potential. Agreement: IDRC – OPS/HEP/CEPIS 2000 – 2002 (In Spanish).

Moscoso-Cavallini, J. and Egocheaga-Young, L. (2000) Wastewater Use in Latin America: Actual situation and Potential. Agreement IDRC – OPS/HEP/CEPIS 2000 – 2002 International Research center for the Development, Canada.

Moscoso-Cavallini, J. and Egocheaga-Young, L. (2004) Integrated systems to treat and use wastewater in Latinoamérica: Present situation and Perspectives. *Hojas de Divulgación Técnica*, **92**:1-8. htpp:www.cepis.ops.oms-org

Muciño, D. (2001) Case Study: Texcoco, México. Regional Project on Treatment and Integral Use of Wastewater in Latin America: Actual situation and potential. Agreement: IDRC –OPS/HEP/CEPIS 2000 – 2002 (In Spanish).

PNUD (2001) Declaración de Quito. La agricultura urbana en las ciudades del siglo XXI. Programa de las Naciones Unidas para el Desarrollo Subprograma de Gestión urbana (PGUALX), coordinación regional para América Latina y el Caribe, Centro de las Naciones Unidas para los Asentamientos Humanos.

Rodrigues-Pimentel, F., Morales-Torres, I., Da Costa de Silva, J., Silveira-Britto junior, A., Bemvindo-Gomes, R. (2001) Case Study: conjunto Renacer, Fortaleza, Estado do Ceará. Regional Project on Treatment and Integral Use of Wastewater in Latin America: Actual situation and potential. Agreement: IDRC – OPS/HEP/CEPIS 2000 – 2002 (In Spanish).

Rosillo, A. (2001) Case Study: Maracaibo, Venezuela. Regional Project on Treatment and Integral Use of Wastewater in Latin America: Actual situation and potential. Agreement: IDRC –OPS/HEP/CEPIS 2000 – 2002 (In Spanish).

Sánchez de León, E. (2001) Case Study: Solola Guatemala Project on Treatment and Integral Use of Wastewater in Latin America: Actual situation and potential. Agreement: IDRC –OPS/HEP/CEPIS 2000 – 2002 (In Spanish).

Shiklomanov, I.A. 1999. World Water Resources and their Use. St. Petersburg, State Hydrogeological Institute, part of the International Hydrological Programme of the United Nations Educational, Scientific and Cultural Organization. St. Petersburg.

Siebe, Ch. and Fisher, W. (1996) Effect of long-term irrigation with untreated sewage effluents on soils properties and heavy metal adsorption of Leptosols and Vertisols in Central Mexico. *Z. Pflanzenernahr. Bodenk.* **159**:357-364.

UN, United Nations (2003) Water for People Water for Life. The United Nations World Water Development Report UNESCO and Berghahn Books. Barcelona.

van der Hoek ,W. (2004) Framework for a global assessment of the extent of wastewater irrigation. Paper presented at the International Water Management Institute workshop on Wastewater Use in Irrigated Agriculture: Confronting the Livelihood and Environmental Realities, in Hyderabad, India, 11-14 November 2002.

Vanegas-Gálvez, M. (2001) Case Study: Ciudad de Ibagué, Colombia Regional Project on Treatment and Integral Use of Wastewater in Latin America: Actual situation and potential. Agreement: IDRC – OPS/HEP/CEPIS 2000 – 2002 (In Spanish)

Von Sperling, C., Chernicharo, C.A., Soares, M.A.E. and Zerbini, A. (2002) Coliform and helminth eggs removal in a combined UASB reactor – baffled pond system in Brazil: performance evaluation and mathematical modelling. *Wat. Sci. Tech.* **45**(10):237–242.

WHO (1989) Health guidelines for the safe use of wastewater in agriculture and aquaculture. Technical Report Series 778. Geneva: World Health Organization.

World Bank (2001). World Development Indicators (WDI). Washington DC. Available in CD-ROM.

World Bank (2002) World Development Indicators (WDI). Washington DC. Available in http://www.rrojasdatabank.org/wdi2002.htm

World Water Forum (2006) Regional Document for the Americas, World Water Forum and Comisión Nacional del Agua, DF, Mexico, 176 pp

WRI, World Resources Institute (2007) Earth Trends: The Environmental Information Portal http://earthtrends.wri.org/index.cfm. Washington D.C.

SECTION TWO

STAKEHOLDERS' POINTS OF VIEW

10

Water reuse practices for agriculture

Valentina Lazarova and Akiça Bahri

10.1 INTRODUCTION

The majority of water reuse projects worldwide are for agricultural irrigation, driven for centuries by increasing water scarcity and agricultural water demand.

It is worth noting that water for agriculture is critical for food security, and for this reason agriculture was and remains the largest water user with about 70% of the world's fresh water withdrawals. In several developing countries (Figure 10.1, adapted from FAO, 2005), irrigated agriculture accounts for more than 80% of the overall water withdrawal with total volume up to 20,000-160,000 million m^3/yr (Mm^3/yr).

Despite the high increase of water usage in agriculture, up to 127% since 1950, the current proportion of irrigated croplands is only 17% of the total cultivated land providing 30-40% of the world's food production (FAO, 2003). In the future, water demand for irrigation will increase, but water availability for agriculture will be threatened by the increasing domestic and industrial demand and by other constraints. For these reasons, the expected growth rate of world water demand in agriculture is lower, about 0.8% per year (Merrett, 2002). In a number of regions however, the annual growth of irrigation demand will remain high and will even increase, as for example in Australia, where the growth rate is estimated at 3% for the irrigation of vegetables and fruits (Ziehrl *et al.*, 2004).

The need for alternative water resources has been emphasized during the last few years by the worst droughts, which occur not only in traditionally arid areas in the United States, the Mediterranean region, the Middle East and South Asia, but also in several temperate states and countries in Europe and North America. For example, it was estimated that the drought of 1999 in the Near East resulted in a relative decline of food production of 51% (FAO, 1999). The economic impacts of the 2003 drought in Europe was estimated at over 15 billion US$, and the new 2005 drought could be the worst for the last 30 years for some regions in Spain, France and England.

In this context, water reuse is becoming more valued, and certain countries are already using a great part of their treated wastewater for irrigation (Argentina, China, Cyprus, Egypt, Israel, Jordan, Kuwait, Libya, Mexico, Saudi Arabia, Spain, Syria, Tunisia, United Arab Emirates). It is important to underline however, that in several developing countries, raw sewage is still used for agricultural irrigation despite demonstrated health problems. The total land irrigated with raw or partially diluted wastewater is estimated at several million ha in fifty countries (20 million ha according to United Nations, 2003, which is a subject of discussion), which is approximately 10% of total irrigated land.

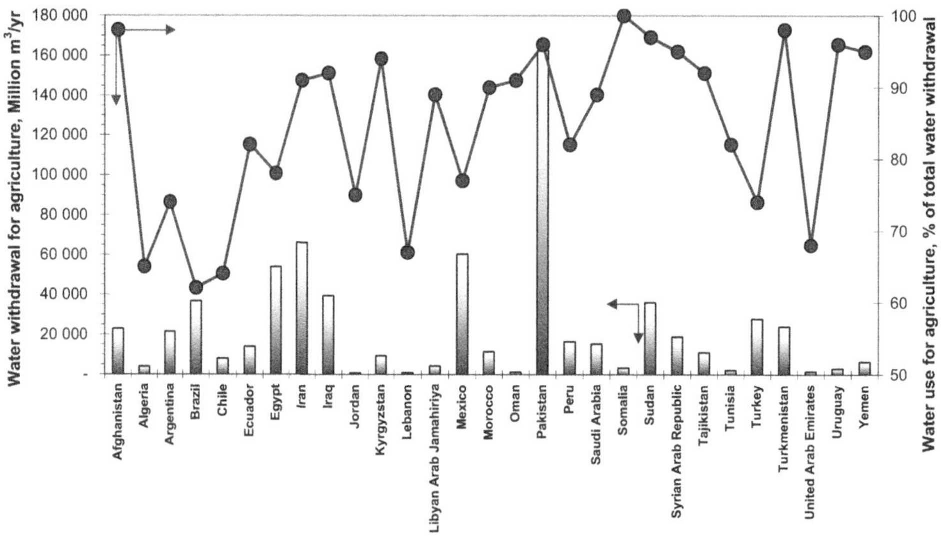

Figure 10.1 Water withdrawal for agriculture for selected countries with high water demand (FAO, 2005).

Other important factors favouring agricultural water reuse are the numerous benefits of irrigation with recycled water in conjunction with relatively less stringent water quality requirements compared to other reuse applications such as urban, industrial and indirect potable reuse (Lazarova and Asano, 2005). Irrigation in agriculture provides several additional health barriers that can increase economic viability of water reuse projects. For these reasons, water reuse for irrigation purposes is becoming a vital resource to enhance agricultural production, providing increased crop yields and decreased reliance on chemical fertilisers. Consequently, in some arid countries such as Israel and Jordan, water reuse provides a large share of irrigation water.

The basic principle of beneficial water reuse in agriculture is that municipal wastewaters may be used for all kinds of irrigation purposes as long as they have been previously treated to appropriate degree to meet specific water quality requirements (Ayers *et al.*, 1985; Lazarova and Bahri, 2005). It is worth nothing that besides the well-recognized benefits of water reuse, the use of recycled water for irrigation may have adverse impacts on public health and the environment, depending on treatment level, local conditions and irrigation practices. In all cases, the existing scientific knowledge and practical experience can be used to lower the risks associated with water reuse by the implementation of sound planning and effective management of irrigation with recycled water.

The most comprehensive book on irrigation with reclaimed water was published by Pettygrove and Asano (1985). Different aspects of water reuse for irrigation were discussed in detail in a number of FAO documents (1985, 1992, 1994, 1997 and 2000). These publications are mainly based on the knowledge gained with traditional irrigation schemes or prepared for specific local conditions and farmers.

Recently, Lazarova and Bahri (2005) published a most up-to-date synthesis of the international expertise on irrigation with recycled water with a large target audience based on a new holistic approach. As shown on Figure 10.2, the holistic approach of water reuse in agriculture includes different institutional, engineering and agronomic practices that could be implemented to improve the health safety, efficiency and competitiveness of irrigation with recycled water. In order to help decision-making, and more precisely estimate the benefits from each specific action, the management practices can be classified in three levels, depending on their main objective, listed in an increasing order of importance:

- health protection;
- improvement of food production and prevention of soil and aquifer degradation;
- improvement of economic competitiveness and public acceptance.

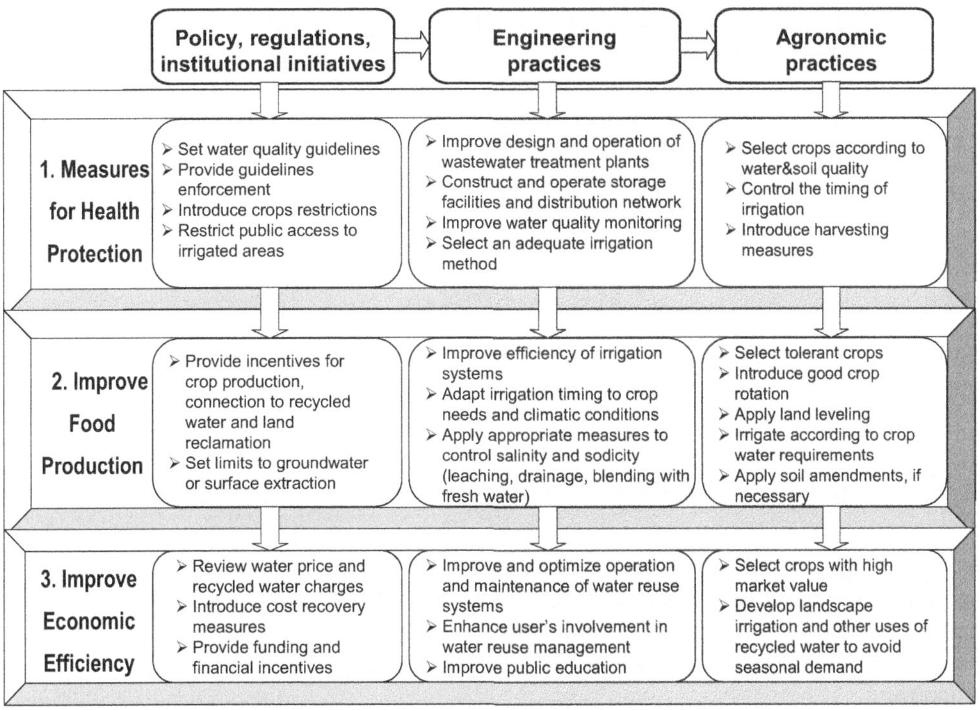

Figure 10.2 Holistic approach to water reuse in agriculture indicating the major management actions to improve the health safety, efficiency and competitiveness of irrigation with recycled water (Lazarova and Asano, 2005).

The main objective of this chapter is to present the major elements of this holistic approach, based on recent and well-recognised experiences worldwide. The described irrigation practices, as well as the terms recycled or reclaimed water are relevant only for irrigation with treated municipal wastewater not contaminated by heavy discharge of high volumes or high strength industrial wastewater.

10.2 RECYCLED WATER AS A VIABLE WATER RESOURCE

10.2.1 Classification of water reuse applications in agriculture

The main use of recycled water in agriculture is irrigation of different crops. It is important to stress that almost all irrigation uses can be satisfied with recycled water when water quality is adapted to the final use.

Two main types of water reuse for agricultural irrigation worldwide may be distinguished: direct and indirect agricultural irrigation. As a rule, direct reuse in agriculture is the use of recycled water directly after treatment and with short-time storage for a wide range of irrigation purposes in commercial and individual farms including:

- cereals such as wheat and maize;
- industrial crops such as soya beans, sugar beet and cane, coconut, oil palm, cotton, sunflower and other crops used in various industries for the production of lubricants, fuels, alcohol, plastics, newsprint and other papers, flavours, fragrances, pharmaceuticals and cosmetics;
- fodder, fiber and seed crops, such as alfalfa, clover, pasture, barley and other crops cultivated for animal feed;
- orchards producing fruits that grow above the ground such as citrus, peach, plum, nuts and grapes;
- vegetables and market gardening produce that are either cooked or processed so that pathogenic organisms are destroyed prior to human consumption, or eaten raw;
- plants grown in nurseries and greenhouses;
- forests, plant barriers against wind, and commercial woodlands;
- landscape irrigation, e.g. irrigation of green areas and golf courses.

Indirect water reuse in agriculture occurs when treated wastewater is used for the various irrigation purposes described above, but after long-term, intermediate or seasonal storage in the following artificial or natural water bodies:

- surface reservoirs;
- aquifers;
- rivers, lakes and wetlands.

Table10.1 illustrates the main categories of crops irrigated with recycled water in selected countries (from literature review). Fodder and industrial crops probably account for the largest area irrigated with wastewater. Irrigation areas of orchards (vineyards and fruit trees) and horticulture are growing in several industrial and developing countries. In a number of arid and semi-arid regions, irrigation of vegetables with raw or recycled wastewater is a concern.

During the last decade, the fastest growth worldwide is reported for landscape irrigation (Lazarova and Bahri, 2005). Compared to agricultural irrigation, this use is associated with numerous additional benefits such as low transportation costs (landscape areas are situated in or near the cities where wastewater is produced), higher absorption of nutrients, tolerance to some chemical compounds and variations of water quality, and less health-related problems (compared to irrigation of food crops).

Table10.1 Example of crops irrigated with recycled water in selected countries (Adapted from FAO 2005)

	Type of irrigated crops with recycled water	Argentina	Australia	Brazil	Chile	Colombia	Costa Rica	Cyprus	Ecuador	Egypt	France	Guatemala	Iran	Jordan	Kuwait	Lebanon	Libya	Malta	Mexico	Morocco	Nicaragua	Pakistan	Peru	Qatar	Saudi Arabia	Spain	Tunisia	Yemen	United States
Main categories	Fodder	●	●	●		●	●	●		●	●	●		●	●		●		●	●		●	●	●	●	●	●	●	●
	Industrial	●	●	●		●		●			●								●	●		●	●		●	●	●	●	●
	Orchards	●	●		●	●			●	●				●		●	●		●	●	●		●		●	●	●	●	●
	Cereals				●									●			●			●							●		
	Nurseries	●	●		●	●		●							●														●
	Vegetables			●					●	●	●	●	●	●				●	●	●		●	●					●	●
	Green areas		●		●			●			●	●			●				●	●				●	●	●	●		●
	Forestry	●	●		●					●	●		●	●	●											●	●		
Individual crops	Alfalfa										●				●				●						●		●		
	Beans					●			●			●							●										
	Cotton		●																●			●	●						
	Maize					●			●		●								●				●		●			●	
	Potatoes				●	●					●	●																	
	Rice					●	●															●							
	Sorghum					●																			●		●		
	Sugar cane		●	●		●	●																						●
	Wheat				●														●			●			●		●		
	Yucca					●																							

10.2.2 Major water reuse developments in agriculture

Agricultural irrigation with municipal wastewater has been practiced for centuries. Nevertheless, a fast and significant development of water reuse has occurred for agriculture and other non-potable reuse practices during the last 20 years, stimulated by increasing water shortages and facilitated by new policies and regulations.

In the Mediterranean region, water is considered as a very valuable resource. Israel, Jordan and Tunisia are the leaders in agricultural water reuse, which satisfy 20, 10 and 1.3% of the total water demand, respectively (Lazarova and Asano, 2005). Cyprus has also developed a sound water reuse strategy and will satisfy 11% of the total water demand with recycled water in the next few years, mainly for irrigation purposes. Another emerging reuse leader in this region is Spain, where 22% of collected wastewater is reused in agriculture, with an overall volume of 340 Mm^3/yr.

The three largest reclamation systems in the Mediterranean region and the Middle East are located in Kuwait, Israel and Saudi Arabia: 1) In Sulaibiya (start-up 2003), near Kuwait City, where 375,000 m^3/d of high quality recycled water are produced by membrane treatment (UF/RO) and used mostly for agricultural irrigation after blending with existing brackish water; 2) in the Dan Region of the Greater Tel Aviv area, where 310,000 m^3/d (95 Mm^3/yr) of high-quality recycled water after soil-aquifer treatment are used for irrigation of various crops in the Negev desert; and 3) in Riyadh, Saudi Arabia, where 674 Mm^3/yr of tertiary disinfected effluents are produced and from which 217 Mm^3/yr are used in agriculture.

In Jordan, about 82 Mm^3/yr of reclaimed water are used directly or indirectly for irrigation. In Tunisia, about 43 Mm^3 of reclaimed water are annually allocated for irrigation of fruit trees, fodder, industrial crops and cereals.

One of the largest reuse projects in Europe (250,000 p.e. (people equivalent)) has been implemented in 1995 in the city of Vitoria, northern Spain with a total volume of recycled water of 20 Mm^3/yr, from which 40% are used for unrestricted irrigation after advanced wastewater treatment and disinfection. The largest water reuse project for agriculture in France started in 1997 in Clermont-Ferrand, where over 10,000 m^3/d of tertiary treated urban wastewater are reused for irrigation of 700 ha of maize. In Italy, one of the largest projects was implemented in Emilia Romagna, where 1250 m^3/d of treated effluents are used for irrigation of more than 400 ha.

Florida and California are the leading US states in water reuse with the greatest volumes of recycled water, with 131 Mm^3/yr (in 2003) and 297 Mm^3/yr (in 2002) respectively. The highest percentage of treated wastewater reused for different purposes are reported for Nevada (80%), California (46%), Florida (16%) and Arizona (35%). Agricultural irrigation is the most common current water reuse practice in the United States. In 1995, 34% (340,000 m^3/d) and 63% (570,000 m^3/d) of the total volume of recycled water in California and Florida, respectively, were used for various agricultural purposes. Agricultural irrigation is often coupled with other uses such as golf course and landscape irrigation.

Driven by water scarcity, the largest water reuse system in South America is located in the arid region of Mendoza, in the western part of Argentina. Over 160,000 m^3/d of urban wastewater is treated by one of the largest lagooning system in the world and used for the irrigation of over 3640 ha of various crops (vineyards, olives, alfalfa, fruit trees).

The use of raw sewage or partly diluted wastewater is common in a number of developing countries such as India, Mexico, Morocco and Pakistan, not without causing health problems. Consequently, enteric diseases and gastrointestinal illnesses have been reported. One of the major examples of agricultural reuse is the Mezquital Valley in Mexico. Almost all collected raw wastewater from Mexico City, 3.9 to 25.9 Mm^3/d dry and wet flows, respectively, is reused for irrigation of over 85,000 ha of various crops. In India, over 73,000 ha of land were irrigated with wastewater in 1985 from at least 200 sewage farms. The use of raw sewage for crop irrigation is also common in Pakistan and it is practiced in 80% of urban settlements. In addition to the poor microbiological quality of raw sewage used for irrigation, a great concern in a number of developing countries (China, India and Pakistan) is the increasing concentration of heavy metals due to industrial wastewater discharge.

10.2.3 Assessment of the feasibility of irrigation with recycled water

The first elements to be considered when planning water reuse projects in agriculture are the background data gathering to assess the feasibility of irrigation with recycled water. The feasibility evaluation must be performed by professionals on the basis of the following elements:

- water quality and quantity considerations;
- site inspection;
- soils characteristics;
- drainage system;
- analysis of the available data on irrigated crops;
- storage system;
- irrigation systems (existing and new);
- operation and management conditions.

Table 10.2 gives the most important information that should be available for decision-making on the feasibility of water reuse. Besides the type of crops and their market value, special attention should be given to soil characteristics, the choice of irrigation systems and the evaluation of potential negative impacts of recycled water quality on crops, soils and groundwater.

10.2.4 Water quality requirements

The quality of irrigation water plays an important role in successful crop production and health protection (Lazarova *et al.* 2005). The three main categories of water quality parameters that are of concern for agricultural irrigation are as follows (Table 10.3):

- biological parameters of major health concern leading to short term diseases associated with direct contacts and/or consumption of water or contaminated crops;
- general parameters;
- chemical inorganic and organic parameters, associated with long-term health effects, as well as with immediate or long-term agronomic and environmental adverse impacts.

Pathogens

Three types of pathogenic organisms are of concern in agricultural irrigation: parasites, mainly helminth eggs, enteric bacteria and enteric viruses. As a rule, water reuse regulations give concentration limits for bacteria content using indicator organisms such as total, fecal coliforms and *E.coli*; WHO and some other guidelines also include helminth eggs. Only few regulations consider virus content.

Table 10.2 Main specific elements to be considered when planning water reuse for irrigation, and factors influencing decision-making (Source: adapted from Lazarova and Bahri, 2005).

Crop type	Soils properties and hydrogeology	Irrigation method	Water quality	Miscellaneous
Crop water requirements depending on climate and type of crops Tolerance to: • salinity • boron • sodicity • chloride Growth characteristics Crop market price and forecast for future needs	Soil profile and slope Soil texture Soil structure Chemical composition (pH, salinity, SAR, Na, Cl, B, etc.) Exchangeable cations (Na, Ca, Mg, K, Al) Physical characteristics (hydraulic conductivity, bulk density, etc.) Depth of the water table (ideally > 10 m) and seasonal variations Drainage requirements Quality of groundwater and use (potable supply, etc.) Proximity to water supply wells (ideally > 250 m)	Surface irrigation systems (flood, border, furrow) Surface trickle irrigation systems (drip, bubbler, micro-sprinkler) Subsurface irrigation systems Sprinklers (central pivot, rolls, stationary, etc.)	Coliforms Helminth eggs pH Total dissolved solids/Electrical conductivity Suspended solids Biochemical oxygen demand/Chemical oxygen demand Forms of nitrogen Forms of phosphorus Bicarbonates Alkalinity Boron Sodium Sodium adsorption ratio Trace elements	Storage of recycled water: • surface reservoirs • aquifers Blending recycled water with other water sources: • groundwater • surface water (lakes, rivers) • drinking water Environmental concerns: • groundwater • surface water (lakes, rivers) • wetlands Health safety aspects Fertilization practices Economic and institutional issues: water rights, charges for water supply, eco and other taxes, financial incentives, etc. Regulatory issues Public acceptance

Table 10.3 Main characteristics and impacts of some important water quality parameters in wastewater (Adapted from EPA, 1992; Metcalf and Eddy, 2003; Lazarova and Bahri, 2005).

Category	Constituents	Parameters of interest	Typical range in wastewater	Characteristics and impacts
Biological parameters (Pathogenic organisms)	Protozoa	*Entamoeba histolytica*, *Giardia lambia*, *Cryptosporidium*, *Cyclospora*	10^{-1}-10^{3}	Infectious dose: 1-20 protozoa High survival time in water: 10-30 days
	Helminths	*Ova*, *Ascaris lumbricoides*,	10^{1}-10^{3}	Infectious dose 1-10 helminth eggs
		Taenia,	10^{-2}-10^{1}	Highly persistent in the environment (many months)
		Hookworm, *Clonorchis*		Considered as the main pathogenic risk in irrigation with recycled water
	Bacteria	Total coliforms, cfu/100mL	10^{5}-10^{9}	Infectious dose highly variable (10 -10^{7})
		Fecal coliforms, cfu/100mL	10^{4}-10^{9}	High survival time in water: 10-60 days
		Streptococci	10^{3}-10^{7}	Total and fecal coliforms are commonly used as indicator
		E.coli, #/100mL	10^{4}-10^{9}	*E.coli* is becoming as more specific indicator
		Shigella, #/100mL	10^{0}-10^{3}	
		Salmonella, #/100mL	10^{2}-10^{4}	
	Viruses	Enterovirus, #/100mL	10^{1}-10^{4}	Detection methods are not sensitive
		Hepatitis, #/100mL		Variable incubation times, very high survival time in water (50-120 days)
		Rotavirus, #/100mL		Person to person contamination is the main mode of transmission
		Adenovirus, Norwalk, #/100mL		Infectious dose: 1-10 viruses
General parameters	Suspended Solids	Total Suspended Solids (TSS), mg/L	100-350	Sorb organic pollutants, trace elements and heavy metals; Shield microorganisms; Plug irrigation systems and soil
	Nutrients	Nitrogen, mgN_{tot}/L	20-85	Induce eutrophication when combined with high concentrations
		Phosphorus, mgP_{tot}/L	4-15	Nitrogen can lead to nitrate build-up in groundwater after leaching
	Hydrogen Ion	pH	6.5-8.5	Impact on coagulation, disinfection, metal solubility, and soils

Category	Constituents	Parameters of interest	Typical range in wastewater	Characteristics and impacts
Inorganics	Dissolved Inorganics	Total Dissolved solids (TDS), mg/L	250-3000	High salinity may damage crops
		Electrical conductivity (EC), dS/m	0-3	Salt can contaminate groundwater
		Boron, mg/L	0-2	High sodicity destabilizes soil structure and affects soil's permeability and structure
		Sodium adsorption ratio (SAR)	0-15	
		Bicarbonate, mg/L	0-600	Sodium can be toxic to plants and can cause soil structure deterioration
		Sodium, mg/L	20-170	
	Trace elements and heavy metals	Specific elements (Cd, Zn, Hg, Ni, Pb, Hg), mg/L	0.001-0.6	Accumulate in certain plants and animals
Organics	Biodegradable organics	Biochemical oxygen demand (BOD_5), mg/L	110-400	Aesthetic and nuisance problems Provide food for microorganisms
		Chemical oxygen demand (COD), mg/L	250-1000	Contribute to chlorine demand
		Total organic carbon (TOC), mg/L	80-290	
	Stable organics	Specific compounds (pesticides, PAH, chlorinated hydrocarbons), μg/L	100-400	Toxic to the environment and public health Limits suitability of recycled water

Numerous research studies have demonstrated that the highest health risks are associated with crops irrigated with raw sewage or poorly treated wastewater and eaten raw (Pearsey *et al.*, 2000). This conclusion was supported by several epidemiological studies (Blumenthal *et al.*, 2000). It was demonstrated that improving recycled water quality to WHO threshold level of <1000 FC/100 mL and <1 helminth egg/L may result in crops with a quality within guideline levels for ready-to eat foods, but this depends on weather conditions, type of crops and other factors (WHO, 2005). The greatest health concerns are associated with crops grown near the soil surface, for example onions, lettuce and zucchini, the latter having sticky surfaces that may concentrate protozoan cysts (Armon *et al.*, 2002). There is no strong evidence that population groups residing near wastewater treatment plants or areas irrigated with recycled water are at greater risk from pathogens resulting from aerosols (USEPA 1998; WHO 2005).

Chemicals

Human health-related issues involving toxic chemicals have been reported only for irrigation with wastewater heavily polluted by industrial waste discharge. Municipal wastewater that has limited industrial wastewater input generally contains concentrations of organic and inorganic compounds that do not present health concerns when the recycled water is used for irrigation. The majority of irrigation water quality criteria give numerical levels only for some potentially toxic elements such as Cu, Cr, Cd, Ni, Pb, Hg and Zn.

Chemical parameters also include several important compounds that are relevant to the yield and quality of crops, maintenance of soil productivity and protection of the environment. The quality of irrigation water is of particular importance in arid zones where extremes of temperature and low relative humidity result in high rates of evaporation, with consequent salt build-up in the soil profile.

Salinity

Salinity is one of the major concerns for agricultural water reuse. Dissolved salts increase the osmotic pressure of soil water, and consequently, lead to an increase of the energy which plants must exert to take up water from the soil. As a result, respiration is increased and the growth and yield of most plants decline progressively as osmotic pressure increases. Compared to many other irrigation waters, recycled water generally has a low to medium salinity with electrical conductivity (EC_w) of 0.6 to 1.7 dS/m. Recently, the classification of saline water has been reconsidered on the basis of research and practical observations of the use of saline water for irrigation (Lazarova *et al.*, 2005). Generally, non-saline water is characterized by TDS<500 mg/L and EC_w <0.7 dS/m, maximum salt content in slightly saline waters is TDS < 2000 mg/L and EC_w <3 dS/m, and water is considered as brine when TDS >30,000 mg/L and EC_w >42 dS/m. It is important to stress however, that recycled water salinity alone cannot be used to assess the suitability of any saline water for irrigation, because a number of other conditions must be taken into account, including crop, climate, soil, irrigation method and management practices.

Toxic ions

The most common phytotoxic ions that may be present in municipal effluents in concentrations high enough to cause toxicity are boron (B), chloride (Cl) and sodium (Na). Each can cause damage individually or in combination. At concentrations of less than 0.5-1 mg/L, boron is essential for plant development, but higher levels can cause problems in sensitive plants.

Chloride concentrations with adverse impacts on irrigation range from 100 mg/L for very sensitive crops to 900 mg/L for tolerant plant species.

Sodium is the most important cation in terms of its effect on soil. When present in the soil in exchangeable form, sodium causes adverse physical-chemical changes, particularly to soil structure, which results in dispersion of particles and consequently, in reduced infiltration rates of water and air into the soil. As a rule, recycled water could be a source of excess sodium (Na) in the soil compared to other cations (Ca, K, Mg), and for this reason it should be monitored. The most reliable index of the sodium hazard of irrigation water is the sodium adsorption ratio, SAR (see Equation 10.1, concentrations in meq/L). The threshold value of SAR of less than 3 indicates no restriction on the use of recycled water for irrigation, while severe damage could be observed when SAR is over 9, in particular for surface irrigation. At a given SAR, the infiltration rate increases as salinity increases, or decreases when salinity decreases. Therefore, SAR and electrical conductivity should be used in combination to evaluate the potential problem.

$$SAR = \frac{Na}{\sqrt{(Ca+Mg)/2}} \qquad \text{(Equation 10.1)}$$

Trace elements

Trace elements are not normally included in routine analysis of regular irrigation water, but attention should be paid to them when using treated municipal effluents, particularly if contamination with industrial wastewater discharges is suspected. These include aluminium (Al), beryllium (Be), cobalt (Co), fluoride (F), iron (Fe), lithium (Li), manganese (Mn), molybdenum (Mo), selenium (Se), tin (Sn), titanium (Ti), tungsten (W) and vanadium (V). Heavy metals also include a special group of trace elements that have been shown to create definite health hazards when taken up by plants. Under this group are included arsenic (As), cadmium (Cd), chromium (Cr), copper (Cu), nickel (Ni), lead (Pb), mercury (Hg) and zinc (Zn). The phytotoxic threshold levels of trace elements are set because of concern for long-term build-up of trace elements in the soil and for protection of agricultural soils from irreversible damage. Under normal irrigation practices, the suggested levels (0.01 to 5 mg/L depending on the trace element) prevent a build-up that might limit future crop production or use of the final product.

Nutrients

The most important nutrients for crops are nitrogen, phosphorus, potassium, copper, manganese, zinc, boron and sulphur. Usually, recycled water contains enough of these nutrients to supply a large portion of a crop's needs.

Nitrogen is a macronutrient for plants that is applied on a regular basis. Nevertheless, at very high concentrations (over 30 mgN_{tot}/L) it can over-stimulate plant growth causing problems such as lodging and excessive foliar growth and also delay maturity or result in poor crop quality. Most plants absorb only nitrates, but other forms such as ammonia and nitrites are easily transformed to nitrates in the soil.

Potassium in recycled water has little effect on crops and phosphorus contained in recycled water is too low to meet crops' needs. Over time, phosphorus can build up in the soil and can reduce the need for supplementation.

In general, irrigation with treated urban wastewater at an application rate of 100 mm/ha would provide fertilizing elements of up to 16-62 kg of total nitrogen (up to 90-300 kg in arid and semi-arid regions), 4-24 kg of phosphorus, 2-69 kg of potassium, 18-208 kg of calcium and 9-110 kg of magnesium (Lazarova *et al.*, 2005).

10.3 HEALTH PROTECTION MEASURES

The management of health risk in water reuse for agriculture includes several types of actions that can be classified in three main groups: (1) policy, regulations and institutional initiatives, (2) engineering actions and (3) agronomic practices.

Water reuse regulations are discussed in more details elsewhere. It is important to stress however, that health safety of irrigation with recycled water could be better guaranteed by the development of appropriate codes of good practices that are as important for farmers and operators as the quality requirements for water reuse (Crook and Lazarova, 2005). Consequently, for each water reuse project, an adequate combination of different health protection measures should be adopted depending on local conditions, type of irrigated crops, irrigation method and exposed groups (Lazarova and Bahri, 2005).

The most important health protection measures are tentatively summarised as follows:

1) Policy, regulations and institutional initiatives:
   - set-up of water reuse regulations or guidelines and their enforcement;
   - crop restriction;
   - human exposure control;
   - immunisation and drug therapy.

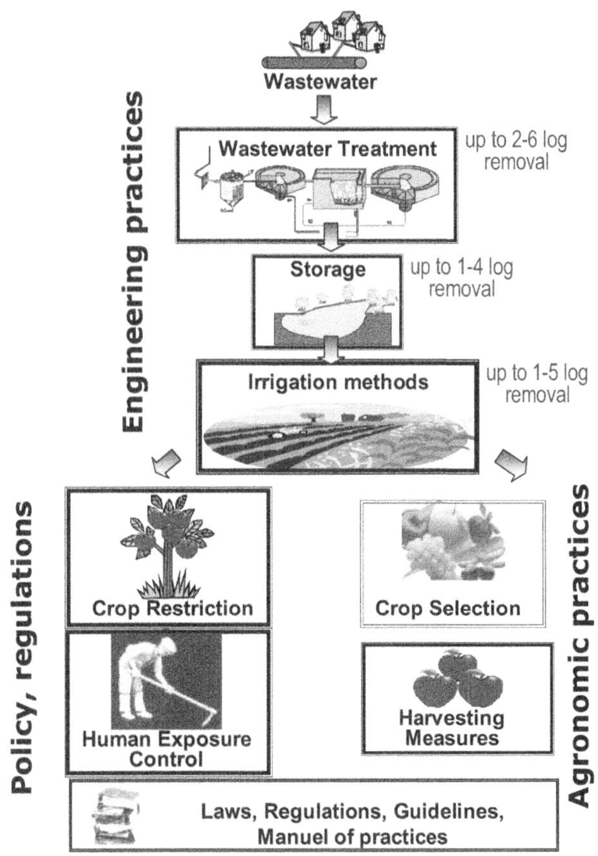

Figure 10.3 Flow-chart of irrigation with recycled water showing different health protection measures that can be implemented to prevent transmission of pathogen bacteria and viruses (Source: Adapted from Lazarova and Bahri, 2005).

2) Engineering practices:
    - wastewater treatment and storage;
    - adequate operation and water quality monitoring of water recycling scheme;
    - control of reclaimed water application: selection of irrigation method.
3) Agronomic practices;
    - crop selection;
    - control of reclaimed water application: timing of irrigation;
    - harvesting measures.

Figure 10.3 illustrates the combination of the major health protection measures that can be applied to irrigation with recycled water providing a suitable level of safety for irrigation with recycled water through a multi-barrier approach (Lazarova and Bahri 2005)

10.3.1 Policy, regulations and institutional initiatives

Historically, agricultural water reuse was the first application for which reuse standards were developed. The most common health concern associated with non-potable reuse of treated municipal wastewater is the potential transmission of infectious diseases by microbial pathogens. Waterborne disease outbreaks have, to a great extent, been controlled, but the potential for disease transmission through the water route has not been eliminated. As mentioned previously, the irrigation of market crops with poorly treated wastewater in developing countries has been and is still associated with enteric and other parasite diseases.

Water reuse regulations

Water reuse standards or guidelines vary with type of application, regional context, and overall risk perception. In practice, these factors are expressed through different water quality requirements, as well as treatment process requirements and criteria for operation and reliability. The major factor in the choice of regulatory strategy is the economics, i.e. costs of treatment and monitoring, as well as the capacity to enforce regulations. Most developed countries have tended to develop conservative low risk guidelines or standards based on a high technology/high cost approach, such as the California Water Recycling Criteria (2000). A number of developing countries advocate another strategy of controlled health risk adopting a low technology/low cost approach based on the WHO recommendations (1989). The new edition of the WHO guidelines (WHO, 2006) proposes a new approach based on risk evaluation allowing ≤103 E.coli/100 mL for unrestricted irrigation and ≤105 E.coli/100 mL for restricted irrigation with the additional barriers as crop cleaning, bacteria die-off, etc.

A comparison of water quality requirements for irrigation with recycled water in some countries (maximum limits for unrestricted irrigation) is provided in Table 10.4.

It is important to have in mind that countries may require microbiological quality standards for imported foods crops irrigated with reclaimed water according to the new recommendations of the International Commission for Microbiological Specifications for Foods (2005) based on the HACCP approach.

Table 10.4 Comparison of water quality criteria (maximum limits) for unrestricted agricultural irrigation (Adapted from Crook and Lazarova 2005).

Parameter	California (2000)	Arizona (2000)	USEPA (2004)	Israel (1978)	WHO[(a)] (1989)	Australia (2000)
	Regulations	Regulations	Guidelines	Regulations	Guidelines	Guidelines
Total coliforms, #/100mL	2.2[b]			2.2 (50%); 12 (80%)		
Fecal coliforms, #/100mL		ND	ND		1000	10 [c]
E.coli, #/100mL						
Salmonella						
Viruses						
Helminths, eggs/L					1	1
BOD_5, mg/L			10	15		
COD, mg/L						
Suspended solids, mg/L				15		
Turbidity, NTU	2	2	2	-		2
Electrical conductivity, dS/m						
Dissolved O_2, mg/L				≥0.5		
Residual Cl, mg/L	[d]		≥1	≥0.5		1
Minimum treatment	Tertiary + disinfection	Tertiary + disinfection	Tertiary + disinfection	Secondary[e]	Lagoons[f]	Tertiary

ND – not detected;
[a] Followed by several developing countries
[b]TC based on running 7-day median;
[c]Median value
[d]dependent on contact time;
[e]Seasonal storage may constitute an equivalent to tertiary treatment;
[f]Stabilisation ponds in series with proper retention time;

Crop restrictions

Water reuse regulations for irrigation specify the type of crops that can be irrigated with recycled water of a given quality, treatment requirements, as well as public access and other restrictions. In a number of countries (Chile, Mexico, Tunisia, etc.), crop restriction was successfully used to protect the health of consumers when recycled water with sufficient quality is not available for unrestricted irrigation. For example, cooking of vegetables results in up to 5-6 log reduction in exposure to pathogens. Consequently, reclaimed water of relatively poorer quality can be used for irrigation of non-food crops or vegetables that are processed before human consumption (e.g. potatoes, egg plants).

Crop restriction is feasible and is particularly facilitated under conditions of strong law enforcement, public body control of water allocation, strong central management of irrigation and adequate market demand for the crops irrigated with recycled water.

It is important to underline, however, that crop restriction is not a sufficient single health control measure, as it does not provide any protection for farm workers.

Human exposure control and immunisation

The objective of this group of measures is to prevent the risk that populations come into direct or indirect contact with pathogens that may be present in recycled water. The main groups of population that may be considered to be at risk as a result of agricultural use of reclaimed water and some recommended measures to protect their health are summarised in Table 10.5.

Health and hygiene promotion are very important and could be associated with other health initiatives such as immunisation. However, it is important to stress that health protection measures should be applied not only by farm workers, crop handlers and other persons concerned with water reuse, but also throughout the chain of food transportation and distribution. For example, numerous cases of contamination of fresh vegetables at the market have been reported due to the use of water with poor quality for washing and "freshening" of crops (Peasey *et al.*, 2000).

Table 10.5 Summary of recommended human exposure control measures for health protection of risk groups in water reuse for agriculture (Adapted from World Bank, 1993; Lazarova and Bahri, 2005).

Risk group	Risks from irrigation with recycled water	Human exposure control
Consumers of crops	• Risk from virus, bacteria, protozoa and nematode infections supported by epidemiological data	• High standards of food hygiene, which should be emphasized in health education associated with recycled water use schemes • Washing and cooking the agricultural produce before consumption
Workers on irrigated sites and crop handlers	• Risk from bacteria infections supported by epidemiological data • Potential risk from protozoa and nematodes infections (no epidemiological data) • No risk from virus infections, confirmed by epidemiological data	• Provision and insistence on the wearing of protective clothing • Maintenance of high levels of hygiene • Immunization against or chemotherapeutic control of selected infections (only if the recycled water is not well disinfected)
People living nearby irrigated areas	• No risk from virus and bacteria infections, confirmed by epidemiological data	• Should be kept fully informed of the use of wastewater in agriculture so that they, and their children, can avoid these areas • Sprinklers should not be used within 50-100 m of houses or roads (even if no evidence exist that those living near irrigated fields are at significant risk)

10.3.2 Engineering practices for health protection

The advance in wastewater treatment technologies, irrigation techniques and analytical methods offers a great potential for the improvement of health safety in water reuse practices.

Wastewater Treatment

Wastewater treatment is considered the most effective way of reducing the health risks associated with use of recycled water for irrigation. The choice of treatment schemes depends on water quality requirements, type of irrigated crops, irrigation method, public access, and potential adverse impacts on soils and crops.

The main technical challenge for the production of reclaimed water for restricted agricultural irrigation is the removal of microbial pathogens, and treatment processes should be selected and designed accordingly. In many countries, the recycled water quality criteria for restricted irrigation are at a minimum equivalent to the quality of secondary treated effluents, while additional tertiary treatment is often required for unrestricted irrigation.

In developing countries, the removal of helminth eggs is reported as a major concern, which represents the most important health risk, in particular for children (Blumenthal *et al.*,

2000). In this case, there may be limited restrictions on BOD removal. Consequently, physical-chemical primary treatment appears as a cost competitive solution that is well adapted to the requirements of restricted irrigation (Jiménez *et al.*, 2001). The main advantage of this treatment is the conservation of the fertilization capacity of wastewater.

As a rule, in the case of unrestricted irrigation, additional polishing steps should be designed for disinfection with or without pre-treatment (e.g., sand or multi-media filtration). The most common processes used for wastewater disinfection prior to reuse for irrigation are chlorination, UV irradiation and maturation ponds. Reed beds have become increasingly used in small treatment plants. For large works, ozonation is an efficient disinfection process, with several additional advantages.

Some typical treatment schemes for production of recycled water for agricultural irrigation according to different water quality criteria are summarized in Figure 10.4.

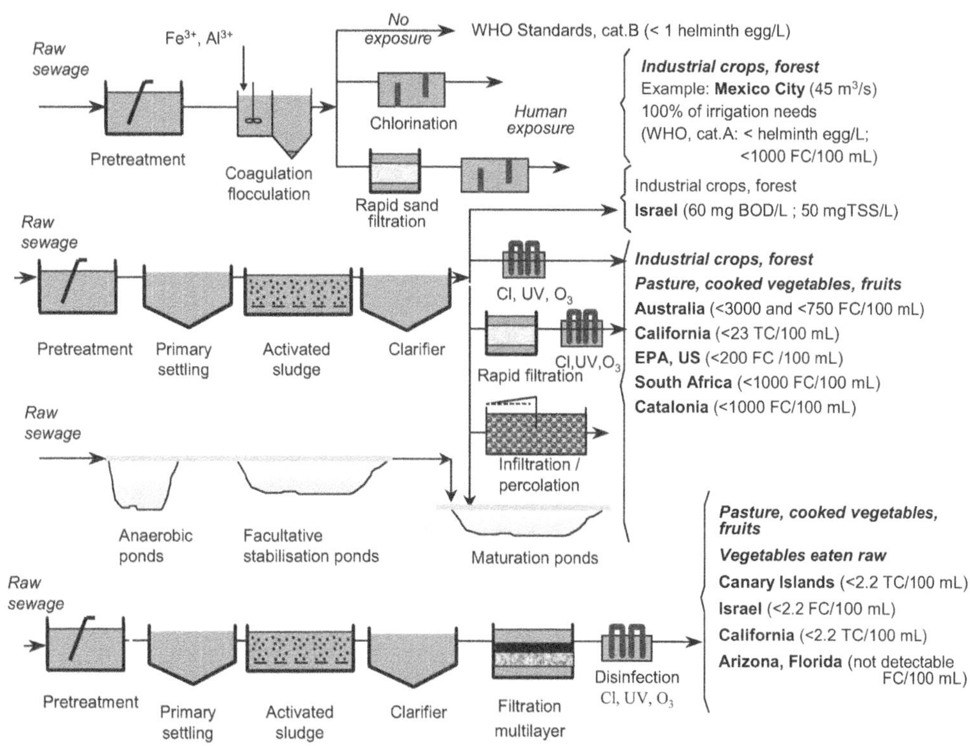

Figure 10.4 Typical treatment schemes for water reuse in agriculture according to some common water quality requirements (Lazarova, 2005).

Advanced treatment to remove wastewater constituents that may be phytotoxic or harmful to certain crops is technically possible, but may not always be justified economically, especially in the case of agricultural irrigation. To use water containing such constituents, farmers should apply appropriate agronomic practices.

Table 10.6 summarizes the main design parameters, advantages and drawbacks of the most common polishing treatments used for unrestricted irrigation.

Table 10.6 Main disinfection processes used for health protection in water reuse schemes for irrigation: design parameters, advantages and drawbacks (Adapted from Lazarova and Bahri 2005)

Process	Design parameters	Advantages	Drawbacks
Polishing (maturation) ponds	High influence of climate Depth 1 to 1.5 m Hydraulic residence time (HRT) 10-90 d, commonly between 15 and 30 d Surface area 2.5 m^2/inhabitant Recommended for small & medium plants (≤20,000-50,000 p.e.)	Provides additional storage capacity Easy to operate Robust to meet WHO guidelines Efficient for removal of bacteria and helminth eggs Cost efficient for small units	High evaporation loss Requires large footprints Effluent quality is a function of the hydraulic loads, climate and season Can cause nuisance (odour, insects) Sludge disposal requires infrequent but intensive work
Constructed wetlands (Reed beds)	Horizontal flux 10 m^2/p.e. Vertical flux 2.5 m^2/p.e. Recommended for small plants (≤100-1000 p.e.)	Simultaneous sludge treatment Simple operation No power consumption if topography allows it Good tolerance of seasonal variations	Large footprint and land requirement Annual removal of vegetation Complete cleaning of the bed every 10-15 years Only removes helminth eggs; does not remove bacteria and viruses
Chlorination	Chlorine dose 5-20 mg/L HRT 20 to 90 min CT 100 – 450 mg.min/L Recommended for all plant sizes	Presence of a disinfectant residual Easy to operate and control Little equipment needed Efficient on nitrified effluents	Formation of potentially harmful by-products Weak efficiency in case of incompletely nitrified effluent Heavy safety regulations in the case of chlorine gas Efficiency depends on water quality Does not remove helminth eggs
UV irradiation	UV dose 25 to 140 mJ/cm^2 Low or medium pressure UV systems	Low footprint No by-products High efficiency for inactivation of viruses, bacteria and cysts (*Cryptosporidium*) Easy to operate Cost efficient (similar to chlorination)	Dose difficult to measure Difficult to assess lamps ageing and fouling No disinfectant residual Efficiency depends on water quality (removal of suspended solids is required for complete disinfection) Does not inactivate helminth eggs
Ozonation	Transferred ozone dose 5 to 20 mg/L Contact time 5 to 10 min Completely mixed contactors	High efficiency for inactivation of viruses bacteria and protozoa Dose easy to monitor Cost efficient for large plants Decreases colour and odour of effluent	Too expensive on a small scale Very efficient for aquaculture Does not inactivate helminth eggs

Maturation ponds are the simplest and least costly technology for disinfection of medium-quality effluent in small communities (≤3000 m^3/d, ≤20,000-50,000 p.e.). This process also provides a storage capacity to accommodate variations in water demand. The major disadvantages of polishing ponds are the low flexibility of the process, high evaporation losses, and the inability to remove or inactivate all pathogens if proper detention times are not applied.

Several studies have shown that maturation ponds can produce effluent with microbiological water quality suitable for unrestricted irrigation according to WHO guidelines (<1000 FC/100 mL and <1 helminth egg/L), but only in conditions of low variation of loads and adequate climate conditions.

Reed beds, or subsurface flow beds with horizontal or vertical flow are the most commonly used constructed wetland systems. Compared to maturation ponds, the disinfection efficiency is lower and the WHO requirements could be achieved only for the removal of helminth eggs i.e. for restricted irrigation.

Chlorination is the most commonly used process for wastewater disinfection. Given the high chlorine doses that may be required, harmful by-products are usually generated. Typical chlorine doses for municipal wastewater disinfection are in the range of 5-20 mg/L with 30-90 min contact time, and usually allow for compliance with regulatory limits for unrestricted irrigation with conventional bacterial indicators such as coliforms. Higher doses or contact times are required for low quality wastewater such as primary or trickling filter effluents, or to meet stringent regulations.

UV disinfection appears as a cost-competitive technology for a broad range of plant capacities (Lazarova *et al.*, 1999). Disinfection studies on various types of secondary and tertiary treated wastewater indicated that 30-45 mJ/cm^2 doses of UV radiation were sufficient to ensure a 3 to 5 log removal of total and fecal coliforms and fecal streptococci. The required UV doses for a given microorganisms log removal are significantly influenced by wastewater quality and especially by suspended matters and transmittance.

In large water reuse projects, the option of ozone disinfection should also be considered.

An increasing number of water recycling schemes for unrestricted irrigation, in particular landscape irrigation, are using other advanced treatment technologies such as membrane bioreactors, tertiary microfiltration or ultrafiltration, and even reverse osmosis when a low water salinity is required (Lazarova, 2005).

It is important to stress that additional removal of pathogens can be achieved in storage reservoirs, which are often required in agricultural irrigation. Disinfection efficiency of storage reservoirs depends on hydraulic residence time and climate conditions and can reach 3-4 log units and more in long-term seasonal storage.

The reliability of operation of water reuse systems and adequate water quality monitoring are great concerns related to health safety of irrigation with recycled water. Many water reuse regulations include specific requirements for water quality monitoring and enhancement of treatment reliability of water reuse schemes, especially for unrestricted irrigation.

A good risk management strategy consists of the definition of key risk points and periodic or on-line measurement of adequate indicators of microbiological water quality, to check the reliability of operation of key treatment processes, storage reservoirs, distribution networks and irrigation devices.

Control of water application: selection of irrigation method

The adequate choice of irrigation method could be an efficient additional barrier for health protection. Significant difference exists between existing irrigation methods in terms of health risk:

a) Low level of health protection:
- associated with flood and sprinkler irrigation;
- recommended health protection measures include protection of farm workers (protective equipment, immunisation, etc.), use of recycled water with better quality, setting-up of minimum distance from roads and human settlements, use of mini-sprinklers, notification of the use of recycled water.

b) Medium level of health protection:
- associated with furrow irrigation;
- recommended health protection measures include protection of farm workers (protective equipment, immunisation, etc.), notification of the use of recycled water.

c) High level of health protection:
- associated with subsurface and drip irrigation;
- no specific health protection measures are required.

Table 10.7 summarizes the main factors affecting the choice of irrigation system in terms of health safety, costs and other specific requirements. All irrigation methods are appropriate to be used for recycled water, which is in compliance with water reuse guidelines or regulations for unrestricted irrigation, provided that agronomic criteria are also met. A number of existing regulations require higher water quality for sprinkler irrigation because of the possible disease transmission with aerosols.

Surface irrigation involves lower investments than other types of irrigation, but exposes field workers to the greatest health risk. If the effluent is not of the required water quality in terms of disinfection, sprinkler irrigation should not be used except for fodder crops, and border irrigation should not be used for vegetables, in particular those eaten uncooked.

Spray irrigation with recycled water that does not meet the health criteria is possible when specific management practices such as crop selection (industrial crops and fodder, for example), irrigation scheduling (irrigation during night) and other restrictions (no irrigation during windy conditions) are implemented.

Table 10.7 Factors affecting the choice of irrigation method and the special measures required for recycled water application (adapted from WHO, 1989).

Irrigation method	Factors affecting choice	Special measures for irrigation with recycled water
Flood irrigation (border)	*Low level of health protection* Lowest cost Exact leveling not required Low water use efficiency	Thorough protection of field workers, crop handlers and consumers (protective equipment, immunisation, drug therapy)
Furrow irrigation	*Medium level of health protection* Low cost Leveling may be needed Low water use efficiency	Protection of field workers, possibly for crop handlers and consumers (protective equipment, immunisation, chemotherapy)
Sprinkler irrigation	*Low level of health protection (aerosols)* Medium to high cost Medium water use efficiency Leveling not required	Water quality restrictions (pathogen removal; avoid use of anaerobic waste) Minimum distance 50-100 m from settlements/roads Use of mini-sprinklers
Subsurface and Drip irrigation	*Highest level of health protection* High cost High water use efficiency Higher yields	No protection measures required Water quality restrictions (filtration) are required to prevent emitters clogging

Subsurface and drip irrigation can provide the highest degree of health protection, as well as using water more efficiently and often producing higher crop yields. However, a highly reliable treatment is required to prevent clogging of emitters.

10.3.3 Agronomic practices for health protection

Crop selection

Crop selection is a specific measure for agricultural practices. Regulatory measures adopted for crop restriction concern only the main categories of crops, such as for example the interdiction of irrigation with reclaimed water of all kinds of vegetables either eaten raw or cooked in Tunisia.

Apart from the usual agronomic factors, crop selection is made according to water and soil properties, as well as to the market value of the crops.

Control of water application: timing of irrigation

In addition to the choice of irrigation method discussed below, another water application measure is the timing of irrigation. For example, this measure is successfully used in landscape and golf course irrigation by scheduling the irrigation only during the night or periods without public access.

Harvesting measures

When poor quality wastewater is used for irrigation, ceasing irrigation for a certain period of time prior to harvesting (crop drying), may help to reduce the risk of infection by allowing time for pathogens to die-off. However, the use of this control measure will depend on the type of crops, the particular pathogens known to be present and their survival times in the environment. For example, crop drying does not result in lowering the health risks associated with helminth eggs and cysts.

10.4 GOOD AGRONOMIC PRACTICES

Agricultural irrigation guidelines, which define the range of variation of irrigation water quality, have been developed to help farmers, operators and decision makers. If recycled water quality does not match these guideline values, customers can take actions either by selecting more tolerant crops, or managing soil characteristics, or changing agronomic practices.

Careful considerations and analysis of the background data, as well as market forecasts, will allow the identification and proposal of the most appropriate management practices for the implementation of irrigation with recycled water, or for the retrofit of existing irrigation systems to recycled water (Table 10.8).

The successful agronomic practices are divided in three main groups: (a) crop selection and management; (b) selection of irrigation method; and (c) management of water application.

Table 10.8 Summary of successful practices of irrigation of agricultural crops with recycled water (Lazarova and Bahri, 2005).

Crop selection and management	Selection of irrigation method	Management of water application
Good practices for salinity management Apply source control (industrial wastes, brines, seawater intrusion in sewers) Reduce evaporation (mulching) Select salt tolerant crops Increase leaching to remove salts from root zone Provide adequate drainage for leaching Maintain adequate soil moisture level Apply more uniform irrigation: localized irrigation, land smoothing Increase irrigation frequency to prevent water stress Irrigate at night or early in the morning at low temperature Apply blending water supply	**Furrow irrigation** No foliar injury Need of land preparation Low cost Possible water stress Salt accumulation in the root zone Fair to medium suitability for saline water Need of good management and drainage practices	**Leaching and drainage** Calculate leaching requirement on the basis of water and soil salinity Apply preliminary soil flushing by means of 10-20 cm of water before planting Adapt leaching frequency during plant growth Apply adequate measures to improve soil permeability Apply adequate drainage: surface or sub-surface (shallow water table)
	Border irrigation Almost no foliar injury (only bottom leaves) Need of land preparation Low cost Possible water stress Good salt flushing past the root zone Fair to medium suitability for saline water Need of good irrigation and drainage practices	***Adjusting fertilized application*** Avoid excess nitrogen application by modification of N-fertilizers, crop rotation, denitrification Excess phosphorus is not a problem: phosphorus content should be adjusted in the first years by additional fertilizers with further reduction due to P build-up in soils Potassium content should be adjusted by additional fertilizers
Good practices to overcome B, Na & Cl toxicity Increase leaching Increase irrigation frequency Avoid foliar wetting and injury using drip irrigation or micro-sprinklers Apply soil amendments such as organic amendments to improve water infiltration	**Sprinkler irrigation** Severe leaf damage can occur when irrigating with salt-affected recycled water Good salt flushing past the root zone Water stress: not possible to maintain high soil water potential Poor to fair suitability for saline water High capital and O&M costs	**Management of soil structure** Deep tillage Application of organic amendments Chemical amendments such as gypsum to counteract sodicity problems Indirect calcium amendment by addition of sulphuric acid requires strict technical supervision
Good practices to overcome trace elements toxicity No specific measures for calcareous soils Apply soil amendments such as liming for acid soils Limit the use of acid fertilizers for acid soils Select tolerant crops Leaching toxic ions past root zone Modification of fertilization Blending of water supplies	**Drip irrigation** No foliar injury Formation of salt wedge Possibility to maintain high soil water potential Excellent to good suitability for saline water High capital and moderate O&M costs	**Management of leakage and clogging problems** Check and repair pipe leakage Apply hydraulic flushing with or without chlorination Install preliminary filtration of recycled water Clean periodically all valves, nozzles and emitters

10.4.1 Crop selection and management

The selection of the crops and plant species to be irrigated with reclaimed water must be done after a complete evaluation of the recycled water quality, not only on the basis of the average values of salinity and other physical-chemical parameters but also taking into account the typical range of variation of each compound of concern, as well as the duration of peak concentrations.

Within this group of practices, salinity control and management appears one of the major issue to be considered when using recycled water for irrigation to avoid any negative impacts on crop yield. For some sensitive crops, boron and trace elements toxicity could be a major concern. The adequate selection of irrigation method and the application of leaching and drainage could reduce negative water quality impacts and prevent salinity build-up. Other cultural practices such as more frequent irrigation, timing of fertilization and seeding procedure, as well as chemical amendment or blending with other water sources may be needed to deal with temporary increase of recycled water salinity. More details on salinity and toxicity management can be found in Lazarova and Bahri (2005).

10.4.2 Selection of irrigation method

It is important to underline that irrigation method is one of the possible health control measures. Even if engineering knowledge is necessary for the choice and design of irrigation devices, the selection of irrigation method is strongly associated with the agronomic aspects such as the type of crops to be irrigated and their water demand, soil characteristics, the ability of farmers to maintain and operate the system, climatic conditions, etc.

In addition to the technical, health and agronomic considerations, socio-economic conditions should also be taken into account. In this context, an irrigation system that is considered as the most appropriate in one country or region may not be so in another one.

Table 10.9 presents an analysis of the most important factors that influence the choice between the four widely practiced irrigation methods, namely border, furrow, sprinkler and drip irrigation. Water application efficiency becomes especially important in areas with water scarcity and high evaporation rates. Ideally, the recycled water should be applied closely to the root zone using micro-sprayers or drip emitters. These techniques have the additional advantages of avoiding foliar wetting and allowing the application of high-frequency irrigation better adjusted to the crop irrigation requirements. In this case, however, adequate engineering management practices must be applied to reduce clogging problems.

Furrow irrigation is suitable when leaching demand is high. To avoid human contact and allow pressurization, recycled water should be conveyed preferably in closed conduits and distributed via resistant plastic tubes. Drip irrigation is particularly suited to sites with water scarcity where a low irrigation rate is desirable: to achieve a given rate of irrigation, the required volume for drip systems is about 4- to more than 10-fold lower than conventional sprinkler, furrow and border irrigation systems, respectively. This irrigation method is recommended for row crops (vegetables, soft fruit), tree and vine crops, where one or more emitters can be provided for each plant. Generally only high value crops are considered because of the high capital costs of installing a drip system.

Table 10.9 Evaluation of common irrigation methods in relation to agronomic and economic aspects of the use of recycled water (FAO, 1992: Lazarova and Bahri, 2005).

Parameters of evaluation	Furrow irrigation	Border irrigation	Sprinkler irrigation	Drip irrigation
Foliar wetting and consequent leaf damage resulting in poor yield	[a]No foliar injury as the crop is planted on the ridge	[a]Some bottom leaves may be affected but the damage is not so serious as to reduce yield	[b]Severe leaf damage can occur resulting in significant yield loss	[a]No foliar injury occurs under this method of irrigation
Salt accumulation in the root zone with repeated applications	[b]Salts tend to accumulate in the ridge which could harm the crop	[a]Salts move vertically downwards and are not likely to accumulate in the root zone	[a]Salt movement is downwards and root zone is not likely to accumulate salts	[c]*Salt movement is radial along the direction of water movement. A salt wedge is formed between drip points*
Ability to maintain high soil water potential	[c]*Plants may be subject to water stress between irrigations*	[c]*Plants may be subject to water stress between irrigations*	[b]Not possible to maintain high soil water potential throughout the growing season	[a]Possible to maintain high soil water potential throughout the growing season and minimize the effect of salinity
Suitability to handle brackish wastewater without significant yield loss	[c]*Fair to medium. With good management and drainage acceptable yields are possible*	[c]*Fair to medium. Good irrigation and drainage practices can produce acceptable levels of yield*	[b]Poor to fair. Most crops suffer from leaf damage and yield is low	[a]Excellent to good. Almost all crops can be grown with very little reduction in yield
Cost of equipments, operation and maintenance	Low costs Land preparation of furrows, 30-450 m long, 20-30 cm deep	Low costs Land preparation of field areas with downslope 0.1-0.4%	Very high cost of equipment Significant O&M costs Need of periodic maintenance	High cost of equipment Moderate O&M costs Need of maintenance

[a]The irrigation method is recommended for this parameter.
[b]The irrigation system is not advisable for this parameter.
[c]*The irrigation method may cause problems for this parameter..*

Mini-sprinklers or micro-sprayers are another type of localized irrigation, which is similar in principle to drip systems where water is applied only to a part of the ground surface ensuring greatest uniformity in effluent distribution. Compared to drip emitters, micro-sprayers can water a larger area. In addition, clogging problems are reduced thanks to the larger nozzle holes.

Subsurface irrigation is another, relatively new, type of localized irrigation where water is applied directly to the root zone via perforated or porous diffusers, placed 10-50 cm below soil surface. This method provides the greatest health safety and is feasible in plantations of fruit trees and other perennial row crops. The major constraint for their implementation is the high capital and operation costs and the high frequency of renewal.

10.4.3 Management of water application

Leaching and drainage are the two most important water management practices to avoid salinization of soils.

Depending on the salinity status, leaching can be carried out at each irrigation, each alternative irrigation or less frequently. With good quality recycled water, the irrigation application level will almost always apply sufficient extra water to accomplish leaching. With high salinity irrigation water, meeting the leaching requirement is difficult and requires large amounts of water.

Soil leaching is needed for almost all crops when electrical conductivities of saturation extracts exceed 10 dS/m and for moderately tolerant crops for values over 3 dS/m.

Certainly, leaching is essential to prevent salinization of the root-zone, but because there can be excess nitrate in this zone, leaching will result in the movement of nitrate and salts to the groundwater. For these reasons, monitoring of groundwater beneath effluent irrigation is an essential indicator of environmental performance.

Salinity problems in many irrigation projects in arid and semi-arid areas are associated with the presence of a shallow water table. In this context, the role of drainage is to lower the water table to a desirable level, at which it does not contribute to the transport of salts to the root zone and to the soil surface by capillarity phenomena.

10.5 SOCIO-ECONOMIC ASPECTS OF WATER REUSE IN AGRICULTURE

The development of sustainable water recycling schemes needs to include an understanding of the social and cultural aspects of water reuse. The drivers, which promote involvement in recycling, may vary between households and cultures, and will certainly be different for the different reuse applications such as irrigation of fodder and industrial crops, orchards or landscape areas. As a rule, the use of recycled waters for irrigation is widely accepted by farmers who believe them to be safer than river waters (Lazarova *et al.*, 2001). Several studies of public perception of water reuse demonstrated that non-potable reuse practices are widely approved by stakeholders and the public (Jeffrey, 2005). Nevertheless, there are strong consumer concerns over the sale of products which have been irrigated with treated wastewater, especially vegetables eaten raw. Farmers may be able to overcome such resistance through positive evidence from consumers and retailers that there will be a market for the products cultivated with recycled water.

In addition, the establishment of standards for reuse and effective management of monitoring programs promote confidence in water recycling schemes. In most countries, setting up or adopting a regulatory framework is an essential step for the development and social acceptance of water reuse. Decision-makers and politicians need clear, sound, reliable standards to endorse reuse projects. Regulations based on internationally acknowledged guidelines are generally preferred. Although an international effort to reduce discrepancies between current standards is highly desirable, regulations must be adapted to suit each individual country's context, health risk and affordability. Regulations have a major influence on the choice of treatment technologies and hence, on the cost of water reuse projects.

Consequently, to gain public acceptance of water reuse project it is necessary to enlarge traditional design and management activities of irrigation to include:

- assessment of cross-cultural barriers and facilitators of water recycling schemes;
- keeping the public informed, by the organisation of forums with local agencies, municipalities, water utilities and legislative officers from the earliest stage of development of water reuse programs. Description of the proposed technologies, their performances, risks associated, costs and benefits to catchment scale water conservation and sustainable development;
- development of public education programs (newsletters, school education programs, open houses and tours, meetings with stakeholders). The public's knowledge and understanding of the safety and appropriate applicability of recycled water is a key component of any successful water reclamation program;
- establishment of new marketing approaches to consider recycled water as a new product for sale.

One of the most critical elements to the success of any water reclamation project is its affordability and financial viability. In general, lack of funding is the major constraint for the implementation of water reuse practices. For this reason most water reuse projects have been developed on the basis of subsidies and grants. The worst situation is reported in developing countries (e.g. India, Pakistan). In all cases, public/private partnerships appear to be an efficient solution to establish appropriate financial plans, cut public sector deficits, promote investments, guarantee reliability of operation and ensure better water resources management with efficient water reuse.

As a rule, the low drinking water rates, which in most cases are subsidised (Figure 10.5), and consequently, the low sale price of recycled water, are the major economic constraints for water reuse development. Moreover, the economic value of recycled water for agricultural irrigation is much lower than in urban and industrial applications, and very often free of charge. Irrigation rates vary from 0.02 to 0.07 €/m^3 in France, Spain, Jordan, Morocco and Tunisia to 0.16-0.23 €/m^3 in Israel and California, with only partial recovery of treatment and distribution costs. This price difference is due to the economic status of the specific countries, different water reuse regulations and related treatment schemes.

In Europe, the EU Water Framework Directive requires EU Member States to ensure, by 2010, that water-pricing policies provide adequate incentives to use water resources efficiently and to recover the true costs of water services in an equitable manner. Most countries are progressing towards water pricing systems. Investing in water supply and sanitation has produced benefits far greater than those directly related to the cost of treating water-related diseases (UNEP, 2005). While there has been a general trend towards higher water prices throughout Europe, they still vary considerably from 0.1 to 2.3 €/m^3 and more. Many of the capitals and major cities in Mediterranean countries have below average prices, as do cities in countries with abundant water supplies. In contrast, water prices are the highest in northern and western European cities.

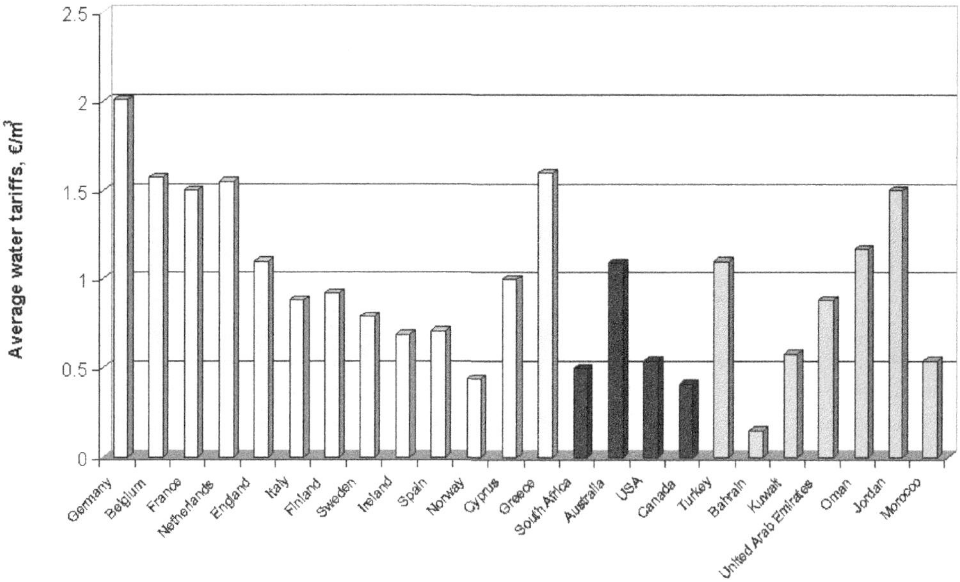

Figure 10.5 Comparison of average water tariffs in some European, Middle Eastern and other countries for the period 1990-2000 (WHO, 2000).

In addition to subsidised water tariffs, another constraint is that sewerage tariffs are lower than water rates even if construction and operation costs of sewers and wastewater treatment plants are very high. The analysis of available information leads to the conclusion that water and sewerage tariffs do not cover the full cost of the provided services (WHO, 2000).

As a rule, charges to users for recycled water for irrigation should not exceed the least-cost alternative, unless water scarcity or rationing exists. A wide variation in recycled water unit pricing is reported depending on the type of reuse, flow rates and local conditions, ranging from 0 to 0.52 US$/m^3 depending on recycled water quality and specific conditions (Morris *et al.*, 2005). Almost 50% of 34 reuse projects recently assessed by WERF (Mantovani *et al.*, 2001) ranged from 0.15 to 0.52 US$/m^3. Among existing water reuse projects, the price of recycled water appear consistently lower than those of potable water with almost all international reuse projects ranging from zero to 25% of potable water rates and only 5 US projects achieving 75 to 100% of potable water rates. Consequently, recycled water revenue appears to recover operating costs, but in most cases tends to rely on some degree of subsidy to recover full costs.

It is important to stress that besides the widely accepted benefits of irrigation with recycled water such as public health protection by eradication of uncontrolled irrigation, water conservation for other uses and environmental protection, water reuse in agriculture is characterised by great economic benefits. The added value of irrigation for food and horticulture production is very high, from 0.12 €/m^3 for fodder crops and up to 1.1 €/m^3 for fruits and vegetables, as reported in Australia (Ziehrl *et al.*, 2004). It is expected that rising irrigation prices are unlikely to lead to major falls in water use as water charges represent only a small proportion of the horticultural costs, between 2 and 7%.

The increase in crop production is another important economic benefit of irrigation with recycled water. In fact, treated wastewater is very effective as a fertilizer for agriculture and can lead to significant increase in crop yield that can be up to 500% higher for some crops (wheat, corn, tomatoes) compared to irrigation with fresh water (Soudi *et al.*, 2000). Table 10.10 gives an example of net economic benefits of agricultural reuse in Morocco. The first benefit is the result of a lower sale price of recycled water, which is 0.05 €/m^3 compared to 0.07 €/m^3 charged for groundwater. In addition to this cost saving, the increased crop yield (from +200 to +500%) allows an increase in farmers' revenues (income) resulting in a net economic benefit in the range of 200-500 €/ha.

Table 10.10 Economic benefits of irrigation with wastewater in Morocco (Adapted from Soudi *et al.* 2000).

Crops	Net benefit from recycled water, €/ha(1)	Net benefit from fertilizing benefits, €/ha(2)	Total benefits, €/ha
Wheat	75	149	224
Grain Maize	159	361	514
Fodder Maize	157	357	514
Zucchini	68	155	222
Marrow	61	122	183
Tomato	155	354	510
Potato	94	214	308

(1)calculated on the basis of the use of recycled water at 0.05 €/m^3 compared to 0.07 €/m^3 of charge for pumping groundwater (1€ = 10 Dhs) and irrigation rate of 100 mm (1000 m^3/ha).

(2)assumption of a total valorisation of fertilizer elements in treated wastewater (40 kgN/ha, 11 kgP/ha, 28 kgK/ha).

10.6 CONCLUSIONS

The review of the existing practices of water reuse in agriculture confirms the needs and general trends for further development and extension of irrigation practices not only in arid and semi-arid regions, but also in temperate climate as an efficient measure for environmental protection and a drought-proof alternative water resource. The successful implementation of water reuse for irrigation requires the application of a holistic planning approach, considering technical, sanitary, agronomic, environmental, institutional and socio-economic aspects.

The safe and beneficial implementation of water reuse schemes could be better guaranteed by the development of appropriate codes of good practice, as described above, which are as important for farmers and operators as quality requirements for water reuse.

10.7 REFERENCES

Armon, R., Dosoretz, C.G., Brodsky, M. and Oron, G. (2002) Surface and subsurface irrigation with effluents with different qualities and presence of Cryptosporidium oocysts in soil and crops, *Wat. Sci. Techn.*, **46**(3), 115-122.

Asano, T. (2002) Water from Wastewater: The Dependable Water Resource, *Wat.Sci.Techn.*, **45**(8), 23.

Ayers, R.S. and Westcot, D.W. (1985) Water quality of Agriculture, FAO irrigation and Drainage paper 29, pp. 99-104.

Blumenthal, U.J., Peasey, A., Ruiz-Palacios, G., and Mara, D.D. (2000) Guidelines for wastewater reuse in agriculture and aquaculture: recommended revisions based on new research evidence, *WELL Study, Task n°68, Part 1. pp.67*, http:/www.lboro.ac.uk/

FAO, Food and Agriculture Organization (1985) Water quality for irrigation, *Irrigation and Drainage Paper n°29*, by Ayers, R. S. and Wescot, D. W., Rome, Italy.

FAO, Food and Agriculture Organization (1992) Irrigation and Drainage Paper 47, *Wastewater Treatment and Use in Agriculture*, by Pescod, M., Food and Agriculture Organization.

FAO, Food and Agriculture Organization (1994) *Guidelines for prediction crop water requirements*, Irrigation and Drainage paper n°24, by Doorenbos J. and Pruitt, W.O., Rome.

FAO, Food and Agriculture Organization (1997) *Quality Control of Wastewater for Irrigated Crop Production*, Water reports 10, by Westcot, D., Rome

FAO, Food and Agriculture Organization (1999) Special report: drought causes extensive crop damage in the Near East raisong concerns for food supply difficulties in some parts, http://www.fao.org/WAICENT/faoinfo/economic/giews/english/alertes/1999/SRNEA997.htm

FAO, Food and Agriculture Organization (2000) *Users manual for irrigation with treated wastewater*, FAO Regional Office for the Near East, Cairo, Egypt, 69 p.

FAO, Food and Agriculture Organization (2003) Managing Water Scarcity for Water Security, Paper prepared by Winpenny, J.T., http://www.fao.org/ag/agl/aglw/webpub/scarcity.htm

FAO, Food and Agriculture Organization (2005) Water Quality and Environment Programme, http://www.fao.org/landandwater/aglw/waterquality/a6

Jiménez, B., Chávez., A., Maya, C., and Jardines, L. (2001) Removal of microorganisms in different stages of wastewater treatment for México City, *Wat. Sci. Techn.*, 43, 10,155.

Lazarova, V. and Asano, T. (2005) Challenges of sustainable irrigation with recycled water. In *Water Reuse for Irrigation: Agriculture, Landscapes, and Turf Grass*, ed. by Lazariva V. and Bahri A., Catalog no. 1649, ISBN: I-56670-649-I, CRC PRESS, 1.

Lazarova, V. and Bahri, A. Ed. (2005) *Water Reuse for Irrigation: Agriculture, Landscapes, and Turf Grass*, Catalog no. 1649, ISBN: I-56670-649-I, CRC PRESS, 456 p.

Lazarova, V., Cirelli, G., Jeffrey, P., Salgot, M., Icekson, N. and Brissaud, F. (2000) Enhancement of integrated water management and water reuse in Europe and the Middle East, *Wat. Sci. Techn.*, **42**(1/2), 193.

Maas, E.V. (1984) *Salt tolerance of plants. The Handbook of Plant Science in Agriculture*, ed. Christie, B.R., CRC Press, Bora Raton Florida.

Merrette, S. (2002) Water for Agriculture: Irrigation economics in international perspective, ed. Spon Press, London, chapter 1.

Metcalf & Eddy (2003) *Wastewater Engineering: Treatment and Reuse*, 4rd Edition, McGraw-Hill Inc.

Pettygrove, and Asano, T. (1985) *Irrigation with reclaimed municipal wastewater – A guidance manual*, Lewis Publishers Inc., Chelsea.

Soudi, B., Kerby M. and Choukrallah R. (2000) Réutilisation des eaux usées en Agriculture. *Transfert de Technologie en Agriculture*, N°67, http://www.iav.ac.ma/pntta/bul67.htm

United Nations (2003) The UN World Water Development Report, Water for People, Water for Life, http://www.unesco.org/water/wwap/wwdr/table_contents.shtml

USEPA, United States Environmental Protection Agency (1992) *Guidelines for Water Reuse*, Manual, EPA & USAID (United States Agency for International Development), 247p.

USEPA, United States Environmental Protection Agency (1998). Indoor Air Facts N°4 (revised). 7 April 1998, revision dare. Sick Building Syndrome (SBS). http://www.epa.gov/iaq/pubs/sbs.htm

World Bank (1983) Sanitation and disease-health aspects of excreta and wastewater management, *World Bank Studies in Water Supply and Sanitation 3*, by Feachem, R. G., Bradley, D. J., Garelick, H. and Mara, D. D., Published by John Wiley & Sons, Chichester, UK.

WHO, World Health Organization, *Health Guidelines for the Use of Wastewater in Agriculture and Aquaculture,* Report of a WHO Scientific Group, Technical Report Series 778, World Health Organization, Geneva, Switzerland, 1989.

Ziehrl, A., Boland, A.M. and Hamilton A.J. (2004) Australian horticulture and water reuse options: summary information to assess, http://www.lwa.gov.au/downloads/final_reports/VPI4_-ResPotential_0312.pdf

11

Wastewater irrigation in urban agriculture

Mark Redwood and Frans Huibers

11.1 INTRODUCTION

Urban agriculture (UA), an activity often considered marginal in an urban economy, has proven to be persistent and common in most cities in developing countries. The growing population in urban areas creates a huge demand for income generation and agricultural produce. Instead of importing food from rural hinterlands, urban agriculture is close to urban markets and thus marketing costs are lower. Apart from providing extra income if it is sold in the market, UA improves household nutrition through food self-reliance and reduces the amount of money spent on food. Aesthetically, UA provides greenery to the city. Research on UA has only really developed in the past decade and some trends are emerging. First, UA is a massive informal enterprise that involves a large number of actors. As city governments recognize its usefulness, it is becoming an increasingly formalized activity. Second, there are many economic benefits that accrue from UA. Finally, UA plays a significant role in making vacant lands productive.

As the recognition of UA has grown, so too has attention on the use of urban wastewater. In many developing countries, particularly those where there is a seasonal water shortage or scarcity, access to fresh water for irrigation is limited or unaffordable and instead both treated and untreated wastewater is used. Wastewater, notwithstanding the risks for negative health implications when used improperly, has huge potential as a source of water. Even when untreated, there is a growing body of evidence that proves that farmers will use wastewater to increase their crop yields and lower their water costs (Faruqui and Al-Jayoussi, 2002; ENDA-IFAN, 2002a and 2002b; Cornish and Lawrence, 2001; Scott *et al.*, 2004). This use of urban wastewater would be straightforward if it were not for the health implications implicit in reusing wastewater.

Where adequate conventional treatment exists, it is common and increasing practice in water scarce regions to use wastewater (California, Tunisia) and there exist guidelines and policies for its management. Where access to conventional treatment is limited, farmers will often use untreated wastewater, or polluted surface water (both domestic and industrial), at least part of the year. Treatment is considered the ideal solution to this problem, but recent experience is showing that conventional treatment, while an important end-goal, is likely far beyond most countries' economic capacity. The Dublin Principles (1992) regarding water promote local, demand-oriented water management as the ideal model for water management. In poorer countries, where a lack of resources impedes infrastructure development, there is an increasing emphasis on low cost treatment systems that allow for water reuse.

11.2 URBAN AGRICULTURE

Rapid population growth, urbanization and the pressure on scant resources in cities and international trade liberalization have all contributed to an increase in food insecurity amongst the urban poor. As Akinbamijo *et al.* (2002) note, food insecurity is the result of either high levels of poverty or the high cost of food. As Table 11.1 shows, food is often the single largest expenditure of poor families – any savings in its purchase can release a significant portion of income for other uses (Egziabher *et al.*, 1994). Meanwhile, urbanization has concentrated food markets in cities. In Havana, a survey of families producing mostly for home consumption reported a 40% drop in household food expenditures (Moskow, 1999). While comprehensive studies of the overall economic contribution that UA makes to the economy are rare, one performed by the Mazingira Institute in Kenya found that in one growing season (1985), urban agriculture contributed 4 million US$ to the Kenyan economy (Mougeot, 2000). An additional advantage is the proximity of food production to its market and thus, lower transportation costs.

The spatial reorganization of people to urban areas (Figure 11.1) has also concentrated food demand in cities. It is common for migrants to improve their situation through the production of food for auto-consumption and sometimes, for the market (Koc *et al.*, 1999). Urban agriculture is not a new phenomenon. What is new, however, is the growing recognition of its practicality and usefulness in easing the burden of poverty and food insecurity. As an adaptive strategy to fight poverty, UA is incredibly malleable to changing economic situations. Its effectiveness is not limited to poverty reduction. UA also increases urban green space, creates economic spin-off industries and employment, as well as improving the urban biophysical environment (Moskow, 1999). This model of agricultural development is well suited to many cities in both the north and the south.

Table 11.1: Percentage of income spent on food by low-income residents in selected cities (source: Akinbamijo, Fall and Smith, 2002).

City	Income spent on food (%)
Bangkok (Thailand)	60
La Florida (Chile)	50
Nairobi (Kenya)	40-60
Dar es Salaam (Tanzania)	85
Kinshasa (Congo)	60
Bamako (Mali)	32-64
Urban USA	9-15

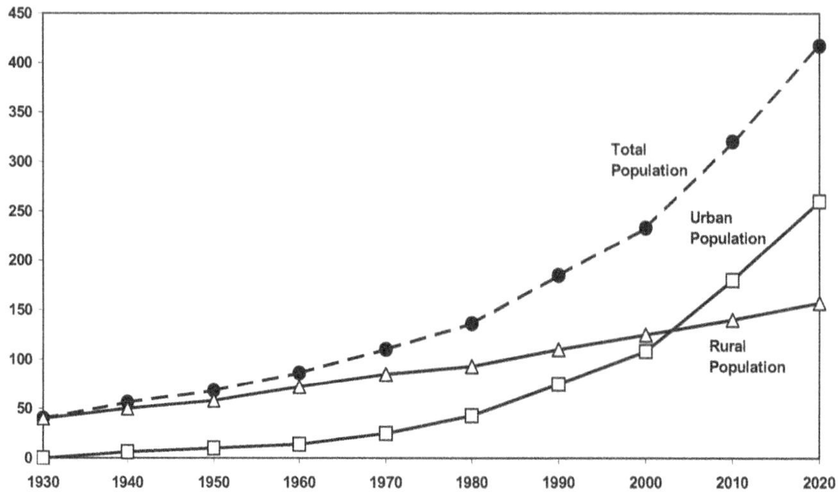

Figure 11.1: Population growth and projection in West Africa in million people (source: United Nations, 2005 Statistics).

Urban agriculture has not been defined in absolute terms. Some define UA as inclusive of agriculture, fish and livestock production within city limits, as well as activities that take place in the peri-urban area. Some urban agriculture occurs in home gardens, a type that is most associated with subsistence living and supplementing incomes. Other types of UA include farming on open spaces such as roadways, public and private lands. Livestock production, common in parts of Africa and in Asia, also occurs in urban areas and comes with associated health risks (Flynn-Dapaah, 2002).

UA also includes spin-off activities such as compost production, small business development, packagers and transportation, for example. A working definition of UA links confined space production, related economic activity, location, destination markets (or home consumption), and the types of products produced in a dynamic interaction that can vary from one urban area to another (Mougeot, 2000). The difficulty in settling on a precise definition of the phenomenon has impacted the extent of knowledge on the subject. Research, and slowly policy, is now acknowledging that peri-urban and urban agricultural systems operate very differently than rural systems. Some examples of this difference are access to land, access to markets, soil quality, urban development pressure, scale, crop diversity and types of production systems.

As the global scope of research into urban agriculture becomes mainstreamed, more and more data on both the positive impacts and what limits UA are being produced. The data that does exist on UA is compelling. The UN estimates that 15% to 20% of the global food output is grown in cities (Smit, 1996). By 2005, estimates suggest that UA will increase its share to 25% to 30% of the global food output. This is based on assumptions that urbanization will increase, technology transfer will continue, there will be an increasing acceptance and reuse of wastes for agriculture, markets will continue to grow, and urban managers will institute supportive policy options. In Africa and Latin America, 40% and 50% (respectively) of urban people are involved in farming in some capacity (Egziabher *et al.*, 1994). One estimate suggests that as many as 800 million people are involved in UA worldwide (Smit, 1996). In Dakar, urban producers meet 60% of urban vegetable demand (de Zeeuw, 1999). Meanwhile, an extensive survey done in Lusaka demonstrated that 50% of respondents use gardens as their primary source of household food (Ogle, 1999).

Considering the contribution of UA to household income generation, the role of women is important. Women often play a central role in both food production and marketing. Still, women are often faced with difficult challenges in accessing services (capital, agricultural extension, technologies, labour), with legal and customary land tenure issues and also, with regard to asset ownership. While inequity is common, urban farming has the potential to afford women more control over household economic decision-making and an independent source of income (Hovorka, 1998).

Other benefits are difficult to quantify and subjective but are nevertheless worth mentioning. Moskow (1999) points out fringe benefits of UA such as neighbourhood beautification, improved safety and an enhanced urban ecology. Economic spin-offs from related industries have yet to be researched thoroughly.

11.3 Constraints on urban agriculture

While the benefits of UA are increasingly evident, the practice still faces constraints that prevent its formal integration into the economy. Urban policy has often been hostile to the practice (Egziabher *et al.*, 1994). Most urban planners view agricultural land uses outside homesteads as inconsistent with their notion of a city. Some municipalities regard this form of UA as a nuisance activity and have rendered it illegal through zoning and health regulations. This contrasts sharply with the view of UA as an activity with positive economic, health and environmental benefits. In open space UA, wastewater use is commonly employed for irrigation and has nutrient benefits for crop production; however, it is often banned outright due to the fear of an outbreak of disease. For example, in Senegal and Ghana, despite officially being banned, treated and untreated wastewater is frequently used by farmers.

Perhaps the greatest impediment to successful integration of agriculture in urban areas is land tenure. For many, urban agriculture is a coping strategy based on their knowledge of farming from rural areas. In Dakar, 60% of those employed in urban agriculture are rural migrants (ENDA-IFAN, 2002b). Because many urban poor have recently migrated they are often without legal title to the land they occupy. Such farmers, in some cases, farm the same land for more than 10 years without significant problems (Amoah *et al.*, 2005). Still, secure land title can be a precondition of housing and infrastructure development (Gilbert and Gugler, 1982). Those with tenuous access to land will be unwilling to support much investment into their households – investments that may be the precursor to on-site sanitation and water reuse systems (Choguill, 1996). Where there is more secure land tenure – and thus a greater sense of ownership – there is interest in improving the household. This is especially true among the urban poor because their home and small plot of land is often the only property they own (Hameed Khan, 1996). Even *perceived* security of tenure and the existence of infrastructure have shown to be important in encouraging incremental household and land improvements.

Beyond land tenure there are also often conflicts in land-use planning that work to disadvantage urban farmers. For example, in Dakar, the 1964 National Land Act, the eminent existing legislation regarding land, grants tenure to "those who cultivate it". However, the same law grants discretionary powers to local community leaders. These community leaders are often not well advised on the advantages of UA. Combined with pressing urbanization and housing needs, valuable land currently farmed is instead being developed (Mbaye and Moustier, 2000; Faruqui *et al.*, 2004). For example, the municipality of Dakar had included a large green area – known as the Niayes – into its development plans for housing and a technological business park. The Niayes happens to be one of the most desirable and productive agricultural areas in Senegal. Only the intervention of the President's Office, and the activism of a hastily developed farmer lobby granted secure rights to the land.

Another result of weak UA policy is the lack of support services existing for its development as a viable economic sector. Most UA occurs outside of the formal economy. Credit programs, one of the most important methods to formalize the practice, are few (Egziabher *et al.*, 1994). Weak policy has also prevented efforts to curb pesticide and raw wastewater use, practices that are widespread and can have serious health impacts when not done properly. Finally, agricultural policy is often directed at medium and large-scale farming and is developed at a macro, or national level. Thus, small landowners, especially urban producers that are not privy to "extension" services, inevitably fall through the cracks (Cornish and Lawrence, 2001).

11.4 URBAN AGRICULTURE AND WASTEWATER USE

In urban areas, where domestic wastewater is readily available, urban farmers will use wastewater for irrigation. Its value is two fold. First, it is a readily available and abundant source of water and second, it contains nutrients that are advantageous for crops. In a city of 500,000 people, assuming an 80% recovery at 80 litres of water use/person/day, a total discharge of 32,000 m^3/day would be generated. With a 5mm application per day, this could theoretically serve to irrigate 640 ha. In Dakar, a city where there is a water shortage, over 100,000 m^3 of wastewater is dumped into the ocean daily amounting to 40% of the city's entire daily water use (Niang, 1999). If there is an adequate strategy to reduce health risks, wastewater use for irrigation of UA is a realistic policy option (Pescod, 1992).

The reasons for wastewater use are numerous and frequently dependant on context; however, there are several recurring themes:

- wastewater is used to reduce the cost of expenditure on piped water;
- year-round availability;
- wastewater provides many nutrients that are not present in water;
- it can be an environmentally safe way to dispose of wastewater.

The nutrients available in wastewater include nitrogen, phosphorus and potassium. While some farmers are familiar with the nutrient benefits of wastewater, evidence from West Africa suggests that most value its reliability as a constant, year-round source of water (Huibers and van Lier, 2005).

In a survey of urban farmers in Dakar in 2002, researchers found that 23% of farming costs were attributed to acquiring a fresh water supply (ENDA-IFAN, 2002b). As a result, 33% of farmers used wastewater primarily for economic reasons in order to avoid paying for potable water. A further 37% cited the nutrient richness as the main reason they used wastewater. Wastewater use was also found to be an activity that is associated with household agricultural production. Approximately 99% of those using wastewater identified urban agriculture as their primary economic activity (ENDA-IFAN, 2002b). In Amman, Jordan, an extensive survey of urban farmers has found that 40% use greywater for their gardens (DOS, 2001).

Socio-cultural elements also impact wastewater use considerably. The WHO guidelines (Mara and Cairncross, 1989) recognize that while the use of human excreta as fertilizer is acceptable in some places (China, Japan), it is looked upon with disaffection or indifference in others (the Americas, Africa), while it has a history of being viewed with hostility in some Muslim countries. Faruqui *et al.*, (2001) challenge the notion that wastewater use is contrary to Islamic principles by arguing that it is actually a natural extension of water conservation, which is an Islamic principle, as long as it is treated to the extent necessary to protect public health. This is especially true in regions such as Saudi Arabia and Tunisia, both very arid, where national wastewater use plans have been developed with the backing of Muslim clerics. A 1978 *fatwa* of the Saudi Arabian clergy, made in concert with government policy makers, decreed that wastewater use was acceptable as long as the water was treated well to protect public health in its intended use. Now, the Kingdom has determined to have 10% of its water demand met by water reuse. Much of the public accept wastewater use, especially when the need for alternative water sources is clear. In Palestine, 80% of people surveyed in a study of wastewater use and urban agriculture were found to be willing to practice wastewater use (Faruqui *et al.*, 2001)

11.5 BUILDING MARKETS FOR WASTEWATER: CONVINCING POLICY MAKERS

As UA becomes more accepted as a legitimate economic sector in urban areas, it is logical that new policies recognizing the usefulness of treated wastewater as an irrigation supplement and attempts at mitigating risk be developed. The economic value of wastewater use is based upon several factors related to water management. First, urban water consumption needs have been met primarily through costly investments into supply infrastructure and water intensive collection, as opposed to demand management or conservation. As a result, water sources have been overexploited, especially in water scarce regions, raising the marginal cost of supply and provision and, eventually, increasing the cost burden on society. These increases have not been reflected properly in pricing and cost recovery policies related to water. To continue this supply-oriented path would exacerbate the cost to already-struggling southern economies. The UN (UNICEF, 2005) estimate that to meet the Millennium Development Goal (MDG) on water and sanitation, Southern countries would need to invest 11.3 billion US$ between now and 2015.

The basic economic premise in support of wastewater use is that using wastewater increases the amount of the available resource and protects valuable potable water supplies. As a demand management strategy, wastewater use conserves existing resources, and reduces the need to invest in finding and exploiting new fresh water supplies. A secondary economic benefit from properly treated wastewater use is its effectiveness as a fertilizer for agriculture. This would reduce fertilizer costs while increasing agricultural output and yield. The economic benefits of wastewater use are particularly significant in urban areas where the cost of water can be up to 10 times higher than in rural areas (Gibbon, 1986).

There is evidence that wastewater is valued by farmers. In Dakar, IDRC-supported research has found that 0.25 US$ per m^3 could be charged for wastewater (ENDA-IFAN, 2002a). Tunisia and Jordan both sell treated wastewater to farmers at heavily subsidized prices. Other evidence emphasises costs saved from wastewater reuse. In Cyprus, a study found that there was a 36% reduction in water bills when household greywater was separated and reused using a simple collection and treatment system in some residences (WHO, 1999). IDRC-supported work on greywater use in Jordan found that the value of greywater used amounted to 27% of the average water bill (Faruqui and Al-Jayoussi, 2002). In the same case study, the average benefit—cost ratio of using greywater was measured to be five. This figure was based on benefits from reduced water bills and increased urban agriculture yields and costs such as the installation of the systems.

Perhaps wastewater's greatest economic contribution is through the improved quantity and value of agriculture. The high nutrient content in domestic wastewater favours the growth of high-value crops such as vegetables. A comparative study of five crops (wheat, mung beans, rice, potatoes and cotton) irrigated with raw wastewater found that when raw wastewater was applied, there was a 36% increase in yield compared to a control group using fresh water (Shende, 1985). Similar increases in yields were found for those crops irrigated with treated wastewater. In Dakar, farmers reported a 37% increase in overall yields of lettuce, tomatoes, eggplant and onion from the use of raw wastewater.

In Mendoza, Argentina, the provincial government has supported widespread irrigation of certain crops using wastewater. An analysis of crop yields has shown a steady increase that has partially resulted from the fertilizer in the wastewater used. In Jordan, household wastewater treatment systems tested by the IDRC have allowed greywater to be used safely on urban agriculture. Among the households reusing their water, an average annual benefit of 376 US$ was reported. This increase is accounted for by increased produce yields, as well as reduced water and fertilizer costs (Faruqui *et al.*, 2001).

Identifying the benefits of wastewater is one thing, but as of yet, the development of markets for wastewater is still young. Are farmers willing to pay for treated wastewater as opposed to using free, raw wastewater – and thus be exposed to health risks? Based on the benefits of high quality wastewater, some research has proposed that farmers would be willing to pay for the resource (Pescod, 1992). In a literature review of existing information on wastewater pricing, Mastenbroek (2001) notes that many studies have not explored willingness to pay for wastewater in enough depth. Instead conclusions are based on their perception of benefits as opposed to sound evidence. Nevertheless, arguments incorporating payment for wastewater have received abundant attention and water management is moving towards incorporating wastewater pricing.

Wastewater or polluted surface water is used for different reasons but often simply because there is no other adequate source of water. Some farmers understand the nutritional value of wastewater use, and use it because it is a cheap, reliable source of water. Most importantly, the choice to use wastewater is frequently related to a tangible increase in income. In Palestine, a survey of farmers found that 53% were willing to pay up to 0.24 US$ per m^3 for wastewater if it were suitable for irrigation (Faruqui *et al.*, 2001). Three-quarters responded that they would pay for the fertilizer sludge produced by the treatment process. In Jordan, where wastewater is considered a legitimate resource, current policy suggests a charge of 0.15 US$/$m^3$ for treated wastewater. The value of wastewater amongst urban farmers is clear from other evidence as well. For instance, in Pakistan, van der Hoek (2002) found that due to an uncertain water supply from canals and the ample available supply of wastewater, urban farmers using wastewater grew five times more vegetables than those using canal water. This higher cropping intensity is reflected in land prices where those plots irrigated with wastewater are, on average, 3.5 times more valuable than areas that are not irrigated with wastewater.

In Tunisia, national policy allows for controlled wastewater use on certain crops. Their national water plan dictates quality standards (they are moving more and more towards WHO standards) as well as directing how much wastewater is used by different economic sectors. Bahri and Brissaud (1996) have noted that the success of their policy is largely influenced by the links between national water planning and agricultural strategies. They also found that price and payment for the resource did not seem to play a large dissuasive role in wastewater use.

Willingness of farmers to pay for treated wastewater and household and collective systems to evacuate and treat wastewater are the basis for improving the cost effectiveness of water and sanitation projects. This suggests that options for creative financing of self-help infrastructure would work (revolving funds, micro-credit etc.). While many diverse methods exist to encourage sustainable treatment and reuse projects, a basic blueprint model is a useful starting point. Instead of prohibiting its use – a strategy that has failed – the challenge for researchers and policy makers is to encourage ways in which wastewater may be used safely. Current research shows that the economic potential of the resource is significant and make the development of infrastructure for its use not only a feasible option, but one that should be encouraged. The question is now how can the benefits of wastewater use be *safely* harnessed for poverty reduction.

11.6 SHIFTING SCALE: ADDRESSING THE NEEDS OF SMALL FARMERS

The emphasis on large-scale projects has distracted from innovation in lower-cost, smaller-scale community and household wastewater treatment systems. Such systems can function well in lower density (*i.e.* peri-urban) areas where reuse is common. Where large-scale treatment is not possible, policy should support the development of decentralized wastewater treatment. Proponents argue that decentralized water management, where local capacity exists, will respond more to local priorities (Bakir, 2001). For wastewater treatment and reuse, Van Lier and Huibers (2004) outline the factors that should influence the selection of an appropriate system: First, **cost-effectiveness** is a priority as it is clear that large-scale infrastructure do not meet the fiscal realities of southern countries. Systems must also be **flexible, compact and robust,** particularly when being adapted for populated areas where land is at a premium. Decentralised technical solutions for wastewater treatment and reuse must be **socially acceptable** in order for them to be sustainable. Finally, a **management plan** is critical to support the implementation of individual systems.

The private sector, from small-scale entrepreneurs to larger firms, already plays an important role in both water provision and sanitation. For instance, in the absence of public services, farmers and residents find ingenious ways to acquire water (through illicit pipes, illegal wells, hoses that bring water from serviced areas etc.). Entrepreneurs will sell truck borne water (at elevated prices) in neighbourhoods where service does not exist. Meanwhile, in sanitation, households and individuals invest in basic pit latrines, wastewater disposal and even, where it can be afforded, septic tanks. Small companies will often provide these services and the materials to build them.

Ownership is a principle concept when developing small-scale infrastructure. When people develop the infrastructure themselves and play a direct role in its operation and management, the likelihood of sustainability of these systems is higher. On the other hand, some detractors of decentralization also note that decentralized systems are difficult first to develop and second, to operate and maintain (Bakir, 2001). Moreover, they argue that some local operators simply would not have the technical capacity to maintain infrastructure even when it is basic. Still, if one considers the failed history of conventional sanitation, the opportunities that exist to harness wastewater, and the value that farmers place on wastewater, there is enough incentive to encourage further research and investment in the development of smaller-scale water and sanitation.

More research is needed to focus on assessing decentralized collection and treatment systems from both a health perspective and also from a cost perspective. A system that proves to be effective at treating wastewater, but can also be cost effective will have a significant impact on policy. The economic incentive to use wastewater should be emphasised. By acknowledging the income and financial benefits of wastewater use, a better understanding of farmers' livelihoods could lead to better policy. For this reason, the importance of community participation in policy development is underscored.

11.7 TOWARDS A MORE RESPONSIVE POLICY

Efforts to boost the acceptance and manage the risks of wastewater use in urban agriculture have been employed at different levels of governance. At the national level, health, agriculture, water, urban development and environmental ministries are all connected to water planning. Moreover, as noted above, water planning and legislation is highly centralized in nature. This highlights the need for a national level intervention to legitimize and establish the safe development of wastewater use in urban agriculture. Wastewater reuse is one component of Water Demand Management – a concept that has proven to be the premise of good water policy, particularly in countries of scarcity (see www.idrc.ca/waterdemand for case studies). Meanwhile, UA is more of a municipal issue. Notwithstanding national government oversight, land-use planning is regarded as a municipal issue and control rests with municipal authorities. Thus, secure access to land, rights and also extension for agriculture tend to be concentrated at this local level.

11.7.1 National level policy considerations

The WHO guidelines on wastewater reuse (Mara and Cairncross, 1989) continue to be the benchmark target for decision makers in the wastewater sector; however, as demonstrated, goals need to be in line with the capabilities of the country in question (Carr, 2005). Considering the needs to first raise local capacity to an appropriate level, moving slowly towards achieving the WHO standards (when wastewater is treated) or minimize health risks (when treatment is deemed unlikely), governments should develop a progressive approach. Such an approach would allow countries to design targets in line with their ability to meet them. A further advantage of such an approach would be the gradual acceptance of wastewater use by the public, and a management of associated risks.

A second national level policy consideration should recognize that the prohibition of wastewater use is ineffective because it is very difficult to enforce. Governments need to lay out a strategy that legitimizes, or at least acknowledges the reality of wastewater use – even in the absence of effective treatment. Such recognition creates a strong basis from which to promote better practices and support initiatives to minimize risks. Such a directive needs to come from the national government as guidelines for all agencies (national and municipal) responsible for urban agriculture and water. Guidelines can then be interpreted and adjusted by local authorities responsible for wastewater UA to create locally appropriate systems.

Third, within guidelines, there needs to be flexibility. Generally, acceptable risk levels in wastewater policy are more stringent than is financially feasible to achieve. Moreover, as shown in developing countries, such standards ignore the reality of the many who use wastewater for its benefits while ignoring (or being unaware of) its risks. Health guidelines must be strong where capacity exists, but in cases where it is unfeasible, governments should not avoid supporting projects that *reduce* health risk. At the moment, the WHO is in the process of revising the WHO guidelines in this sense.

Decentralization of water management is an important element in achieving a better understanding of local needs and priorities with regard to water. Municipalities should be able to support the development of options for wastewater. It is this level of government that is often best positioned to organize local groups of producers, support safe reuse options as well as dialogue between the different stakeholders.

With such a variety of interests at stake, coherent policy-making that encourages the appropriate use of wastewater in agriculture, and that also responds to related health risks has been elusive. Not surprisingly, it is usually the health authorities that demonstrate the most reluctance to press for change or promote a more flexible policy on wastewater. This is still the case despite the enormous evidence that wastewater use is commonly employed and contributes tremendously to food supply and raising incomes. The result is a policy framework completely detached from the reality on the ground. Considering the varied impact of the use of wastewater, a responsible policy needs to come from coordination between different ministries.

11.7.2 Municipal level policy considerations (UA oriented)

With regard to UA, municipalities should emphasize public involvement, the development of agricultural markets and education. Generally, it is also here that land use planning and strategies related to UA can be developed. Urban planners, with their mandate to use land as a resource and tool for municipal development, are an essential part of the solution. Planning should be based on the principal that income constraints are probably easier to overcome than technological constraints. Therefore, linking community development with the installation of low-cost on-site and off-site water treatment technologies is a good approach. Planning needs to also enable livelihood activities (relative to health risk) instead of constrain them. Ensuring land tenure and protecting agricultural areas (with access to various sources of water often resulting in higher land values) through zoning are one possibility.

Innovative work on low-cost wastewater treatment technologies at the community level also needs to be supported. NGOs, research and academic institutes are at the forefront of research into low-cost treatment. Since investment in such systems is so far minimal, donor organizations (along with governments) need to play a facilitating role in supporting the development of such research. Individual groups can be helped in designing projects, developing networking and dissemination activities for their work. CBOs and NGOs must be involved to some degree as they represent the best means to involve water users: a key principle needed to ensure success of local water and sanitation work.

Where wastewater treatment is not possible, local authorities cannot expect that farmers will cease using wastewater. Instead, non-treatment management options are necessary (Scott *et al.*, 2004). Where this is not possible, alternative strategies at the farmer level, market level and consumer level that mitigate negative health risks are necessary.

11.8 CONCLUSIONS

This chapter has focused on introducing urban agriculture and the link between UA and wastewater use. When used as an input into the income-generating activities of UA, wastewater use has clear economic potential. Still, the policy environment remains fairly hostile to the concept out of concern for associated health risks. As the UA sector develops, the more it becomes clear that there are technologies and policy options that hold promise. In the right socio-economic and institutional setting, such options can work. Several things are clear: (a) in the absence of alternatives, farmers will use wastewater including raw wastewater; (b) if regulations are excessively strict and not enforced, they will be ignored for the most part; (c) when wastewater is used raw, any treatment and risk mitigation possibility, even those that do not meet international health guidelines, are better than no treatment; and finally (d) for poor urban areas, low-cost conveyance and treatment is the preferable solution to the urban infrastructure dilemma.

There is some movement towards revising the internationally-used WHO guidelines towards a more appropriate balance between water and sanitation capacity, health risk and allowing for an incremental, "step by step", improvement in treatment. The WHO assumes responsibility to assess current epidemiological evidence and revise guidelines based on this evidence. In response to criticism that the WHO guidelines are too strict and unattainable, the WHO has indicated that guidelines for wastewater use will become more flexible by using the *Stockholm Framework*. This framework examines health risks in the context of other potential risks (drinking water, poor hygiene practices, contaminated food) thus ensuring that international policy guidelines are not a catch-all and in fact can be adapted to local context. This allows for authorities to prioritize and develop a locally appropriate response to the most significant health problems (Carr, 2005). A phased approach to developing guidelines, updated every few years as technological and sanitation options are improved, may be recommended.

In order to respond to the growth of UA and the problem of access to safe water for UA production, there is a need for a coordinated research effort towards new technological and policy options. Decision-makers consistently cite the need for more information on the effectiveness and capability of low-cost methods to treat wastewaters to internationally recognised standards while retaining its reuse potential. At the moment, the WHO, IDRC, the International Water Management Institute (IWMI), and the Resource Centre on Urban Agriculture and Forestry (RUAF) represent some international institutions attempting to develop a unified global research and action agenda on the topic. The Hyderabad Declaration represents a benchmark containing key issues and research needs (www.iwmi.org 2003).

This chapter has highlighted some examples of where demand management and cost-recovery strategies are being employed, but more emphasis can be placed on wastewater use as an integral component of a sound water management plan. The evidence cited has also shown that there is a sound economic argument to do so. There are many models whereby costs associated with inexpensive treatment can be recuperated. Systems at the household level can be supported through simple building code adjustments and design guidelines. Training programs, through agricultural extension services, could be developed focusing on decentralized wastewater treatment and wastewater use. Moreover, such extension should include the development of micro-credit facilities to allow farmers access to credit for basic sanitation and irrigation equipment as well as business development services. Extension would also help in the development of markets and an entrepreneurial sector interested in promoting the use of wastewater for urban agriculture.

On a final note, three basic concepts that are necessarily the starting point in order to guide policy development in this area at both national and municipal levels; first, it is necessary for governments to accept that wastewater use for irrigation is a common reality and cannot be simply outlawed or ignored; second, where health standards are adopted, the minimum health guidelines should be appropriate for the local context and not based on standards that are difficult, or too strict, to achieve, otherwise, the gap between treatment options, water and sanitation capacity and the reality "on the ground" of UA will remain too large; finally, the need to shift policy towards developing appropriate technologies – technologies that are not capital intensive and large-scale – is key. Focusing an increasing amount of attention on decentralized treatment is not ignoring centralized and extensive wastewater treatment, but simply acknowledging that in the vast majority of developing country urban areas, large-scale systems are just not viable.

11.9 REFERENCES

Akinbamijo, O., Fall S. and Smith, O. (2002). *Advances in Crop-Livestock Integration in West Africa*, IDRC, Ottawa.

Amoah, P., Dreschel P., Abaidoo R.C., and W.J. Ntow. (2005). *Pesticide and pathogen contamination of vegetables in Ghana's urban markets in Environmental Contamination and Toxicology*. Springer-Verlag, New York.

Bahri, A. and Brissaud F. (1996) Wastewater Reuse in Tunisia: Assessing a National Policy. *Water Science and Technology*. **33** (10-11), 87-94.

Bakir H. (2001). Sustainable Wastewater Management Strategies in Middle East and North Africa. *The Journal of Environmental Management*, (**61**), 319-328.

Carr, R. (2005) WHO guidelines for safe wastewater use: More than just numbers. In Special Issue of Irrigation and Drainage. Managing Water for Sustainable Agriculture. *The Journal of the ICID*, **54** (Supplement 1): S103-S111.

Choguill, C. (1996). Ten Steps to Sustainable Infrastructure. *Habitat International* **20**(3): 389-404, Pergammon Press, Oxford.

Cornish, G.A, and Lawrence P. (2001). Informal Irrigation in Peri-urban Areas : A summary of findings and recommendations, Report OD 144, HR Wallingford, London.

de Zeeuw, H. (1999). Some More Urban Agriculture Case Studies: Dakar. *GATE: Technology and Development*, no.2, GTZ, Bonn.

Department of Statistics (2001). Jordan in Figures, Issues 4, Hashemite Kingdom of Jordan.

ENDA-IFAN. (2002a). Politiques d'assainissement au Sénégal et implications des populations dans la prise en charge de leur besoins de bases. Scientific report 1, ENDA-IFAN, Dakar.

ENDA-IFAN. (2002b). Valorisation des eaux usées domestiques dans l`agriculture urbaine à Dakar. Scientific report 2, ENDA-IFAN, Dakar.

Egziabher, A., Lee-Smith D., Maxwell D., Ali Memon P., Mougeot L., and Sawio, C. (1994) *Cities Feeding People*, IDRC Books, Ottawa.

Faruqui, N., Niang, S. and Redwood M. (2004). Untreated Wastewater Reuse in Market Gardens: A Case Study of Dakar, Senegal. In *Wastewater Use in Irrigated Agriculture: Confronting the Livelyhood and Environmental Realities* (eds. Scott, C., Faruqui, N. and Raschid L.), IWMI-CABI-IDRC, Nairobi.

Faruqui N., Biswas A. and Bino M.J. (2001) *Water Management in Islam*. United Nations Press, Tokyo.

Faruqui, N. and Al-Jayoussi O. (2002) Greywater Reuse in Urban Agriculture for Poverty Alleviation: A Case-Study in Jordan. *Water International*, **27**(3)

Flynn-Dapaah, K. (2002) Land Negotiations and Tenure Relationships: Accessing Land for Urban and Peri-Urban Agriculture in Sub-Saharan Africa, Cities Feeding People Report 36, IDRC, Ottawa.

Gibbon, D. (1986) Evaluating Market-Oriented Water Policies in Jordan: A Comparative Study. *Water International* **20**(2): 88 – 97.

Gilbert, A. and Gugler, J. (eds.) (1992) *Cities, Poverty and Development*, 2nd Edition, Oxford University Press, Oxford.

Hameed Khan, A. (1996) *Orangi Pilot Project*, Oxford University Press, Oxford.

Hovorka, A. (1998). Gender Resources for Urban Agriculture Research: Methodology. Cities Feeding People Report 26, IDRC, Ottawa.

Huibers, F. and van Lier, J. (2005) Use of wastewater in agriculture: The water chain approach. Wastewater Irrigation. In Special Issue of Irrigation and Drainage. Managing Water for Sustainable Agriculture. *The Journal of the ICID*, **54** (1): S3-S10

Koc, M., McRae, R., Welsh, J., and Mougeot L. (1999) *For Hunger Proof Cities*, IDRC, Ottawa.

Mara, D. and Cairncross, S. (1989) *Guidelines for the Safe Use of Wastewater and Excreta in Agriculture and Aquaculture*, United Nations Environmental Programme/World Health Organization, Geneva.

Mastenbroek, A. (2001) Treated Wastewater Pricing. Unpublished Master's thesis, Irrigation and Water Engineering Group, Wageningen University, Netherlands.

Mbaye, A. and Moustier P. (2000) Market-Oriented Urban Agricultural Production in Dakar. In *Growing Cities, Growing Food* (ed. Bakker, N.), DSE, Feldafing, Germany.

Moskow, A. (1999) The Contribution of Urban Agriculture to Gardeners, Their Households and Surrounding Communities: the Case of Havana, Cuba. In *For Hunger Proof Cities* (ed. M. Koc *et. al.*), IDRC, Ottawa.

Mougeot, L. (2000) Urban Agriculture: Definition, Presence, Potentials and Risks. In *Growing Cities, Growing Food: Urban Agriculture on the Policy Agenda* (Bakker et al., 2000). Food and Agriculture Development Centre, Feldafing, Germany.

Niang, S. (1999) Utilisation des eaux usées brutes dans l'agriculture urbaine au Sénégal : Bilan et perspectives. In *Urban Agriculture in West Africa*, (ed. O. Smith), IDRC, Ottawa.

Ogle, B. (1990) *Traditional Vegetables in Zambia : A study of procurement, marketing and consumption of traditional vegetables in urban areas.* Sveriges Lantbruksuniversitet, Uppsala, Sweden.

Pescod, M. (1992) Wastewater Treatment and Reuse in Agriculture. *FAO Drainage and Irrigation Paper 47*, FAO, Rome.

Scott, C., Faruqui N. and Raschid-Sally L. (2004) *Wastewater Use in Irrigated Agriculture: Confronting the Livelyhood and Environmental Realities*, IWMI-CABI-IDRC, Nairobi.

Shende, G.B. (1985) *Status of Wastewater Treatment and Agricultural Reuse with Special Reference to Indian Experience and Research and Development Needs.* Paper presented to the FAO Regional Seminar on the Treatment and use of Sewage Effluent for Irrigation, 7-9 October, Nicosia, Cyprus.

Smit, J. (1996) Urban Agriculture Progress and Prospects: 1975-2005, IDRC Reports, Cities Feeding People Series Report 18, Ottawa.

UNICEF (2005) *Meeting the MDG Drinking Water and Sanitation Target.* Posted at www.unicef.org/wes/mdgreport/disparities1.php

Van der Hoek, W., Hassan, MU., Ensink, J., Feenstra, S., Rschid-Sally, L., Munir, S, Islam, R., Ali, N., Hussain, R., Matsuno, Y. (2002) Urban Wastewater: A Valuable Resource for Agriculture: A Case Study from Haroonabad, Pakistan. IWMI Research Report 63, IWMI, Colombo. www.iwmi.cgiar.org/pubs/pub063/Report63.pdf

van Lier J. and Huibers, F. (2004) Agricultural use of treated wastewater: the need for a paradigm shift in sanitation and treatment. Joop Steenvoorden and Theodore Endreny (eds), Wastewater Re-use and Groundwater Quality, Wallingford, UK, IAHS Publ. 285: 5-18

World Health Organization (1989). *Health Guidelines for the Use of Wastewater in Agriculture and Aquaculture.* WHO Technical Report no. 778, World Health Organization, Geneva.

World Health Organization (1999) *Wastewater Management in EMR Countries.* Conference, Amman, Jordan, 1999.

12

Municipal water reuse

John Anderson

12.1 INTRODUCTION

The world's supply of fresh water is finite and is threatened by pollution. Urban development has had a significant impact on the natural water cycle. Water drawn for urban water supply reduces streamflows in rivers. At the same time, stormwater runoff and wastewater discharges often carry high levels of pollution that cause a decline in the water quality of rivers. The economic, social and environmental impacts can be severe.

In many countries, the available fresh water resources are already heavily committed and in some cases perhaps already over-committed. To avoid a water crisis, many countries must use water more efficiently, manage supply and demand, pollute less and reduce the environmental impacts of growing population.

Municipal and public uses of water are often a significant component of urban water use. However, many municipal water uses do not need high quality drinking water. Using recycled water for public and municipal water needs can contribute to reducing demands on high quality freshwater sources, make urban water use more efficient, and help to achieve sustainable water management.

12.2 MUNICIPAL WATER REUSE AND INTEGRATED WATER MANAGEMENT

Water efficiency measures and water recycling projects can help to overcome water shortages and reduce the need for expenditure on new water sources and transfer schemes. A

©2008 IWA Publishing. *Water Reuse - An International Survey of current practice, issues and needs* by B. Jimenez *et al.* ISBN: 9781843390893. Published by IWA Publishing, London, UK.

municipal water efficiency and water reuse program should be included in all urban water efficiency programs. Elements of an urban water efficiency program should include:

- mandatory water efficiency targets for new buildings;
- water efficiency labelling for appliances;
- household tune-up and retrofit programs to reduce losses and install water efficient fittings and appliances;
- financial incentives to encourage the purchase of water-efficient household appliances;
- commercial and industrial audits to reduce losses, install water efficient processes and fittings, and identify opportunities for internal recycling;
- a municipal water efficiency and water reuse program;
- a public education program.

12.3 MUNICIPAL REUSE APPLICATIONS

In recent years local water authorities in many countries have implemented successful municipal water reuse projects. This experience has demonstrated the feasibility of municipal water reuse on a large scale and its role in achieving efficiency of urban water use. The following is a listing of the municipal water reuse applications that have been implemented by water authorities.

Public facility applications:

- irrigation of municipal landscaping and public open space areas;
- irrigation of sporting fields and golf courses;
- toilet flushing in public buildings, sporting facilities, camping grounds, etc.;
- cooling tower make-up water.

Municipal service applications:

- street cleaning;
- roadmaking;
- dust control;
- washing of municipal trains and buses;
- process water in wastewater collection and treatment systems;
- heat recovery and district heating systems;
- fire fighting.

Environmental applications:

- urban water features;
- water features in zoological gardens;
- restoration of urban streams;
- snow making.

12.4 MUNICIPAL REUSE CASE STUDIES

12.4.1 Public facility applications

Community-wide municipal reuse systems: The City of St Petersburg in Florida has constructed an extensive urban water reuse scheme which has been in operation since 1977. Recycled water uses include urban and residential landscape uses, industrial uses, air conditioner chiller water and a backup source for fire protection. The scheme supplies an average of about 80,000m^3/d of recycled water. The quantity used depends on weather conditions. In 1993, more than 100,000m^3/d of recycled water was supplied to consumers. A further 70,000m^3/d is used for deep well injection to prevent saline intrusion into the city's drinking water aquifers (Johnson and Parnell, 1998).

The Irvine Ranch Water District (IRWD) commenced construction of a dual reticulation recycled water scheme in 1977. Recycled water is used for landscape irrigation including residential gardens (2000 ha), irrigation of food crops (400 ha), ornamental lakes, car washes and industrial uses including a carpet mill. IRWD has mandated the use of recycled water for toilet flushing in new high-rise office buildings. The IRWD dual reticulation system supplies 57,000m^3/d to 1750 customers and delivers more than 41,000 m^3/d of recycled water (Young *et al.*, 1998).

Abu Dhabi in the United Arab Emirates has grown rapidly from less than 5000 people in the 1950s to more than 650,000 people today. Abu Dhabi and the adjoining mainland satellite towns consume more than 350,000 m^3/d of desalinated water. All wastewater from the city is pumped to a single treatment plant at Mafraq, about 40 km from the city centre. Over the last 25 years, Abu Dhabi Municipality has implemented a strategy to treat 100% of its wastewater to a standard suitable for the irrigation of public parks, roadway landscaping and fodder farms. This has allowed Abu Dhabi to develop a garden city aspect despite high temperatures and minimal rainfall. The recycled water receives supplementary treatment by sand filtration and chlorine disinfection. About 200,000m^3/d of recycled water is delivered to the city for green space irrigation.

Ornamental lakes: The Mawson Lakes housing development in Adelaide, South Australia will house 10,000 people in 3700 houses and also serves a university and a commercial and industrial estate. Stormwater from the site is collected, treated and recycled to provide lakes and water features within the estate and supplementary water for landscape irrigation. Aquifer storage and recovery will be used to store surplus winter stormwater flows to supply peak irrigation needs during summer.

Sporting facilities: In Australia, a water recycling scheme has been installed at the Sydney Olympic Park where the 2000 Olympic Games were staged. Up to 7000m^3/d of recycled water from stormwater and treated wastewater sources is used for toilet flushing in sporting venues and for irrigation of sporting venues and open space areas. Recycled water is also supplied to 2000 residential houses for gardens and toilet flushing. Microfiltration and reverse osmosis treatment processes are used to achieve the required water quality. The scheme reduces demands on Sydney's freshwater supplies by about 2 300 m^3/d (Chapman, 2005).

Caravan parks and camping grounds: Recently as a drought management measure, the local authority at Wyong in Australia has reduced potable water consumption at the municipal caravan parks and camping grounds by about 50% by using rainwater as a supplementary source for showers, and by using recycled water for toilet flushing and landscape irrigation.

12.4.2 Municipal services

Roadmaking, street cleaning and dust control: As a drought management measure, a number of Australian local authorities have installed water tanker filling stations using stormwater, water treatment plant backwash water or recycled water as the water source for road construction, dust control, street cleaning, and for the watering of municipal gardens and street landscaping where it is uneconomic to construct recycled water pipelines. Appropriate regulations are needed to separate water tankers being used for these purposes from tankers being used to supply drinking water to unreticulated properties.

Washing of municipal trains and buses: An urban redevelopment project on the site of a former tram depot in Newcastle, Australia uses rainwater, stormwater and recycled water to supply all the residential garden irrigation and toilet flushing needs and delivers surplus water to meet all the bus-washing needs of the adjacent Hamilton bus depot. (Coombes *et al.*, 2000).

The Seoul Metropolitan Government in Korea uses recycled water for washing the exteriors of its municipal train fleet.

Process water in wastewater treatment plants: The amount of water used in wastewater treatment plants for washing down structures, screens, sludge handling equipment, odour control, filter backwashing, lubricating and cooling pump seals, cooling water for equipment such as pump motors, compressors, blowers, chillers, cleaning and site landscaping, can be as high as 1% to 2% of total urban water use. Recycled water is commonly used for these purposes. Knight and Sokol (1991) reported that the Greater Chicago Metropolitan Water Reclamation District uses 42,000 m^3/d of recycled water in its water reclamation plants, just under 1% of incoming wastewater flows.

Power plants: At the Eraring Power Station at Lake Macquarie, about 100 km north of Sydney, Australia, recycled water which has received dual membrane treatment is supplied as boiler feed to provide steam for the power station turbines. This recycled water replaces 2,740 m^3/d of potable water previously supplied from the urban water supply system (Cole and Deans, 1994).

Heat recovery and heating systems: In cold climate countries, wastewater and recycled water streams are warmer than ambient temperatures during the winter months. Funamizu (2001) describes opportunities to recovery heat energy from wastewater and recycled water streams for various municipal service applications. Also in cold-climate countries, there is potential to use recycled water in district heating networks if the recycled water is adequately treated and disinfected, and suitable biofilm control measures are used.

Fire fighting: In emergency situations, water for fire fighting is often sourced from secondary water sources including untreated water from natural water bodies. High grade recycled water can be used safely for fire fighting in place of potable water provided fire crews are aware of the water source and adjust their fire-fighting practices to suit. Using recycled water systems as the principal system for fire fighting and locating fire hydrants on the recycled water network potentially may allow savings in the sizing of potable water networks and water treatment plants.

12.4.3 Environmental reuse applications

Urban water features: Kurayabashi (1991) has described a project to create an artificial stream (seseragi) in the Kawasaki aqua-promenade in the dried up riverbed of the Egawa River in Kawasaki City, Japan.

Restoration of urban streams: There has been extensive use of recycled water to restore degraded urban streams in Japan. Ohgaki and Sato (1991) have described a project in Tokyo where 43,300 m^3/d of recycled water is being used to restore flows in the Nobidome and Tamagawajousi channels. River water which previously flowed in the channels was diverted for public water supply in 1965. Ogawa (2001) has described municipal reuse projects in Nagoya where 7,400 m^3/d of recycled water from the Uchide treatment works has been used for restoration of flows in the Arako River.

River restoration: In Australia, the New South Wales state government proposes to use recycled water to restore environmental flows in the Hawkesbury-Nepean river system downstream of the Sydney water supply storages. In the first stage of the project it is proposed to use up to 49,300 m^3/d of recycled water for this purpose. In the longer term, up to 110,000 m^3/d of recycled water may be used for this purpose (up to 6% of the total urban supply).

Zoological gardens: A small water recycling scheme (250m^3/d) has been installed at the Taronga Zoological Gardens in Sydney. Wastewater from the zoo is treated in a conventional activated sludge process followed by microfiltration and disinfection. The recycled water is used for landscape irrigation, to fill ornamental water features, for water barriers around animal enclosures, and for washing down of animal enclosures.

Snow making: Recycled water has been used as a supplementary water source for artificial snow making at a ski resort in Australia. When natural snowfalls are low, water is sprayed into the air over the ski-slopes at night to produce extra snow cover. Recycled water used for this purpose receives tertiary filtration and disinfection.

12.5 System management and operation

12.5.1 General

Recycled water should be treated and distributed so that it is safe for its intended use. In many municipal water reuse applications, there are medium to high levels of human exposure to recycled water so the management and operation of municipal reuse systems should be similar to those used for third-pipe urban and residential water reuse systems.

12.5.2 Recycled water quality

Water recycling requires effective measures to protect public health and the environment. The use of filtered and disinfected tertiary-treated recycled water is desirable for municipal reuse systems to ensure low public health and environmental risks. Appropriately disinfected secondary-treated recycled water may be suitable for other municipal applications where systems are closed or public access is controlled.

Care is required to ensure that recycled water used for high-grade municipal applications and the production of food crops has low levels of heavy metals and pesticides. If practical, recycled water for these applications should be sourced from wastewater flows which do not contain heavy industrial wastes. Where practical, industrial trade wastes should be treated separately from residential wastes or pretreated and recycled within each industrial plant.

For reuse in municipal landscaping, it is also necessary to limit salinity and heavy metals, and to ensure that the nitrogen and phosphorus loadings do not exceed the long-term sustainable capacity of the soils. Recycled water often has higher levels of salts and nutrients than river water or drinking water. Application of recycled water for irrigation in excess of the needs for vegetation growth may lead to long-term adverse environmental impacts from the accumulation of salts and nutrients in the soil. Some plants are sensitive and suffer growth impairment at quite low salt levels. If salt levels exceed 1000mg/L great care is required to manage irrigation in way that results in the salts being leached to below the soil profile.

Care is also required to avoid over-irrigation. Rising groundwater levels can have very serious consequences if saline groundwater is lifted in to the soil profile and causes soil salinisation. There can also be adverse consequences if groundwater tables are already high, and irrigation with recycled water leads to excessive BOD load, waterlogging and anaerobic conditions in the soil. Careful planning should be undertaken for municipal landscaping projects to ensure that they will be sustainable.

12.5.3 Water reclamation plant operation

Water Quality: Good treatment plant management, to maintain process stability and reliability, are important for higher-grade water recycling applications. Parameters such as coliform counts provide assurance of process integrity but the time taken to obtain test results means that online monitoring of additional parameters such as turbidity and disinfection residuals is needed for effective operational control.

Water reuse guidelines should nominate the frequency of recycled water quality testing required to meet regulatory requirements. The prime focus in the California and Australian guidelines is on ensuring that the required recycled water quality is achieved at a designated point at the end of the water reclamation process. In some guidelines there is some supplementary testing of water within the recycled water distribution. The purpose of this supplementary testing is not compliance monitoring but as a check for any failure of disinfection residuals or recontamination of the recycled water within the network such as can occur if cross-connections occur.

Monitoring: For high-grade recycled water applications, on-line monitoring of turbidity and chlorine residual is required. The plant should automatically cease to deliver recycled water to users if these parameters are outside the allowable limits. The plant should not be capable of automatically resetting after an automatic shutdown without the intervention of an operator. Automatic alarm devices should be installed to warn of loss of power or failure of any process unit and include operator call-out alarms that are on-line whenever the water reclamation plant is running unattended.

Commissioning: The operation of all equipment and the water reclamation plant as a whole should be fully tested during commissioning. Following commissioning the water reclamation plant should undergo one month of quality assurance proof testing to ensure that the plant meets specifications. Recycled water should be diverted and not be supplied to consumers until the regulator has certified compliance with water quality specifications.

Diversion: Facilities should be provided for the diversion or storage and re-treatment of any recycled water which fails to meet the required water quality standards in the guidelines. The system should be designed so that bypassing of untreated or partially treated wastewater direct to the point of use is not possible.

12.5.4 Storage and distribution of recycled water

Extensive summaries of recycled water storage and distribution requirements can be found in NSWRWCC (1993), Asano and Levine (1998) and Holliman (1998). Operational storage may be needed to balance the differences between daily demands for recycled water and the outputs of the water reclamation plant. Unless recycled water has been treated to a drinking water standard, the recycled water storage should be closed to public access and preferably be enclosed in a fenced area. The storage should have appropriate warning signs that the recycled water is not suitable for swimming or drinking. Open reservoirs may be the most economical form of storage for irrigation uses. However algal growth and suspended solids from open reservoirs may clog distribution and sprinkler systems. It may be necessary to install screens, strainers or filters to remove solids before recycled water from storage enters a distribution system. Recycled water, which is intended for high-grade uses, should be stored so that it is not subject to recontamination by surface runoff or by water birds.

A fundamental principle in the design of recycled water distribution systems is to maintain the separation of the recycled water and drinking water systems to avoid the potential health risks which may result from cross-connection of the two supplies. No connection of the recycled water supply into the drinking water system shall be permitted in any circumstances and penalties should apply for any breach of this rule. If the local authority supplies drinking water into the recycled water system as make-up water in peak demand periods, an approved air gap meeting international standards must be installed in the drinking water supply at the point where it enters the recycled water system.

Experience with dual reticulation urban and residential recycling schemes shows that distinctive marking of pipes is desirable to reduce the risk of accidental cross-connections. Recycled water pipes should be a different material or colour to drinking water pipes or should be laid with a distinctive marking tape. All above ground taps and fittings should be appropriately colour-coded and taps should have signs identifying that they supply recycled water, which is not fit for drinking. If practical, taps and hose fittings should be a different pattern to drinking water taps and hoses. In the USA and Australia, recycled water pipework, fitting and taps are coloured a distinctive purple or lilac. In Australia, recycled water pipework within houses has a distinctive lilac-coloured plastic jacket.

Where both drinking water and recycled water pipes enter private property, backflow prevention devices should be installed on the drinking water pipe so that if a cross-connection occurs within the property, other consumers are not affected. The local authority should undertake a systematic and ongoing inspection procedure to detect any accidental or deliberate cross-connections. The inspections should cover:

- all new services at installation;
- any service at change of ownership;
- any service following completion of building extensions or plumbing modification;
- a sample of other services every year.

In some systems, the recycled water pipe network is operated at a lower pressure than the drinking water network to reduce the chance of recycled water entering the drinking water network.

12.5.5 System management and quality control

Supply agreements: For large users of recycled water it is appropriate to establish supply agreements between the supplier and the user of recycled water setting out the responsibilities of each party.

Conditions of supply: For reticulated recycled water to a large number of small users such as occurs in the Australian residential reuse schemes, the local authority should adopt appropriate regulations to control the use of recycled water and give effect to these regulations in a set of mandatory conditions of supply. The conditions of supply should include provisions covering:

- permitted uses;
- the responsibilities of users to use recycled water safely and avoid cross-connections;
- the responsibility of users to allow only licensed plumbers to undertake modifications;
- the right of the authority to enter premises during reasonable hours for the purpose of testing water plumbing for cross-connections.

Community information: The acceptance and understanding of recycled water by users is vital. The local authority should provide users with education about system operation and instructions on the proper use of the recycled water system. Users should be made aware of their responsibilities under the conditions of supply. The authority should maintain an ongoing community information program once the scheme has been commissioned to keep users aware of system performance and user responsibilities. Appropriate signs should be erected in areas where recycled water is in use to alert visitors and trades people to the presence of recycled water.

Quality management: Recycled water systems require a high standard of operation, control and monitoring to protect public health. The local authority should implement a detailed quality and risk management system conforming to either an ISO9000 Quality Management Standard or an ISO14000 Environmental Management Standard. The local authority shall designate a recycled water supervisor with appropriate qualifications to maintain an effective quality management system. The recycled water supervisor should be responsible for compiling and updating quality procedures. The quality procedures shall include procedures for recycled water quality monitoring, diversion of recycled water not conforming to guidelines, mains tapping and service installation, cross-connection control and incident response. The recycled water supervisor should maintain written records and logs of system operation, recycled water quality, all audits and inspections and records of any system incidents, failures or violations and corrective actions undertaken during the incident or afterwards to prevent recurrences. Quality records should be made available for audit by the regulatory authorities at any time.

Reporting: Local authorities should provide an annual report to the regulatory authorities summarising system performance and compliance with water quality and other required system parameters. Annual reporting should not preclude the conduct of audits by the health or environmental regulatory authorities at any time and the local authority should provide access to the water reclamation plant and monitoring records for auditing when requested.

Incident management: In the event of any major incidents including supply of out-of-specification recycled water or detection of any cross-connection, the local authority should inform the local health authority of the incident within 24 hours and supply details of the incident and the corrective actions being taken. Where necessary, the General Manager of the local water authority and the Director of the local health authority should consult and necessary public information or warnings should be issued to consumers under agreed local emergency management protocols.

12.6 CONCLUSIONS

Municipal and public uses of water are often 5% or more of total urban water use. As most municipal water uses do not need high quality drinking water, using recycled water for municipal water can make a worthwhile contribution to reducing demands on high quality freshwater sources.

There is considerable experience worldwide in water reuse for municipal and related purposes. However there are surprisingly few case studies reported in the literature, possibly because many of the individual applications are relatively small and "unglamorous" compared to the landmark urban, industrial and agricultural reuse projects.

Where recycled water is reticulated through urban areas for municipal uses, recycled water quality, storage, distribution, cross-connection control and system management should be at a sufficient standard to protect public health, especially for applications where there is public access.

12.7 REFERENCES

Asano, T. and Levine, A.D. (1998). Reclaimed Water distribution and storage. In T Asano Ed., Wastewater Reclamation and Reuse, Ch 1. Technomic Publishing, Lancaster, PA

Chapman, H. (2005), WRAMS, Sustainable Water Recycling, in Khan, S.J., Muston, M.H. & Schafer, A.I. (eds), Integrated Concepts in Water Recycling, Wollongong, February 2005.

Cole, C. and Deans, G (1994) From Effluent to Power: A Major Recycling Initiative by Hunter Water and Pacific Power, Proc RWCC 4th NSW Recycled Water Seminar, Newcastle May , pp 28-35.

Coombes, P.J., Argue, J.R. and Kuczera, G. (1999) Figtree Place: a case study in water sensitive urban development (WSUD), *Urban* Water, **1**:335-343.

Funamizu, N., Iida, M., Sakakura, Y and Takakuwa, T. (2001). Reuse of heat energy in wastewater: implementation examples in Japan. *Water Science and Technology* **43** (10): 277-285

Holliman, T.R. (1998). Reclaimed Water distribution and storage. In T Asano Ed., Wastewater Reclamation and Reuse, Ch 9. Technomic Publishing, Lancaster, PA

Johnson, W.D. and Parnell, J.R. (1998). Wastewater Reclamation and Reuse in the City of St Petersburg, Florida. In T Asano Ed., Wastewater Reclamation and Reuse, Ch 22. Technomic Publishing, Lancaster, PA

Knight, E.W. and Sokol, R 1991, *Reuse of Treated Wastewater in Metro Chicago*, Water Science & Technology 24 (9) 143-152.

Kuribayashi, S 1991, *Reuse of Treated Wastewater in an Artificial Stream in Kawasaki City in Japan*, Water Science & Technology **23**: 2209-2214 .

NSWRWCC, (1993), *NSW Guidelines for Urban and Residential Use of Reclaimed Water,* NSW Recycled Water Coordination Committee, 1st Edition, May 1993.

Ogawa, T, (2001), *Ensuring new water sources for urban area – effective use of treated wastewater*, AWA 19th Federal Convention, Canberra, April 2001.

Ohgaki, S. and Sato, K (1991), *Use of Reclaimed Wastewater for Ornamental and Wastewater Purposes*, Water Science and Technology 23: 2109-2117 1991.

Young, R.E. Thompson, K. A., McVicker, R. R., Diamond, R. A., Gingras, M. B., Ferguson, D., Johannesse, J., Herr, G. K., & Parsons, J. J. (1998). Irvine Ranch Water District's Reuse Today Meets Tomorrow's Conservation Needs. In T Asano Ed., Wastewater Reclamation and Reuse, Ch 21. Technomic Publishing, Lancaster, PA.

13

Current practices of water reuse in industry

In S. Kim, Saravanamuthu Vigneswaran and Namjung Jang

13.1 INTRODUCTION

The term wastewater is defined, according to the Food and Agriculture Organization (FAO) of the United Nations, as "the spent or used water of a community or industry that contains dissolved and suspended matter." About 99% of most wastewater is water, and only 1% is solid waste. Therefore, natural shortage of water can be overcome by reuse of wastewater. Wastewater is a preferable unconventional water source, since the supply is increasing because of industrialisation and population growth.

The United Nations World Water Development Report (UN WWDR, 2003) indicates that global annual industrial water use is expected to rise from an estimated 725,000 Mm^3 in 1995 to about 1,170,000 Mm^3 by 2025, when industrial water usage will represent 24% of all world water usage. The UN WWDR cited the World Bank (2001) that the percentage of water used by industry was related to country income: 10% for low- and middle-income countries (who used 82% on agriculture) as opposed to 59% for high-income countries (30% on agriculture).

The water needed for industrial purposes is not required to be of the same quality as that needed for drinking. Therefore, much of the wastewater generated can be used again in the same location (usually referred to as recycling), or collected from one or more utilities that generate wastewater for use elsewhere (referred to as reuse). By using water several times, societies can get more production out of each liter, thereby lessening the need to develop new water supplies and hence, conserving water resources. Furthermore, the recycling and reuse of wastewater results in the containment of polluted waters, preventing their contamination of other freshwater sources, thus increasing the availability of these freshwater sources for more beneficial uses. By taking measures to improve the quality of water resources that have been polluted, the source can be reclaimed (called reclamation) to further supplement the availability of freshwater.

Almost all industries are large consumers of water and also generators of huge quantities of wastewater. There is potential here to recycle wastewater generated in industry and also to reuse wastewater generated in the community, after the appropriate level of treatment.

13.2 POTENTIAL FOR INDUSTRIAL WATER REUSE

UN WWDR (2003) estimated world average water usage, by country income, for agricultural, industrial and domestic use to be as shown in Table 13.1. A simple calculation shows that 110,000 Mm^3/yr can be newly generated for industrial water reuse if only 10% of industrial water is recycled [12,500,000 Mm^3 (total available water resources for human use) x 0.5 (water usage rate for humans: 50%) x 0.22 (world average industrial usage from Table 13.1: 22%) x 0.8 (wastewater production rate: 80%) x 0.1 (reuse rate: 10%) = 110,000 Mm^3/yr]. This quantity of water, 110,000 Mm^3, is about three times bigger than the storage capacity of the Hoover Dam (32,000 Mm^3) on the Colorado River, in the southwest of the USA.

Table 13.1 Competing water uses for main income groups of countries (UN WWDR, 2003).

	Agricultural Use (%)	Industrial Use (%)	Domestic Use (%)
World	70	22	8
Low Income	87	8	5
High Income	74	13	12
Lower middle income	75	15	10
Upper middle income	73	10	17
Low & Middle Income	82	10	8
East Asia & Pacific	80	14	6
Europe & Central Asia	63	26	11
Latin America & Caribbean	74	9	18
Middle East & North Africa	89	4	6
South Asia	93	2	4
Sub-Saharan Africa	87	4	9
High Income	30	59	11
Europe Economic and Monetary Union (EMU)	21	63	16

13.3 MAIN INDUSTRIES FOR WATER REUSE

Water reuse for industry must consider distance for any transport system as a critical factor for cost-effectiveness. Asano (1998) reported that three categories were particularly interesting for municipal wastewater reclamations: (1) recirculating cooling tower makeup, (2) once-through cooling, and (3) process water. Also, he reported that industrial recycling applications in commercial laundries, car and truck washing establishments, pulp and paper industries, steel production, textiles, electroplating and semiconductor industries, boiler-feed water and water for stack gas scrubbing.

Water reuse and recycling are practiced to different degrees in industries both in developing and developed countries. For example, in the car industry where cathodic painting is used, a significant amount of paint goes with water to waste. Both the paint and water can easily be recovered by using ultrafiltration. An automobile manufacturing plant in Europe, which installed a UF system to recycle paint cleaning solvents and water, reported a payback period of only 5 months due to savings in waste disposal and fresh paint and water costs (Roulet, 1984).

Meat processing industry is very common both in developing and developed countries. For example, in South Africa, there are more than 300 registered abattoirs which consume a significant quantity (7 Mm^3) of potable quality water every year and generate 6 Mm^3 of effluent. This water can easily be processed by membrane processes to the required raw water quality for the industry. This technology was found even cheaper than the conventional anaerobic process; the anaerobic process can only treat the water to the discharge quality to municipal sewers. In contrast, UF followed by reverse osmosis remove salt, phosphate and organic matter which make the water reusable in the industry. A cost comparison made, based on a daily flow of 820 m^3, indicated that UF and RO treatment (capital cost of \$1.05 million and operating and maintenance costs of \$0.95 million) is more economical compared to the conventional treatment of anaerobic digestion (capital cost of \$1.22 million and operating and maintenance costs of \$1.56 million). Apart from the cost considerations, membrane processes lead to higher quality effluent (Cowan *et al.*, 1992).

Paper industries both in developing and developed countries are reusing wastewater from paper mills, after simple decantation process. An economical feasibility study conducted in a major paper mill in Bangkok indicated that the installation cost of the settling process can be recovered with the first 20 months of operation (Vigneswaran *et al.*, 1998).

In breweries and beverage, more than 70% of water is used for bottle cleaning and related activities and the resultant wastewater is less contaminated, compared to process wastewater. The former can easily be reused after it undergoes simple separation processes to remove suspended matter.

Metal plating is often a small-scale industry and the wastewater containing valuable metals creates a major hazardous pollution which cannot be treated by conventional processes. Processes, such as ion exchange in developing, and membranes in developed countries, can both handle this wastewater and recover the water as well as wasted heavy metals (Vigneswaran and Sundaravadivel, 2001).

Metal pickling is a process of removing scales, rusts (metal oxides) and other foreign materials from the surface of metal by treating it with acids. In India, pickling of stainless steel plates is carried out in small-scale units, which process about 50 to 100 tonnes of stainless steel per year. In the Capital Territory of Delhi, many such pickling units are located in the 70 industrial estates. A simple neutralization process with lime can easily recover most of the water. For a plant discharging 10 m^3/day of effluent, the cost of providing a wastewater reclamation arrangement has been estimated to be 250-375 US\$, and the cost of lime is approximately 1.5 US\$ per day. The quantity of freshwater saved will be at least 2,500 m^3 per year.

Power plants require a huge quantity of demineralised water. Even the water used for drinking has to undergo a softening process prior to use in power plants. In countries like Australia, Hungary and USA, effluent from sewage treatment plants is treated by microfiltration and reverse osmosis and the reclaimed water is used in power stations. A similar system adapted in a power plant in Newcastle, Australia indicated a payback period of 5 years for the installation and operation of the membrane system. This benefit is in addition to a number of other environmental benefits. Similarly, Singapore, through their NeWater project, is reclaiming a significant quantity of water from sewage for industrial uses.

Although a number of case studies have indicated the economic and environmental benefits of water reuse and recycling, the reluctance of top management within the industry remains a major hurdle. Better cooperation between environmental authorities and industries, and the effort taken to conduct waste audits can promote wastewater recycling and reuse.

13.4 TECHNOLOGIES FOR WATER REUSE IN INDUSTRY

Quality requirements for different industrial waters vary with their use. For example, boiler feed water must typically be of very high quality, to avoid corrosion and scale problems. The advanced treatment processes utilized in tertiary wastewater treatment systems are mostly physico-chemical in nature. Coagulation-flocculation, filtration, and sedimentation followed by chlorination are the standard form of advanced treatment scheme that is mostly utilized in water reuse applications.

However, with technological advances and the ever-increasing stringency of water quality criteria, membrane processes are becoming a more attractive solution to the challenge of water reuse. The use of membrane technology particularly in wastewater treatment and reuse has received increased attention since the early 1990s. Membrane technologies currently being used in different industries include microfiltration (MF), ultrafiltration (UF), nanofiltration (NF), reverse osmosis (RO), pervaporation, dialysis and electrodialysis. In this chapter, some technologies and their practices are introduced below.

13.4.1 Membrane bioreactor for industrial reuse

One of the most promising water reuse technologies is the combined process of biological treatment and membrane separation, called "membrane bioreactor (MBR)". MBR has gained considerable attention due to its potential advantages over those of conventional biological treatment processes. In comparison with the activated sludge system, MBR has merits for the complete removal of solids from an effluent, superior nutrient and organic removals, higher loading rate capability, reduced sludge production, and smaller footprint (Jang *et al.*, 2004).

Table 13.2 The industrial applications of MBR in the Netherlands (Brons, 2005).

Company	Location	Sector	Year	Capacity (m^3/hr)
Noviant	Nijmegen	Coatings	1996	20.0
SCA	Suameer	Paper	1998	0.7
Driessen	Dongen	Leather	1998	10.0
Rendac	Bergum	Rendering	1999	40.0
VWS	Broek op Langedijk	Flowerbulb cleaning	1999	3.0
VHP	Ugchelen	Paper	2000	12.0
Platvis	Volendam	Food	2000	2.0
Cerestar	Bergen op Zoom	Wheat refinery	2001	35.0
Rentex Floron	Bolsward	Textile cleaning	2002	35.0
Astra Faam	Harlingen	Food	2002	2.5
Rendac	Son	Rendering	2003	100.0
Akzo Nobel	Oss	Phamaceutical	2003	1.2
Du Pont/Invista	Dordrecht	Chemical	2003	60.0
Fuji	Tilburg	Photo/film	2004	35.0
Bavaria	Eemsmond	Malt	2005	55.0
Noviant	Nijmegen	Coatings	2005	3.0
Cargill	Amsterdam	Soja bean creshing	2005	3.0

Industrial process water, such as water used for cleaning, usually does not have strict water quality requirements. In cases where low/medium water quality is acceptable, MBR directly, or MBR with disinfection process could be applied for water reuse. If higher water quality is necessary, reverse osmosis or nanofiltration should be considered after the MBR (Jang *et al.*, 2005).

Table 13.2 shows the industrial applications of MBR in the Netherlands (Brons, 2005). For example, in 2001 the Cerestar Company built a 35 m^3/h capacity MBR plant to provide process water for cooling purposes. The plant saved 270,000 m^3/yr, 20% of the total water usage.

13.4.2 Wafer fabrication (Singapore)

The Singapore NEWater project started as a joint initiative between the Public Utilities Board (PUB) and the Ministry of the Environment and Water Resources (MEWR) in 1998. The NEWater Factories at Bedok and Kranji Water Reclamation Plants were commissioned at the end of 2002. Since February 2003, NEWater has been supplying to wafer fabrication plants at Woodlands and Tampines/Pasir Ris, and to other industries for non-potable use. In January 2004, the third NEWater Factory at Seletar Water Reclamation Plant began supplying NEWater to the wafer fabrication plants at Ang Mo Kio. The total capacity of the 3 NEWater factories is 92,000 m^3/day (http://www.pub.gov.sg/NEWater).

Figure 13.1 shows the process of NEWater Factory plants, comprising of membrane and ultraviolet treatment processes. The quality of NEWater exceeds drinking water standards stipulated by WHO and USEPA.

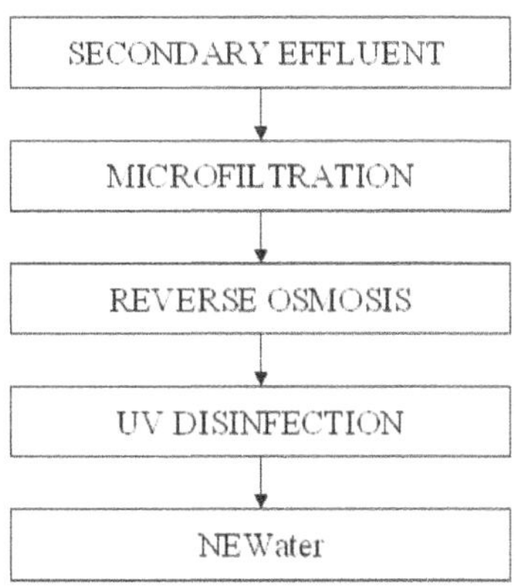

Figure 13.1. Process diagram of NEWater in Singapore.

13.4.3 Steel industry (South Korea)

The INIsteel Company obtains its cooling water by treating the secondary effluent of the Gajwa Wastewater Treatment Plant (West Inchen). The cooling water requirement is 70,000 m^3 and the circulation flow is 1 Mm^3/day. The reclaimed water makes up the loss by evaporated cooling water (12,500 m^3/day), saving about US$3 million/yr. Figure 13.2 shows the water reuse process. The technology train uses flocculation and media filtration as pretreatment for reverse osmosis (RO). Microfiltration is applied for the protection of RO in case of faults or process breakdowns.

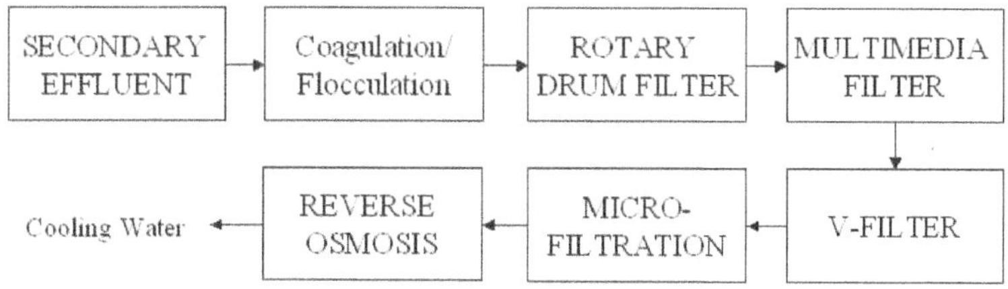

Figure 13.2. Process schematic for wastewater reuse in the steel industry.

13.4.4 Aluminium can manufacturer (USA)

Continental Can Company has applied UF/RO process for the reuse of rinse water. As the aluminium can manufacturer is located far from sewer access, it had to haul all wastes by truck (http://www.gewater.com). Figure 13.3 shows the UF/RO process for rinse water reclamation. The can rinse wastewater contains metal fines, cutting oils and dissolved salts. The bag filter removes the metal fines. Ultrafiltration and reverse osmosis remove oils and low molecular weight organics, and salts, respectively. The system is designed to treat 90 m^3/d influent.

The system can save over 400,000US$/yr in haulage costs (assuming 90 m^3/d and 250d/yr operation, and given that UF/RO treated water costs 1.32US$/$m^3$, which is less than 10% of haulage charges of 19.8US$/$m^3$).

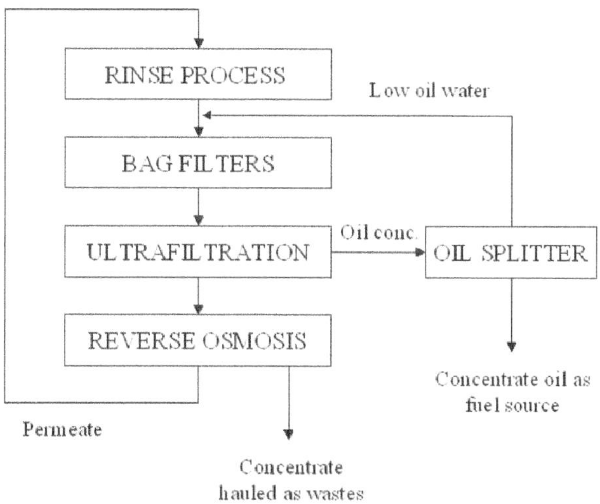

Figure 13.3. Process schematic for the can rinse water reuse.

13.4.5 Precision Glass (South Korea)

A water reuse program was implemented at the Gumi plant of Samsungcorning which produces glass for display units (Braun tube, LCD). Through the wastewater recycle program, 12,500 m^3/day of recycle water is produced to save about 3 million US$/yr of water supply costs. Wastewater production was reduced from 20,000 m^3/day to 12,500 m^3/day, which also saves 850,000 US$/yr treatment costs. The program included three phases as detailed below.

- Phase 1 (1998): The wastewater treatment process was modified. A cost-effective direct sand filtration process was applied for wastewater reuse. The grind process water for Braun tube manufacturing was replaced with recycled water (8,000 m^3/day, 1.46 million US$/yr).
- Phase 2 (2000): All processes for Braun tube manufacturing and fire service water were connected to recycled water (10,000 m^3/d, 2 million US$/yr).
- Phase 3 (2005): 2,500 m^3/d of wastewater effluent was recovered as industrial water by a packed bed type ion exchange resin process. Figure 13.4 shows the wastewater recycle system after the 3 steps of modification.

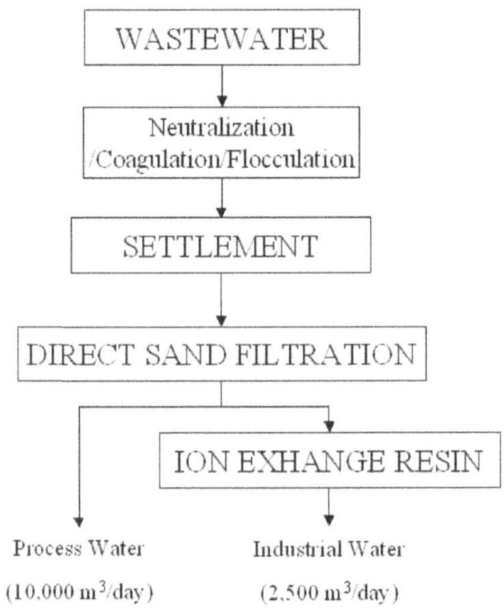

Figure 13.4. Process schematic for wastewater reuse in the Precision Glass plant.

13.4.6 Refinery (Australia)

Brisbane City Council (Queensland state) has provided 14,000 m^3/day of treated secondary effluent from the Luggage Point wastewater treatment plant for industrial use by the BP Amoco Bulwer Island Refinery.

The reclamation process is a dual membrane plant of microfiltration (PVDF membrane, 0.1 μm pore size) followed by reverse osmosis (DOW thin film composite polyamide membranes). The reuse water is chlorinated before the 4 kilometre transfer to the refinery, and stored in 2 x 6,000 m^3 polyethylene lined and covered reservoirs.

13.4.7 Power station (Boiler feed water, Hungary)

The water quality of the cooling lake, which is the source of the boiler feed water to the Bokod Power Station, Hungary, gradually deteriorated due to pollutant discharges over a decade. Suspended solids, TDS, and algal counts levels caused a dramatic increase in the chemical usage of the deionizer. The regeneration periods of the deionizer became very short and operation was troublesome. To resolve the problems, various treatment methods were evaluated, and RO with continuous microfiltration (CMF) was selected as the technically best and economically most viable option.

The CMF system includes self-cleaning screens to remove large suspended solids at the inlet of the raw water feed tank. Two CMF units (44 kL/h each) can supply 44 kL/h in duty/standby mode, or double flow in parallel mode. The filtrate produced has < 3 Silt Density Index with particle removal down to 0.2 micron size. The backwash uses compressed air operating every 30 minutes, while periodic cleaning is accomplished with a clean-in-pipe (CIP) system. The pretreated water is pumped via cartridge (safety) filters into the RO units, and dosed with anti-scalant and acid to prevent membrane fouling. The membranes are sanitized every 4-8 weeks. The RO system has two units in parallel, with a total 52 kL/h capacity, and the reject volume is 34-36 kL/h.

Key benefits achieved with the RO system are:

- minimal chemical consumption;
- guaranteed treated water quality regardless of poor and varying raw water quality;
- compact plant requiring much less land area;
- lower power consumption;
- significant increase of deionising capacity with reduced chemical usage;
- dramatic reduction of operational and maintenance labour due to automatic operation;
- payback period under 4 years due to high cost savings.

13.4.8 Piggery Farm, Victoria, Australia (UNEP, 1994)

In a pig farm in Victoria, Australia, liquid manure composed of a suspension of faecal and urinary waste including spilled feed and hair bedding, was sized and segregated by screening. Liquid manure was sized and segregated by screening (Figure 13.5). The retained solids were spread over the farmland and the liquid was held in a large reservoir for summer irrigation. The 21 kL/d liquid slurry had 1.7% of organic solids content and caused odour problems.

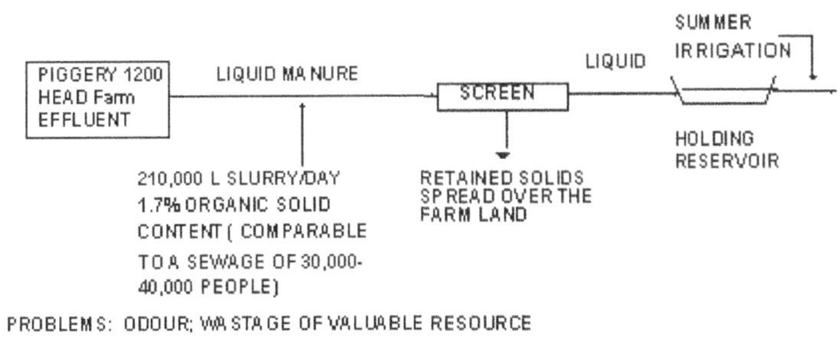

Figure 13.5. Earlier water management system in the piggery farm.

In a waste minimization program, anaerobic digestion was introduced for the treatment of effluent. The new technology involved:

- an automatic flushing system to clean the effluent from the pig sheds;
- a grit removal machine;
- a high-rate dissolved-air flotation unit to concentrate the solids and produce biogas;
- a cogeneration plant to produce electricity and hot water;
- development of an organic soil conditioner.

Figure 13.6 shows the improved treatment process of the pig farm waste. The advantages due to the application of waste minimization techniques were:

- separated grits from the effluent sold to worm farmers;
- 70% reduction in fresh water usage due to water recovery by dissolved air flotation;
- 1,700 m^3 of biogas production per day, which was converted to electricity (3,840 kWh/d) and 28,800 MJ/d energy as hot water;
- production of 4 ton of humus solids per day;
- production of 100 kL of nutrient-rich liquor everyday for irrigation or fertilization;
- far fewer flies and rodents around the farm;
- total elimination of odor which made the neighbours much happier.

The economic analysis showed an approximate 5-year payback period on the capital investment of 1.36 US$million.

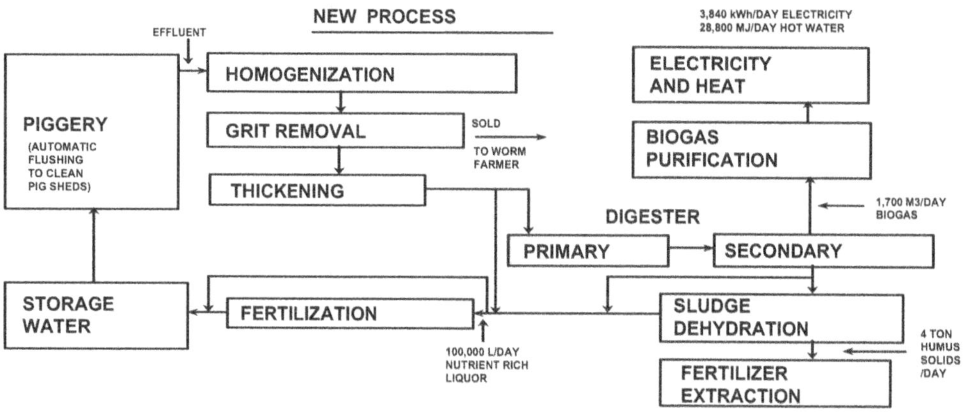

Figure 13.6. Improved treatment at the piggery farm.

13.5 CONCERNS ASSOCIATED WITH WATER REUSE IN INDUSTRY

With the development of technologies for water reuse and the economical benefits from them, many industries are under increasing opportunities to use reclaimed water within their facilities. They have tried to use recycled and reclaimed industrial wastewater as well as reclaimed municipal wastewater as an extra water source for industrial applications. Some water quality parameters in wastewater reuse must be considered to minimize negative effects on industrial process performance. Table 13.3 shows negative effects of water quality parameters for facilities (Asano, 1998). It is necessary to control and monitor these parameters to reduce obstructions for long-term operation; if not properly managed, they will cause corrosion, foaming, scaling, biological growth and fouling of the process. These are closely related to quality requirements for the specific industry and applied technologies for water reuse.

Another concern is very low concentration of specific compounds such as persistent organic pollutatns (PCPs) in the reclaimed water, even if it does not directly affect the water usage for industry. Monitoring biological pathogens, toxic volatile organic compounds and micropollutants is to protect worker's health in industries. Biological pathogens in aerosols can cause diseases, and toxic volatile organic compounds also induce toxicities if workers are exposed to aerosols in the water reuse process. Micropollutants such as pharmaceuticals and hormones which are identified as an emerging class of potential pollutants for the aquatic environment may affect human health (Gagne *et al.*, 2005).

Table 13.3 Summary of negative effects of water quality parameters for the facilities.

Water quality parameter	Negative effects for the process
Residual organics	Bacterial growth, microbial fouling on surfaces, foaming in process waters
Ammonia	Forms combined chlorines with lower disinfection effectiveness, causes corrosion, promotes microbial growth
Phosphorus	Scale formation, algal growth, biofouling of process equipment
Total suspended solid (TSS)	Deposition in materials, microbial growth
Total dissolved solid (TDS)	Corrosion, scale formation
Dissolved minerals: Calcium, magnesium, iron, and silica	Scale formation

REFERENCES

Brons H. (2005) Industries is ready for MBR. *H2O*, April, 27-29.

Cowan,J.A.C, MacTavish, F.,Brouckaert, C.J. and Jacobs, E.P., (1992) *Wat.Sci. Technol.*, **25** (10), 137-148.

Gagne, F., Blaise, C. and Andre, C. (2005) Occurrence of pharmaceutical products in a municipal effluent and toxicity to rainbow trout (*oncorhynchus mykiss*) hepatocytes, *Ecotoxicology and environmental safety*, in press.

Jang, N.J., Hwang, M.H., Yeo, Y.H., Shim, W.G., Vigneswaran, S. and Kim, In S. (2004) The kinetics on the biological reaction in membrane bioreactor (MBR) with gravitational and transversal filtration, *Environmental Engineering Research*, **9**, 238-247.

Jang, N, Ren, X., Moon, J., Cho, J. and Kim, In S. (2005) Water Reuse Technology Train for Medium-level Water and Industrial Cooling Water in South Korea, *Proceeding of Wastewater Reclamation and Reuse for Sustainability Conference (WRRS2005)*, Jeju Island, South Korea.

Roulet M. (1984) *Les applications industrielles de l'ultrafiltration*, *Techniques Separatives sur Membranes*, Cycle ICPI, Lyon, France, October 22–26.

Takashi A. (1998). *Wastewater reclamation and reuse*, Water quality management library, volume 10, CRC Press, Boca Raton, FL.

The United Nations World Water Development Report (2003) *Water for people Water for life*, www.unesco.org/water/wwap

Two case histories of industrial reuse via membrane technology, http://www.gewater.com/library/tp/761_Two_Case.jsp

UNEP, Cleaner Production in the Asia Pacific Economic Cooperation Region (1994).

US Filter Report – ACT Electricity and Water, Application Bulletin No. A2049, U.S Filter Australia Pty. Ltd., Australia

Vigneswaran, S., Ngo, H.H., Chaudhary, D.S. and Hung, Y.T. (2005) Physico-chemical processes for water reuse, Book Chapter in "Physicochemical treatment processes", Editors: Wang, L.K, Hung, Y.T., Shammas, N.K, P 635-675, Ch 18,Vol. 3, Humana Press Inc., NJ, USA.

Vigneswaran, S. and Sundaravadivel, M. (2001) *Water pollution, recycling and reuse*, *Our Fragile World: Challenges and Opportunities for Sustainable Development*, Editor: M. K. Tolba, ELOSS Publishers Co. Limited, UK, Vol. 2, 1381-1406

Vigneswaran, S., Visvanathan, C., Jegatheesan, V. 1998. *Industrial Waste Minimization.* Malaysia, Environmental Management and Research Association of Malaysia

World Bank (2001) *World Development Indicators* (WDI)

World Resources 2000-2001. *People and ecosystems*, WRI

14

Water reuse via aquifer recharge: intentional and unintentional practices

Peter J. Dillon and Blanca Jiménez

14.1 INTRODUCTION

Managed aquifer recharge (the intentional recharge of aquifers) is not itself a use of reclaimed waters but is a step that can lead towards reuse, such as for agricultural, industrial and drinking water supplies and for protection of groundwater-dependent ecosystems. Traditionally urban aquifers have been sources of high quality water supplies for cities and often inadvertently also received urban wastewaters. Increasingly, intentional replenishment of aquifers by highly treated reclaimed waters is being practiced in developed countries with the full support of communities, and health and environmental regulators, for aquifers that are under stress through imbalances between rates of extraction and natural recharge. In addition, for water-short cities, recharge of treated stormwater and reclaimed water to brackish aquifers is occurring to develop new water supplies from otherwise untapped resources. In developing countries, cases of water reuse through aquifer recharge are also increasing, but in contrast, they involve unplanned recharge with polluted water or effluents.

It is therefore timely to look at the lessons learned through these practices and from cities where unintentional recharge of effluents have taken place over many years with a view to understanding the potential and limitations for contaminant attenuation in the subsurface. Taking account of the natural treatment that the subsurface can provide will allow better design of wastewater treatment and management methods that ensure that desirable end points are reached for water quality, well yields and groundwater levels. Acknowledging the linkages that aquifers provide between wastewater streams and drinking water supplies is important in managing existing risks and in justifying investment in treatment, monitoring and research necessary to allow planned partial closure of the urban water system in ways that provide the necessary level of water security and safety at an affordable cost and with acceptable greenhouse gas emissions.

In this chapter examples are presented of intentional and unintentional recharge of aquifers in a variety of settings, with a range of purposes, methods and outcomes. As wastewater treatment processes and water quality issues are comprehensively addressed in other chapters, it may be helpful here to provide some background information on hydrogeology that will explain the different capabilities for subsurface storage and treatment of reclaimed water and wastewater in the main types of aquifers that are commonly encountered.

14.2 AQUIFER PROPERTIES THAT INFLUENCE RECHARGE WITH RECLAIMED WATERS

Aquifers are those parts of the subsurface that contain water that can be extracted at useful rates. They can range from unconsolidated sands and gravels, through to consolidated sandstone and limestone and to hard rocks that can convey water through cracks. In general aquifers can be divided into two broad classes; unconfined, where water can infiltrate from the ground surface to a water table below which all pore space is saturated, and confined, where locally the roof of the aquifer is relatively impermeable to water, *e.g.* clay, and infiltration from overlying land does not reach the aquifer. Hence confinement is the most important determinant of the impacts of unintentional recharge of wastewater on water quality within an aquifer. It also defines the types of recharge enhancement that are viable. Table 14.1 summarises the various characteristics of aquifers and identifies their influence on potential for managed aquifer recharge using reclaimed waters.

Beneath a city there may be no aquifer, or one or more aquifers, *e.g.* in Bangkok six distinctive aquifers have been identified in a layer cake formation with lower permeability formations between them. Without a permeable formation there is no opportunity for subsurface storage. In some arid locations permeable but dry formations suggest the possibility for managed aquifer recharge. However, sometimes those formations are dry because groundwater can discharge so quickly that storage does not accumulate, and opportunities for managed recharge here are only illusory, unless this is regarded only as a filtration step prior to discharge. If there are several aquifers available, depending on their degree of inter-connection, it may be possible to store water of different qualities in different aquifers. Table 14.1, along with sound local hydrogeological knowledge, may help in identifying the most suitable aquifers for use in storing reclaimed waters of different quality for different purposes.

In Table 14.1 the term "recovery efficiency" refers to the volume of water that can be recovered at a suitable quality for its intended use as a proportion of recharged water. The table does not cover all possibilities as there are many subtleties. For example, many aquifers are semi-confined with characteristics that are intermediate between those shown in the table. Semi-confined aquifers can have other effects such as reducing recovery efficiency if the leaky confining bed contains more saline water than the aquifer.

Because of the range of characteristics that come into play, it is difficult to extrapolate from one site to predict performance of managed aquifer recharge (MAR) with reclaimed water at another. Unfortunately a site-by-site approach will be necessary until the knowledge base has extended sufficiently to facilitate reliable predictions, such as currently occurs in groundwater geochemical modelling, and is being developed holistically in the EU project RECLAIM WATER (http://www.reclaim-water.org).

Table 14.1 Characteristics of aquifers and their influence on potential for managed aquifer recharge (MAR) using reclaimed waters.

Characteristic	Feature and influence on MAR	
Confinement	Unconfined - surface infiltration methods viable - unprotected from surface contamination - storage capacity depends on depth to water table	Confined - well injection methods only - protected from surface contamination - storage capacity depends on aquifer thickness
Permeability	Moderate to High - high rates of recharge possible - recharged water can be dispersed - lower capital and energy costs per unit of water recovered	Low to Moderate - lower rates of recharge possible - recharged water more localised - greater capital and energy costs per unit of water recovered
Thickness	Thick - high storage potential - more sensitive to salinity stratification if native groundwater is brackish	Thin - low storage potential - may limit rate of recovery by wells
Uniformity of hydraulic properties	Homogeneous - minimal mixing and higher recovery efficiencies if native groundwater is brackish	Heterogeneous - lower recovery efficiencies if native groundwater is brackish - in karstic and fractured rock systems limited ability to contain recharged water
Salinity of groundwater	Fresh - recovery efficiency is not limiting - requirement to protect wider range of beneficial uses of aquifer (higher treatment costs)	Saline - recovery efficiency can limit effectiveness - less beneficial uses to protect so treatment requirements less onerous
Lateral hydraulic gradient	Gentle - recharged water contained closer to point of recharge	Moderate to Steep - recharged water dispersed down-gradient and lower recovery efficiencies if native groundwater is brackish
Consolidation	Consolidated - simpler well completions - easier to maintain recharge wells to prevent irrecoverable clogging	Unconsolidated - screens required for injection and recovery wells - land subsidence a consideration
Aquifer mineralogy	Unreactive with recharge water - recovered water quality unaffected by geochemical reactions with aquifer matrix - likelihood of clogging of injection wells is sometimes increased	Reactive with recharge water - consider metal (*e.g.* As) mobilisation, F, H_2S effects on recovered water and groundwater - in carbonate aquifers less onerous treatment required to avoid clogging of injection wells
Redox state of native groundwater	Aerobic - higher rates of inactivation of pathogens and biodegradation of some endocrine disrupting chemicals (EDCs)	Anaerobic - higher biodegradation rates for trihalomethanes (THMs)

14.3 INTENTIONAL RECHARGE WITH RECLAIMED WATERS

Intentional recharge with reclaimed water can play a role in providing a balancing storage and supplemental treatment for either indirect non-potable reuse, such as irrigation of agriculture fringing a city or landscape irrigation and industrial use within a city, or for return to the raw water supply for further treatment and indirect potable reuse (Figure 14.1). Subsurface storage has the advantage of significantly extending the time period between uses giving an opportunity for inactivation of pathogens and biodegradation of a number of trace organics within the aquifer (Bouwer, 2002; Dillon and Toze, 2005). It also provides a storage that occupies a minimum of valuable urban land, is protected from evaporation, and will not produce algae or mosquitoes (Pyne, 2006). Capital costs of subsurface storage are also frequently significantly less than for surface storages. However there are considerations that need to be made, such as that investigations are required to determine the characteristics and extent of the storage zone (EWRI/ASCE, 2001) (issues to check off are in Table 14.1), there are energy costs associated with recovering water and in some cases with recharging it, maintenance of recharge facilities is required, and monitoring of the operation is needed to ensure that the system behaves as expected. Adverse impacts of subsurface storage on water quality may include recovery of metals such as arsenic at concentrations that exceed those of native groundwater and recharged water, and increased concentrations of fluoride, or radioactive elements through mixing with groundwater. Effects of elevated water tables and pressures must also be taken into account.

In some cases the quality of reclaimed water that is fit for direct non-potable reuse is not satisfactory for recharge. This is particularly true where well injection is the method of recharge and clogging of the interface between the well and the aquifer by particulates and biofilms can reduce the injection rate to an uneconomic level, even with good well maintenance practices. In these cases supplemental treatment may be required to reduce turbidity and nutrient levels in water to be injected. If an unconfined aquifer is available for storage and there is adequate land available, soil aquifer treatment (SAT) may be viable and that has much less stringent water quality requirements than well injection.

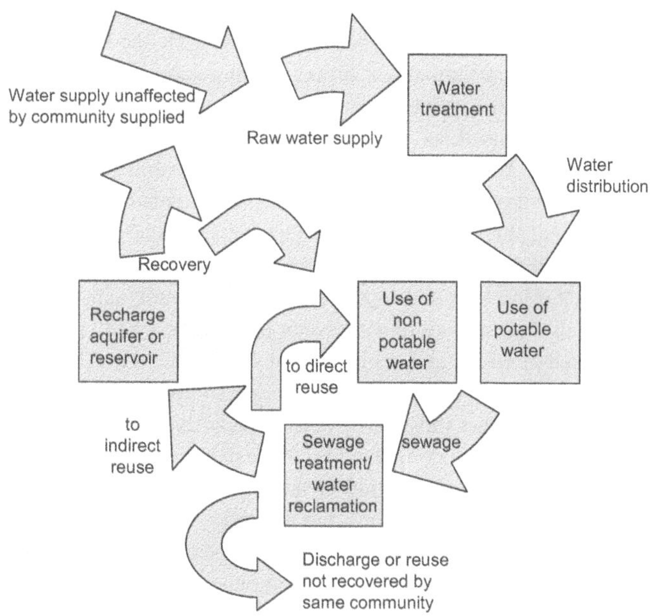

Figure 14.1 Intentional reuse and roles for subsurface storage and recovery within a city water supply.

Commonly used classification systems for reclaimed water refer only to health aspects of water reuse and not to environmental aspects, and therefore are inadequate on their own to define whether a specific water type is suitable for aquifer recharge by wells, soil aquifer treatment basins or other means for a particular aquifer. A more detailed understanding of aquifer-specific clogging and in particular the effect of nutrients and labile organic matter is needed. This information base is slowly developing.

Water quality requirements for recharge also need to address the end uses of water on recovery. In recent years the World Health Organization (2003), and in Australia the National Water Quality Management Strategy (2004, 2005), have encouraged a risk management approach to water quality management for drinking water and non-potable recycled water. The same hazard assessment and critical control points (HACCP) approach has been applied to planning for urban storm water to be recharged into a brackish aquifer with the intention of recovering water suitable for drinking water supplies (Swierc *et al.*, 2005; Rinck-Pfeiffer *et al.*, 2006). This approach is equally well suited to manage aquifer recharge with reclaimed water for indirect non-potable and potable uses.

14.4 EXAMPLES OF INTENTIONAL RECHARGE WITH RECLAIMED WATERS

A number of examples of intentional recharge with reclaimed water are given in Table 14.2 and more detailed descriptions of two projects are presented in Box 14.1 and in Chapters 23 and 24. The earliest reported intentional indirect potable reuse commenced in Orange County near Los Angeles in 1954 with pond infiltration of Santa Ana River water, which in dry months is composed entirely of treated sewage effluent discharges, to replenish depleting aquifer drinking water supplies. Soil Aquifer Treatment commenced in 1962 at Montebello Forebay, near Los Angeles and Aquifer Storage Transfer and Recovery (separate injection and recovery wells) commenced at Orange County, in 1976. Long before this, in the 1870s, Berlin had adopted bank filtration and subsequently pond infiltration for drinking water supplies with water derived from lakes downstream of a wastewater treatment plant. It is not known at what stage Berlin's MAR projects became overtly recognised as indirect potable water reuse as opposed to being unintentional. However in 2002 research was undertaken by several organisations led by the Berlin Water Competence Centre, to explore the fate of constituents of effluent in groundwater and the water supply system (Table 14.2). This water now satisfies a large proportion of the city's drinking water supply. SAT in the Dan Region of Israel also satisfies a very large and increasing demand for irrigation water. However some projects are quite small, satisfying localised needs only. In general, economy of scale applies to wastewater treatment capacity more so than recharge and recovery capacity, which often requires replication of small-scale units (basins, wells).

For the cases reported in Table 14.2 the majority of aquifers used for storing reclaimed water have been unconfined, and of these most have been alluvial and suitable for drinking water supplies. Recharge methods have been dominated by those that involve passage of water through the unsaturated zone, *e.g.* SAT, pond infiltration and infiltration galleries, where additional treatment can occur and maintenance procedures are simpler (Bouwer, 2002). In Berlin there is some evidence to suggest that groundwater extraction, combined with the schmutzdecke (organic rich layer on the base of the lake) may create an unsaturated zone through which bank filtrate must pass. Only a relatively small number of cities have used confined aquifers and well injection methods and these include indirect potable and non-potable cases. However there is an increasing awareness of the potentially valuable role of confined aquifers that initially contain non-potable native groundwater in integrated urban water management.

Table 14.2 Examples of intentional recharge of reclaimed water.

City	Aquifer confinement & material	Ambient ground-water uses	recharge method	Recharged water and treatment	Residence time in aquifer	Proportion of reused water in recovered, and type of reuse	Average annual recharge volume (Mm^3)	Year commenced	Issues addressed	References
Dan Region, Israel	unconfined sandstone	drinking and agricultural irrigation	SAT	secondary effluent	> 6 months	100% agricultural irrigation (unrestricted)	120-140	~1989 distribution system3rd line	metals, nutrients, organics, salinity, clogging	Guttman *et al.* 2002, Banin *et al.* 2002 Icekson & Blanc 1998 Icekson Tal *et al.* 2003 Idelowitch *et al.* 2003
Atlantis, Sth Africa	Unconfined sand	drinking	pond infiltration	stormwater & secondary effluent	> 6 months	25-40% drinking	~4	1979	salinity, pollution	Tredoux *et al.* 2002 and 2003
Montebello Forebay, Cal, USA	Unconfined sand	drinking	SAT	tertiary effluent		0-31% drinking	62	1962	health effects studies	Crook 2004 Sloss *et al.* *1999*
Orange County, Florida, USA	unconfined		infiltration basins	Tertiary effluent		irrigation	30	1986		Crook 2004
Berlin, Germany	unconfined alluvial	drinking	bank filtration	river and tertiary effluent	<1 year	20-70% drinking	~160	~1870	organics, hydrogeochem, algal toxins, pathogens	Grutzmacher *et al.* 2006 Knappe *et al.* 2006 Grunheid & Jekel 2006 Massmann *et al.* 2006 Heberer 2006

City	Aquifer confinement & material	Ambient ground-water uses	recharge method	Recharged water and treatment	Residence time in aquifer	Proportion of reused water in recovered, and type of reuse	Average annual recharge volume (Mm^3)	Year commenced	Issues addressed	References
Orange County, Cal, USA	confined alluvial	drinking	ASTR	Advanced treatment of effluent incl. RO	>2 years	50% drinking	6	1976	small organic molecules, sea water intrusion barrier	Mills 2002 Crook 2004 Wehner (pers. comm.)
Orange County, Cal, USA	unconfined alluvial	drinking	pond infiltration	river and tertiary effluent	>6 months	65% drinking	250	1954	clogging management	Mills 2002, Clark 2002 Crook 2004, Wehner (pers. comm.)
Sun Lakes, Phoenix Arizona, USA	unconfined alluvial	irrigation	ASR	tertiary effluent plus filtration	<6 months	golf course irrigation	0.78	1998	clogging management	Toy *et al.* 1999 Unangst *et al.* 1999 Herman Bouwer (pers. comm.)
Tucson (Sweetwater), Arizona, USA	alluvium – sand, gravel, clay	drinking	SAT & wetlands	secondary & tertiary effluent	<1 year		3-4	1990	pathogens, organics, clogging	Wilson *et al.* 1995 Fox *et al.* 2001
Chandler (Tumbleweed), Arizona, USA	alluvium – sand, gravel, clay	drinking	ASR	tertiary effluent	<6 months		7	1999	pathogens, organics, clogging	David Pyne (pers. comm.)
Torrele, Flemish dunes, Belgium	unconfined sand	drinking	pond infiltration	secondary effluent plus UF and RO	>40 days	60-70% drinking	2.5	2002	ecology, salinity	Van Houtte & Verbauwhede 2006
Sabadell, Spain	Unconfined alluvium	n.a.	riverbed infiltration	secondary effluent, (UV on recovery)	n.a.	100% park irrigation	9		ecology	www.reclaim-water.org
Bolivar, South Australia	confined limestone	brackish - unused	ASR	tertiary effluent	< 3 months	>90% irrigation	0.1	1997	pathogens, organics, clogging	Dillon *et al.* 1999, 2003 and 2006
Halls Head, Mandurah, Western Australia	Unconfined karstic limestone	irrigation	pond infiltration	secondary effluent	1-2 months	~15% urban green space irrigation	0.006	2000	nutrients, salin-ity, pathogens, organic carbon, heavy metals	Toze *et al.* 2004

City	Aquifer confinement & material	Ambient ground-water uses	recharge method	Recharged water and treatment	Residence time in aquifer	Proportion of reused water in recovered, and type of reuse	Average annual recharge volume (Mm^3)	Year commenced	Issues addressed	References
Floreat, Western Australia	unconfined sand	irrigation	infiltration galleries	secondary effluent plus rapid sand filtration	2-3 months	irrigation	0.02 pilot	2005	pathogens, nutrients, organics clogging	Toze *et al.* 2005
Windhoek, Namibia	Unconfined-semi-confined fractured rock	drinking	injection wells (ASTR)	tertiary plus granular activated carbon	<1 month to 5 years	0-30% drinking	0.4 (to 5.3)	1999	salinity, iron, sulphate	Murray and Tredoux 2002 Van der Merwe *et al.* 2005
New Delhi, India	unconfined	irrigation	filtration ponds with wells	urban stormwater		irrigation	0.02	2001	declining ground-water levels	Central Groundwater Board 2005
Nardò, Salento Region, Italy	Confined fractured and karstified	Potable and non-potable	aquifer injection by sinkhole	secondary effluent and untreated waters due to rainfall	2 months	5-10%, irrigation and domestic supplies	3.2 + 0.6	1991	saltwater intrusion	Masciopinto *et al.* 2000 Masciopinto & Carrieri 2002
Gaobeidian, Beijing, China	Clay soil and sandy aquifer	Drinking	wells and recharge basins	tertiary effluent plus slow sand filtration			0.23 (max capacity)			www.reclaim-water.org

Generally reclaimed water has been through secondary or tertiary treatment prior to recharge, and there appears to be no consistent alignment of treatment class with recharge method, ambient groundwater quality or end use of the recovered water. This is possibly due to contrasts in regulations between states and countries reported here and dates of project implementation.

The proportion of recharged reclaimed water in recovered water and residence time in the aquifer varies from site to site. Typically the proportion is low where water is recovered for drinking supplies and native groundwater is fresh. That is, managed aquifer recharge simply augments natural recharge. However, in brackish aquifers, dilution using ambient groundwater is not a viable option and proportions of recharge water in recovered water of necessity are high. Measurement methods for the proportion of reused water in recovered water are well advanced (*e.g.* gadolinium, Massmann *et al.*, 2006; carbon isotopes, Le Gal La Salle *et al.*, 2002) but infrequently applied. Residence times in aquifers were not reported for many reuse projects, and this will become a very important parameter as greater reliance is placed on the aquifer as part of the treatment train in water recycling. Tracer techniques are available (*e.g.* sulphur hexafluoride, Jordan, 2002) to enable residence time to be quantified even in complex hydrogeological systems.

Box 14.1 Case description of reclaimed water ASR at Bolivar, South Australia

Buoyed by the success of Andrews Farm stormwater ASR (see Box 25.2) and the coming availability of large volumes of reclaimed water which is unused in winter but for which summer demand is forecast to grow, a research project was established at Bolivar to determine the technical, environmental and economic viability of ASR with reclaimed water in a brackish aquifer. Between October 1999 and July 2005, in the first three ASR cycles, 520,000m^3 water was injected and 370,000m^3 was recovered from a single ASR well. Intensive monitoring with 8 piezometers and 8 observation wells within a 300m radius enabled an understanding to be developed on water movement, mixing and quality changes in the aquifer. The ASR trial used water from the nearby Bolivar water reclamation plant. Secondary treatment involved trickling filters until January 2001 when these were replaced by activated sludge reactors. The treated effluent was stored in stabilisation ponds then passed through a water reclamation plant involving coagulation, dissolved air flotation and filtration (DAF/F) followed by chlorination. The reclaimed water was delivered to a balancing storage pond before pumping via the Virginia Pipeline Scheme (VPS) for horticultural irrigation on the Northern Adelaide Plains or for storage in an aquifer via a single ASR well.

The target aquifer was a confined limestone aquifer from 100m to 160m below ground surface and contained brackish native groundwater which was too saline for irrigation. It was confined above by 7m of clay which isolated the ASR operation from a local drinking water supply aquifer. Travel time to the well at a radius of 4m was 1-2 days and to 4 of the 8 piezometers at a 50m radius was 90 to 120 days. Injectant was not detected in the other piezometers at 50m completed in lower permeability horizons. Detections were observed at fully penetrating wells at radii of 75m and 120m but not at 300m (Pavelic *et al.*, 2006).

Oxidation of organic matter was evident within 4 metres of the ASR well during the injection phase and was responsible for removal of up to 4mg L^{-1} DOC, or 20% of that injected. In contrast, strongly reducing conditions were evident around the ASR well during the storage phase, when sulphate reduction and methanogenisis locally occurred

(Vanderzalm *et al.* (in press), Greskiowak *et al.* (2005)). Dissolved organic compounds with larger molecular weight and more reactive functional groups were adsorbed or decomposed close to the ASR well. Recalcitrant organic matter having the same characteristics as natural deep groundwater was persistent in the aquifer (Skjemstad *et al.*, 2002). Trihalomethanes (THMs) and haloacetic acids (HAAs) (both of which are disinfection by-products) degraded before recovery with chloroform, the most persistent of the THMs, with degradation rates shown to be dependent on redox status of the aquifer (Pavelic *et al.*, 2005 and 2006). The microbial pathogens tested were Poliovirus, Coxsackievirus, *Salmonella typhimurium*, the opportunistic pathogen *Aeromonas hydrophila*, as well as the indicator microorganisms *Escherichia coli* and the coliphage MS2. Results indicated that the bacteria had the fastest attenuation rates followed by the coliphage MS2, with the enteric viruses inactivating slowest of all the organisms tested. In all cases, the time for one $\log_{10}$ removal was less than one month (Dillon and Toze, 2005).

Laboratory column studies were used to investigate the combined effect of physical, biological, chemical and mechanical factors on clogging and to quantify calcite dissolution processes that would offset this on the perimeter of an injection well (Rinck-Pfeiffer *et al.*, 2000). Initial teething problems and infrequent redevelopment produced clogging at the beginning of the trial; however, periodic pumping of the ASR well was effective in restoring injection rates which have been maintained over the duration of the trial. The relationship between water quality and well clogging was evaluated from the field and laboratory results, allowing water quality criteria to be developed for this aquifer.

The recovered water quality was found to satisfy the requirements for use as irrigation water according to Australian guidelines. Salinity remained below the locally permissible limit for irrigation of 1500 mg/L TDS until the volume of recovered water reached 66% of recharge in the first cycle and 80% in the second and third cycles. Subsurface storage reduced the concentrations of pathogenic micro-organisms, suspended solids, organic carbon and some metals, as predicted from the *in-situ* and laboratory pathogen survival experiments by Toze and Hanna, 2002.

Table 14.3 Quality of injected water, local ambient groundwater and recovered water.

Parameter (mg/L)	Irrigation Guidelines (NWQMS)	Ambient Groundwater (n=17)	Recharge water (cycle 1) (mean, n=24)	Recovered water (cycle 1) (final sample)
Electrical Conductivity (S/cm)		3592 ± 326	2265 ± 191	2470
Temperature (°C)		25.9 ± 1.0	20.4 ± 4.5	22.7
pH	4.5-9	7.3 ± 0.1	7.1 ± 0.4	7.06
Dissolved Oxygen		0.77 ± 0.79	4.4 ± 3.4	0
Total Suspended Solids		12 ± 14	14 ± 13	1
Total Dissolved Solids (by EC)	1500	2006 ± 188	1267 ± 58	1470
Total Nitrogen		0.08 ± 0.03	19.9 ± 10.8	15.6
Total Phosphorus		0.02 ± 0.001	0.72 ± 0.65	0.24
Dissolved Organic Carbon		0.3	16.7 ± 2.1	10.5
Total Organic Carbon		0.3	18.2 ± 2.3	10.6
Chlorine residual - total		0	0.7 ± 0.4	0
E Coli. (cells/100mL)	1000	0 ± 0	42 ± 113	0

Neglecting treatment and distribution costs, the cost of ASR with reclaimed water was found to be substantially cheaper than for stormwater ASR and similar to groundwater extraction by individual irrigators for typical annual production volumes taking into account capital and operating costs and the expected lifetimes of wells and pumps. Surface storage of large volumes of reclaimed water is not considered an acceptable option on the Northern Adelaide Plains due to lack of topographic relief and high costs.

The study has demonstrated that the recovered water met the guidelines for unrestricted irrigation. The quality of the water improved during ASR, particularly with respect to pathogens, disinfection by-products, suspended particles, organic carbon and most metals. The anticipated clogging was found to be manageable using simple methods and the cost of the operation was found to compare favourably with conventional alternatives.

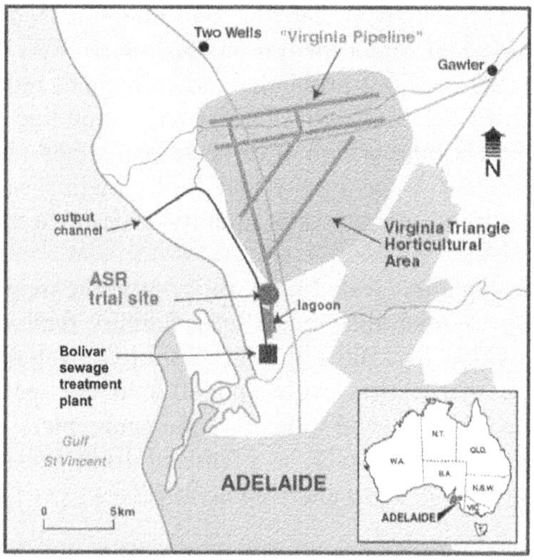

Figure 14.2. Location of the Bolivar Reclaimed Water ASR trial site.

The Bolivar Reclaimed Water ASR Research Project has been the first reclaimed water ASR project in Australia, and is the first known successful ASR trial with nutrient-rich irrigation water. Further details are found in Dillon *et al.* (1999) and Dillon and Toze (2005). The project partners were CSIRO, SA Department of Water Land and Biodiversity Conservation, United Water International Pty Ltd, SA Water Corporation and SA Department of Administrative and Information Services. The American Water Works Association Research Foundation also contributed through Research Project No. 2618 'Water Quality Improvements During ASR'.

14.5 UNINTENTIONAL RECHARGE WITH RECLAIMED WATERS

Traditionally, water reuse has been considered only as a managed activity where wastewater is intentionally treated to be used once again. Nevertheless, unintentional or 'natural' reuse occurs much more widely within the urban hydrological cycle, but is rarely acknowledged. Such reuse occurs because water supplies (both surface and groundwater) are not isolated in the environment but are linked with each other and with extractive and in-stream uses, exchanging water and constituents. This has been happening to such an extent and for so long now that most water supplies contain water that has been used previously. This is especially so for multiple towns that draw water from, and discharge wastes to, the same river, or for towns that are unsewered or have combined stormwater and sewage canals, that take their drinking water from underlying unconfined aquifers. Although water is being reused, wastewater treatment is generally performed to protect human health and the environment but is rarely done for the purpose of water reuse. Often the subsurface pathways by which reuse occur bypass wastewater treatment, and in some cases with local well or informal water supplies, may also bypass water treatment (Figure 14.3).

Surprisingly little is known about incidental water reuse, simply because it is not an accepted concept. Available information is dispersed in literature, which often links it with pollution problems. Nevertheless, some recent works have begun to talk about "unintentional", "unplanned" or "incidental" water reuse (Jiménez, 2005). These cases, in general, describe situations where used water is mixed with (or becomes) part of the water supply. Most of these cases deal with groundwater, not just because it constitutes an important source of water but also because decontamination during passage of treated or untreated sewage through soils and aquifers makes its origin and history more difficult to identify. In surface water bodies pollution and even used water discharges are more easily detected. Unintentional water reuse cases reported for drinking water supplies are presented in Table 14.4. A detailed case study is presented in Chapter 24.

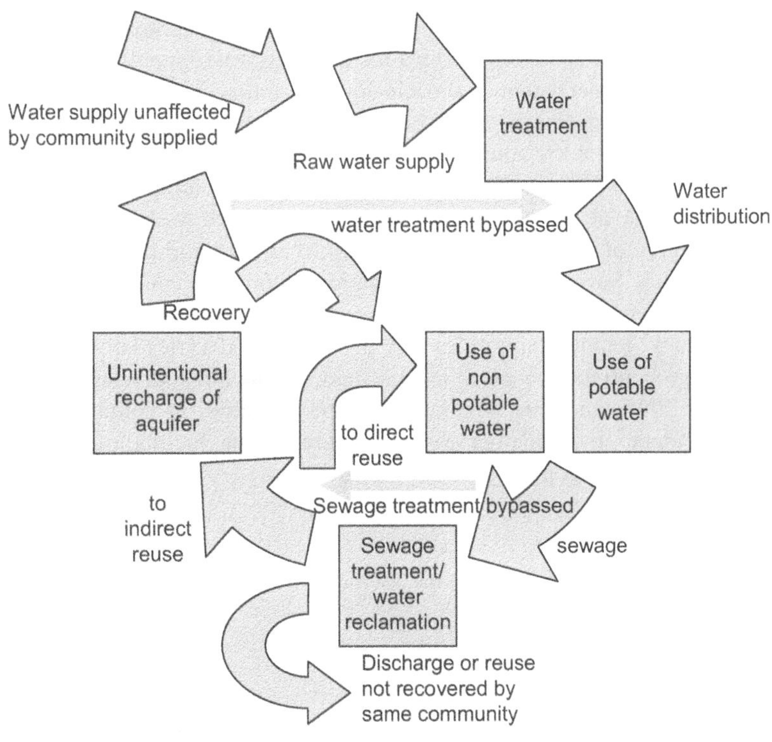

Figure 14.3 Unintentional reuse via subsurface storage and recovery within a city water supply and semi-formal sanitation system.

It is evident that pollution of groundwater drinking supplies by urban sewage varies from negligible to severe, and that recognition of unplanned reuse is needed in order to advance to an understanding of how to manage the risks. This may involve continuing groundwater recharge with water of improved quality and/or separating the recharge areas further from points of water abstraction. Activities or lack of capacity that cause water or wastewater to bypass treatment plants are obvious targets for early intervention. Risk-based approaches, as previously discussed, with appropriate monitoring information, will allow the most cost effective investments to be identified. These may include development of trade/industrial waste strategies for sewer systems, disinfection for water supply wells, deepening of water supply wells and lining of canals, installing sewage collection systems for permeable areas with dense on-site sanitation, upgrading wastewater treatment processes and capacity, and establishing early warning groundwater monitoring programs. The City of Mount Gambier in South Australia has undergone a transition from 100 years of unintentional recharge of their drinking water supply from urban stormwater drainage wells, to planned recharge with a risk abatement strategy in place (Gorey and King, 2006). This includes measuring travel time to the water supply with a tracer study, performing a risk assessment and developing a risk management plan in partnership with stakeholders, providing guidance to developers on improved management of stormwater quality prior to recharge, and continuing a monitoring program to validate or refine the risk assessment (Vanderzalm *et al.*, 2004).

It can be seen from Table 14.4 that pathogens and nitrogen are the most common water quality issues. Recent work by Toze *et al.* (2004) among others suggests that with sufficient residence time, particularly in warm aerobic aquifers, pathogen inactivation is adequate to reduce numbers of viable pathogenic viruses, protozoa and bacteria of wastewater origin to below guideline values for drinking water supplies. Nitrogen, on the other hand, may be quite persistent in aerobic aquifers and acceptability of drinking water supplies would depend on source water concentrations, wastewater treatment processes and the degree of dilution in the aquifer prior to recovery. Mixing and dilution also applies to conservative solutes, such as chloride. Metals may have longer travel times due to sorption, and the time at which breakthrough in recovered water could be expected depends on the degree of retardation which relates to aquifer and groundwater properties and the metal species. Sorption and biodegradation of organic constituents of reclaimed water in aquifers depend on the specific organic compound and on aquifer characteristics including mineralogy, temperature, redox and pH status, and varies between species. The body of knowledge continues to grow in this field, particularly in relation to MAR (*eg.* Dillon and Toze, 2005) and this is expected to be of value, in conjunction with local observations and data, in validating the sustainability, or facilitating refinements, of initially unintentional indirect potable reuse via aquifers.

Table 14.4 Examples of unintentional indirect potable reuse via aquifers.

City	Aquifer confinement and material	Ambient groundwater uses	Recharged water and treatment	Issues	References
Hanoi, Vietnam	unconfined alluvial	irrigation and drinking	sewer, storm water	NH_4 , metals, organics	Chilton and Kinninburg, 2003
Leon, Mexico	semi-confined	irrigation and drinking	industrial effluent	NO_3 , TDS, chromium, organics	Foster *et al.*, 1994 and 2000
Tula Valley, Mexico	unconfined	irrigation and drinking	untreated effluent	phthalates phenols	Cisneros and Ordonez, 1999 Jiménez and Chavez, 2004 Jiménez, 2005 BGS, 1998
Merida, Mexico	unconfined karstic limestone	drinking	sewer, storm water		Foster *et al.*, 1998
Mexico City (Southern part), Mexico	unconfined	drinking	on-site sanitation	faecal coliforms	Jiménez *et al.*, *2004* Mazari *et al.*, 2000 and 2001 Cifuentes *et al.*, 2002 & 2004
Hat Yai, Thailand	semi-confined alluvium	drinking	drainage canals, on-site sanitation	NH_4, Cl, SO_4	BGS, 1995-1998 Foster *et al.*, 1998
Ica Valley, Peru		drinking	primary effluent	pathogens	Foster *et al.*, 1994
Santa Cruz, Bolivia	semi-confined alluvium		on-site sanitation	NO_3, CL, Mn	Foster *et al.*, 1998
Sana'a, Yemen	unconfined alluvium	drinking	cess pits	NO_3, Cl, g/w levels	Foster *et al.*, 1998
Nottingham, UK	unconfined sandstone	drinking	sewer, storm water	NO_3 pathogens	Wakida and Lerner, 2005
Gabel El Aswar, Egypt	unconfined alluvium	saline	primary effluent canal seepage	NO_3, NO_2, dilution of salt	Troeger, 2002
Mt Gambier, South Australia	unconfined limestone	drinking	untreated urban stormwater	organics, nutrients	Gorey and King, 2006 Vanderzalm *et al.*, 2005

14.6 CONCLUSIONS

With climate change and increasing urban populations, water reuse for non-potable purposes will increasingly substitute for drinking water supplies, particularly for large and centralised uses such as industrial supplies and public open space irrigation. However, water scientists and engineers recognise that indirect potable reuse may also be an economic option in the longer term, due to the relatively high costs of second distribution systems in relation to costs for advanced water treatment. It is ironic that there is currently more unintentional potable reuse via groundwater in developing and developed countries than there is planned potable reuse. Studies of these unplanned systems that acknowledge reuse and consider adaptations to manage risk in reuse could be expected to facilitate transition to planned systems.

Reuse via managed aquifer recharge has potential as an adaptive strategy for cities with existing unplanned recharge and reuse, to increase safety and reliability of water supplies in water-stressed regions. Subsurface storage can offer economic, environmental and health benefits over above-ground urban water storage where suitable aquifers are available. The characteristics of aquifers that favour subsurface storage of reclaimed water have been identified. Public confidence in water recycling is known to increase when reclaimed water is put back into natural systems such as streams, lakes and aquifers before recovery for reuse. Natural systems are generally robust with respect to removal of human pathogens, the dominant concern to human health. Putting reclaimed water into the natural environment increases the residence time before reuse and so assists in biodegradation of contaminants that degrade more slowly. Sorption of metals and some organics may also provide effective removal from the water, if sorbed concentrations do not overwhelm processes by which such substances are inactivated, or if sorbent is removed. In streams, sometimes there is dilution by mixing with other flow. In aquifers, while this is also true, often the ambient groundwater quality is brackish, so unacceptable for the intended reuse, and mixing needs to be restricted.

Recognizing unintentional water reuse via aquifers will not only change human perception of planned reuse but will also increase awareness of the importance of properly closing the hydrological cycle and facilitating the natural treatment processes that occur in aquifers. Cities in different hydrogeological settings in developed and developing countries may adopt different balances between engineered and natural treatments, and knowledge of aquifer-specific treatment capabilities and limitations will be vital for wise selection of strategies. For managed aquifer recharge to be effective in integrated water resources management and water reuse requires its value and limitations to be understood by water utilities, urban planners, water resources managers, environmental and health regulators, and the public. Transition from unplanned to planned reuse via aquifers will demand a greater knowledge of the residence time and sustainable treatment capabilities of aquifers as a common component in wastewater treatment and water supply systems.

14.7 ACKNOWLEDGEMENTS

This paper was prepared with the support of CSIRO Land and Water, the Water for a Healthy Country Program of CSIRO, and by the Salisbury ASTR storm water to drinking water research project, a component of the European Union project RECLAIM WATER. The ASTR project is proudly supported by the Department of Education, Science and Training International Science Linkages Program established under the Australian Government's innovation statement, Backing Australia's Ability (Grant number: CG080091), and by the Government of South Australia, Minister for Science and Information

Technology, Hon. Mike Rann, under the Premiers Science and Research Fund. The authors gratefully acknowledge contributions to information in tables of case studies by Avi Aharoni, Herman Bouwer, Constantino Masciopinto, Paul Pavelic, David Pyne, Simon Toze, Gideon Tredoux, Emmanuel Van Houtte, Mike Wehner and Thomas Wintgens, and reviews of the draft manuscript by Paul Pavelic and Rudi Regel.

14.8 REFERENCES

Banin, A., Lin, C., Eshel, G., Roehl, E.K., Negev, D., Greenwald, D., Shachar, Y., Yablekovitch, Y. (2002) Geochemcial processes in recharge basin soils used for municipal effluents reclamation by the soil-aquifer treatment (SAT) system. In: Dillon, P (ed) Management of Aquifer Recharge for Sustainability. Swets and Zeitlinger, pp. 327-332.

BGS, CNA, SAPAL, WAJ, DMR and PSU (1998) Protecting groundwater beneath wastewater recharge sites. Final Report. BGS Technical Report WC/98/39.

Bouwer, H. (2002). Artificial recharge of groundwater: hydrogeology and engineering. Hydrogeology Journal **10**: 121-142.

Capella, A. (2002) Situación del Agua en el Valle de México. World Bank Report [In Spanish]

Central Ground Water Board (2005). STAP/UNESCO Workshop on managing the subsurface environment, 19-22 Sept 2005 – Field guide to sites of rain water harvesting and artificial recharge to ground water in NCT of Delhi. Ministry of Water Resources, New Delhi.

Chapman, D. (1992). Water Quality Assessments. Ed Chapman Hall, Great Britain

Chilton, J. and Kinniburgh, D. (2003) Soil and Groundwater Protection in the South-East Asia Region, British Geological Survey. www.unescap.org/esd/water/publications/cd/escap-iwmi/keynote.pdf

Cifuentes, E., Suárez, L., Espinosa, M., Juárez, L, Martinez-Palomo, A.. (2004). The risk of Giardia intestinalis infection in children from an artificially recharged groundwater area, Mexico City. *American Journal of Tropical Medicine and Hygiene* **71**(1):65-70

Cifuentes, E., Suárez, L., Solano, M.. and Santos, R. (2002). Diarrhea in children from water reclamation-site, Mexico City. *Environmental Health Perspective* **110** (9):219-624 .

Cisneros Jiménez , B..and Ordonez, J.E. (1999). The Mezquital valley aquifer recharge due to massive reuse of municipal wastewater for agricultural irrigation. p15-24 In Proc. 9th Biennial Symp on Artificial Recharge of Groundwater, Tempe, Arizona, 10-12 June, 1999.

Crooke, J. (2004) Innovative applications in water reuse: ten case studies. WateReuse Association. Report, May 2004, 41pp..

Dillon, P., Pavelic, P., Toze, P., Rinck-Pfeiffer, S., Martin, R., Knapton, A., Pidsley, D. (2006). Role of aquifer storage in water reuse. *Desalination*, **188**, 123-134.

Dillon, P.J. (2005). Future management of aquifer recharge. *Hydrogeology Journal*, **13** (1): 313-316.

Dillon, P. and Toze, S. (eds) (2005). Water Quality Improvements During Aquifer Storage and Recovery. AWWARF Report 91056F (286p+2CD).

Dillon, P., Martin, R., Rinck-Pfeiffer, S., Pavelic, P., Barry, K., Vanderzalm, J., Toze, S., Hanna, J., Skjemstad, J., Nicholson, B. and Le Gal La Salle, C. 2003. Aquifer storage and recovery with reclaimed water at Bolivar, South Australia. In: Proc. 2nd Australian Water Recycling Symposium, Brisbane 1-2 Sept 2003. Australian Water Association, Artarmon, NSW, Australia.

Dillon, P. Pavelic, P. Toze, S. Ragusa, S., Wright, M., Peter, P. Martin, R., Gerges, N., and Rinck-Pfeiffer, S. (1999). Storing recycled water in an aquifer: benefits and risks. Aust Water & Wastewater Assoc. *J. Water* **26** (5) 21-29.

Dillon, P., Pavelic, P., Sibenaler, X., Gerges, N., and Clark, R. 1997. Aquifer storage and recovery of stormwater runoff. Australian. Water & Wastewater Assoc. *J. Water* **24** (4) 7 - 11.

Dillon, P.J. and Pavelic, P. (1996). Guidelines on the quality of stormwater and treated wastewater for injection into aquifers for storage and reuse. Urban Water Research Assoc of Aust., Research Report No 109, July 1996, 48p.

Dillon, P.J. (ed) (2002). Management of Aquifer Recharge for Sustainability. Proc.4th Intl Symp on Artificial Recharge. A.A. Balkema Publishers.

EWRI/ASCE (2001) Standard Guidelines for Artificial Recharge of Ground Water. American Society of Civil Engineers, 106p.

FID BGS British Geological Survey (1995-1998) DFID Protecting Groundwater Beneath Wastewater Recharge Sites R Number R6231 http://www.dfid-kar-water.net/projects/files/R6231.html

Foster S., Garduño, H., Tuinhof, A.., Kemper, K. and Nanni M. (2000) Urban Wastewater as Groundwater Recharge Evaluating and Managing the Risks and Benefits Briefing Note 12

Foster, S., Chilton, J., Moench, M., Cardy F., and Schiffler, M. (2000) Groundwater in rural development. Facing the challenges of supply and water resource sustainability. The World Bank Ed., Washington. D.C., USA

Foster, S., Gale, I., Espanhol, I. (1994) Impacts of wastewater use and disposal of groundwater British Hydrogeological Survey and WHO Technical report 94/55

Foster, S., Hirata, R., Gomes, D., Delia, M. and Paris, M. (2002) Groundwater Quality Protection: A Guide for Water Utilities. Municipal Authorities and Environment Agencies. The World Bank, Oxford, UK

Foster, S., Lawrence, A. and Morris, B. (1998) Groundwater in Urban Development Assessing Management Needs and Formulating Policy Strategies World Bank technical paper No. 390 The World Bank Washington, D.C.

Fox, P. et al, (2001) An investigation of soil aquifer treatment for sustainable water reuse. AWWA Research Foundation and American Water Works Association Report.

Gallop, G. (2003). Securing Our Future – A State Water Strategy for Western Australia. Office of Premier, Govt. of Western Australia.

Gelt. J., Henderson. J., Seasholes. K., Tellman. B., Woodard. G., Carpenter. K., Hudson. C. and Sherif, S.(1999) In search of Adequate Water Supply, in Water in the Tucson Area seeking sustainability in A status report by the Water Resources Research Centre, College of Agriculture, The University of Arizona® Water Resources Research Centre

Gorey, P.G. and King, H.J. (2006). Developing regulatory controls for stormwater discharge to a potable aquifer in regional South Australia. In: Proceedings ISMAR 5 International Symposium on Management of Aquifer Recharge, Berlin, June 10-16, 2005. UNESCO (in press)

Greskiowak, J., Prommer, H., Vanderzalm, J., Pavelic, P., Dillon, P. (2005). Modeling of carbon cycling and biogeochemical changes during injection and recovery of reclaimed water at Bolivar, South Australia. Water Resources Research, 41, W10418.

Grünheid, S. and Jekel, M. (2006) Fate of bulk organics during bank filtration and groundwater recharge of wastewater-impacted surface waters. In: Proceedings ISMAR 5 International Symposium on Management of Aquifer Recharge, Berlin, June 10-16, 2005. UNESCO (in press)

Grützmacher, G., Wessel, G., Bartel, H., Chorus, I. and Holzbecher, E. (2006) On the behavior of microcystins in saturated porous medium. In: Proceedings ISMAR 5 International Symposium on Management of Aquifer Recharge, Berlin, June 10-16, 2005. UNESCO

Guttman, Y., Sellinger, A. and Bein, A. (2002) Simultaneous freshwater production and wastewater reclamation in a coastal aquifer at the Dan Plant, Israel. In: Dillon, P (ed) Management of Aquifer Recharge for Sustainability. Swets and Zeitlinger, pp. 321-326

Heberer, T. (2006) Occurrence, Transport, Attenuation and Removal of pharmaceutical residues in the aquatic environment and their relevance for drinking water supply in urban areas. In: Proceedings ISMAR 5 International Symposium on Management of Aquifer Recharge, Berlin, June 10-16, 2005. UNESCO

Herczeg, A.L., Dillon, P.J., Rattray, K.J., Pavelic, P. and Barry K.E. 2004. Geochemical processes during five years of aquifer storage recovery. *Ground Water* **42**(3):438-445.

Hodgkin, T. in press. Aquifer storage capacities of the Adelaide region. South Australia Department of Water, Land and Biodiversity Conservation Report 2004/47.

Icekson, N. and Blanc, R (1998) Wastewater treatment and groundwater recharge for reuse in agriculture: Dan Region reclamation project, Shafdan. In: Peters, et.al (eds) Artifical Recharge of Groundwater. Balkema, Rotterdam, p 99-103

Jiménez, B. (2005) Treatment Technology And Standards For Agricultural Wastewater Reuse: A Case Study in Mexico. *Irrigation and Drainage Journal* **54**:1-11

Jiménez B. (in press) UNESCO EOLSS Water Encyclopaedia

Jiménez B. and Chavez A. (2004) Quality assessment of a potential use of an aquifer recharged with wastewater: "El Mezquital case" *Water Science and Technology* **50** (2): 269-273

Jiménez, B., Mazari, M., Cifuentes, E. and Domínguez, R. (2005) The water in the Mexico City Valley. In Jiménez B. and Marin L. The water view from the Mexican Academy of Science: 15-32. Mexico City, Mexico [in Spanish]

Knappe, A., Massmann, G., Richter, D., Ohm, B. and Pekdeger, A. (2006) Exploring surface and groundwater interactions with help of environmental tracers and sewage indicator in Berlin. In: Proceedings ISMAR 5

International Symposium on Management of Aquifer Recharge, Berlin, June 10-16, 2005. UNESCO (in press)

Le Gal La Salle, C., Vanderzalm, J.L., Hutson, J.L., Dillon, P.J., Pavelic, P. and Martin, R. (2005). Isotope evolution and contribution to geochemical investigations in aquifer storage and recovery: a case study using reclaimed water at Bolivar, South Australia. *Hydrological Processes*, **19**, 3395-3411.

Martin, R. and Dillon, P. 2002. Aquifer storage and recovery – future directions for South Australia. Dept. for Water Land and Biodiversity Conservation Report. DWLBC 2002/04. Aug 2002. http://www.dwlbc.sa.gov.au/water/ publications/pdfs/reports/asr_future_directions_2002_04.pdf

Masciopinto, C., Benedini, M., Troisi, S., and Straface, S. (2000) Conceptual models and field test results in porous and fractured media. p245-270 in 'Groundwater Pollution Control'. K. L. Katsifarakis (ed.), WIT Press, Boston.

Masciopinto, C. and Carrieri, C. 2002. Assessment of water quality after 10 years of reclaimed water injection: the Nardò fractured aquifer (Southern Italy). Ground Water Monitoring & Remediation, Winter, 22 (1), 88-97.

Massmann, G., Greskowiak, J., Dünnbier, U. and Zuehlke, S. (2006) The influence of variable redox conditions on the behavior of pharmaceutically active compounds during artificial recharge in Berlin. In: Proceedings ISMAR 5 International Symposium on Management of Aquifer Recharge, Berlin, June 10-16, 2005. UNESCO (in press)

Mazari-Hiriart, M., López Vidal, Y., Castillo-Rojas, G., Ponce de León, S. and Cravioto, A. (2001). Helicobacter pylori and Other Enteric Bacteria in Freshwater Environments in Mexico City. *Archives of Medical Research* **32** (5):458-467.

Mazari-Hiriart, M., Cifuentes, E., Velázquez E, and Calva J. (2000). Microbiological Groundwater Quality and Health Indicators in Mexico City. *Urban Ecosystems* **4** (2):91-103.

Mills, W.R. (2002). The quest for water through artificial recharge and wastewater recycling. p3-10 in Management of Aquifer Recharge for Sustainability (ed P.J. Dillon) A.A. Balkema Publishers.

Nadebaum, P., Chapman, M., Morden, R. and Rizak, S. 2004. A guide to hazard identification and risk assessment for drinking water supplies. CRC for Water Quality & Treatment Research Report 11.

National Water Quality Management Strategy (2004). Australian Drinking Water Guidelines. National Health and Medical Research Council and Natural Resource Management Ministerial Council.

National Water Quality Management Strategy (2005). National Guidelines for Water Recycling: Managing Health and Environmental Risks. Natural Resource Management Ministerial Council and Environment Protection and Heritage Council (draft for public consultation).

Oliver, Y..M., Gerritse, R.G., Dillon, P.J. and Smettem, K.R.J. (1996). Fate and mobility of stormwater and wastewater contaminants in aquifers. II. Adsorption studies for carbonate aquifers. Centre for Groundwater Studies Report No. 68.

Pavelic, P., Dillon, P.J. and Simmons C.T. (2006) Multi-scale characterization of a heterogeneous aquifer using an ASR operation. *Ground Water* **44**(2), 155-164.

Pavelic, P., Dillon, P.J., and Nicholson, B.C. (2006). Comparative evaluation of the fate of disinfection by-products at eight aquifer storage and recovery sites. *Environ. Sci. & Technol.*, **40**, 501-508.

Pavelic, P., Nicholson, B.C., Dillon, P.J., and Barry, K.E. (2005). The fate of disinfection by-products in groundwater: A case study from the Bolivar aquifer storage and recovery site. *J. Contaminant Hydrology* **77**, 351-373.

Peters, J. (ed) (1998). Artificial Recharge of Groundwater. Proc 3rd Intl. Symp. on Artificial Recharge of Groundwater. Amsterdam. A.A. Balkema Publishers.

Pitman, C. (2003). Stormwater harvesting and utilization. In: Proc. Water Recycling Aust. 2nd National Conf., Brisbane, 1-3 Sept, 2003. Australian Water Association, Artarmon, NSW, Australia.

Pyne, R.D.G. (2006). Aquifer Storage Recovery: A guide to groundwater recharge through wells. ASR Systems, Gainesville, Florida, 620p.

Radcliffe, J. 2004. Water Recycling in Australia. Review for the Australian Academy of Technological Sciences and Engineering. www.atse.org.au

Rinck-Pfeiffer, S., Pitman, C. and Dillon, P. (2006). Stormwater ASR in practice and ASTR (Aquifer Storage Transfer and Recovery) under investigation in Salisbury, South Australia. Proc. ISMAR5, Berlin, June 2005. UNESCO (in press)

Sloss, E., Geschwind, S.A., McCaffrey, D.F. and Ritz, B.R. (1996). Groundwater recharge with reclaimed water: an epidemiological assessment in Los Angeles County, 1987-1991. RAND Corporation, Santa Monica, California.

Sloss, E., McCaffrey, D.F., Fricker, R.D., Geschwind, S.A. and Ritz, B.R. (1999). Groundwater recharge with reclaimed water: birth outcomes in Los Angeles County, 1982-1993. RAND Corporation, Santa Monica, California.

Skjemstad, J., Hayes, M.H.B. and Swift, R.S. (2002) Changes in natural organic matter during aquifer storage. In: Management of Aquifer Recharge for Sustainability, P.J. Dillon (Ed.) Proceedings of the 4th International Symposium on Artificial Recharge (ISAR4), Adelaide Sept. 22-26, 2002, Swets & Zeitlinger, Lisse, ISBN. 90 5809 527 4, pp 149-154.

South Australia Environment Protection Authority (2004). Code of Practice for Aquifer Storage and Recovery. Jan 2004. www.epa.sa.gov.au/pdfs/cop_aquifer.pdf

Swierc, J., Page, D., Van Leeuwen, J. and Dillon, P. (2005). Preliminary Hazard Analysis and Critical Control Points Plan (HACCP) - Salisbury Stormwater to Drinking Water Aquifer Storage Transfer and Recovery (ASTR) Project. CSIRO Land and Water Tech Report 20/05, Sept 2005. www.clw.csiro.au/publications/technical2005/tr20-05.pdf

Tejada-Guibert, J. And Maksimovic, C. (2001) Outlook for the 21st Centry. In Maksimović, C and Tejada-Guibert, J (eds) Frontiers in Urban Water Management: Deadlock or Hope. IWA Publishing, 2001, pp.399-409.

Toy, D., Goldman, F., Walker, J. and Voorhees, D. (1999) Aquifer recharge and recovery: a case study of the Sun Lakes effluent recharge/recovery pilot project – first year of operation. In: Artificial Recharge and Integrated Water Management", Proc 9th Biennial Symposium Artificial Recharge of Groundwater, Tempe, Arizona 1999, pp. 81-90.

Toze, S., Hanna, J., Smith, T., Edmonds, L. and McCrow, A. 2004. Determination of water quality improvements due to the artificial recharge of treated effluent. Wastewater Reuse and Groundwater Quality *IAHS Publication* **285**:53-60.

Toze, S., McFarlane, D. and Turner, N. (2005). Water reuse technologies. Research partnerships providing solutions and technologies for water reuse.
www.cmis.csiro.au/healthycountry/brochures/water_reuse_factsheet.pdf

Tredoux, G., Cavé, L.C. and Bishop, R. (2002) Long-term storm water and wastewater infiltration into a sandy aquifer South Africa. In: Management of Aquifer Recharge for Sustainability. Dillon, P (ed), Swets and Zeitlinger, pp. 35-40.

Tredoux, G, Murray, E C, and Cavé, L C, (2003). Infiltration Basins and other Recharge Systems in Southern Africa. p67 – 77 In: Tuinhof, A., Heederik, J.P. (eds) Management of Aquifer Recharge and Subsurface Storage. Netherlands National Committee of IAH, Utrecht Netherlands. NNC-IAH Publicn. No. 4.

Tuinhof, A., Heederik, J.P. (eds) (2003). Management of Aquifer Recharge and Subsurface Storage. Netherlands National Committee- Intl Assoc Hydrogeol., NNC-IAH Publicn. No. 4.

Unangst, L., Fisher, A. and Swieczkowski, D. (1999) Trends and issues in effluent recharge for the Phoenix and Tucson active management areas. . In: Artificial Recharge and Integrated Water Management", Proc 9th Biennial Symposium Artificial Recharge of Groundwater, Tempe, Arizona 1999, pp. 67-77.

UK Water Industry Research Ltd (in press). Framework for developing water reuse criteria with reference to drinking water supplies. UKWIR Report.

UN (2003) Water for People Water for Life. The United Nations World Water

Van der Merwe, B., Tredoux, G., Johansson, P-O. and Jacks, G. (2005). Water quality requirements for artificial recharge of the Windhoek Aquifer. City of Windhoek, Tender No INF 260/2003, Final Report of Windhoek Aquifer Joint Venture Consultants.

Vanderzalm, J.L., Le Gal La Salle, C. and Dillon, P. J. (in press). Fate of organic matter during Aquifer Storage and Recovery (ASR) of reclaimed water in a carbonate aquifer. *Applied Geochemistry*.

Vanderzalm, J.L, Schiller, T., Burn, S., and Dillon, P. (2004). Impact of stormwater recharge on drinking water supply for Mount Gambier. Proc. Intl. Conf. on Water Sensitive Urban Design, Adelaide, Nov 2004.

Van Houtte, E. and Verbauwhede, J (2006). Artificial recharge of treated wastewater effluent enables sustainable groundwater management of a dune aquifer in Flanders, Belgium. In: Proceedings ISMAR 5 International Symposium on Management of Aquifer Recharge, Berlin, June 10-16, 2005. UNESCO (in press)

Wakida, F. and Lerner, D. (2005) Non-agricultural sources of groundwater nitrate: a review and case study *Water Research* **39:** 3–16.

Wilson L G, Amy G L, Gerba C P, Gordon H, Johnson B and Miller J, 1995. Water quality changes during soil aquifer treatment of tertiary effluent. *Water Environ. Res.* **67** (3), 371-376.

World Health Organization (2003). State of the Art Report Health Risks in Aquifer Recharge Using Reclaimed Water. Eds. R. Aertgeerts and A. Angelakis, Water Sanitation and Health Protection and the Human Environment WHO Geneva, and WHO Regional Office for Europe Copenhagen, Denmark. SDE/WSH/03.08.

SECTION THREE

EMERGING TOPICS

15

Ethical dilemmas in water recycling

Cheryl K. Davis

15.1 INTRODUCTION

As professionals in the water industry, we often perceive ourselves as upright people seeking to serve the common good, producing high-quality water to support commerce and protect public health. Although not generally a flamboyant or self-dramatizing group, sometimes those of us who support increased use of recycled water may even detect a quiet heroism in our battle against public ignorance, government inefficiency (or even corruption), and/or corporate greed. We may be more inclined to congratulate than doubt ourselves when rallying behind a cause like water recycling which seems to offer so many benefits for both humanity and the environment. Recycling advocates, whose competence and motivations are often questioned by others, can end up feeling defensive or even self-righteous; this does not contribute to a spirit of open inquiry about possible ethical dilemmas.

But in a complex world rife with competing values, only a zealot could avoid noticing that the right path sometimes becomes unclear – not because one wishes to do wrong, but because there are so many competing goods. The technical, legal, political, economic, regulatory and environmental challenges of water recycling are such that there is a tendency to ignore or oversimplify ethical issues. We are tempted to pretend (both to ourselves and the public) that there is a scientific, legal, or economic answer to every question. We are often more comfortable speaking in technical terms (where we are the experts) than in terms of values, which put us on a more level playing field with the public and other stakeholders. As a result, the values imbedded in our recommendations may not be clearly articulated or openly discussed.

Ethical issues in recycling can't be meaningfully discussed in isolation. They will be discussed here as they must be addressed in real life – in the context of a menu of options

 Water Reuse - An International Survey of current practice, issues and needs by B. Jimenez *et al.* ISBN: 9781843390893. Published by IWA Publishing, London, UK.

available for meeting water demand. At the same time that growing populations are increasing demand, many supplies are being degraded. Thomas Terne *et al.* (2006) cite projections that by 2025, two-thirds of the world's population will face conditions of moderate to high water stress, and about half the population will be confronted by real constraints in their water supply.

The intent of this chapter is to illustrate the number and complexity of the ethical issues associated with water resource planning in general and water recycling in particular, and to suggest approaches for analysing the costs and benefits of our choices. These trade-offs will be discussed using the Triple Bottom Line analytical framework defined by John Elkington (1998). In *Cannibals with Forks*, he recommends that businesses develop strategies to address three bottom lines: economic prosperity, environmental quality, and social justice. Elkington acknowledges that the three bottom lines are not stable and that they move independently, making it difficult to address the three simultaneously. In reality, trade-offs have to be made among competing bottom lines, as well as different approaches to addressing each one of them. The ethical dilemmas associated with water recycling reside in the ambiguous territory where many values have merit, and none can be optimised.

15.2 ECONOMIC BOTTOM LINE

The most basic responsibility of water utilities is to support public health, safety, and economic well-being by providing an adequate supply of safe water. On a global level, water supply remains an unsolved problem. UNICEF's Joint Monitoring Program for Water Supply and Sanitation estimates that 1 billion people (in Africa, two out of five) are currently without access to adequate water supplies. Although climate is an obvious factor, availability of funding for investment can also be decisive.

In *The Value of Water: Concepts, Estimates, and Applications for Water Managers*, the authors Robert S. Raucher *et al.* (2005) point out that water managers try to minimize the costs to their customers; therefore, utilities deliver first the water which costs least. If demand grows, a utility will typically add its next lowest cost option to meet the next increment of demand. This point is illustrated (Figure 15.1) with a water supply curve that shows increasing costs by source of supply. In this example, local groundwater, treated local surface water, and conservation are all less expensive than reclaimed water.

This reality is reflected in San Francisco's local water supply options study, in which five potential water supply strategies for meeting increased demand in the future (water conservation, increased use of groundwater, recycled water, desalination, and rationing during droughts) were analyzed in terms of three criteria: affordability; reliability; and responsible management of entrusted resources (Figure 15.2). Water conservation, groundwater, and rationing received more favorable marks on the affordability scale than recycled water, although recycled water received a positive score for reliability and environmental stewardship.

The cost differences between these sources can be expected to change over time. A report by the American Water Works Association Research Foundation on "Managing the Future: Trends in Drinking Water" indicates that changing treatment technologies will narrow the cost gap. Ed Means *et al.* (2005) predict that "as the population in general continues to grow, and traditional water supplies are less available, we'll need to develop more marginal supplies. Thirty years ago we would not have dreamed of drinking recycled water . . . because it was so far out of reach technologically and other supplies were available. That's simply not the case today. . . as technology becomes cheaper, the cost of marginal supplies continues to shrink until they are nearly the cost-effective source of expanded supplies."

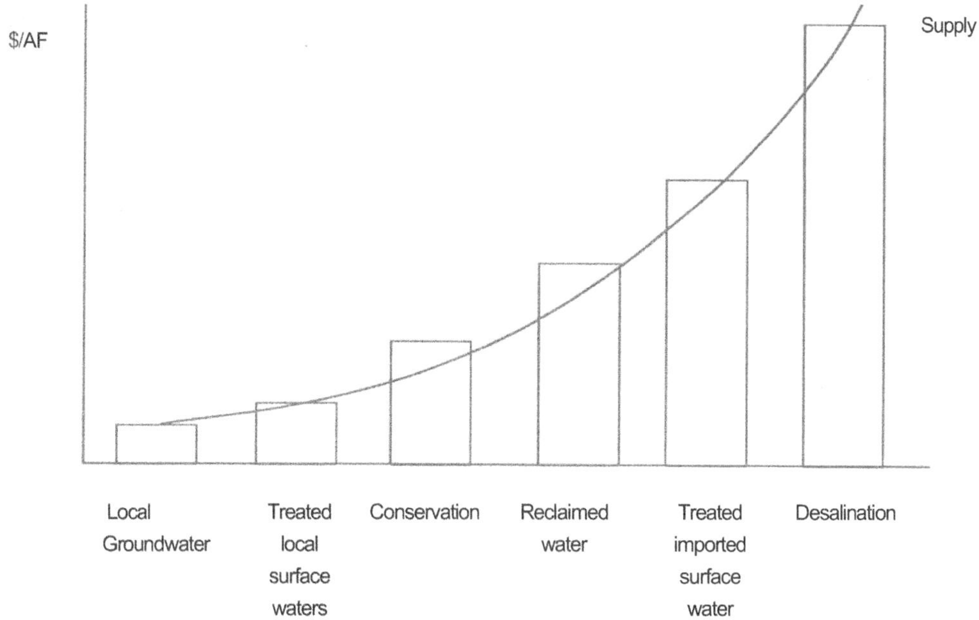

Figure 15.1. The cost of options for delivering potable water shapes a utility's supply curve (Raucher, *et al.*, 2005).

	Water Conservation	Groundwater	Recycled Water	Desalination	Rationing During Drought
Affordability	+	+	−	−	✓
Reliability	+	✓	✓	✓	−
Responsible Management of Entrusted Resources	+	✓	+	−	+

+ More Favorable
✓ Moderately Favorable
− Less Favorable

Figure 15.2. Water supply planning section, Water Enterprise of San Francisco Public Utilities Commission, 2005.

In considering water supply alternatives, the economic bottom line is sometimes defined by decision-makers as the immediate costs and benefits for current customers. However, this analysis should also consider benefits to the economy which the utility supports, as well as future rate-payers. Where recycled water is a more reliable source due to its location, its accessibility, its availability during droughts, or the risk that other sources may not be available in the future, the economic bottom line of a community may be better served by development of recycled water than by purchase of other supplies that are currently less expensive.

It can also be argued that higher water rates have the advantage of discouraging water waste. In developed nations, water is often under-priced, with rates that fail to take into account long-term infrastructure maintenance, the marginal costs of new water sources as demands increase, or environmental costs of producing the water. In countries where many consumers can afford to pay high prices for bottled water, it seems feasible to raise water rates to cover the cost of more expensive supplies; this allows utilities to remain financially viable while providing financial incentives for water conservation.

However, the situation in developing countries can be very different. It should be noted that the assumption that recycled water is an expensive water source reflects the water treatment standards of developed nations. In reality, wastewater is used in some parts of the world for irrigation with little or no treatment, making it an inexpensive and economically valuable source of supply. In "Social, Political, and Scientific Dilemmas for Massive Wastewater Reuse in the World," Blanca Jiménez and Hector Garduño (2001) state that "Wastewater use is essentially ethical, because it is an economical option – and perhaps the only feasible one – for coping with developing countries' water scarcity. The massive reuse of water can be achieved only through agricultural irrigation." They point out that use of untreated sewage effluent in poor countries has increased agricultural productivity (and poor farmers' incomes) at a relatively low cost because no synthetic fertilizers are required. They also acknowledge (Table 15.1) that this practice has contributed to the spread of waterborne disease (citing an irrigation district downstream of Mexico City where children exposed to untreated wastewater experienced *Ascaris Lumbricoides* infection rates 16 times as high as children from an area irrigated with potable water). Because of the equipment, expertise, energy, and treatment chemicals required to produce recycled water which meets the effluent standards set by wealthy countries, highly-treated recycled water may not be a commodity that poor countries can afford. They argue that treatment technology and standards should be tailored to specific needs (e.g. the desirability of retaining nutrients in water used for agricultural purposes), so that recycled water can be both affordable and safe.

The water industry actively supports advancements in technology in water and wastewater treatment. For example, IWA's World Water Congress in Beijing (2006) included discussion of emerging biological and nano-technologies. Such technologies will be invaluable in expanding our options for safely producing additional water from new sources. However, no ladder has been built to help developing countries move incrementally from food crop irrigation with untreated wastewater to levels of treatment that would produce water safe for multi-purpose use. The NEWater Project in Singapore, for example, which treats recycled water for potable use, using complex technology and multiple treatment barriers, is one vision of the cities of the future. However, many cities do not currently have sufficient resources to realize this vision.

High-tech solutions, even if affordable in initial construction due to private investment or subsidized loans, may not be affordable or maintainable long-term. Complex multi-barrier systems with diverse treatment processes and automated quality control require operations and maintenance by staff with sophisticated skills in water treatment and computer technology. Where financial resources are limited, water resource planning should address not only up-front investments but also the life cycle costs of treatment technologies.

While cost issues are more pronounced in developing countries, water rate increases can also have a negative impact on low-income rate-payers in more wealthy countries. Where all consumers pay the same water rate, a choice that benefits the economy as a whole may be disproportionately burdensome for low-income rate-payers. For example, when Poland was governed by a Communist regime, water infrastructure was reportedly inadequately maintained, but water rates were relatively low. When a private water company began providing water service in Gdansk, both the treatment and distribution systems were upgraded; however, one manager noted that some people could no longer afford to bathe as often as before. While the water industry has tended to simplify its billing by charging the same rates for all customers, the flat rate approach that has been both administratively efficient and justifiable as fair in the past may not be socially defensible when rates rise to cover the costs of sound asset management and more expensive supplies.

Table 15.1. Comparison of the frequency of waterborne illness in the Mezquital Valley and an area that uses clean water (Jiménez, B. and Garduño, 2001).

		Death Rate		
Species	Affected Population by Age	Irrigated With Wastewater (A)	Irrigated With Clean Water (B)	Relation Ratio (A/B)
Ascaris lumbricoides	0 to 4	15.3	2.7	5.7
	5 to 14	16.1	1.0	16.0
	15+	5.3	0.5	11.0
Giardia lamblia	0 to 4	13.6	13.5	1.0
	5 to 14	9.6	9.2	1.0
	15+	2.3	2.5	1.0
Entamoeba hystolitica	0 to 4	7.0	7.3	1.0
	5 to 14	16.4	12.0	1.3
	15+	16.0	13.8	1.2

Janice Beecher (2001) has observed that there is a "clear tension between the efficiency goals of markets and the equity goals of government." She points out that an emphasis on cost recovery can make water less accessible and affordable for low-income communities. Although the frame of reference for her remarks is privatization, publicly-owned utilities which intend to produce more expensive water supplies may also need to consider discounts for low-income consumers.

In wealthy communities, analysis of the economic bottom line generally means the rate of return for the individual utility; in such cases, the community can afford to pay for the services it needs. Finding a way to a viable economic bottom line can be difficult in developing nations. Jerry Delli Priscolli and M. Ramón Llamas (2001) take the position that "although water pricing needs to take into account water costs, we must take into account the role that subsidies have played and are realistically going to play in the future."

Jeremy Pelczer (2005), while serving as President of American Water and Deputy CEO of RWE Thames Water, acknowledged that unrestrained market processes often leave the poor under-served. He noted that in many developing countries, water policies have given insufficient attention to ensuring that the needs of the poorest in society are met. While reasons for this are various (including political unwillingness to give any priority to those who are unlikely to vote, and fear of legitimising informal settlements), the consequences are clear: middle-class consumers pay lower prices for adequate service than the poor pay street vendors for inferior water. He suggested formation of partnerships between private companies, international NGOs, willing municipalities, and community-based organizations, with each partner doing what they do best. For example, a private company might provide expertise and human resources, with a local entity providing billing and collection.

In reality, many private companies have found it more difficult than expected to achieve an acceptable rate of return while upgrading service levels in developing countries. The attempt to address social inequities while providing an adequate rate of economic return to stimulate investment is an example of what John Elkington describes as a "sustainability challenge."

15.3 ENVIRONMENTAL BOTTOM LINE

Jerry Delli Priscolli and M. Ramón Llamas (2001) have described environmental stewardship as a moral imperative: "The ethical principle of stewardship teaches respect for creation and moral responsibility to that creation." Similarly, Eric Rosenblum has proposed that we make decisions based on an awareness of the interdependence of life on earth. He believes that adherence to environmental ethics implies an ethical consideration of all living beings (biospherical egalitarianism) and respect for the rights of future generations (intergenerational equity). He argues that we should accept responsibility for long-term environmental impacts, while acknowledging the limitations of our knowledge about the consequences of our actions. He argues for an expanded scale of responsibility, both in space and in time. He notes in this discussion that recycled water can be environmentally preferable to tapping new supplies.

From an environmental perspective, recycled water offers many potential benefits:

- protecting aquatic species by avoiding additional diversions from streams and rivers;
- restoring wetlands and other natural habitats;
- eliminating or reducing overdraft of groundwater supplies; and
- avoiding degradation of receiving waters (e.g., pollution of streams, or freshwater intrusion into saltwater habitats).

On the other hand, environmentalists sometimes prefer water conservation to water recycling, for reasons which include the following:

- development of any new water supply is sometimes viewed as a stimulus to further population and commercial growth, addressing one resource need while increasing pressure on the underlying ecosystem. Robin Saunders (2001) has noted that "In regions where development and growth are pressing issues, limiting the water supply is often construed as a means to control growth."
- recycled water projects involve construction of treatment and distribution facilities, as well as ongoing use of equipment, power, and chemical supplies. As a result, environmentalists often promote water conservation as a less intrusive alternative.
- wastewater treatment (including treatment expansions needed to support water recycling) is disproportionately located in low-income neighborhoods, raising environmental justice issues. Although one approach to water supply planning is to try to achieve the greatest good for the greatest number, Peter Wenz (2004) has noted that "in utilitarianism, there is no guarantee that the good of some will not be sacrificed for the greater good of all." With water recycling (as with any form of water treatment), location has both economic and ethical implications.
- concerns have been expressed about the potential environmental impacts of pharmaceuticals and chemicals used in personal care products on organisms in receiving waters, particularly in environmentally sensitive areas (e.g., for wetlands restoration).

These environmental concerns can be proactively addressed by water managers. For example, while water professionals have historically preferred to avoid involvement in land use planning, some have come to the conclusion that there is an unavoidable nexus between water supply and population growth. Randele Kanouse (2001), in "Water Supply Planning and Smart Growth," recommends that utilities inform decision-makers and the public about the level of growth current supplies will support, as well as any additional supplies that will be required if there is new development.

Impacts associated with construction and operation of water recycling facilities can be reduced if environmental considerations play a key role in selection of treatment technologies and design of treatment facilities. In their article on "Implementing Sustainability in Water Recycling," authors Lydia Holmes, Michael Ban, Tom Fox, Jim Hagstrom and Susan Stutz-McDonald (2005) describe the use of a number of environmental analysis tools to encourage environmentally sound implementation of recycling projects:

- the Natural Step^{TM}, a framework for evaluating sustainability and establishing goals;
- the Ecological Footprint, an analytical framework which compares alternatives in terms of material and energy consumed and disposed of, using a common unit of comparison (acres); and
- the Leadership in Energy and Environmental Design (LEED^{TM}) Green Building Rating System, a performance-based certification system for buildings that meet environmental performance standards.

These tools were used by the City of Petaluma, California, in planning for a project to replace an existing water recycling plant with an ecologically and economically sustainable facility. The Natural Step^{TM} process was used to develop these project goals:

- meet discharge standards for effluent into the Petaluma River;
- expand the use of recycled water for irrigation of crops, parks, schools, and open space, to reduce use of potable water for non-potable purposes; and
- construct a sustainable facility to serve community educational and recreational needs.

Treatment alternatives were identified and screened with assistance from experts in pond systems, wetland systems, and conventional systems (Figure 15.3). Five treatment alternatives (ranging from conventional activated sludge to natural land-based systems, such as aerated lagoons) were evaluated using the Ecological Footprint process. This meant that alternatives were compared in terms of the materials and energy used for construction, as well as the chemicals and energy that would be required for operation.

This analysis resulted in selection of an extended aeration process using oxidation ditches; use of vegetated wetlands in the algae removal process; and use of ultra-violet rather than chlorine disinfection. In design of their facility, they considered the source of their materials (e.g., whether the materials were new or recycled); selected a sod-vegetated roof to reduce energy use; required waterless urinals; and used native plants for landscaping.

The City also included neighborhood quality criteria in their consideration of environmental impacts. These included odor control, mosquito abatement, visual appeal, noise control, and construction impacts on traffic patterns. In considering the educational and recreation potential of new facilities to be constructed, they evaluated alternatives in terms of their consistency with the City's river enhancement plan, bike plan, and marsh plan. Social benefits were taken into account, but did not override environmental stewardship commitments.

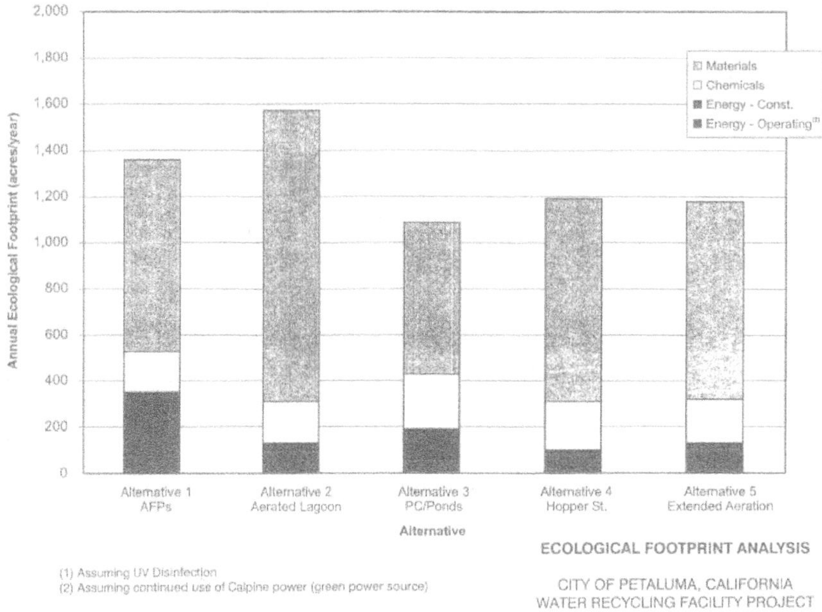

Figure 15.3. Ecological footprint analysis (City of Petaluma, 2000).

As illustrated by the above example, some treatment approaches to production of recycled water are environmentally preferable to others. One approach which offers potential benefits in terms of both construction impacts and environmental justice is decentralization of water recycling facilities. If, for example, small water recycling facilities can be placed at a site where new residences or commercial sites are being developed, those benefiting from the use of additional water will also make some contribution to replenishing the supply. Whereas centralized wastewater treatment facilities are often located in low-income neighborhoods where treatment loads can increase odors and the potential for sewer back-ups, placement of decentralized recycling facilities in new developments can distribute impacts more fairly. Decentralization of treatment and on-site use of recycled water can also minimize the environmental disruption associated with construction of distribution lines from centralized wastewater treatment facilities to sites where recycled water will be used.

The potential environmental impacts of treated wastewater on the environment are currently receiving intense scrutiny in the water industry, particularly in relation to pharmaceuticals and personal care products. Thomas Knacker *et al.* (2006) note in "Environmental Risk Assessment" that substances which remain in sewage treatment effluent will eventually turn up in rivers or lakes, posing a potential problem for the receiving environment. They cite research on various test organisms (e.g. zebrafish), test systems (mortality and fertilization rates), and concentrations required to demonstrate negative effects. They also discuss a risk assessment scheme which includes analysis of terrestrial toxicity, fish biocentration, and sediment toxicity.

In industrialized countries with high levels of pharmaceutical and personal care product use, the underlying problem can't be addressed by avoiding use of recycled water. These compounds can already be found in surface water, groundwater, and sludge. From a sustainability perspective, the ethical response to this potential environmental risk is not to dismiss it because it is already widespread, but to continue research on its implications for all water sources and uses.

15.4 SOCIAL BOTTOM LINE

Social issues which can generate ethical dilemmas for water professionals include the following:

- communication and stakeholder input;
- water quality;
- religious, cultural, and aesthetic values.

15.4.1 Communication and stakeholder input

The most basic ethical issue relating to communication with the public and stakeholders on water recycling issues is intent: whether or not policy-makers and water professionals actually wish to keep the public and other stakeholders informed and involve them in decision-making in a meaningful way. Bonnie Nixon (2001) recommends well-planned community involvement: "I believe people genuinely want to do the right thing. Too often we don't provide them with enough information . . . In fact, I often hear clients or other consultants say that stakeholders are dangerous if we arm them with enough information." She believes that water professionals have an ethical obligation to provide people with the right information and trust them to make the right decisions for themselves and society. However, her experience is that clients often ask her to help them use a public participation process to reach a predefined outcome. Water professionals often believe that they have already defined both the right questions and their answers.

Some problems which can undermine meaningful communication with stakeholders are discussed below:

(1) *Providing data which is incomprehensible to the public.*
Nixon notes that we frequently do not commit the resources needed to translate scientific jargon into plain English, so that the public can understand the criteria and tradeoffs involved in complex public decisions.

(2) *Superficial communication that does not address valid concerns*
Because of the complexity of the issues involved and the difficulty of discussing them effectively, there is a temptation to stress marketing rather than two-way communication. Advertising campaigns which over-simplify issues and gloss over issues of legitimate concern do not help consumers understand the implications of decisions which are being made on their behalf. Marketing is a valid activity for water providers, performed most effectively when the specific attributes of water are appreciated, so that water is not marketed like an ordinary consumer product. Marketing is different from public education, which is also an important component in promoting understanding and support of water recycling programs. Both marketing and public education may be informed and shaped by customer surveys, focus groups, and other forms of information-gathering on stakeholder questions and interests. However, in both marketing and public education, messages are crafted with the intention of achieving predefined outcomes. Involvement of the public and other stakeholders in the decision-making goes further. Stakeholders are not simply informed of the results of the utility's analysis, but invited to participate in the process, which requires some understanding on their part of the costs, benefits, and risks associated with various alternatives.

(3) *Inadequate listening skills*
Sometimes public hearings become pro forma, where comments are recorded without being addressed. Sometimes water professionals provide venues for input, but do not really listen to or empathize with public concerns.

(4) *Questionable timing*
Stakeholders are sensitive to whether their participation is requested at a point when questions are being defined or when proposed solutions are being offered.

(5) *Unwillingness to acknowledge or express values*
Jack Ward Thomas (1998) has commented on the aversion of resource professionals to acknowledging their own values in policy discussions with the public. He believes there is no way to avoid emotions in important resource management decisions, and that "avoidance of expressing values – other than to indicate unmitigated reliance on dispassionate science – results in persistent miscommunication between professionals and the public." He observes that "We speak calmly in science. The public speaks passionately in values." He believes that when professionals express their own values, they use a language that is more widely understood, resulting in communication that is not only more effective but more ethical: "Personal and professional integrity are more fully aligned when passions are expressed alongside science."

(6) *Limiting avenues for input to those which are easily provided, but not easily accessed by a wide range of stakeholders*
The time and travel constraints of the general public and important stakeholders are not always taken into account. Traditional public meetings and hearings are often dominated by special interest groups and vocal minorities who have the ability and motivation to attend meetings at inconvenient times and locations. Majority opinion may be under-represented, and the opportunity may be lost to gain the benefit of a wide range of perspectives. Overcoming this limitation requires finding the right tools and techniques to broaden opportunities for input, including community surveys, customer surveys, focus groups, media outreach, and web-based communication tools.

Maintaining effective communication with the public is not only ethically right but often a practical necessity in terms of program implementation. Ed Means (2005) has noted that the dynamics of communication with public and policy-makers have been radically altered by the formation of thousands of non-governmental organizations (NGOs) in recent years. They are largely single issue focused advocacy groups, whether they are protecting wet-lands or endangered species. In many cases they have built very effective grass roots communications networks by leveraging the potential of the Internet to disseminate information and mobilize action (e.g. letter-writing campaigns). He points out that highly focused, technologically sophisticated use of communication techniques can "run circles around a water utility's traditional means of managing information." Roger Patrick (2005) has noted that the disparity between the communication skills of NGOs and those of water utility professionals is exacerbated by the fact that many members of the public are not scientifically literate: "The lack of scientific and critical thinking and the proliferation of sensational reports is a dangerous mix."

Ed Means suggests that water managers, who typically have management backgrounds, need to become more adept at keeping the public informed, and should make the effort to find out, on any given issue, where the public stands: "Use survey information, focus groups, modern communication tools so you're never surprised." Water professionals are often aware that their technical skills exceed their skill in public education and stakeholder involvement. Competence can become an ethical issue when professionals assume responsibility for tasks they are unprepared to handle. Competence as an ethical issue has been addressed by David Schwartzel (2001) in relation to the skills of engineers who determine specifications for water treatment equipment, noting that the engineers have a "moral obligation, a duty of care, to understand the issues," and to spend the time that may be needed to educate themselves to adequately handle their responsibilities. Many water managers were educated at a time when public participation in water resource planning was limited. However, water professionals who wish to support water recycling efforts today must be prepared to address communication issues, both by improving their own skills and obtaining assistance as needed.

15.4.2 Water quality

Communication about the treatment and quality of recycled water is particularly challenging. The technical language and tools used by the water industry to evaluate, quantify, and describe health risks is often neither comprehensible nor credible to the public. The early socialization of human beings to keep waste out of their mouths is often more persuasive than scientific analysis published by experts. This learned aversion can be exacerbated by stakeholders and decision-makers who (for reasons ranging from paranoia and ignorance to self-interest) complicate public debate by distortion of the facts. The ability of water professionals to make responsible decisions is complicated by conflicting regulatory standards and, in some cases, a cultural climate that encourages litigation and claims for damages (Gaston, 2001).

Water quality can be perceived as a health risk by consumers even when the planned use is non-potable (e.g. use to irrigate turf areas used by school children for sports). It has been argued by Robin Saunders that there is a greater (although small) public health risk associated with distributed use of recycled waster for irrigation than with direct injection into a potable water reservoir: "As a matter of risk, the single controlled stream of recycled water being blended into a reservoir allows a lower public risk than potential contamination issues in the more usual distribution of recycled water for irrigation. For an equal volume of water being recycled, the non-potable water distribution system would be a greater risk for cross-connection and/or public exposure (as low as that is) in the hundreds of individual irrigation sites."

Whether the intent of a wastewater project is to serve non-potable purposes, replenish a drinking water reservoir, recharge groundwater as a barrier to seawater intrusion, or recharge groundwater for drinking water purposes, public concern often focuses on the danger that the water will be ingested. With irrigation of food crops, the possibility exists that contaminants in effluent will not be removed prior to use of the food. With use of recycled water for urban irrigation, firefighting, or irrigation of green spaces, there is a concern that fire-fighters or children may drink the water. Water intended to recharge aquifers for the sake of preventing seawater intrusion may end up in portions of the aquifer where groundwater is being extracted for potable purposes.

In some cases, public concerns seem to reflect a kind of water innocence. In reality, many drinking water supplies already contain wastewater effluent from upstream sources. Ternes *et al.* (2006) point out that "although not admitted, many cities in the world undergo incidental, unplanned, indirect potable reuse: in France (Aubergenville, Le Pecq, Rouen), the UK (London), the USA (Indianapolis, New Orleans, Philadelphia, etc.), where up to 70-95% of low flow surface water during summer periods is wastewater." From the perspective of scientific measurement, many surface water and groundwater sources are inferior in water quality to highly treated wastewater.

It is sometimes annoying to water professionals that the public lives so calmly with familiar risks and is so averse to changes which professionals perceive as having acceptable (or even lower) levels of risk. However, water professionals bear some responsibility for this, in that historically utility officials have not been eager to publicize or even investigate the wastewater contribution to public drinking water supplies. While changes inspire scrutiny, risks which are familiar generally do not. Peter Sandman (1996) in *Responding to Community Outrage: Strategies for Effective Risk Communication* has pointed out that people generally underestimate familiar risks and respond with more fear and outrage to unfamiliar ones. He says that the tendency of most managers to try to soothe and reassure the public on the risks of new projects tends instead to convince the public that managers are not taking the risks seriously enough – or worse yet, that the reality is so bad managers refuse to admit it. He suggests that it is more beneficial for managers to indicate that they share a concern than to downplay it. Sandman also cites memorability as a factor which tends to increase public concern, and mentions symbolism as an important source of memorability. An illustration of Sandman's point is a public relations campaign which effectively prevented implementation of an indirect potable reuse project in Southern California. This campaign focused on a "Toilet to Tap" visual which the public found much more compelling than the statistics provide by water professionals.

It is important for water professionals to investigate and report fully on the risks associated with all water sources (including those that have been in long-term use). It is also important to acknowledge the specific risks associated with inadequately treated wastewater. The nature of wastewater, which has higher concentrations of nitrite, ammonia, and organic nitrogen than most drinking water supplies, increases the likelihood of certain water quality problems. This is complicated by the addition of waste products associated with industry, agriculture, personal care, and human and veterinary medicine. With a continually expanding list of potential contaminants, the scientific community is continually behind in terms of being able to accurately measure their presence and potential health risk.

Use of recycled water, both intentionally for non-potable purposes and on an unplanned basis in drinking water, has an extensive history. However, current circumstances are unparalleled in terms of both the amount and type of pollutants entering the water stream. For example, global consumption of pharmaceuticals has been cited by Alfreda Ader *et al.* (2006) as 100,000 metric tons per year. Advances in medical technology have resulted in an alphabet soup of antiphlogistics, antibiotics, anti-diabetics, anti-epileptics, beta blockers, antihistamines, calcium antagonists, psychotropics, muscle relaxants, diuretics, decongestants, and anti-gout medications which enter the waste stream from sources which include factories, hospitals, and homes. These are modified to different degrees by human digestion and wastewater treatment. However, the wastewater treatment processes now most commonly in use were not designed to address these influents. As a result, these substances are sometimes found in surface waters receiving discharge from wastewater plants, as well as groundwater where crops have been irrigated with recycled water. Chemicals can also reach groundwater in connection with groundwater recharge downstream from municipal treatment plants. Ader *et al.* cite studies in which pharmaceuticals and x-ray contrast media have been found in treated wastewater, the aquatic environment, and drinking water.

Harold Muckter (2006) has noted that both the pharmaceuticals used in human and veterinary medicine and the copious use of personal care products have resulted in the delivery of "a respectable annual load of xenobiotics to the environment." Personal care products reach the environment mostly unchanged, whereas pharmaceuticals may reach the environment in modified form. Neither degradation in soil nor wastewater treatment have affected these substances to the degree earlier expected. He identifies three drug classes as most relevant for environmental contamination:

1. hormonally active compounds, because effective concentrations that will affect aquatic fauna are within ranges that have been found in wastewaters;
2. cytostatic and antineoplastic compounds, which carry some genotoxic material; and
3. anti-infective agents that help spread antimicrobial resistance and pose a risk for the human population, which is continuously struggling with pathogenic microbials.

Our knowledge has grown to the extent that we now realize that about 3000 different pharmaceutical ingredients are in common use, reach the waste stream in a combination of forms, and are not totally removed by conventional systems of wastewater treatment. What we are still investigating is the potential impact of pharmaceuticals and personal health care products in wastewater on the environment and on the health of humans who may subsequently ingest them. Consistent with these unknowns, Ternes *et al.* (2006) recommend vigorous, multiple barrier treatment requirements for planned indirect potable reuse. Even with a high level of treatment and reliability, they recommend natural polishing of finished water in aquifers and reservoirs prior to human consumption.

Whether the water under discussion is surface water, groundwater, brackish water, or recycled water, a basic reality of water treatment today is that research continually unveils new potential health threats. Meanwhile, water utilities must continue to meet water demand. This has been discussed by Mills (2001) in relation to N-nitrosodimethylamine (NDMA), which is considered a probable human carcinogen based on animal testing. NDMA was initially discovered in California groundwater in 1998. Water Factory 21, operated by Orange County Water District (OCWD) in Southern California, treats wastewater to drinking water standards and injects it into the groundwater basin; this creates an underground freshwater pressure ridge to repel seawater that would otherwise intrude into the groundwater basin. Water Factory 21 has been in operation since 1975 and produces about 15 million gallons a day of water that meets drinking water standards.

Since monitoring was not required, the first ethical dilemma faced by OCWD was whether to test for NDMA. Wastewaters contain relatively higher concentrations of the nitrogen compounds that are believed to be precursors to the formation of NDMA. OCWD had established more than 100 water quality monitoring wells in the aquifer, so a number of measuring points were available. OCWD decided to test its drinking water wells and the aquifer monitoring wells, and later expanded its testing to include the product water from Water Factory 21.

OCWD initially discovered no NDMA in the drinking water wells or in product water from Water Factory 21. However, subsequent testing with a more sensitive testing technique reflected NDMA at a level that exceeded 20 parts per trillion (ppt), a level which would have required notification of the governing board if OCWD had been serving Factory 21 effluent as drinking water.

OCWD then began an intensive effort to develop a better testing method. By January of 2000, the district's laboratory had developed the capability to detect NDMA down to 1 ppt. Further testing using this improved method revealed NDMA levels which exceeded 20 ppt in several monitoring wells and two drinking water wells near the seawater intrusion barrier. Water Factory 21 product water was discovered to contain substantial amounts of NDMA (up to 400 ppt).

At this point, OCWD staff recognized that there was a potential for public concern regarding the discovery of NDMA, a probable carcinogen, in the drinking water supply. This raised a number of disturbing questions:

- how would the media characterize the contaminant – would Water Factory 21 be characterized as a "Cancer Factory"? Would the public understand that NDMA was a long-term threat rather than an immediate public health crisis?
- would the discovery undermine confidence in wastewater projects throughout the nation?
- how long could the District afford to shut down Water Factory 21, since the seawater intrusion barrier was necessary to protect the water quality in drinking water wells? If drinking water wells were taken out of production, was there enough supply to meet demand?

To address these communication, water quality, and water supply issues, OCWD took the following actions:

- OCWD disclosed test findings to its Board, the two retail agencies using the affected water wells, regulatory agencies, community leaders, and the public (through the media). Their public education effort provided information on the nature of the potential long-term chronic health threat of NDMA.
- OCWD recommended that the water retailers using the two drinking water wells take the wells out of service until the problem was addressed.
- OCWD designed, constructed and put into operation ultraviolet (UV) oxidation systems at Water Factory 21 and at one affected drinking water production well. The other affected well was converted to a monitoring well. The UV destruction systems were costly but effective.
- OCWD began source control efforts to address possible sources of DNMA (e.g., the local metal plating industry).

Orange County Water District was dealing with a specific chemical, a specific set of regulations, a specific use of the recycled water (groundwater recharge), and a level of expertise and funding that allowed for a very proactive response to the problem. The water quality challenge of other water recycling projects will be different. However, it can be expected that recycling projects throughout the world will experience their own ethical dilemmas in relation to communication, water quality, and water supply.

15.4.3 Religious, cultural and aesthetic values

No substance in the world is endowed with more cultural and religious significance than water. Jerry Delli Priscolli and M. Ramón Llamas (2001) point out that Hindu tradition considers water a medium of purification and a source of energy. The Sagrada offers a water prayer: "The waters in the sky, the water of the rivers, and waters in the well whose source is the ocean, may all these sacred waters protect me." In Islamic tradition, the Shariah, which is literally translated as "the source of water", contains legal rules and principles. One of the principles is that water is proof of God's existence, proof of God's care, and proof of resurrection, as water restores life every day.

The deities worshipped in the Candomblé religion (practiced in the Yoruba region of Africa) include Yemanjá, who reigns over the ocean. Yemanjá is the mother of many other Orixás (deities); she represents motherly love, fertility, and feminine beauty. The goddess of fresh water is Oxum, who reigns over the rivers. African slaves from the Yoruba region brought their beliefs to Brazil, Cuba, and Puerto Rico where the two water goddesses are still worshipped through song, dance, and gifts. Both the Jewish and Christian traditions use water as a symbol of cleansing. Baptism has been a religious ritual since ancient times, as a symbol of forgiveness, redemption, and renewal. The words of a popular Protestant hymm say "Wash me, and I shall be whiter than snow."

No substance in the world has deeper emotional resonance or aesthetic appeal than water. Throughout history, water has inspired art, architecture, poetry, song and dance. Garcia Lorca, one of Spain's greatest poets, wrote:

> water that is the whole soul of a thousand ruined fogs
> that turns the stones into irises and jasmines

Garcia Lorca's family home was in Granada, Spain, where the Moors had in earlier centuries built the Alhambra, with its gorgeous gardens, fountains and pools. A series of channels, fountains, and pools allowed the Moors to revel in the luxury and beauty of water.

The value assigned to water has been reflected in different ways over time. An example is the Basilica Cistern, constructed in Istanbul in the 6th Century by Emperor Justinian. The 336 columns with Doric and Corinthian capitals supporting the reservoir were brought from various locations in Asia Minor for the construction of the Cistern, making the reservoir as beautiful as a temple. Two of the columns stand on carved heads of Medusa brought from the Temple of Apollo in Didyma. Because of its stunning architecture, the Turkish name for the Basilica Cistern was Yerebatan Saray ("Underground Palace"). Similarly, in the early 1900s, San Franciscans expressed their appreciation of water by constructing a water temple (Figure 15.4) to mark an important site in their water system.

A vivid sense of the beauty of water is still part of our world today. *Aquas de Sierra Nevada*, published by EMASAGRA of Granada, contains seven poems and countless photographs devoted to water. In Portugal, Lisbon's infrastructure includes the reservoir Maé de Agua, constructed in 1834 with an internal "waterfall" that water cascaded over as it entered the reservoir. Today, the reservoir is used for art exhibitions, concerts, ballet, and theatre, bringing together love of beauty with appreciation for water. California's *River of Words* project encourages children to write poetry and produce artwork relating to water. Hundreds of communities participate in their exhibitions, creek clean-ups, watershed festivals, poetry readings, award ceremonies, and celebrations. Heart-felt love of water is expressed by a painting (Figure 15.5) produced by Parkpoom Poonpana, aged 15, for the River of Words Project, entitled "My precious water, I kiss you."

In *Water Music*, a publication which combines photographs of water with writings by sixty-six musicians from around the world, Wesley Jefferson (2003), a blues bassist and songwriter says, "So it's pleasure, water is, it's everything."

Unfortunately, modernization of the water industry has too often meant an attempt to eliminate or ignore the religious, cultural, and aesthetic values of water. Now we talk in numbers, calculations, and slogans; the facilities we construct rarely resemble temples. In recent years, environmental stewardship has had less to do with loving water than doing without: replacing the flower with a cactus, the grass with rock; emptying the fountain; taking a shower that is not warm and cleansing, but standing naked in the cold while you soap up.

Figure 15.4. Historical Photograph of Sunol Temple, constructed by Spring Valley Water Company in 1910 and restored by San Francisco Public Utilities Commission in 2002.

Figure 15.5. "My precious water, I kiss you." (Parkpoom Poonpana, aged 15) © 1996 River of Words, www.riverofwords.org

Few water professionals would dispute the religious, cultural, and aesthetic values that have been associated with water over time. However, in our work, we often relate to water as a product. Our focus on physical properties and economic values generally serves our objectives. However, sometimes the invisible cost of what we learn is what we forget. The sensitivity to values which exist in the general public may actually be dulled in the professional. Aldo Leopold (1949) observed this phenomenon in relation to something he defined as a 'land aesthetic': "Let no man jump to the conclusion that [the ordinary man] must take his Ph.D. in ecology before he can 'see' his country. On the contrary, the Ph.D. may be as callous as an undertaker to the mysteries at which he officiates."

Water professionals (who often have strong personal feelings about non-quantifiable values associated with water) may see those values as irrelevant to their professional lives and the decisions made by the public about water. However, the impression that values and emotions can or should be separated from public policy-making may reflect a narrow bias. We may be naïve to imagine that the factors which are ultimately decisive in public decision-making are the facts, figures, and analytical tools which are dear to the hearts of the managers, engineers, scientists, and consultants of the water industry.

Again, an analogy to the effect of 'land aesthetics' on land management may be instructive. J. Baird Callicott acknowledges that aesthetics is "generally a poor step-sister, a despised and neglected subject of theoretical discussion in the general field of philosophy." Nonetheless, he is convinced that aesthetic appreciation of nature has played a decisive role in American conservation policy and land management: "Many more of our conservation and management decisions have been motivated by aesthetical rather than ethical values, by beauty instead of duty." We may be able to work more effectively with the public if we take cultural and aesthetic values into account, rather than assuming that they are extraneous to water resource planning.

Jerry Delli Priscolli and M. Ramón Llamas (2001) propose a new ethics based on finding a balance between the sacred and the utilitarian, the rational and the emotional, through the wisdom encoded in traditional and secular symbols and rituals. They advocate an appreciation of water that "is not captured in our traditional utilitarian calculus of transactions." They believe this can be accomplished by combining traditional values regarding conservation with the use of technological advances.

Recycled water is exactly the place we can combine cultural, spiritual, and environmental values with the use of advanced technology. By expanding our water supply through reuse, we can experience the joy and solace of water without sacrificing other inhabitants of nature. We can expect that increased use of recycled water will become a clear necessity, given population growth, the degradation of other sources, and possible impacts of climate change. However, we can probably develop this resource best if we appreciate it more broadly. Robert Raucher (2005) acknowledges cultural water values, the emotional and aesthetic appeal of water, the benefits of fountains, and spiritual and ceremonial uses of water. "Water soothes and relaxes, inspires reflection, is a source of beauty and has spiritual qualities recognized in religion and ritual, from baptism to death." Recycled water offers an ethical way to claim for ourselves the pleasures of abundance without encouraging water waste. The challenges posed by water recycling are significant, but the potential benefits are literally incalculable.

15.5 REFERENCES

Beecher, J. (2001), The Ethics of Water Privatization, in *Navigating Rough Waters: Ethical Issues in the Water Industry*, p. 247, American Water Works Association, Denver, CO.

Callicott, J. Baird (2004), The Land Aesthetic, in *Environmental Ethics: Divergency and Convergence*, Third Edition, p. 135, McGraw-Hill, New York, NY.

City of Petaluma (2000) City of Petaluma Water Recycling Facility Project Report, p. V1-I through VIII-6, unpublished

Elkington, J. (1998) The Triple Bottom Line: Sustainability's Accountants. In *Cannibals with Forks*, p. 69-73, New Society Publishers, Gabriola Island, BC, Canada and Strony Creek, TC, USA.

Gaston, J. (2001), Environmental Fear Mongering: An Ethical Dilemma for the Water Industry: An Ethical Dilemma for the Water Industry, in *Navigating Rough Waters: Ethical Issues in the Water Industry*, p. 117, 127, American Water Works Association, Denver, CO.

Holmes, L., Barr, M., Fox, T., Hagstrom, J., Stutz-McDonald, S. (2005) in Implementing Sustainability in Water Recycling, delivered at American Water Works Association's Annual Infrastructure Management & Technology Conference in Denver, CO.

Jefferson, W. (2003) in *Water Music,* p. 103, The University of Michigan Press, USA.

Jiménez, B. and Garduño, H. (2001) Social, Political, and Scientific Dilemmas for Massive Wastewater Reuse in the World, in *Navigating Rough Waters: Ethical Issues in the Water Industry*, p. 142-144, 148, American Water Works Association, Denver, CO.

Kanouse, R. (2001), Water Supply Planning and Smart Growth, in *Navigating Rough Waters: Ethical Issues in the Water Industry*, p. 90, American Water Works Association, Denver, CO.

Knacker, T., Liebig, M., and Moltmann, J. (2006), Environmental Risk Assessment, in *Human Pharamaceuticals, Hormones and Fragrances: The Challenge of Micropollutants in Urban Water Management*, p. 121-135, IWA Publishing, London.

Leopold, A. (1949), *A Sand County Almanac: And Sketches Here and There*, p. 174, Oxford University, New York.

Lorca, Garcia (11/03), quoted in *Aguas de Sierra Nevada*, Emasagra-Granada.

Means, E., and Patrick, R. (2005), Managing the Future: Trends in Drinking Water, p. 2-4, AwwaRF's Publication, U.S.A.

Mills, William R., Jr., (2001) Ethical Dilemmas in Converting Wastewater into Drinking Water, in *Navigating Rough Waters: Ethical Issues in the Water Industry*, p. 277-284, American Water Works Association, Denver, CO.

Muckter, H. (2006), Human and Animal Toxicology of Some Water-Borne Pharmaceuticals, in *Human Pharamaceuticals, Hormones and Fragrances: The Challenge of Micropollutants in Urban Water Management*, p. 149-153, IWA Publishing, London.

Nixon, B. (2001) Top Ten Ethical Challenges in the Public Arena, in *Navigating Rough Waters: Ethical Issues in the Water Industry*, p. 336, 345-346, American Water Works Association, Denver, CO.

Oliveira, J., Website, http://www.sobresites.com/candomble/orixas2.htm

Pelczer, J. (2004), Water Partnership and Social Responsibility. National Council for Science and the Environment, *Water for a Sustainable and Secure Future: A Report of the Fourth National Conference on Science, Policy and the Environment*, p. 16-17, Washington, D.C.

Poonpana, P., My Precious Water, I Kiss You, in *River of Words: Images and Poetry in Praise of Water*, p. 49, Heyday Books, Berkeley, CA.

Priscolli, D. Jerry, Llamas, M. Ramón, (2001), International Perspective on Ethical Dilemmas in the Water Industry, in *Navigating Rough Waters: Ethical Issues in the Water Industry*, p. 41, 59, American Water Works Association, Denver, CO.

Raucher, R., Chapman, D. Henderson, J., Hagenstad, L. Marca., Rice, J., Goldstein, J., Huber-Lee, A., DeOreo, W., Mayer, P., Hurd, B., Linsky, R., Means, E., Renwick, M. (2005). In *The Value of Water: Concepts, Estimates, and Applications for Water Managers*, p. 36-37, p. 101, American Water Works Association Research Foundation, Denver, CO. Reprinted with permission. Copyright 2005 AwwaRF.

Rosenblum, E. (2005) House Rules: Environmental Ethics for a Sustainable World, Australian Water Association Annual Conference, Brisbane.

Sandman, P. (1996), *Responding to Community Outrage: Strategies for Effective Risk Communication*, Third Edition, p. 3-73, American Industrial Hygiene Association, VA.

Saunders, R. (2001), Recycled: The Other Good Water, in *Navigating Rough Waters: Ethical Issues in the Water Industry*, p. 95, 100, American Water Works Association, Denver, CO.

Schwartzel, D. (2001), Ethical Dilemmas From an Equipment Manufacturer's Perspective and Practical Advice From Front Lines, in *Navigating Rough Waters:Ethical Issues in the Water Industry*, p. 367-379, American Water Works Association, Denver, CO.

Ternes, T., Kreuzinger, N., and Lazarova, V. (2006), in "Indirect Potable Reuse," in *Human Pharmaceuticals, Hormones and Fragrances: The Challenge of Micropollutants in Urban Water Management*, p. 323, IWA Publishing, London.

UNICEF Website, *http://*www.wssinfo.org

Ward Thomas, J. Ward (1998), Integrity as Professionalism: Ethics and Leadership in Practice, in *Environmental Ethics: Divergency and Convergence, Second Edition*, p. 44, McGraw-Hill, New York, NY.

Wenz, P. (2004) Just Garbage: Environmental Injustice, in *Environmental Ethics: Divergency and Convergence*, Third Edition, p. 110, McGraw-Hill, New York, NY.

16

The economic dilemmas of water management and reuse

Darla Hatton MacDonald and Wendy Proctor

16.1 INTRODUCTION

Wastewater reclamation and reuse is an increasingly studied alternative for urban and rural areas as rainfall diminishes, water storage levels in dams fall and populations increase. This has been particularly the case in Australia over the last ten or so years with the effects of persistent droughts leading to acute water shortages. City water supplies, dams and rivers used for irrigation have come under severe threat. Apart from the need to assess the biophysical and technical aspects of water recycling for effective use and successful implementation, the economics of water management and reuse also needs to be assessed. The economics of water can prove to be a binding constraint on reclamation and reuse of wastewater.

In this chapter we look at several aspects of the economics of water and water reuse and discuss why these features in combination make for an unusual and complex resource management problem. We discuss the economic dilemmas from a largely Australian perspective and consider reuse case studies to highlight some of the practical dilemmas. First there is the issue of pricing, laden with conflicting multiple objectives. Second, there is the underlying cost structure of different water delivery system configurations that varies with the scale of infrastructure. Third, there can be a series of environmental costs associated with water harvesting, use, reuse and disposal that may not be accounted for in the full cost of water.

These features are embedded in a set of institutional arrangements and social perceptions that may restrict a society's set of choices. On the demand side, water is essential for life, serves as an input to a multitude of production processes, is central to all ecosystems and finally, water is often a key amenity to human recreational experience of the landscape.

©2008 IWA Publishing. *Water Reuse - An International Survey of current practice, issues and needs* by B. Jimenez *et al.* ISBN: 9781843390893. Published by IWA Publishing, London, UK.

These uses, as well as perceptions of quality, shape the demand for water. These factors, especially as they are often present in combination, require careful analysis to ensure that the full implications of particular approaches are understood.

Only on rare occasions will planners or policy advisors face a blank surface to begin building infrastructure. Exceptions include planned cities such as Islamabad, Brasilia and Canberra, new urban developments *e.g.* subdivisions, or, on a bleak note, areas devastated by war or natural disasters. Groundwater bores, water diversions and storages and some provision for reuse/wastewater will often exist. UN (2003) reports that 94% of households in major cities are connected to piped water (and 86% connected to sewers). These figures do not suggest that these households have continuous access to good quality, safe drinking water or improved sanitation. It only highlights the presence of existing infrastructure that may be augmented where feasible.

The water industry (whether public or private) is typically characterized by large scale infrastructure designed to deliver potable water to the urban population and then dispose of wastewater. One of the more striking features of the megacities of the developing world is the state of wastewater services. Water and wastewater services often do not reach the urban poor. In many developing-country megacities, wastewater may not be treated prior to disposal, and urban streams and rivers have become little more than sewage canals (National Research Council, 1995). However with technologies changing rapidly and allowing for smaller scale options with respect to reuse, there is considerably more choice in options for diverting wastewater to agriculture, to irrigated green space or potentially extending systems to rapidly growing urban areas.

Future investment paths will be highly dependent on previous decisions and investments in infrastructure. Experience from the developed and developing world, regarding what has worked and not worked, is useful for considering the opportunity cost of particular choices and what options are being inadvertently shut down by particular choices.

Australia has opted for a system where the potable water supply and wastewater are largely separate though some indirect potable reuse does exist (see Chapter 1 for definitions). Stormwater in urban areas is largely a potential hazard (*e.g.* flooding) to be managed. Wastewater and sewage are something to be disposed of rather than treated and reused as a resource. This is not the case throughout the world where numerous examples can be cited where reuse water is an integral part of the water resource system (*e.g.* Singapore, Mexico City, Windhoek, etc). With increasing competition for water resources, this separation is being considered and reconsidered.

The planning landscape is further shaped by the maturity of the water economy. Quiggin (2001) summarizes the early stages of developing water resources for irrigation as a case where the demand for water is low but expanding, competing uses such as those for urban or industrial areas are minimal and the impacts on third parties, including the environment, are low. As a water economy matures, the demand for water is high and increasing, competition for use becomes intense and the environmental consequences can be quite severe. This setting, reflecting an evolving set of resource conditions, will motivate periodic reviews of the use and reuse of water and wastewater resources. Many of these economic dilemmas emerge as governments attempt to rebalance the competing objectives as priorities of a society change over time. As well, the full cost of past choices may only become apparent after the investment process is complete.

16.2 WATER PRICING ARRANGEMENTS

The dilemmas of wastewater reclamation and recycling become apparent with a cursory glance at the costs and prices charged for potable water and recycled water. We use the Australian water industry and some international examples to illustrate. The cost of recycled wastewater can often be considerable higher than first use potable water. The reasons often relate to the nature of economies of scale of existing systems, patterns of past financing of infrastructure and the nature of the external costs to the environment by different systems. Further water pricing arrangements are often embedded in shifting institutional structures (the set of formal and informal rules, policies and regulations) which govern the use of resources. In the case of Australia, wastewater reclamation is becoming feasible as costs of recycling come down and the water reform process resulting in higher prices being charged for potable urban water and rural irrigation water.

Dinar (2000) suggests that successful implementation of pricing reform is often embedded in larger reform processes. In some cases, new governments/regimes may introduce wide ranging reforms across sectors as a change in direction or the reforms may be a response to a financial crisis. Water reform is a slow and politically charged process. Australia started down the path of water reform in 1994 with the Council of Australian Governments issuing a Water Reform Framework. Water reform became part of the National Competition Policy and non-compliance was discouraged through the withholding part of tranche payments to a State if agreed reforms were not implemented. Commitment to the reform process was renewed with the 2004 National Water Initiative (NWI). The NWI is a comprehensive reform agreement containing eight inter-related elements of water management: water access entitlements and planning, water markets and trading, best practice water pricing, integrated management of water for environmental and other public benefit outcomes, water resource accounting, urban water reform, knowledge and capacity building, and community partnerships and adjustment. The overall objective of the NWI is to achieve a nationally compatible market, regulatory and planning based system of managing surface and groundwater resources for rural and urban use that optimises economic, social and environmental outcomes.

The NWI will bring about a mixture of administrative pricing of water along with water trading. State and Territorial governments have committed to achieving lower bound pricing (full cost recovery including the cost of operations, maintenance, externalities, taxes or equivalencies and future asset refurbishment) and movement towards upper bound pricing (which includes full costs plus a weighted average cost of capital) for all rural systems. Urban systems are to move to upper bound pricing and develop pricing policies for recycled water and stormwater that are congruent with pricing policies for potable water.

A high level national commitment to reform is necessary as state/territory governments, local governments and 'independent' regulators tend to benchmark off one another (AWA, 2003). Political pressure makes it difficult for jurisdictions to put their populations and economies at a disadvantage relative to others. However, price bands have narrowed considerably and most States have opted for these prices to apply regardless of the amount of water consumed.

The Independent Pricing and Regulatory Tribunal (IPART) in the State of New South Wales in Australia recently announced that the price of potable water will be increased over the next few years to better encourage users to save water (Table 16.1). Fully costing potable water would add around \$0.76-1.44/m^3 to current prices bringing the true price of water to around \$1.90/m^3 at September 2005 levels (figures adjusted from Perkins and MacCormick, 1998).

The cost of wastewater recycling systems has been slowly coming down. White and Howe (1998) estimated the cost of supply of greywater reuse systems in Australia to be on average around \$1.85/m^3. AATSE (2004) reports that the cost per m^3 of recycled water with some of the more recent greenfield development (e.g. Springfield, QLD) are in the range of \$1.10 per m^3. Therefore based on these figures, on some sites, recycled water is becoming more competitive with potable water, particularly if the cost to the environment of both potable and recycled water was considered and incorporated in pricing.

Table 16.1. Sydney water prices. The prices are not applicable under certain circumstances e.g. for large, low income families etc. (Source: Independent Pricing and Regulatory Tribunal, 2005).

	Sept 2005 \$/m^3	Oct 2005 \$/m^3	2008/9 \$/m^3
Up to 400 m^3/year	0.78	0.86	1.00
Over 400 m^3/year	0.78	1.09	1.52

It should be noted that water resources can also be allocated via market mechanisms or through administrative processes across and/or within sectors of the economy. Market mechanisms are often designed to enhance economic efficiency and move resources to their highest and best use. Markets can be constrained in any number of ways to serve social or environmental objectives.

Administrative processes can also be used to address economic efficiency or alternatively to address goals and objectives such as cost recovery, demand management, regional or social equity considerations or environmental quality. However, these goals are unlikely to be achieved simultaneously (Hatton MacDonald *et al.,* 2005) by either process. Public policies need to be designed to address each objective separately (Tinbergen, 1952; Young and McColl, 2003).

16.2.1 Setting Charges through Administrative Processes

Governments, local associations or water utilities can charge for water through non-volumetric means such as a general tax based on area, use or crop type or set the charges according to any number of rules associated with the volume (Cornish *et al.,* 2004). Water charges refer to the payments made for the delivery of water. The charging system may discriminate among residential, industrial or agricultural uses, *i.e.* charge different customers different prices based on demand for water.

There can be considerable differences in the willingness to pay for water among sectors of the economy. With increasing urbanization, the demand for water increases and governments face the dilemmas of the political economy of reallocating water from agriculture to urban areas. Historically, in Australia, the price of rural irrigation water was set quite low to encourage the development of agriculture or industry in an area.

In developing economies, the ability to pay for water is often limited by income and/or by the lack of investment in institutions at the community level to collect fees and to maintain services. Therefore the provision of water must be placed in the context of the limited resources of these governments to address the many competing demands on resources as well as the fundamental importance of food production for these economies.

Experience from developed countries shows that an administrative process can be used to recover the full suite of costs associated with the varying quality of water. Ideally, charges for the disposal of wastewater are also worth consideration. Renzetti (2000), reviewing the structure of water supply, notes the many difficulties in detailing the cost of water for urban, irrigation or industrial uses. For wastewater, there are similar difficulties associated with establishing appropriate charges. In addition there is the issue of metering disposal to compound the difficulties.

If a critical objective of a charging system is to improve economic efficiency, then it is often suggested that a volumetric charge based on the marginal costs of supply or disposal (incremental cost of supplying/disposal of a m^3 of water) needs to be estimated. The indivisibility of capital, however, makes these calculations complicated (Munashinghe, 1992; Russell and Shin, 1996) but better water pricing practices have the potential to bring about significant changes in water conservation (Rogers *et al.,* 2002). By accounting for the full cost of potable water, including the unintended impacts on third parties and the environment, potable water will be used more efficiently. This argument is driven by the 'polluter pays' principle which suggests that a given society holds the property rights to a particular state of the environment. Water users facing higher charges, whether urban and rural, will internalise the full cost of water in their decision making. The result will be a balancing of the incremental benefits of the water use with the cost to the environment.

While the cost structure of production processes will be altered with costs increasing in the short run, technological adaptations will occur over time. Substitutes, such as reuse water and desalinisation (ideally including the cost of carbon) will become relatively more attractive, all other things staying the same.

The alternative is to subsidise practices which encourage behaviour in the direction that a given society wishes to move. For instance, to encourage the use of recycled wastewater, a per unit subsidy based on environmental costs avoided to a receiving environment might be used. For instance, the provision of 'cheap' recycled water for increased use in agriculture may discourage the use of potable water supplies for agricultural purposes. Greater in-situ treatment of wastewater to reduce the health and pollution-related problems in the developing world, may similarly be encouraged through subsidies of simple, effective technologies. The difficulty is that there are often limits to the fiscal resources available to developing country governments and without movement to financial sustainability of projects, maintaining services becomes increasingly problematic.

Two-part tariffs (that consist of a fixed cost plus a cost per volume of water used) have become increasingly widespread in Australia and parts of the EU. The approach allows for the recovery of the costs of infrastructure, while the volumetric charge portion sends important conservation messages to users. The fixed cost side incorporates costs that occur regardless of the volume of water or sewage involved. Fixed costs include the expenses relating to the existing infrastructure including some accounting for the capital costs and scheduled and unplanned maintenance relating to the existing capital. The fixed cost component ensures that costs are covered and that system assets are not simply run down in the short term.

The volumetric charge includes all volume related costs including the obvious expenses relating to the operating side such as pumping, labour, treatment and future capital costs. In essence, the volumetric charge approaches a long run marginal cost. Ideally future expansions and upgrades are to be included in the calculation of long run marginal costs for the volumetric charge. The closer the volumetric charge is to the long run marginal costs, the more efficient the charge will be. The fixed charge is designed to ensure that all costs are covered and be roughly related to costs that do not vary in the short term. Care, however, must be taken to avoid double-counting.

The volumetric side would include an externality[1] charge reflecting the externalities directly linked to volumes of water. It is important that the externality charge relates as closely as possible to the incremental damage associated with volumes of potable water, reuse water or wastewater being extracted and returned to the environment. This means that, in many cases, it will be necessary to set different charges for extraction and return via a sewage treatment process or equivalent. For the externality charge to be effective, it must retain its incremental properties to balance the value of the water use against the incremental damage at each point in the water cycle. Ideally, the externality charge does not attempt to recover costs of infrastructure.

For a developing country, the investment in research, institutions and governance may be onerous. The principles of full cost pricing and establishing a path towards full cost recovery and accounting for the environment represents an ambitious, long term goal. Moving in the direction of cost recovery to cover operating and maintenance costs, however, is an important first step in addressing the requirements of the urban populations in the world's megacities.

16.2.2 Setting Prices through Market Mechanisms

An alternative to administrative processes is to allow markets to set prices. In order for a market to function, buyers and sellers, preferably a large number of buyers and sellers, must be able to trade the good. In the case of water, trade is facilitated by the separation of title to land and water and the development of tradeable entitlements to water (Brennan and Scoccimarro, 1999). In Australia, the development of tradeable entitlements for ground and surface water has gained momentum as a result of the water reform agenda. In particular, the prospect of payments being withheld from the States and Territories has encouraged reform across a number of sectors of the economy.

There are examples of groundwater trading systems (Young and Hatton MacDonald, 2003), trade in surface water within jurisdictions (Brennan, 2004) and interstate trading in surface water (Young *et al.,* 2000; Young and McColl, 2003). With reductions in transaction costs, markets have emerged as an efficient mechanism for allocating the resource among competing uses. To accommodate other objectives of society, constraints on trade can be introduced. For example, the requirements for environmental flows in the River Murray have emerged as a concern (Wentworth Group, 2003) as well as surface and groundwater connectivity in Australia. Young and McColl (2003) have suggested that an allocation for the environment could be set aside as a means of addressing environmental requirements. An environmental trust would then buy and sell environmental water where environmental requirements allowed.

Using markets to allocate water in ways that are sustainable through time requires extensive research with respect to sustainable yields of water resources through varying climatic conditions, the requirements of water dependent ecosystems as well as the supporting regulations and the administrative structures to record trades in entitlements. Experience in the Australian context suggests that these investments are necessary, if communities are to avoid 'trading into trouble' (Young and Hatton MacDonald, 2003).

For developing countries, the choice of administrative pricing versus markets is not straight forward. Markets are an efficient means of allocating water across sectors of the economy. Within sectors such as irrigated agriculture, irrigators can choose to enter the market and purchase additional water (of varying quality). Using markets requires large

---

[1] A consequence of an economic activity that is experienced by unrelated third parties.

upfront investments of human capital which can be onerous for a developing nation. Administrative pricing allows for a more incremental approach as part of developing a price path. Administrative pricing may represent a solution to water scarcity issues that can be implemented in a relatively short time frame with fine-tuning of environmental externalities implemented over longer time frames. Administrative pricing approaches for potable water and reclaimed wastewater can be used to reduce distortions in the market or essentially "level the playing field".

16.3 THE ECONOMIC DILEMMAS

Regardless of whether an administrative process or a water market is used, supply and demand considerations become important in the analysis. In the case of administrative processes, the pressures of supply and demand considerations will continue to exist and shape the actions of households, firms and irrigators.

It is important to note that there may be some reasons why markets may fail to allocate water resources efficiently (in the short term or dynamically through time). These factors enter on either the supply side or the demand side. On the supply side there may be a number of institutional impediments (regulations, legislation, formal and informal policies of agencies) which significantly increase the cost of moving away from the status quo. For instance, the nature of past investments may also impede the development path for water and recycled water. There is also the potential for the cost of third party impacts of water and water reuse which are not taken into account in pricing (administratively or through markets).

The demand for different qualities of water is driven by the economic value a user derives from the product (and its attributes), the cost of managing water supplies of differentiated qualities, the price of substitutes and the social perceptions of the product. The economic value of water will depend largely on its use in industry, agriculture, household applications (consumptive and non-consumptive uses), and so on. Higher value uses will have a much higher willingness to pay for this input. The set of available substitutes will depend on local conditions such as alternative surface water or groundwater sources, desalination, or recycled versus potable sources and aside from physical constraints, institutional impediments will enter as a significant constraint on supply. Social perceptions, integral to the acceptability of recycled water, will be critical in consumptive applications.

16.3.1 Supply side - Economies of Scale

Water and wastewater infrastructure have traditionally been characterised by economies of scale. For instance the construction of a dam to supply a city is a major investment and usually built to supply water for residents and industry far into the future. The economies of scale imply a decreasing cost structure where the incremental cost of delivering water decreases for each additional unit until some capacity regarding volumes of water delivered is reached. If the dam is built with excess capacity, new suburbs can be added to the system at a relatively low incremental cost. Often it is only as the dam reaches capacity will new technologies or approaches become feasible.

Wastewater recycling projects in the Australian economy have been successfully implemented where the capacity of a system has been reached and the demand for water high and increasing. An example is the Willunga Basin Scheme in South Australia where treated wastewater is used in an important viticulture area producing premium wines. The groundwater resources were considered to be overallocated and recycled water was deemed to be one of the few options available.

In the absence of capacity constraints particularly in separate water/wastewater systems, wastewater recycling is likely to remain as pilot projects. Further, the economies of scale may be lost with respect to recycled water when there is a need to reconstruct infrastructure in already established housing and industrial areas.

Retrofitting/reconstructing infrastructure as part of a pilot project is unlikely to be a viable option for many developing country contexts. However with the pace of urbanisation and the scarcity of freshwater, new developments will need to be well planned if the development is to serve the requirements of new arrivals. There may be opportunities to design greenfield sites and divert wastewater for food production and outside use in safe, cost effective ways. The use of urban wastewater in agriculture can be expected to increase in arid and semi-arid areas. In many cases, this source of water represents a means of increasing food security and a way for households to increase their farm income. UNESCO (2006) recommends a risk-benefit approach to assess the benefits of increased food security nutrition and local economic development against the potentially negative health impacts. Key to managing risk will be wastewater treatment and health protection measures.

16.3.2 Supply Side - Risk

Recycling technology is subject to continuous innovation. New analytical methods for detecting pathogens in water and recycled water are being developed. Regulators have to decide how to develop strategies for optimal levels of regulation. Technologies for filtering particles and eliminating pathogens are also under continuous development (EPA, 2003; EPA, 2005). From a public health perspective, regulators must balance public health benefits against cost. Finally a water recycling pilot project carries a risk associated with technological development. The potentially long life span of projects combined with quickly evolving technology, presents a risk that projects may be rendered uncompetitive or economically obsolete with the development of new technologies. Innovation risk can be managed, in part by trialling promising technologies and working out the potential problems on a limited scale. Implementation on a larger scale might go ahead when, for instance, capacity constraints are reached on existing systems. In the context of a developing country, recycled wastewater technologies will have to be fit for purpose and the maintenance plan worked out in advance. A vicious cycle can emerge if operating and maintenance expenditures are not sufficient to maintain the provision of services and customers across sectors fail to see the benefit of payments.

16.3.3 Supply side - Distortions of Past subsidies

There are also a number of pre-existing distortions in the way infrastructure has been delivered and paid for over this century (AATSE, 2004; Young and McColl, 2005). In the Australian context, much of the urban infrastructure relating to water supply was built to address public health issues or to drive economic expansion. In the case of irrigation, systems throughout the Murray Darling Basin were built to drive agricultural expansion and thereby fuel economic growth and development. While attempts are being made to move both urban and irrigation systems to operate on a cost recovery basis, progress in urban areas has been more successful. The political economy of agriculture and the reality of the cost-price squeeze of agricultural production places real limits on how quickly, real cost recovery can occur.

Developing countries face some of the same dilemmas with respect to subsidised infrastructure. When priorities change to the extent that funding for major infrastructure may not be available, governments face difficult decisions regarding system expansions. As well the same issues emerge with respect to decisions regarding potable versus recycled water options. Further, where donors have built infrastructure, a consumer culture of "user pays" may not be present.

16.3.4 Supply side - Externalities

Externalities are the external benefits or costs of goods produced that are not reflected in the item's price. These external benefits or costs are not taken into account or internalised by those immediately involved in the transaction. For example, with regard to water use, negative externalities might include the damage caused to the environment and the possible health effects on swimmers of allowing untreated sewage waste to flow into the ocean. For the producer of that waste, the full cost to society is not taken into account in this process and therefore is not reflected in the price. A positive externality could be one where irrigators in a catchment, through the use of improved technology (for example, a water recycling system), take less water out of the river system and improve environmental flows having positive effects on the river ecosystems (such as improved fish breeding areas) which in turn flow through as benefits to humans. One reason why externalities exist is because of the lack of ownership or property rights associated with the area in which the positive or negative externality manifests itself.

Spatial and temporal issues also surround the nature of externalities. Actions leading to externalities in one part of the water catchment may have downstream effects on other parts. Similarly actions occurring at one point in time may not be fully felt until some years later.

Young (2000) concentrates on the spatial aspects related to urban water externalities with water supply, water use and wastewater return externalities occurring with:

- Dams, streams, rivers and ecosystems: these include externalities related to where the water supply comes from e.g. the effects on the environment from diverting water to the urban area and away from ecosystem uses or agricultural uses.
- The built environment where water is consumed: these are called in-situ externalities and result from water use e.g. fear of health impacts from residents near to a grey water irrigated oval (negative externality) or the amenity value to nearby residents from the watering of a community garden (positive).
- The return of contaminated wastewater: wastewater and sewage returned to the environment may have negative effects depending on how the water is treated and where it returns to e.g. the externalities of treated wastewater returning to an ocean outfall may be less than those of untreated sewage returning to a river.
- Stormwater flows which have in situ and downstream effects related to contaminants picked up, for example, from motor vehicle or industrial pollution.

Externalities are an important issue to take into account when considering urban water reuse systems. Externalities affect the well being and welfare of individuals but this is not necessarily taken into account in the market system for recycled water. Rosenblum (2003) argues that the value of positive externalities should be taken into account in water reuse projects. Following this line of logic, the benefits of positive externalities will be internalised by the proponents of a recycled water project. A co-contribution by the beneficiaries to the project proponents may be required to achieve the optimal level of investment.

Given the many priorities of governments in the area of health, education and infrastructure, it can be argued that the environment, despite its importance, is one area of investment in a set of priorities. Most countries, but particularly developing countries, face difficult challenges with the competing demands on public sector funds. The ability to provide subsidies to encourage wastewater recycling may be limited. Subsidies may also limit the incentive to bring costs down over the longer term.

Hatton MacDonald (2004) argues that incorporating the full cost of all qualities of water through the urban water cycle is an alternative approach. The argument rests on the logic that water users faced with the full cost, including externalities imposed on third parties including the environment, will incorporate these costs in their decision making. Wastewater recycling will tend to occur where the externalities associated with either extraction/storage/diversion of water or the disposal of wastewater are large.

16.3.5 Supply Side - Institutional impediments

Existing institutional arrangements (including policies, laws, regulations, organisations and responsibilities) have been cited as possible impediments to water recycling schemes. Hatton MacDonald and Dyack (2004) summarise the Australian context and state that some of the key impediments include:

- The uncoordinated nature of regulations and policies in the area of environmental protection, health, planning and property rights regarding infrastructure, wastewater and water resources. In the Australian context, the responsibilities for these regulations and policies are vested in various local, state and federal government bodies;
- The need to articulate the property rights of all parties with respect to stormwater and sewage for the purposes of developing aquifer storage and recovery projects.

Toze (2005) suggests that there is also a need for a better system to account for the real and perceived human health risks with using recycled water. The risk to human health increases with direct household use, in particular, with the potential for cross connections in plumbing. Risk can be managed but not entirely eliminated.

Institutional impediments can also result from the lack of investment in the institutions, basic research and human capital development. Where developed countries struggle with policies that overlap due to conflicting environmental, health and economic efficiency considerations, developing countries can be faced with models of wastewater reclamation without sufficient home-grown human resources to manage large scale projects and the development and enforcement of regulations to protect public health and the environment.

16.3.6 Demand side - Social impediments

Consumer attitudes and preferences are also an important consideration in the success or failure of a recycled water scheme. Consumer resistance to recycled water, particularly in developing countries, is often due to the perceived health risks that may accompany it, even for recycled water used to irrigate parklands and gardens or to flush toilets.

Po *et al.* (2005) state that, internationally, recycled water schemes, mainly in developed countries, have been known to fail because of a failure of consumers to accept and use them. They conducted an extensive three year project in Australia analysing consumer behaviour and attitudes to the reuse of different wastewaters for different uses. Two issues that they looked at included peoples attitudes to eating produce grown using recycled water and factors affecting their attitudes to drinking recycled water. They found that "...knowledge did not emerge as a factor in people's decisions to buy the produce. Anyone considering using communication and education as the main feature of a program to obtain community acceptance of a recycling scheme should take note of this finding. However, the provision of comprehensive and open information is a factor in engendering trust, so the role of knowledge should not be totally discounted" (Po *et al.,* 2005).

As well they found that people's trust in such schemes was dependent on their faith in the relevant recycled water authorities and so their aversion to such schemes, because of the risks involved, could be directly affected by increasing their trust in the providing authority. A number of other conclusions were also drawn from the study including:

- people with lower levels of education were less trusting of the authorities than those with higher levels of education, and
- females were more inclined to hold more negative emotions about the schemes than were men.

In addition they concluded that, the more people felt an obligation to protect the environment, the more positive were their attitudes to buying vegetables grown with recycled wastewater.

Marks *et al.* (2002) also conducted extensive research into community perceptions of recycled water in both Australia and the United States. They found that people value reclaimed water for the benefits that accrue from water conservation but they are more enthusiastic about the personal costs and benefits involved if using the recycled alternative is cheaper than potable water. This in turn depends on the quality of information provided to users. However, with regard to reclaimed water for toilet flushing, they found that there was a strong indication that the water conservation and cost saving benefit is comparatively small with regards to the initial cost and the increased 'cross connection' risks.

Both studies therefore point to the need for targeted marketing and well designed communication and educational campaigns in gaining consumer acceptance of recycled water schemes. In Germany and Scandinavian countries, recycled water projects have been undertaken by local communities who initiate, design and manage these cooperative projects. The results have shown much greater acceptance of recycled water by communities when they are directly involved from the outset (AATSE, 2004).

In the context of developing countries, the importance of community participation in the development of potable water, recycled water and wastewater facilities is even more acute. Evidence from the condominial sewerage system in Brazil demonstrates how the same technology can succeed in one community and subsequentally fail in others (World Bank, 1992) with the crucial difference being the degree of community participation in the project development.

16.3.7 Demand side – mis-match of conditions

Wastewater recycling projects in the Australian setting have been successfully implemented where supply has matched demand conditions. The supply of recycled wastewater from urban settings will be largely constant whereas the demand by agriculture and the residential outdoor use and urban amenity space application may be seasonal. AATSE (2004) identified industrial applications as having the greatest potential for matching supply with demand.

This section has highlighted the difficulties and complex trade-offs associated with using markets and pricing policies to manage resources efficiently. All countries face binding constraints on public sector resources and there are always competing demands across health, education, infrastructure development, etc. The main problem with subsidizing water is that the utilities have little incentive to build a culture of quality service delivery and may not have the resources to maintain the assets. Subsidies are often proving ineffective as a means of addressing the needs of the poor.

According to the US National Research Council: "Historically, high subsidies in many countries have allowed large numbers of people to have access to drinking water and wastewater services at artificially low cost. While subsidies for these services are often promoted as antipoverty measures, the beneficiaries tend to be the wealthier residents of the city. Residents in low-income areas generally do not have in-home or on-property access to drinking water and must make do with high-cost and/or labor-intensive substitutes, such as carrying water long distances or paying high prices to water vendors." (NRC, 1996, p.44).

Australia, countries in the EU, metropolitan areas in Canada and the United States are moving towards water pricing arrangements which encourage efficiency. Developing countries may not be in the position to move rapidly in this direction but there is considerable merit in establishing longer term pricing paths that move in this direction. The most compelling arguments relate to the needs of all by addressing the interests of sustainability of infrastructure and the water resources.

16.4 CASE STUDIES OF WATER REUSE

Severe droughts in Australia over the last fifteen years, coupled with the increasing acceptance for conservation of water for environmental purposes and a growing population has seen an increasing number of recycled water schemes being implemented in this country in both rural and urban areas. Internationally, many countries are also under increasing pressure to address dwindling water supplies as populations increase. In this section, we review case studies of water reuse from the developed country perspective of Australia and also some developing country projects that are growing in number.

Some examples of Australian experience with water recycling include cities such as Wagga Wagga which have had urban dual pipe recycling systems for garden watering in place for over thirty years now. The New Haven Village in Adelaide uses treated wastewater mixed with stormwater for use in open space and garden watering as well as toilet flushing. The new housing estate of Mawson Lakes in Adelaide uses an ecosystem services approach with a series of wetlands used to capture and recycle water naturally. The Rouse Hill development in Sydney's north west is a large dual reticulation scheme that serves 15 000 properties at present with work on a further 10 000 to be completed in 2006. Recycled water is used for garden watering, washing cars, toilet flushing, park and golf course irrigation and industry in the area. As with many recycled water projects, taps, pipe-work and plumbing fittings are coloured lilac to ensure that recycled water is not confused with drinking water.

One of the most innovative and advanced recycling schemes in Australia was built as a result of Sydney's successful bid for the 2000 Olympic Games. The following gives a brief overview, including the key economic and institutional aspects, of the Water Reclamation and Management Scheme (WRAMS) operating at Sydney Olympic Park and the nearby Newington Estate.

The WRAMS began operation in July 2000 and is capable of servicing a population of 20,000 people. It sources its water from local sewerage (75%) and locally collected stormwater (25%). Its main features include:

- the collection and treatment of sewage and stormwater;
- supply of recycled water for non drinking uses to all residents, commercial premises and sporting venues; and
- irrigation of parklands and playing fields.

The system itself is comprised of a water reclamation plant that removes water from sewage and this water is then fed into a series of water storage reservoirs. The reservoirs were abandoned quarries that were originally set up to supply material for the NSW government brickworks. Prior to the Olympic Games site being developed, an endangered species of frog, the Green and Golden Bell Frog, was discovered in the brick pits and so the reservoirs now serve two purposes: one being to store water and the other as a habitat for the endangered frogs. A water treatment plant on the site filters and disinfects water from the reclamation plant and storage reservoirs and the water is then piped through the Olympic Park and the Newington Estate via a dedicated pipe system.

The recycled water is then used for:

- toilet flushing;
- gardens lawns and parks irrigation;
- clothes, car, windows and brickworks washing;
- ornamental ponds; and
- firefighting.

The WRAMS is intended to save around 800 000 m^3 of drinking water per year and this is expected to increase to over 1 million m^3 per year over the next two years. In addition, around this amount of sewage is also saved from being discharged into the Pacific Ocean. Thus the savings in potable water use, the lack of effluent discharge into the ocean, the pollution benefits of capturing the stormwater runoff as well as the provision of native habitats for local animals are all examples of positive externalities provided by this recycling scheme.

The recycling scheme has benefited from strong community involvement from the outset. This has been in the form of:

- customer meetings and community representation on advisory panels;
- education in schools;
- information stands at Olympic Park events;
- Sydney Olympic Park Authority (SOPA) presentations to council, community and catchment groups; and
- information pamphlet distributions.

Originally the scheme was intended to eventually provide recycled water for potable use but this was not achieved due to negative feedback from the potential users via a survey on the issue. Consumers do benefit financially though from the recycled water system as at present, the recycled water price is kept at $0.11/$m^3$ below Sydney's drinking water price.

The implementation of appropriate institutional and regulatory arrangements for the water recycling scheme have been an ongoing concern to SOPA. SOPA has been given a license to act as a water authority which gives it legal status to operate a water treatment and reticulation system. Under this license, SOPA is allowed to collect waste water, treat this waste water and distribute the treated water to households, businesses and venues. It has also been required to adhere to strict regulations including those related to health standards by continually monitoring to meet the New South Wales Department of Health water quality standards and also regulations related to endangered species. It has met the endangered species requirements through the redesign of the original brickpit with the dual objective of both water storage and habitat. Despite these detailed arrangements, there are still institutional impediments that have acted to discourage the further expansion of the WRAMS. The major one is that SOPA does not receive any returns for the provision of recycled water services from household rates which accrue to both Sydney Water and the local councils. As well, there has been no mechanism to ensure that developers provide adequate capital contributions to fund ongoing growth of the scheme. Both of these impediments now mean that further maintenance and, in particular, extension of the WRAMS scheme, is now financially unviable (SOPA, 2005).

Detailed information on the economics of recycled water case studies in developing countries is often difficult to obtain although some information is now becoming available as these schemes become increasingly necessary due to rising populations and declining water supplies in some countries. AATSE (2004) provides a good overview of water recycling in developing countries and reports that little treatment of water is undertaken in South American countries and the Asia Pacific where water supplies in tropical regions are more plentiful. West Asian countries such as Yemen, Syria, Lebanon, Palistine Territories, Egypt, Iran and Bahrain have extreme water shortages and have increased recycled water supplies in recent years mainly for agricultural purposes. The exceptions in the Asia Pacific countries include India, Vietnam and China where the demand for water due to increasing populations has been exacerbated by increasing occurrences of drought.

In developing countries, agriculture is by far the biggest user of water supplies and so recycling of wastewater for irrigation purposes has become an important issue in diverting scarce potable water supplies more suited for drinking. Untreated wastewater is often used for irrigation purposes but this also has health and environmental problems if such uses are not monitored and regulated. Both Scott *et al.* (2004) and Lahlou (2005) regard a lack of suitable institutional structures in developing countries as a major problem in the enhancement and increased production of recycled water. Such effective institutions are necessary to ensure compliance with quality standards as well as the application of cost recovery and pricing mechanisms.

Some exceptions to this however include Tunisia, Morroco and Jordon where the 'polluter pays' principle is enforced to recover costs of reuse, for example, in the form of sewerage tariffs and where progressive tariffs are applied according to the volume of water consumed (Lahlou, 2005). Some Latin American countries have also embarked on formal reuse programs using similar institutional and regulatory structures (Scott *et al.*, 2004). Without such structures, financing and cost recovery of effective recycling systems cannot be achieved and the high establishment costs of recycling systems can be prohibitive to their uptake. Scott *et al.* (2004) conclude that for recycled water to be successful in developing countries there is a desperate need for a 'coherent legal and institutional framework with formal mechanisms to coordinate the actions of multiple government authorities, sound application of the 'polluter pays' principle, conversion of farmers towards more appropriate practices for wastewater use, public awareness campaigns to establish social acceptability for reuse, and consistent government and civil society commitment over the long term with realisations that there are no immediate solutions'.

Interestingly, another outcome of the 2008 Olympic Games are plans by the Beijing municipal council to build ten 'neutral water plants' to supply non drinking water with a daily capacity of 590,000 m^3 by 2008 to address the increasing demand by this time (Beijing Daily, March 22, 2006). This will increase the total usage rate of recycled water in Beijing to 50 per cent of supplies by 2008 which is a significant increase on the rate of 30 per cent in 2005. Although no detailed information on the economic and institutional arrangements for the Beijing system is available yet, it may provide an important precedent for other cities in China and other Asian countries to follow.

16.5 CONCLUDING REMARKS

Most reuse schemes involve subsidies from governments in order for them to be initiated and viable (see for example Section 4 and the Sydney Olympic Park reuse scheme where the Sydney Olympic Park Authority subsidised the establishment of the recycled water and stormwater scheme by approximately $1.5m over five years). With the exception of the Australian Capital Territory (ACT), regulations and policies in Australia related to recycled water are spread between the State and local governments. Also the risk of having diseases spread through recycled water use is usually borne by local governments. Therefore the enthusiasm often shown by state governments to introduce recycled water schemes is often dampened by local governments not wanting to take such risks.

In Australia, most states have water recycling plans or strategies (*e.g.* Victorian Water Recycling Actions Plan 2002, Queensland Water Recycling Strategy, Western Australia State Water Strategy, NSW Water Conservation Strategy) which set state water reuse targets and encourage reuse usually with financial and non financial assistance available. However, the implementation of specific reuse systems is usually carried out at the local government or agency level.

The policy making and responsibilities of water recycling in the ACT however is carried out within the one organization (ACTEW AGL) and this has led to a much more coordinated approach than in the states and the Northern Territory. In a study on the economic aspects of water recycling in Queensland, (Price Waterhouse Coopers, 2000) it was noted that water recycling schemes are predominantly implemented by local governments but funded by state governments. They list a number of subsidy schemes available in various states but comment that many are still not enough to cover the initial costs and risks borne by the local authorities. In the ACT, subsidies are not provided but rather capital initiative expenditures are used to encourage recycling schemes.

At present there is only one privately owned and operated wastewater recycling project that does not involve a subsidy in Australia. A particular set of conditions, over-allocated groundwater in an area of high demand for water (an important viticulture area), gave rise to this private project. Generally, experience in Australia suggests that the economics of scale of the current infrastructure for the potable water system, distortions of past subsidies, the lack of externality pricing, a number of sources of risk to project proponents and resistance by consumers has served to limit the opportunities to utilise wastewater. In developing countries there are additional issues related to institutional structures as well as the effective involvement of communities in such schemes that need to be addressed. Resolving these economic dilemmas requires targeted tools depending on the competing objectives. In the case of external costs imposed of third parties, incorporating the cost of these externalities presents an opportunity to send economic signals to all water users to internalise the full cost of water in their decision-making. While there are obvious limits to the degree to which this can occur in developing countries, the many competing demands on all countries natural and fiscal resources suggests that there is merit in encouraging efficient use of water and recycled water.

16.6 REFERENCES

ACIL Tasman 2005, Research into Access to Recycled Water and Impediments to Recycled Water Investment.

Australian Academy of Technological Sciences and Engineering (AATSE) 2004, ‘Water Recycling in Australia’, AATSE, Burwood, Victoria.

Australian Water Association 2003, 'Submission to the Sustainable Cities 2025 Inquiry', Commonwealth Standing Committee on Environment and Heritage, http://www.urbanecology.org.au/articles/sustainablecities2025.html

Brennan, D. and M. Scoccimarro, M. 1999, Issues in defining property rights to improve Australian water markets. Australian Journal of Agricultural and Resource Economics 43 (1), 69-89

Brennan, D 2004, "Price formation on the Victorian Water exchange", Presented at the Australian Agricultural and Resource Economics Society Annual Conference, Melbourne, February 2004.

Cornish, G., B. Bosworth, C. Perry and J. Burke (2004). Water charging in irrigated agriculture: An analysis of international experience. Food and Agricultural Organization of the United Nations, Rome, 2004

Dinar, A. 2000, Political Economy of Water Pricing Reforms. The Political Economy of Water Pricing Reforms. Oxford University Press.

EPA 2003, Managing Urban Watershed pathogen Contamination. EPA/600/R-03/111 http://www.epa.gov/ORD/NRMRL/Pubs/600r03111/600r03111.pdf

EPA 2005, Guidelines Establishing Test Procedures for the Analysis of Pollutants; Analytical Methods for Biological Pollutants in Wastewater and Sewage Sludge (August 16, 2005) http://www.epa.gov/waterscience/methods/

Hatton MacDonald, D. 2004, 'The Economics of Water: Taking Full Account of First Use, Reuse and Return to the Environment', CSIRO Land and Water Client Report, No. S/03/1474.

Hatton MacDonald, D. S. Lamontagne and J. Connor 2005, "The Economics of Water: Taking Full Account of First-use, Reuse and the Return to the Environment". Irrigation and Drainage 54: S93-S102.

Hatton MacDonald, D. and Dyack, B. 2004, 'Exploring the Institutional Impediments to Conservation and Water Reuse- National Issues', CSIRO Land and Water Client Report, No. S/04/221. Adelaide.

Independent Pricing and Regulatory Tribunal 2005, 'Sydney Water Corporation, Hunter Water Corporation and Sydney Catchment Authority – Price of water supply, wastewater and stormwater services: Final Determination and Report.' IPART, Sydney.

Lahlou, A. 2005, 'Wastewater Reuse' in Baroudy, E., Lahlou, A. and Attia, B. (Eds.) Managing Water Demand: Policies, Practices, And Lessons From The Middle East And North Africa Forums', IWA Publishing.

Marks, J.S.. Cromar, N.J Fallowfield, H.J. and Oemcke. D.J. 2003 Community experience and perceptions of water reuse. Water Science & Technology: Water Supply, 3(3):9–16,.

Musgrave, 2000, The Political Economy of Water price Reform in Australia. The Political Economy of Water Pricing Reforms. Oxford University Press.

Munashinghe, M. 1992 Water Supply and Environmental World Applications. Westview Press.Boulder, Colorado

National Water Commission NWI (2004). Intergovernmental Agreement on a National Water Inititiave. http://www.coag.gov.au/meetings/250604/iga_national_water_initiative.pdf

National Research Council (1996). Meeting the Challenges of Megacities in the Developing World: A Collection of Working Papers. National Academy of Sciences. Washington, DC.

Perkins, P.J. and MacCormick, A.B. 1998, 'Water reuse – A pricing or technology challenge', AWWA WaterTECH Conference, Brisbane.

Po, M., Nancarrow, B. Leviston, Z., Porter, B., Syme, G. and Kaercher , J. 2005 'Predicting Community Behaviour in Relation to Wastewater Reuse: What Drives Decisions to Accept or Reject?' Water for a Healthy Country National Research Flagship. CSIRO Land and Water: Perth.

Price Waterhouse Coopers 2000, 'Economic Aspects of Water Recycling in Queensland', Study Report Five prepared for the Queensland Water Recycling Strategy, Department of Natural Resources, Queensland.

Quiggin J. 2001, Environmental economics and the Murray–Darling river system. Australian Journal of Agricultural and Resource Economics, 45 (1): 67-94.

Randall, A. 1981, Property Entitlements and Pricing Policies for a Maturing Water Economy. Australian Journal of Agricultural and Resource Economics 25: 195-220.

Renzetti, S. 2000, An empirical Perspective on Water Pricing Reforms. The Political Economy of Pricing Reforms. Ed A. Dinar. Oxford University Press.

Rogers, P., de Silva, R. & Bhatia, R. 2002, 'Water as an Economic Good: How to Use Prices to Promote Equity, Efficiency, and Sustainability', Water Policy, Vol. 4, pp. 1-17.

Rosenblum, E. 2004, 'House Rules: Environmental Ethics for a Sustainable World' Ozwater 2005 Conference, Brisbane, 8-12 May 2005.

Russell, C. and B. Shin, 1996 "Public Utility Pricing: Theory and Practical Limitations. Marginal Cost rate Design and Wholesale Water Markets". Advances in the Economics of Environmental Resources. Vol. 1 Greenwich, Connecticut: JAI Press.

Scott, C.,. Faruqui, N. and Raschid, L. 2004, 'Wastewater Use in Irrigated Agriculture: Confronting the Livelihood and Environmental Realities, CABI/IWMI/IDRC.

SOPA 2005, Sydney Olympic Park Authority Submission to IPART Review of Water and Wastewater Industry Structure and Pricing in the Greater Metropolitan Area, June 2005.

Tinbergen, J. (1952). On the Theory of Economic Policy. Amsterdam: North Holland.UN 2003. UN World Water Development Report. Water for People, Water for Life. UNESCO/Berghahn Books.

UNESCO (2006). Water: a shared responsibility. The United Nations Water Development Report 2, UNESCO, Paris.

Wentworth Group of Concerned Scientists 2003, Blueprint for a Living Continent. http://www.wwf.org.au/publications/blueprint_for_a_living_continent/

White, S. and Howe, C. 1998, 'Water efficiency and reuse: a least cost planning approach', Procedings of the Sixth NSW Recycled Water Seminar, November.

World Bank . 1992 . Development and the Environment: World Development Report 1992 . New York: Oxford University Press .

Young, M.D. 2000, 'Managing Externalities: Opportunities to Improve Urban Water Use', CSIRO Urban Water Program, No. 01/1324.

Young, M. and D. Hatton MacDonald 2003, "An Opportunity to Improve Water Trading in the South East Catchment of South Australia". Water Policy 5(2): 127-146

Young, M.D. and McColl, J. C. (2003) Robust reform. The case for a new water entitlement system for Australia. Australian Economic Review 36(2):225-34.

Young, M., D. Hatton MacDonald, R. Stringer and H. Bjornlund 2000. Inter-State Water Trading: A Two Year Review, CSIRO Land and Water Client Report, December 2000

17

Public policy and institutional capacity building: opportunities for innovation in recycling

John C. Radcliffe

17.1 INTRODUCTION

There has been increasing interest in the recycling of waste and storm water, both for productive purposes rather than discharging it to the environment, and also as a response to increasing difficulties in accessing any new surface and groundwater resources. This has resulted in innovative changes to policies for the management of water resources. A survey as part of the AQUAREC project (an international project *Integrated Concepts for Reuse of Upgraded Wastewater,* supported by the European Commission under the 5th Framework Programme and contributing to the implementation of the Key Action 'Sustainable Management and Quality of Water') established over 3,300 water recycling projects globally. In some cases, these are to use the nutrients remaining in the recycled water beneficially for agricultural production as an alternative to installing further nutrient removal facilities at the behest of environment protection agencies. However, many communities are realising that catchment and ground water resources are limited, threatening the ability to cater for population growth and economic development. Changing climate trends, such as the measured 20% reduction in European annual precipitation, reinforce this recognition (Durham *et al.*, 2005). Recycled water from treatment plants can be used to offset or defer the need for new water resources, and is becoming a saleable product in its own right. Consumption of recycled water can be anticipated to increase. Provision should be made for such water within the developing water rights and water titles frameworks being established in many countries.

Recent innovations including improvements in membrane technologies have facilitated water recycling. Sometimes unplanned, recycling is managed in widely different ways. Almost always there is a role for government in the processes, and indeed, many will argue that there must be a role for government. To evaluate what it is and how it should be managed, one needs to review the breadth of water recycling, the recycled water sources, the options for managing water recycling systems and how these fit into the entire hydrologic system. There is a need for complementary innovation in the policies for management and regulation of recycled water use and assignment of rights to use it. This chapter discusses the issues and particularly reviews and draws on lessons from the policy changes that have emanated in Australia since 1994.

17.2 ACCESS TO WATER

Traditionally, water has been accessed on the basis of customary rights which provided the basic needs of individuals and their families. As water use has developed, legal frameworks have arisen that allow recognition of legitimate private use of what has been considered as a public resource. Access has been managed on several alternative bases. These have included on a basis of "riparian rights" from land adjacent to rivers, a policy usually adopted in higher rainfall areas, for example in Europe and Eastern United States, or by priority rights ("first in time – first in right"), in which those with prior applications had prior access, a policy often adopted in more arid environments, for example in the Western United States. Both of these rights approaches were adopted in some countries, such as in Japan (Liu, 2005).

Four alternative philosophies have developed with regard to how rights regimes should be implemented:

- the *"state" approach,* advocating governmental control over water management and over the allocation and adjustment of water rights;
- the *"market" approach,* seeking to decentralize management and to allocate or adjust water rights by means of market forces, using the "rational decisions of individual stakeholders";
- the *"consensus-based management" approach (*or *"concertation" approach),* seeking to decentralize management and to allocate or adjust water rights through regulation by consensus-based platform entities; and
- the *"empowerment" approach,* advocating the strengthening of local organizations (grassroots groups and marginalized sectors) to improve equitable resource distribution and to generate a balance between the power and opportunities of stakeholders involved in water management decentralization (Boelens *et al.*, 2005).

As water recycling increases, how this additional water resource is to be managed, owned and accessed, and within what alternative policy frameworks, will become increasingly important.

Within Australia, it was recognised from the late nineteenth century that water resources were limited and varied widely between seasons, so a policy of sharing available water proportionately has long been adopted (Haisman, 2005).

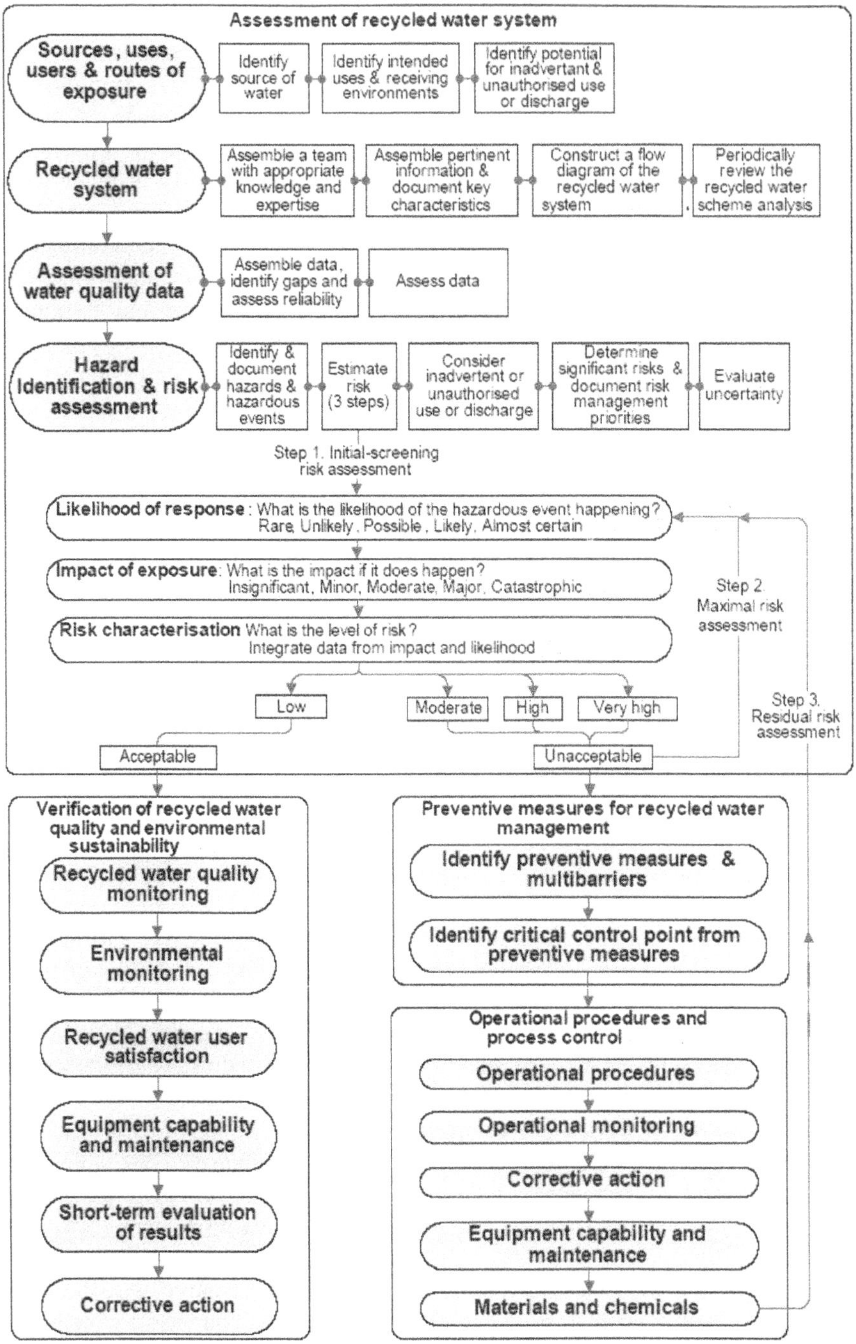

Figure 17.1 Details of analysis and management of recycled water systems recently drafted for Australian use (NRMMC/EPHC, 2005).

17.3 PROCESSES TO BE MANAGED

The human health and environmental risks to be managed in water recycling are complex. The assessments that must be considered are summarised in Figure 17.1 (NRMMC/EPHC, 2005). Traditionally, these issues have been initiated and managed by a single service-providing entity, usually a government agency, though increasingly subject to independent regulation by other agencies of government. The approach taken in dealing with the assessment and integration of these issues has already been extensively adopted by the food industry – the Hazard Analysis and Critical Control Point (HACCP) strategy (Table 17.1), a process widely recognised but adopted by very few wastewater projects recently surveyed in Europe (Bixio *et al.*, 2005). Supporting measures which are useful pre-requisites to more effective process management include trade waste provisions to improve effluent quality, and streetscape management to improve stormwater quality through pollution abatement plans.

Public policy and institutional capacity must be available to deal with:

- the receiving capacity and quality of the water resource (sewage, stormwater etc);
- the transmission of the water resource for remediation;
- the remediation process including disposal of unwanted discharges;
- the infrastructure security for and the transmission of the recycled water to the point of use;
- the approval of the use and the fitness of the recycled water for the purpose;
- the management of its use;
- the control and management of any health outcomes;
- the control and management of any environmental outcomes;
- economic viability of recycled water production without excessive profits; and
- the integration of the use of recycled water with other water resources, principally surface catchment water and groundwater.

Water recycling has usually involved a singular focus on water quality parameters. Innovative new policies are required to integrate recycled water regulation into broader economic and environmental policies and the social needs and perceptions of the community. Water provision should encompass safety, reliability, and effective sharing of the resource, with regulatory enforceability when necessary, able to be applied with consistency. Policies should be built upon innovation in water resource technology and innovation in regulatory management.

Table 17.1 Hazard Analysis and Critical Control Point steps (Cunliffe and Stephens, 2003).

Steps in the HACCP Process

Step 1 Assemble HACCP Team

The HACCP Team plan, develop, verify and implement the HACCP Plan. Generally comprises multi-disciplinary people (maintenance, operations, sanitation, quality control, marketing, chemists, microbiologists, etc.) knowledgeable of the process and product

Step 2 Describe the Product

A full description of the product is documented. This description may include water source, treatment processes, storage and distribution, and any special considerations to maintain product safety

Step 3 Identify Intended Use

The expected use of the product is documented, including how the product is to be used, consumer instructions for product use, and for whom the product is intended for.

Step 4 Construct Flow Diagram

Flow diagram must clearly indicate all process steps in the operation. The flow chart must state when the company's responsibility starts (bulk treated water, raw source water) and ends (at the meter box, at consumer tap). Steps prior to and after the organisation's direct responsibility should also be included.

Step 5 Confirmation of Flow Diagram

The HACCP Team confirms that the flow diagram is both complete and accurate as it is used in the hazard analysis. The best validation is to walk through and verify the set up of the system and processes. If this is not possible, those with operational knowledge of the system can validate the flow diagram.

Step 6 Conduct a Hazard Analysis **Principle 1**

A significant hazard is one that must be prevented, eliminated or reduced to an acceptable level to produce a safe drinking water. Significant hazards, associated control measures, and modifications to improve safety are identified as a basis for determining which control measures become critical control points.

Step 7 Determine Critical Control Points (CCP) **Principle 2**

A CCP is a point, step or procedure at which control can be applied and a hazard can be prevented, eliminated or reduced to acceptable levels. The decision tree recommended by Codex Aliamentarius may be used to determine if a process step is a CCP.

Step 8 Establish Critical Limits **Principle 3**

Critical limits are assigned to each control measure at a CCP. A critical limit defines the cut-off to ensure product safety. If a critical limit is not met, then the hazard is not controlled and corrective action is taken.

Step 9 Monitoring **Principle 4**

Monitoring is planned observations or measurements to provide a record. All critical limits have associated monitoring activity to ensure that the critical limit is met. If monitoring indicates that the critical limit has not been met, then corrective action must be taken.

Step 10 Establish Corrective Action **Principle 5**

Corrective actions ensure that the CCP is brought under control. Corrective actions can include immediate action, responsibility for corrective action, disposition of product and root cause of the problem.

Step 11 Establish Verification Procedures **Principle 6**

Verification is the use of methods, procedures or tests to determine if the HACCP system is in compliance and confirms that the system follows the HACCP plan and that identified hazards, CCPs and critical limits are appropriate.

Step 12 Establish Documentation and Record Keeping **Principle 7**

Documentation is required as proof of compliance to the HACCP plan and provides a legal defence for due diligence. HACCP records should be dated and signed. Records should provide product traceability.

17.4 THE CHANGING POLICY FRAMEWORK FOR WATER AND WASTEWATER

In recent years, particularly in Europe and Australia, many of the service functions of water supply and sewage treatment previously operated directly by arms of government have been transferred to newly-established companies. In some cases, the shareholding has been wholly retained by government. In other cases it has been sold to private sector investors. A number of private sector companies have become global in their operations.

Although such companies should have an obvious incentive in their own self-interest to adopt management practices that minimise risks and hence liabilities arising from potential hazards within their services, governments have adopted policies of ensuring regulators are available and empowered to oversee the operations of water service companies and are transparently independent from them. Furthermore, policies for service standards are now being increasingly managed within the broader compass of public policies. These policies may extend well beyond the direct management of water resources to encompass privately-held and jurisdictionally-held titles to water and obligations to ensure adequate provision of water resources for the environment. Broader economic policies adopted by governments and designed to minimise costs to consumers through encouraging competition in the provision of services can also influence the way in which water and wastewater treatment services are provided. However, within developing countries, water resources and water services are generally managed by government or public bodies. Where privatisation of water supplies has been introduced, there has been some evidence of growing discontent, conspicuously so in South America (Boelens *et al.*, 2005). This is only likely to be overcome if there are mechanisms in place that ensure the transparent and equitable sharing of access to water.

17.4.1 Jurisdictional collaboration

Catchments, rivers, lakes and aquifers, either as initial water sources or as resources to which treated effluents might be added, are not necessarily congruent with national, state or local government boundaries. Mechanisms are essential to establish agreements on how these resources are to be managed and to what standards. For example, in Osaka Prefecture, Japan, local water supply agencies express concern about the quality of water they withdraw from the Yodo River, into which discharges are made upstream from sewage treatment plants (STPs) in Kyoto Prefecture (Radcliffe, 2004). Although countries within the European Union have developed national measures to ensure sustainable water management through development of the Water Framework Directive, there appear to be no EU-wide technology-based regulations or guidelines for tertiary treatment, such as for instance the California Title 22 regulation. Despite this, several member states or autonomous regions have now published their own technical standards or regulations for wastewater reclamation and reuse based on California Title 22 or Australian standards (Bixio *et al.*, 2005).

Australia presents an interesting microcosm of the issues to be confronted globally. Though it is an island nation, it faces the same jurisdictional difficulties as are faced between countries sharing water resources internationally as responsibility for water resources is constitutionally a matter for State governments, not the Australian (Commonwealth) government. Water policy issues are addressed nationally through six-monthly meetings of the Ministers responsible for water resources from the six State governments and two Territories, along with the responsible Australian (Commonwealth) Government Minister who chairs the meetings of what is called the Natural Resource Management Ministerial Council. Policy papers and proposals are prepared for the Ministerial Council by the Natural

Resource Management Standing Committee comprising the heads of the respective Commonwealth, States and Territories Departments encompassing water resources, and also including representatives of the Commonwealth Scientific and Industrial Research Organisation (CSIRO) and the Bureau of Meteorology. Subordinate bodies and working parties are commissioned to address specific issues. Examples include the *National Guidelines for Water Recycling - Managing Health and Environmental Risks* endorsed in 2006 (EPHC, 2006) and the draft *National Guidelines for Water Recycling - Managing Health and Environmental Risks (Phase 2) - Augmentation of Drinking Water Supplies* (EPHC, 2007). These guidelines are also being extended to encompass stormwater and recharge management of aquifers.

The complexity of policy changes faced by governments is illustrated below through a consideration of the "water reform" processes that have been introduced into Australia's water resources management since 1994. These were established at the highest level through the Council of Australian Governments (CoAG) comprising the Prime Minister, State Premiers and Territory Chief Ministers. Since 2004, these governments have signed an Intergovernmental Agreement on a National Water Initiative, leading to the establishment of a National Water Commission with an independent Chairman, three members nominated by the Commonwealth Government and three members nominated by the States and Territories governments. The Commission is responsible for accrediting the States/Territories plans to implement the agreement and monitoring their progress in doing so.

17.4.2 Title – the concept of "ownership" of water

Property rights are a bundle of entitlements that define the owner's right to the use of a resource or asset. For a market for these rights to be effective, the property rights should be well specified, exclusive, transferable and enforceable. Australian governments have agreed to the ownership of water, whether from surface or groundwater sources, being separated from the ownership of land, each having a separate title. Access to water has become a recognised property right. Consequently water is now tradeable. This has required governments to establish titles registers for water assets and to note on the water title any encumbrances, such as a mortgage, over the water resource. The water titles have recognised historical legal access to water and in essence, the recipient of a title to existing resources has generally received a "free good" which has acquired a capital value. This may be at the expense of the capital value of the owner's land which no longer automatically commands access to water. "Unallocated" water is generally perceived as belonging to the State governments. Where there is unallocated water available, access may be granted to it by the state on a basis of competitive bids. Owners of water have the option of selling or leasing some or all of their water in any given season for "temporary transfer" to a buyer at a mutually negotiated market price. Alternatively, the owner of the water can sell the title outright (a "permanent transfer"), in which case his land may no longer have access to water. Market experience suggests that the price of a "permanent" sale is roughly five to ten times that of a "temporary" sale of one season's water.

The Australian States are establishing statutory regional Catchment Management Boards or similar bodies responsible for developing water resource management and water allocation plans. These plans must encompass both surface and groundwater resources, recognising their inter-connectivity where necessary, though historically they have been managed as though they were separate, unrelated resources. However, the total availability of water from a resource, particularly a catchment, may vary from year to year depending on seasonal conditions. It therefore becomes necessary to define a number of terms to manage the water represented by the title, and how the water may be used.

These terms have been established within the Australian Intergovernmental Agreement on the National Water Initiative (CoAG, 2004). The owned water asset is defined as a ***water access entitlement***, being a perpetual or ongoing entitlement to exclusive access to a share of water from a specified ***consumptive pool*** as defined in the relevant ***water plan*** accredited by the appropriate level of government. The ***consumptive pool*** is the amount of water resource that can be made available for ***consumptive use*** in a given ***water system*** (catchment and/or groundwater basin) under the rules of the relevant ***water plan***, while ensuring sustainable protection of the natural environment. The ***water allocation*** is the specific volume of water allocated to a water access entitlement in a given season, defined according to rules established in the relevant water plan. The allocation reflects the seasonal availability of water in that year to be shared in proportion to holders' entitlements within a catchment or groundwater management unit. A ***water use licence*** defines the purposes for which the water can be used.

Water entitlements and allocations, as well as being held by water supply service providers and private owners (usually irrigators or industrial users), can also be provided to environmental managers, with scope to hold the water within an Environmental Trust structure (Young and McColl, 2003). The managers potentially have the right to make temporary sales of water when it is not needed for environmental purposes, or they can buy additional water, if, for example, the ecosystems they manage require an occasional flood to maintain them – a process with obvious public benefits and liabilities. In times of scarcity, the water allocation to the environment will be reduced in the same proportion as it is to private water owners, reflecting natural drought conditions.

Urban stormwater has traditionally been seen as a hazard to be disposed of, but is now being recognised as a potential resource capable of remediation for a variety of uses. Entitlement and allocation regulations are required to clarify the rights of local government and individual landholders to intercept overland stormwater flows which have potential to be brought directly into remediation systems to improve water quality for specific purposes, effectively adding value to the resource. In some Australian states, the private interception of overland flows and their capture in dams is already controlled in urban and rural areas.

If urban stormwater flows are to be captured for further use, there is likely to be a need for storage as supply and demand are unlikely to match. If urban stormwater is being harvested, the provision of urban real estate for remediation and storage dams is likely to be prohibitively expensive except in certain circumstances such as on the edges of airfields, as has been undertaken in Adelaide (Hains, 2003). In this project, which is in some senses ahead of the regulatory framework for secure title of the water, the Salisbury City Council is marketing recycled stormwater to a wool scouring firm. When necessary, additional storage capacity is provided through aquifer storage and recovery (ASR). A subsequent similar project supplies the nearby plant of one of the world's largest car firms while minimising the deleterious impact which the stormwater previously induced on a local estuary. Golf courses and recreation parks also provide opportunities, but there needs to be a trade-off between benefits and costs. An advantage in the use of aquifer storage and recovery where there are appropriate confined aquifers available is the scope for lateral transfer within the aquifer, where further remediation is likely to be achieved (Dillon and Toze, 2005) with potential to upgrade the groundwater quality. An example is given by Swierc *et al.* (2005) of a HACCP plan prepared for the recovery of injected stormwater from a previously saline aquifer that is intended to become a resource for drinking water.

Similarly, the recycling of sewage has been expanding in dry environments or where access to water sources is not secure. The recent developments in Singapore which now has several wastewater treatment plants consistently producing recycled water to indirect potable standards, albeit mainly used for industrial use, are well known (Ong *et al.*, 2002). Within Australia, over 500 wastewater treatment plants are engaged in recycling at least some of their treated wastewater (Radcliffe, 2004). Industrial use where there is a steady demand, is particularly attractive for matching to the supply from the wastewater treatment plants, examples being the BP-Amoco petroleum refinery in Brisbane, served with up to 13ML/day from the Luggage Point STP (Hopkins and Barr, 2002) and the Bluescope Steel Port Kembla steelworks, provided with up to 20ML/day from the Wollongong STP (Hird, 2005). Agricultural use of recycled water from STPs presents the same problems as industrial use of stormwater but in the reverse relationship. The supply is relatively constant and the demand may be seasonal. Thus either storage facilities are required, or there are appropriate discharge arrangements available when demand is low. A broader description of current Australian projects is given elsewhere in this volume (Anderson *et al.*, 2007).

If greater remediation and use of recycled water is to occur, investors will require adequate provisions of security and freedom to operate. The same principles of entitlement, allocation and use licensing should be adopted as already apply to suppliers and users of water from surface catchments and groundwater basins. The issue of title in the management of sewage and recycled water has been given little attention. However, Spain under its 1999 Water Law. has provided for a legal entitlement for reuse waters arising from exploitation processes including desalination. This law authorises contracts between the holder of an effluent discharge licence and another party that obtains an entitlement to reuse waters. In this way, recycled water gains importance (Garrido, 2005). Similarly, drainage water rights, which refer to the right to polluted drainage water (including contaminated water), have come to the forefront, and could be a pilot area for water rights transfers. Cases here include the wastewater from Shanghai City and Jiaxing City in Zhejiang Province (Liu, 2005). However, some countries such as Japan and Indonesia explicitly prohibit trading of water rights (Kobayashi, 2005; Sarwan *et al.*, 2005). Transfer out of individual Western states of the United States is also prohibited (Bruns *et al.*, 2005).

Whether the title to potable water passes from the water supply service provider (which will have secured a bulk entitlement to run its business) to the consumer and then, after it has been "used", passes back to the service provider in the form of sewage has been given little published consideration. Similarly, the position of an industrial or agricultural recycled water user, who may have an entitlement for surface or groundwater as an alternative to recycled water, is probably unclear. The industrial or agricultural user may just be purchasing a "product" (recycled water) from a supplier where contractual arrangements between suppliers and users of reclaimed water are well established (Di Carlo and Sherman, 2004). The supplier may be a recycled water distribution company which has acquired title to the water from the STP operator. If the consumer is building a business, should the notion of entitlement continue down the line to the user to ensure security of allocation, and if thought appropriate, can the user sell that entitlement as is possible with groundwater and surface water allocations? These questions may be of little import where the recycled water is coming from a monopoly supplier, but if competition is introduced, circumstances may change.

17.4.3 Impact of competition policy

Governments of various persuasions in Britain, California, Australia and elsewhere have introduced competition into the provision of services previously supplied by monopoly service providers. Initially these may have involved the private sector tendering for provision of services for the monopoly provider, such as build, own and operate (BOO) facilities. These have commonly included sewage treatment plants which provided a contracted service. More recently, competition has been directly introduced to some services previously supplied by monopoly service providers. These have included electricity, railways and gas services, with various providers having access to transmission infrastructure, often originally established by governments, but also sometimes provided by monopoly private sector service providers.

The issue by the Australian National Competition Council in December 2004 of a declaration on the application by a private sector firm, Services Sydney Pty Ltd, for competitive access to sewage transmission and interconnection services provided by the Sydney Water Corporation (NCC, 2005) has introduced a completely new aspect into the potential for water recycling in Australia. This could open sewage services to retail competition in a similar way to those for electricity, water, railway and urban public transport services. The Services Sydney proposal envisaged accessing part of the sewage effluent stream currently flowing to Sydney Water's North Head, Bondi and Malabar STPs, and remediating the accessed flow at a new facility which it would build. Pursuant to section 44H(9) of the Trade Practices Act 1974 (Cwlth), the Premier of New South Wales was deemed to have made a decision not to declare the services open for competition, as he had not made a decision on declaration within a 60 day period following receipt of the National Competition Council's final recommendation. Services Sydney sought review of the Premier's decision by the Australian Competition Tribunal. On 21 December 2005, the Tribunal handed down its decision to set aside the deemed decision of the Premier and to declare the services open for competition. In the meantime, the Premier had announced on 29 November 2005 that the Government would embark on significant competition reforms for the water industry—reforms that "will encourage growth in water recycling over the next few years, creating a dynamic water reuse industry". These reforms "will create a level playing field for competition, enabling the private sector to compete with Sydney Water for the efficient reuse of effluent. Both Sydney Water and all new entrants into this field will have stringent obligations on public health, consumer protection and the environment." (Iemma, 2005). The New South Wales parliament subsequently passed the Water Industry Competition Act 2006.

The issue of title could ultimately become important as homeowners, depending on whether they choose to have their sewage services undertaken by the government-owned Sydney Water Corporation or a private sector entrant such as Services Sydney Pty Ltd, could be transferring title of their effluent to one or the other. Services Sydney anticipates developing its own sewage remediation and water recycling plant, with the aim of then on-selling the recycled water. One anticipated market would be for "environmental flows" which, if introduced below water catchment reservoirs, would allow a greater proportion of the water held in the reservoirs to be used for consumptive purposes, a potential benefit to Sydney Water as a drinking water service provider.

An aspect to be resolved would be operation of the existing sewerage infrastructure which is owned and managed by Sydney Water Corporation from individual homes to its STPs. One option would be for Services Sydney to pay a transport fee to Sydney Water Corporation for access to its infrastructure. Another might be for the infrastructure to be spun off into a separate company, as has been done in a number of locations for railway tracks, with each service provider paying access rights. Alternatively, Sydney Water could sell some of its sewage to Services Sydney at the cost of transport to the point of access. The NSW Independent Pricing and Regulatory Tribunal (IPART) has set a current price for the sale of raw sewage to a sewer mining operator at zero. More recently, in a comprehensive investigation into water and wastewater services for the Greater Sydney Region (IPART, 2005), it was hypothesised that where open access was given to the sewerage system, new entrants would be permitted to seek access to wastewater infrastructure to compete to collect wastewater from customers, transport it across the network and withdraw it for treatment. There could be circumstances in which the price might be a negative sum – i.e. giving a net benefit to the owner of the infrastructure. The Tribunal believes that:

- the price of services to both small and large customers should continue to be regulated, and the need for this regulation should be reviewed when an open access framework is established and competition in the provision of services to customers emerges;
- the legal and regulatory framework should be reviewed to ensure appropriate obligations are placed on incumbents and new entrants in relation to a range of non-price matters, including security of supply, water quality and public health, environmental impacts and customer contracts.

In its earlier Draft Report, the Tribunal recommended that the Government establish property rights for sewage and stormwater, and consider establishing property rights for the injection and withdrawal of water from storage facilities such as natural aquifers. However, the Tribunal now believes that comprehensive property rights regimes for sewage and stormwater are not required. The development of such regimes would require significant policy work, and it is unlikely that the benefits of such work would justify its costs. This is because urban stormwater and sewage are not currently scarce resources (in many cases, they are still considered wastes to be disposed of or managed at minimal cost), and in some cases are managed by different authorities. This attitude unfortunately reinforces the long-held view that sewage and stormwater are hazards rather than resources of potential value. The Tribunal considered that the rights to sewage were vested in the owner of the pipes through which the sewage flowed – effectively the Sydney Water Corporation beyond the boundaries of the land from which it emanated.

Returning to the example of the recycled stormwater provided by the Salisbury City Council, this water is demonstrably competing with potable water from Adelaide's monopoly service provider, the South Australian Water Corporation. Despite the costs of retrofitting plants for its use, the recycled stormwater has been attractive to industry, both in terms of quality (salinity of 200-250 ppm, substantially lower than the potable water it replaced), and price.

17.4.4 Costing and pricing

In many environments, the provision of urban water supplies and the treatment of the subsequent wastewater stream is managed by the same service provider. The STP operator has generally been faced with discharging the treated effluent into rivers, lakes and estuaries. Environment protection agencies now impose quite demanding discharge standards on such disposal. This has led to the effluent being treated to an appropriate standard, usually at the expense of the consumer on the basis of "the polluter pays". The treated effluent may then be provided at little or no charge to irrigation users, with the cost of supply being borne in a price to the recycled water consumer that reflects little more than the transmission costs.

A recent survey of Australian recycled water prices (Figure 17.2) showed the majority of treatment plants supplied recycled water without charge while eight supplied recycled water at a charge unrelated to the volume consumed. Twenty-five schemes did not provide a price (Woolston and Jaffer, 2005). However, charging a premium price for recycled water is possible. In 2007, the Western Australian Water Corporation was selling high quality reverse osmosis recycled water to industry at a 26% premium over that charged to industry for Perth drinking water.

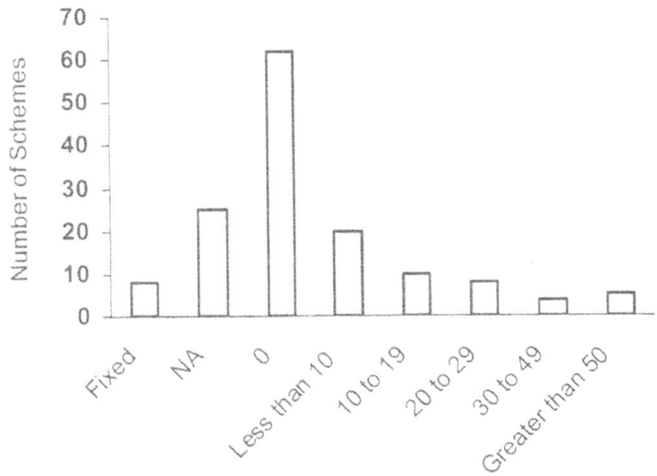

Figure 17.2 Distribution of prices of recycled water from major Australian cities, measured in cents per KL; currency = Australian $ (Woolston and Jaffer, 2005).

The issue of costing and pricing of potable water, sewage treatment and water recycling is complicated by the capital values of the infrastructure used in providing the services. When most of the government water instrumentalities in Australia were corporatised, the infrastructure was considered a "sunk cost" and valued at zero for accounting purposes, and has been so treated by pricing regulators and advisers. Prior to this time, capital expenditure and borrowings were serviced from the consolidated revenues of the governments. Domestic and industrial water rates were set on a basis of the capital value of the property being serviced. In some cities, supply was not even metered. Since that time, water rates have been generally amended to include a fixed service connection fee and a consumption-based tariff. However, there is an aspiration by most governments to provide an initial water allocation to domestic households as a "community service obligation" at a low tariff, then incremental increases in one or two "blocks" with the ultimate aim of discouraging wasteful consumption at the highest level. Nevertheless, the price of water to domestic consumers in Australia is about half that of much of Western Europe (Barton Group, 2005). Similarly, the development of water recycling schemes has generally involved government subsidies, some from water utilities and some from Australian or State Governments. Until 2005, only one of the many completed schemes had been funded on a fully commercial basis by its shareholders – that by the South Australian Willunga Basin Water Company, which provides recycled water, class B/C to grapegrowers. It accesses its effluent without charge from the SA Water Corporation, but prices its recycled water at $0.59 AUD/KL, delivered pressurised to growers. It can be seen from Figure 17.2 that this is at the upper end of the price range for recycled water in Australia.

Further issues arise with water pricing. The true impacts on the environment (environmental externalities) must be recognised in water pricing to send price signals to consumers of first use and recycled water that support environmental objectives on which each have differentiated impacts (Hatton MacDonald, 2004). The scarcity factor has also been little recognised in water pricing, but is likely to be increasingly important as further pressure develops on access to limited water resources. Although some water recycling projects have been instituted without establishing the size of the market (Mantovani *et al.*, 2001), market demand and water scarcity will ultimately influence recycled water prices.

17.4.5 Recycled water for the environment

Water resource managers should include recycled water in their consideration of the entire hydrologic cycle. Recycled water can be used to offset the non-sustainable use of surface or groundwater resources attributable to historical over-allocation, sometimes driven by political opportunism. For recycled water to be accepted for environmental flows, environment protection agencies are likely to require the recycled water to be matched to the quality of that in the original stream or aquifer, and managed so that flow regimes, including wetting and drying cycles, are congruent over the seasons with what would have happened under "natural" conditions. These conditions may vary from one locality to another and require significant research, monitoring and management planning before the recycled water is likely to be accepted as contributing to environmental flows. Where this occurs, the title of the water would pass from the water recycler to the environmental manager. Though the cost of this transfer may be paid for by the environmental management agency, it could also be recouped from the beneficiaries of the original over-allocation of the water resource. Recycled water can also be used to offset increasing salinity in surface water bodies due to over-allocation or pollution or due to over-extraction from groundwater allowing ingress of saline water to an aquifer. These management decisions are now increasingly being delegated to local community-based statutory committees and boards which are backed up by the legislative provisions of government.

As the provision of water services, wastewater services and recycling becomes transferred to corporatised bodies or the private sector, governments have an obligation to set the best possible standards based on sound science to ensure the effective and sustainable management of all available water resources. Yet the chosen processes must provide opportunities for innovation and rewards as well as regulatory sanctions. The bottom line in all this is that Australia has chosen "the market" along with a defined ownership and regulated trading framework to manage the nation's water resources demand. The approach appears to be in contrast with many other countries such as the EU which, for example, is using a pressure and impact risk assessment regulation policy derived from the Water Framework Directive, within a priority mechanism aimed at water quality, run-off and flooding (UKTAG, 2006) This mechanism gives priority to the environment, then to urban use rather than addressing the need for conservation and adopting a sharing mechanism for the more efficient use of water selected on a basis of being "fit for purpose", while providing greater opportunities to productively use recycled water and giving responsibility for water management to community groups and local environmental managers.

17.5 COMMUNITY PERCEPTIONS

Despite the effectiveness or otherwise with which governments develop public policies and institutional capacity to facilitate the greater use of recycled water, the ultimate acceptance of its introduction is likely to be expressed in the ballot box, at least in democratic societies. Bruvold (1975) pointed out that when contemplating a recycling project, there is a need for research on public perceptions to the actual usage of recycled water and to relate all stakeholders' attitudes, beliefs and behaviour in a realistic manner. There have been numerous reports since then reinforcing those views (Po *et al.*, 2003), followed more recently by the Redwood City experience (Ingram *et al.*, 2005), a questionnaire survey reported by

Baggett *et al.* (2005) and a comprehensive review of the topic of stakeholder communication in water reuse operations (Khan and Gerrard, 2005). These issues are all the more important in communities which may be contemplating incorporating an indirect potable water recycling scheme into their water resource management, especially if the proposal is in advance of the regulatory framework to manage it. Four Australian cities have considered incorporating recycled water into their drinking water supplies. Toowoomba voters rejected it in a referendum, Goulburn eventually achieved an alternative pipeline from a distant reservoir, and Canberra after a community consultation is seeking to build a pilot plant and an interpretive centre similar to that at Bedok, Singapore. The Queensland government is building a very large scheme to take recycled water from three STPs to the Wivenhoe Dam, which is Brisbane's principal drinking water reservoir. Not only are community perceptions important in the acceptance of recycled water, but also in the acceptance by consumers of foodstuffs, such as salad vegetables, produced with recycled water (Marks and Boon, 2005).

17.6 CONCLUSION

Many of the world's cities remain without any sewage treatment system, discharging the raw sewage directly to land or water. This may pose serious health risks if these locations are immediately accessed for agriculture, aquaculture or direct human activities. However, water reuse is becoming an increasingly common component of water resource planning as opportunities for conventional water supply development dwindle and the costs of wastewater disposal climb. The greatest water recycling occurs in world regions suffering water scarcity, such as the Middle East, Australia or the US south-west, or in regions with severe restrictions on disposal of treated wastewater effluents, such as Florida, coastal or inland areas of France and Italy, and densely populated European countries such as Britain and Germany. Although Japan has a mean annual precipitation of 1714mm, population density is very high in the settled areas, some regions can suffer from water shortages, and urban wastewater reuse has become common.

As experience with water recycling spreads to new environments, governments will need to ensure that the regulatory framework is in place. The responsible water treatment and remediation service agencies will need to have the expertise to undertake the production of recycled water and manage its use safely and effectively within an appropriate risk assessment framework. The regulators must have the positive vision and scientific skills and integrity to underpin that safety. To be successful, the receiving communities must have sufficient confidence and trust in their service providers and government regulators to ensure acceptance of recycled water. The technical and policy fields are continuing to evolve. There is scope for further innovation in the development of policies to encourage and manage increased water recycling, and to do so within a consideration of the entire hydrologic cycle and associated ecosystems. Policy frameworks must be in place to not only manage the technical quality of recycled water, but also to provide security of access to the asset so produced, integrated into the entire framework for the management of water resources. Industry and government must fulfil their obligations to remain at the leading edge of improved approaches for the production, use and regulation of recycled water and to do so with the confidence of consumers and other stakeholders.

17.7 REFERENCES

Anderson, J, Bradley, J and Radcliffe J (2007) Water Reuse in Australia and New Zealand. In *International Survey of Wastewater Reclamation and Reuse Practice* (Ed Jiménez, B and Asano, T.) IWA Publishing, London, UK

Asano, T (1998) *Wastewater reclamation and reuse.* Water Quality Management Series Volume 10, CRC Press, Boca Raton, FL, USA.

Baggett, S., Jeffrey, P and Jefferson, B. (2005) Identification and management of risk perception in participatory planning for water reuse. In *Integrated Concepts in Water Recycling 2005* (Ed Khan, S.J., Muston, M.H. and Schäfer A.I.) University of Wollongong, Wollongong, Australia.

Barton Group (2005) *Australian Water Industry Roadmap - A Strategic blueprint for sustainable water industry development*. Report. May 2004 http://www/bartongroup.org.au

Bixio, D., De Koning, J., Savic, D., Wintgens, T., Melin, T. and Thoeye, C. (2005) Wastewater reuse in Europe. In *Integrated Concepts in Water Recycling 2005* (Ed Khan, S.J., Muston, M.H. and Schäfer A.I.) University of Wollongong, Wollongong.

Boelens, R; Dourojeanni, A. and Hoogendam, P. (2005) Improving Water Allocation for User Communities and Platforms in the Andes. In *Water Rights Reform: Lessons for Institutional Design* (Ed Burns, R., Ringler, C. and Meinzen-Dick, R) IPFRI, Washington, DC, USA

Bruns, B.R., Ringler, C. and Meinzen-Dick, R. (2005) Reforming water rights: Governance, tenure and Transfers. In *Water Rights Reform: Lessons for Institutional Design* (Ed Burns, R., Ringler, C. and Meinzen-Dick, R) IPFRI, Washington, DC, USA

Bruvold, W.H. (1975) Human perception and evaluation of water quality. *CRC Critical Reviews in Environmental Control* **5**:153-231

CoAG (2004) Council of Australian Governments, Intergovernmental Agreement on the National Water Initiative, June 24 2004, http://www.coag.gov.au/meetings/250604/iga_national_water_initiative.pdf

Cunliffe, D. and Stevens, M. (2003) *Success Of HACCP In The Drinking Water Industry – Can It Be Adapted To Reuse Schemes?* . CD-ROM, Water Recycling Australia, 2nd National Conference 1-3 September, 2003 Brisbane. Australian Water Association, Sydney, Australia.

Di Carlo, A. and Sherman, A. (2004). The reclaimed water agreement manual. Australian Water Conservation and Reuse Program Report No 8. CSIRO and Australian Water Association, Australia. http://www.clw.csiro.au/awcrrp/stage1.html

Dillon, P. and Toze, S. (eds) (2005). Water Quality Improvements during Aquifer Storage and Recovery. American Water Works Assoc. Research Foundation Report 91056F, 286p + 2CDs.

Durham, B., Angelakis, A.N., Wintgens, T., Thoeye, C. and Sala, L. (2005) *Water recycling and reuse in Eureau countries, trends and challenges.* Presented at "The integration of reclaimed water in water resource management: fostering the role in the territorial region" Conference, Consorci de la Costa Brava, Lloret de Mar, Spain, 19-20 October 2005

Garrido, A. (2005) Analysis of Spanish Water Law Reform. In *Water Rights Reform: Lessons for Institutional Design* (Ed Burns, R., Ringler, C. and Meinzen-Dick, R) IPFRI, Washington, DC, USA

EPHC (2006) National Water Quality Management Strategy - Australian Guidelines for Water Recycling: Managing Health and Environmental Risks (Phase 1) - November 2006, http://www.ephc.gov.au/pdf/water/WaterRecyclingGuidelines-02_Nov06_.pdf

EPHC (2007) National Water Quality Management Strategy - Australian Guidelines for Water Recycling: Managing Health and Environmental Risks (Phase 2): Augmentation of Drinking Water Supplies - Draft for Public Comment July 2007 - http://www.ephc.gov.au/pdf/water/AugmentationofDrinkingWaterSupplies__ConsultationDraft_July07.pdf

Hains, S. (2003) *Environmental diversity and economic opportunities – the Salisbury wetland experience.* Planning Institute of Australia Congress, Adelaide, Australia, March 31 2003

Haisman, B. (2005) Impacts of Water Rights Reform in Australia. In *Water Rights Reform: Lessons for Institutional Design* (Ed Burns, R., Ringler, C. and Meinzen-Dick, R) IPFRI, Washington, DC, USA

Hatton-MacDonald, D. (2004). The economics of water: Taking full account of first use, reuse and return to the environment. Australian Water Conservation and Reuse Program Report No 1D. CSIRO and Australian Water Association, Australia. http://www.clw.csiro.au/awcrrp/stage1.html

Hird, W. (2005) Recycled water case study: BlueScope Steel Port Kembla steelworks, In *Integrated Concepts in Water Recycling 2005*, (Ed Khan, S.J., Muston, M.H. and Schäfer, A.I.), University of Wollongong, Wollongong, Australia.

Hopkins, L., and Barr, K. (2002) *Operating a water reclamation plant to convert sewage effluent to high quality water for industrial use.* 3rd World Water Congress of the IWA, Melbourne, Australia, 7-12 April 2002, paper e21595.

Iemma, M. (2005) *Sydney Water Supply.* Reply by Premier of New South Wales to Ms Virginia Judge, Questions without notice, New South Wales Legislative Assembly 29 November 2005

Ingram, P, Young, V.J., Millan, M., Chu Chang, P.E. and Tabucchi, T (2005) From Controversy to Consensus: the Redwood City Recycled Water Experience. In *Integrated Concepts in Water Recycling 2005* (Ed Khan, S.J., Muston, M.H. and Schäfer A.I.) University of Wollongong, Wollongong, Australia.

IPART (2005) *Investigation into water and wastewater service provision in the Greater Sydney Region.* Independent Pricing and Regulatory Tribunal of New South Wales, Sydney, Australia, October 2005 http://www.ipart.nsw.gov.au

Khan, S.J. and Gerrard, L.E. (2005) Stakeholder communication for successful water reuse operations. In *Integrated Concepts in Water Recycling 2005* (Ed Khan, S.J., Muston, M.H. and Schäfer A.I.) University of Wollongong, Wollongong, Australia.

Kobayashi, H., Japanese Water management Systems from an Economic Perspective: The Agricultural Sector. In *Proceedings of OECD Workshop on Agriculture and Water: Sustainability, Markets and Policies*, Adelaide and Berri, Australia, 14-18 November 2005

Liu, B. (2005) Institutional Design Considerations for Water Rights Development in China. In *Water Rights Reform: Lessons for Institutional Design* (Ed Burns, R., Ringler, C. and Meinzen-Dick, R) IPFRI, Washington, DC, USA

Mantovani, P., Asano, T., Chang, A. and Okun, D.A. (2001) *Management practices for nonpotable water reuse.* Project 97-IRM-6, Water Environment Research Foundation, Alexandria VA

Marks June S. and Boon, Katherine F. (2005) *A social appraisal of the South Australian Virginia Pipeline Scheme: Five years' experience.* Flinders University of South Australia, Adelaide

NCC (2005) *Application by Services Sydney for declaration of sewage transmission and interconnection services provided by Sydney Water - Final Recommendation.* National Competition Council, Australia, 1 December 2004 http://www.ncc.gov.au/pdf/DeWaSSFR-001.pdf

Ong, C.N., Rose, J., Lauer, W., Ng, W.J., Chew, S.K.,Tam, J. P., Singh, M., Bllodworth, B.C. and Lee, H.K. *et al.* (2002). Singapore Water Reclamation Study – Expert Panel Review and Findings, Singapore, June 2002, http://www.pub.gov.sg/NEWater_files/download/review.PDF

Po, M., Kaercher, J.D. and Nancarrow, B.E. (2003). Literature Review of Factors Influencing Public Perceptions of Water Reuse. CSIRO Land and Water Technical Report 54/03, Australia, http://www.clw.csiro.au/publications/technical2003/tr54-03.pdf

Radcliffe J. C. (2004) – *Water Recycling in Australia.* Australian Academy of Technological Sciences and Engineering: Melbourne, Australia, 253pp. http://www.atse.org.au/index.php?sectionid=597

Sarwan, S., Subijanto, T.W., and Rodgers, C. (2005) Development of Water Rights in Indonesia. In *Water Rights Reform: Lessons for Institutional Design* (Ed Burns, R., Ringler, C. and Meinzen-Dick, R) IPFRI, Washington, DC, USA

State of California (2003a). California Codes – Water code Section 13050, subdivision (n). California, USA, (http://www.leginfo.ca.gov/)

Swierc, J., Page, D., Van Leeuwen, J. and Dillon, P. (2005). Preliminary Hazard Analysis and Critical Control Points Plan (HACCP) - Salisbury Stormwater to Drinking Water Aquifer Storage Transfer and Recovery (ASTR) Project. CSIRO Tech Report 20/05, Australia, Sept 2005.

UKTAG (2006). United Kingdom Technical Advisory Group – Water Framework Directive site http://www.wfduk.org/about_wfd/

Woolston, M and Jaffer, S. (2005). Pricing for Recycled Water. Occasional Paper no 12, Water Services Association of Australia, Melbourne

Young, M. D. and McColl, J. C. (2003) Robust Reform: Implementing robust institutional arrangements to achieve efficient water use in Australia. CSIRO Land and Water Policy and Economics Research Unit, Australia, November 2003; http://www.clw.csiro.au/research/society/peru/publications.html

18

Public acceptance of water reuse

John Anderson, Sue Baggett, Paul Jeffrey, Linda McPherson, June Marks and Eric Rosenblum

18.1 INTRODUCTION

There are many factors affecting public acceptance of water reuse. Water reuse projects, like other water and sewerage projects, are capital intensive, impact on community finances and affect environmental outcomes. Members of the community may also be concerned about health issues. For theses reasons, community consultation and engagement in developing and managing water reuse projects is highly desirable.

Social and environmental trends have encouraged a social climate that makes better communication absolutely essential. Social shifts have given rise to a more active and concerned community, and regulatory developments have invested the public with more opportunities to participate. These trends are changing the face of public communication. Old forms of public information and public relations – which are primarily one-way forms of communication – are giving way to community involvement, which is a two-way process.

This chapter presents experience on public acceptance of water reuse and community consultation drawn mainly from project experience in the USA, UK, Australia and Singapore. Unfortunately experience in other countries, particularly those from the developing world, is scarce. The information provided in this chapter may be useful in framing public education and consultation processes for water reuse projects in other regions.

©2008 IWA Publishing. *Water Reuse - An International Survey of current practice, issues and needs* by B. Jimenez *et al.* ISBN: 9781843390893. Published by IWA Publishing, London, UK.

18.2 FACTORS INFLUENCING PUBLIC PERCEPTIONS

18.2.1 Previous research

Understanding public perceptions and the determinants of acceptance of water reuse are now recognised as one of the main ingredients of success for any reuse project. There has been only limited social research into the different factors that influence public perceptions of water reuse and how these factors mediate people's decision-making processes. Decisions are often conditioned by local circumstances. Social studies conducted so far have mainly been initiated by water agencies considering the possibility of using recycled water, or by their professional associations.

The US WateReuse Foundation sponsored a project to examine how people perceive the value of indirect potable reuse, and how the messages and management practices of sponsoring utilities have affected these perceptions. The purpose of the research was to understand the key ideas of value related to the project, how communication of the project was managed, and how key people involved in the project perceived the benefits and risks. The project report, entitled *Best Practices for Developing Indirect Potable Reuse Projects* (2004) describes 25 best practices to help ensure project acceptance, ranging from "Create a Perception of Improvement" to "Establish the Water Agency as the Source of Quality."

Po *et al.* (2004) reviewed the literature to identify factors that may influence the behavioural acceptability of a reuse scheme to the general community and identified the following factors:

- perceptions of recycled water risks;
- disgust or "yuck" factor;
- the specific uses of recycled water;
- the sources of water to be recycled;
- the issue of choice;
- trust and knowledge;
- attitudes toward the environment;
- environmental justice issues;
- socio-demographic factors;
- the cost of recycled water.

The remainder of this section is an abbreviated summary of key points from the Po, Kaercher and Nancarrow (2004) assessment of each of these factors, with some additional commentary from some other authors including Baggett *et al.* (2003).

18.2.2 Recycled water risks

The perceived risk of using recycled water is a key factor that influences public acceptance of water reuse. Table 18.1 summarises some of the key public perceptions of risks associated with the use of recycled water. The risk perception is often related to the public health risks of using recycled water. In interviews with representative focus groups in Australia, participants consistently reported concerns about the safety of using recycled water, considering the potential lethality of pathogens in the water and the unknown impact of chemicals used to treat the water (Melbourne Water, 1998; Sydney Water, 2002b; Kaercher *et al.*, 2003). The majority of respondents (89%) in a study conducted by Jeffrey and Jefferson (2002) in the United Kingdom agreed with the statement *'I have no objection to water recycling as long as safety is guaranteed'*. Concerns about the safety of using recycled water on children were especially pronounced. Ninety-two percent of respondents in the Sydney Water study (1999) agreed or strongly agreed with the statement *"people will worry about the safety of recycled water for their children."*

The public often perceives risks in using recycled water despite constant assurances from experts and the authorities that the water is safe for the intended uses. In the general risk literature, it is evident that risk perceptions are different between the experts and lay people.

The public tends to capture a broader conception of risk, incorporating attributes such as uncertainty, dread, catastrophic potential, controllability, and equity into their risk equation (Slovic, 1998). The experts, on the other hand, define risk in terms of event probabilities and treat subjective factors as "accidental" dimensions of risk. These "accidental" dimensions of risk may however play an important role in forming people's attitudes towards a risky situation. The Sydney Water study (1999) found that just over half of the participants (59%) agreed or strongly agreed with the statement *"no one can guarantee the safety of recycled water."*

If the use of recycled water is considered risky, people are inclined to avoid using the water in order to reduce the potential of regret (*i.e.* anticipated regret). Empirical evidence has shown that anticipated regret is an important determinant of people's behavioural choices (*e.g.* Bakker *et al.*, 1997; Zeelenberg and Beattie 1997). People often prefer the existing service and known experience rather than a new innovation because the potential of feeling regret and disappointment is lower.

From the general risk literature, Frewer *et al.* (1998) stated that people used their social and moral values to evaluate situations, especially if the situation is poorly defined. If a situation threatens to undermine one or more of these values, they may become outrage factors, which are those public concerns most likely to inspire opposition.

Many studies have indicated that people do not use the same underlying frames of reference when making judgements of risky situations (*e.g.* Beckwith, 1996; Syme and Bishop, 1992). Beckwith (1996) found that people who supported the introduction of a risk were more willing to consider the individual and societal benefits of the risk as part of their decision-making strategy. Those who opposed did not seem to factor these benefits into their thinking.

Knowledge of the judgment strategies people use is therefore crucial in framing effective risk communication strategies to match the decision strategies adopted by different people.

Table 18.1. Public values associated with perceived risks.

Public Values	Perceived Risks
Cleanliness is of prime importance to public health; so pristine public water supplies must be kept free of all pollutants, even those not yet identified.	• Recycled water is unsafe because it may contain trace amounts of pollutants present in wastewater even after treatment. • Recycled water is unsafe because it contains pollutants that may be found harmful in the future.
Nature is safer and more reliable than technology. Natural water from rivers and springs is preferable to recycled water from treatment plants.	• Recycled water is unsafe because it is unnatural. • The public lacks adequate control over the quality and safety of recycled water.
The community shouldn't put "all its eggs in one basket" by making expensive and irreversible decisions.	• Water reuse schemes are more costly than tapping "natural" supplies. • The decision to use recycled water is irreversible, since once it enters the water supply it may not be readily removed. • Recycled water use will result in unknown negative future consequences.
Although public officials should be honest, but they often lie to maintain their positions of authority.	• Utility managers have a vested interest in reuse schemes, so the quality of recycled water cannot be trusted.

18.2.3 Disgust – the "yuck" factor

Disgust - the "yuck" factor - has been cited in the literature since the beginning of public attitudes studies towards reuse back in the 1970s. On many occasions, the general community openly acknowledged there was a psychological barrier when it came to using recycled water (Melbourne Water, 1998; Kaercher *et al.*, 2003). This psychological barrier appears to be the disgust emotion derived from the thought of using recycled water. Some people reported that they were caught up in their own mental imagery associated with raw sewage and could not move beyond it (Melbourne Water, 1998).

The disgust emotion is defined as the emotional discomfort generated from close contact with certain unpleasant stimuli (Angyal, 1941). An early reuse study in Australia by Hamilton and Greenfield (1991) indicated that the psychological rejection of potable reuse as *filthy* and *unclean* accounted for the majority of respondents who totally rejected the reuse scheme. From the perception of filthy and unclean, the disgust reaction was generated and served to tip the balance, motivating people to stay away from using recycled water in order to prevent illness and disease. The relationship between one's perception of contamination and avoidant behaviour has been demonstrated in past research (see Rozin *et al.*, 1986).

One possible mode discussed in the psychology literature of how people associate recycled water and the emotion of disgust, is through the law of contagion (Rozin and Fallon, 1987). This law suggests that a neutral object may acquire disgusting properties from another object through brief contact (*e.g.* hair in the soup). So regardless of the fact that recycled water has been treated to highest standards, people may still perceive the water to be disgusting because it has been in contact with disgusting stimuli, in this case human wastes. The large-scale community survey research by Sydney Water (1999) had indicated that individual perception of disgust was a good predictor of his or her support for various uses of recycled water.

18.2.4 Recycled water uses

The disgust emotion and risk perceptions in using recycled water can further explain the research results which have consistently shown that the way recycled water is used affects people's perceptions and acceptance of the water. Specifically, the closer the recycled water is to human contact or ingestion, the more people are opposed to using the water (Marks *et al.*, 2006).

The focus groups conducted by Melbourne Water (1998), Water Corporation of Western Australia (2003) and Australian Research Centre for Water in Society (ARCWIS) (Kaercher *et al.*, 2003) also indicated that the acceptability of water reuse decreased substantially as the use moved from public areas to inside the home, and from toilet flushing, laundry, bathroom and kitchen uses to drinking.

Bruvold (1988) proposed two major factors that influenced public perception to reuse: (1) degree of human contact, and (2) the five factors (*i.e.* health, environment, treatment, distribution, and conservation). He asserted that the first factor only took greater effect when people were asked about general reuse options. When salient reuse options were used, the second factor had greater impact on people's perceptions.

18.2.5 Sources of recycled water

The source of water to be recycled, or 'use history' of the water, was also found to affect the acceptability of recycled water (Jeffrey, 2002; Nancarrow *et al.*, 2002; Kaercher *et al.*, 2003). Specifically the reuse of greywater or treated wastewater from one's own household was more acceptable than water obtained from other public or secondary sources (Jeffrey, 2002; Nancarrow *et al.*, 2002). This finding can be associated with the disgust emotion in using recycled water. People may perceive using their own waste to be less 'disgusting' than other people's. In contrast to this, participants in the ARCWIS focus group (Kaercher *et al.*, 2003) indicated a preference towards a more public source of recycled water. In the preliminary study, the participants indicated preferences towards reusing greywater from their neighbourhood rather than their own household. They also preferred using treated wastewater from the whole city, rather than their own neighbourhood. Further examination of these studies revealed that the differences man be attributed to two underlying factors: (a) the perceived quality of the recycled water and (b) the perceived control over the quality of recycled water received. In the Jeffrey (2002) and Nancarrow *et al.* (2002) studies, participants were reassured of the safety of the recycled water quality regardless of their source. In fact, when given a chance to comment on their acceptability of a particular water source, participants in the Nancarrow *et al.* (2002) study consistently mentioned that they could accept the source if it was safe and treated to appropriate health standards. Meanwhile, participants in Jeffrey's study were told that the greywater quality had been assured by the two institutions that they trusted. Participants in Sydney Water (2002a) and ARCWIS (Kaercher *et al.*, 2003) studies expressed concern about their ability to manage and monitor on-site treatment systems. As a result, the quality of the water might be compromised. The citywide wastewater treatment plant was therefore preferred.

Studies have also found that the acceptability of recycled water is dependent upon the type of water to be recycled (ARCWIS, 2002; Nancarrow *et al.*, 2002; Kaercher *et al.*, 2003). Rainwater harvesting from one's household roof was considered to be more acceptable than greywater and wastewater reuse. Reusing greywater was on the other hand more acceptable than treated wastewater. Nancarrow *et al.*, studies (2002, 2003) found aquifer recharge of recycled water to be less acceptable than using greywater for home gardens and wastewater for parks and gardens. Focus group participants often queried the levels of recycled water treatment before they could decide upon the different uses of recycled water (*e.g.* Kaercher *et al.*, 2003).

18.2.6 The issue of choice

The issue of choice is also an important determinant to the public acceptance of water reuse. In places where there were water shortage issues, people were reported to readily accept water reuse because of the heightened awareness of the need to conserve water (Dishman *et al.*, 1989). Rebhun (as cited in Dishman *et al.*, 1989) reported that in Israel, it had not been necessary to convince the public of the need for water reuse as everyone was aware of the country's water shortage problem. In instances where an alternative water source is available, people question the need for reuse. When the option of water reuse was discussed, the Melbourne Water focus group participants (1998) stated that there had to be a genuine need for using recycled water. It should only be considered if other solutions were impractical and economically unfeasible. This finding is also cited in Bruvold and Crook (1981).

A heightened need for other water sources, however, does not necessarily warrant public acceptance of water reuse. The San Gabriel Valley Groundwater recharge project is an example of a failed project despite being conceived during the 1987-92 droughts. Acceptance of reuse may be affected by other factors such as trust.

18.2.7 Trust in authorities and scientific knowledge

Trust in the authorities to provide safe recycled water, and in the scientific investigations and technologies could play a crucial role in determining public acceptance of water reuse. Scientists from universities or the CSIRO are most trusted by the community to provide information about recycled water (Sydney Water, 1999). Trust in the Water Corporation of WA to provide safe recycled water was said, by focus group participants to be the main reason they would be willing to use recycled water (*e.g.* Kaercher *et al.*, 2003). Factors in this acceptance were that it had been established for many years, had a good safety record, and its intention was not monetarily or politically driven. These characteristics are similar to those found in a food risk perception study by Frewer *et al.* (1996). For communicating food-related hazards, people trusted institutions that they perceived to be truthful, trustworthy, responsible, accurate, concerned with public welfare, factual and had a good track record. Despite public trust in certain institutions to set and monitor the quality of recycled water, Jeffrey and Jefferson (2002) found that people might remain unwilling to use recycled water for higher risk applications because perceptions of water quality (mainly water turbidity and the content of suspended solids) played a bigger role in deciding whether they could accept water reuse.

Overall, the relationship between trust in the authorities and attitudes towards policies has been consistent in the general empirical investigations. However, the relationship between trust and attitudes towards policies was often found to be either weak or moderate (Bord and O'Connor, 1992; Biel and Dahlstrand, 1995). In the siting of a local high-level nuclear waste repository, Drottz-Sjoberg (as cited in Sjoberg, 2001) found that people trusted agencies to make decisions that protected public safety, but they still might oppose a siting proposal. Frewer *et al.* (2003) comment on how the relationship between trust in the source of information and the impact that information has on the recipient's perception of that risk is not clear. Therefore it is crucial to understand what the person receiving the information feels is important. Lack of trust was the reason most frequently noted by people who were in opposition to using recycled water for agriculture, watering parks or drinking (Sydney Water, 1999).

18.2.8 Attitudes towards the environment

Attitudes toward the environment also play an important in determining public acceptance of recycled water. This topic was explored in some depth at the Mawson Lakes development in South Australia where a dual water supply scheme was proposed. Residents who favoured the scheme also agreed that the potential to do something positive for the environment had motivated them to live in a greenfield development (Hurlimann and McKay, 2002). However, none of the 20 households interviewed cited recycled water as a reason for choosing to live at Mawson Lakes (Marks, 2002b).

Sydney Water (1999) reported that people who supported water conservation and recycled water use for all purposes except cooking and drinking believed strongly about the importance of environmental issues and also in their own ability to make a difference.

It is clear that an awareness of environmental attitudes and other commonly held social values is critical to understanding how a community will perceive the risks associated with water reuse. In fact, various authors have identified public values as the key drivers that influence public perception of risk (Frewer, 1998) and also determine public willingness to adopt water reuse projects (Rosenblum and Anderson, 2004). In their discussion of political considerations relating to water reuse, Rosenblum and Anderson suggest that political decision makers take into account environmental, technical or economic factors, but that the direction of influence is primarily form the top down, since "with sufficient political will, economic resources can be found to purchase technical solutions to produce and distribute recycled water for a sustainable environment." Rosenblum (2005) presented a hierarchical model of decision-making in which ethical values determine political motivations. Rosenblum (2005) also cites the diagram by Norgaard (see Figure 18.1) in which values, technology and even social organization are presented in an evolutionary relationship.

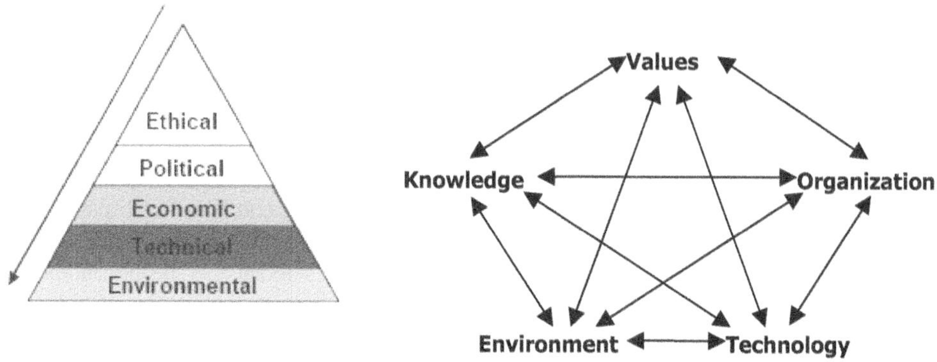

Figure 18.1. Two models relating ethical values and environmental issues in the decision-making process (Rosenblum, 2005).

18.2.9 Environmental justice issues

Environmental justice issues can also influence how people perceive the acceptability of a reuse project. One of the strong opponents of the San Diego Repurification project, Herman Collins, stated that his opposition to the project was due to the perceived injustice to the low and medium income communities as they were deemed the major recipients of recycled water (Recycled Water Task Force, 2003). A politician promoted this information and it resulted in strong community resentment and subsequently the demise of the project.

The perceived fairness in the decision making process is also important (Syme *et al.*, 1999, 2000 and 2001). The Maroochy and Caloundra councils failed to initiate a reuse project because the lead up process was seen to be unfair by some community groups. The Rivermouth Action Group stated that they were not consulted and involved at the conception of the project.

Aesthetic concerns over neighbourhood treatment plants have been expressed by participants in Sydney Water (2002b) and ARCWIS (Kaercher *et al.*, 2003) focus groups. People expressed little opposition to having either estate- or neighbourhood-based treatment plants, but expected the treatment plants to be located away from main residential areas. The idea of having on-site sewage treatments system was completely rejected for fear of unpleasant smells and contamination (Sydney Water, 2002b). In addition, there is a perception amongst community members that they should not be targeted for water reuse initiatives. Any water reuse projects should start with big water users such as industries before domestic householders. The comparative amount of water that could be saved from householders was said to be minimal (Kaercher *et al.*, 2003).

18.2.10 Socio-demographic factors

Some demographic factors have been identified in reuse studies to be influential in public perception of water reuse. For example, McKay and Hurlimann (2003) predicted that the greatest opposition to water reuse schemes would be from people aged 50 years and over. As a result, they recommended education and information campaigns to target this specific age group. Some surveys in California and Colorado (cited in Hartley 2003) further indicated that "older" women tended to be less supportive of potable water reuse and hazardous waste siting. In contrast, Jeffrey (2002) found no significant variation in the public support of greywater reuse across gender, age or socio-economic groups. The Sydney Water (1999) study indicated differences in the responses of participants from different genders, levels of education, place of residence, and language spoken. These inconsistent findings may in turn suggest that demographic factors alone were not able to explain the individual differences in the perceptions and acceptance of water reuse. Research in the area of risk perceptions suggested that 80-90% of the variation in risk perceptions across individuals resulted from attitudinal and contextual factors other than demographics (Savage 1993).

18.2.11 Cost of recycled water

Cost consideration has been suggested in the National Water Quality Management Strategy (NWQMS) as an important determinant for community acceptance. People generally expect to pay less for using recycled water as it is considered to be of lower quality. In a survey of residents who were living in a dual supply development, Marks *et al.* (2002a) found that the majority of people expected to pay less for using recycled water because of the water quality and restrictions on people's use of this resource. Some residents thought the lower price was necessary to encourage acceptance and investments in the up-front costs. This was also indicated by the focus group participants (*e.g.* Kaercher *et al.*, 2003). As an incentive to encourage industrial reuse, Gagliardo (2003) further asserted the need to show the potential users economic advantages in recycled water.

18.3 CONSULTATION

18.3.1Community consultation approaches

Involving the community in the project planning and development process requires a commitment by the local water authority to consult in a meaningful way. Consultation is a process by which parties can interact. The process can involve the giving, receiving and sharing of information. Clear communication is essential in consultation so that all parties understand each other.

Effective consultation is essential to gain a proper understanding of community needs and requirements. For a water authority, effective consultation is essential to good management of the business and it will enhance the quality of service it provides to its customers and community.

Lines of communication need to be established between the local authority and those affected by its proposals. This will require the development of a community consultation process with clearly defined objectives. Objectives of a community consultation process and the reasons for involving the community may include:

- to gain a full understanding of the community's needs and community knowledge;
- to develop mutually agreed objectives;
- to identify concerns about social, environmental and economic issues;
- to foster better understanding of the proposed project in the community;
- to develop ownership of the underlying problems which are driving the project and agreement on the most appropriate way to solve those problems;
- to enable emerging issues to be resolved at an early stage in a non-threatening way;
- to develop trust between the parties and a cooperative approach;
- to allow the community to propose options it wants the authority to evaluate;
- to satisfy the community that final decisions will be made after proper evaluation of the issues that the community has raised.

18.3.2 Identifying stakeholders

The community is made up of many diverse groups, each with its own objectives, motivations and problems. In planning for meaningful community consultation it is necessary to identify the community groups that will be affected by the proposal or are likely to have an interest. When identifying community groups and stakeholders, it is important to recognise that an individual shares interest with others and may belong to more than one stakeholder group. It is a mistake to categorise any member of the community as belonging to a perceived interest group. Positions may shift over time in response to new information, re-assessment of the facts, changes in interests or changing alliances.

There are several different approaches to identifying community groups and stakeholders. One approach is to group stakeholders according to specific individual interests. Brotherton and Owens (1991) proposed the stakeholder categories in Table 18.2.

Table 18.2: Categories of Stakeholders (Brotherton and Owens, 1991).

Category	Description	Consultation Issues
Broad Interest Groups	Examine issues from the public interest and environmental perspectives. The acceptance of a proposal by broad interest groups can be earned on the basis of merit. Opposition may occur if authorities do not evaluate projects objectively or negotiate in good faith.	Will examine need for the proposal, alternative ways of satisfying the need, and how project impacts can be mitigated or eliminated. The processes that were used to devise the solution and the conduct of the proponent will also be thoroughly worked over.
Vested Interest Groups	Segments of the community that may be advantaged or adversely affected by a proposal. Concerned with the outcome as it affects them. Their support or opposition depends on how they are affected by the project.	They resist the imposition of costs on them from proposals that are designed to serve the public interest. Essentially, they are advising the proponent (and others) of their valuation of the private costs that the proposal imposes on groups and individuals.
Affected Individuals	Some individuals may be adversely affected by a proposal. The personal costs they are asked to bear are their concern, and they are thus identifying the costs in much the same way as vested interest groups do.	Individuals want the costs they bear removed or ameliorated to tolerable levels. Affected individuals may appear weak in the face of the proponent's power and expertise. It is a serious mistake to disregard their concerns.
General Public	The general public usually wants to be left alone to get on with their lives and do not want to be involved in complex projects which are the responsibility of the proponent. This does not mean that the general public does not care.	While the general public does not expect proponents to be perfect, it does expect a reasonable level of competence. Evidence of incompetence, unfairness or favouritism may convince the general public to oppose a project.
Champions	Some individuals participate in public issues on the basis of championing perceived public interests. Personal interests may also be involved.	Champions may support the cause of special interest groups or affected individuals.
Media	The media have difficulty reporting the facts of complex issues. What they do well is to report conflicts and differences no matter what the proponent says or does. Media acceptance of a proposal can be earned on the basis of merit, and the reverse also applies.	If possible, avoid conflict. When conflict occurs, focus on matters of real consequence. Assessment and rational response to the media on issues is preferable to over-reaction in response to the discomfort of media scrutiny.

18.3.3 Involving the community in projects

There is no foolproof formula that will guarantee successful community involvement. It is important to recognise that different consultation techniques are appropriate to different types of communities and different types of issues. For example, simple informal techniques may work well in a small close-knit community, but more formal techniques may be required in a large community or where a project has regional implications.

Finding the right approach will depend on the consultation objectives, which may be to seek information, to share information or to share the decision-making process. Generally the consultation process will involve sharing information, obtaining feedback and openly discussing the options and issues presented by the authority and the stakeholders.

Communication and outreach methods

In all contact with the community, whether in meetings or through the media, it is important to provide high quality information to effectively communicate project issues and objectives. Forms of communication include:

- ***Public Meetings*** – Public meetings can be a useful way of information sharing with a large audience and of allowing the audience to comment on or ask questions about project options. There is a risk that public meetings may become a vehicle for conflicts so meetings must be well planned and chaired by an experienced person who understands the issues but has no personal involvement. The meeting attendance may not represent majority community opinion, so meetings should not be used to determine consensus about a preferred option.
- ***Scoping Meetings*** – Scoping meetings provide information on the options and the proposed decision making process. They also help to identify stakeholders, local expertise and issues.
- ***Forums*** – Meetings at which stakeholders and interest groups can present their views and discuss issues of concern.
- ***Workshops*** – Sessions in which professionals and community members work together in groups to share information and identify solutions to specific issues and problems.
- ***Community Advisory Committees*** – Advisory committees are a more formal method of involving community representatives in the planning process. They can be useful in providing community groups with a window into the decision making process.
- ***Personal Presentations*** – In personal presentations to community groups, presenters can link their presentation to the interests of the audience. Such presentations provide an opportunity for interaction and to clarify misunderstandings that may have arisen
- ***Site Inspections*** - Site inspections and open days at facilities such as water reclamation plants and reuse schemes provide members of the public with the opportunity to see first-hand the processes involved and the quality of the recycled water. They can be even more effective if part of a wider educational program about water and environmental management.
- ***Displays*** – Displays are a way of making information available to the public and providing feedback opportunities. Displays of project options with photos and diagrams at shopping centres, libraries and other public places are an effective educational tool, particularly if staff are available to discuss the project with interested parties.
- ***Printed Matter and Fact Sheets*** – These can convey information to a large number of people and include the use of press releases, brochures and fact sheets. Printed material needs to be factual and interesting. Special supplements in newspapers and magazines

are a way to tell the community about project objectives, project options and decision-making processes.

- ***Video and Audio Messages*** – A video is a very effective way of presenting information about options and getting a message across provided it is well scripted, well made and targeted to the intended audience. A program of audio messages (TV and radio) can be useful if it seeks to promote information about the available options. However, it may be counter productive if seen as a political campaign to sell a particular option.
- ***TV, Radio, Speeches*** – TV, radio and public meeting presentations are the hardest forms of communication because they are immediate and direct to the audience. Everything presented in public needs to be said in a way such that it will not be misconstrued or misquoted.
- ***Market Research*** – Market research including the use of surveys can provide information on attitudes, preferences and issues of concern to the community.

Wegner-Gwidt (1998) presents further information about communication and outreach programs that have been utilised by water authorities in California.

18.3.4 Communicating about risks

Hazard and outrage

Many people have adapted to live with risks in their communities and manage the risks to minimise adverse consequences. To avoid being confronted by community members who demand zero risk in a project, it helps to involve them in decisions that affect them. By and large, people insist on zero risk when they feel that they are being treated unfairly or that their concerns are being ignored.

According to Hance, Chess and Sandman (1990), the public perception of risks is often greater than indicated by scientific assessment. To scientific experts, risk means the annual risk of incidents, accidents, illness or death.

To the public (and even the experts when they go home at night), risk means more than the scientific definition and includes perceptions about real and imagined adverse consequences. We can call the risk as defined by the experts as "hazard". We can call the other factors "outrage". Public perception of risk is the sum of hazard plus outrage.

The public pays too little attention to hazards; the experts pay too little attention to outrage. While proponents should aim to make information about hazards available and easy to understand, they need to work equally hard at understanding and dealing with the other dimensions of public concerns. Working to keep outrage low is as much a part of risk management as working to minimise and mitigate the hazards.

Effective communication about risks can help people to understand and accept risks that they might otherwise find unacceptable. However, simply talking openly and honestly about risks is not enough – risk communication is not a way to paper over ineffective risk management. The proponent needs to demonstrate that it is taking positive steps to manage, mitigate and reduce risks.

Hance, Chess and Sandman (1990) have distilled risk communication guidelines and provide detailed checklists in their Industry Risk Communication Manual for the implementation of each strategy.

Table 18.3. Characteristics that influence the acceptability of risk (after Hance, Chess and Sandman, 1990).

More Acceptable	Less Acceptable	Comments
Voluntary	Involuntary	People feel less at risk when the choice is theirs. Find ways to share decisions with the community.
Natural	Man-made	Man-made risks may be seen as a form of coercion; acknowledge such risks and show how they will be avoided or mitigated.
Fair	Unfair	People who bear the most risk often reap the least benefit; such unfairness naturally provokes outrage. Risks should be reduced or compensation provided in a form the community finds most appropriate.
Familiar	Exotic	Familiar risks and surroundings diminish outrage. Facility tours, displays, school programs and the like can increase familiarity if presenters are honest about the tough issues.
Unmemorable	Memorable	Whether through personal experience or media presentation, memorable incidents and images of risks exacerbate outrage. When such images are present in people's minds, ignoring them just makes the problem worse.
Trivial	"Dread"	Some risks are more dread than others (e.g. cancer versus asthma). Acknowledge an appropriate level of concern and mitigation actions for risks that cannot be removed.
Knowable	Unknowable	Risks may be considered "unknowable" when there are large areas of uncertainty or experts disagree about the extent of the hazard. Giving citizens access to all pertinent data may increase the degree of "knowability".
Morally Acceptable	Immoral	Moral relevance enters the picture when a risk (e.g. pollution) is considered not only harmful but also wrong. Rather than suggest tradeoffs, sponsors should demonstrate their intention to get as close to zero risk as possible.
Individual Control	System Control	Most people feel safer when they are in control. Being at someone else's mercy provokes outrage—especially when that someone is a faceless corporation. Responsibility for the project needs to be made local, visible and accessible.
Trustworthy	Untrustworthy	Build trust by making actions public, collaborative and accountable.
Open Process	Closed Process	Elements of an open process include prompt release of information, ready admission of errors and their correction, and courteous response to expressed public concerns.

Improving dialogues with communities

Sandman gives five guidelines to assist water utilities plan community involvement in projects:

- recognise the importance of community input;
- try to involve the community in the decision-making process;
- identify and respond to the needs of different audiences;
- develop alternatives to public meeting and public hearings;
- recognise people's values and feelings.

Wegner-Gwidt (1998) and Jansen *et al.* (2005) have developed similar lists of actions needed in dialogue with the community, based on a survey of successful and unsuccessful projects to supplement drinking water sources:

- involving the stakeholders before project conception;
- educating and cooperating with the local media;

- developing an education and information campaign to increase public awareness;
- using pilot demonstration projects;
- using credible third party testimonies;
- showing successful projects elsewhere;
- communicating risks and management measures to stakeholders;
- encouraging stakeholders to communicate with each other.

18.3.5 Identifying and dealing with issues

Identifying issues

In any project which can affect the environment, there are issues that may lead to conflicts. By involving the community during planning, decision-making and project implementation, conflicts can often be avoided or resolved with results acceptable to most of the community. It is useful at the outset to identify issues that might arise during a project. By anticipating and acknowledging that there are issues to be resolved, it may be possible to defuse potential conflict in the community. Types of issues include:

- ***Planning Issues*** – These can relate to the planning of the project and the adequacy of local planning instruments and approvals to protect public safety, health and well-being.
- ***Environmental Issues*** – These are often the most difficult to deal with in project planning and implementation. They are often complex and the assessment of impacts cannot usually be made with full certainty.
- ***Social Issues*** – These relate to the quality of people's lives and their perceptions about how the project will change them for better or worse.
- ***Economic Issues*** – These include project funding and the impact on rates and services charges. They also include impacts on property values and advantages or disadvantages to local businesses.

Evaluating issues

Once the issues relating to a project have been identified, it will be necessary to evaluate them. The validity of the issues should be checked and their importance to the project assessed. Some of the costs and benefits may not be readily quantifiable in economic terms. The local authority and the community need to arrive at a decision making process which can weigh up both economic and non-economic factors and make sensible decisions about the option which achieves the best balance between decision making factors and preferences.

Resolving conflict

Debate about the issues arising from a particular project can create conflict. An early assessment of the issues will increase the local authority's chances of being able to defuse potential conflicts before the protagonists become entrenched in positions from which they cannot retreat. The political nature of conflict situations requires information sharing and negotiation between the parties for their resolution. It is necessary to listen to the views of other parties and a willingness to accept them if they are right. Coming up with acceptable solutions to conflicts is possible. It often requires diverse thinking, inventiveness and a willingness to consider alternative solutions. Goal directed problem-solving approaches could be helpful in resolving disputes or defusing potential disputes. In some situations, mediation may be appropriate. An experienced and independent third party acceptable to all parties should conduct mediation.

18.4 COMMUNITY EDUCATION

Water reuse will become increasingly important in the years ahead as a means of achieving sustainable use of the world's water. Advances in treatment technologies now make it possible to recycle water of a quality which is fit for all public and personal uses.

However, public perceptions have been affected by views that reused water is a spoiled or unclean resource, unsuitable for public or personal uses. Such concerns relate to long-standing inhibitions about bodily functions and community concerns about health risks. Negative language used in the water industry e.g. sewage effluent, pollutant discharges, adds to the concerns.

It has also been observed that there are gaps in community knowledge of human interaction with the water cycle. These gaps include an almost total lack of awareness of how water reuse systems work. Much of the available information on water issues is too technical for the average person in the community. Community consultation processes on water reuse projects are often delayed and sometimes frustrated because of this lack of knowledge. There is more likely to be informed and rational debate about proposals if the community is knowledgeable and well informed on water issues before the commencement of any community consultation process (Katz, 1997).

Macpherson and Simpson (2004) have proposed a new approach to public education and communication about water recycling to break down the perception barriers to greater reuse. The steps proposed include:

- developing a new *lingua franca* for the water industry to change the focus from used water being a problem to recycled water being a new water source and an opportunity to achieve sustainable use of the world's water;
- developing a clear description of the water quality needed for various uses: *the right water for the right use*;
- creating value propositions that describe the benefits of water recycling;
- building an on-line community clearing-house of water reuse success stories;
- creating a public outreach program that delivers water recycling case studies and testimonials to industry and to popular publications.

An Australian water education project to improve community understanding on water issues (Bovill and Simpson, 1998) has become the 'We All Use Water' suite of documents and education aids which have been published by the Australian Water Association.

18.5 CASE STUDIES

18.5.1 Urban reuse

The Irvine Ranch Water Recycling Program is said to be one of the longest and most successful multi-use recycling projects in California (Redwood City, 2003). Irvine Ranch Water District (IRWD) imports 50% of its water to meet the district irrigation and domestic requirements. In order to reduce the dependency on this imported water, IRWD first introduced recycled water to the local agricultural sector in 1967 (D'Angelo Report, 1998). Recycled water has since been widely used in the district for the irrigation of local crops, golf courses, parks, school grounds, greenbelts, street medians, freeway landscaping, other industrial uses and even commercial toilet flushing. The IRWD has also supplied recycled water for non-potable uses to individual homeowners, through a dual distribution system

(Holliman, 1998). Currently, recycled water supplies about 15% of the annual water needs of the District (Young *et al.*, 1998).

The success of the Irvine Ranch Water Recycling Program can be credited to their commitment to inform and educate the local community about efficient water use and reuse, thus creating a greater awareness of the water shortage issues. Every member of the community was told about the value of water by the IRWD *for decades* (Young *et al.*, 1998). Water reuse was promoted to the public as a means of protecting the environment, saving money and energy, and providing a drought resistant supply (D'Angelo Report, 1998). Intensive water conservation programs combining public tours, in-school education, outreach and community education programs were fully utilised to promote water reuse.

18.5.2 Agricultural reuse

The Monterrey County Water Recycling Project was designed to minimise seawater intrusion into the local groundwater aquifers by using recycled water for irrigation rather than groundwater (Sheikh *et al.*, 1999). Currently, over 145,000m^3/d of recycled water is produced for irrigation of high quality food crops, such as artichokes, lettuce, cauliflower, celery, and strawberries. The success of the project is credited to years of careful planning with great emphasis on public involvement. It took *almost 20 years of planning* before the project was fully operational in 1998 (Recycled Water Task Force, 2003). In the early planning stage, public discussion with the local growers suggested a field study was needed to look at the safety and marketability of produce irrigated with recycled water. A five-year health study was consequently conducted and the results showed that produce irrigated with recycled water was safe to consume (Sheikh *et al.*, 1998). The market study also returned a positive note for the project (Sheikh *et al.*, 1998). Buyers, shippers and other intermediaries were accepting of recycled water for their irrigation needs, providing that the practice was endorsed by regulatory agencies and that their competitors and the media would not target them with negative publicity. More importantly, the labelling of the products as being irrigated with recycled water was not required since the water had been treated to the acceptable standard. Overall, support for this project was well received from different groups in the County (Recycled Water Task Force, 2003). The environmental community and local people endorsed the project for reducing wastewater discharge to the Monterey Bay Sanctuary. Actively involving the growers in the health study and providing safe, reliable and inexpensive water for irrigation obtained the support by the agricultural industry. However, two years before the project was due to start, concerned growers formed a committee to express their concerns about the project design and operation (Recycled Water Task Force, 2003). The County took these concerns seriously and subsequently a supplemental food safety study was conducted to investigate potential pathogen growth in the irrigation water. The study again failed to find any pathogens and thus the project commenced (Sheikh *et al.*, 1999). To assure growers of the plant's operating safety, the County provided continual information, both directly to them, and via the Monterrey Regional Water Pollution Control Agency – the agency which manages the reuse scheme website. The project was said to have a strong support at the community level because of the extensive educational programs introduced (California Recycled Water Task Force, 2003). These programs involved using school presentations, tours to treatment plants, project exhibitions at local community events, and providing pamphlets to consumers with their water bills outlining the project.

18.5.3 Indirect reuse

There is little research performed on this subject in developing countries, therefore the Singapore case will be the only one cited here. Singapore is a small, wealthy island that depends heavily on its neighbouring countries for their natural resources, including water. In fact, half of the country's water supply is imported from Malaysia (Seah, 2002). Singapore recognised a need to diversify its water sources to cater for its future water supply needs (Kyodo News International, 2003).

The recycled water project, commonly known in Singapore as NEWater, was seen as a strategic option as it reuses the available water and it is cheaper than other options such as desalinated water. In 2001, NEWater was introduced to the industries for non-potable applications such as wafer fabrication processes and cooling towers. The scheme has since been extended to include indirect potable uses for the public. Currently, about 9,000 m^3/d of highly treated wastewater is mixed with the reservoir water before it undergoes the conventional water treatment (Collins, 2003). This figure represents about 1% of the daily water consumption by the country and it is expected to increase to 2.5% (about 45,000 m^3/d) by 2011.

To raise people's awareness of NEWater, Singapore's Public Utilities Board (PUB) launched intensive education campaigns, using a documentary feature film, media exposure, information briefings at community centres and schools. The PUB also created the NEWater visitor centre where visitors can view the water recycling processes in action and participate in a wide range of interactive educational displays (Collins, 2003).

In relation to the quality of NEWater, the PUB reassured the public by citing that there had been no ill-health impacts to US citizens who had consumed recycled water throughout the past 20 years. The government supported a two-year study by a panel of local and international experts to look at the different aspects of introducing NEWater to supplement the country's drinking water supply. The study concluded that the quality of NEWater was well within standards specified by the World Health Organization and the US Environmental Protection Agency and that it was considered to be safe for potable uses (PUB, 2002). Using the study results, PUB promoted NEWater as cleaner than their tap water in terms of colour, clarity, organic substances and bacteria count. The government also distributed 1.5 million bottles of NEWater for the public to see and try for themselves (Kyodo News International, 2003).

A few newspaper articles indicated signs of public hesitation to using the NEWater (see Seah, 2002; Kyodo News International, 2003). Two leading newspapers in Singapore further reported their studies which found the majority of respondents did not feel comfortable with drinking the water (Seah, 2002). Some respondents even considered paying more for imported water rather than having to drink NEWater (Seah, 2002). These findings are, however, in contrast to an independent poll by Forbes Research (Kyodo News International, 2003) which the government cited. The poll indicated that 82% of people surveyed were prepared to drink the NEWater directly and 16% were prepared to drink it indirectly through mixing with reservoir water. With little public controversy, NEWater was introduced in February 2003. The government hopes to use NEWater to meet 20% of the country's needs by 2015 (Collins, 2003).

18.6 DISCUSSION

Although the importance of community acceptance for a successful reuse program is widely acknowledged, further research is needed to understand public perceptions of water reuse and the psychological factors governing decision-making processes. Po *et al.* (2004) have proposed a list of the areas for further investigations including:

- judgment strategies;
- factors influencing risk perceptions;
- the role of trust and scientific knowledge;
- sensitivity to disgust factors;
- why different recycled water sources influence decisions;
- perceived economic and environmental benefits of water reuse;
- environmental justice issues; and
- behavioural issues affecting willingness to use recycled water.

Perceptions are also influenced by social and economic factors and better understanding is needed of the differences in perception between developed and developing countries.

18.7 CONCLUSIONS

Public acceptance of water reuse is dependent on a wide range of factors. A proper understanding of the factors that determine public acceptance can assist in framing suitable communication and education programs.

Communities increasingly demand a say in the development of projects, particularly where environmental issues are involved. Communities want to know how a project will affect them and the local environment. They want to know how any perceived negative impacts of a project will be mitigated and how the decisions will be made.

Community views should be sought and potential conflicts should be identified and resolved as early as possible in the project development process. Community involvement in development of project options increases the chance of shared ownership of both the process and the outcomes.

18.8 REFERENCES

Agence France Presse. (2003, March 12). *Singapore wages high-tech campaign to promote recycled water.* Retrieved June 10, 2003 from Dialog Newsroom database.

Angyal, A. (1941). Disgust and related aversions. *Journal of Abnormal and Social Psychology*, **36**, 393-412.

Australian Research Centre for Water in Society (ARCWIS). (2002). Perth Domestic Water-Use Study Household Appliance Ownership and Community Attitudinal Analysis 1999-2000. Sydney: CSIRO Urban Water Program.

Baggett, S., Jeffrey, P., & Russell, R. (2003). Stakeholder perceptions of water reuse and participatory planning models. AQUAREC Integrated Concepts for Reuse of Upgraded Wastewater, Project EVK1-2001-00213, Sept 2003.

Bakker, A. B., Buunk, B. P., & Manstead, A. S. (1997). The moderating role of self-efficacy beliefs in the relationship between anticipated feelings of regret and condom use. *Journal of Applied Psychology*, **27**: 2001-2014.

Beckwith, J. A. E. (1996). Judgement Strategies in Determining Risk Acceptability. Unpublished doctoral dissertation, Curtin University of Technology, Perth, Western Australia.

Biel, A., & Dahlstrand, U. (1995). Risk perception and the location for a repository of spent nuclear fuel. *Scandinavian Journal of Psychology*, *36*, 25-36.

Bord, R. J., & O'Connor, R. E. (1990). Risk communication, knowledge, and attitudes: Explaining reactions to a technology perceived as risky. *Risk Analysis*, *10*, 499-506.

Bovill I & Simpson J (1998), Water Education Project, Proc AWWA/RWWC, *6th NSW Recycled Water Seminar*, Sydney November 1998, pp126-131

Brotherton, P.D. & Owen, D.K. Sustainable Development: The Public Consultation Imperative. International Hydrology and Water Resources Symposium 1991, IE Aust Conf Pub 91/22.

Bruvold, W. (1988). Public Opinion on water reuse options. Journal *WPCF*, *60*(1), 45-49.

Bruvold, W., & Crook, J. (1981). What the public thinks: reclaiming and reusing wastewater. *Water Engineering and Management*, 65-

Collins, R. (2003). Hard sell for new water. *Water and Wastewater*, *14*(4), 39-40.

Crook, J. (1999). Indirect potable reuse – Reclaimed water can be used to augment drinking water sources if engineers and operators are aware of the uncertainties and proceed with caution. *Water Environment & Technology*, May 1999.

D'Angelo Report. See Using Reclaimed Water to Augment Potable Water Resources. (1998). *Public Information Outreach Programs* (Special Publication, Salvatore D'Angelo, Chairperson). Publishers: Water Environment Federation & American Waterworks Association.

Dishman, C. M., Sherrard, J. H., & Rebhun, M. (1989). Gaining public support for direct potable water reuse. *Journal of Professional Issues in Engineering*, *115*(2), 154-161.

Frewer, L. J. Scholderer, J. Bredhal, L. (2003). Communicating about the risks and benefits of genetically modified foods: The mediating role of trust. Risk Analysis 23 (6): 1117-1133.

Frewer, L. J., Howard, C., Hedderley, D., & Shepherd, R. (1996). What determines trust in information about food-related risks? Underlying psychological constructs. *Risk Analysis*, *16*(4), 473-486.

Frewer, L. J. Howard, C., & Shepherd, R. (1998). Understanding public attitudes to technology. *Journal of Risk Research*, *1*, 221-235.

Gagliardo, P. (2003, April). Use of Reclaimed Water for Industrial Applications. Paper presented at the Ozwater 2003 Convention and Exhibition, Perth, Australia.

Hance, B.J., Chess, C. & Sandman, P.M. "Industry Risk Communication Manual – Improving Dialogue with Communities". Lewis Publishers 1990.

Hamilton, G. R., & Greenfield, P. F. (1991). Potable Reuse of Treated Wastewater. Proc. Australian Water and Wastewater Association 14th Federal Convention (Vol.1, pp. 497-506). Perth, 17-22 March.

Hartley, T. W. (2003). Water Reuse: Understanding Public Perception and Participation. Virginia: Water Environment Research Foundation.

Holliman, T. R. (1998). Reclaimed water distribution and storage. In T. Asano (Ed.), *Wastewater Reclamation and Reuse* (pp. 383-436). Pennsylvania: Technomic Publishing

Hurlimann, A., & McKay, J. (2002, April). Community attitudes to an innovative dual water supply system at Mawson Lakes South Australia. Paper presented at the Enviro 2002 Convention and Exhibition, Melbourne, Australia.

Jansen, H.P., Senstrom, M.K. and de Koning, J. (2005), Development of Indirect Potable Reuse in Impacted Areas of the United States, Proc. IWA Spec. Conf. WRRS2005 Water Reclamation and Reuse for Sustainability, 8-11 Nov 2005, Jeju, Korea.

Jeffrey, P. (2002). Public attitudes to in-house water recycling in England and Wales. *Journal of the Chartered Institution of Water and Environmental Management*, *16*, 214- 217.

Jeffrey, P., & Jefferson, B. (2002, April). Public receptivity regarding 'in-house' water recycling: Results from a UK survey. Paper presented at the Enviro 2002 Convention and Exhibition, Melbourne, Australia.

Kaercher, J. D., Po., M., & Nancarrow, B. E. (2003). Water Recycling Community Discussion Meeting I (Unpublished Manuscript). Perth: Australian Research Centre for Water in Society (ARCWIS).

Katz, S. M., & Tennyson, P. (1997). Public education is the key to water repurification's success. A paper presented at Beneficial Reuse of Water and Biosolids April 6-9, Malaga, Spain.

Kyodo News International (2003, February 21). Singapore starts pumping reclaimed water.

Retrieved June 10, 2003 from Dialog NewsRoom database.

Macpherson, L. and Simpson, J. (2004), Communicating Rightly in a Left-Brained Industry - Promoting Public Understanding about Water Use and Reuse Workshop, Hatfield School of Government Portland State University, 27 Oct 2004.

McKay, J., & Hurlimann, A. (2003). Attitudes to reclaimed water for domestic use: Part 1. Age. *Journal of the Australian Water Association*, *30*(5), 45-49.

Marks, J., Martin, B., & Zadoroznyj, M. (2006). Acceptance of water recycling in Australia: national baseline data. Jnl AWA Water 33 2, March 2006, pp151-157.

Marks, J., Cromar, N., Howard, F., Oemcke, D., & Zadoroznyj, M. (2002a, April). *Community experience and perceptions of water reuse.* Paper presented at the Enviro 2002 Convention and Exhibition, Melbourne, Australia.

Marks, J. (2002b). Community Perceptions of Recycled Water: Mawson Lakes. Unpublished report, The Flinders University of South Australia, Adelaide.

Melbourne Water (1998). Exploring Community Attitudes to Water Conservation and Effluent Reuse. A consultancy report prepared by Open Mind Group. St Kilda, Victoria.

Nancarrow,, B. E., Kaercher, J. D., Po, M., & Syme, G. J. (2003). Social Values and Impact Study: South West Yarragadee Blackwood Groundwater Area. Perth: CSIRO Land and Water.

Nancarrow, B. E., Kaercher, J. D., & Po, M. (2002). Community attitudes to water restrictions policies and alternative water: A longitudinal analysis 1988-2002. Perth: CSIRO Land and Water.

Po, M., Kaercher, J. & Nancarrow, B.E. (2004), Literature review of factors influencing public perceptions of water reuse. Australian Water Conservation and Reuse Research Program, CSIRO Land and Water April 2004, ISBN 0 643 09175 0.

Public Utilities Board. (2002). Singapore water reclamation study: Expert panel review and findings. Retrieved June 10, 2003 from http://www.pub.gov.sg/NEWater_files/download/review.pdf.

Recycled Water Task Force (2003). White paper of the public information, education and outreach workgroup on better public involvement in the recycled water decision process Website: http://www.owue.water.ca.gov/recycle/docs/PubInfoDraftPaper.pdf. California Department of Water Resources.

Redwood City. (2003). Water Recycling in California. Retrieved June 30, 2003, from http://www.ci.redwood-city.ca.us/water/pdf/Water_Recycling_in_CA_9_09_02.pdf.

Rosenblum, E. (2005) House Rules: Environmental Ethics for a Sustainable World. Proceedings, Australian Water Association OzWater 2005 Annual Conference, May 7-11, 2005 (Brisbane, Australia). o5285.

Rosenblum, E., & Anderson, J. (2004). The Water Reclamation Matrix: A Framework for Sustainable Urban Water Use. IWA Leading Edge Conference on Sustainability, November 7-11, 2004 (Sydney, Australia)

Rozin, P., & Fallon, A. E. (1987). A perspective on disgust. *Psychological Reports*, *94*, 23-41.

Savage, I. (1993). Demographic influences on risk perception. *Risk Analysis*, *13*(4), 413-420.

Seah, C. N. (2002, July 21). Media blitz on the yuck factor. *The Star*. Retrieved June 13, 2003 from http://www.singapore-window.org/sw02/020721st.htm.

Sheikh, B., Cooper, R. C., & Israel, K. E. (1999). Hygienic evaluation of reclaimed water used to irrigate food crops – A case study. *Water Science Technology*, *40*(4-5), 261-267.

Sheikh, B., Cort, R., Copper, R. C., & Jaques, R. S. (1998). Tertiary-treated reclaimed water for irrigation of raw-eaten vegetables. In T. Asano (Ed.), *Wastewater Reclamation and Reuse* (pp. 779-825). Pennsylvania: Technomic Publishing

Sjoberg, L. (2001). Limits of knowledge and the limited importance of trust. *Risk Analysis*, *21*(1), 189-198.

Slovic, P. (1998). The risk game. *Reliability engineering and system safety*, *58*, 73-77.

Sydney Water (2002a). WaterPlan 21: Concept Testing Customer Research. An internal report prepared by Eureka Strategic Research. Sydney.

Sydney Water (2002b). Doing Things Differently, Edmondson Park Customer Reactions to Service Delivery. An internal report prepared by Eureka Strategic Research. Sydney.

Sydney Water (1999). Community Views on Recycled Water. Sydney, Author.

Syme, G.J. & Nancarrow, B.E. (Eds.) (2001). Applying Social Justice Research to Environmental Decision-Making. *Social Justice Research*, *14*(4).

Syme, G.J., Kals, E., Nancarrow, B.E. and Montada, L. (2000). Ecological Risks and Community Perceptions of Fairness and Justice: A Cross-Cultural Model. *Risk Analysis*, 20(6), 905-916.

Syme, G.J., Nancarrow, B.E., & McCreddin, J.A. (1999). Defining the components of fairness in the allocation of water to environmental and human uses. *Journal of Environmental Management*, *57*, 51-70.

Syme, G.J. and Bishop, B.J. (1992). Community Perceptions of Dam Safety Issues: A Preliminary Study. CSIRO Division of Water Resources. DWR Consultancy Report No. 92/32.

Timko, S. (2002, September 14). Officials consider replenishing ground water with recycled sewer water. *Reno-Gazette-Journal*. http://www.rgj.com/news/stories/html/2002/09/14/23791.php Retrieved June 13, 2003.

Upper San Gabriel Valley Municipal Water District (2003). Recycled Water Program – Fiscal Year 2002-03. Retrieved December 12, 2003 from http://www.usgvmwd.org/capital.htm.

Water Corporation of Western Australia (2003). Community Attitudes and Public Perceptions. Paper presented at the Water Recycling Workshop 25-26 June 2003, Perth, Australia.

Wegner-Gwidt, J. (1998). Public support and education for water reuse. In T. Asano (Ed.), *Wastewater Reclamation and Reuse* (pp. 1417-1462). Pennsylvania: Technomic Publishing

Young, R. E., Thompson, K. A., McVicker, R. R., Diamond, R. A., Gingras, M. B., Ferguson, D., Johannesse, J., Herr, G. K., & Parsons, J. J. (1998). Irvine Ranch water district's reuse today meets tomorrow's conservation needs. In T. Asano (Ed.), *Wastewater Reclamation and Reuse* (pp. 941-1036). Pennsylvania: Technomic Publishing

Zeelenberg, M., & Beattie, J. (1997). Consequences of regret aversion 2: Additional evidence for effects of feedback on decision making. *Organisational Behaviour and Human Decision Processes*, *73*, 63-78.

19

Water reuse criteria: environmental and health risk based standards and guidelines

Alan Godfree and Samuel Godfrey

19.1 INTRODUCTION

In those regions of the world that operate centralised systems for the collection and treatment of sewage, discharges of treated wastewater often act to augment river flows and sustain aquatic habitats. Subsequent abstraction, whether it be for irrigation, potable supply, or industrial applications, constitutes indirect reuse that can best be described as unconscious. Unplanned reuse has been the prevailing situation for many years. Recognition that reuse can address aspects of water scarcity, and play a positive role in terms of sustainability, is encouraging greater use of wastewater. Reclaiming water is seen as a sustainable practice, particularly in the urban environment (UKWIR, 2004) and the use of treated wastewater to irrigate urban landscapes, parks and recreational areas is increasing. Other uses for reclaimed water include irrigation of golf courses, creation of amenity water bodies and the maintenance of wetland habitats. Reclaimed water may be used to augment drinking water supplies by means of aquifer recharge either through direct injection or infiltration. Finally, and most controversially of all, a small number of potable water supplementation schemes involve the direct addition of reclaimed water to open storage reservoirs; an example is Singapore where surplus reclaimed water from the NEWater factories (approx. 6%) is used for indirect potable use, contributing 1% of Singapore's daily potable water requirements of 1135 million litres.

World wide, the single largest user of freshwater is agriculture, on average accounting for almost 70% of total usage, rising to in excess of 90% in some arid countries (FAO, 2005). Much of this irrigation water is drawn from freshwater groundwater and surface water sources. However, water scarcity coupled with the demands of agricultural production has resulted in wastewater being increasingly used to irrigate crops and non agricultural land, and it is estimated that more than 10% of the world's population eat food crops produced on land irrigated by wastewater (Smit and Nasr, 1992). This is likely to increase as population growth contributes to additional water scarcity and stress. Wastewater has a number of positive attributes that favour its use over other sources of irrigation water. In several regions of the world the supply of wastewater is more dependable than other sources of freshwater. In addition, wastewater contains plant nutrients, particularly nitrogen and phosphorus, which can reduce or even eliminate the need for additional fertilisation, thereby reducing farmers' costs and mitigating the environmental impacts of eutrophication and fertiliser production. The total area of land irrigated with raw or partially diluted wastewater is estimated at 20 million hectares, which is approximately 10% of total irrigated land (United Nations, 2003).

AQUAREC, a project funded by the European Commission, identified and characterised water reclamation schemes around the world (http://www.aquarec.org/). The uses of the water reclaimed were categorised into five types: a) irrigation of agricultural land, b) urban, recreational and environmental (including aquifer recharge, c) process water for industry, d) potable water production (indirect and direct), and e) mixed use (Bixio *et al.*, 2005). Based on the number of schemes, agricultural irrigation was the predominant application in Latin America and the Mediterranean, but in Japan (a highly industrial country) urban use was the major application. The situation in North America and Europe is that the number of schemes involving agricultural irrigation were similar to those for urban applications.

Against this background most countries where reuse is practiced have developed standards or guidelines to address the public health risks arising from the use of wastewater in food production. Often these were based on, or adapted from, guidance produced by the World Health Organization (WHO, 1989). The stringency of these guidelines varied from region to region, reflecting local circumstances and the affordability of treatment. The WHO guidelines took into account the circumstances in developing countries and were established using data obtained from epidemiological studies that estimated the burden of disease attributable to wastewater use in agriculture. Industrialized countries favour stringent water quality criteria for reuse based on the application of advanced technologies that are capable of achieving essentially pathogen-free water that has also undergone treatment to remove organic contaminants. It could be argued that stringent standards provide for the stakeholder reassurance that is requisite for the successful implementation of reuse schemes in these regions.

The situation in developing countries is characterized by limited access to advanced treatment or no treatment at all, often allied with water scarcity particularly for the irrigation of agricultural land. Under these circumstances the primary concern is balancing the need for water, and increased agricultural output, with the disease risk. This requires the utilization of reused and reclaimed wastewater for domestic as well as agricultural purposes. Due to rapid demographic growth and increased urbanization, reused wastewater is rapidly becoming an essential source of additional water supply for many drought affected and water scarce environments within developing countries. In this respect the WHO guidelines have been seen as particularly relevant.

The World Health Organization has recently completed a revision of the 1989 guidelines, embracing the risk management approach espoused in the Stockholm framework (Fewtrell and Bartram, 2001). This accords with the risk management paradigm underpinning the third edition of the Guidelines for Drinking Water Quality (WHO, 2004) and which is being adopted in the Australian national guidelines for water recycling (Anon, 2005a). This risk management approach facilitates a flexible approach to defining a level of health protection and control measures that take into account local circumstances. Health protection measures are framed within the concept of multiple barriers, the application of each will sequentially reduce the risks to health of exposed populations.

19.2 WATER REUSE AND ENVIRONMENTAL HEALTH

Wastewater contains pathogens (viruses, bacteria, protozoa, and helminths) and chemical constituents that are of concern if the wastewater is to be used beneficially (WHO, 2006). The nature and amounts of these will vary widely depending on the inputs to the sewerage system. There may also be positive benefits to human health in terms of improved nutrition as a result of increased production of vegetables or other food crops, whether for household use or sale which in turn generates higher levels of economic activity.

Inappropriate use of wastewater may have adverse effects on human health, either directly or indirectly. Direct contact may be an important route of exposure in regions that practice widespread irrigation of agricultural land with reclaimed water. Agricultural workers are likely to be directly exposed to reclaimed water, the extent of exposure being related to the mode of irrigation. In many countries it is not unusual for children to accompany adults working in agriculture or to eat uncooked field vegetables that have been irrigated with wastewater. A number of studies have shown that the risk of gastrointestinal infection is greatest in this group, exceeding that of adults (Blumenthal and Peasey, 2002). The use of reclaimed water for landscape irrigation is another potential route of direct exposure, particularly where the area under irrigation includes communal facilities such as parks or sports fields.

Routes of indirect exposure include the consumption of food crops irrigated with reclaimed water and the use of drinking water sources augmented with wastewater, whether planned or unplanned. Another route for exposure is occupational or recreational contact with ornamental water features that utilise or receive treated or untreated wastewater. The number of people who may be exposed indirectly is significantly greater than those who could conceivably come into direct contact with reclaimed water because of the multiple pathways involved.

19.2.1 Pathogens

The principal concern associated with the reuse of municipal wastewater is the possibility of infectious disease transmission. Pathogens potentially present in untreated wastewater have been extensively documented in the literature (Feacham *et al.*, 1983; Geldreich, 1990; NRC, 1996; Mara and Feacham, 2003). Table 19.1 provides a summary of the range of concentrations of pathogens and indicator organisms that may be present in wastewater.

The actual pathogens present and their numbers will depend on the pattern of endemic disease in the communities served by the wastewater collection system. Infections caused by helminths (*Ascaris*, hookworks, tapeworms) are endemic in the developing countries of Africa, South America and South East Asia. It is estimated that worldwide, a billion people suffer from intestinal helminth infection (WHO, 2006). This is mirrored by the numbers of helminths present in wastewater from these regions (Jiménez, 2005). Conversely these diseases are rare in the developed world and numbers of eggs in wastewater are correspondingly low (typically <10/L).

Table 19.1 Levels of pathogens and indicator organisms present in wastewater (adapted from WHO, 2006).

Organism	Number per litre of wastewater
Bacteria	
Theromtolerant coliforms	10^8 - 10^{10}
Campylobacter jejuni	10 – 10^4
Salmonella spp.	1 – 10^5
Shigella spp.	10 – 10^4
Vibrio cholerae	10^2 – 10^5
Helminths	
Ascaris lumbricoides	1 – 10^3
Ancyclostoma duodenale / Necator americanus	1 – 10^3
Trichuris trichuria	1 – 10^2
Protozoa	
Cryptosporidium parvum / Cryptosporidium hominis	1 – 10^4
Entamoeba histolytica	1- 10^2
Giardia intestinalis	10^2 – 10^5
Viruses	
Enteric viruses	10^5 – 10^6
Rotavirus	10^2- 10^5

Despite advances in molecular techniques that permit culture-independent methods of detecting the presence of pathogens, their use in determining the safety of wastewater reuse is limited and likely to be so for some time. Indicators of the presence of waterborne pathogens are employed, including coliform organisms, thermotolerant coliforms, *Escherichia coli*, enterococci, Clostridium, and bacteriophages (viruses that infect bacteria, considered to mimic the fate of enteric viruses (Havelaar *et al.*, 1993)). The use of indicators for assessing safety is predicated on the assumption that the indicator organism mimics the occurrence and fate of a pathogen or category of pathogens (e.g. protozoan parasites). Unfortunately none of the recognized indicators, either singly or in combination, is able consistently to predict the presence of pathogens in wastewater or reclaimed water. There is also concern about the relationship between these indicators and emerging pathogens such as *Helicobacter pylori*, verotoxigenic *Escherichia coli* (VTEC), Cryptosporidium, and Cyclospora (Huffman *et al.*, 2003; NRC, 2004).

19.2.2 Heavy metals

Wastewater may contain a wide range of chemicals of anthropogenic origin, including heavy metals. Elements of concern include lead, cadmium, mercury, chromium, molybdenum, copper and nickel. The concentrations of these and other metals will depend on the amount and content of any industrial discharges to the wastewater collection system. Faeces contribute 60–70% of the load of Cd, Zn, Cu and Ni in domestic wastewater and >20% of the input of these elements in mixed wastewater from domestic and industrial premises.

Other sources of copper and zinc present in domestic wastewater are plumbing materials and personal care products/cosmetics respectively. Actual amounts of metals in urban wastewater vary, often by orders of magnitude depending on a number of factors including rainfall and seasonal influences. Table 19.2 shows the range of concentrations reported from industrialized countries. Metals present in wastewater used for irrigation will tend to accumulate in the soil. If the concentration reaches a threshold it is possible that metals will be taken up by crop plants giving rise to a potential health risk to those consuming them.

Table 19.2 Concentrations of metals present in urban wastewater (from Thornton *et al.*, 2001 and UKWIR, 2003).

Metal	Concentration (ug/L)
Cadmium	0.4 – 30
Mercury	0.1 – 3.5
Lead	<30 – 535
Nickel	<50 – 72
Copper	<100 – 266
Chromium	<20 - 7100
Zinc	835 - 1000
Arsenic	2.2
Selenium	0.3 – 0.4

19.2.3 Nutrients

Nitrogen, phosphorus and potassium are present in wastewater in agronomically beneficial amounts (Table 19.3). Unlike other wastewater constituents, the presence of N, P, and K is advantageous when irrigation is the end use. At an application rate of 5000 m^3 ha $year^{-1}$, the fertiliser contribution per year of the effluent would be: nitrogen, 250 kg ha^{-1}; phosphorus, 50 kg ha^{-1}; and potassium, 150 kg ha^{-1}. Thus all the nitrogen and much of the phosphorus and potassium normally required for agricultural crop production over the growing season would be supplied by the effluent (FAO, 2002). Micronutrients and organic matter contained in the effluent may also provide agronomic advantages.

A positive environmental benefit accrues since most of these nutrients are absorbed by the crop and therefore are removed from the water cycle, hence playing no further role in the eutrophication of rivers, lakes and coastal waters (Anon, 2005b).

19.2.4 Salinity

The land application of wastewater will inevitably result in an increase in soil salinity because wastewater contains higher concentration of salts (sodium, chlorides, boron) compared with freshwater. Over time this will reduce yields and limit the range of crops that can be grown. Although not a direct human health impact, salinization of soil is the most important negative environmental consequence of agricultural wastewater use, affecting about 20-30 million ha of the world's 260 million ha of irrigated land (FAO, 2002).

Table 19.3 Typical composition of wastewater (adapted from Metcalf and Eddy, 2003).

Constituent	Concentration (mg/L)		
	Low strength	Medium strength	High strength
Total solids	390	720	1230
Suspended solids	120	210	400
Biochemical oxygen demand	110	190	350
Total organic carbon	80	140	260
Chemical oxygen demand	250	430	800
Nitrogen (total as N)	20	40	70
Phosphorus (total as P)	4	7	12
Chloride	30	50	90
Sulfate	20	30	50
Oils and grease	50	90	100
Volatile organic compounds	<100	100-400	>400

19.2.5 Organic matter

The organic content of wastewater (as measured by BOD) can enhance soil fertility by increasing humic content, aiding moisture retention, and enhancing microbial biomass. Decomposition of organic matter in soil is rapid, particularly under aerobic conditions. Soil clogging is unlikely to occur unless untreated wastewater containing high levels of suspended solids and BOD is land applied (Siegrist and Boyle, 1987).

19.2.6 Toxic organic compounds

Domestic wastewater is unlikely to contain toxic organic substances. In most instances however, wastewater collection systems receive discharges from industrial facilities or run off that may contain a wide range of organic compounds, often in relatively high concentrations. Inputs of the main persistent organic pollutants of concern, including polycyclic aromatic hydrocarbons (PAHs), polychlorinated biphenyls (PCBs), Polychlorinated dibenzo-p-dioxins (PCDDs) and polychlorinated dibenzofurans (PCDFs), are principally from atmospheric deposition onto paved surfaces and runoff. These groups of chemicals are hydrophobic in nature and tend to bind to sewage solids or soil particles, limiting their subsequent transport in the environment. The most commonly found organic compounds in wastewater are detergent residues (e.g. nonyl phenol), surfactants (e.g. linear alkyl benzene sulphonates), plasticising agents (e.g. di-(2-ethylhexyl)phthalate) and polyacrylamide compounds (used as sewage sludge dewatering aids) (Thornton *et al.*, 2001).

19.2.7 Endocrine disrupting substances and pharmaceutical residues

Wastewater effluent is known to contain a wide number of potential oestrogenic chemicals including oestrogens which are naturally secreted by humans and other animals (oestrone and 17β-oestradiol), synthetic oestrogens such as those found in the contraceptive pill (ethinylestradiol) and alkylphenols used in plasticizers (e.g. nonylphenol, octylphenol and their bisphenols and ethoxylates) (Desbrow *et al.*, 1998). Another group of substances about which concern has been raised is residues and metabolites of pharmaceutically active compounds (Halling-Sørensen *et al.*, 1998). Some over-the-counter (notably acetaminophen, caffeine, ibuprofen) and prescription medicines (diltiazem, sulfamethoxazole, and trimetheprim) have been detected in wastewater and receiving waters (Tixier *et al.*, 2003).

19.2.8 Application of greywater use

Uses of greywater are becoming increasingly important due to the scarcity of freshwater and increase in anthroprogenic pollution of ground and surface water bodies (Ottoson *et al.*, 2003). Greywater is defined as *wastewater without input from toilets (i.e. wastewater from laundries, showers, bathtubs, handbasins and kitchen sinks)* (Ottoson *et al.*, 2003). Due to the absence of faecal waste in greywater, the potential environmental health risk is low. However, the level of the risk associated with greywater is dependent on the intended usage. This may include the following:

- direct–potable, where wastewater is treated to potable standards;
- indirect–potable, where treated effluent enters a river or lake or aquifer that is also used as a source of drinking water;

- direct–non-potable, where treated wastewater is used for irrigation and toilet flushing;
- indirect–non-potable, where effluent goes into a water source that is also used for irrigation.

To ascertain an appropriate level of health risk derived from greywater it is critical to determine the following:

1. hazards associated with greywater;
2. level of exposure of target population to greywater;
3. dose-response of the population to greywater;
4. characterisation of risk level based on intended use (Westrell *et al.*, 2004).

The sections below will discuss in more detail the application of each of these criteria in relation to both the establishment of greywater and wastewater guidelines and standards.

19.3 INTERNATIONAL, NATIONAL AND REGIONAL REGULATIONS

A number of countries used the World Health Organization Guidelines published in 1989 as the scientific point of departure for establishing national wastewater standards. For example, France and Mexico have retained the recommended microbial guideline values and enforce them using crop restriction or site-specific anti-helminthic campaigns (WHO, 2006). However, some countries view the WHO guidelines as too stringent (Tables 19.4 and 19.5). California has developed its own set of public health laws pertaining to recycled water (namely Titles 22 and 17 of the California Code of Regulations, State of California). These are commonly known as the *Purple Book* standards and are widely adopted elsewhere as the basis for forming standards.

Within the European Union no regulation for wastewater reuse exists. The only reference is Article 12 of the European Wastewater Directive (91/271/EEC) which states that *treated wastewater shall be reused wherever possible* (Angelakis *et al.*, 2001). Table 19.6 outlines examples of parameters of concern and their application to wastewater reuse standards or guidelines in selected countries of the world. There are minimal guidelines for wastewater reuse for non-potable purposes.

Table 19.4: Comparison of microbiological quality guidelines for irrigation (Angelakis *et al.*, 1999).

Agency	Reuse Conditions	Intestinal nematodes*	Thermotolerant or Total Coliforms	Wastewater Treatment Requirements
WHO	Irrigation of cereal crops, industrial crops, fodder crops, pasture and trees	<1/L	No standard recommended	Stabilisation ponds with 8-10 day retention or equivalent removal
WHO	Irrigation of crops likely to be eaten uncooked, sports fields, public parks	<1/L**	<1000/100mL	A series of stabilisation ponds or equivalent treatment
USEPA	Irrigation of pasture for milking animals, fodder, fibre and seed crops and landscape improvements	No standard recommended	200/100mL	Secondary treatment followed by disinfection
CA	Irrigation of pasture for milking animals, landscape impoundment	No standard recommended	<23/100mL	Secondary treatment followed by disinfection
WHO	Landscape irrigation where there is public access such as hotels	<1/L	<200/100mL	Secondary treatment followed by disinfection
USEPA	Surface or spray irrigation of any food crop including crops eaten raw	No standard recommended	Not detectable	Secondary treatment followed by filtration (with prior coagulant and/or polymer addition and disinfection)
CA	Spray and surface irrigation of food crops high exposure landscape irrigation such as parks	No standard recommended	<2.2/100mL	Secondary treatment followed by filtration and disinfection
Mexico	Irrigation of cereals, animal fodder, fruits & vegetables (unrestricted)	<1/L	<1000/100mL	Not specified
Mexico	Irrigation of agricultural crops excluding vegetables to be eaten raw (restricted)	<5/L	<1000/100mL	Not specified; treatment must be economically viable***

* Ascaris, Tricuris species; hookworms

** WHO (2005) recommends additional health protection measures for children <15 years, including antihelminthic chemotherapy campaigns, washing field vegetables to reduce numbers of helminth eggs, or alternatively additionally treating wastewater to achieve $\leq$0.1 eggs per litre

*** Jiménez (2005)

Table 19.5 USEPA guidelines for water reuse (from Crook and Surampalli, 2005).

Type of Use	Treatment	Reclaimed Water Quality
Urban uses, crops eaten raw, recreational impoundments	Secondary Filtration Disinfection	pH = 6 – 9 $\leq$ 10 mg/L BOD $\leq$ 2 NTU[a] No detectable fecal coli/100 mL[b] $\geq$ 1 mg/L Cl_2 residual[c]
Restricted access area irrigation, processed food crops, non-food crops, aesthetic impoundments, construction uses, industrial cooling[d], environmental reuse	Secondary Disinfection	pH = 6 – 9 $\leq$ 30 mg/L BOD $\leq$ 30 mg/L TSS $\leq$ 200 fecal coli/100 mL[e] $\geq$ 1 mg/L Cl_2 residual[c]
Groundwater recharge of non-potable aquifers by spreading	Site specific and use dependent Primary (minimum)	Site specific and use dependent
Groundwater recharge of non-potable aquifers by injection	Site specific and use dependent Secondary (minimum)	Site specific and use dependent
Groundwater recharge of potable aquifers by spreading	Site specific Secondary & Disinfection (minimum)	Site specific Meet drinking water standards after percolation through vadose zone
Groundwater recharge of potable aquifers by injection, augmentation of surface supplies	Includes the following: Secondary Filtration Disinfection Advanced wastewater treatment	Includes the following: pH = 6.5 – 8.5 $\leq$ 2 NTU[a] No detectable fecal coli/100 mL[b] $\geq$ 1 mg/L Cl_2 residual[c] $\leq$ 3 mg/L TOC $\leq$ 0.2 mg/L TOX Meet drinking water standards

a Should be met prior to disinfection. Average based on a 24-hour time period. Turbidity should not exceed 5 NTU at any time.

b Based on 7-day median value. Should not exceed 14 fecal coli/100 mL in any sample.

c After a minimum contact time of 30 minutes.

d Recirculating cooling towers.

e Based on 7-day median value. Should not exceed 800 fecal coli/100 mL in any sample.

Table 19.6: Comparison of legislation criteria (adapted from Angelakis et al., 1999).

Parameter	California (1978)	USEPA (1992)	WHO (1989)	Israel (1978)	Tunisia (1975)	Cyprus (1997)	France (1991)	Italy (1977)
Type of regulation	law	guidelines	guidelines	law	law	Provisional standards	guidelines	Law
Minimum treatment required	Advanced treatment	Advanced treatment	Stabilisation ponds	Secondary treatment	Stabilisation ponds	Tertiary treatment		Secondary treatment
Total BOD_5 (mg/l)		10		15	30	10		
Dissolved BOD_5 (mg/l)				10				
Suspended solids (mg/l)		5		15	30	10		
Turbidity (NTU)	2	2						
pH					6.5-8.5		As WHO	
Conductivity (Ds/M)					7			
Dissolved O_2 (mg/l)	present			0.5				
Total Colif	2.2 (50%)			2.2 (50%)				
Faecal Colif			1000			50		
Helminths			1		<1	0		
Res Cl_2	present	1		0.5				SAR < 10
Salinity								Yes
Main treatment processes	Oxidation, Clarification, Filtration, Disinfection	Filtration. Disinfection	Stabilisation ponds or equivalent	Long Storage Disinfection	Stabilisation ponds or equivalent	Filtration Disinfection		

19.4 APPLICATION OF REGULATIONS AND GUIDELINES FOR HEALTH PROTECTION

Health and aesthetic concerns related to the reuse of wastewater dictate the need for stringent guidelines and regulations, which provide guidance on reduction of exposure to toxic chemicals and microbes through the adoption of a risk based assessment and management approach. Central to the establishment of these guidelines are two key scientific resource documents published by the World Health Organization: "Guidelines for Drinking Water Quality - 3rd edition" (WHO, 2004), and "Guidelines for the safe use of wastewater in agriculture" (WHO, 2006).

Following an expert meeting in Stockholm, Sweden, the WHO proposed a harmonised framework for the development of health-based guidelines for both water and wastewater (Fewtrell and Bartram, 2001). (See Figure 19.1).

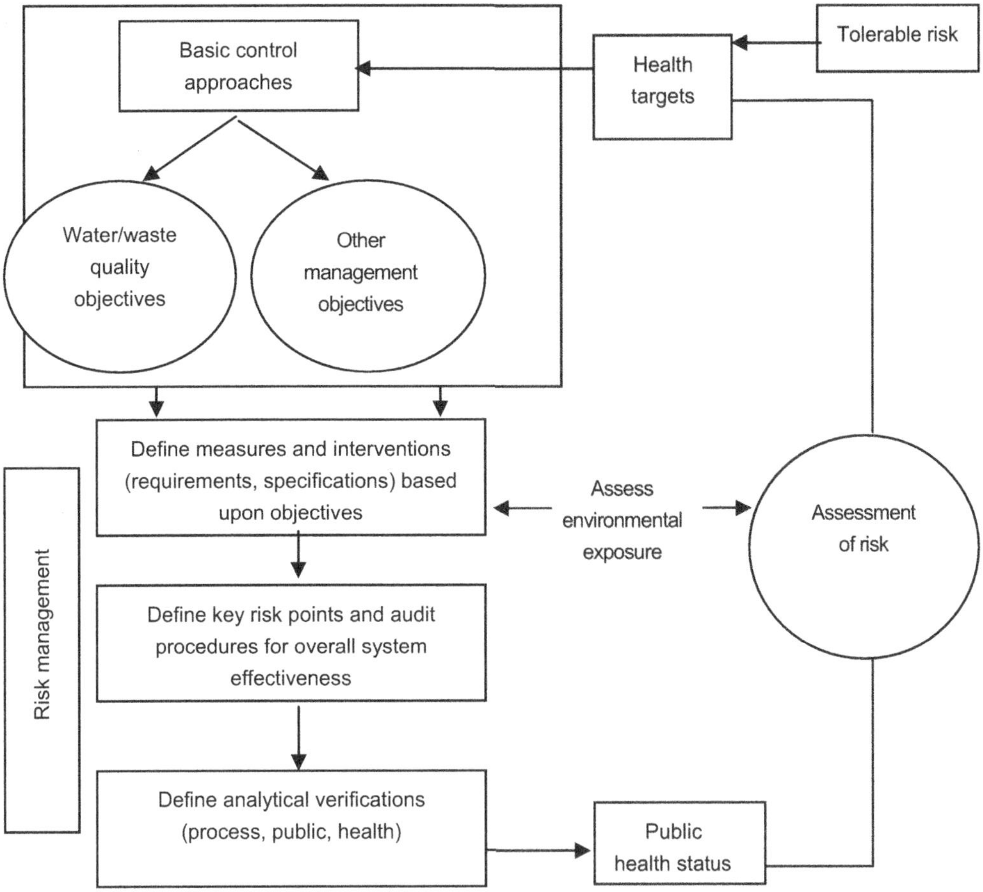

Figure 19.1: The Stockholm framework for developing harmonised guidelines for the management of water related infectious disease (WHO, 2006).

The guidelines provide the basis for the establishment of national and intra-national standards, whilst allowing for local social, economic and environmental considerations. A practical illustration of this harmonised framework is the WHO third edition (2004) *Guidelines for Drinking Water Quality (GDWQ)* (WHO, 2004). The GDWQ marked a fundamental departure from conventional methods of water quality control indicating a move away from reliance on end product testing towards risk assessment and risk management processes. The guidelines note that *"The most effective means of consistently ensuring the safety of a drinking-water supply is through the use of a comprehensive risk assessment and risk management approach that encompasses all steps in water supply from catchment to consumer"* (WHO, 2004).

To establish the risk based approaches, a reference level of risk is defined for all constituent risk parameters. These are based on Quantitative Chemical and Microbial Risk Assessments (Godfrey *et al.*, 2005). The risk assessment defines a tolerable level of risk for selected microbial indicators and toxic chemicals based on their prevalence and exposure to the population. Risk management is then performed based on multiple barrier principles such as HACCP (Hazard Assessment Critical Control Points) used in the food industry (Havelaar, 1994). This comprehensive risk assessment and risk management approach is termed in the WHO GDWQ as *Water Safety Plans (WSPs).*

The principles of *WSPs* when applied to drinking water are:

- to prevent contamination of source waters;
- to treat the water to reduce or remove contamination that could be present to the extent necessary to meet the water quality targets; and
- to prevent re-contamination during storage, distribution and handling of drinking-water (Davison *et al.*, 2004, Godfrey *et al.*, 2004).

Adaptation of these principles for greywater systems are outlined in Figure 19.2.

Following progress in the drinking water sector with the GDWQ being established for river, lake and well drinking water sources, application of the risk management approach to wastewater and greywater was a natural progression and addresses concerns regarding the initial WHO guidelines for reuse (Asano *et al.*, 2004). Indeed, many nations including South Africa, Japan and Australia deferred from using the 1989 WHO guidelines for wastewater quality believing that they were too lenient for industrialised countries (Angelakis *et al.*, 2001). Therefore to establish risk based approaches for wastewater, an expert group convened by WHO was established in 2001. The group proposed greater reliance on establishing reference levels of risk and risk management principles for a) microbiological quality, b) total mineral content (total dissolved solids) c) presence of heavy metal toxicants and d) the concentrations of stable and potentially harmful organic substances (Asano *et al.*, 2004). Figure 19.2 outlines recommended steps for the application of the health risk based approaches outlined in the WHO guidelines for individual scenarios.

The revised wastewater guidelines have been published (WHO, 2006) and mark a radical departure from the 1989 document, much in the same way as the third edition of GDWQ did from previous editions. The risk management approach facilitates and enables the setting of national standards and regulations that are protective for public health. Central to this is the derivation of health-based targets using the metric Disability Adjusted Life Years (DALYs) that define the tolerable level of additional disease burden attributable to the use of wastewater. In the GDWQ, WHO established a tolerable burden of waterborne infectious disease from drinking water of $\leq 10^{-6}$ DALY per person per year. This value is recommended by WHO for application to health risks associated with the agricultural use of wastewater on the basis that consumers expect the same level of safety when consuming ready to eat crops that have been irrigated with wastewater as when drinking potable water. This is an important consideration given the global nature of trade in vegetables, fruits, and other ready to eat products, which are an increasingly valuable source of income to developing countries. Increasing recognition of food poisoning outbreaks linked with globally traded fresh produce has brought this issue to the fore (Döller *et al.*, 2002, Ho *et al.*, 2002).

WHO Wastewater Guidelines	
Assess wastewater use activities and policies	Crop restriction
Assess health implications	Waste application
Develop national or local action plans for safe wastewater use	Human exposure control
	Treatment
Involve stakeholders	Microbial wastewater quality standards
Strengthen national/local capacity	Other health interventions
Health and hygiene education	Industrial effluents

Figure 19.2: Sample action plan for incremental adoption of WHO Guidelines (WHO, 2006).

The risk management framework is illustrated in Table 19.7. It can be used to derive limits for pathogen content and application controls for wastewater taking into account local conditions. The key metrics of DALYs and tolerable additional disease burden facilitate the derivation of guidelines or regulations that encompass the interest of the various stakeholders.

Table 19.7. Risk management framework (WHO, 2006).

Step	Comments
Step 1 Define Tolerable Disease Burden	The metric unit for disease burden is the DALY. WHO suggest $\leq 10^{-6}$ DALY per person per year, consistent with risks from drinking water. This value may be tailored (increased or decreased) to local circumstances.
Step 2 Derive Tolerable Risk of Infection	Tolerable risk of infection per person per year is derived from knowledge (or assumption) of the proportion of those infected that are ill. For individual pathogens the risk of infection is calculated from the following equation: "disease risk ÷ disease/infection ratio"
Step 3 Conduct Quantitative Microbiological Risk Assessment (QMRA)	QMRA is employed to estimate the maximum tolerable pathogen load (the number of pathogens necessary to give rise to the target additional disease burden e.g. $\leq 10^{-6}$ DALY per person per year) ingested during a single exposure event. Exposure may be through direct contact (e.g. field workers) or indirect (e.g. consumption of crops grown on land irrigated using wastewater or consumption of drinking water drawn from an aquifer receiving wastewater). QMRA cannot be used for helminthic infections since no credible data exist on infection risks and severity of health outcome. For these pathogens, epidemiological data are used instead of QMRA (Blumenthal and Peasey, 2002).
Step 4 Estimate pathogen reduction target Tolerable Risk of Infection Burden	This step requires information on a) numbers of pathogens in untreated wastewater and b) the volume of treated wastewater remaining on crops following irrigation or the dilution afforded by aquifer storage. The estimate is described in terms of $\log_{10}$ reduction (Table 19.8).
Step 5 Establish health protection measures	Achieving the $\log_{10}$ pathogen reduction targets previously established (Step 4) are described. This may involve a single measure e.g. wastewater treatment or a combination of measures (multiple barriers) e.g. treatment, method of irrigation, protective clothing, pathogen die-off prior to harvesting, and washing of produce prior to consumption.
Step 6 Verification monitoring	Verification monitoring is carried out to confirm that the required pathogen reduction is being achieved by wastewater treatment. For most pathogens this can be achieved by measuring levels of indicator bacteria (*E. coli* or thermotolerant coliforms). For helminths this entails direct enumeration of eggs. Hydraulic retention time is an appropriate surrogate measure where waste stabilisation ponds are used.

Table 19.8 Health-based targets for wastewater use in agriculture (From WHO (2006).

Exposure	Health-based target (DALY per person per year)	Log_{10} pathogen reduction	Helminth eggs per litre
Unrestricted irrigation	≤10^{-6} [a]		
Lettuce		6	≤1 [b,c]
Onion		7	≤1 [b,c]
Restricted irrigation	≤10^{-6} [a]		
Highly mechanised		3	≤1 [b,c]
Labour intensive		4	≤1 [b,c]
Localised (drip) irrigation	≤10^{-6} [a]		
High-growing crops		2	No recommendation[d]
Low-growing crops		4	≤1 [c]

a Rotavirus reduction. The health-based target can be achieved, for unrestricted and localised (i.e. drip) irrigation, by a 6-7 log_{10} pathogen reduction (obtained by a combination of wastewater treatment and other health protection measures); for restricted irrigation, it is achieved by a 2-3 log_{10} pathogen reduction.

b When children under 15 are exposed, additional health protection measures should be used (e.g. treatment to ≤0.1 egg per litre, protective equipment such as gloves or shoes/boots or chemotherapy)

c An arithmetic mean should be determined throughout the irrigation system. The mean value of ≤1 egg per litre should be obtained for at least 90% of samples in order to allow for the occasional high-value sample (i.e. >10 eggs per litre). With some wastewater treatment processes (e.g. waste stabilisation ponds), the hydraulic retention time can be used as a surrogate to assure compliance with the target of ≤ 1 egg per litre.

d No crops to be picked up from the soil.

19.5 WATER REUSE, WATER SCARCITY AND DEVELOPING COUNTRIES

The Pan-American Health Organisation (PAHO) estimates that only 14% of wastewater in Latin America receives treatment before discharge into surface water courses. Of this, only 6% is treated to within acceptable limits (Cavallina and Young, 2002). In Chile, for example, 97% of wastewater is not treated prior to disposal and is significantly contributing to environmental pollution. Additionally, in water scarce environments in Africa and Asia, wastewater reuse and reclamation is highly under-utilised due to the lack of available technologies, limited operation and maintenance systems and minimal affordability. In India, for example, the utilisation of wastewater through reuse and recycling has become very important due to the nitrogen, phosphorous and potash content that may be reused for agricultural purposes. However, despite this, water reuse is still uncommon.

In contrast, Mediterranean countries are investing in wastewater reuse and reclamation due to high transpiration, low rainfall and increased demand for water for irrigation and tourism (Angelakis *et al.*, 2001). Equally, in water scarce developing countries, greywater reuse in schools, hospitals and government institutions is proving to be an essential alternative water resource to fresh ground, surface or rainwater supplies. Studies from the Middle East and India, for example, indicate that greywater systems have a water saving of between 3.4% to 33.4% per annum (Al-Jayyousie, 2003; Al-Jayyousie, 2005).

BOX 19.1 GREYWATER REUSE AND RISK IN DEVELOPING COUNTRIES: APPLICATION OF QMRA IN INDIA

UNICEF aided a local NGO to construct a greywater reuse system with a production capacity of 2000 lts ($2m^3$) serving a residential school population of 50 children in Jhabua, Madhya Pradesh, India. Acute water scarcity in the district results in limited water availability for sanitation. Greywater from bathrooms is recycled and used for sanitation using the multiple barrier principles of sedimentation and filtration.

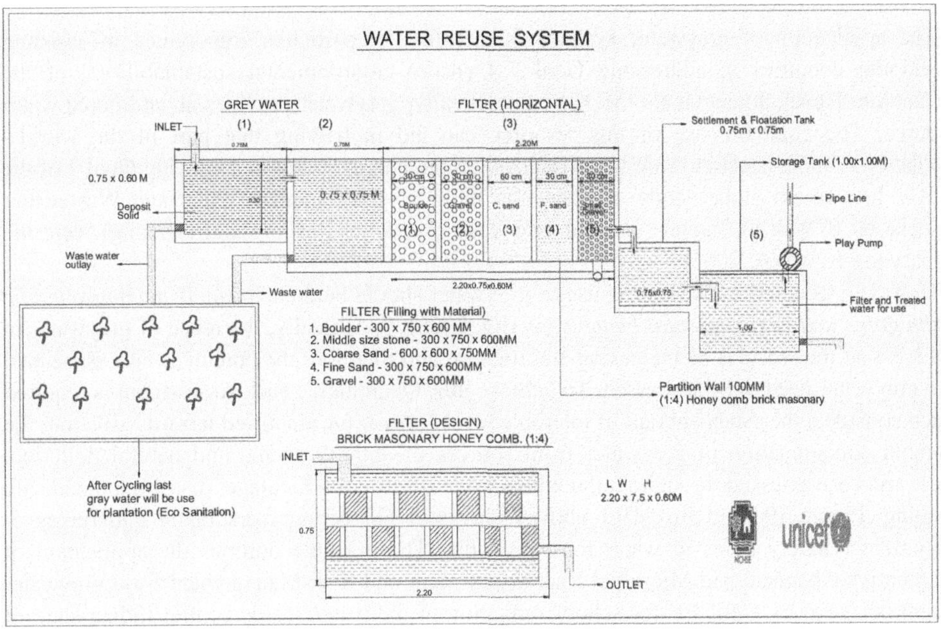

Figure 19.3: WHO risk based approach for greywater in India.

To ascertain the health risk associated with the supply, a QMRA was undertaken in four stages:
Stage 1: Hazard Identification – Analysis of prevailing microbial hazards in the system
Stage 2: Exposure Assessment – See example below of assessment for 100 children

Type of Exposure	Volume ingested (mL or g)	Frequency (times per year)	Number of persons affected
1. (Un)intentional ingestion of greywater during hand washing	30	1	50
2. Child playing in greywater tank	1	2	30
3. Child drinking greywater	100	10	50
4. (Un)intentional ingestion of greywater during tooth brushing	30	1	2

Stage 3: Dose response – Use of available health data to establish DALY targets combined with microbial verification of functioning of system using selected pathogenic organisms (*Ecoli* 0157:H7, Rotavirus, Clostridia Perfringens)
Stage 4: Risk Characterisation – Characterisation of risks based on frequency of occurrence of hazard and its severity of consequence (Westrell *et al.*, 2004).

The QMRA concluded that the highest individual microbial risk came from *Rotavirus*. Microbial results indicate tolerable level of risk $<10^{-3}$, a target within the acceptable risk benchmark (Regli *et al.*, 1991). Risk management reduced risk through implementation of HACCP/Water Safety Plan principles where control points were identified, monitored and corrected within the greywater multiple barrier cycle.
(See Godfrey, S., Kumar, P., Swami, A., Wate, S., Dwivedi, H., (2005) *Water safety plans for greywater and rainwater*, XII International Rainwater Catchment Systems Conference, New Delhi, India (Godfrey, S., *et al.*, 2005)).

The application of greywater systems is therefore of particular importance in assisting developing countries in addressing Goal 7 ("ensure environmental sustainability") of the Millennium Development Goals (MDGs). Specifically, greywater provides an additional water resource. The equitable use of this resource can aid in halving that part of the world's population without access to safe water and sanitation, and therefore in achieving Goal 7 of the MDGs. In a recent global study of greywater reuse by the Canadian Water and Wastewater Association (CWWA), sanitary uses of greywater (i.e. toilet flushing and cleaning) were the primary use (CWWA, 2005).

However, as noted by WHO, the use of greywater should be done to *safely maximise public health gains and environmental benefits* (WHO, 2006). Specifically, the reuse of greywater in arid areas of the world is of increasing significance and therefore the potential risks associated with greywater need to be addressed. To achieve this, Quantitative Risk Assessment is required which considers the establishment of tolerable levels of risk for identified hazards affecting the potential contamination of greywater from sources of anthroprogenic and natural pollution. These risks are assessed throughout the whole cycle of greywater storage, distribution and safe handling. Figure 19.3 (in Box 19.1 above) illustrates pioneering work being undertaken to apply Water Safety Plans to water reuse in India. The example outlines the application of Quantitative Chemical and Microbial Risk Assessment and Risk Management for a greywater system designed by UNICEF for school sanitation in the water scarce central Indian state of Madhya Pradesh (Godfrey *et al.*, 2005).

19.6 CONCLUSIONS

When carried out properly, the practice of reclamation and reuse can bring significant environmental and health benefits, including:

- increased agricultural productivity through irrigation;
- augmentation of potable water supplies through aquifer recharge;
- recycling plant nutrients thereby reducing eutrophication;
- reserving drinking water supplies by substituting with reclaimed water e.g. landscape irrigation, toilet flushing, industrial uses, and cooling water.

There is a clear need to establish and document Good Practice as it applies to water reuse. Central to this is the application of water quality criteria, allied with appropriate controls on their use and post application activities. This is essential to protect public health and maintain ecosystems. There is a wide degree of consensus regarding the physico-chemical quality of reclaimed water (and natural surface waters) that is necessary to maintain soil quality and sustain agricultural production (FAO, 1992).

It is not unusual for freshwater, used as the source of potable water, to receive inputs of wastewater. This is particularly the case for surface waters where the reuse is not the primary intention and is often referred to as 'unplanned'. There are numerous examples of 'planned' indirect potable reuse involving both surface and ground water. Aquifer storage and recovery

(ASR) is an important component of resource management in a number of countries, including Australia and USA. In most countries wastewater discharges are subject to quality standards with the intention of protecting the ecological status of the receiving water. In the situation that the source is used incidentally for drinking water production, public health is protected by appropriate water treatment that is capable of achieving specific drinking water guidelines. The factors that contribute to a reuse scheme being successful are economic, financial, regulatory, psychological, organisational and technical (Lazarova *et al.*, 2000). Of these the psychological factor is the most difficult to plan for. Surveys of the public's attitude toward various wastewater reuse projects indicate that public acceptance tends to increase with income and education. Planned potable reuse receives the greatest opposition even in developed countries.

The area of greatest debate has been in relation to wastewater use for non-potable applications and in particular microbiological considerations. Historically, the argument revolved around the applicability of the values suggested by WHO in their 1989 guidelines (<1000 faecal coliforms/100 ml; ≤1 intestinal nematodes/L) and whether these were sufficiently stringent to safeguard health when crops that are eaten raw are irrigated.

These divergent views can be reconciled by the application of a Risk Management approach, which is now considered the most effective means of ensuring the safety of water supplies (WHO, 2004). This approach is equally applicable to reclaimed water and recreational waters (WHO, 2001). It entails the identification and management of risks proactively rather than reacting when problems become evident by which time it is often too late to intervene to protect health. The process involves a number of sequential steps:

- systematic identification of all the hazards that could potentially affect human or ecosystem health;
- assessment of the risks posed by each hazard characterised in terms of its likelihood and severity;
- identification of control measures capable of controlling the risk or reducing it to acceptable limits;
- establishing monitoring to ensure that the control measures are in place and working effectively.

The most recent WHO guidelines for wastewater use in agriculture adopt this risk management approach that will facilitate development of national or regional practices of safe wastewater use (WHO, 2006). Local economic, environmental and cultural conditions can be taken into account during the development of Best Management Practices for the use of reclaimed water. In developing countries, the health gain (of adopting wastewater criteria) can be enhanced when it is integrated with other sanitary measures such as access to safe water, sanitation and programmes of health education and promotion.

19.7 REFERENCES

Al-Jayyousie, O.R., (2003) Greywater reuse: towards sustainable water management. *Desalination* **156**(1), 181-192.

Angelakis, A.N, Marecos Do Monte, M.H.F., Bontoux, B. and Asano, T. (1999) The status of wastewater reuse: practice in the Mediterranean basin: need for guidelines. *Water Res.* **33**(10), 2201-2217.

Angelakis, A.N. and Bontoux, L. (2001) Wastewater reuse and reclamation in Eureau countries. *Water Policy* **3**, 47-59.

Anon (2005a) National guidelines for water recycling: Managing health and environmental risks. Australian Environment Protection and Heritage Council and Natural Resources Management Ministerial Council. Available from: URL: http://www.ephc.gov.au/ephc/water_recycling.html

Anon (2005b) Closing the loop on phosphorus. EcoSanRes factsheet. Available from: URL: http://www.ecosanres.org/

Asano, T. and Cotruvo, J.A. (2004) Groundwater recharge with reclaimed municipal wastewater: health and regulatory considerations. *Water Res.* **38**(8), 1941-1951.

Bixio, D., De heyder, B., De koning, J., Savic, D., Wintgens, T., Melin, T. and Thoeye, C. (2005) Wastewater reuse in Europe. In Proc. Intl. Conf. Integrated concepts in water recycling, Wollongong, NSW Australia, 14-17 February 2005.

Blumenthal, U.J. and Peasey, A. (2002) Critical review of epidemiological evidence of the health effects of wastewater and excreta use in agriculture. Available from: URL: http://www.who.int/water_sanitation_health/wastewater/gsuww/en/index.html

Cavallina, J.M. and Young, L.E. (2002) Integrated systems for the treatment and recycling of wastewater in Latin America: Reality and potential. Urban Agriculture **8**: 18-19. Available from URL: http://www.ruaf.org/node/276

Crook, J. and Surampalli, R.Y. (2005) Water reuse criteria in the United States. *Water Supply* 5(3-4), 1–7

CWWA (2005) Rainwater harvesting and greywater reuse.

Davison, A., Howard, G., Stevens, M., Fewtrell, L., Deere, D., Callan, P. and Bartram, J. (2004). Water Safety Plans: Managing drinking-water quality from catchment to consumer. WHO, Geneva

Desbrow C., Routledge E.J., Brighty G.C., Sumpter J.P. & Waldock M. (1998) Identification of estrogenic chemicals in STW effluent 1. Chemical fractionation and in vitro biological screening. *Environ. Sci. Technol.* **32**(11), 1549-1558.

Döller PC, Dietrich K, Filipp N, Brockmann S, Dreweck C, Vonthein R. Cyclosporiasis outbreak in Germany associated with the consumption of salad. *Emerg. Infect. Dis.* **8**(9) 2002 Sep [date accessed 7/12/05]; Available from: URL: http://www.cdc.gov/ncidod/EID/vol8no9/01-0517.htm

FAO (1992) Wastewater treatment and use in agriculture. FAO Irrigation and Drainage Paper No. 47. Food and Agriculture Organization of the United Nations, Rome

FAO (2002) Crops and drops: Making the best use of water for agriculture. Food and Agriculture Organization of the United Nations, Rome

Available from: URL: http://www.fao.org/ag/agl/aglw

FAO (2005) Water use in agriculture. Food and Agriculture Organization of the United Nations, Rome. Available from: URL: http://www.fao.org/ag/magazine/0511sp2.htm

Feacham R.G.A., Bradley D.J., Garelick H. and Mara DD (1983) Sanitation and disease. In: *Health aspects of excreta and wastewater management.* Chichester, Wiley, 375-399.

Fewtrell, L. and Bartram, J. (2001) Water quality - Guidelines, standards and health: Assessment of risk and risk management for water-related infectious disease. IWA Publishing, London

Geldreich (1990) Microbiological quality of source waters for water supply. In: *Drinking water microbiology: Progress and recent developments* Ed. McFeters, G.A. Springer-Verlag, New York 3-33.

Godfrey, S. and Howard, G. (2004) Water Safety Plans (WSP) for urban piped water supplies in developing countries. WEDC, Loughborough, UK

Godfrey, S., Kumar, P., Swami, A., Wate, S. and Dwivedi, H. (2005) Water Safety Plans for greywater and rainwater. AFPRO 12th International Rainwater Catchment Systems Conference, New Delhi, India

Godfrey, S., Saxsena, A., Kumar, P., Swami, A. and Joshi, N. (2005) A preliminary study on potential of greywater reuse system in water scarce and fluoride contaminated rural areas of Jhabua, Madhya Pradesh, India Group, I. W. R. S. Water Reuse and Reclamation for Sustainability (WRRS) South Korea IWA

Halling-Sørensen, B. Nors Nielsen, S. Lanzky, P.F., Ingerslev, F., Holten Lützhøft, H.C. and Jørgensen S.E. (1998). Occurrence, fate and effects of pharmaceutical substances in the environment - A review. *Chemosphere* **36**(2), 357-393.

Havelaar AH, van Olphen M, Drost YC. (1993) F-specific RNA bacteriophages are adequate model organisms for enteric viruses in fresh water. *Appl. Environ. Microbiol.* **59**(9),2956-2962.

Havelaar, A.H., (1994) Application of HACCP to Drinking Water Supply *Food Control* **5**, 145-152.

Ho A.Y., Lopez A.S., Eberhart M.G., Levenson R., Finkel B.S., da Silva A.J., *et al.* Outbreak of cyclosporiasis associated with imported raspberries, Philadelphia, Pennsylvania, 2000. *Emerg. Infect. Dis.* [serial online] 2002 Aug [date cited];8. Available from: URL: http://www.cdc.gov/ncidod/EID/vol8no8/02-0012.htm

Huffmann, D.E., Quinter-Betancourt, W. and Rose, J.B. Emerging waterborne pathogens. In: *The handbook of water and wastewater microbiology.* Eds. Mara D.D. and Horan N. Academic Press, London 193-208.

Jiménez, B. (2005). Treatment technology and standards for agricultural wastewater reuse: A case study in Mexico. Irrig. and Drain **54**: S23-S33.

Lazarova V., Cirelli G., Jeffrey P., Salgot M., Icekson N. and Brissaud F. (2000). Enhancement of integrated water management and water reuse in Europe and the Middle East. Wat. Sci. Technol. **42**(1-2),193-202.

Mara, D.D. and Cairncross, S. (1989) Guidelines for the safe use of wastewater and excreta in agriculture and aquaculture. World Health Organization, Geneva

Mara, D.D. and Feacham, R.G.A. (2003) Unitary environmental classification of water- and excreta-related communicable diseases. In: *The handbook of water and wastewater microbiology*. Eds. Mara D.D. and Horan N. Academic Press, London 185-192.

Metcalf & Eddy Wastewater Engineering, Treatment and Reuse. 4th Edition, McGraw-Hill 2003

NRC (1996) Use of Reclaimed Water and Sludge in Food Crop Production. National Academy Press, Washington DC

NRC (2004) Indicators for waterborne pathogens. National Academy Press, Washington DC

Ottoson, J. and Stenstrom, T.A., (2003) Faecal contamination of greywater and associated microbial risks. *Water Res.* **37**(3), 645-655.

Regli, S., Rose, J.B., Haas, C.N. and Gerba, C.P. (1991) Modeling the risk from giardia and viruses in drinking water *Journal American Water Works Association* **83**, 76-84.

Siegrist R.L. and Boyle W.C. 1987. Wastewater induced soil clogging development. *J. Environ. Eng.* ASCE. **13**(3), 550-566.

Smit, J. and Nasr, J. (1992) Urban agriculture for sustainable cities: Using wastes, idle land, and water bodies as resources. *Environment and Urbanization* **4**(2), 141-152.

Thornton, I., Butler, D., Docx, P., Hession, M., Makropoulos, M., McMullen, M., Neuwenhuijsen, M., Pitman, A., Rautiu, R., Sawyer, R., Smith, S., White, D., Wilderer, P., Paris, S., Marani, D., Braguglia, C. and Palerm, J. (2001) Pollutants in urban wastewater and sewage sludge. Report to DG Environment, Brussels. Available from: URL: http://europa.eu.int/comm/environment/sludge/sludge_pollutants.html

Tixier, C., Singer, H.P., Oellers, S. and Müller, S.R. (2003) Occurrence and fate of Carbamazepine, Clofibric Acid, Diclofenac, Ibuprofen, Ketoprofen, and Naproxen in surface waters. *Environ. Sci. Technol.* **37**(6), 1061-1068.

United Nations (2003) The UN World Water Development Report: Water For People, Water For Life. Available from: URL: http://www.unesco.org/water/wwap/wwdr/table_contents.shtml

UKWIR (2003) Priority hazardous substances, trace organics and diffuse pollution (Water Framework Directive): Screening study and literature review of quantities in sewage sludge and effluents. Report Ref 04/WW/17/2. UK Water Industry Research, London

UKWIR (2004) Framework for Developing Water Reuse Criteria with Reference to Drinking Water Supplies. Report Ref 05/WR/29/1. UK Water Industry Research, London

Westrell, T., Schonning, C., Stenstrom, T.,A. and Ashbolt, N.J. (2004) QMRA (quantitative microbial risk assessment) and HACCP (hazard analysis and critical control points) for management of pathogens in wastewater and sewage sludge treatment and reuse. *Wat. Sci. Technol.* **50**(2), 23-30.

WHO (2004) Guidelines for Drinking Water Quality 3rd Edition. WHO, Geneva

WHO (2006) Guidelines for the safe use of wastewater, excreta and greywater: Vol 2 Wastewater use in agriculture. WHO, Geneva.

SECTION FOUR

CASE STUDIES

20

Water reuse in Japan

Naoyuki Funamizu, Takuya Onitsuka and Shigeki Hatori

20.1 MAIN TYPES OF WATER REUSE IN JAPAN

20.1.1 Municipal water reuse

Wastewater reuse has a long history in Japan. The earliest planned wastewater reuse started in 1955 for supplying reclaimed wastewater to industries in Tokyo. The Tokyo Metropolitan Government promoted the supply of reclaimed wastewater to industry to prevent over-drafting of groundwater in the Tokyo Bay area. In the 1970s, Japan experienced severe droughts in wide areas of the country. For example, in 1978, the prolonged drought conditions in Fukuoka forced citizens to accept serious water supply limitations for 283 days. These drought experiences let people recognize reclaimed wastewater as dependable water in urban areas. In the 1980s, local government undertook most planned wastewater reuse projects and national government supported these projects by subsidy. In cities having no drought problem, wastewater reclamation was promoted with the positive image of environmental protection provided by the construction of sewerage and wastewater treatment facilities. Reclaimed wastewater has, then, been promoted by municipalities as a safe, dependable and esthetically acceptable "new" water resource for the environment (Ogoshi *et al.*, 2001).

Table 20.1 shows the current status of wastewater reclamation and reuse in Japan. In 2003, 367 wastewater treatment plants (WWTPs) in Japan supplied approximately 486,200 m^3/d (180Mm^3/y) of reclaimed water. In addition, 1,058 on-site individual building and block-wide wastewater recycling systems generated toilet flush water in commercial buildings and apartment complexes. Figure 20.1 shows that, contrary to many other countries where agricultural irrigation is a dominant water reuse application, water reuse in Japan has decisively been for non-potable urban water applications. Also, providing in-stream flow needs and environmental water to restore "aquatic amenities" in the urban environment characterizes recent trends in large-volume reclaimed water use.

Table 20.1 Current status of wastewater reuse in Japan.*

Application	Number of WWTPs and on-site treatment facilities	Volume of use (m^3/d)
Toilet flushing:		
From WWTPs	40	15,900
From on-site	1,058	87,600
Environmental water	102	289,300
Snow melting	20	80,300
Tree planting	76	500
Cleansing water	58	1,100
Agricultural irrigation	22	35,300
Industrial water	49	63,800
Sub-total		
WWTPs	367	486,200
On-site	1,058	87,600
Total	1,425	573,800

* Compiled from the published data from the Japan Sewage Works Association (2004) and the National Land Agency (2004).

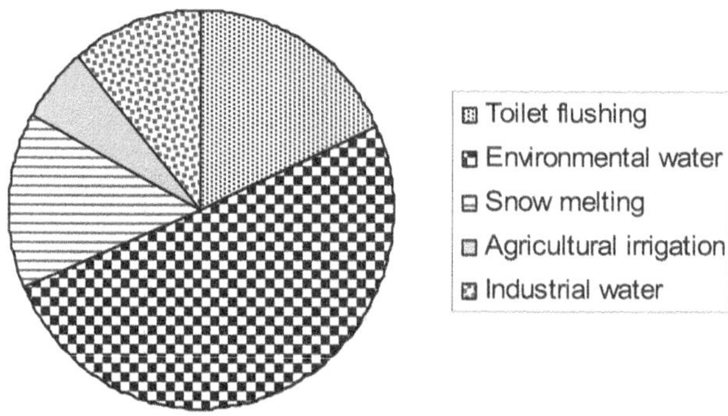

Figure 20.1 Reclaimed wastewater application in Japan.

20.1.2 Wastewater reuse schemes

a) Individual building/block-wide wastewater reuse systems

Individual wastewater reuse (Figure 20.2(a)) mainly comprises toilet flushing in large office buildings or apartment complexes with an on-site wastewater treatment system. Some cities set a guideline to request building owners to install a dual distribution system for newly constructed buildings of certain floor space and/or water demand, e.g.:

- buildings with floor space greater than 5000m^2 or installed water supply pipe diameter of greater that 50mm (Fukuoka city);
- buildings with floor space greater than 3000m^2 or potential demand of reclaimed wastewater greater than 100m^3/day.

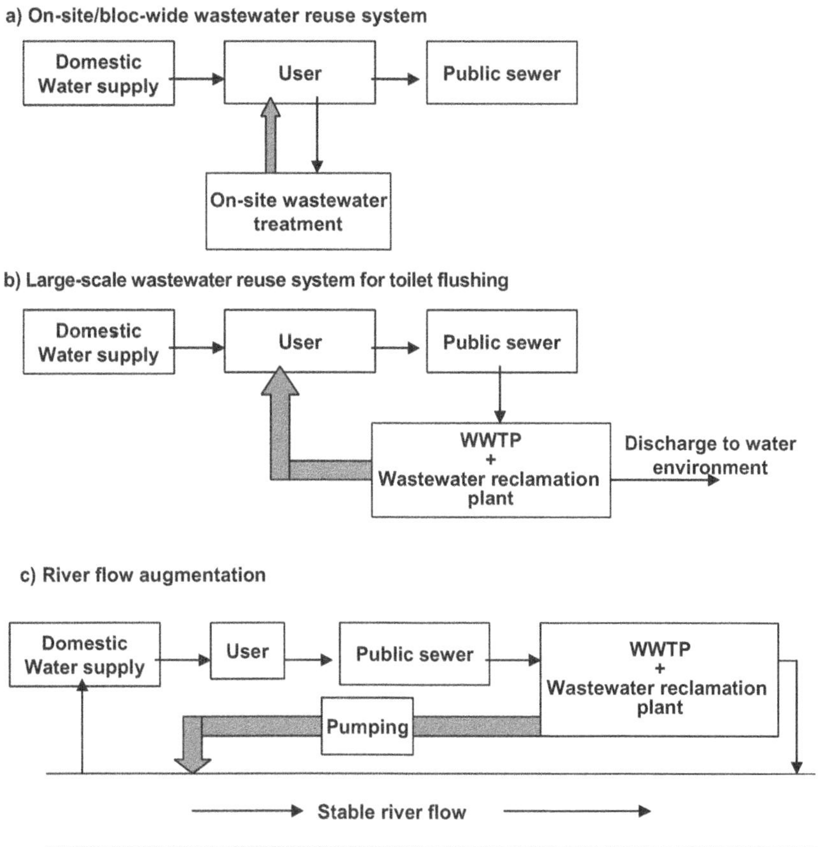

Figure 20.2. Wastewater reuse system schematics used in Japan (adapted from Ogoshi *et al.*, 2001).

Several buildings are connected together to a block-wide wastewater treatment facility and their reclaimed water is distributed back to the buildings mainly for toilet flushing.

b) Large-scale wastewater reuse system

As shown in Figure 20.2(b), reclaimed wastewater is distributed through a pipe network to large service area. Tertiary or advanced wastewater treatment processes are employed for further treatment. The main uses of reclaimed water are for toilet flushing and environmental water, but there are other uses such as landscape irrigation, garden watering and melting and transporting snow. This type of wastewater reuse has been increasing and Tokyo has four big projects: Shinjuku Sub-Center; Ariake District; Shinagawa Station East Side District; and Osaki Station East District. There are other large-scale wastewater reuse systems constructed in Fukuoka City, Nagoya City, Makuhari New Town in Chiba Prefecture, Hamamatsu City, and Kobe City.

c) Wastewater reuse for in-stream flow augmentation

In the scheme (Figure 20.2(c)) reclaimed water is pumped from a wastewater treatment plant (WWTP) to the point of introduction to a river to augment various stream flow needs. Examples include the Shibuya River, the Meguro River, the Nobu River and the Tamagawa Josui in the Metropolitan Tokyo, the Naka River and the Mikasa River in Fukuoka

Prefecture, the Yasuharu River in Sapporo City and the Ina River boarding Hyogo and Osaka Prefecture.

Large-area wastewater reuse and in-stream augmentation are mainly related to municipal wastewater reuse in urban areas.

20.1.3 Reclaimed water quality criteria

Table 20.2 shows current reuse water quality criteria in the wastewater reuse manual published by the Japanese Ministry of Land, Infrastructure and Transport in April 2005. In the criteria, water quality parameters and minimum requirement for tertiary treatment system are specified for toilet flushing water, garden watering, landscape and recreational waters. In the new criteria, *E.Coli.* instead of total coliform counts is used as a parameter.

Table 20.2. Criteria for reuse water quality and reclaimed treatment system.

Parameter	Toilet flushing	Garden watering	Landscape water	Recreational water
E.coli.	Not detected	Not detected	See foot note[1)]	Not detected
Residual chlorine, (mg/L)	0.1 (free chlorine) or 0.4 (combined chlorine)	0.1 (free chlorine) or 0.4 (combined chlorine)	-	0.1 (free chlorine) or 0.4 (combined chlorine)
Turbidity, (NTU)	Less than 2	Less than 2	Less than 2	Less than 2
Appearance	Not unpleasant	Not unpleasant	Not unpleasant	Not unpleasant
Color(color unit)	-	-	Less than 40	Less than 10
Odor	Not unpleasant	Not unpleasant	Not unpleasant	Not unpleasant
pH	5.8 – 8.6	5.8 – 8.6	5.8 – 8.6	5.8 – 8.6
Minimum requirement for tertiary treatment system	Granular-medium filtration	Granular-medium filtration	Granular-medium filtration	Chemical coagulation and Granular-medium filtration

1) 1000CFU/100mL as total coliform counts is used tentatively now

20.2 LARGE-SCALE WASTEWATER REUSE IN TOKYO: APPLICATION OF THE NEWLY DEVELOPED OZONE-RESISTANT MF MEMBRANE FOR WASTEWATER REUSE

20.2.1 The current status of wastewater reuse in Tokyo (Sone, 2004)

The population of the Tokyo Metropolitan Area is approximately 33 million and its water is supplied from the dam lakes built in three watersheds: the Tone river, Ara river and Tama river. All domestic wastewater from this region is treated and discharged to Tokyo Bay and the Pacific Ocean through rivers. In spite of 1,500mm of annual rain fall, the available water resource *per capita* counts only 900m^3/y because of high population density, and this value is one third of the average value of Japan. This situation has led to the promotion of rain water harvesting and wastewater reuse as water sources. Wastewater reuse in Tokyo has a long history, starting in 1955. Today, the Bureau of Sewage Works, Tokyo, operates twenty wastewater treatment plants which treat 550 Mm^3/d from 2,187km^2 of service area, and its wastewater reuse percentage reached about 9% in 2002.

The Tokyo Metropolitan Government supplys reclaimed wastewater for several purposes with the required quality of water by using advanced treatment systems. The use of reclaimed wastewater to sustain stream flows is attractive in Tokyo and the Ochiai wastewater treatment plant supplied approximately 85,000m^3/d (in 2002) of advance treated water by sand filter to "Johnan" three rivers. The Bureau of Sewage Works, Tokyo is also promoting wastewater reuse for toilet flushing by a large-scale water reuse system. A tertiary or advanced wastewater treatment process is employed for further treatment before water reuse. As shown in Table 20.3, Sand filters, biofilm filters and ozonation have been conventionally used as the treatment methods. 50m^3/d of reclaimed wastewater by RO treatment is used for recreational purposes at the park adjacent to the wastewater treatment plant. Supplying secondary effluent to the railway company for cleaning purposes and to the domestic waste incineration plant is another application of reclaimed water in Tokyo.

Table 20.3. Wastewater reclamation systems in Tokyo.

Wastewater treatment plant	Reclamation system	Amount of supply water in 2002 (m^3/d)
Shibaura	Rapid sand filter	1,375
Ochiai	Rapid sand filter	3,267
Ariake	Biofilm filter + ozone	2,175

20.2.2 Development of a new wastewater reclamation system

The Bureau of Sewage Works, Tokyo, had researched the new wastewater reclamation system for 1998-2000 in cooperation with the Water Re-use Promotion Center, Japan. The purpose of this project was developing new reclamation systems to distribute reclaimed water which was more aesthetically pleasing, especially improving the quality in color and odor, and with higher hygienic safety. In addition, the new wastewater reclamation system was desired to have a relatively low cost.

(1) The new wastewater reclamation system

To meet the above mentioned needs, a reclaimed water system with biofilter, ozone and MF membranes was developed as shown in Figure 20.3. The pre-ozone unit oxidizes ammonia and nitrite and a part of organic matter in secondary effluent by dosing a low level of ozone. The biofilter is a granular particle bed filter with anthracite, which removes organic matter and nitrite by microorganisms on the surface of anthracite particles. The purposes of the ozone are to treat color and odor and inactivate pathogenic micro-organisms (virus, bacteria, and protozoa) by strong oxidation. The MF membrane unit removes micro-particulate matter including bacteria and pathogenic protozoa such as *Cryptosporidium*.

(2) Features of MF membrane

The MF membrane used in the system is called an "ozone resistant membrane" and is a hollow fiber with pore size of 0.1μm and made of poly-vinyli-dene-fluoride (PVDF). It had been impossible to connect ozone treatment units and membrane units directly because ozone broke up the membrane by ozone's oxidation power. The newly developed MF membrane, however, is tolerant to ozone because of a membrane made of fluorine resin, PVDF. The merit of direct connection of the ozone and membrane units is that ozone can decompose organic matters on the membrane surface and reduce fouling of the membrane.

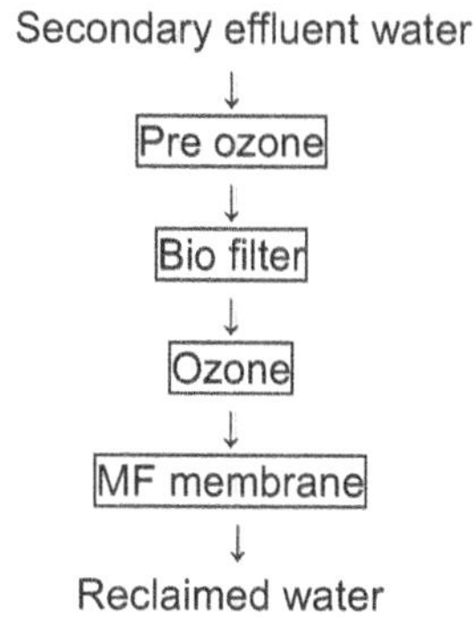

Figure 20.3. A schematic description of the new reclamation system.

(3) Pilot plant study

To confirm the treatment efficiency and to obtain the suitable operation conditions, several tests were performed in the pilot plant. The treatment capacity of the pilot plant was approximately 30 m^3/d. The parameters examined in the study included BOD, COD, heterotrophes, Escherichia coliform, fecal coliform, ammonia nitrogen, nitrite nitrogen, nitrate nitrogen, color, and *Cryptosporidium*. The trans-membrane pressure (TMP) at different dissolved ozone concentrations on the surface of the membrane was monitored and the effect of ozone concentration on the operation of MF unit was examined. Then, the continuous operation of the system was carried out at the adequate dissolved ozone concentration by controlling the dose of ozone automatically.

Table 20.4 illustrates a typical performance of the pilot plant study, showing that:

1) heterotrophes, *E.coli* and fecal coliform are made inactive in the ozone treatment process;
2) color is reduced to 8 (cu) by the ozone treatment and furthermore reduced to 1(cu) by the MF membrane;
3) most of ammonia nitrogen is nitrified by the biofiltration process and converted to nitrate nitrogen;
4) MF membrane can remove not only suspended matters but also ova or larva of organisms.

Table 20.4. Treated water quality in each treatment process.

Item	**Raw water (Secondary effluent)**	**Treated water by each treatment process**		
		Bio-filter	**Ozone**	**MF membrane**
T-BOD (mg/L)	16	4.4	2.3	2.1
COD (mg/L)	11	8.2	5.9	5.7
NH_4-N (mg/L)	10.8	2.97	3.34	3.55
NO_2-N (mg/L)	1.66	0.84	0.44	<0.03
NO_3-N (mg/L)	4.9	14.7	16.3	16.3
Color (CU)	31	17	8	1
Odor (TON)	10	-	-	1
Heterotrophes (CFU/mL)	4.4E+04	2.6E+03	0	0
E.Coli (CFU/100mL)	1.8E+05	1.5E+04	0	0
Fecal coliform (CFU/100mL)	6.5E+04	1.2E+03	0	0
Cryptosporidium Oocysts (count/L)	2	-	-	0

It is also found that keeping the dissolved ozone concentration on the surface of the membrane at 1.0mg/L can prevent the rise of trans-membrane pressure (TMP). The continuous filtration test showed that TMP increased very slowly and reached a plateau, 120kPa, after 3,800 hours (about 5 months) operation at the membrane flux, $5.0m^3/m^2.d$. Since this TMP value, 120kPa, is far less than the TMP level at which chemical cleaning is necessary (200kPa), it is confirmed that the control of the dissolved ozone concentration at 1.0 mg/L on the surface of the membrane ensures stable membrane operation at high flux for a long time.

(4)Cost estimates

Initial cost, and maintenance and operation costs in the new reclamation system are shown in Table 20.5. In addition to the newly developed system, cost estimates for two other systems are also listed: a system with sand filter + Ozone and a system with MF+RF. It should be noted that the new reclamation system gives a reasonable cost performance compared with other systems.

Table 20.5. Estimated cost of the new system and other reclamation systems* (unit: USD/m^3).

	New reclamation system	Sand filter + ozone	MF + RO
Initial cost	0.23	0.19	0.50
M & O cost	0.23	0.16	0.61
Total cost	0.46	0.35	1.11

[*Estimate conditions]

1) Treatment scale : 3,000 m^3/d

2) Initial cost : All units such as bio filter, ozonizer, membrane filtration unit and electrical control panel. The building of a concrete house, storage tank for inlet and outlet water are not included. Repayment duration: 20 years.

3) M & O cost : Electrical power cost, membrane changing cost, chemical cleaning cost for membrane and payment for operator.

(5) Application of the new reclamation system at a wastewater treatment plant

Based on the results of the above-mentioned pilot plant experiment and cost estimation, the Bureau of Sewerage of Tokyo installed a reclamation plant (treatment capacity of 4,300 m^3/d) with the newly developed system at Shibaura water reclamation center in April 2004. The flow chart of the reclamation system constructed is shown in Figure 20.4. Although the reclaimed water is stored in a reclamation water tank, no re-growth of microorganisms has been observed in the tank. The Shibaura Water Reuse Center is supplying reclaimed water for toilet flushing to the Shinagawa, Shiodome and Oosaki areas. In the Shiodeme area, the reclaimed water is used for landscape and garden watering as well as for street spraying (it is reported that spraying water on asphalt roads has the effect of mitigating heat island phonomena in urban areas).

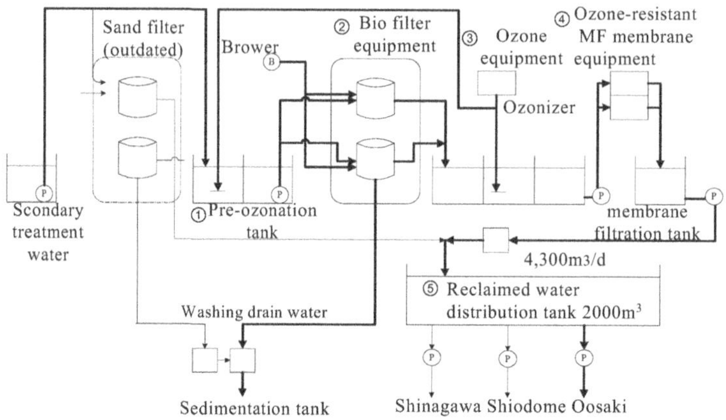

Figure 20.4. Flow chart of the new reclamation system in Shibaura Water Reuse Center.

20.3 LARGE-SCALE WASTEWATER REUSE FOR NON-POTABLE PURPOSES IN FUKUOKA CITY

Fukuoka City experienced very severe drought conditions twice, in 1978 and in 1994. Citizens had to accept serious water supply limitations for 283 days in 1978 and for 295 days in 1994. Since 1979, a large-scale wastewater reuse project has been running to supply toilet flushing water to large buildings with floor space greater than 3000m^2 and for distributing reclaimed water for garden watering and landscaping.

In 2005, the wastewater reuse project covered five districts with an area of 994ha and the wastewater reclamation plant supplies 9600m^3 of water daily. The wastewater reclamation plant in Fukuoka has the treatment system as shown in Figure 20.5.

Fukuoka City sets three levels of price for reclaimed water, depending on the total amount of water consumption, e.g. 1.36 USD/m^3 for a total monthly water consumption of less than 100m^3, 2.73 USD/m^3 for between 100m^3 and 300m^3, and 3.18 USD/m^3 for total water consumption of more than 300m^3.

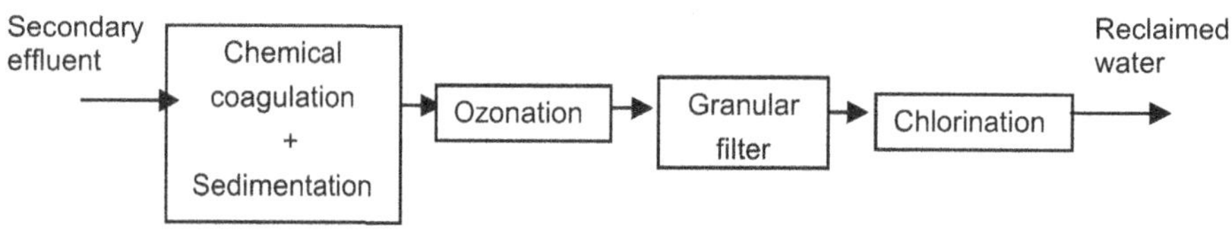

Figure 20.5. Wastewater reclamation system in Fukuoka City.

20.4 LARGE-AREA WASTEWATER REUSE FOR MELTING SNOW IN SAPPORO CITY (FUNAMIZU *ET AL.*, 2001)

Sapporo City has a so-called "snow problem," having to remove snow from streets and then dumping that snow. The problem has given rise to a multi-purpose use of the existing sewerage system to transport and melt snow. Figure 20.6 shows the new strategy of Sapporo City for its combined sewer system. There are three projects depicted in Figure 20.6. The first is a multi-purpose use of basins and reuse of heat energy from treated water for melting snow. The second project is the construction of channels for transporting snow by supplying reclaimed water. The third is transporting snow by a combined sewer. The facilities for throwing snow into the combined sewer were constructed in the central part of Sapporo. Dirty snow from the downtown area is transported and melted in the sewer, then the snowmelt is treated at the wastewater treatment plant (WWTP).

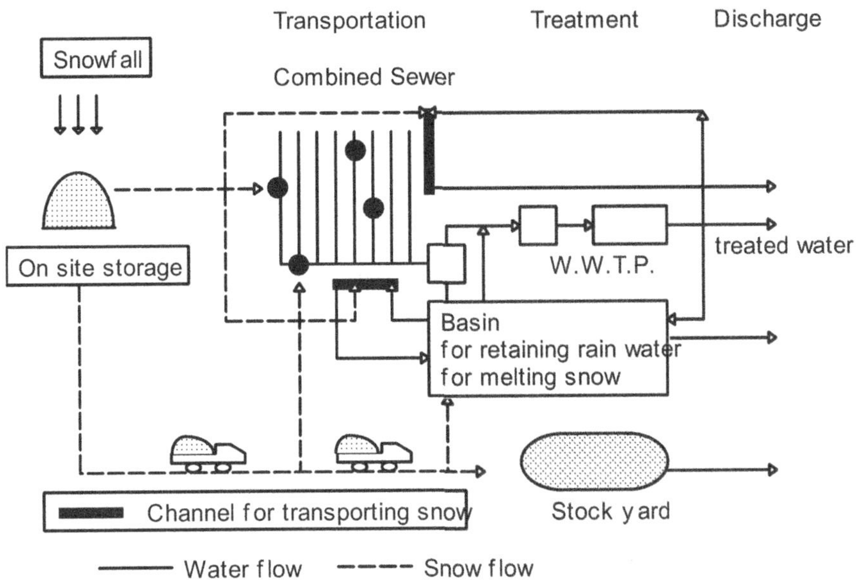

Figure 20.6. Sketch of the sewage system used to transport and melt snow in Sapporo (Funamizu *et al.*, 2001).

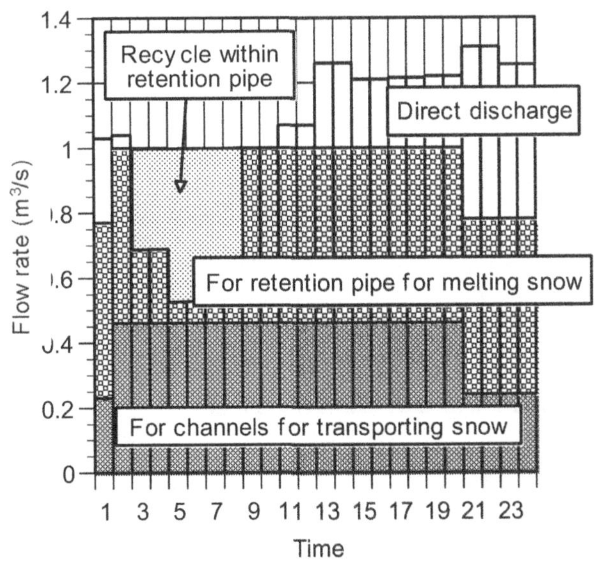

Figure 20.7. Distribution plan of treated water of SOSEI WWTP for transporting and melting snow (Funamizu *et al.*, 2001).

The SOSEI wastewater treatment plant treats approximately 120,000m^3 of wastewater daily by multi-stage nitrification and de-nitrification process (secondary treatment) followed by a granular media filter system and a disinfection system (tertiary treatment).

The distribution plan of the effluent in winter is given in Figure 20.7. The reclaimed water is used for melting snow at the combined sewer overflow (CSO) retention pipe and for transporting snow. The CSO retention pipe, 5m in diameter and 2495m in length, can store 46,400m^3 of water. This pipe has two functions: retaining storm water in the rain season and melting snow in winter. Snow is thrown into the retention pipe after being reduced in size by a rotary grinder. Snowmelt is pumped up and discharged to the river.

In summer, SOSEI WWTP supplies the 28,330m^3/d of reclaimed water for four rivers in the north part of Sapporo City to augment river flow.

20.5 INDIVIDUAL BUILDING WASTEWATER REUSE SYSTEMS

20.5.1 An example of individual building wastewater reuse

Figure 20.8 shows water economy of three buildings with an individual wastewater reuse system (Tazawa and Kurihara, 1995 and Funamizu *et al.*, 1998). These buildings have three zones: offices, restaurants and shops with 168,000m^2 of total floor area. Figure 20.8 shows that:

1. 490m^3 of drinkable tap water is supplied daily and is used at the restaurants and offices for hot-water supply and wash basins;
2. graywater from the restaurant and from the offices is collected separately by individual pipes;
3. graywater from the restaurants is treated and treated wastewater is diverted to reclamation units and to the public sewer;
4. graywater from the office and treated graywater from the restaurants are mixed and treated by the reclamation unit and then reclaimed water is pumped to toilets for flushing purposes.

It should also be noted from Figure 20.8 that the input of water to the buildings is 490m^3/d of drinkable tap water, and the output is 490m^3/d of black water to public sewer; however, the total water demand for the buildings is 880m^3/d.

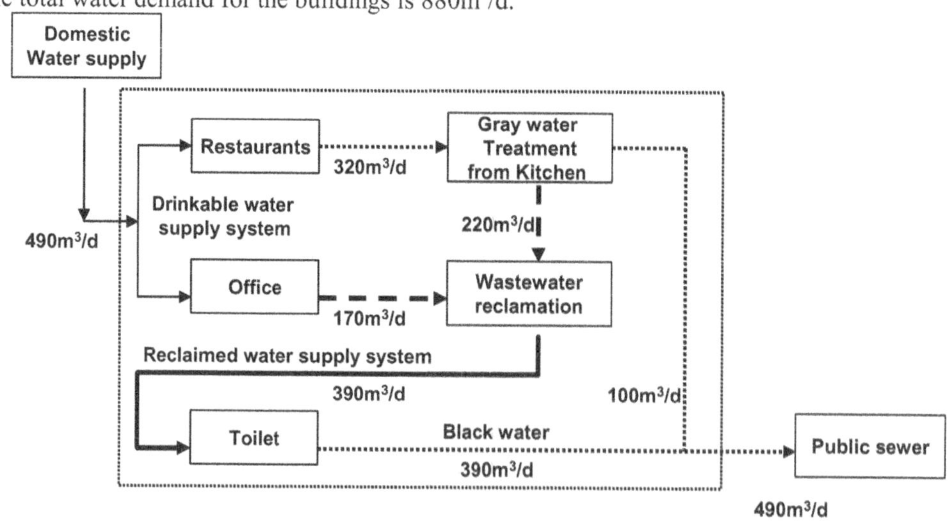

Figure 20.8. a water economy of the three building with individual wastewater reuse system (Tazawa and Kurihara, 1995 and Funamizu et al., 1998).

20.5.2 Demand for reclaimed water and production of wastewater in buildings

In an individual building water recycling system, the reclaimed water demand of the building and wastewater production from the building need to be harmonized. The reclaimed water demand and the quantity and quality of wastewater depend strongly on the size of the building and its use. In this section, the typical values used for planning and design of an individual wastewater recycling system in Japan are presented.

(1)Reclaimed water demand in a building

Table 20.6 summarizes the demand for reclaimed wastewater in office buildings, hotels and other types of buildings.

Table 20.6. Demand of reclaimed wastewater.

(a)Office buildings

Building size (Floor space: m^2)	Volume of reclaimed wastewater (m^3/d)	Volume of reclaimed wastewater per square meter, (L/m^2,d)
12,000	25 to 30	2.0 to 2.5
36,000	75 to 90	2.0 to 2.4
55,000	110 to 130	2.0 to 2.4
87,000	165 to 190	1.9 to 2.2
150,000	270 to 310	1.9 to 2.2

(b) Hotel

Building size (Floor space: m^2)		Number of guest rooms	Volume of reclaimed wastewater (m^3/d)	Volume of reclaimed wastewater per square meter (L/m^2.d)
Resort-hotel	43,000	450	180 to 300	4.2 to 8.0
	67,000	700	300 to 500	4.5 to 7.5
	73,000	800	300 to 550	4.1 to 7.5
City hotel	48,000	500	250 to 305	5.0 to 6.4
	87,000	800	445 to 50	5.1 to 6.3
	100,000	1,000	500 to 650	5.0 to 6.5

(c) Station buildings and other buildings

Building size (Floor space: m^2)	Volume of reclaimed wastewater (m^3/d)	Volume of reclaimed wastewater per square meter, (L/m^2.d)
40,000	160 to 190	4.0 to 4.8
53,000	200 to 250	3.8 to 4.7
68,000	270 to 350	4.0 to 5.1

(2)Wastewater production from several appliances in a building

Since reclaimed water is produced from wastewater, it is important know the characteristics of wastewater from several different appliances in a building. Especially in planning and designing a low cost reclamation plant, with less sludge production and no odor emission, decisions on which wastewaters should be used for reclamation should be made by considering the volume and quality of wastewater and the total demand of reclaimed water. In other words, the balance of the volume of wastewater to be reclaimed as well as the balance of the plumbing costs for collecting different types of wastewater separately, and the cost for treating collected wastewater, should all be considered.

Table 20.7 lists the volumes of wastewater from several appliances in a building. The characteristics of these wastewaters are summarized as follows:

a) wastewater from washing bowls and hot-water service rooms (lower loaded graywater): wastewater from washing bowls and hot-water service rooms account for about 10% of total wastewater from a general building. Since it contains little suspended particles it is very suitable for reclamation.
b) wastewater from bathrooms (lower loaded graywater): this wastewater is mainly discharged from accommodation facilities such as hotels and Japanese-style inns. It contains relatively low levels of suspended particles. In addition to the advantage of low loaded quality, volume of this graywater is large and it is considered a suitable source for reuse.
c) wastewater from cooling towers: this wastewater is discharged from air-conditioning systems. Although this water has some rather high conductivity, it is also considered as a suitable source for reclamation and reuse.
d) wastewater from a kitchen (higher loaded wastewater): this wastewater is discharged from cafeterias and restaurants in a building, and contains organic matter with high concentration of BOD, oil contents (normal hexane extracts) and surface-active agents. The volume of the wastewater is larger than the other wastewater. Since Japanese regulations prohibit the direct discharge of wastewater to public sewers from a large restaurant, large restaurants have to install a treatment unit of kitchen wastewater and this treated wastewater from kitchens becomes a suitable raw water for reclamation and reuse.
e) Black water: Since black water contains faeces and urine, this means that this wastewater contains concentrated organic matter, ammonia nitrogen, bile-pigment and pathogenic micro-organisms, and it is rarely used as a raw water for reclamation and reuse in individual building water recycle systems.
f) Other water sources for reclamation and reuse: rain water and infiltrated groundwater to an underground pit can be a source for reclamation and reuse.

Table 20.7. Volume of wastewater (m^3/d)*.

		Size of building		
		6,000 m^2	15,000 m^2	30,000 m^2
Appliance	Toilet flush	11.4 ±2.1	28.5 ±5.5	56.9 ±11.1
	Wash bowls	1.9 ±0.1	4.8 ±0.4	9.7 ±0.7
	Water heating	3.0 ±1.5	7.5 ±3.8	15.0 ±7.5
	Kitchen	8.0 ±3.2	19.9 ±8.0	39.8 ±15.9

(*Source) Construction Minister's Secretariat Government Buildings Department, 1991. Standards for wastewater reuse and rainwater use system design and their descriptions, edited by Public Buildings Association (1991 edition.) p.18.

20.5.3 Examples of treatment system for individual building recycling

(1) Treatment process of lower loaded graywater

The treatment system for lower loaded graywater consists of a biological reactor (membrane bio-reactor) and disinfection unit. The lower loaded wastewater is treated biologically by activated sludge in the aeration tank and then the water is sucked and filtered by the submerged membrane unit followed by a disinfection unit.

The membrane unit is a combination of a cartridge membrane (flat membrane) made from chlorinated polyethylene with nominal diameter of 0.4 μm. Because the membrane unit is used, the quality of treated water is very stable and good. Table 20.8 shows the operation results of BOD, suspended solids (SS) and normal hexane extracts (n-Hex).

Table 20.8. Operation results of lower loaded wastewater treatment.

	Design water quality		Operation Data	
	Raw water	Treated water	Raw water Average (max - min)	Treated water
BOD (mg/L)	50	10	41 (22 – 61)	Less than 1
SS (mg/L)	50	5	39 (25 – 59)	Less than 1
n-Hex (mg/L)	10	1	1.5 (1 - 5)	Less than 1

(2) Treatment process of higher loaded graywater

The typical treatment system for higher loaded graywater is a treatment train with contact aeration + fixed bed biological filtration + sand filtration/activated carbon + disinfection. This system is designed to treat wastewater from the kitchen, wash bowls and hot-water service rooms.

Since wastewater from kitchens contains emulsified oil contents as well as suspended and dissolved organic matter in high concentration, kitchen wastewater should be treated biologically by oil-decomposing fungi in the contact aeration unit separately. Then the effluent of the contact aeration unit is mixed with wastewater from wash bowls and hot-water service rooms and treated biologically by fixed bed biological filtration, followed by a sand filtration unit. Then, the effluent from the sand filter is treated by an activated carbon unit before disinfection. Because the volume of lower loaded graywater can not meet the demand of reclaimed water, the effluent of the contact aeration unit which treats wastewater from the kitchen is mixed with lower loaded graywater to compensate the lack of raw water for reclamation of water. The addition of oil-decomposing fungi contributes to reduce the emission of odor but leads to increased sludge generation. The fixed bed biological filter has advantages in terms of stable treatment and low operation cost. Table 20.9 shows the operation results of the system.

Table 20.9. Operation results of higher loaded wastewater treatment.

	Design water quality		Results	
	Raw water	Treated water	Raw water Average (max - min)	Treated water Average (max - min)
BOD (mg/L)	600	10	475 (242 – 688))	2.2 (1 – 4)
SS (mg/L)	600	5	108 (44 - 168)	1.4 (1 - 2)
n-Hex (mg/L)	150	1	115 (60 – 160)	Less than 1

(3) Treatment process of wastewater from a cooling tower

Wastewater from a cooling tower is treated by the system of filtration and disinfection. The filter media of the system is sand and anthracite particles, and this filter is used to remove particulate inorganic and organic matter in the wastewater. Table 20.10 summarizes the operation results of the system.

Table 20.10. Operation results of cooling tower wastewater.

	Design water quality		Results	
	Raw water	Treated water	Raw water	Treated water
BOD (mg/L)	-	-	4.6	4.5
SS (mg/L)	10	5 or less	3	Less than 1
Conductivity (μs/cm)	-	-	1210	1200

(4) Treatment system of other wastewater

If rain water and/or infiltration water to the underground pit is reused, the system with sand filtration + disinfection can be used for its treatment. In this operation, a flow meter is required to measure the volume of water.

20.6 REFERENCES

Construction Minister's Secretariat Government Buildings Department (1991) *Standards for wastewater reuse and rainwater use system design and their descriptions*, edited by Public Buildings Association Tokyo, Japan, p. 18.

Funamizu, N., Ohgaki, S. and Asano, T (1998) Wastewater reclamation and reuse, In *Iwanami lecture series on earth environment, vol.7 Water environment and recycle of water* (eds. By Y.Takahashi and K.Kawata), pp.211-240, (Iwanami Shoten, Japan, 1998) (in Japanese).

Funamizu N. Iida M., Sakakura Y. and Takakuwa T: Reuse of heat energy in wastewater: implementation examples in Japan, *Water Science and Technology*, **43**(10), 277-286, 2001.

Japan Sewage Works Association: *Sewage Works in Japan,* Tokyo, Japan, 2004

National Land Agency: *Japan's Water Resources,* Tokyo, Japan, 2004.

Ogashi, M., Suzuki, Y. and Asano, T.: Water reuse in Japan, *Water Science & Technology,* **43** (10), 17-23, 2001.

Sone K. (Bureau of Sewage Works, Tokyo Metropolitan Government): Wastewater reuse in Tokyo, Japan, Newsletter October, 2004, pp2-5, International Water Association, Specialist Group on Water Reuse

Tazawa, R. and Kurihara, T.: A study on design and operation of wastewater recycling system – Analysis water demand and water economy of buildings and their optimization, Japan journal of Heating, Air-conditioning and sanitary engineering, **58**, 163-174, 1995 (in Japanese)

Acknowledgement

The authors gratefully appreciate the Bureau of Sewerage, Tokyo Metropolitan Government and Water Re-Use Promotion Center for providing the experimental data and the information on the reclamation plant of Shibaura water reuse center.

21

Livelihoods from wastewater: water reuse in Faisalabad, Pakistan

Jeroen H. J. Ensink, Wim van der Hoek and Robert W. Simmons

21.1 INTRODUCTION

As a result of population growth and land degradation over the past decades, the irrigated area per person has declined to such an extent that currently in a growing number of countries and regions of the world, water has become the single most important constraint to sustainable increases in food production (Seckler *et al.*, 1998). In the next 24 years the world population is expected to grow by 1.7 billion; this growth will almost exclusively take place in cities in the developing world. In absolute and relative numbers, growth in Pakistan will be among the highest in the world and Pakistan's urban population is expected to more than double in the next quarter of a century (United Nations Population Division, 2005).

The Indus Basin Irrigation System is the largest contiguous irrigation system in the world, irrigating over 16 million hectares of agricultural land, and is a vital supplier of domestic water in those areas where groundwater is unfit for consumption (van der Hoek *et al.*, 2001). Rapid population growth will put irrigated agriculture in Pakistan under severe pressure and the use of (treated) wastewater in agriculture has been suggested as a possible alternative water source. A nation-wide survey in Pakistan revealed that over 32,500 hectares of agricultural land were irrigated with untreated wastewater and that as much as 26% of Pakistan's vegetables were grown on untreated wastewater (Ensink *et al.*, 2004).

The use of wastewater poses health and environmental risks but also provides important benefits by facilitating a source of livelihood to its users. Benefits include savings in chemical fertilizers, higher cropping intensities and the ability to cultivate high value crops (van der Hoek *et al.*, 2002). This chapter presents the findings of a study on the risks and benefits of wastewater use in Faisalabad, Pakistan. For the study, a water and nutrient balance was set-up, water quality and soil quality were assessed, and a large cross-sectional health survey into the prevalence of intestinal parasites was undertaken.

21.2 FAISALABAD-CITY

With a population of over 2 million, Faisalabad is the third largest city in Pakistan and one of the main industrial centres (Population Census Organization, 2001). The city is located in a semi-arid zone in the heart of Punjab province where temperatures range from -4ºC in January to 48ºC in May. The climate is arid with an annual rainfall of 190 mm and a total annual evaporation of 1,992 mm. The local Water and Sanitation Agency (WASA) estimated that approximately 550,000 m^3 of wastewater is disposed of daily. A survey in and around the city identified nine different wastewater use sites, with a total area of over 2,200 hectares. The research presented in this chapter will focus mainly on one of the largest wastewater irrigated sites which is located right next to the city's only wastewater treatment plant.

At this site, wastewater is pumped on a 24-hour basis from the main sewerage network into a primary drain bringing wastewater to the waste stabilization ponds (WSP). Local farmers have organised themselves in to a water user association, which successfully established their right to use untreated wastewater for irrigation. They even won legal proceedings on the argument that the farmers had no access to any alternative source of water. At the moment, on average 42,200 m^3 of untreated wastewater is used by approximately 290 farming households. These farmers paid a combined annual fee of Rs 440,000 (US$ 7,500) to the WASA to use this untreated wastewater. The main crops cultivated with untreated wastewater were fodder, wheat and vegetables. The vegetables included: spinach, cauliflower, aubergine, chillies and tomatoes. Surprisingly, farmers were unwilling to use final effluent from the WSP, which (likely as a result of high evaporation) was considered too saline. Farmers also appreciated the higher nutrient value of untreated wastewater, thereby reducing their reliance on chemical fertilizers.

With regards to potential heavy metal contamination of soil and crops associated with the long-term use of untreated industrial wastewater, an additional site, namely 'Channel 4', will be discussed. Faisalabad is a major industrial centre, where over 150 industrial units have been identified by WASA, most of which are involved in textile processing. The Channel 4 wastewater irrigation area is approximately 500 ha and utilizes a mixture of industrial and domestic wastewater. Crops and cropping systems at the Channel 4 site are similar to those at the wastewater treatment plant site.

21.3 WATER QUALITY

Regular irrigation water and untreated wastewater was sampled on a monthly basis during the period from September 2001 till August 2002 for key water quality parameters. Untreated wastewater showed very high concentrations of helminth eggs and *E.coli*, the key indicator organisms for faecal contamination of water (Table 21.1). Based on World Health Organization guidelines, this water was unfit for unrestricted use in agriculture (WHO, 1989). Although wastewater showed higher concentrations for most metals as compared to regular irrigation water, all metals with the exception of cobalt were within guidelines set by the Food and Agricultural Organisation (FAO) (Pescod, 1992). For currently unexplained reasons, cobalt in regular irrigation water was also found to have exceeded the guideline value.

Table 21.1 Untreated wastewater and regular irrigation water quality (values in parenthesis represent one standard deviation).

	Unit	Guideline value	Untreated Wastewater	Regular irrigation water
Helminth eggs	Ova/L	<1[a]	700 (677)	0 (0)
E.coli	CFU/100mL	<1000[a]	$5.3\ 10^7$ ($7.7\ 10^7$)	246 (154)
BOD_5	mg/L		394 (73)	6 (6)
NO_3^-	mg/L	< 30[b]	184 (32)	0.5 (1.2)
PO_4^{-3}	mg/L	-	18.4 (4.0)	0.2 (0.1)
EC	dS/m	< 3.0[b]	2.0 (0.3)	0.3 (0.1)
pH		6.5 – 8.4[c]	7.0 (0.1)	7.6 (0.2)
SAR		<9[b]	12.9 (7.0)	1.9 (1. 5)
Cadmium	mg/L	0.01	0.004 (0.010)	0.010 (0.025)
Cobalt	mg/L	0.05	0.104 (0.067)	0.060 (0.040)
Copper	mg/L	0.2	0.11 (0.05)	0.12 (0.12)
Chromium	mg/L	0.1	0.02 (0.07)	0.01 (0.02)
Iron	mg/L	5.0	1.18 (1.06)	0.93 (1.38)
Manganese	mg/L	0.2	0.20 (0.08)	0.12 (0.13)
Nickel	mg/L	0.2	0.07 (0.05)	0.08 (0.02)
Lead	mg/L	5.0	0.13 (0.27)	0.08 (0.02)
Zinc	mg/L	2.0	0.44 (0.38)	0.24 (0.33)

[a] For unrestricted irrigation; produce consumed uncooked (WHO, 1989); [b] The upper limit of the medium restriction of use classification in order to maintain maximum yield; [c] Normal range (Pescod, 1992); EC: Electric conductivity; SAR: Sodium absorption ratio.

Wastewater was found to have a relatively high salinity, however, salinity levels were still found to be below salt tolerance levels of the most commonly irrigated crops in Pakistan, like wheat (6.0 dS/m), cotton (7.7 dS/m), rice (3.0 dS/m) and spinach (3.2 dS/m). Farmers, however, did report that the use of untreated wastewater limited their choice in crops, with root crops like potatoes, carrots and onions deemed unsuitable for cultivation with wastewater.

21.4 FARMER CHARACTERISTICS AND PERCEPTIONS OF WASTEWATER USE

The reluctance of farmers to use final effluent from the WSP and the high prices they paid for untreated wastewater prompted an investigation into farmer perceptions regarding wastewater use. A group of 100 farmers (50 wastewater farmers and 50 farmers using regular [non-wastewater] irrigation water) were randomly selected following an agricultural census at the WSP site and were asked about their views on the health risks associated with untreated wastewater, reasons why they used or would use untreated wastewater and whether they felt it was acceptable to use wastewater from a religious point of view. Basic farmer characteristics were also collected. Regular farmers were selected from a community of farmers that lived close to the wastewater irrigated sites but did not have access to wastewater. Potentially they could get access to final effluent if a distribution canal from the WSP was constructed.

In censuses taken during colonial rule, farmers were typically divided into functional castes or agricultural tribes, or in the case of Muslim farmers, qaums. Tribal affiliation, descent and occupational specialisation merge into a qaum identity. The traditional occupation gives the qaum its name and until today distinctions based on qaum remain significant social markers, particular in rural areas. There was a distinct difference between wastewater farmers and regular farmers with regard to qaum (Table 21.2). Regular farmers more often belonged to the land-owning, politically well-connected and financially strong Arrain and Rajpoot qaums, while the majority of the wastewater farmers belonged to qaums that traditionally work the land but often as share croppers or land renters. This was confirmed by a significantly larger percentage of regular farmers that owned land, as compared to wastewater farmers. The rent paid for land with access to wastewater was found to be more than 66% higher as compared to land without access to wastewater, which would indicate that wastewater was a well sought after resource for which farmers were willing to pay. This would be supported by the fact that almost all wastewater farmers grew vegetables, which brought a good price at the local markets, while only a small percentage of regular farmers grew vegetables. No significant difference in socio-economic status, following a wealth ranking exercise, was found between the two farmer groups.

Table 21.2 Characteristics of wastewater and regular farmers.

	Wastewater Farmers n= 50	Regular Farmers N = 50	Test statistics[a]
Tribe/*qaum*			
Rajpoot/Arrain	10%	32%	
Jutt	66%	44%	
Gujjars	10%	0%	χ^2=28.5, P < 0.001
Doggar/Baluch	2%	12%	
Malik	12%	0%	
Sheik	0%	12%	
Wealth Ranking of Households			
Poor	28%	20%	
Intermediate	48%	58%	χ^2=1.2, P = 0.55
Rich	24%	22%	
Landowner	68%	96%	χ^2=12.6, P< 0.001
Average farm size (ha)	2.5 (0.3)	2.1 (0.3)	P = 0.24
Land rent (US$/ha/year)	171 (11.5)	113 (4.1)	P < 0.001
Vegetables cultivated	88%	22%	χ^2=44, P < 0.001

[a] Differences between the two farmer groups for each variable were tested with analysis of variance (t-test) for continuous data and chi-squared tests for proportions.
Values in parentheses are one standard deviation

When asked why wastewater farmers were using wastewater, a combination of three answers emerged: they lacked access to fresh water, wastewater was more reliable as it was not subjected to a rotational schedule, and the nutrients in wastewater made it unnecessary to use additional chemical fertilizer (Table 21.3). Nutrients were identified by both groups of farmers as the key benefit of wastewater. All regular farmers used chemical fertilisers at some stage during the cropping season, while 70% of the wastewater farmers used no chemical fertilizers at all, with the remaining 30% only applying a limited amount of chemical fertilizer during specific growth stages, like flowering or grain setting.

A small majority of regular farmers saw health problems associated with wastewater use as its greatest risk, in contrast to a large majority of the wastewater farmers that felt that there were no health or other risks associated with the use of wastewater.

Table 21.3 Perceptions of wastewater quality; risks, benefits and its acceptability.

	Wastewater Farmers (n= 50)	Regular Farmers (n = 50)	Test statistic
Benefits of wastewater use			
No benefits	4%	12%	
Nutrients	72%	86%	$X^2=15.6$
Nutrients + reliability	24%	0%	P= 0.001
I don't know	0%	2%	
Risk of wastewater use			
No risks	74%	36%	
Bad for health	20%	56%	$\chi^2=16.1$
Bad for soil	6%	6%	P= 0.001
I don't know	0%	2%	
Would you use final effluent			
No (< nutrients)	48%	12%	$\chi^2=41$
No (< nutrients & >saline)	52%	28%	P < 0.001
Yes	0%	60%	
Is wastewater use acceptable			
No	2%	0%	$\chi^2=1.7$
Yes	72%	66%	P= 0.44
I don't know	26%	34%	

Most wastewater farmers claimed that when they started using wastewater they had suffered from some initial health problems. However, with the passing of time, these problems had all disappeared. The large majority of wastewater farmers claimed not to have suffered from health problems. However, 18% had suffered from diarrhoea, 12% from itching and cracking of the soles of the feet and 6% from more general skin problems. A surprisingly large number of wastewater farmers (56%) claimed that use of untreated wastewater could cure skin problems.

Untreated wastewater farmers were unanimously unwilling to use final effluent from the WSP for irrigation because they considered it too saline and because it had a lower nutrient value than untreated wastewater and would therefore result in a greater need for chemical fertilizer. Regular farmers saw final effluent as an additional source of water and were willing to use final effluent even though many acknowledged the higher salinity levels. The large majority of farmers considered the use of wastewater religiously acceptable.The main reason for its acceptability was the fact that, according to farmers, the soil and the roots filtered the wastewater and that therefore there was no contact between wastewater and edible produce. The study yielded no significantly different response on religious acceptability between untreated wastewater and final effluent.

All farmers in both groups classified the wastewater as very dirty and foul smelling but a large majority in both groups considered (untreated) wastewater, for a number of reasons, acceptable from a religious point of view for the use in agriculture. In contrast, a *fatwa* issued by Muslim scholars in Saudi Arabia considered only treated wastewater acceptable for the use in agriculture, and then only on the condition that it would not lead to adverse health effects (Faruqi *et al.*, 2001). Farmers did acknowledge that wastewater was not fit for *wudu* (ablution) and that they needed to wash and change clothes before being able to attend mosque.

21.5 HEALTH IMPACTS

To access the health impact of untreated wastewater irrigation, a cross-sectional health study was conducted from July 2002 till June 2003. Fresh stool samples were collected from adult wastewater and regular farmers and their children aged from 2 to 12 years. Stool samples were analysed, using the formal-ether concentration technique, for intestinal helminths and protozoa.

Table 21.4 Prevalence of helminth and protozoa infections in wastewater farmers, regular farmers and their children (2-12 years old).

	Regular Farming Household		**Wastewater Farming Household**	
	Adults (n = 167)	Children (n = 309)	Adults (n =176)	Children (n =310)
Helminths				
Ascaris lumbricoides	0.6%	1.0%	4.0%	1.6%
Hookworm	0.0%	0.6%	13.6%	6.1%
Trichuris trichiura	0.0%	0.0%	0.0%	1.6%
Hymenolepis nana	1.8%	2.9%	11.9%	13.5%
Taenia saginata	0.0%	0.6%	0.6%	0.3%
Protozoa				
Giardia intestinalis	49.1%	49.2%	80.1%	75.8%
Entamoeba histolytica	19.2%	10.7%	40.9%	39.4%

The survey showed a relatively low prevalence of intestinal nematode infections (Table 21.4), though the prevalence of infection was significantly higher in wastewater farmers and their children than in regular farmers and their children. This difference proved to be significant for hookworm infection after controlling for other risk factors (Ensink *et al.*, 2005). A very high prevalence of protozoal infections was observed in wastewater farmers but also in regular farmers. Wastewater farmers and their children had a significantly increased risk of *Giardia* infection as a result of exposure to untreated wastewater (Ensink *et al.*, 2006).

21.6 WATER APPLICATIONS

During the period from September 2001 till August 2002, six agricultural fields (approximately 1 hectare each) at the WSP site and three at a neighbouring village were selected for an irrigation and nutrient application evaluation. Out of these nine fields, three fields were irrigated exclusively by untreated wastewater, three fields used only regular (non-wastewater) irrigation water, while the remaining three fields relied on the conjunctive use of both untreated wastewater and groundwater.

Water delivery to the nine selected fields was monitored with the help of a cut-throat flume (Siddiqui *et al.*, 1996) installed at the water entry point of each field. Two research assistants recorded water head and time during the full length of each irrigation event.

During the period of observation, fodder and wheat were grown in all fields. Cauliflower was grown with wastewater and regular irrigation water, while tomatoes were only grown with wastewater. During the study period there were disputes over wastewater charges, which resulted in prolonged closure of irrigation canals, shifts from day to night irrigation, and fallow periods. Consequently, on average only nine months of data were collected for each field.

Wheat and cauliflower cultivated with wastewater were more frequently irrigated than the same crops irrigated with regular water (Table 21.5). Fodder fields irrigated by wastewater showed significantly higher irrigation application than both other types. A significant difference in irrigation application was also observed between wheat fields irrigated by wastewater and regular irrigation water. No difference was observed for cauliflower irrigated by wastewater, regular irrigation water or conjunctively, and for wastewater irrigated wheat. Wastewater was able to meet crop water demand for three out of four crops, while regular irrigation water was never able to meet crop water demands.

Variations between wastewater irrigation applications were large, with monthly irrigation applications ranging from 45 mm $month^{-1}$ to 175 mm $month^{-1}$ (Mean 114 mm $month^{-1}$, Standard Deviation (SD) 62 mm $month^{-1}$). In contrast, monthly irrigation applications for regular irrigated land ranged from 68 mm $month^{-1}$ to 71 mm $month^{-1}$ (mean 69 mm $month^{-1}$, SD 1.5 mm $month^{-1}$).

Table 21.5 Irrigation applications for (untreated) wastewater, conjunctively and regular irrigated fields.

Crop	Water source	Total irrigation application (mm)	Crop water requirement[a] (mm)	Irrigation application (n)	Average irrigation application mm(SD)
Fodder	Wastewater	914		8	113 (77)
	Conjunctive	583	558	8.5	71 (18)
	Regular	432		8	54 (20)
Wheat	Wastewater	630		7.5	84 (57)
	Conjunctive	404	570	7	58 (21)
	Regular	351		6	55 (13)
Cauliflower	Wastewater	321	344	10	32 (8)
	Regular	256		8	32 (9)
Tomato	Wastewater	585	557	25	21 (9)

[a]Crop water requirements obtained with Cropwat 4.

21.7 NUTRIENT APPLICATIONS

Monthly water quality measurements were taken in the main wastewater irrigation canal. The timing of these monthly assessments was determined on the basis of two (Summer and Winter season) twenty-four hour survey's in which water quality and flow measurements were taken every second hour.

Water samples were analyzed following standard methods (APHA, 1998) for three macro nutrients (total nitrogen (TN), total phosphorus (TP), and potassium (TK)), for pH and Electro-conductivity (EC) and Sodium Absorption Ratio (SAR). Nutrient applications were obtained by multiplying irrigation applications with average monthly water quality. Farmers were also asked to report additional chemical fertilizer application.

As compared with nutrients applied through chemical fertilizers and farm yard manure at regular irrigated fields, nutrient application through wastewater for all three nutrients was at least double, and for potassium in fodder even fifteen times as high (Table 21.6). With the exception of cauliflower, the exclusive and conjunctive use of wastewater was able to supply sufficient TN, TP and TK for all crops without the use of supplementary chemical fertilizer. However, it should be noted that the use of wastewater resulted in an over application of N, P and K, which ranged from 25% to over 500% in fields exclusively irrigated by wastewater (Ministry of Food, Agriculture and Livestock, 1997). Conjunctive use of wastewater led to a much lower over-application of the aforementioned nutrients. Chemical fertilizer and farm-yard manure application in regularly irrigated fields was almost able to meet crop nitrogen demand but was not sufficient for phosphorous and potassium demand (Table 21.6).

Table 21.6 Nutrient application for (untreated) wastewater, conjunctively and regular irrigated fields.

Crop	Water source	TN applied (kg/ha)	TN advised (kg/ha)	TP applied (kg/ha)	TP advised (kg/ha)	TK applied (kg/ha)	TK advised (kg ha^{-1})
Fodder	Wastewater	306		44		152	
	Conjunctive	254	180	40	15	144	45
	Regular	145		0		10	
Wheat	Wastewater	336		54		126	
	Conjunctive	187	100	43	13	51	20
	Regular	98		28		12	
Cauliflower	Wastewater	159	125	18	105	60	100
	Regular	71		1		7	
Tomato	Wastewater	257	100	40	60	111	30

The use of chemical fertilizer on wastewater-irrigated land was on average 65 kg per hectare per year, while on regular irrigated land it was 530 kg per hectare per year. With an average chemical fertilizer input reduction of 465 kg per hectares per year for wastewater irrigated fields, farmer perceptions on the fertilizer value of wastewater were confirmed. For the complete WSP site, this would have resulted in a saving of almost 300,000 kg of chemical fertilizer, equivalent to approximately 3,000,000 Rupees (51,100 US$), (equivalent to seven times the annual wastewater fee).

In Haroonabad, a small town in the Punjab, wastewater irrigated crops carried a high cost for land preparation, seeds and pesticides (van der Hoek *et al.*, 2002). Savings from chemical fertilizers accounted for approximately 30% of total gross margin for wastewater farmers in Haroonabad, with higher cropping intensities and the cultivation of higher valued crops making up the remaining 70%. If these values were translated to Faisalabad this would mean that wastewater farmers could earn approximately 35,000 RS (600 US$) per year more than regular farmers.

21.8 SOIL QUALITY

In 2002, composite wheat grain and concurrent soil samples were collected from 20 fields at 1 km intervals along the length of Channel 4 (an industrial wastewater irrigation channel). Concurrent soil samples were taken at 0-20cm depth. Further, in 2003 rice grain (8 fields) and sorghum fodder (12 fields) samples and concurrent soil samples were collected from fields along the length of Channel 4 previously sampled for wheat. Wheat and rice grain samples were collected at physiological maturity. Thirty randomly selected ear and panicle samples were taken from the pre-selected wheat and rice fields respectively. In addition, in 2003 composite spinach (30 plant composite per plot) and concurrent soil samples (0-20cm) were collected from the 0-3km 'urban' stretch of Channel 4. In addition, as a control, soil samples were collected from 2 fields that were irrigated by regular irrigation water.

To determine soil total Cd, Pb, Zn, Ni, Cr and Cu air-dried and ground (<1.0mm) soil samples were digested in aqua regia (3:1 $HCl:HNO_3$) using an open tube digestion method (McGrath and Cunliffe, 1985). Soil pH_w and EC were determined on a 1:5 soil:water suspension (Rayment and Higginson, 1992) and organic carbon determined following the Walkley-Black Method (Walkley, 1947; Nelson and Sommers, 1996). Plant samples were digested in 2:1, $HNO_3:HClO_4$ using an open tube digestion technique. To assess within-batch and between-batch precision, two reagent blanks, three replicates of an appropriate

In-House Standard Reference Material (IH-SRM), and three duplicates from the previous analytical batch were included in each batch of samples analyzed. In addition, test portions were analyzed in triplicate. Element concentrations were determined using a Perkin Elmer Analyst 3000 Atomic Adsorption Spectrophotometer with deuterium lamp background correction. Analytical accuracy was further assessed through the 'real time' comparison of IH-SRM results with element-specific control charts and the use of Continuous Verification Standards (CVS) at regular intervals.

The results indicated that irrespective of type of irrigation and sampling location, soil Cd, Zn, Pb, Cu, Cr and Ni levels were well below the levels established for agricultural soils by the EU (European Economic Commission, 1986) (Table 21.7). The results also indicated that soil Pb, Cu, Cr and Ni concentrations and Cu and Ni concentrations were significantly (ANOVA $p=<0.001$) lower at the wastewater treatment plant site in 100% wastewater and conjunctively-irrigated plots as compared with all other plots (Table 21.7). This may in larger part be due to inherent differences in soil mineralogy as indicated by the significantly lower silt and clay fractions (ANOVA $p=<0.001$) associated with the 100% and conjunctively-irrigated wastewater plots (Table 21.7).

With the exception of the regular water irrigated plots at the site adjacent to the WSP, organic carbon (Org-C) levels in the soils evaluated were relatively low, irrespective of irrigation source. The significantly higher Org-C levels associated with the regular irrigated plots at the WSP site (Table 21.7) may be indicative of the significantly higher soil EC 5.20 (±0.66) and associated impacts on microbial organic matter decomposition. Further, the level of soil EC associated with the regular water irrigated plots at the Narwala Road site would be inhibitive to the cultivation of rice, spinach, tomato and pepper which experience 10% yield reductions at 3.0, 3.2, 3.5 and 2.2 dS m^{-1}, respectively.

Table 21.8 shows that fields adjacent to the WSP irrigated by wastewater, either exclusively or conjunctively had higher concentrations of N, P and K than regularly irrigated soils. These differences were only found to be significant for the 0-20 cm and not for the 21-40 cm layer. The impact of wastewater was particularly clear for phosphorus (ANOVA, $p< 0.01$). Conjunctively-irrigated soil showed significantly increased levels for all nutrients when compared with regular water irrigated soils (*t-test*, $p< 0.05$), while soils exclusively irrigated by wastewater showed only a significantly increased level for phosphorus when compared to regular water irrigated soil (*t-test*, $p= 0.02$). No significant difference was found between soils exclusively and conjunctively irrigated by wastewater.

After thirty years of wastewater irrigation N, P and K levels in the first 20 cm of the soil profile were found to be slightly higher as compared with regular water irrigated soil, whilst the reverse can be seen for soil EC.

Salinization and sodification of soils is a common problem in Pakistan, where large areas of agricultural land have been taken out of production as a result of rising (saline) groundwater tables and poor drainage. The significantly lower soil EC values associated with the wastewater and conjunctively-irrigated fields (Table 21.7) may indicate that the observed high wastewater applications are beneficial by promoting the leaching of salts to lower soil layers, but could, in the longer term, lead to a more rapid rise of groundwater tables.

Table 21.7. Mean soil Cd, Pb, Zn, Ni, Cr and Cu concentrations (mg/kg) and selected soil characteristics in relation to irrigation source.

	Mean Element concentration (mg/kg)						**Soil Characteristics**					
Irrigation	Cd	Zn	Pb	Cu	Cr	Ni	pH	EC (dS m^{-1})	Org-C (%)	Sand (%)	Silt (%)	Clay (%)
Narwala Road												
100% Wastewater	/	42.37**a** (±1.020)	7.66**a** (±0.524)	12.98**a** (±0.083)	14.63**a** (±0.064)	18.68**a** (±0.152)	8.50**a** (±0.122)	2.26**a** (±0.151)	0.485**a** (±0.190)	92.8**a** (±1.67)	4.17**a** (±1.428)	3.02**a** (±1.00)
Conjunctive water use	/	39.98**a** (±2.57)	7.11**a** (±0.051)	12.35**a** (±0.080)	20.37**b** (±0.613)	24.26**b** (±0.907)	8.56**a** (±0.05)	2.96**a** (±0.527)	0.450**a** (±0.339)	78.86**b** (±5.24)	11.42**b** (±4.35)	9.71**b** (±1.22)
100% Non-wastewater	/	57.82**a** (±0.870)	10.96**b** (±0.004)	22.35**b** (±0.563)	24.45**b** (1.25)	38.99**c** (±1.25)	8.50**a** (±0.122)	5.20**b** (±0.655)	1.03**b** (±0.487)	58.59**c** (±2.65)	23.62**c** (±1.95)	17.58**c** (±0.843)
Channel 4												
100% Wastewater	0.227**b** (±0.130)	51.38**a** (±14.63)	10.72**b** (±1.59)	21.36**b** (±2.53)	26.33**b** (±3.56)	33.82**c** (±4.11)	8.23**b** (±0.123)	2.41**a** (±0.529)	0.441a (±0.144)	60.63**c** (±6.58)	21.84**c** (±3.90)	17.51**c** (±3.69)
Dhudi Wala Minor												
100% Non-wastewater	0.217**b** (±0.015)	44.52**a** (±4.73)	11.57**b** (±0.677)	22.76**b** (±0.412)	24.11**b** (±1.51)	35.22**c** (±0.966)	8.3**b** (±0.00)	2.07**a** (±0.150)	0.369**a** (±0.060)	54.55**c** (±0.55)	24.92**c** (±0.415)	20.53**c** (±0.409)
EU MP levels (mg/kg)	1.0-3.0	150-300	50-300	50-140	100-150	30-75	/	/	/	/	/	/

EU Maximum Permissible (MP) levels for sludge amended soils (Directive 86/278/EEC). Values indicate the arithmetic mean. Values in parentheses represent ±1 STDEV

On a heavy metal and soil characteristic basis, mean values followed by a different letter are significantly different (ANOVA $p=<0.001$).

Table 21.8. Soil quality at (untreated) wastewater, conjunctively and regular irrigated fields.

	Available P (mg/kg)		Organic matter (%)		Total K (mg/kg)		Total N (mg/kg)	
	0 – 20	21-40	0 -20	21-40	0 - 20	21-40	0 -20	21-40
Untreated	14.9	9.6	0.86	0.69	168	152	251	99
	(3.1)	(3.5)	(0.18)	(0.10)	(34)	(34)	(140)	(35)
Conjunctive	20.0	11.9	1.16	0.72	210	183	273	120
	(7.5)	(5.8)	(0.14)	(0.14)	(55)	(61)	(108)	(50)
Regular	8.7	6.4	0.91	0.70	145	116	147	114
	(4.1)	(3.1)	(0.16)	(0.11)	(41)	(19)	(87)	(52)

21.9 CROP QUALITY

Internationally recognized Maximum Levels (ML) for contaminants in foods are established by the Joint FAO/WHO Expert Committee on Food Additives (JECFA). Specifically, MLs are established by the Codex Committee on Food Additives and Contaminants (CCFAC) under the Joint FAO/WHO Food Standards Programme, Codex Alimentarius Commission. In terms of human dietary exposure pathways, MLs have been established for Cd and Pb. JECFA MLs for Pb and Cd in leafy vegetables, wheat and rice grain are indicated in Table 21.9.

The results indicate that for wheat grain, rice and spinach, Pb levels were non-detectable. Further, rice grain Cd concentrations are below recommended levels. In contrast, 60% of wheat grain samples and 100% of spinach samples contained Cd at levels that exceed the respective JECFA recommended value (Table 21.9). This may in part be due to the high Cl^- levels in irrigation water (data not shown) which has been shown too promote the uptake of Cd to edible portions of crops (McLaughlin *et al.*, 1994; Li *et al.*, 1994).

Table 21.9 Range and mean heavy metal concentrations (mg/kg) for selected crops irrigated with Channel 4 Wastewater.

Element Concentration (mg/kg)	*Wheat Grain (n=20)	*Rice Grain (n=8)	**Sorghum (n=12)	**Spinach (n=15)
Cd	0.137 – 0.331 **0.228** (±0.061)	0.03 – 0.124 **0.079** (±0.032)	0.03 – 0.137 **0.093** (±0.042)	0.555 – 0.806 **0.694** (±0.080)
Zn	15.07 – 48.00 **28.16** (±9.38)	11.94 – 30.14 **23.57** (±6.50)	22.36 – 47.35 **30.21** (±8.75)	32.43 – 136 **67.28** (±26.71)
Pb	nd	Nd	Nd	nd
Cr	<0.01 – 0.373 **0.192** (±0.116)	0.02 – 0.596 **0.192** (±0.186)	3.11 – 8.69 **6.07** (±1.89)	0.37 – 1.29 **0.932** (±0.243)
Ni	nd	0.621 – 1.15 **0.830** (±0.183)	1.87 – 4.44 **3.10** (±0.980)	0.622 – 4.44 **2.05** (±1.36)
Cu	5.33 – 9.93 **6.32** (±0.116)	4.72 – 8.78 **6.37** (±1.18)	/	10.4 – 25.8 **18.65** (±4.49)
Cd:Zn Ratio	0.004 – 0.013 **0.008** (±0.002)	0.001 – 0.005 **0.003** (±0.001)	na	0.005 – 0.017 **0.011** (±0.003)
[1]Critical Cd:Zn Ratio	<0.015	<0.015		<0.015
[2]JECFA MP Level of Pb (mg/kg)	0.2	0.2	na	0.1
[2]JECFA MP Level of Cd (mg/kg)	0.2	0.4	/	0.2

Values in bold indicate the arithmetic mean. Values in parentheses represent ±1 STDEV
na = not applicable; nd= non-detectable; * 30 ear/panicle composite per field; **30 point composite (10 leaves per plant) per field; [1]Chaney *et al.* (1996); [2]JECFA (2005).

However, and critically, Chaney *et al.* (1996) suggested that a Cd:Zn ratio of <0.015 effectively provides protection against Cd-induced human health impacts. For the wheat and spinach samples the Cd:Zn ratio ranges from 0.004 – 0.013 and 0.005 – 0.017, respectively. For wheat, this indicates that the Cd:Zn ratio is acceptable (Table 21.9). Further, only 30% of spinach samples have a Cd:Zn ratio >0.015. In addition, it should be noted that spinach makes a small contribution to daily calorific intake with the majority of daily intake in the study area derived from wheat and rice. Further, it should also be noted that potential Cd-induced human health impacts result from prolonged dietary exposure in the absence of adequate dietary Zn and Fe (Reeves and Vanderpool, *1998*; Berglund *et al.*, *1994;* Reeves and Chaney, 2001). Further studies will be undertaken to evaluate potential Cd-related health risks in the study area.

21.10 GROUNDWATER

Depth to groundwater was measured bi-weekly in 19 piezometers which were installed for this study and evenly spread over the untreated wastewater, conjunctive and regular irrigated areas. Out of these 19 piezometers, 9 were installed in the untreated wastewater, 7 in the conjunctive and 3 in the regular irrigated sites. Groundwater samples were taken in March 2002 on a single occasion. Samples were tested for *E.coli*, EC, Nitrate (NO_3^-) and Nitrite (NO_2^-).

At the start of the study the average depth to groundwater at the study site was 5.19 (SD = 0.83) meters, while at the regular water irrigated sites this was 4.98 (SD = 0.23) meters. After a year, groundwater levels at both sites had dropped respectively to 5.83 (SD = 0.75) meters and 5.81 (SD = 0.19) meters below the surface level.

Groundwater at all three sites was unfit for consumption because of the high salinity concentrations. However, groundwater salinity at the wastewater irrigated sites was higher as compared to the regular irrigated sites, this likely as a result of prolonged wastewater use. This was further confirmed by high nitrate and *E.coli* concentrations (Table 21.10).

Table 21.10. Groundwater quality at the three different sample sites.

	Unit	Guideline value [a]	Wastewater irrigated	Conjunctive use	Regular irrigated
E.coli	CFU 100/mL	0	78 (198)	5 (7)	0 (0)
EC	dS/m	1.8	4.6 (1.6)	3.8 (1.1)	2.1 (0.4)
NO_2^-	mg/L	3	0.17 (0.24)	0.05 (0.06)	0.03 (0.02)
NO_3^-	mg/L	50	56 (61)	10 (19)	0.33 (0.58)

[a] WHO, drinking water guidelines (1993)

21.11 CONCLUSION

In Faisalabad, wastewater irrigation seems to be a sustainable and profitable affair for all those involved. Daily availability of (waste) water and as result the more flexible rotational schedule allows farmers to grow vegetables and meet crop water demands. In addition, a major benefit for farmers is the more than adequate supply of N, P and K, which resulted during the study period in an 85% reduction in chemical fertilizer application. Savings in fertilizer costs and the cultivation of vegetables resulted in an annual income which was estimated to be 600 US$ higher than that of farmers using regular irrigation water. The local water utility, by auctioning wastewater, supplemented its scarce resources, thereby allowing it to operate, maintain and upgrade its current drinking water and sewerage collection system. This seems to be a win-win situation for all involved actors with little or no motivation for change.

Heavy metal concentrations in water and soils were found to be within critical levels set by the FAO, and the EU. However, the elevated levels of Cd in wheat and spinach samples irrigated by Channel 4 wastewater are of concern. In addition, a further and negative impact of wastewater irrigation comes from human pathogens. The majority of the crops cultivated at the WSP-site are cooked before consumption and therefore pose little to no risk to humans. Therefore the greatest health risk comes from direct contact of farmers with untreated wastewater. The cross-sectional health study conducted showed a much higher prevalence of hookworm and *Giardia* infections in wastewater farmers as compared to regular farmers. However these negative impacts on a relatively small group are easy to control through regular anthelmintic treatment programmes.

It is important to acknowledge the clear benefits that wastewater irrigation can bring to the urban and peri-urban poor and how the cultivation of fresh cheap vegetables can benefit the health of urban residents and producers alike. As a result of rapid urbanization and uncontrolled urban sprawl, most municipalities in Pakistan and many other countries in Africa and Asia will remain unable to collect and treat all wastewater for the foreseeable future. To ban untreated wastewater irrigation is neither desirable nor achievable according to sanitation officers in Pakistan (Ensink *et al.*, 2004). Consequently, there is a need to look at other options to reduce health and environmental risks.

21.12 ACKNOWLEDGEMENT

International Water Management Institute (IWMI) gratefully acknowledges the financial support from the German Federal Ministry for Economic Cooperation and Development (BMZ) for its research on the use of wastewater in Pakistan. We would like to thank Safraz Munir, Tipu Naveed, Tariq Mahmood, M. Asghar and M. Rizwan Aslam for their dedication to the field work which involved flow and water quality measurements at odd hours.

21.13 REFERENCES

APHA, AWWA and WPCF (1998). Standard methods for the examination of water and wastewater. 20th edition. American Public Health Association, Washington D.C., USA.

Berglund, M., Akesson, A., Nermell, B. and Vahter, M. (1994). Intestinal absorption of dietary cadmium in women depends on body iron stores and fiber intake. Environ. Health Perspect. 102, 1058-1066.

Chaney, R.L., Ryan, J.A., Li, Y.M., Welch, R.M., Reeves, P.G., Brown, S.L. and Green, C.E. (1996). Phyto-availability and bio-availability in risk assessment for cadmium in agricultural environments In Sources of Cadmium in the Environment, pp 49-78. OECD, Paris, France.

Ensink J.H.J., Mahmood, T., van der Hoek, W., Raschid-Sally, L., and Amerasinghe, F.P. (2004). A nation-wide assessment of wastewater use in Pakistan: an obscure activity or a vitally important one? Water Policy. 6, 197-206.

Ensink, J.H.J., W. van der Hoek, M. Mukhtar, Z. Tahir & F.P. Amerasinghe. (2005). High risk of hookworm infection among wastewater farmers in Pakistan. *Transactions of the Royal Society of Tropical Medicine and Hygiene,*99, 809-818.

Ensink, J.H.J., W. van der Hoek & F. P. Amerasinghe (2006). *Giardia duodenalis* infection and wastewater irrigation in Pakistan. *Transactions of the Royal Society of Tropical Medicine and Hygiene,* 100, 538-542

European Economic Commission. (1986) Council Directive 86/278/EEC on the protection of the environment, and in particular of the soil, when sewage sludge is used in agriculture, 12 June 1986.

Faruqi, N.I., Biswas, A.K., and Bino,.M.J. (2001) Water Management in Islam, IDRC, Ottawa, Canada

JECFA, (2005). Report of the 64th Meeting of the JECFA Rome, 8-17 February 2005

Li, Y-M, Chaney R.L., and Schneiter, A.A. (1994) Effect of soil chloride level on cadmium concentration in sunflower kernels. Plant and Soil. 167, 275-280.

McGrath, S.P., Cunliffe, C.H. (1985). A simplified method for the extraction of the metals Fe, Zn, Cu, Ni, Cd, Pb, Cr, Co and Mn from soils and sewage sludges. J. Sci. Food and Agric. 36, 794-798.

McLaughlin, M.J., Maier, N.A., Correll, R., Smart, M.K., Sparrow, L.A., and McKay, A. (1994) Increasing soil salinity causes elevated cadmium concentrations in field-grown potato tubers. Journal of Environmental Quality, 23, 1013-1018

Ministry of Food, Agriculture and Livestock, Federal water management cell. (1997). On farm water management field manual, volume VI; irrigation agronomy. Government of Pakistan, Islamabad, Pakistan.

Nelson, D.W. and Sommers, L.E. 1996. Carbon and organic matter. In Methods of Soil Analysis Part 3, Chemical Methods Ed. D.L. Sparks, pp 961-1010. Soil Sci. Soc. Am. and Am. Soc. Agron. Madison, Wisconsin.

Pescod, M.D. (1992). Wastewater Treatment and Use in Agriculture. Irrigation and drainage paper 47, Food and Agricultural Organization (FAO), Rome, Italy.

Population Census Organization. (2001). 1998 Provincial census report of Punjab. Statistics Division, Population Census Organization, Islamabad, Pakistan.

Rayment, G.E. and Higginson, F.R. (1992). Electrical conductivity and soil pH. In Australian Laboratory Handbook of Soil and Water Chemical Analysis Eds. G.E. Rayment and F.R., Higginson, pp15-23. Inkata Press, Melbourne.

Reeves, P.G. and Chaney, R.L. (2001). Mineral nutrient status of female rats affects the absorption and organ distribution of cadmium from sunflower kernels ((Helianthus annuus L.). Environ. Res. 85, 215-225.

Reeves, P.G., and Vanderpool, R.A. (1998). Organ content and fecal excretion of cadmium in male and female rats consuming variable amounts of naturally occurring cadmium in confectionery sunflower kernels (Helianthus annuus L.). J. Nutr. Biochem. 9, 636-644.

Seckler, D., Amarasinghe, U., Molden, D., de Silva, R., and Barker, R. (1998). World water demand and supply, 1990 to 2025: scenarios and issues. Research report 19, International Water Management Institute, Colombo, Sri Lanka.

Siddiqui, R., Lashari, B., and Skogerboe, G.V. (1996). Converting a fabricated cutthroat flume into a discharge measuring instrument. Report No. T-5, IIMI, Lahore, Pakistan.

United Nations Population Division (2004). http://esa.un.org/unup/

van der Hoek, W., Konradsen, F., Ensink, J.H.J., Mudasser, M., and Jensen, P.K. (2001). Irrigation water as a source of drinking water: is safe use possible? Trop.Med.Int. Health. 6 (1), 46-54.

Van der Hoek W, ul Hassan, M., Ensink, J.H.J., Feenstra, S., Raschid-Sally, L., Munir, S., Aslam, R., Ali, N., Hussain, R., and Matsuno. Y. (2002). Urban wastewater: a valuable resource for agriculture. A case study from Haroonabad, Pakistan. Research report 63, International Water Management Institute, Colombo, Sri Lanka.

Walkley, A. (1947). A critical examination of a rapid method for determining organic carbon in soils: Effect of variations in digestion conditions and of inorganic soil constituents. Soil Sci. 63, 251-263.

WHO. (1989). Health guidelines for the use of wastewater in agriculture and aquaculture. Technical report series 778. World Health Organization, Geneva, Switzerland.

World Health Organization (WHO). 1993. *Guidelines for drinking-water quality, volume 1; Recommendations. Second edition.* WHO, Geneva, Switzerland, 188pp.

22

Indirect water reuse for human consumption in Germany: the case of Berlin

Martin Jekel, Steffen Gruenheid

22.1 INTRODUCTION AND OBJECTIVES

In the city of Berlin, 56% of drinking water is derived from bank filtration and 14% from artificial recharge (BWB, 2003). The water utility is using well-operated bank filtration and artificial recharge facilities to provide high quality water. With the growth of the city, a semi-closed urban water cycle has been established in parts of the city area and its surroundings. At some bank filtration sites the surface water is strongly influenced by treated domestic waste water (*e.g.* 15–50% in Lake Tegel) (Ziegler, 2001). Figure 22.1 shows the location of Lake Tegel, with inflows and sources, including the well-treated wastewater treatment plant effluent of Schoenerlinde, contributing about 86400m^3/d. The lake water is used for bank filtration and artificial groundwater recharge. Specific investigations of the processes during both these natural treatment methods were done during the years 2002 to 2005 on 2 major transects (linear clusters of monitoring wells) depicted in Figure 22.1.

The respective inflows to Lake Tegel are shown in Fig. 22.2, including the wastewater treatment plant effluent of Schoenerlinde, the surface water of Nordgraben and Tegeler Fliess and the late pipeline delivering Havel river water in the dry summer periods. The wastewater portion is clost to 50% in the winter period and about 33% in the summer half year.

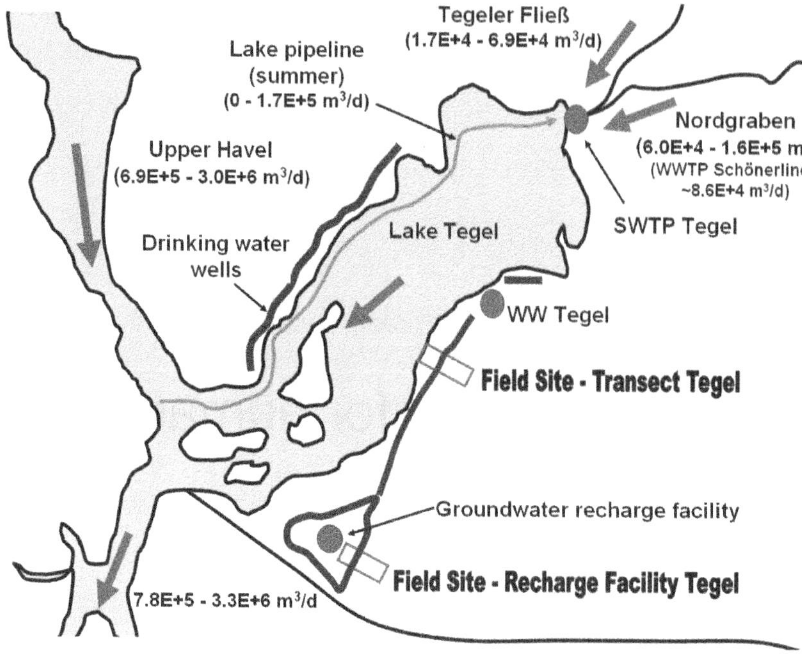

Figure 22.1. Lake Tegel in Berlin, showing water flows and locations of drinking water well fields.

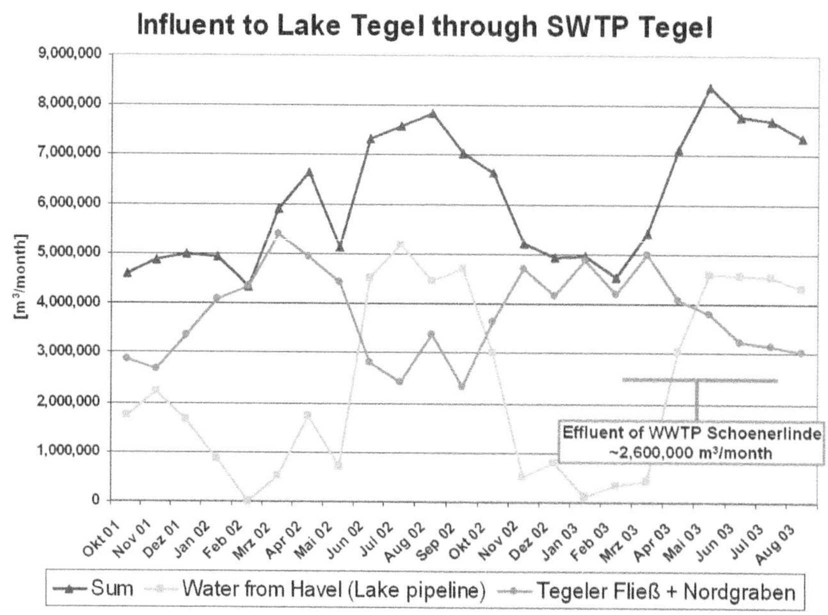

Figure 22.2. Monthly flows of incoming water to Lake Tegel, including a submerged pipeline for Havel river water, the inflow via Tegeler Fliess and Nordgraben (treated in a phosphorous removal plant) and the flow of the effluent of a wastewater treatment plant in the water shed of Nordgraben.

Bank filtration and groundwater recharge are traditional methods for drinking water treatment and extraction in central Europe, but the processes for removal of dissolved organic substances are not yet well understood, especially for emerging compounds of various origin, like industrial chemicals and pharmaceutical substances. This research study is part of the NASRI-project of the Center of Competence for Water Berlin (NASRI, 2003). As part of this project, the Department of Water Quality Control at the Technical University of Berlin (DWQC) investigates the removal of bulk organics (dissolved organic carbon. DOC), effluent organic matter (EfOM)) and selected trace organics at various field sites with different characteristics in the Berlin, Germany area and in soil column systems.

Despite this indirect potable reuse, the bank filtration system is providing high quality water which needs little further treatment and is distributed without chlorination. Since the processes during infiltration are very complex, it is difficult to predict bulk organic composition in the bank filtrate or to estimate important factors of influence for the degradation. There are indications that bank filtration under anoxic/anaerobic conditions provides different removal of organics than aerobic bank filtration (Gruenheid *et al.*, 2004). In addition to the redox state, factors such as retention time, temperature, initial degradable carbon concentration, soil properties, and hydrogeological conditions may affect the final concentration. The factors of influence and the fate of these bulk organics are studied in the field and in long term experiments.

22.2 METHODS

The major focus of the study was the fate of bulk and trace organics at a bank filtration site in Berlin-Tegel, where the highest percentages of treated domestic wastewater are found in dry periods (up to 50%). The wastewater collected receives secondary and tertiary treatment (removal of nitrogen and phosphorous) in the Schönerlinde sewage treatment plant plus coagulation/flocculation, settling and deep bed filtration for P-removal down to 0.02 mg/L in the Tegel surface water treatment plant before discharge to the north end of Lake Tegel.

Additionally, results of experiments on a long retention column system which was operated with the same water under aerobic and anoxic conditions will be presented. The monthly analytical program of field and column samples is comprised of DOC, UV-Absorption, LC-OCD (Organic Carbon Detection), differentiated Adsorbable Organic Halogens AOX (AOI, AOBr)-analysis and solid phase extraction (SPE)/ HPLC-analysis of three groups of trace compounds. The groups are X-ray contrast media (Iopromide); bacteriostatica (Sulfamethoxazole) and the industrial chemicals naphthalenedisulfonates (1,5-NSA; 1,7-NSA; 2,7-NSA). Additionally, all samples were analyzed for general hydrochemistry parameters to characterize the infiltration process more detailed.

22.2.1 Field site

The field site is operated with water from Lake Tegel which is influenced by discharge from the Schönerlinde wastewater treatment plant (Figures 22.1 and 22.2). Based on LC-OCD analysis, the DOC-composition of Lake Tegel is characterized by high values (~7.5 mg/L), of mostly natural origin: fulvic acids associated with the background groundwater (4–5 mg/L DOC), effluent organic matter with a contribution of 1–2 mg/L DOC and algal cellular products. The dominant soil type at the Lake Tegel bank filtration site is sand. In the direct infiltration zone near the lake, the hydraulic permeability of the clogged sand (k = 5*10-6 m/s) (Fritz *et al.*, 2002) is lower than in the deeper aquifer ($k = 2 - 8*10^{-4}$ m/s) (Pekdeger *et al.*, 2004). The sampling points for the bulk organic analysis at the field site were selected to reproduce the flow path of the infiltrating water to the production well. The individual retention times for each monitoring well were determined by the shift of the seasonal variation of the ratio of oxygen isotopes (18 O) in the surface water (Pekdeger *et al.*, 2004). At the bank filtration site, some dilution with deeper groundwater was observed and quantified with the help of boron data (Pekdeger *et al.*, 2004). Data for retention times and dilution with deeper groundwater are included in Figure 22.3. A detailed description of the field site and a distribution of the used monitoring wells are available at Pekdeger *et al.* (2004).

At the bank filtration site it was found that most of the surface water is infiltrating under oxic conditions. Because of mineralization of DOC and sedimentary bound particulate OC (~0.5% w/w) during infiltration, oxygen is used up quickly and most of the 4–5 month long infiltration (100 m) to the production well usually takes place under anoxic and anaerobic conditions (iron and manganese reduction). The extension and position of the redox zones varies seasonally and is horizontally stratified (Pekdeger *et al.*, 2004). The average nitrate concentration in Lake Tegel is 1.83 mg/L.

During the first sixteen months (05/02 – 09/03) of the observation period the nitrate (NO_3-N) level was significantly reduced during infiltration (Average NO_3-N in Well 3302 of 0.07 mg/L and in Well 3303 <Limit of Quantification, LOQ).

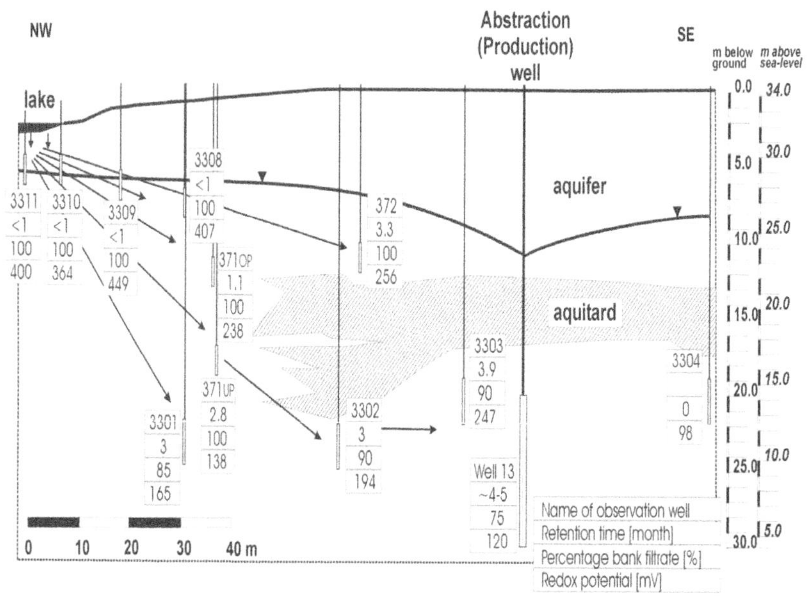

Figure 22.3. Cross section of bank filtration site at Lake Tegel.

The summer of 2003 was very dry in Berlin, and the combination of the low water level in Lake Tegel and the heavy pumping of the production well pumps led to an expanded zone of unsaturated infiltration under the lake. After passage of the biologically very active water-sediment interface, the bank filtrate was aerated again during a short unsaturated infiltration. Beginning in October 2003 until June 2004, elevated oxygen and nitrate concentrations were observed in the deeper monitoring wells and the production well (NO_3-N content in Well 3302 of 1.16 mg/L and in Well 3303 of 0.47 mg/L). For nine months of the observation period, the dominant redox conditions at the field site changed from anoxic/anaerobic with Fe/Mn-reduction to oxic. The differences in the fate of bulk and trace organics between the two redox conditions were investigated.

22.2.2 Column system

A long retention soil column system was used to simulate a one-dimensional aquifer. The study was conducted on a 30 m column system, which was operated with spiked water from Lake Tegel and had a retention time of 30 days (Figure 22.4). The columns were filled with natural fine sand from the Berlin area. Flow and hydraulic loading rate were selected to fit the conditions at the field site. The columns were sampled at different depths to investigate the degradation kinetics.

An online dosage system for five trace compounds was installed to assure a constant concentration of these five compounds in the influent. The spiking concentrations were selected to be approximately ten times higher than the concentration in Lake Tegel. Table 22.1 gives an overview of the dosages of trace compounds to the lake water.

Table 22.1. Spiking concentrations for the long retention columns in Marienfelde.

Trace Compound	Iopromide	Sulfamethoxazole	1,5-NSA	1,7-NSA	2,7-NSA
Spiking Concentration	10 µg/L	2.5 µg/L	2.5 µg/L	2.5 µg/L	2.5 µg/L

During the first part of the experiment oxic conditions were established in the soil columns. The oxic period lasted twelve months until May 2004. During this period oxygen was present in the whole column and no denitrification was observed. From May 2004 the redox status of the column system was switched to anoxic (NO_3-reducing) by sparging the influent with nitrogen (O_2 in influent: ~1mg/L). Denitrification was observed but nitrate was not completely depleted from the system. No oxygen was detectable in the sampling ports. Monitoring of the DOC, UVA and redox-conditions ensured stable conditions over the time. Results from both phases will be presented.

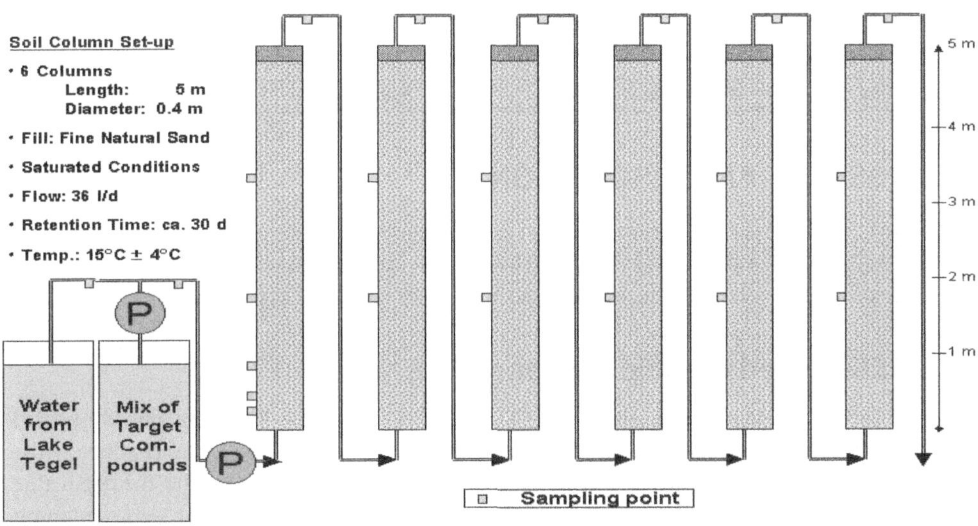

Figure 22.4. Experimental set-up of the long retention columns.

22.2.3 Analytics

At the field site the surface water, monitoring wells, and abstraction (production) well were sampled monthly over a period of 27 months for the bulk organic parameters and for 16 months for trace pollutants. The wells were purged for 30 minutes at pumping rates of 1–2 m3/h before sampling to ensure stable conditions (on line measurement of temperature, redox potential, dissolved O_2, pH, and conductivity). Afterwards the samples were stored in glass bottles at 4°C in the dark and if necessary filtered through 0.45 μm cellulose nitrate membrane filters. DOC was determined after 0.45 μm filtration with an ELEMENTAR HighTOC analyzer (Hanau, Germany). UV absorbance at 254 nm (UVA254) was measured using a PERKIN-ELMER Photometer Lambda 12 (Berlin, Germany). For the LC-OCD, a liquid chromatography method that quantitatively distinguishes between different fractions of DOC was used, a system from the DOC-Laboratory Dr. Huber (Karlsruhe, Germany), (Huber and Frimmel, 1996). Furthermore, the analytical program comprised differentiated AOX-analysis and trace organic analysis. The protocol for the differentiated analysis of the organohalogens consisted of coupled combustion and ion chromatography and is described in Oleksy-Frenzel *et al.* (2000).

The trace organic compounds Iopromide, Sulfamethoxazole and the isomers of naphthalenesulfonic acids were all extracted by different solid phase extractions (SPE) and measured with standard addition in high performance liquid chromatography with MS/MS- and FLD-detectors. Table 22.2 shows more details on the trace compound analysis.

Table 22.2. Details of trace organic compound analysis.

	Iopromide	Sulfamethoxazole	Naphthalenesulfonic acids
SPE-Cartridge	Lichrolute EN	Lichrolute EN	Lichrolute RP-18
Liquid chromatography	H_2O/MeOH with Gradient	H_2O/MeOH with Gradient	Ion pair chromatography
Detector	MS/MS (Micromass)	MS/MS (Micromass)	Fluorescence (Gynkotek)
Limit of detection	5 ng/L	5 ng/L	10 ng/L
Limit of quantification	20 ng/L	20 ng/L	30 ng/L
Method described in	Putschew *et al.*, 2001	Hartig, 2000	Storm *et al.*, 1999

22.3 RESULTS AND DISCUSSION

Results of 2¼ years of field monitoring at Lake Tegel and more than 2 years of column studies are summarized in this chapter. The results are presented with the focus on the influence of the redox conditions on the removal of bulk and trace organics.

22.3.1 Dissolved organic carbon (DOC)

The monthly DOC-monitoring (05/2002–08/2004) for the bank filtration site confirmed stable surface water concentrations in Lake Tegel of 7–7.5 mg/L. The results of the monitoring well sampling are plotted in Figure 22.5a as box plots with the median value (horizontal centre line), the 25%- and 75%-quartiles (box) and the minimum and maximum values (vertical line). The removal rates are calculated from the arithmetic means of all measurements. Under oxic and anoxic/anaerobic conditions the DOC is reduced by 2.8 and 2.6 mg/L from the surface water to monitoring Well 3303, respectively.

The degradation kinetic indicates a more rapid removal of DOC under oxic conditions and a slower but continuing removal under anoxic/anaerobic conditions. Under oxic conditions, 35% of the DOC is degraded during infiltration from the surface water to the first deep monitoring well (well 3301). During further infiltration, the fraction of degraded DOC increases by only 5% (3303) or 9% (Well 13, including some dilution). Under anoxic/anaerobic conditions 24% of the initial DOC is degraded in monitoring Well 3301. The continuing removal of DOC leads to reduction rates of 34% (Well 3303) and 42% in the production well (Figure 22.5). Comparing the concentrations of DOC during oxic and anoxic/anaerobic infiltration reveals that, under both conditions, 3.1 mg/L DOC were removed between the surface water and the production well.

The dilution in the production well is assumed to be constant (25% background groundwater) and the background monitoring Well 3304 shows low DOC-concentration during the observation period (oxic: 2.5 mg/L; anoxic/anaerobic: 2.9 mg/L). Under anoxic/anaerobic conditions the slower process of DOC-mineralization demands the entire retention time, whereas under oxic conditions the efficient removal during the initial infiltration is followed by a plateauing of the DOC levels and a slower removal. Under both conditions, a similar residual DOC concentration can be achieved if sufficient retention time is allowed under anoxic/anaerobic conditions.

The field results are consistent with findings from the soil column experiments (Figure 22.5b). The simulation experiments allowed a more precise investigation of the fate of DOC during the first month of infiltration depending on the dominant redox conditions. In the oxic soil columns, 47% of the DOC was removed. Under anoxic conditions, the initial oxic phase which invariably occurs in the field because of oxygen saturated surface water was reduced to the lowest practicable dimension (<0.21m). Only 31% of the source DOC was removed during 30 days of infiltration. The results support the field results that the DOC removal under oxic conditions is faster than under anoxic or anaerobic conditions. The removal rates are within the range of the rates observed in the field. A slightly higher removal can be explained by a higher average water temperature in the soil columns (15°C; field: 12°C). Because of their limited retention time, the column experiments could not demonstrate that the residual not-degradable DOC-concentration is similar under oxic and anoxic/anaerobic conditions.

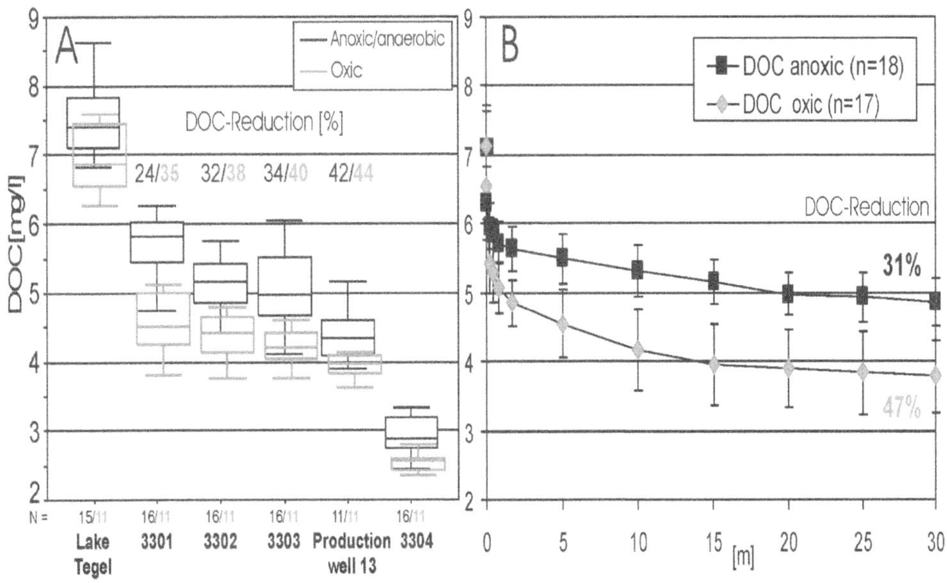

Figure 22.5. DOC for two observed redox conditions at (a) the bank filtration site and (b) in the column experiment.

22.3.2 LC-OCD

Figure 22.6 shows LC-OCD chromatograms of samples from the soil column experiment, and indicates that the character of the DOC partly changed after infiltration. The first peak corresponds to the polysaccharide (PS; elution time at 35–45 min) peak, the second corresponds to the humic substances (HS; elution time 52 min) and HS building blocks (secondary peak; 57 min), and the third to low molecular weight acids (LMA; 62 min).

LC-OCD measurements of samples from the bank filtration site (not shown) clearly indicated that the fraction of polysaccharides is completely removed very fast under both redox conditions, whereas other fractions (humics, humic hydrolysates) exhibited only partial removal. At no time were the polysaccharides detected in the first observation well (Well 3311). It is assumed that this fraction is mineralized instantly in the oxic infiltration zone. The results of the soil column experiments support the assumption of a rapid biodegradation for this DOC-fraction, because in an abiotic column only minor filtration effects were observed. Generally, LC-OCD chromatograms of the field site were comparable to the chromatograms of the oxic period of the long retention column system and showed the same trends (Figure 22.6a).

Both, in the field and in the columns the most fundamental change during infiltration was observed for the fraction of the polysaccharides. The largest molecular weight fraction, interpreted as polysaccharides, was efficiently removed in the first 0.21 m of oxic infiltration in the soil columns (Figure 22.6a).

Furthermore, the fractions of humic substances (HS), HS building blocks and LMA were only partially degraded. It is noteworthy that even a part of the HS peak was removed, generally assumed to be mostly nondegradable. Under anoxic conditions in the column system a slower degradation of the polysaccharide fraction (PS) was observed. PS were still detectable after 30 days of anoxic infiltration (Figure 22.6b). The fractions of HS, HS building blocks and LMA showed a similar partial removal as under oxic conditions. Because of the slower DOC degradation kinetics, the chromatograms are more pooled together in the anoxic period. However, except for the PS-fraction, no difference in the degradability of the single fractions was observed under oxic or anoxic/anaerobic conditions.

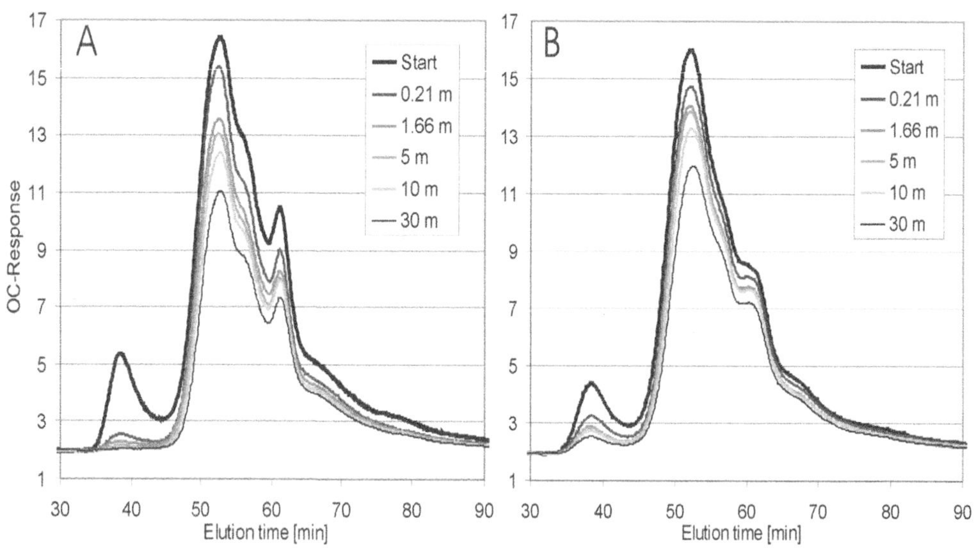

Figure 22.6. LC-OCD chromatogram of soil column experiment: (a) oxic and (b) anoxic.

22.3.3 Adsorbable organic iodine (AOI)

Adsorbable organic halogens (AOX) have been an issue for the aquatic environment for several years. Triiodinated X-ray agents, which are very stable and hydrophilic, contribute to large amounts (>50%) of the total AOI in Berlin's municipal wastewater and are introduced to the surface waters. Long-term AOI-monitoring has confirmed the seasonal changes of AOI-concentration in Lake Tegel due to variations in the dilution of sewage treatment plant effluents. Additionally, a general increase in AOI concentration in spring 2003 was observed, which was due to an extension of the Schönerlinde treatment plant that discharges into Lake Tegel.

Therefore, the AOI surface water concentrations were considerably higher during oxic conditions at the bank filtration site (oxic: 13.3 μ/l; anoxic: 9.6 μg/l) (Figure 22.7a). The comparison of the removal rates of the anoxic/anaerobic infiltration and the oxic infiltration at the same field site points towards a more efficient removal of AOI under anoxic/anaerobic conditions. During the anoxic/anaerobic part of the observation period, 60–70% of the source water AOI were removed during infiltration. The AOI of the background groundwater at the bank filtration site is very low (1 μg/L) and the mixture in the production well leads to an AOI-concentration in the extracted raw water of 1–2 μg/L. The time series of the AOI measurements (not included) shows the reaction of the monitoring wells to the change in redox conditions in Autumn 2003. The deep monitoring wells (Wells 3301, 3302, and 3303) reacted with an increase in AOI-concentration and significantly lower reduction rates. During the oxic period, only 40–50% of the surface water AOI was removed during infiltration. It was found that under oxic conditions the AOI level of the extracted raw water was slightly increased by 1.5 μg/L. The results give evidence that there is a more efficient reduction of AOI during infiltration under low redox potentials. These results are consistent with findings at other field sites in Berlin. A similar dependency of the AOI-concentration from the dominant redox conditions was observed at a bank filtration site at Lake Wannsee. AOI reduction rates of approximately 65% were observed during one month of anaerobic infiltration, whereas under oxic conditions only 30–40% of the AOI was removed.

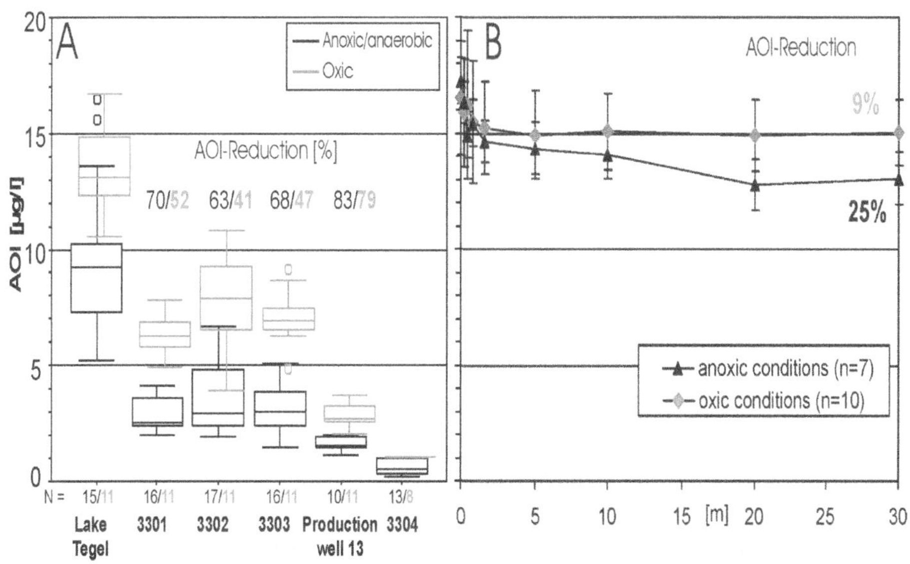

Figure 22.7 AOI for two observed redox conditions (a) at the bank filtration and (b) at the column experiment.

At the oxic artificial recharge facility in Berlin-Tegel the AOI-reduction rate was 30% (retention time: 50 days). This inverse correlation between AOI-removal and redox potential most probably originates from the initial step of AOI-mineralization: reductive dehalogenation. It is known that the degradation of halogen-substituted organics is more effective in soil passages with low redox potential (Mohn and Tiedje, 1992). The results from the field site at Berlin-Tegel evidently show that low redox potentials are essential for an efficient degradation of AOI. In the case of a permanent transition of the redox conditions at the Tegel bank filtration site to oxic, considerably higher AOI-concentrations can occur in the extracted raw water.

Figure 22.7b displays the fate of AOI in the long retention column system under two different enforced redox conditions. The high standard deviation of the measurements originates from the seasonal variations in the used Lake Tegel water. Under oxic conditions only 9% of the source water AOI is dehalogenated, but under anoxic conditions 25% is degraded. These results support the field studies indicating a better removal under anoxic/anaerobic conditions. A more efficient reduction of AOI in 30 days might be possible under strictly anaerobic conditions which were not tested. Under anoxic conditions, some iron and manganese reduction occurred towards the end of the column, but nitrate was still present and the redox potential remained high (180–240 mV).

22.3.4 Trace pollutants

The single compound analysis for the trace pollutants Iopromide, Sulfamethoxazole and different naphthalenesulfonic acids was started in spring 2003, providing data for the time period May 2003 to August 2004. The monitoring showed that these trace organic compounds which stand for different groups of persistent pollutants behave differently during infiltration. For some compounds, an influence of the changing redox conditions on the degradability in the field and in the columns could be detected.

Iopromide, the X-ray contrast agent, occurred in highest concentrations in the surface water (970 ng/L, n=16, σ=270 ng/L). Due to the fact that Iopromide, as a triiodinated benzene derivate, is part of the bulk parameter AOI (share in surface water ~5%), it was expected that the removal mechanisms for both parameters would be similar. This could not be confirmed. Contrary to the fate of AOI, the degradation of Iopromide showed no dependency on the dominant redox conditions. At the bank filtration site in Tegel, Iopromide was very efficiently degraded under both redox conditions (oxic and anoxic/anaerobic: 99% removal). This was consistent with findings at other field sites. In the soil column experiments 99% and 98% of the initial Iopromide concentration were removed under oxic and anoxic conditions respectively. The measurement of AOI parallel to the single compound analysis revealed that Iopromide is not mineralized but metabolized during infiltration. In the soil column experiment, the spiking of 10 μg/L Iopromide increased the AOI-concentration in the feed by ~4.81 μg/l (~28%). But only a removal of 1.5 μg/L (9%, Figure 22.7b) was observed under oxic conditions. This proves the assumption that the Iopromide molecule is rapidly metabolized, but not mineralized during infiltration. It still remains an iodinated organic molecule and is detectable as AOI. The chemical structure of the metabolite/s is unclear. Under anoxic conditions, 4.3 μg/L AOI (25%, Figure 22.7b) were removed during 30 days of infiltration in the column system. The share of Iopromide-originated AOI on this removal remains unclear.

Figure 22.8a presents data on the fate of Sulfamethoxazole during infiltration at the bank filtration site. The effluent concentration of the bacteriostatic in Berlin treatment plants varies between 370–1200 ng/L. Because of its high stability, it is also found in the surface waters at bank filtration sites. The degradation of Sulfamethoxazole seems to be redox-dependent as well. In Berlin-Tegel, Sulfamethoxazole shows a better removal under anoxic/anaerobic conditions where removal rates of 80–90% were observed. Under oxic conditions the removal decreased to rates of 50–70% and considerably higher concentrations were found in the production well.

The better removal of Sulfamethoxazole under anoxic/anaerobic conditions is consistent with findings of Schmidt *et al.* (2004) at other bank filtration sites in Germany. The results of the soil column experiment did not support these findings. Under anoxic conditions a removal of 65% Sulfamethoxazole was observed, whereas under oxic conditions 95% were removed (Figure 22.8b). The different behavior of Sulfamethoxazole in the column system might originate from higher spiking concentrations, higher average temperatures or a higher portion of cometabolic degradation. This effect is presently being investigated.

Naphthalenedisulfonates are well known polar contaminants of treated wastewater and surface water. The different isomers behave similarly in the environment but show different degradation properties. Consistent with findings of Stüber *et al.* (2002) (who reported the 1.5-NSA as very stable in wastewater treatment plants) the 1,5-NSA-concentration remained nearly constant during oxic and anoxic/anaerobic infiltration in the field and the column system. For the 1,7- and the 2,7-NSA isomers, the degradation was more efficient under oxic conditions. These results show that some naphthalenedisulfonic acids (1,5-NSA) are very poorly biodegradable. An elimination of these isomers can not be expected during bank filtration. Other biodegradable isomers (1,7- and 2,7-NSA) display a more efficient reduction during oxic infiltration.

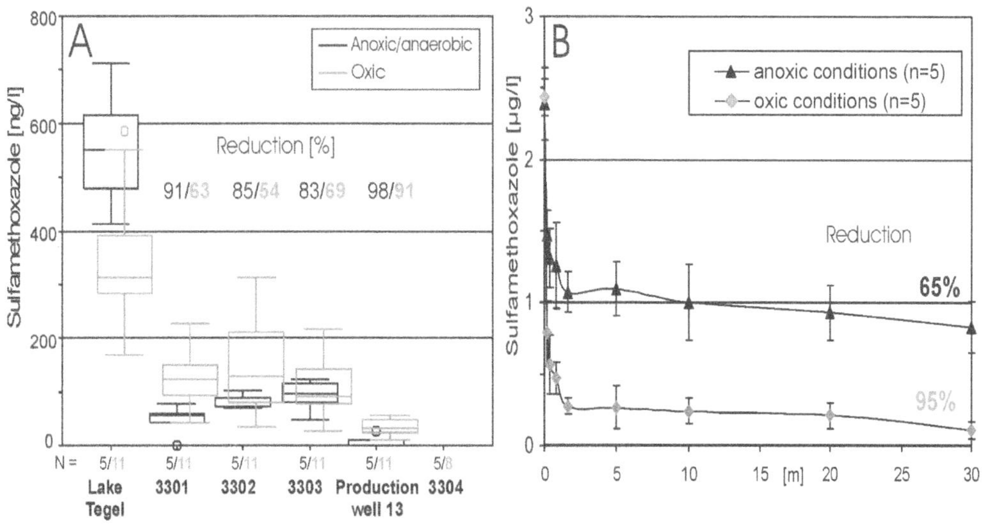

Figure 22.8 Sulfamethoxazole for two observed redox conditions (a) at bank filtration site and (b) column experiment.

22.4 CONCLUSIONS

The results of the field monitoring program give new insights into the mechanisms of bulk and trace organic removal. Aerobic and anoxic subsurface conditions during infiltration can lead to approximately the same residual DOC but the study indicates significant differences in the kinetics of DOC-removal depending on the redox conditions. Aerobic conditions result in a fast degradation of biodegradable DOC within a month, while anoxic conditions appear to require 3–6 months. LC-OCD analysis reveals that the change in character is comparable. Under oxic conditions the fraction of polysaccharides is removed more efficiently than under anoxic conditions. The fractions of humic substances, building blocks and low molecular weight acids are degraded partially. Significant differences were observed in AOI removal during aerobic and anoxic/anaerobic bank filtration. Dehalogenation of AOI occurred under reducing conditions, whereas during oxic infiltration no efficient dehalogenation was observed. Trace compound monitoring revealed different behavior by the examined trace compounds during infiltration. For some compounds (Sulfamethoxazole) an influence of redox conditions was observed. In addition to rapidly removed compounds (e.g. Iopromide) and very persistent compounds (1,5-NSA), the study determined a group of medium persistent compounds (Sulfamethoxazole, 1,7-NSA, 2,7-NSA) that require special conditions for optimal degradation.

The overall conclusion, out of all results of the NASRI-project (including findings by other researchers on complete removal of algae toxins, pathogenic bacteria and pathogens and most pharmaceutical residues) confirms the multi-barrier function of the underground passage for an indirect potable reuse system. The main requirements are long travel and residence times and appropriate distances, in the range of 5–6 months during anoxic conditions. It emerged that reducing conditions may be favorable for the degradation of halogenated trace pollutants and some pharmaceuticals, while aerobic conditions seem to speed up biodegradation of DOC (especially the polysaccharide fraction) and some other trace organics. The long-term use of bank filtration and recharge (for about 100 years) is based on sustainable biodegradation and reliable efficiencies.

When considering the application of indirect water reuse for potable supplementation, underground passages are a very feasible approach, including storage and treatment plus loss of identity of wastewater during its residence and in conjunction with the mixing of surface and other groundwater. A number of other field sites, such as in the South-West of the United States or in Israel, support these conclusions fully. The only real alternative to natural underground treatment would be a treatment with high-pressure membranes. But for these approaches the disposal problem of the concentrate needs still to be solved. If desalination is a major issue, membrane based treatments, like nanofiltration or reverse osmosis are indispensable, but in the case of Berlin and other indirect reuse systems, the salt concentrations are acceptable.

22.5 REFERENCES

Berliner Wasser Betriebe, (2003): http://www.bwb.de.

Fritz B., J. Sievers, S. Eichhorn and A. Pekdeger. Geochemical and hydraulic investigations of river sediments in a bank filtration system. In: Dillon, P. (ed.): Management of Aquifer Recharge for Sustainability, Balkema, Adelaide, Australia (2002) 95-100.

Grünheid, S., Schittko, S. and Jekel, M. Behaviour of bulk organics and trace pollutants during bank filtration and groundwater recharge of wastewater-impacted surface waters. Proceedings of the annual meeting of the Water Chemical Society, Bad Saarow, pp. 75-80 (2004).

Hartig C. (2000) Analytik, Vorkommen und Verhalten aromatischer Sulfonamide in der aquatischen Umwelt. (Analysis and behaviour of aromatic sulfonamides in the aquatic environment). Doctoral Dissertation accepted by: Technical University of Berlin, School of Process Sciences and Engineering, 2000-08-29, http://edocs.tu-berlin.de/diss/2000/hartig_claudia.htm.

Huber, S. and Frimmel, F. (1996): Gelchromatographie mit Kohlenstoffdetektion (LC-OCD): Ein rasches und aussagekräftiges Verfahren zur Charakterisierung hydrophiler organischer Wasserinhaltsstoffe, Vom Wasser, Vol. 86, pp.277-290.

Mohn W. W. and J. M. Tiedje (1992) Microbial reductive dehalogenation. Microbiol Rev. 1992; 56(3): 482–507.

NASRI, (2003): http://www.kompetenz-wasser.de.

Oleksy-Frenzel J., S. Wischnack and M. Jekel (2000) Application of ion-chromatography for the determination of the organic-group parameters AOCl, AOBr and AOI in water. Fresen. J. Anal. Chem., 366, 89-94.

Pekdeger A., G. Massmann, B. Ohm (2004) 2nd Periodic Report: Hydrogeological-hydrogeochemical processes during bank filtration and groundwater recharge using a multitracer tracer approach. NASRI-Projekt, 80 pages.

Putschew A., S. Schittko, M. Jekel (2001) Quantification of triiodinated benzene derivatives and X-ray contrast media in water samples by liquid chromatography-electrospray tandem mass spectrometry. J. Chromatogr. A, 930, 127-134.

Schmidt C.K., F.T. Lange; H.-J. Brauch. (2004) Assessing the impact of different redox conditions and residence times on the fate of organic micropollutants during riverbank filtration. Proceedings "4th International Conference on Pharmaceuticals and Endocrine Disrupting Chemicals in Water", 13.-15.10.2004, Minneapolis, Minnesota.

Storm T., T. Reemtsma, M. Jekel (1999) Use of volatile amines as ion-pairing agents for the high-performance liquid chromatographic-tandem mass spectrometric determination of aromatic sulfonates in industrial wastewater. J. Chromatogr. A, 854, 175-185.

Stüber M., T. Reemtsma, M. Jekel (2002) Determination of naphthalene sulfonates in tannery wastewater and their behaviour in a membrane bioreactor. Vom Wasser, 98, 133-144.

Ziegler D. (2001) Untersuchungen zur Nachhaltigkeit der Uferfiltration und künstlichen Grundwasseranreicherung in Berlin. (Investigations on the sustainability of bank filtration in Berlin's water cycle), Doctoral Dissertation accepted by: Technical University of Berlin, 2001-04-24. http://edocs.tu-berlin.de/diss/2001/ziegler_doerte.htm.

23

Unplanned reuse of wastewater for human consumption: The Tula Valley, Mexico

Blanca Jiménez

23.1 INTRODUCTION

In irrigation, water is frequently used with low efficiencies (< 50%), seldom realizing that the "lost" water often recharges aquifers that are being used for several purposes. This results in unplanned water reuse together with concerns that depend on the quality of the irrigating water. This situation is illustrated by the Tula Valley case study that shows how an "inefficient" use of wastewater turned out to be a successful, though unplanned, example of water reuse in a semi-arid area. Environmental and economic conditions were dramatically improved while a new drinking water source was provided. However, in order to maximize the advantages while reducing future risks, special management – described in this chapter – is required.

Payne was the first to report, in 1975, that 90-100% of the aquifer in Tula Valley was formed by Mexico City's wastewater. Later, in 1995, the British Geological Survey and the National Water Commission (BGS-CNA, 1998) quantified the phenomenon as at least 2 194 560 m^3/d (25.4 m^3/s). It was then realized that more than 400,000 people were using the infiltrated wastewater as a water supply. Several projects were launched to assess the potability of the aquifer water, to find proper potabilization methods and even to look for new water supply uses. This chapter describes the results of some of those projects.

©2008 IWA Publishing. *Water Reuse - An International Survey of current practice, issues and needs* by B. Jimenez *et al.* ISBN: 9781843390893. Published by IWA Publishing, London, UK.

23.2 DESCRIPTION OF THE PROBLEM

Mexico City has around 21 millions inhabitants and is formed by the Federal District (Mexico's capital) and 18 municipalities in the State of Mexico. Mexico City uses 6 652 800m^3/d (69 m^3/s) of water, of which 5 356 800 m^3/d (62 m^3/s) are supplied through the water mains, 604 800 m^3/d (7 m^3/s) are pumped on site from the local aquifer (for agricultural irrigation) and 691 200 m^3/d (8 m^3/s) come from treated wastewater. Reclaimed water is used for recreational purposes, lawn irrigation, industries (cooling mainly), car washing and to fill the Texcoco lake for recreational and environmental purposes.

Because Mexico City is located in a closed basin at 2,240 meters above sea-level (masl), three collectors were built to get rid of the wastewater and avoid floods caused by storm water: the Central Collector (conveying 55% of the wastewater); the El Gran Canal (30%) and the Western Interceptor (the remaining 15%). These collectors discharge their contents to the Tula, El Salado and El Salto rivers respectively, all located in the Tula Valley. In total, 518,400 m^3/d (60 m^3/s) are being disposed of, 75% of which is combined industrial and municipal wastewater and 25% excess rain water. In the Tula Valley 4492,800 m^3/d (52 m^3/s) of wastewater are used to irrigate and the rest (691 200 m^3/d or 8 m^3/s) is sent out of the Valley through the Tula River (Figure 23.1).

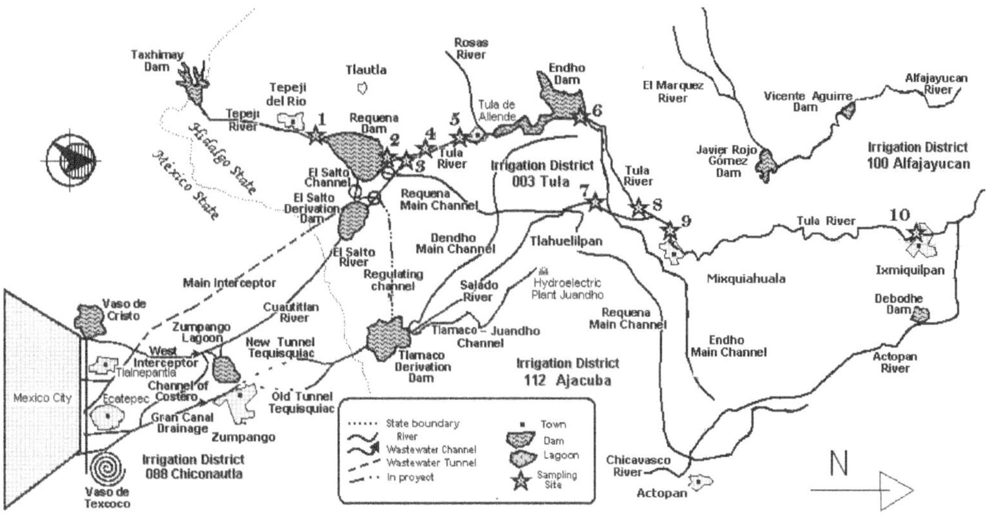

Figure 23.1 Sewerage conduits and disposal sites for Mexico City wastewater.

23.3 STUDY AREA

The Tula Valley (also colloquially known as the Mezquital Valley) is located north of Mexico City with an area of around 4 100 km^2. In the south of the valley (near Mexico City) the altitude is 2 030 masl, while to the north it is 1 990 masl. The Tula Valley is semi-arid to arid, with a mean annual temperature of 17°C, precipitation (from June to September) of 450mm and evapotranspiration of 1750mm. More than 400 000 people live in the valley, 43% in three cities (Tezontepec de Aldama, Actopan and Tula) and the rest in 294 localities. The economy is mainly based on agriculture and commerce. Industrial activity is limited to 5 industrial plants (one power generation plant, one refinery, two cement plants and one pheno-chemistry plant). As mentioned above, irrigated agriculture in the valley is based on wastewater. At the beginning of the 1990s a maximum wastewater irrigated area of 90,000 ha was registered but in 2004 the area had seemed to have decreased to 76,119 ha owned by 73,632 farmers (CNA, 2004).

Figure 23.2 Flood irrigation in the Tula Valley.

Farmers resort to furrow and flood irrigation (Figure 23.2) and thanks to wastewater, Tula soils (originally poor in organic matter and nutrients) receive 56 kg P/ha.yr, 1,200 kg N/ha.yr and 5,200 kg of organic matter measured as BOD/ha.yr, thus increasing productivity by 150% for maize, 100% for oats, 94% for tomatoes, 71% for alfalfa and chilli, and 67% for wheat. The reliability of the wastewater means that 2-3 crops per year can be grown instead of 1. For this reason, land with access to wastewater is rented at 455 USD/ha.yr instead of 183 USD/ha.yr in areas using just rain water (Jimenez, 2005). The main crops are alfalfa and maize (60% of the total) followed by oats, barley, wheat and beans. Small quantities of vegetables, such as chilli pepper, Italian squash and tomatoes are also produced, mainly for local consumption (Siebe, 1994). Health effects caused by the wastewater used to irrigate have been documented in several papers (Blumenthal *et al.*, 1991; Siebe and Cifuentes, 1995; Cifuentes, 1998 and Blumenthal *et al.*, 2001) and refer to a considerable increase in helminthiases diseases, especially in children under 15 years.

23.3.1 Development of the irrigation area

Wastewater was sent to the Tula Valley for the first time in 1789 and it first began being used for irrigation in 1896 in a small region near Tlaxcoapan and Tlalhuelilpan and expanded towards Mixquiahuala. At that time, wastewater was conveyed by the El Salado River. The wastewater's economic impact soon became evident and the government decided to officially acknowledge it in 1920. The implementation of a complex irrigation system ensued. The Tequisquiac tunnel was built to convey most of the wastewater through the El Gran Canal to the El Salado River. Nowadays the irrigation system encompasses 9 dams (3 with "first-use" water from perennial rivers and irrigation drainage water, and 6 storing wastewater), 575 km of primary channels, 283 km of secondary ones and thousands of kilometres of interconnections into parcels. One of the wastewater dams is Endhó, with a capacity of 50 Mm^3. Sixty two percent of the primary channels and almost all the secondary ones (Figure 23.3) are unlined, allowing the infiltration of 25% of the conveyed wastewater. Using field data, the BGS (1998) estimated that the Requena and the Endho main channels have an infiltration rate in the unlined sections of 0.4 and 1.4 m/d, respectively, while in the lined parts it is only 0.1 m/d. Wastewater infiltration from channels is equivalent to 8 219m^3/km.d.

(a) Primary Requena canal near Doxey, lined (b) Secondary canal near Tlaxcoapan, unlined

Figure 23.3 Irrigation channels in The Tula Valley.

Field irrigation also produces aquifer recharge because most of the crops have a water demand of less than 1 $m^3/m^2 \cdot yr$, while the applied rate to wash soil salinity is 1.5-2.5 $m^3/m^2 \cdot yr$. It is estimated that both phenomena, wastewater transport and excess irrigation rates, recharge the aquifer to at least 2 194 560 m^3/d or 25.4 m^3/s (Figure 23.4).

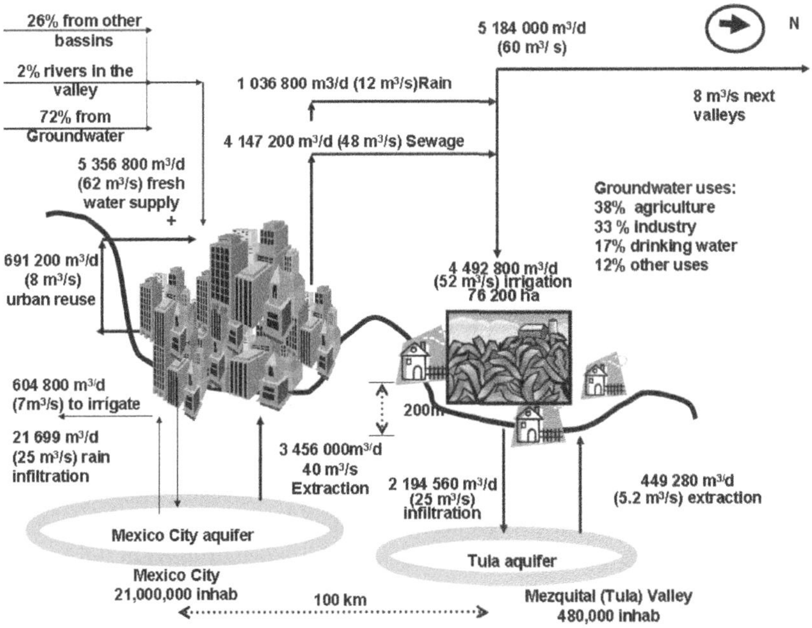

Figure 23.4 Water Balance in the Mexico and Tula Valleys.

23.3.2 Groundwater

The Mezquital valley is part of the Mexican plateau. It is surrounded by igneous extrusive rocks mountains at the south east and by limestone marine sediments and intensely folded limolites at the northwest. The central part of the valley is formed by erosion deposits that originate from the surrounding mountains and volcanic layers. Soil layers are randomly arranged with an undefined stratigraphical order, due to the non homogenous erosion and deposit periods as well as to important orogenic and tectonic land movements. For this reason, hydrogeology is very complex. In general, and according to the BGS-CNA 1998, there are three aquifers (a) Superior (b) Tarango and (c) El Doctor. The Superior aquifer is shallow, unconfined and of a variable depth. It is located irregularly in alluvial deposits. It is permeable and recharged with rainwater, wastewater and lateral groundwater flow. The Tarango aquifer (or Inferior aquifer) is the most important due to its storage and abstraction. It is recharged mainly with wastewater and some rainwater, stored in its deepest parts. The Tarango aquifer is located in volcanic ash sediments with variable granulometry and basaltic layers. Its permeability varies from 0.1 to 50 x 10^{-2} m^2/s and has a mean value of 0.015 m^2/s. In some parts the aquifer is semi-confined with groundwater depth of only 0.3 m, while in others the depth is 268 m. Several new springs have been formed from this aquifer. It is believed that the Tarango formation is interconnected with the Superior Aquifer. Finally, the third aquifer, El Doctor, is also recharged with wastewater and a significant volume is abstracted from it. Due to the complexity of the geological structure, in the Tula Valley there are also thermal springs in the lower (mainly) and upper regions. The groundwater flow is, in general, from south to north in the Tula area and from south to east in the Actopan river basin. Due to wastewater recharge for more than one hundred years, the aquifer level rose 15–30 m from 1938 to 1990 and dozens of new springs have appeared with flows varying from 8 640 to 51 840 m^3/d (0.1 to 0.6m^3/s). As mentioned, recharge was estimated to be at least 2 194 560 m^3/d (25.4 m^3/s), a value 13.4 times the natural recharge value (BGS-CNA, 1998).

23.3.3 Hydrology

The biggest rivers in the area are the Actopan, Tula and Salado (Figure 23.5). The Tula River is the main one and is known in its upper part as Tepeji River. It discharges to the Requena Dam, which also receives wastewater from the Central Collector through the El Salto River. Depending on the season of the year, water from the Tula River is either used to irrigate or is partially stored at the Endhó Dam, located west of Tula City. After Tezontepec de Aldama City, the Tula River merges with the El Salado River which contains, almost exclusively, wastewater collected by El Gran Canal. In the north and near Ixmiquilpan, the Tula River has the Actopan River as an influent. At the end of the valley, the Tula River joins the San Juan and the Hondo river taking the name of Moctezuma River which is one of the main tributaries of the Panuco River that discharges into the Gulf of Mexico.

All along their course, the Tula, Actopan and Salado rivers receive wastewater discharges from the 297 localities of the Tula Valley (sewer coverage is only 30% and wastewater treatment capacity almost nil) as well as water from the irrigation drainage channels (which is in fact wastewater used to irrigate and thus treated through the soil) as well as water from the newly formed springs. Downstream, irrigation drainage water is mixed with river water; the local population perceives the mixed water as first use or clean water.

In the past, the rivers did not receive water from aquifers due to the arid conditions. Nowadays, however, the Tula and the Actopan rivers receive 164 160 m^3/d (1.9 m^3/s) from groundwater plus another large volume from drainage channels. For all these reasons the base flow of the Tula River increased from 138 240 m^3/d (1.6 m^3/s) in 1945 to 1 097 280 m^3/d (12.7 m^3/s) in 1995. The spatial variation of the water quality in the Tula River based on all the aforementioned interactions is shown in Table 23.1.

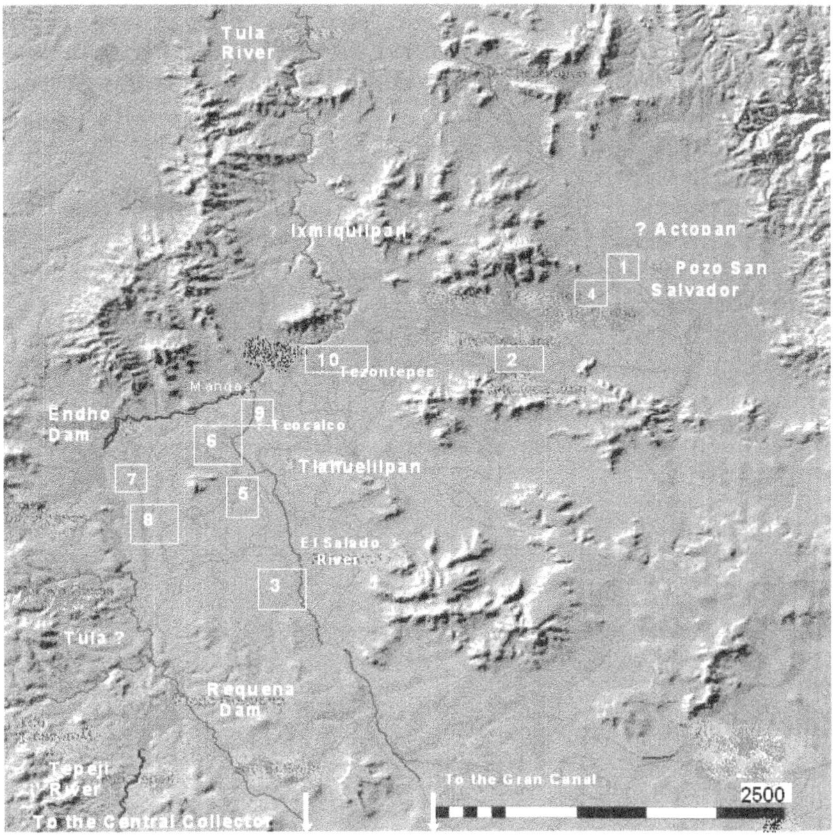

Figure 23.5 Wells and springs monitored at the Tula Valley.

Table 23.1 Water quality of the Tula River and some of its tributaries, with information from BGS-CNA, 1995.

Parameter mg/L unless indicated	Tepeji River	Central Collector	Tula River	Endhó Dam	Salado River	Gran Canal	Tula River Near Cfe	Tula River Near Teoacalco
pH[1]	7.0	7.0	7.0	7.1	7.6	7.7	7.4	7.5
Conductivity[2]	431	1114	995	1136	1590		1673	4092
O2	2.5	0.6	1.0	2.2	1.0		1.8	4.3
BOD	53	131	77	57	78	460	67	7
COD	186	315	256	204	295	210	192	208
Faecal coliforms[3]	$28x10^5$	$6.5x10^7$	$6x10^7$	$7.8x10^4$	$5.2x10^6$		$6.4x10^6$	$2.01x10^4$
ABS	3.4	6.7	5.2	4.2	2.6	13	1.4	0.5
Boron	1.1	1.6	1.3	0.9	1.0	25	1.9	1.8
Organic nitrogen	6.1	14.7	4.6	2.4	3.8		5.9	5.0
$N\text{-}NH_3$	2.9	8.5	8.2	10.9	18.5	23	3.4	25.9
$N\text{-}NO_2$	3.4	0.0	0.0	0.0	0.1		0.8	0.8
$N\text{-}NO_3$	0.4	0.2	0.3	0.1	0.3		12.4	8.0

[1] no units, [2] μS/cm, [3] MPN/100mL

23.4 DRINKING WATER QUALITY

Before it was realized that water used in the region was infiltrated wastewater, a change in groundwater quality was noticed in 1938. First, it was thought to be a pollution problem, and even though its chemical composition (salts mainly) was evidently very similar to that of wastewater, in 1985 the groundwater was still considered to be of "good quality". Only more recently, in 1995, was it realized that the difference was not due to a pollution problem but to the replacement of the original groundwater by infiltrated wastewater. At the present time there are 283 groundwater wells or springs in the area producing 449,280 m^3/d (5.2 m^3/s) of water for domestic (17%), agricultural (25%), industrial (33%) and other purposes (25%). Since 1995, several studies have been performed to asses the water quality. The first covered the whole valley and was performed by the BGS-CAN. Hundreds of compounds were analyzed. Results are shown in Figure 23.6. No major problems besides faecal pollution (controlled by disinfection) and high nitrate contents were found.

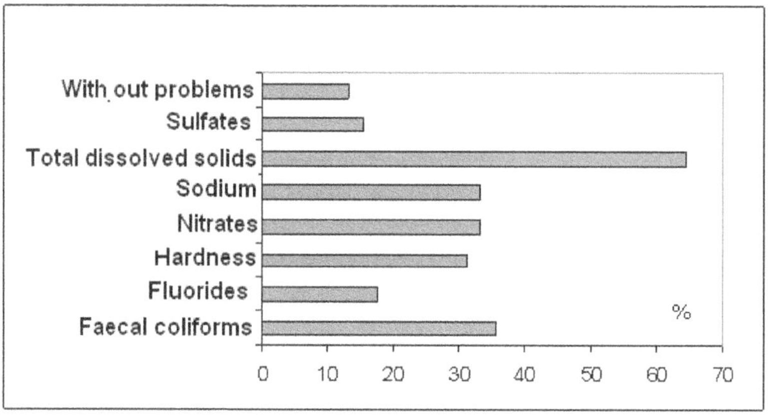

Figure 23.6 Percentage of sites with a problem in the Tula Valley by BGS-CNA, 1998.

Later, Jimenez *et al.* (1997) performed a study limited to the irrigation district 03 Tula (the oldest district in the Tula Valley – irrigated with wastewater and where the aquifer has been recharged the most). The study area was 45,215 ha. It has several springs and artesian wells (mainly near Chilcualtla, Tezontepec and San Salvador) but also some wells (near Ajacuba and Actopan) with a water level up to 90 m deep. In total, 128 abstraction sites produced 164,160 m^3/d (1.9 m^3/s) of water from springs (74%), wells (23%) and boreholes (3%). Seventeen sites were located inside the irrigated area and 50 close to the wastewater distribution channels, the rest were distributed in the irrigation fields. Prior to supply, 96% of the water was disinfected with chlorine. The study was performed in three phases.

First Phase. In 36 sites, three sampling campaigns were performed, one in the dry season and two during the rainy one, measuring the 45 parameters considered in the Mexican drinking water standard. Results showed that 37% of the supplies did not fulfil the standard due to their chloride, iron, nitrate, sulfate and/or faecal coliform content (Table 23.2). All other parameters complied with the regulation. These results were similar to those found by BGS-CNA in 1998.

Table 23.2 Number of sites that did not meet the Mexican drinking water standard (Jimenez *et al.*, 2003).

Parameter	Sites not complying with drinking water standards		Mexican drinking standard
	% Number	% Volume	
Dissolved solids	64	95	1000, mg/L
Sodium	38	73	200, mg/L
Faecal coliforms	42	25	0, MPN/100 mL
Nitrates	31	12	10, mgN/L
Chlorides	24	10	250, mg/L
Hardness	31	9	500, mg$CaCO_3$/L
Sulfates	18	2	400, mgSO_4/L
Fluorides	18	1	1.5, mgF/L
Without problem	13	5	

Second phase. To perform a more detailed analysis, 10 sites were selected: 6 supplying a large part of the population and located near the irrigation channels or inside the irrigated area, 2 with the influence of wastewater and 2 considered as future "clean" water supplies (Table 23.3). The selected sites supplied 39% of the population. The parameters analyzed were the pesticides used in the region for maize and alfalfa (atrazine, 2-4D, parathion methyl and permethrine), helminth ova and enteric viruses. Results showed absence of these compounds but chromatograms performed during analysis showed small peaks of non identified organic compounds.

Table 23.3 Selected sites for detailed analysis within the Irrigation District 03 Using information from: (Jimenez *et al.*, 1999).

Site	Municipality	Situation with respect to the irrigation channel or irrigation area	Flow, m^3/d	Served population
1. Bothibaji No.1;	Actopan	50 m from a wastewater distribution channel	2592	30000
2. Pozo Grande, la Noria	Actopan	200 m from a wastewater distribution channel	8640	4,000
3. El Capulín	Atotonilco	Out of the irrigated area	950	583
4. San Salvador	San Salvador	Out of the irrigated area. The site is flooded.	1728	4,000
5. Cerro Colorado	Tezontepec	1.5 km from a wastewater irrigation channel and an irrigated area. The site is flooded.	28512	72413
6. Puedhe	Tezontepec	Infiltration gallery 500 m from the Tula river. Water is used to irrigate vegetables	2333	8645
7. San Fco. Bojay	Tula de Allende	50 m from a wastewater irrigation canal	5011.2	6,000
8. El Llano 2a Sec.	Tula de Allende	20 m from the unlined Requena channel and 5 m from a wastewater channel	3024	5946
9. La Cueva 500	Tezontepec	At the bottom part of a hill	ND	-
10. El Géiser	Tezontepec	In a gorge, 500 m from the Tula River.	ND	-

ND: No data available.

Table 23.4 Monitored compounds to assess the feasibility of using the Tula Valley aquifer as a water supply for Mexico City.

Organoleptic	**Microbiologic**	**Aromatic**
▪ Colour	▪ Total Coliform	▪ Benzene
▪ Taste	▪ Faecal coliform	▪ Ethyl benzene
Nutrients	▪ Faecal Streptococci	▪ Toluene
▪ NTK	▪ *E. histolytica*	▪ m – Xylene
▪ Nitrates	▪ Helminth Ova	▪ o – Xylene
▪ Nitrites	▪ *Salmonella spp.*	▪ p – Xylene
▪ Ammonia nitrogen	▪ *Shigella*	▪ pyrene
▪ Phosphates	**Organic matter**	**Polynuclear aromatic**
Physical	▪ COT	▪ Naphthalene
▪ Turbidity	▪ COT total	▪ Fluorene
▪ Conductivity	▪ COT soluble	▪ Criseno
▪ Total dissolved solids	▪ BOD total	▪ Pyrene
▪ Total suspended solids	▪ BOD soluble	▪ Anthracene
▪ Total solids	▪ MBAS	▪ Benzo (a) Anthracene
▪ Redox potential	**Pesticides**	▪ Benzo (k) fluorantene
▪ Temperature	▪ Aldrine	▪ Benzo (b) fluorantene
▪ pH	▪ Chlordane	▪ Benzo(g,h,i) pyrilene
Non metals and other	▪ Chlordane A	▪ Dibenz(a,h) Anthracene
compounds	▪ Chlordane G	▪ Indene (1,2,3-cd) pyrene
▪ Bore	▪ Lindane	▪ 2 – metilnaphthalene
▪ Dissolved oxygen	▪ Heptachlor	▪ Acenaphtylene
▪ Selenium	▪ Heptachlor epoxy	▪ Acenaphthene
▪ Carbon dioxide	▪ Metoxychlore	▪ 2- chloronaphthalene
▪ Chlorides	**Aromatic halides**	**Halogenated ethers**
▪ Total Hardness	▪ 1,2 dichlorobenzene	▪ 4 chlorophenyl phenyl ether
▪ Cyanides	▪ 1,3 dichlorobenzene	▪ Ether bis 2-chloroetthyl
▪ Fluorides	▪ 1,4 dichlorobenzene	▪ 4 bromo phenyl phenyl ether
▪ Sulphates	▪ 1,2,4 dichlorobenzene	▪ Ether bis (2-chloroisopropyl)
▪ Sulphides	▪ 1,2 dichloropropane	**Phenols**
▪ Total Alkalinity	▪ Hexachlorobenzene	▪ Phenol
Fenoftaleina alkalinity	▪ 1,3 dichloropropane	▪ 2,4 dimethyl phenol
▪ Carbonates	▪ 2,2 dichloropropane	▪ Pentachlorophenol
▪ Bicarbonates	▪ Methylene chloride	▪ m-cresol
▪ Hydroxides	**Organic nitrogen compounds**	▪ 2-chlorophenol
Metals	▪ N – nitrosodimethylamine	▪ 2,4 dichlorophenol
▪ Aluminium	▪ Nitrobenzene	▪ 2,4,5 triclorophenol
▪ Arsenic	▪ 2,4 dinitrotoluene	▪ 2,4,6 triclorophenol
▪ Barium	▪ 2,6 dinitrotoluene	▪ p-cresol
▪ Cadmium	▪ Bencidine	**Phtalates**
▪ Calcium	▪ 1,2 diphenilhydrazine	▪ Bis 2 ethylhexyl-phtalate
▪ Cobalt	▪ N – nitrosodiphenilamine	▪ Di-n-octyl phtalate
▪ Cupper	▪ 2- nitrophenol	▪ di-n-butyl phtalate
▪ Chrome (Total)	▪ 4 nitrophenol	▪ Dimetil phtalate
▪ Iron	▪ 2,4 dinitrophenol	**Other organic compounds**
▪ Manganese	▪ 3,3 dichlorobencidine	▪ Fluorantene
▪ Mercury	**Aliphatic halogenated**	▪ Isoforone
▪ Nickel	▪ Chloroform	▪ BHC a
▪ Potassium	▪ Bromoform	▪ BCH b
▪ Silver	▪ Dichlorobromometane	▪ BCH d
▪ Lead	▪ Diclorodibromometane	▪ BCH g (Lindane)
▪ Zinc	▪ Hexachlorobutadiene	▪ Toxaphene
▪ Sodium	▪ Hexaclorociclopentadieno	**Radioactivity**
	▪ 1,1 dichloro ethylene	▪ Alfa radioactivity
	▪ Trans - 1,2 dichloro ethylene	▪ Beta radioactivity
	▪ Triclorofluro methane	**Toxicity test**
	▪ Tetrachloro ethylene	▪ Microtox
	▪ Trichloroethylene	
	▪ Vinyl chloride	
	▪ Tetrachloride carbon	

Third phase. Based on the previous results, it was decided to include more sites (22 in total to cover 88% of the supplied population) and measure first the organic matter content (as TOC and COD). Subsequently, in those sites with high organic contents (6 sites with vertisol or with reduced compounds such as iron, manganese or nitrites), semi-volatile organic compounds were measured using the EPA SW-8270 method. No semi-volatile organic compound was detected above the detection limit (> 5ppb). Samples were also tested for toxicity using the Microtox test, with negative results.

23.5 WATER QUALITY IN THE TULA VALLEY AQUIFER

Given that no urgent or evident problems were found and that the area had plenty of water of an apparently good quality, a study was performed (Jimenez *et al.*, 1999) to determine the feasibility of using the new aquifer formed as a possible future water source for Mexico City. In this third phase, 276 parameters (Table 23.4) were analyzed simultaneously by 5 laboratories (4 nationally certified, 1 also certified by the USEPA). Parameters to be analyzed were defined based on international criteria for drinking water and water reuse, the national standard for drinking water and the results of previous studies. Three sites, from the Tarango aquifer were selected (Teoacalco, Tezontepec and El Salvador, Figure 23.7a). Teocalco is a 11 232 m^3/d (130 L/s) well used to supply water to an oil refinery plant. Tezontepec is an infiltration gallery of 51 840 m^3/d (600 L/s) supplying water to 8645 people in three cities and also for recreational purposes (Figure 23.7b). San Salvador is a 3024 m^3/d (35 L/s) well that supplies 4000 people. Table 23.5 shows the results and compares them with the Mexican drinking water standard, WHO criteria (WHO, 2004), the European Union (Council Directive, 1998) and the USEPA guidelines (USEPA, 1992).

(a) (b)

Figure 23.7 Wastewater and water in excess from the aquifer: (a) The Central Collector with wastewater; and (b) recreational use of the Tezontepec water.

Table 23.5 Parameters above the national or the international criteria for drinking water.

Parameter	Criteria or Standard	Teocalco,	Tezontepec	San Salvador
Total Coliforms MPN/100 mL	MEXST = 2 USEPA(1)WHO & EU = 0	16± 21	27± 27	228± 222
Faecal Coliforms, MPN/100 mL	MEXST, USEPA, WHO, EU(2) =0	1.3±1.3	4.2±9.0	88±88
Boron, mg/L	WHO = 0.5 EU= 1	0.55 ±0.14	0.63±0.1	0.2+0.2
Lead, mg/L	MEXST = 0.025 USEPA = 0.015 WHO & EU = 0.01	0.02±0.02	0.02±0.03	0.014±0.014
Mercury, mg/L	MEXST, WHO & EU = 0.001 USEPA= 0.002	0.001±0.0009	0.001±0.002	0.0004±0.0004
Chlorides, mgCl⁻ /L(3)	MEXST, WHO = 250	149±16	179±17	264±63
Sodium,mg/L (3)	MEXST, WHO & EU = 200	176±98	168± 73	215± 105
Total hardness, mg $CaCO_3$/L(3)	MEXST = 500	324±37	452±15	492±19
Nitrates mgN-NO_3/L	NOM-127 & USEPA = 10 WHO & EU = 11(4)	24±24	17±17	19±18
Ammonia, mgN-NH_4/L	MEXST = 0.5	0.7±0.7	0.08±0.06	0.07±0.07
Sulfates, mg $SO_4^{=}$/L	MEXST = 400, WHO = 500	109±8.0	130±55	147±78
MBAS, mg/L	MEXST = 0.5	0.2±0.2	0.14±0.3	0.2±0.7
Total dissolved solids,	MEXST = 1000 EU = 1500	945±100	1038±188	1179±125

MEXST: Mexican drinking water Standard or NOM 127 SSA1USEPA: Environmental Protection Agency of USA
WHO: World Health Organization EU: European Union
1) 5% positive for samples during one month (2) as *E. Coli*
(3) Criteria or standard due to esthetical or operational nuisances 4)11 mgN-NO_3/L = 50 NO_3

23.5.1 Comparison with wastewater

To really appreciate the aquifer water quality, it is useful to compare it with: (a) the original quality of the wastewater, (b) the quality of the water supplied in Mexico City, considered as one of the best in the country, and (c) the water quality produced by biological secondary treatment. Table 23.6 compares the mean characteristics of Mexico City's wastewater with that of Teocalco, Tezontepec and San Salvador sites in the Tula Valley. Table 23.6 also shows the difference between values in the wastewater and groundwater expressed as a percentage. It can be seen that in general, during the passage of wastewater through soils, organic matter, metals and nutrients are considerably reduced while ion salts (like Ca, Mg and Sulfates) are considerably increased. The soils of Tula were found to be acting as an unplanned Soil Aquifer Treatment or SAT (Bower, 1989).

At this stage aromatics, chlorinated benzenes and nonylphenols were considered the more conspicuous compounds found, both in wastewater and groundwater samples, as some of these classes have been reported as endocrine disrupters in the literature. Quantitative estimates for these are shown in Table 23.7, and even though their concentration in groundwater is low and in most cases below the detection limits, they were nevertheless present. The logical question then is: how will the groundwater quality evolve with time?

Table 23.6 Characterization of Mexico City's wastewater and percentage (%) difference with water in three sites of the Tarango aquifer.

Parameter in mg/L unless indicated	WHO criteria	Mexico City Wastewater	Teocalco		Tezontepecc		San Salvador	
		Value range	Mean value	D,%	Mean value	D, %	Mean Value	D, %
Fecal Coliforms, MPN/100 mL	0	10^{04}-10^{8}	2	99.9	4	99.9	88	99.9
Salmonella (3 varieties), CFU/mL	0	ND-positive	ND	100	ND	100	ND	100
E. histolytica, cysts/L	0	0-1.5	ND	100	ND	100	ND	100
Shigella, CFU/mL	0	0 - positive	ND	100	ND	100	ND	100
Helminth ova, ova/L	0	12-90	0	100	0	100	0	100
Turbidity, NTU	0.1[(1)]	100 - 249	1	99	1	99	1.5	99
Total suspended solids	-	83-153	3.8	97	3	97	4.1	97
Total dissolved solids	1000[(1)]	758-860	945	**-11**	1038	**-22**	1179	**-39**
Conductivity, μmhos/cm	-	1437-1689	1577	**-1**	1698	**-9**	1918	**-23**
BOD	-	200-451	4	99	8	98	5	98
Total organic carbon	-	35-188	18	84	28	75	11	90
Total COD	-	450-496	11	98	23	95	14	97
Soluble COD	-	274	9	97	4	99	15	95
MBAS	-	5.9-6.2	0.2	97	0.15	98	0.2	97
Aluminium	-	1.3-5.5	0.08	98	0.09	97	0.07	98
Arsenic	0.01	ND-0.008	0.002	50	0.004	0	0.002	50
Boron	0.5[(2)]	1-1.2	0.5	55	0.6	45	0.29	74
Cadmium	0.003	0.0030	0.00015	95	0.00015	95	0.0015	50
Calcium	-	41-44	80	**-88**	73	**-72**	110	**-159**
Chromium	0.05	0.042	0.004	90	0.004	90	0.004	90
Copper	2	0.05-0.07	0.015	75	0.02	67	0.01	83
Iron	0.3[(1)]	1-1.2	0.04	96	0.15	86	0.09	92
Lead	0.01	0.09-0.1	0.02	98	0.02	98	0.01	99
Magnesium	-	24-29	30	**-15**	64	**-146**	46	**-77**
Manganese	0.4	0.03-0.2	0.004	97	0.01	91	0.004	97
Mercury	0.001	0.001	0.0006	40	0.001	0	0.0004	60
Nickel	0.02	0.08-0.15	0.015	87	0.01	91	0.01	91
Potassium	-	25-41	35	**-6**	31	6	22	33
Sodium	-	198-206	175	13	167	17	215	**-6**
Chlorides	250[(1)]	155-248	149	26	179	11	264	**-31**
Cyanides	0.07	0.005-0.01	0.008	**-7**	0.006	20	0.008	**-7**
Fluoride	1.5	0.7-4	0.5	79	1.1	53	0.3	87
Sulfates	-	21-51	109	**-203**	130	**-261**	147	**-308**

Parameter in mg/L unless indicated	WHO criteria	Mexico City Wastewater	Teocalco		Tezontepecc		San Salvador	
		Value range	Mean value	D,%	Mean value	D, %	Mean Value	D, %
Sulfides	-	3-3.5	1.2	63	1.0	69	1.7	48
Total Kjeldahl Nitrogen	-	37-38	1.5	96	1.6	96	1.3	96
Ammonia nitrogen	-	24-32	0.7	98	0.08	100	0.07	100
Nitrates	11	ND -1	24	**-47900**	18	**-35900**	19	**-37900**
Nitrites	0.9	ND -0.001	0.008	**-1500**	0.006	**-1100**	0.01	**-1900**
Phosphorus	-	2.7-3	0.1	96	0.2	93	0.2	93
Bicarbonates, mg /L	-	485	585	**-22**	642	**-34**	572	**-19**
Total Hardness, mg /L	-	210-220	324	**-51**	452	**-110**	492	**-129**
Ethylbenzene	0.3	1.2	ND	100	ND	100	ND	100
p. cresol (methyl phenol), µg/L	-	46.5	ND	100	ND	100	ND	100
Chloroform, µg/L	0.2	0.2-0.8	ND	100	ND	100	ND	100
Tetrachloroethylene µg/L		2	ND	100	ND	100	ND	100
o – Xylene, µg/L	0.5	3.8-4	ND	100	ND	100	ND	100
m – Xylene, µg/L	0.5	9.2	ND	100	ND	100	ND	100

(1) Value set on the basis of acceptability aspects

(2) Provisional guideline value because calculated guideline value is below the level that can be achieved through practical treatment methods, source protection, etc,

ND: Not detected

Table 23.7 Selected organic compounds in wastewater and groundwater at the Tula Aquifer (µg/L), with data from Capella, 1997.

Compound	Central Collector	Teocalco	Tezontepec	San Salvador
Methyl (1-methyl-ethyl)-benzene	5-10	< 5	< 5	< 5
1,1-oxy-bis-benzene	10-50	< 5	< 5	< 5
4-nonyl phenol	1000	5-10	5-10	10-50
1,2,4, trichlorobenzene	5-10	< 5	< 5	< 5
Benzenes	100	< 5	< 5	< 5
Phenols	1500	10	10	50
PAH	25	< 5	< 5	< 5

23.5.2 Comparison with Mexico City´s water supply

For this comparison, two wells located in Mexico City's aquifer were selected. One is considered of good quality (Ciudad Universitaria) while the other (La Noria) is known for its faecal pollution (Cifuentes *et al.*, 2002). Figures 23.8 a, b and c show the comparison. Mexico City's groundwater has a quality similar to that of the Tula Valley, the difference being that the latter is more saline. Results concerning the microbial content are not shown but water from La Noria had a faecal content similar to that of the other sites studied in the Tula Valley (< 100 MPN/100 mL).

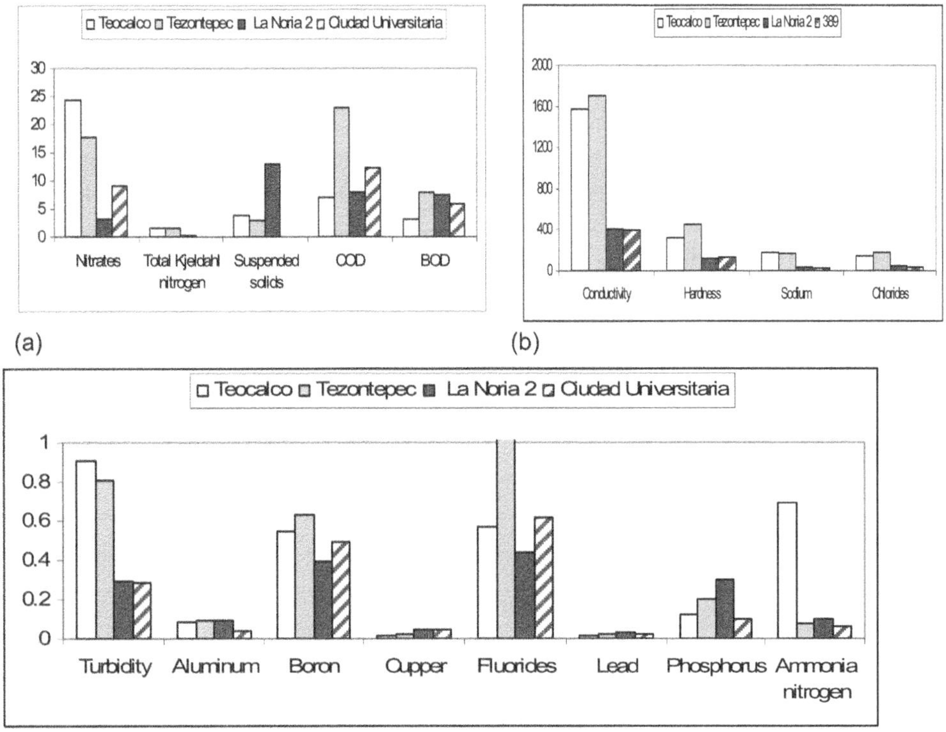

(c)

Figure 23.8 Mexico City and the Tula valley water supply quality. Content expressed in mg/L.

23.5.3 Comparison with a secondary effluent

Mexico City has several wastewater treatment plants that permit onsite water reuse. One such plant is Cerro de la Estrella which treats 172 800 m^3/d (2m^3/s) with activated sludge, filtration and chlorination. This plant is considered one of the best operated in the city. Table 23.8 compares the effluent quality with the water from Tezontepec spring (Figure 23.9). It is evident that, with the unplanned SAT in the Tula Valley, almost all the wastewater produced by the city (4 492 800 m^3/d or 52m^3/s) is treated to a better degree than at the El Cerro de la Estrella wastewater treatment plant.

Figure 23.9 Tezontepec spring.

Table 23.8 Comparison with the effluent of an activated sludge plant.

Parameter, mg/L unless indicated	Secondary effluent	Tezontepec	Parameter, mg/L unless indicated	Secondary effluent	Tezontepec
Turbidity, NTU	4.5	1.0	Copper	0.05	0.02
Total suspended solids	23	3	Chromium	0.008	0.004
Total COD	45	23	Iron	0.13	0.02
Soluble COD	24	7	Mercury	0.001	0.001
BOD	14	8	Nickel	0.025	0.02
Faecal coliforms, MPN/100mL	$9.2x10^5$	4	Lead	0.02	0.01
Total coliforms, MPN/100 mL	$1.1x10^6$	27	Selenium	0.009	0.002
Alkalinity, mg $CaCO_3$/L	174	511	Cyanides	0.024	0.006
Conductivity, µmhos/cm	726	1699	Chlorides	68	179
Hardness, mg/L($CaCO_3$)	164	452	Fluorides	0.8	1.0
Sodium	82	168	Sulfates	67	130
Dissolved solids	496	1038	Sulfides	0.3	1
Aluminium	0.1	0.09	Nitrates	12	18
Barium	0.03	0.01	Nitrites	0.40	0.006
Boron	0.61	0.63	Ammoniacal nitrogen	1.6	0.08
Cadmium	0.002	0.002			
Calcium	27	73	TKN	3.0	1.9
Cobalt	0.05	0.01	Phosphorus	2.8	0.2

23.6 ENVIRONMENTAL EFFECTS

The use of wastewater certainly has a big impact on Tula soils. According to Siebe (1994), using FAO classification system, soils in the Mezquital Valley are of three types: (a) rendzic and melanic Leptosols, (b) calcic and haplic Phaeozems; and (c) eutric Vertisols. In general, soil pH is 6.7–8.4, conductivity is 0.41–2.22 µS/cm (with values of up to 40 µS/cm where the aquifer level is very high), clay content is 11–60%, and texture tends to be clay-loamy. Because the area irrigated with wastewater gradually increased with time as the wastewater flow from Mexico City intensified, it is possible to find areas in the Tula Valley with varying numbers of years under wastewater irrigation. This has permitted some interesting comparisons. Concerning metals, although Mexico City's wastewater contains very low levels (below the Mexican irrigation standard which is similar to that of USEPA, 1992), irrigation annually adds small amounts of metals to soils. Thus, sites irrigated for more than 80 years show a metal accumulation in the arable layer (up to 15 cm) around 3 to 6 times more than the regional baselines. For instance, Pb content has increased from 0.15 to 0.28 g/m^2, Cd from <0.009 to 0.011 g/m^2, Cu from 0.19 to 0.40 g/m^2, and Zn from 0.49 to 1.13 g/m^2. Despite this accumulation, international criteria have not been exceeded (Siebe and Cifuentes, 1995). Wastewater has also contributed to increasing the available phosphorus from low-medium levels (2–9 g P/m^2) to medium-high ones (14–25 g P/m^2) after 80 years under irrigation. For nitrogen, a slight increase has been observed from 0.22 to 0.8 kg N/m^2 (Siebe and Cifuentes, 1995; Siebe, 1998), depending on the period for which soils have been irrigated. However, unlike phosphorus, this element is easily leached, which explains its low accumulation in soils and the high nitrate content in the aquifer and the irrigation drainage (Jimenez and Chavez, 2004).

Wastewater also adds organic matter to soil to such an extent that in soils irrigated for more that 65 years it has increased from 1.6–3.6% to 3.1–6.4%. Additional organic matter increases moisture content, diluting salts, and also limits metal mobility (Siebe, 1998). Effects on crops have also been analyzed by Siebe, 1998, and Herre *et al.*, 2004, reporting no yield changes on crops although a slight salinity increase in plant tissues was observed.

23.6.1 New ecosystems

Due to the lack of water, the original vegetation in the Tula Valley was limited to Xerophila scrubs, such as mesquite, sweet acacia, yucca and a wide variety of cactus (Siebe, 1994). But nowadays, in the lower parts of the valley, where the new springs have appeared, new ecosystems have been formed. The Cerro Colorado spring is one of these springs with a flow of 51 840 m^3/d (0.6 m^3/s). It appeared in 1964 and since then new aquatic flora and fauna have sprung up (Figure 23.11). Cerro Colorado spring is used not only to supply 70,000 people, but also to wash clothes, cars and for recreational purposes. Water from the spring is even bottled with no further treatment to be sold as "spring water". Because there were some concerns about endocrine disruptors found in the water, 4 monitoring campaigns in 5 sites and at 2 depths of a small lake formed by a spring were performed to find the percentage of male and female fish in *jonesi and regalis* species (Figure 23.12). The mean distribution was 42% male and 58% female, and it was concluded that no statistically significant effects had occurred.

Heterandria jonesi (gupi) Fish

Cambarrellus sp. (acocil mexicano)

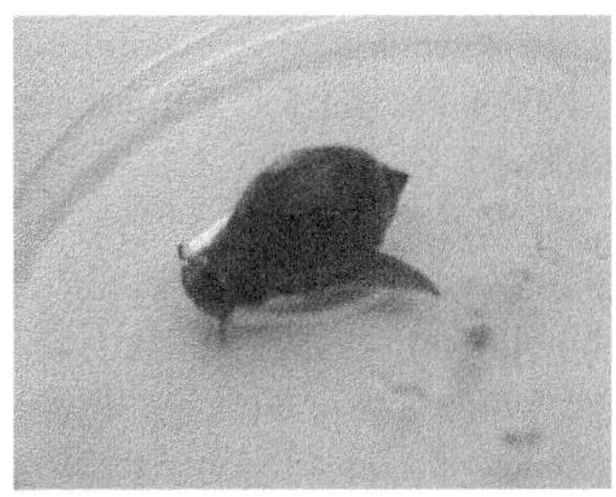

Artropoda Physella sp Mollusk

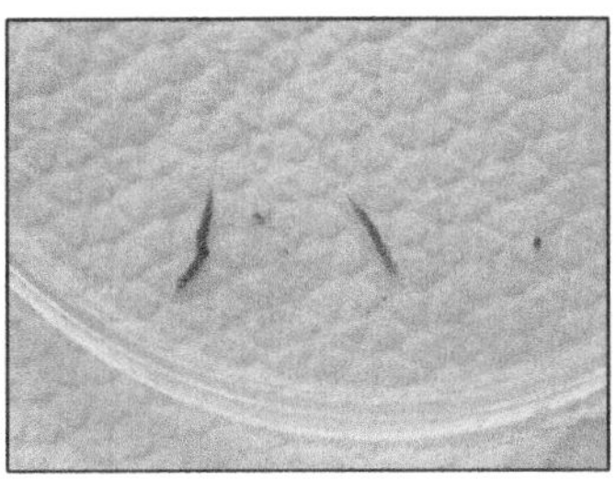

Planaria (*Dugesia*) Platelminth

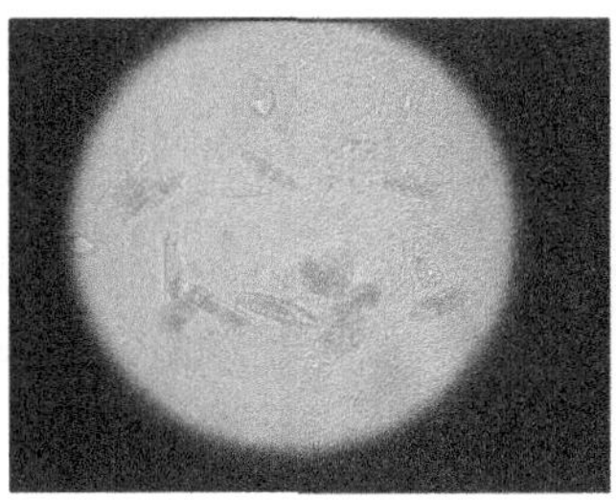

Diatomeas Algae

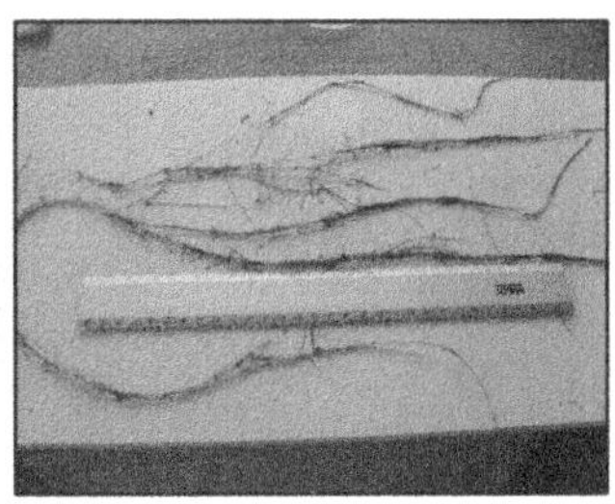

Aquatic grass *(Palustris)* Plant

Figure 23.11 New flora and fauna in the Cerro Colorado spring.

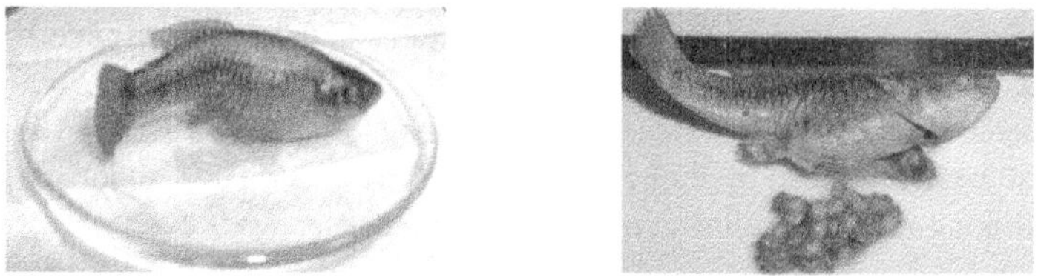

Figure 23.12 (a) Poecilidae, Heterandria, H. jonesi and (b) Goodeidae, Allotoca, A. regalis.

23.7 Water potabilization

Based on criteria set for drinking water it could not be concluded that the water represented severe health risks. But considering the origin of the water and the presence of unidentified and identified compounds in chromatograms (even at very low concentrations) it was decided to define a way to treat the water prior to its consumption in order to remove organic compounds and salts. To investigate this, membrane pilot plants were operated, one with nanofiltration membranes and the other with reverse osmosis ones, using water from Cerro Colorado spring and Teocalco well. Pre-treatment before the pilot plants consisted of sand and anthracite filtration plus a 5 μm filtration step. During the 1-year study, the treated water consistently met drinking water standards using both processes, but reverse osmosis produced an effluent with very low TDS (40–57 mg/L) and an acid pH (Jimenez and Chavez, 2004). In both cases, organic matter removal was high, producing an effluent with < 1–2 mg COT/L, with no bacteria or coliphages.

In wastewater, 60 compounds were detected (Figure 23.13a), notably 4, nonyl phenol (500–1000 ppb), 1,1,3,3 tethramethilbutylphenol (100–500 ppb), 2,3 dihydro-1,2 dimethil-1 H-Indene (10–50 ppb), 1,1 –oxyibis-benzene (10–50 ppb) and 1,1 biphenil (5 ppb), according to Capella and Pegueros, 1998. In groundwater some volatile halo compounds (alkyl benzene and chlorobenzenes) were found in very low concentrations; nanofiltration considerably removed nonylphenols and all the halo compounds, while reverse osmosis removed almost all compounds. Considering these results and based on a benefit/cost analysis, nanofiltration was selected as an adequate treatment process. Studies are still being carried out to optimize the membrane selection, complete the treatment scheme and further reduce the treatment cost.

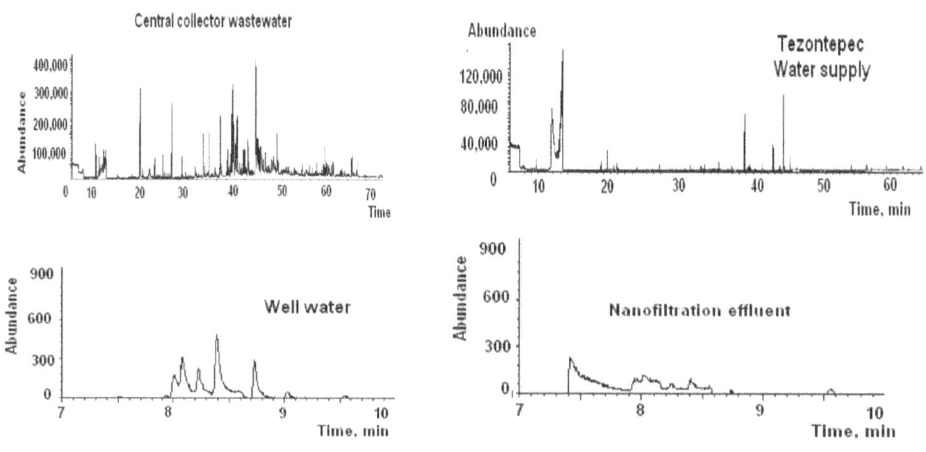

Figure 23.13 Chromatograms in wastewater, groundwater and nanofiltration effluent With information from: Capella and Pegueros, 1998 and García *et al.*, 2000.

23.8 USE OF THE TULA VALLEY AQUIFER TO SUPPLY MEXICO CITY'S WATER SUPPLY

In Mexico City, around one million people lack an adequate formal water supply, and it is estimated that due to population growth an additional 864,000 m^3/d (10 m^3/s) will be needed for around 3.5 million people over the next 10 years, mainly in the northern part of the City. Due to groundwater over-exploitation, Mexico City is subsiding at rates of up to 40 cm/yr in some areas. Thus, additional water abstraction from soil is prohibited – relying on water rights markets to reallocate water from irrigated agriculture or livestock uses to the water supply of new urban developments – and a large number of wells are to be closed in order to restore hydrological equilibrium. Other options are to bring water from other valleys, some of them located 300 km away and 1700 m below Mexico City. Besides this, the Temascaltepec option, considered the most favourable one, had to be stopped due to social pressure from the regional population that rejects sending "their water to the capital", despite water being national property. Thus, new alternatives are being taken into account. One is to potabilize wastewater *in situ*, as is done in Windhoek, Namibia, while another is to bring back the water from the Tula Valley aquifer. Geohydrological studies in the Mezquital Valley have shown that there is an excess of water of at least 864,000 m^3/d (10 m^3/s). This excess is causing groundwater levels to rise, leading to soil salinization problems, flooding of the irrigation areas and 86 400 m^3/d (1 m^3/s) of water loss from evaporation. In an initial stage, 518 400 m^3/d (6 m^3/s) would be extracted for drinking purposes. This amount was set using a cautious approach in order to study the possible effects of extraction on the local groundwater availability. This volume can be produced using 90 wells (6048 m^3/d or 70 L/s each) over a 322 km^2 area near Tezontepec, Cerro Colorado and San Salvador. In this scenario the total water level depression in the Tula Valley would be around 30 m. Compared to the option of potabilizing wastewater *in situ*, this option has the advantage of enabling agriculture activities in the Tula Valley to continue and thus increase the added value of water through its successive reuse. Table 23.9 shows the comparative costs. As reference, the present cost of supplying water from the local aquifer in Mexico City is 0.02 USD/m^3 (just to chlorinate groundwater) while the cost of bringing water from a site located 151 km away and 1 100 m below Mexico City is 0.75 USD/m^3.

Table 23.9 Cost of future water supply options for Mexico City.

PROJECT	USD/m^3 (2000)
Amacuzac	2.36
Tecolutla	2.13
Temascaltepec	0.75
Tula	0.72
Potabilization in situ of the wastewater	1
Potabilization in situ, reinjection and extraction for water supply	1.3

23.9 CONCLUSIONS

It is evident that infiltration through the Tula Valley soil is acting as an unintentional SAT system that efficiently depollutes Mexico City's wastewater. While the wastewater origin of the aquifer water is not causing urgent problems for Tula Valley inhabitants, it is certainly a source of great concern, especially as it is not known how long the treatment capacity of the soil will last. For this reason studies are still being carried out and additional support is required to determine the fate of pollutants and quantify soil behaviour. Also, additional studies should be done to fully determine the long-term effects of the aquifer water supply using toxicological tests in larger species, such as fish. Also, population health surveys should be carried out. With respect to water treatment, besides optimizing the process, a safe disposal method needs to be defined for the brine produced from the membrane process.

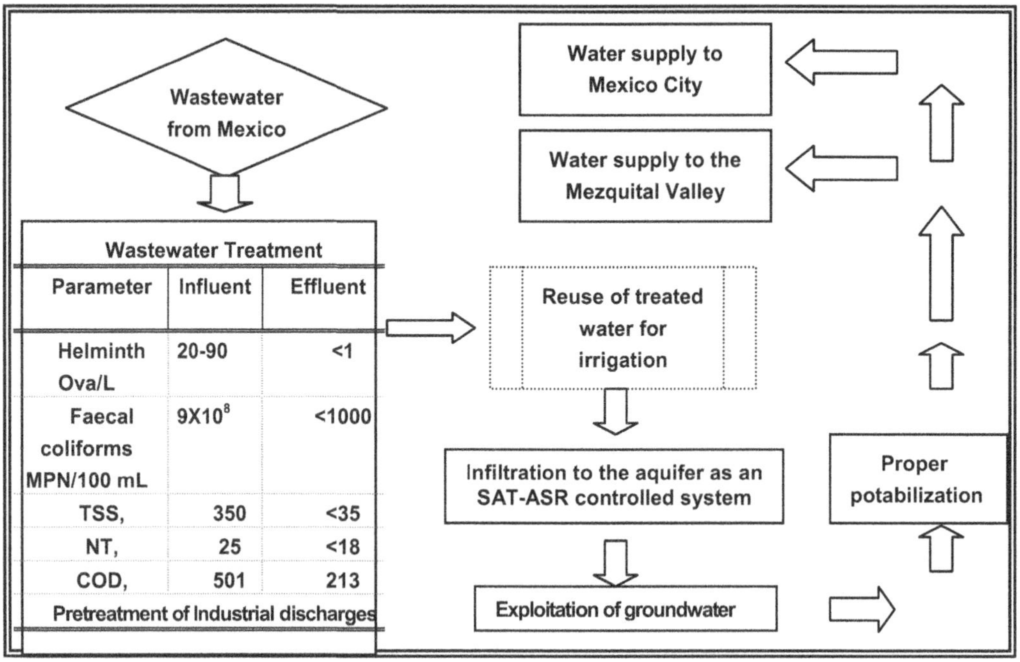

Figure 23.14 Management proposal.

Because any study will take time, it is recommended that high priority be given to controlling industrial discharges to Mexico City's sewage system, as well as treating wastewater in accordance with irrigation norms prior to its use in the Mezquital Valley; that way, the soil's treatment capacity will last longer. As defined by national standards since 1995, Mexico City's wastewater treatment should be implemented independently of the Mexico City's government decision to reuse the water from the Tula Aquifer as supply in order to protect the health of the Tula Valley inhabitants. Also, independently of Mexico City's decision, the Hidalgo state government should be reviewing the potabilization process applied in the area (chlorination) or, as proposed by other researchers, extracting water from parts of the aquifers where wastewater has no or less influence. The water management proposal for the valleys of Mexico and Mezquital is presented in Figure 23.14.

23.10 REFERENCES

BGS and CNA (British Geological Survey & Comisión Nacional del Agua) (1998), "Impact of wastewater reuse on groundwater in the Mezquital Valley, Hidalgo state, Mexico", Final report, CNA, DF., Mexico (In Spanish)

Blumenthal U., Abisudjak B., Cifuentes E., Bennett S., Ruiz-Palacios, G. (1991) Recent epidemiological studies to test microbiological quality guidelines for wastewater use in agriculture and aquaculture. Public Health Reviews 19: 237-242.

Blumenthal U., Cifuentes E., Benett S., Quiley M., Ruiz-Palacios G. (2001) The risk of enteric infections associated with wastewater reuse: the effect of season and degree of storage of wastewater. Transactions of the Royal Society of Tropical Medicine and Hygiene 95: 1-7

Bouwer H. (1989). Groundwater recharge with sewage effluent. Water Science and Technology. 23: 2099-2108.

Capella S. and Pegueros A., (1998). Analysis by solid Phase Micro-extraction (□-SPE) and Capillary Gas Chromatography-Mass Spectrometric Detection (CGC-MSD) of Chlorinated Benzenes and Phenols in treated Wastewater from Mexico's City Municipal Effluent. Proceedings of the 20th International Symposium on Capillary Chromatography, 4 pp.

Cifuentes E. (1998) The epidemiology of enteric infections in agricultural communities exposed to wastewater irrigation: Perspectives for risk control. International Journal of Environmental Health Research 8 (3): 203-213.

Cifuentes E., Suárez L., Solano M, and Santos R. (2002) Diarrheal Diseases in Children from a Water Reclamation Site in Mexico City Environmental Health Perspectives 110 (9):619-624

CNA, Comisión Nacional del Agua (2004) Water Statistics CNA Ed., DF, Mexico (in Spanish)

COUNCIL DIRECTIVE 98/83/EC of 3 November 1998 on the quality of water intended for human consumption Official Journal of the European Communities L 330/32L 330/53

Herre A., Siebe C and Kaupenjohann (2004) Effect of irrigation water quality on organic matter, Cd and Cu mobility in soils of central Mexico Water Science and Technology 50(2): 277–284

García V, Peña A and Capella S. (2000). Impact of Wastewater Irrigation on the Quality of Groundwater. Preliminary Evaluation of Organic Compounds by SPME-CGC-MSD. Proceedings of the 23th International Symposium on Capillary Chromatography, Edition in CD, 4 pp.

Jiménez B., Cruickshank C., Capella S., Chávez A., Palma A., Pérez R. y García V. (Diciembre 1999) Feasibility study for the Mezquital Valley aquifer use as water supply for Mexico City. Institute of engineering, UNAM. Mexico Project 8384, 1500 pp. (In Spanish)

Jiménez, B., Barrios J.A. and Cruickshank, C. (5-7 Junio 2003), Evaluation of a Wastewater Recharged Aquifer as a Source of Water Supply. 11th Biennial Symposium of Groundwater Recharge. Arizona Hydrological Society, Arizona Department of Water Resources, Salt River Project, U.S. Water Conservation Laboratory of USDA-ARS, Memorias en disco, JIMENEZE. Tempe, Arizona, U.S.A.

Jiménez B. and Chavez A. (2004) Quality assessment: a potential use of an aquifer recharged with wastewater: "El Mezquital case" Water Science and Technology 50 (2): 269-273

Jimenez B. (2005) Treatment Technology And Standards For Agricultural Wastewater Reuse: A Case Study in Mexico. Irrigation and Drainage Journal 54:23-35

Payne B.(1975) Interaction between the water to irrigate and the groundwater and the Tula River in the El Mezquital Valler. Final report. Hydrology isotopic section, pp. 9.

Siebe, Ch. (1994) Heavy metal accumulation in soils from the Irrigation District 03, in Tula, Hidalgo irrigated with wastewater. International Magazine of Environmental Pollution, 10:15-21 (in Spanish).

Siebe, Ch. (1998) Nutrient inputs to soils and to their uptake by alfalfa through long-term irrigation with untreated sewage effluent in Mexico. Soil Use and Management 13: 1 -5.

Siebe Ch. and Cifuentes E. (1995) "Environmental Impact of wastewater irrigation in central Mexico: an overview" International Journal of Environmental Health Research, 5:161-173

USEPA (1992) Guidelines for Water Reuse Report EPA/625/R-92/004 Agency for International Development, Washington, D.C., 186 pp

WHO (2005) Guidelines for Drinking Water Quality, Third Edition. WHO Ed. Geneva.

24

Water reuse in Windhoek, Namibia: 40 years and still the only case of direct water reuse for human consumption

Ben van der Merwe, Piet du Pisani, Juergen Menge and Erich König

24.1 INTRODUCTION

Namibia is located in the south-western part of Africa and is the most arid country south of the Sahara Desert. Windhoek, the capital of Namibia, is situated in the Central Highlands of Namibia, approximately 1 600 m above mean sea level. The average annual rainfall is 360 mm, while the average evaporation is 3 400 mm/yr. Namibia's water resources are unevenly distributed over the country and there are no perennial rivers within the borders of Namibia, as indicated in Figure 24.1 below. The nearest perennial river, the Okavango, is 700 km from the city on the north-eastern border of the country.

The estimated population of Windhoek in mid 2001 was 250 000. According to the Residents Survey Report of 1995, the average population growth rate of Windhoek was 5.44%/yr from 1991 to 1995. The annual population growth rate (birth rate less deaths) in the city over the same period was only 1.52%/yr, which results in a net annual migration of 3.92% from rural areas to the city (TRP Associates, 1995). It is expected that the annual population growth rate will be 3.25% from 2005 to 2015 as a result of the devastating effect of Aids (WCE, 2000).

 ISBN: 9781843390893. Published by IWA Publishing, London, UK.

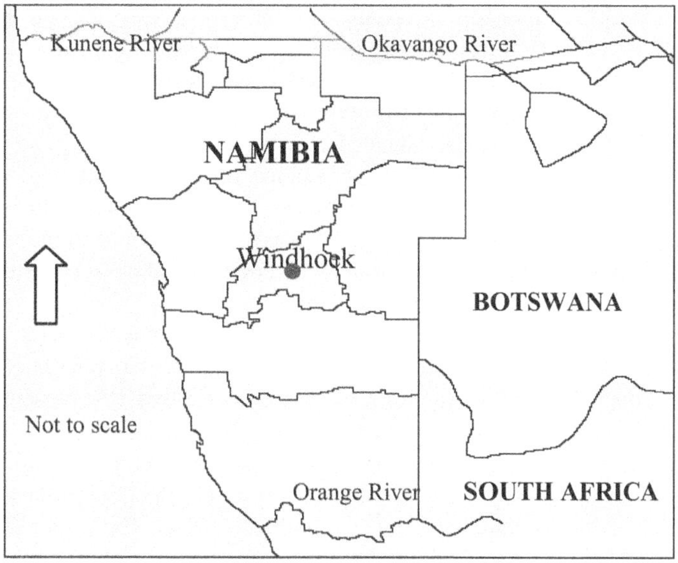

Figure 24.1. Map of Namibia.

According to the Namibian Economic Policy Research Unit (1996) Windhoek has 51% of manufacturing, 96% of utilities, 56% of construction and trade, 94% of transport and communications, 82% of finances and business services and 68% of community and social services in the country. The city produces 47% of value added, and private consumption expenditure in Windhoek comprises 35% of the national total. In contrast only 8% of production comes from the north where 60% of population lives and where 33% of private consumption expenditure takes place.

24.2 HISTORIC DEVELOPMENT OF WATER SOURCES IN WINDHOEK

The first settlers stayed in Windhoek because of the abundance of water from the hot springs in the areas of Windhoek and neighbouring Klein Windhoek. The first recorded formal settlement in Windhoek was built towards the end of 1840 when Jonker Afrikaner settled at the hot springs of //Ai//gams ("Fire water" in Nama) or Otjomuise ("Place of steam" in Herero) known today as Klein Windhoek. On 1st September 1911, a state owned water scheme was developed for Windhoek. The government did not favour the sinking of boreholes as they feared the springs might dry up, and so they bought a site from August Schmerenbeck with a borehole yielding 300 m^3/d. The provisional pump installation was replaced by a permanent one. A second borehole on private property in Windhoek was drilled in 1913. Groundwater remained the sole source of water for Windhoek until 1933 when the Avis Dam, with a capacity of 2.4 Mm^3, was constructed. This dam has a small catchment area and therefore a very small assured yield; often, it could not supply any water at all. The Avis Dam is currently used exclusively for recreational purposes. After various extensions to the wellfield there are 50 municipal production boreholes (2005) in Windhoek with a combined safe yield of 1.73 Mm^3/yr. Except from some augmentation from the Avis Dam, Windhoek relied almost solely on groundwater until 1960.

Water supply remained a challenge, and on 19 April 1956, the Mayor of Windhoek made a submission to the Administrator-in-Executive Committee, where three proposals were made as an interim solution for water supply to the city. These included the building of the Goreangab Dam, the use of groundwater from the empty Avis Dam, and the reuse of sewage effluent. This submission, signed by Town Clerk Conradie, appears to be the first mention of reuse (Wipplinger, 1956). In a paper presented to the Institute of Water Pollution Control, dated March 1970, it is however indicated that "As early as 1954, it was realized that the future economic progress of Windhoek depended on the utilization of reclaimed water" (Stander and Clayton, 1970) The Goreangab Dam (3.6 Mm^3), located in an ephemeral river near the city, was completed in 1959. The 95% assured safe yield of the Goreangab Dam is approximately 1.1 Mm^3/yr. A conventional treatment plant was constructed to treat the surface water from this dam to potable standards.

During the water supply crisis in 1957, the idea of reclaiming water for industrial use was again broached. On 18 November 1957, the Municipality of Windhoek reported to the Secretary of South West Africa (as Namibia was then known): "At present the reservoirs in Windhoek are emptied completely each day except Sundays, Mondays and Tuesdays. Sundays the draw-off is comparatively low and allows the reservoirs to build up. By Tuesday afternoon this advantage is lost and the draw-off is higher than the production. All borehole pumps are operating continuously for 24 hours per day. The water-table has dropped alarmingly… the most alarming drops have been at Kessler, 7.4 of 52.0 metres, and Aub, 10.4 of 44 metres." (Brand, 1957). These drops were recorded over a period of 4 months. At that time, it was estimated that the underground source was used at a level some 57% above the safe yield, estimated at 1.6 Mm^3/yr. At the same time, the city experienced a population growth rate of 8.5%/yr. In a report to the Honourable Administrator in Executive Committee, the building of two additional dams was proposed (excluding Goreangab Dam, at that time under construction), for the harvesting of the runoff in ephemeral rivers near Windhoek. Significantly, in the same submission, Wipplinger (1957) states: "The reuse of suitably treated effluent for certain special purposes, such as the power station, cemetery, etc. would appear to be a measure which is likely to yield positive results by September 1958 as any other. The source of supply for this measure is available, whereas other sources of supply still have to be developed."

During the early 1960s, the Council of Windhoek Municipality resolved to run a pilot plant for the reclamation of treated sewage effluent and a pilot scale plant was erected at the Gammams Waste Water Treatment Plant (Clayton, 2005). A pilot plant was also installed at Pretoria, called the Stander Plant, after the pioneer Dr G.J. Stander who was also intimately involved in the Windhoek project. As South West Africa was then administered as an integral part of South Africa, part of such research was conducted and paid for by the National Institute for Water Research of South Africa. During the mid-1960s, upon favourable results from the Gammams pilot plant, the Council of the Windhoek Municipality took a decision to implement potable reclamation. Consistent pressure on water supply led to a decision to extend the Goreangab treatment plant. Work commenced in 1967 and was completed at the end of 1968. The conventional Goreangab water treatment plant was extended with the addition of a second treatment train to treat not only the surface water from the Goreangab Dam but also the final effluent from the Gammams Waste Water Treatment Plant. Gammams had been constructed to treat predominantly Windhoek's domestic effluents while industrial effluents were diverted for separate treatment at a different plant, situated to the north of the city. Thus, the Goreangab Water Reclamation Plant (GWRP) was born. It had an initial capacity of 3 287 m^3/d (1.2 Mm^3 /yr). The reclamation of potable water from sewage at the GWRP has been ongoing for the past 40 years and is well publicised and recognized as the only plant of its kind in the world. The process was improved and the plant extended on various occasions over a period of

30 years and ultimately to a capacity of 7 500 m^3/d (2.7 Mm^3/yr) during the drought of 1996/1997.

In 1969 the Government started to develop bulk infrastructure in ephemeral rivers further from Windhoek to provide water to the city. The Von Bach Dam (48.6 Mm^3) was added in 1970, the Swakoppoort Dam (63.5 Mm^3) was completed in 1977, while the Omatako Dam (43.5 Mm^3) was completed in 1982. The distances from Windhoek to the various dams are 70, 100 and 200 km respectively. The safe yield, based on 95% assurance of supply, of the three dams is approximately 20 Mm^3. Of this volume, approximately 17 Mm^3 is available for supply to Windhoek. During a severe drought in 1996 the Berg Aukas Mine in the North was added as an emergency groundwater supply to the Central Area of Namibia.

During discussion of the Central Area Water Master Plan of 1992 it was concluded that Water Demand Management and extension of the Goreangab Water Reclamation Plant with a capacity to supply approximately 35% of the total potable supply to Windhoek were the most feasible interim measures for implementation. Technically the extension of water reclamation was seen as a readily available source of water for the city of Windhoek. The New Goreangab Water Reclamation Plant (NGWRP) was completed in August 2002 with a capacity of 21 000 m^3/d (7.6 Mm^3/yr).

During 8 years of water shortages from 1968 to 2000 the old plant produced from 12% to 18% of the total potable water supply to Windhoek with highest production in 1997 of 7 650 m^3/d (2.8 Mm^3/yr) (18%) of the total demand of 42 565 m^3/d (15.5 Mm^3/yr). The new plant will be operated continuously to provide potable water to the city including water for the planned artificial recharge project for the Windhoek Aquifer. During 2003 the plant produced 14 330 m^2/d (26%) of the supply of 55,340 m^3/d (20.2 Mm^3/yr) to Windhoek. The lower production rate resulted mainly from the restricted availability of raw water for reclamation, technical problems on the plant and raw water supply infrastructure, and the blending ratio restriction of 35% reclaimed water in the final blended water distributed within the city.

Figure 24.2. New Goreangab Water Reclamation Plant under construction (2000) with the old plant and the Goreangab Dam on the right.

Other initiatives to cope with limited water supply augmentation options included:

- Conjunctive use of water based on the premise that surface water is used during periods of ample supply while groundwater is used as a backup system during drought.
- Implementation of Water Demand Management which lowered the unrestricted water demand by approximately 30% and lowered the annual increase in consumption to below 3%.
- Installation of a dual pipe system to distribute treated (filtered and chlorinated) maturation pond effluent for landscaping in the City, contributing to annual savings of 6-7% in potable water supply.
- Research and development to establish the technical viability of artificial recharge of the Windhoek Aquifer (1996 to 1999).

24.3 FUTURE WATER SUPPLY AUGMENTATION TO WINDHOEK

A recent study (Central Area Joint Venture Consultants, 2004) determined that artificial recharge of the Windhoek Aquifer (water banking) is the "best next" supply option to the Central Area of Windhoek. It contemplated the storage of excess treated water from surface dams (periods of high runoff) including reclaimed water, in the Windhoek Aquifer for abstraction during periods of drought. In a study on the injection water quality, it was determined that a blend of 3 to 1 of surface water with reclaimed water will be suitable for injection.

The long-term solution to secure the water supplies for central Namibia is to link the Eastern National Water Carrier to the Okavango River. The Okavango Delta in Botswana is an important international wetland that is very sensitive from an ecological perspective and abstraction from the Okavango River would require negotiation with neighbouring Botswana. It is essential to develop all local sources to their maximum potential and to consider alternative water supply sources which could be developed to replace or postpone, for as long as possible, the further development of the Eastern National Water Carrier to the Okavango River.

24.4 VARIOUS PROCESS MODIFICATIONS FROM 1968 TO 1995

The Reclamation Plant has undergone a process of evolution and improvement since 1968. The different configurations are summarised in Table 24.1 and have been described in greater detail in an earlier publication, (Haarhoff and Van der Merwe, 1995).

Modifications to Configuration Four (adapted in 1995) entail the use of ferric chloride in the place of alum, carbon filtration moved to immediately follow sand filtration, and with breakpoint chlorination after carbon filtration. Since 1996 a filter-to-waste facility was added to the sand filters to eliminate possible breakthrough of Giardia and Cryptosporidium after filter back-wash.

Table 24.1. Process configurations and modifications since commissioning of the Old Goreangab Water Reclamation Plant

Configuration One (1969)	Configuration Two (1977)	Configuration Three (1980)	Configuration Four (1986)
GAMMAMS WASTEWATER TREATMENT PLANT			
primary settling	primary settling	primary settling	primary settling
biological filters	activated sludge	activated sludge	activated sludge
secondary settling	secondary settling	secondary settling	secondary settling
maturation ponds	maturation ponds	maturation ponds	maturation ponds
GOREANGAB WATER RECLAMATION PLANT			
carbon dioxide alum	lime settling	chlorine alum and lime	alum
autoflotation	ammonia stripping	settling	Dissolved air flotation
foam fractionation	primary carbon dioxide		
breakpoint chlorination	chlorine	breakpoint chlorination	chlorine* lime
settling	settling	settling	settling
sand filtration	secondary carbon dioxide	sand filtration	sand filtration
	sand filtration		
carbon filtration	breakpoint chlorination	chlorine	breakpoint chlorination
	chlorine contact	chlorine contact	chlorine contact
	carbon filtration	carbon filtration	carbon filtration
chlorine	chlorine	chlorine	chlorine
blending	blending	blending	blending

Notes: Process modifications in bold boxes.
Chemical addition in shaded blocks.
* only for intermittent shock dosing.

24.5 PROCESS DESIGN FOR THE NEW GOREANGAB WATER RECLAMATION PLANT

The process design of the New Goreangab Water Reclamation Plant summarised in this section was described in detail in a publication by Haarhoff, Van der Walt and Van der Merwe (1998).

24.5.1 Summary

The sensitive nature of wastewater reclamation demands unusual attention to the process selection criteria for such a plant. The recent upgrading of the Goreangab Reclamation Plant necessitated a complete re-analysis of the process needs and the treatment objectives in the light of new technologies and ever-tighter water quality guidelines. The process selection methodology is summarised below:

- establishing a raw water quality profile in probabilistic terms;
- defining final water quality objectives, based on own experience and recent international developments;
- developing the vague "multiple barrier" concept into a quantitative tool for comparison of process options;
- critical analysis of the process train successfully used in the past;
- evaluation of numerous experimental and pilot trials performed during the period from 1997 until 2000; and
- the finalisation of the process train for the new reclamation plant.

24.5.2 Raw water quality profile

The raw water data records over the period 1997 to 2000 were statistically analyzed and the most important results are reflected in Table 24.2. Sixty different parameters were measured, of which forty were also duplicated by external laboratories as part of the ongoing quality control programme.

The two raw water sources are chemically remarkably similar, both in terms of their mean values as well as the 95th percentile values. Physically they are slightly more dissimilar, with the natural surface water from Goreangab Dam more turbid and algae-infested than the maturation pond effluent from Gammams Water Care Works. This dissimilarity, however, is small enough to allow essentially the same process design for both sources. The variability of the raw water quality is higher for Goreangab Dam water than Gammams raw water, as would be expected. Once again, these variabilities are not excessive and do not, superficially, pose any special problems in terms of process design. In terms of phase separation, the algal and turbidity concentrations are well within the ranges very commonly encountered for Southern African raw waters, and do not pose special problems. The high levels of iron and manganese, and ammonia do necessitate an early oxidation step in the process train. The occasional high level of nitrate in the water from Gammams is a problem that is not treated by conventional physical treatment processes, but which requires ion exchange or low cut-off membrane treatment. The occurrence of nitrate is a special problem and should be dealt with at the Gammams Water Care Works.

Due to the high organic loading of the raw water, organic removal processes should form part of the treatment process, as has been the case since the inception of the GWRP. The organic material from Gammams is the refractory material left after biological treatment and is not very amenable to precipitation during normal coagulation. This is supported by the specific UV absorbance (SUVA) values of 2.94 for Goreangab raw water and 1.87 for Gammams raw water, suggesting that only about 20% of the organics would be precipitable.

Table 24.2. Raw water quality.

Physical and Organic	Units	GOR 95thtile	GAM 95thtile
Chemical Oxygen Demand	mg/L	43.0	43.6
Colour	mg/L Pt	80.0	63.8
Dissolved Organic Carbon	mg/L	13.6	16.6
Total Dissolved Solids	mg/L	344.5	781.0
Turbidity	NTU	102.2	3.7
Alkalinity	mg/L	153.1	282.3
UV_{254}	abs/cm	0.4	0.3
Inorganic	Units	GOR 95thtile	GAM 95thtile
Aluminium	Al mg/L	1.7	0.9
Ammonia	N mg/L	1.7	2.9
Chloride	Cl mg/L	54.9	142.
Fluorine	F mg/L	0.8	1.0
Iron	Fe mg/L	5.3	0.38
*Manganese	Mn mg/L	1.6	0.17
Nitrate and Nitrite	N mg/L	0.1	15.3
Microbiological	Units	GOR 95thtile	GAM 95thtile
Heterotrophic Plate Count	per 1 mL	399,100	265,200
Total Coliforms	Per 100 mL	464,000	26,250
Faecal Coliforms	Per 100 mL	37,400	7,183
Chlorophyll a	µg/L	46.86	24.11
Giardia	per 100 L	276.00	152.50
Cryptosporidium	per 100 L	200.00	468.00

Notes: GOR = Goreangab raw water
GAM = Gammams raw water

24.5.3 Determination of treatment objectives

Table 24.3 summarizes the finally adopted set of water quality treatment objectives, as determined from an analysis of various documents and the experience gained on the old GWRP. The guidelines used included the Namibian Guidelines (1991), Rand Water (1994), World Health Organisation (1993), National Drinking Water Standards and Health Advisories issued by the United States Environmental Protection Agency (1996) and the Water Research Commission Guide (RSA, 1982) for the planning, design and implementation of a water reclamation scheme.

Table 24.3 reflects the current USEPA "treatment technique" philosophy that treatment plant operation should be directed towards a log reduction approach when routine measurements of contaminants cannot be quickly or easily measured. By maintaining a residual concentration C after time t, the link between the Ct-product (concentration in mg/L x time in minutes) and the log-reduction of certain microbiological contaminants will ensure compliance with the required log-reduction. The 5-log requirement for the protozoan oocysts is based on the restraint of no more than 1 infection per 10 000 people per year.

Table 24.3 Final water quality criteria and operational results of the New Goreangab Water Reclamation Plant.

Parameter	Units	Final Water Specification	Actual Operational Results	
Physical and Organic			**50%tile**	**95%tile**
Chemical Oxygen Demand	mg/L	10 - 15	6.6	11
Colour	mg/L Pt	8 - 10	0.5	0.5
Dissolved Organic Carbon	mg/L	3*	1.7	2.8
Total Dissolved Solids	mg/L	1000 max or 200 above incoming	838	938
Turbidity	NTU	0.1 - 0.2	0.05	0.10
UV_{254}	abs/cm	0.00 - 0.06	0.015	0.027
Inorganic				
Aluminium	Al mg/L	0.15	0.005	0.05
Ammonia	N mg/L	0.1	0.05	0.18
Iron	Fe mg/L	0.05 - 0.10	0.01	0.03
Manganese	Mn mg/L	0.01 - 0.025	0.005	0.015
Microbiological				
Heterotrophic Plate Count	per 1 mL	80 - 100	0	4
Total Coliforms	per 100 mL	0	0	0
Faecal Coliforms	per 100 mL	0	0	0
Chlorophyll a	µg/l	1	0.27	2.58
Giardia	per 100 L	0 count/100 L or 5 log removal	0	0
Cryptosporidium	per 100 L	0 count/100 L or 5 log removal	0	0
Disinfection by products				
Trihalomethanes	µg/L	20 - 40	35	57

Note *A target was set for 3 mg/L DOC (with a maximum of 5), based on the premise that in the final blended water not more than 1 mg/L DOC originating from sewage water should be present. A maximum blending ratio of 35% to 65% water from other sources was specified.

24.5.4 The multiple-barrier concept

The term "multiple barriers" is frequently encountered in relation to wastewater reuse, but is often loosely used as an approximate equivalent of "safety factor" without clear meaning. It is necessary to differentiate between non-treatment barriers, treatment barriers and operational barriers. Three significant non-treatment barriers have been present in the past, and will continue to be present in the future:

- the diversion of industrial effluents to different municipal drainage areas, and the continuous policing of discharges into the sewerage system draining to the Gammams Wastewater Treatment Plant.
- rigorous, continuous quality monitoring of raw and treated water, which will detect problems as they arise and allow corrective action before the public can be exposed to any prolonged risk.
- blending of reclaimed water with water from conventional sources to limit reclaimed water to a maximum of 35% of the blended water supplied to residents.

Operational barriers are treatment processes which are not normally used, but which provide backup or standby for other essential treatment processes. In the case of the NGWRP, for example, a powdered activated carbon facility was installed to provide additional adsorption capacity if the water quality deteriorates during periods of drought to such an extent that the final water quality criteria could not be met.

Treatment barriers are barriers always present in the treatment plant against specific contaminants, such as the barrier posed by chlorination against bacteria, for example. It should be noted that "barriers" in this sense mostly cannot imply absolute dead-stop barriers. The use of settling and filtration, for example, is regarded as a barrier against turbidity, although it does not completely remove all turbidity. The precipitation of DOC during coagulation, as another example, is only partial and therefore far from a complete "barrier". Different contaminants respond differently to different treatment methods. Likewise, the consequences of failure are different for each contaminant. By considering all these often confounding factors, the following criteria for treatment barriers were eventually adopted:

- for aesthetic parameters such as turbidity or colour, two complete barriers are required;
- for microbiological parameters, three complete barriers are required;
- as an additional precaution against protozoan oocysts, filters with slow-start and filter-to-waste facilities are required to ensure the required removal or better. (The combination of DAF with dual-media filtration will give 5 log removal if properly operated);
- for organic parameters, two partial barriers are required;
- for the parameters which determine stability (with respect to corrosion), one barrier is required;
- the minimization of disinfection by-products will be ensured by adherence to practical guidelines such as using a chemical dosing strategy which maximizes turbidity as well as organic removal ("enhanced coagulation" in current US vernacular), moving chlorine dosing as far downstream as possible, and using alternative oxidants as far as possible.

The performance data of each individual process on the old Goreangab Reclamation Plant was scrutinized over a three-year period, and the main conclusions were:

- the flocculation, DAF, settling and filtration provided one consistent barrier against the aesthetic parameters. The GAC filtration did provide a small secondary filtration effect, but a more formidable second barrier against the aesthetic parameters was required;
- the settling step contributed very little to overall process performance and could be omitted without compromising the existing barrier;
- the three breakpoint chlorination steps provided three complete barriers against microbiological contaminants;
- there was only one barrier against Cryptosporidium (DAF with filtration). The Ct products (concentration, C, in mg/L multiplied by the time, t, in minutes) after the first and second breakpoint chlorination points were adequate to provide two additional barriers against Giardia;
- the organic removal in the GAC filters was good with 35% COD removal, 36% DOC removal and 40% UV254 removal. A filtering effect was also obtained by the GAC with 16% removal of turbidity.

- the high dosage of chlorine could be reduced or delayed to minimize DBP production. The THM concentration on the existing plant was consistently below 70 µg/L, but even lower values should be ensured where they could be easily attained.
- the occasional spikes of iron and manganese were removed with reasonable efficiency, to the extent that it did not lead to consumer complaints. For consistent compliance with the adopted treatment objectives, however, additional measures were required.

The deficiencies in the Old Goreangab process train could theoretically be rectified by any number of additional processes. Before the final selection was made, numerous experiments were performed at full-scale, at pilot scale and at bench scale. The most important experimental results are summarized below.

24.5.5 Experiments and pilot studies to determine process design criteria

Two full-scale experiments were performed with activated carbon on the Old Goreangab Water Reclamation Plant. In the first, a GAC layer of 1200mm was placed on top of the filter sand of two filters, and a layer of only 500mm on the other four filters remaining as a control. The performance of the GAC filters was excellent, leading to DOC removals down to levels of below 3 mg/L. The breakthrough time, however, was about three months after which the carbon had to be replaced. This rendered it impractical for continuous use and frequent regeneration.

A second experiment was performed to assess the efficiency of powdered activated carbon (PAC) adsorption added to the raw water, and to determine whether the deliberate lowering of the pH with acid improved the rate of adsorption. PAC was dosed at different dosages (mg/L) while also varying the pH of the water. In general it was found that the addition of PAC did reduce the organic load substantially, but samples taken after the DAF and sand filter process units show that there is little difference between the addition of 20 mg/L or 60 mg/L PAC. The lowering of the pH had a positive effect on the organic removal efficiency of both the PAC and the DAF process, and in general better results were obtained with 20 mg/L PAC with acid than with 60 mg/L PAC without acid. The DOC levels were reduced to a low value of less than 4 mg/L after sand filtration.

An ozone/granular activated carbon (GAC) pilot plant was operated for 1 year from November 1995 and consisted of an ozonated train and a non-ozonated train. Two different carbons were tested and the ozone dosage was regularly adjusted with varying organic content of the incoming water.

Evaluation of the results concluded that:

- ozone dosage improved the adsorption of organic material on the GAC;
- ozone enhanced biological activity and adsorption on the GAC. After an empty bed contact time of 28 minutes, biological activity on the ozonated stream accounted for 40% of the nett removal of dissolved organic carbon, while it was only 11% in the non-ozonated control;
- ozone dosage should be a function of the dissolved organic carbon (DOC) content of the water;
- the carbons tested in the pilot plant performed significantly different which influenced the final selection for the full scale plant;
- the results allow an estimate of the regeneration frequency on the full-scale plant;

- the best results were obtained with one biological activated carbon (BAC) adsorber followed by a conventional GAC adsorber;
- such a system should comfortably reach the treatment objective for organic contaminants.

Four different ultrafiltration membranes were tested for varying periods over more than one year on a pilot scale. These provided useful results on allowable flux rates and backwash frequency. Moreover, it showed that:

- membrane filtration reduced turbidity to very low levels and provided a physical, dead-stop barrier against protozoan oocysts;
- membrane filtration did not lower the organic content of the water;
- some of the membranes were susceptible to premature fouling by manganese and iron and indications were that the previously stated treatment objective for manganese and iron might have to be lowered to ensure smooth membrane operation.

A bench-scale study was performed on the raw water from Gammams in order to investigate the beneficial effect of lower pH on organic removal. Where normal optimum dosage requirement for this water is 35 mg/L of ferric chloride at pH 7.4, it was found that by increasing the dosage to 44 mg/L and lowering the pH to 6.8, additional organic precipitation (measured as DOC) of about 12% can be achieved. The benefits of enhanced coagulation should however be measured against TDS increase of the water and stabilization costs.

24.5.6 Selection of final process train

The resulting treatment train, after consideration of all the foregoing, is shown in Figure 24.2. The accepted process train provided for the following:

- two barriers against turbidity (DAF/filtration and membrane filtration) as well as the secondary filtration effect of GAC;
- three complete barriers against microbiological contaminants (ozone, membrane filtration, chlorination);
- four complete barriers against Giardia (DAF/filtration, ozonation, membrane filtration and sufficient chlorination);
- two complete barriers and two partial barriers against Cryptosporidium (DAF/filtration and membrane filtration as complete barriers, ozonation and chlorination as partial barriers);
- four partial barriers against organic material (enhanced coagulation, ozonation, GAC adsorption);
- pre-ozonation is used on a permanent basis for the beneficial effect on coagulation/flocculation;
- the use of ozone with GAC on a pilot scale has reduced the THM formation potential to less than 20 µg/L (compared to the less than 70 µg/L THM's actually measured during the past few months), showing the beneficial effects of using ozone in lieu of later chlorination with respect to DBP's;
- the PAC facility will be used only when the ozone/GAC facility is not operating at full capacity for a short period or during quality deterioration during periods of drought;
- the membrane choice between micro- and ultrafiltration was only made during tender adjudication stage.

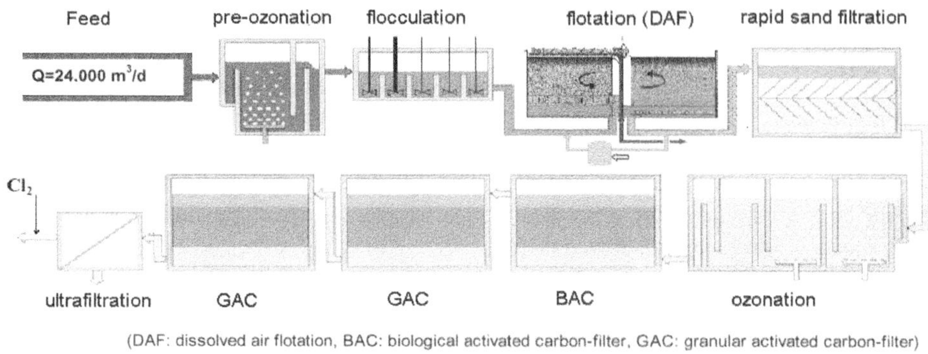

Figure 24.2 Final treatment train for the New Goreangab Water Reclamation Plant.

24.6 OPERATION, WATER QUALITY AND COMPARATIVE COSTS OF THE NEW PLANT

As summarised in Table 24.3 (above) the new plant produced good quality water that complied to a large extent with the required final water specification.

24.6.1 Operational experience to date

Enhanced coagulation is practiced (considered as best practice) where pre-ozonation is a critical component for floc development and DOC removal, typically 40% at the dissolved air flotation process.

With effective iron and manganese oxidation across the sandfilter media, no significant DOC removal is experienced. The main ozonation process ensures the fractionisation of organic substances. Biological GAC filters as well as the GAC filters account for 20% DOC removal respectively, which highlights the necessity of the BAC filters as an effective process that extends GAC saturation and reduces operational cost.

24.6.2 Water quality and monitoring

From the experience gained over the past 18 years since 1990 with the operation of the old reclamation plant, and during the research for the design and operation of the NGWRP, the decision was taken that instead of testing for numerous individual parameters, testing would instead be done on a more frequent basis for certain key parameters at process points that represent a barrier for that particular parameter or would contribute towards its removal.

The monitoring is divided into three parts. First is the on-line monitoring for key parameters at different processes. Second is the process monitoring part, which is vital in the control of the proper operation of a process unit; for representative sampling, automatic samplers are used. Third is the health monitoring program, which is aimed at parameters that are still in a state of research or that are mostly measured in concentrations where it does not constitute an immediate health concern with the current available knowledge but which are still deemed to be important for inclusion in the monitoring program. To ensure that the treatment unit would be operated according to specification, heavy penalties are levied if certain parameter values are not met at critical control points.

On-line monitoring of Giardia/Cryptosporidium and viruses: on-line sampling devices have been developed in-house to sample one litre every 15 minutes over a period of seven days (filtration volume: 762 litres) for Giardia and Cryptosporidium, (Testing methodology CSIR RSA Method PMP 1 May 2003 as adapted from USEPA [1999] Method 1622 and 1623). These have been proven to be very effective and give a clear indication if a plant failure or breakdown of a barrier occurred in a certain period. The ultra filtration device will also be used for the concentration of viruses onto a cartridge.

On-line monitoring for chemical parameters: pH, turbidity, conductivity, dissolved organic carbon (DOC), UV254 absorption, dissolved oxygen (DO), particle counters, ozone concentration and free chlorine concentration are measured at various positions in the process.

Automatic composite samplers are placed at critical control points in the process to collect the desired volumes of samples over a 24-hour period for chemical analysis.

The operational monitoring program includes:

- physical and organoleptic: Turbidity, colour, pH, conductivity, TDS, hardness, corrosiveness;
- macro elements: Potassium, sodium, chloride, sulphate, fluoride, iron, manganese, silica;
- organics and nutrients: Dissolved organic carbon (DOC), Chemical oxygen demand (COD), UV254 absorption, nitrate, nitrite, ammonia, total kjeldahl nitrogen (TKN), ortho-phosphate and trihalomethanes (THM);
- microbiology, biology and parasites: Heterotrophic plate count (HPC), Total coliform (Tc), Faecal coliform (Fc), E coli (Ec), Faecal streptococci, Clostridium perfringens (viable and spores), Coliphage (Somatic), Giardia and Cryptosporidium, chlorophyll a.

Monitoring (Research) for health aspects includes:

- micro elements: silver, argon, arsenic, gold, boron, barium, cadmium, copper, chromium, cobalt, mercury, nickel, lead, selenium, tin, zinc;
- toxicity and mutagenicity: Waterflea lethality (24 + 48 hours), Urease enzyme assay, Bacterial growth test and Ames Salmonella mutagenicity;
- virus: Enteric virus by cytopathogenic effect and Polymerase chain reaction (PCR), f-RNA phages;
- other tests that are or have been part of a research program or that are conducted at irregular intervals: Bromate, estrogens: estrogenic activity, inflammatory activity, medical substances, organic pollution index (OPI).

24.6.3 Quality concerns with the present process configuration

The following concerns needs to be addressed with the operation of the NGWRP:

- the build-up of TDS in the distribution system. If not addressed the water quality of the NGWRP will soon exceed the stipulated TDS value of 1 000 mg/L. This is aggravated during periods of drought with the higher production from boreholes with higher TDS values;
- another concern is the high concentrations of nitrates emanating from the wastewater treatment plant. This will have to be addressed at the wastewater treatment plant by creating longer retention times in the anoxic basins;
- during droughts Windhoek experiences increasing DOC concentrations due to the deterioration of the water quality from surface sources. An investigation is underway to address this at the Von Bach Water Treatment Plant that supplies water from surface sources to Windhoek with relatively high DOC values.

24.6.4 Cost considerations

The cost of water produced by the NGWRP versus other supply sources are summarised in Table 24.4. The cost for new supply augmentation options were calculated according to the Unit Reference Value based on the estimated capital investment as well as the estimated costs based on 2007 cost estimates with an 8% return on investment. All the alternative supply augmentation options are based on an accepted level of security of supply. The project design goal is based on a 1 in 100 year risk that shortfall magnitudes will not exceed a 13% annual shortfall in supply over the 18 years planning period for the next scheme.

Table 24.4. Cost comparison of existing and future water sources.

Supply Source	N$/m^3	U$/m^3
Groundwater (Borehole supplies)	1.75	0.25
NamWater Supply (Surface sources)	5.75	0.82
New Goreangab Water Reclamation Plant Supply	7.35	1.05
Estimated Artificial Recharge Supply (Unit Reference Value)	10.60	1.51
Estimated Tsumeb Karst Aquifer Supply (Unit Reference Value)	22.50	3.21
Okavango River Supply (Unit Reference Value)	188.00	26.86
Irrigation Water Production Costs	3.65	0.52

Note: The exchange rate was accepted as N$7.00=U$ 1.00

24.7 PUBLIC ACCEPTANCE OF DIRECT POTABLE REUSE

On 24 November 1968, the Sunday Tribune carried a headline: "WINDHOEK DRINKS SEWAGE WATER: a purified world first". The city Engineer of Johannesburg made the following statement in 1974: "It had been predicted that by the year 1985 we would all be drinking sewage and by the turn of the century there would not be enough to go around". In practice, this prediction has not been realized with respect to direct water reuse. However the question is, was it realised in certain places without the consent of the consumers through indirect reuse of water and even with conventional water treatment processes in place? The debate on direct versus indirect potable reuse is not actively pursued today, but possibly only for reasons of expediency. It is anticipated that the 21st Century consumer would not summarily accept their water if he/she knew fully the path travelled by the individual water molecules in the water cycle, from precipitation up to consumption and back to nature.

The US Environmental Protection Agency investigated the number of domestic water supplies that contain wastewater. The study examined 525 towns and cities with populations over 25000 and showed that 15 million people were supplied from surface supplies containing at least 10% wastewater and 4 million people by sources that contained 100% wastewater during low flow conditions. The study concluded that the indirect, unplanned reuse of wastewater for domestic purposes was widespread, and that wastewater sometimes represented a significant portion of the total flow in many receiving waters (USEPA 1980 as Reported by Salmon).

In the American Waterworks Association Journal of January 1968, Dwight F. Metzler and Heinz B. Russelmann, reported on how the city of Chanute, Kansas, survived the severe drought of 1956, on water from the virtually dry Neosho River. They reported: "The 4,000 sq mile drainage area yielded only the wastes of upstream communities. Much of the river flow past Chanute during the drought was made up of sewage plant effluent, and the city had been using this water for several months without known ill effects."

Today, 40 years later, Windhoek is still the only example of direct potable reuse. The question should be asked why the people of Windhoek have come to accept and are indeed proud of still being the only place in the world where direct potable reuse is practiced on a commercial scale. In the case of Namibia and Windhoek in particular, the water supply situation described at Chanute, where effluent is the only reliable water source, occurs almost every 5 years. The main reason for water reclamation in 1968 was related to fact that there was no available supply alternative.

24.7.1 Historic reasons for acceptance

Perhaps the most important cornerstone of potable reclamation is public acceptance and trust in the quality of this water. The most difficult task facing any town or city wanting to emulate what is done at Goreangab in Windhoek would be to overcome the obstacle of the consumers' psychological barriers to the principle of direct reuse for potable purposes.

When reclaimed water was first introduced in Windhoek in 1968, a local daily newspaper "The Suidwester" interviewed a large number of residents and widely published the facts. Public opinion ranged from very positive, to the point that the water was said to be better than what people were used to, to people indicating that they felt a little nervous drinking the water. Rumours that the water had caused an outbreak of typhoid were investigated by the same newspaper; A Dr Hitzeroth of the Department of Health gave the assurance that no such outbreak had occured (Die Suidwester, 1968).

In a letter dated 6 December 1968, the Chief Scientist of the Windhoek Municipality writes to Dr Lucas van Vuuren at the National Institute for Water Research in Pretoria: "A press release was made on 22/11/68. It actually revealed little. In general, the news was received very calmly. No single complaint has been lodged with me in 14 days. The public has definitely accepted it very well. The Southwester' handled the matter very well and carried a very good main article on Monday 25/11/68. The Department of Health (Administration) also had no objection. I had invited their Heads (Dr Oosthuizen and Dr Hitzeroth and others) to come and visit (the plant) and they were all satisfied – after a large number of questions from Dr Hitzeroth." (Engelbrecht, 1968).

Dr Lucas van Vuuren, a pioneer of water reclamation in South Africa during the 1970s, coined a phrase: "Water should be judged not by its history, but by its quality." To get public acceptance of this viewpoint, however, is not easy; it places a severe responsibility on the city to exercise the required level of ongoing control over water quality. For this purpose, the city has over the years invested substantially in laboratory facilities and manpower. The Gammams Laboratory of the city's Scientific Services Division boasts state-of-the-art facilities and analytical equipment to ensure customers' continuing comfort with using the treated water. This is backed up with a rigorous monitoring program covering all the supply sources, the reservoirs and the consumer points in the distribution network.

The fact is that reclaimed water is widely used for aquifer recharge, but often in such a way that it loses its identity as sewage water. Water Factory 21 in Orange County, California, is a prime example, where reclaimed water is injected into the coastal freshwater aquifer to prevent intrusion of saline seawater into the aquifer. This reclaimed water ultimately re-enters the drinking water stream. Also, treated sewage is often discharged to rivers and lakes. It is not uncommon that downstream users reuse such releases, which are then treated as natural water through conventional treatment methods. These are in fact instances of indirect reclamation or reuse, even though the water in question has lost its "sewage identity." At Goreangab, the history of the feed water is specifically recognized as treated sewage and treatment is designed to cope with just that. In that sense, the very aspect of this water that initially made people reluctant to use it, is in fact the aspect that leads to its being adequately treated to ensure its quality.

24.7.2 Acceptance of the latest increased production

Over the years, the Goreangab Reclamation plant has been widely publicized locally and internationally. Reclaimed water has become a fact of life in Windhoek, and the average citizen drinks water from the tap on a daily basis. Very few if any water quality complaints ever linked bad quality to Goreangab.

About 15 articles per year are carried in the local newspaper on water issues. The reports range from bad water quality in the distribution system to announcements on water savings measures and treatment plant upgrades to Windhoek citizens. One third deal with national issues on water supply in different regions of the country, the Water Act and international water issues. The water quality problems are mainly caused by the bulk supplier's inability to handle diurnal taste and odour problems caused by algae blooms on the Von Bach Dam. This leads to many consumer complaints about the foul smell of the water; many cartoons have appeared demonstrating citizens' dislike of consuming such water. These problems have placed a heavy burden on the city of Windhoek to embark on an intensive monitoring program covering all supply sources, reservoirs and consumer points in different residential areas and supply zones to be able to link quality problems to a particular supply source. When a consumer complaint is being dealt with, the consumers are fully informed about the water situation, a sample is taken at the consumers property and a report on the water quality is issued. In 99% of the cases the consumers accept the reasons given and the remedial measures taken by the city of Windhoek. When such incidents occur the city of Windhoek informs the public through a press release.

Since 1995 the city of Windhoek has embarked on an active campaign to inform the public about water saving measures. This includes inviting schools and different public target groups including various people from different diplomatic corps to visit the wastewater treatment plant, the laboratory facilities and the reclamation plant. Water pollution, water saving and environmental protection issues have also been introduced into the school curriculum at primary and high school levels.

In 1992 a technical committee of experts in the different disciplines of water and wastewater treatment and health aspects was selected to provide advice during the planning and design stages of the reclamation plant. A group of knowledgeable people in the public health sector and water affairs were also invited to attend meetings or read the reports from the technical committee. This kept key persons informed about the technical soundness of the reclamation process and efforts to cover the important health issues with the monitoring program.

When the new Goreangab plant was inaugurated by His Excellency the President of Namibia in December 2002, the event was televised and widely reported in the local media.

To retain public confidence, water quality at the Goreangab Treatment Plant is monitored on an ongoing basis through on-line instrumentation as well as through collection of composite water samples after every process step. In the event of any quality problems (specified target values that are not met), the plant goes into recycle mode and water is not delivered. The final water product is also continuously sampled by way of online instrumentation and composite sampling and is analyzed for the full range of currently recognized contamination parameters, inclusive of pathogens such as viruses, Giardia and Cryptosporidium.

The city recognizes that viruses also constitute a real threat, and these are equally monitored and studied as an ongoing research project; however, virus studies world-wide are mostly in the research phase and much work needs to be devoted to this subject before more specific treatment barriers can be designed. Currently ozone treatment, ultrafiltration and chlorination combined are deemed to form an effective barrier for viruses.

In its own news publication, the city of Windhoek makes sure that these facts are well publicized and that the citizens are informed. In eight years, no single consumer has complained about the fact that reclaimed water was supplied. Inquiries received centred more around the possibility of contracting HIV Aids through the medium of reclaimed water. With the aid of science and knowledge of the virus, it has always been possible to satisfactorily deal with such enquiries.

During 1996 a local newspaper enquired about the possibility of medicinal substances being unintentionally distributed to consumers via the water supply through direct potable reuse. At that time this subject became a big issue in Europe. The City of Windhoek was unable to respond to the article, but in 1998 samples from the Goreangab treatment train were analysed by a European laboratory for certain groups of medical substances. The results confirmed that through the maintaining of sound process protocols, the reclamation plant successfully removed all the substances in these groups.

The citizens of Windhoek have over time become used to the idea that potable reuse is included in the water provision process. In fact, they have grown to harbour a fair amount of pride in the fact that their city in many respects leads the world in direct reclamation.

The distribution layout of the bulk supply system prior to 2002 was such that reclaimed water could only be distributed to the main supply zones such as the CBD and high-income areas with high water consumption within the city. A new pump station was designed to distribute blended water (surface water and reclaimed water 2:1 ratio) to all domestic and the central business areas, which account for approximately 93% of the daily water consumption. Two industrial areas receive surface water only, for the production of consumer goods for export.

24.8 NEW RESEARCH AND DEVELOPMENT OPTIONS

24.8.1 Process related refinements

From a process optimisation point of view, various projects have been identified and are currently in progress to enhance process efficiency and to reduce operational cost. The evaluation of alternative chemicals for coagulation and flocculation is continuously performed such as the use of polymers and ferric salts.

The sensitivity of the ultra filtration membranes to DOC increases and accompanied rapid membrane fouling have resulted in the continuous evaluation and adjustment of the chemical cleaning regime. This includes various blends of caustic soda and sodium hypochlorite or the respective singular use as dictated by the quality of the feed water.

Effective wastewater treatment is one of the corner stones of a successful reclamation scheme. The city of Windhoek will invest considerable amounts for the upgrade of its Gammams Wastewater Treatment Plant. The primary process which includes effective digestion of sludge will be upgraded in the next three years. An additional settling tank is currently being built to operate the nutrient removal activated sludge process at a longer sludge age or higher mixed liquor suspended solids concentration. It has been found that lower DOC and UV254 can be achieved. Nitrification and phosphorus removal is also enhanced.

In future the use of Membrane Bioreactors (MBR) will be tested to convert the current wastewater treatment plant to improve the effluent water quality especially through the biological reduction of organic carbons.

24.8.2 Quality control

The QC system in place can be divided into three categories: four-hourly sampling and analysis, on-line sampling and analysis, and weekly composite sampling and analysis. QC is comprehensive by any measure and results are used for performance measurement as well as to demonstrate that the risks involved in direct reclamation are well managed.

The microbiology and virus monitoring program has been changed to include critical control points according to the Hazard Analysis and Critical Control Points (HACCP) principle where certain relevant parameters are tested on a more frequent basis to establish direct relationships between different microbiological indicators and virus detected by means of the Polymerase Chain Reaction (PCR) method. From this, the risk and removal efficiency will be determined. A new development is the continuous sampling for the concentration of virus on a suitable medium. Another development is the continuous sampling and concentration of parasites on a filter. This has also proven over several years to be a good visual indicator of certain treatment plant mishaps.

The HACCP or critical control point principle has also been introduced in the monitoring of the distribution system. This has helped considerably to improve the water quality in the distribution system.

24.8.3 Health

Important issues related to health especially with the use of ozone are bromate and THM formation. This research is nearing completion and early indications are that the multiple barrier approach reduces the risk of THM in the final water to the extent that European Water Quality Standards are consistently met. The determination of bromates is part of a current study.

Research on the prevalence and occurrence of endocrine disruptive substances as identified and prioritised by the Global Water Research Coalition in the process water is in progress. Due to the fact that Windhoek does not have a significant industrial sector and that industrial wastewater is diverted from domestic wastewater, the components identified for observation are minimised.

From 2001-2003 estrogenic screening tests and inflammatory activity tests, developed during a research project at a South African University, were conducted on Windhoek's supply system. Findings suggest that estrogenic activity was removed during reclamation. Inflammatory screening tests suggest a seasonal pattern, which is higher during the rain season. The New Goreangab Water Reclamation Plant is effective in reducing it. Indications are that if very high inflammatory activity is present in the raw water, breakthrough could occur in the final water. This work will be followed by direct determination of estrogenic substances. Further research on additional screening tests is needed as this can be used as a cheaper indicator test than testing directly for hormones.

Research on the quantitative characterisation of organic substances in the raw water sources and process water, the removal by each process and the use of the information for process optimisation was initiated recently.

24.9 CONCLUSION

The experience of Windhoek with the direct reclamation of domestic wastewater to a potable standard over 40 years is important in view of water shortages in certain parts of the world. The reclamation of water can increase the total quantity of water substantially, at an affordable price, without impinging on natural water sources. The cost of reclaimed water in the case of Windhoek is approximately 30% of the cost of additional augmentation schemes from conventional supplies over long distances.

Reclamation should not be treated as a technical process on its own but forms part of a holistic approach in respect of the water supply strategy, collection of waste water, prevention of pollution of waste water, optimal treatment of waste water through biological processes, public participation and water demand management.

With the development of new technology as well as the refinement of test methods for water quality, the process is dynamic and should be continuously scrutinised and upgraded to ensure safe potable water.

The new reclamation plant by far exceeds the performance of the previous reclamation plant. It is good to know that the targets that had been defined for each treatment unit can consistently be reached after one and a half years in operation. It can truly be called advancement in reclamation technology. Although many operational challenges remain to be solved, the new plant has proved to produce water of potable quality to national and international drinking water specifications. Therefore the plant can be seen as a success to augment the limited resources of a city on a cost-competitive basis compared with conventional water supply sources.

This again begs the question, why is Windhoek still the only place where direct potable reclamation is practised on a commercial scale? Are the people of Windhoek sufficiently ignorant to accept the fact that treated sewage effluent is treated with appropriate technology, or are consumers elsewhere sufficiently ignorant of the fact that treated sewage water might be indirectly reclaimed, probably treated conventionally and distributed as drinking water?

The sage words of Dr. Lucas van Vuuren, one of the pioneers of the Windhoek reclamation system, have successfully withstood the test of the past 40 years:

"Water should be judged by its quality; not its history"

24.10 ACKNOWLEDGEMENTS

Appreciation is recorded for the numerous individuals (and organisations) who have been involved with the Windhoek Water Reclamation Plant since the early 1960s; for their vision, perseverance and sterling work. The authors acknowledge the enormous contribution made during the past years by diligent individuals who performed innumerable analyses, who operated pilot plants under trying conditions and who analyzed reams upon reams of data. These efforts have provided the scientific base necessary for the critical selection process for the new process design. Many good friends in their special fields of expertise from abroad have contributed with their invaluable advice to help finalise the design of the plant.

The German KfW (Kreditanstalt für Wiederaufbau) and the European Investment Bank financed the construction of the New Goreangab Water Reclamation Plant through favourable loan conditions. The city of Windhoek made all information available for the preparation of this chapter.

24.11 REFERENCES

Brand JG, (1957) *On the Water Position in Windhoek during November 1957 and the Expected Consumption Trend for The Future* Memorandum of the Town Engineer, Windhoek, Dated 16/11/1957 and Submitted to The Secretary of South West Africa on 18 November 1957.

Central Area Joint Venture Consultants, (2004), *Feasibility Study on Water Augmentation to the Central Area of Namibia*, Volume 6, Report No. NWPC-IP-Tsumeb97-1, Client: Namibia Water Corporation Ltd.

Clayton AJ, (April 1974), *The Engineering Economics of Water Reclamation* Paper delivered to the Fifty First Conference of the Institution of Municipal Engineers of Southern Africa,

Clayton AJ, Personal communication, July 2005.

"Die SuidWester" Windhoek daily newspaper 22 November 1968.

CSIR (2003), Method PMP 1 Revision 2, "Determination of Cryptosporidium Oocysts and Giardia Cysts in Water"

Dwight F Metzler and Heinz B. Russelmann: "Wastewater Reclamation as a Water Resource"; American Waterworks Association Journal January 1968.

Engelbrecht, A.J.,(1968) Chief Scientist Windhoek Municipality: An official communication from the Chief Scientist to the Director of the National Institute for Water Research, CSIR, Pretoria 6 December 1968)

Haarhoff, J., Van der Merwe, B. F. (1996) "Twenty-five Years of Wastewater Reclamation in Windhoek, Namibia" Water Science and Technology, Volume 33, No 10-11.

Haarhoff, J, Van der Walt, C.J, and Van der Merwe, B.F., (1998), "Process Design Considerations For Windhoek Water Reclamation Plant" Paper presented at the biannual Conference of the Water Institute of Southern Africa, Cape Town May 1998.

Namibian Guidelines, 1991, Guidelines for the Evaluation of Drinking Water for Human Consumption with Regard to Chemical, Physical and Bacteriological Quality, Department of Water Affairs, Republic of Namibia.

Namibian Economic Research Unit (NEPRU), 1996: "Projections and Opinions on Economic and Business Prospects in Windhoek", 1996.

Rand Water, 1994, "Potable Water Quality Guidelines", Scientific Services, Rand Water, Johannesburg, Republic of South Africa.

Stander, G.J, Clayton, A.J. (1970), "Planning and Construction of Waste Water Reclamation Schemes as an Integral Part of Water Supply": Paper presented to the 1970 Conference of the Institute of Water Pollution Control, Cape Town March 1970

TRP Associates, 1996: "1995 Residents Survey Report". Report prepared for the Municipality of Windhoek, 1996.

USEPA (1980), "Wastewater in Receiving Waters at Water Supply Abstraction Points"; EPA-600/2-80-044, U.S. Environmental Protection Agency, Cincinnati, Ohio.

USEPA, 1996, "National Drinking Water Standards and Health Advisories", United States Environmental Protection Agency, Washington DC, USA.

Water Research Commission, Republic of South Africa, 1982, "Guide for the Planning, Design and Implementation of a Water Reclamation Scheme"

Windhoek Consulting Engineers, 2000, "Analyses of Present and Future Water Demand in Namibia", Client; Namibia Water Resources Management Review.

Wipplinger O, (1957) "Memorandum on Windhoek Water Supply", Addressed to the Honourable the Administrator-in- Executive Committee dated 18 December 1957 and referenced w.w 80/1.

Wipplinger O., (1956) – "Report to The Administrator in Executive Committee on Interim water supply Scheme for Windhoek"

World Health Organization (1993), "Guidelines for Drinking Water Quality", World Health Organization, Geneva, Switzerland.

25

Industrial water resource management and recycling in Germany: case studies from the food and beverage industry

Karl-Heinz Rosenwinkel, Jörg Brinkmeyer and Maike Beier

25.1 INTRODUCTION

Industrial water resource management and recycling are interesting alternatives to end-of-pipe-technology and for saving water resources. Water is needed in industry for products, for cleaning and different processes, so that a water management system is necessary to determine the part streams that can be recycled and the technology required for part stream treatment. A decision support system is required for water resource management that takes into account the economic and ecological components of the treatment and recycling technology. In every case, possibilities for internal measures like recycling, product recovery, reduction of loads, nutrient content, temperature, cleaning and disinfecting agents have to be verified. Normally it is much more efficient to reduce the waste water load by in-plant measures than by the handling of organic or inorganic loads in municipal or industrial waste water treatment plants. Within the scope of long-term half- and full-scale experiments in different branches of industry, information about costs, practicability and effectiveness from closed loop systems has been brought together. All experiences were collected in close

©2008 IWA Publishing. *Water Reuse - An International Survey of current practice, issues and needs* by B. Jimenez *et al.* ISBN: 9781843390893. Published by IWA Publishing, London, UK.

collaboration with industry partners and this case study shows a review of an assortment of the results. The compiled experiences could be applied to new developments and refits, either at home or abroad, to reproduce the potential benefits under local conditions.

25.2 IN-PLANT MEASURES AND INTEGRATED ENVIRONMENT PROTECTION TO REDUCE WATER VOLUMES FOR INDUSTRIAL PROCESSES

In Germany, there is long-standing experience including substantial data available for different production technologies and involving specific water volumes and waste water loads. These data compare water volumes and waste water loads for different industries and different technologies, and may be used to identify the best available technologies for a production process. The reasons for saving water in industries in Germany are normally cost-oriented, but in some cases they are image-oriented.

German industry used approximately 10 billion m^3 of water during 1995, most of it being surface water and 27% groundwater. When looking at the tobacco and food industry at a total volume of approximately 432 Mm^3/d, we see that most of the water volume for the food industry is taken from groundwater and from the public water system (Figure 25.1).

The utilisation factor for water is defined as the quotient of utilisation divided by the water provided. In terms of the food industry, Germany had a stable utilisation factor of approximately 3.5 from 1980, though it increased again in 2000 (Figure 25.2). The reason for this is to be found in the requirements for the quality of water in the food industry, combined with the costs of extensive measures for treatment to achieve a high hygienic standard for recycling in the food-industry.

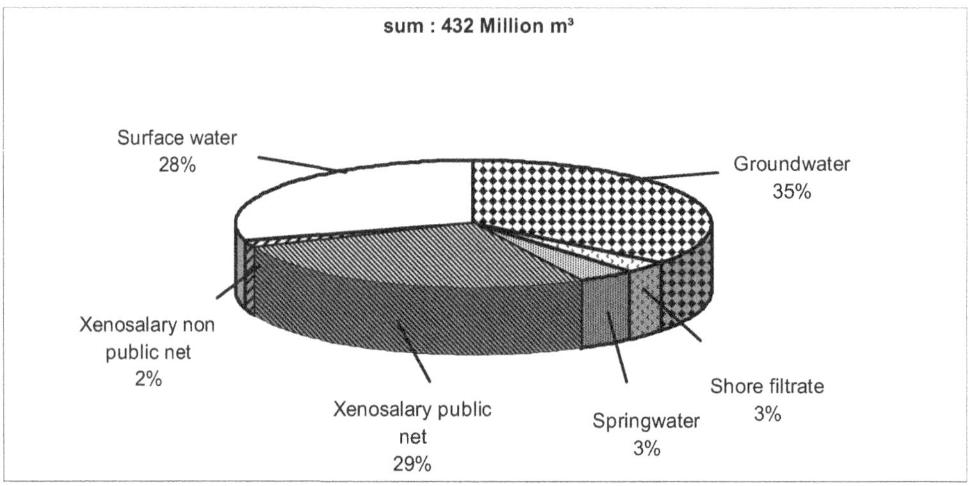

Figure 25.1. Water consumption in the food and tobacco industry 2001.

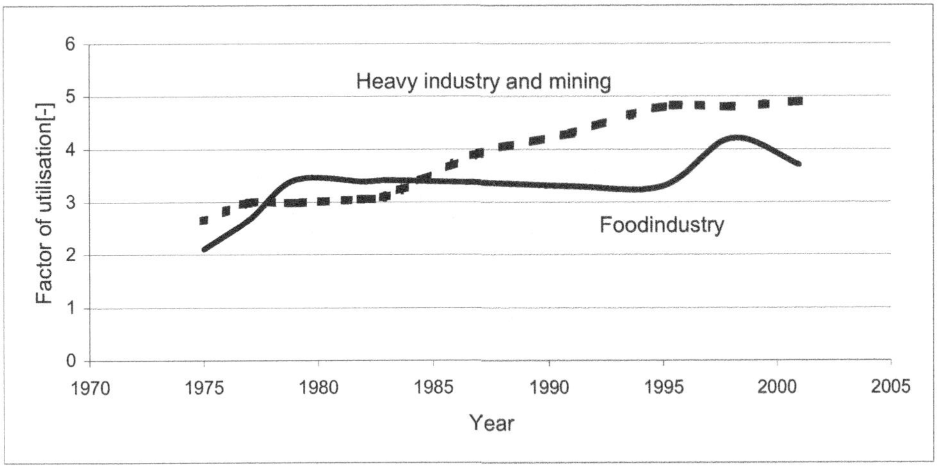

Figure 25.2. Development of the utilisation-factors.

25.3 IN-PLANT MEASURES

The method for reducing water demand and load through in-plant measures must be divided into three main steps.

The first step is a detailed determination of production demand and load in terms of the production scheme, the water demand in different part streams, entailing the calculation of specific demands and loads. These calculations and measures have to be divided into different load components, like carbon, nitrogen, phosphorus, suspended solids, heavy metals and so on. After calculation of the specific demands and loads for comparison with other examples of this branch of industry, a proposal for in-plant measures has to be prepared and, as a third step, the calculation of efficiency of the extensive measures for recycling has to be carried out in order to determine the efficiency of the measures.

In Germany, there is substantial specific data regarding different sectors of the food and non-food-industry. Table 25.1 shows the specific volume of waste water in m^3/unit-production and the wastewater load in kg BOD/unit-production, and also the population equivalent for the BOD load per unit-production for several branches of the food industry. The figures have been stable in most branches over the last few years. The cost of further measures is in most cases not efficient. However, technological advancements continue and in some cases – when not enough fresh water is available – the situation changes, and it would be possible to develop technologies to reduce the given specific data in terms of load and volume.

In general, there are a lot of proposals for reduction measures, such as training personnel, as well as daily measurements and cost calculations. However, in addition, there are many possibilities that could be integrated into the process to reduce water, like changing transport systems or changing the cleaning system and integrating water cycles, or changing cooling systems. In terms of load reduction in the case of the food industry, the recovery of products from waste is doubly efficient. Through this measure, it would be possible to increase production and to decrease the waste load.

Table 25.1. Specific waste water loads in relation to several branches of the food industry (Rosenwinkel 1991).

operating mode	unit	specific wastewaterloads		population equivalent
		m^3/unit	kg BOD/unit	
dairy	t	1 - 2	0.8 - 2.5	13 - 42 /t
brewery	hl	0.25 - 0.6	0.3 - 0.6	12 - 15 /hl
wine	ha	0.04 - 0.3	0.32 - 0.97	5 - 16 /ha
distillery	d	0.5 - 0.8	6 - 35	100 - 500 /d
beverages	m^3	1.4 - 2.8	1.7 - 4.5	20 - 60 /m^3
fruit juice	m^3	1.8 - 2.8	1.7 - 4.5	28 - 75 /m^3
slaugtherhouse				
cattle	C	0.5 - 1.0	1.0 - 3.5	28 - 85 /C
pigs	P	0.1 - 0.3	0.2 - 0.3	5 - 9 /P
tinned food				
peas	t	12 - 30	18 - 30	300 - 500 /t
carrots	t	19 - 30	25 - 30	415 - 500 /t
beans	t	15 - 35	10 - 22	165 - 365 /t
fastfood	10^3 piece	2.5 - 3.6	1.5 - 6.4	38 - 160 /10^3 piece
Sauerkraut	t	5 - 9	4.2 - 9.2	70 - 150 /t
Potato processing	t	5 - 8	5 - 10	85 - 170 /t
sugar factory	t	0.5 - 1.0	0.8 - 1.6	13 - 27 /t
leaven	t	10 - 80	140 - 250	2330 - 4170 /t
margarine	t	1 - 3	(0.5 - 3.0)	(8 - 50) /t
cooking oil	t	10 - 25	(3.0 - 7.0)	(50 - 115) /t

t - [ton] hl - [hectoliter] ha - [hectare] d - [day] m^3 - [cubic meter] C - [Cattle] P - [Pig]

25.4 QUALITY ASSURANCE IN FOOD AND BEVERAGE INDUSTRY DURING REUTILISATION AND RECIRCULATION

Water recycling continues to gain in significance and is also, for economic reasons, gaining entry into more sensitive areas of production. Treated water is fully recirculated and is maintained at the same level of quality through permanent treatment. Biological contamination is only evident after a period of at least 48 hours through conventional processes; germfree water which contains an obtainable substrate can be stored for this time before bacterial contamination starts. Preventative measures like chlorinating or the addition of biocides, when permitted under law, are often undesirable since these substances adhere to the product and in addition substantially contaminate the waste water streams. Additional components that possibly kill germs that are present entail substantial additional costs, and the measurement of the functionality of these systems involve further costs. In particular, UV-lamps in the medium- and high-pressure range, as well as oxidation (advanced oxidation process) are used. Unfettered application is, however, not possible since for example with much of the wash water the transmission values of the filtrate are so poor, even after pore filtration, that an extremely high UV dose is required. During oxidation, operating costs also rise considerably in line with increased COD content.

25.5 EXAMPLES FROM DIFFERENT GERMAN INDUSTRIES

The following sections outline examples from German industries. The tables and graphs indicate some full-scale practical examples from different branches of German industry.

25.5.1 Sugar industry

Water demand in the sugar industry has decreased in Germany in the last 20 years almost to zero. The only waste water volume emanates from sugar beet, which today is approximately 0.8 m^3/tonne sugar beet.

During sugar production, however, up to 17 m^3 of water per tonne of beets is required. This total can be divided into the following types as indicated in Table 25.2.

Table 25.2. Water and waste water types in the sugar industry (Rüffer and Rosenwinkel 1991).

Water type	Consumption (L water/tonne beets)
Flume water	500 – 800
Sweet water	150 – 200
Hot well water	400 – 600
Fresh water for juice extraction	30 – 40
Cooling and sealing water	20 – 100
Total consumption	600 – 1740

This water consumption is, however, not covered through fresh water, but through recirculation of process water and supplementation of losses with process condensates.

Since the reduction of water consumption cannot be further optimised, future task areas lie in the reduction of waste water contamination with organic substances. Product losses are to be kept to a minimum, both from a waste water technical viewpoint and economic perspective.

25.5.2 Malting industry (Ahrens 1998)

A variety of soaking processes are utilised at malt factories. Depending on the soaking process, the volume of waste water varies and contamination exhibits considerable differences (Table 25.3).

Table 25.3. Contamination of soaking water (Ahrens 1998).

Parameters	Immersion soaking process		Spray soaking process
	Box	Germinating line	
Spec. volume (m^3/t barley)	1.8 to 2.5	1.73	0.2 or 0.2
BOD (mg/L)	600 to 880	410	1800
COD (mg/L)	925 to 1732	823	2720
Sediments (ml/l)	1.8 to 2.9	2.9	13

In terms of the malting industry, an integrated process water treatment system can be illustrated. The integrated system (Kraft 1997) uses a microfiltration unit as its treatment technology. The treated water, virtually disinfected through membranes, can be returned into the production process so that only a small input of fresh water is necessary for evaporation from the product. The BOD from the effluent is approximately 5 mg/L and the ammonia concentration lower than 1 mg/L. Further advantages are the high reduction rate of germs, a low footprint of the technology, and low sludge-production (Figure 25.3).

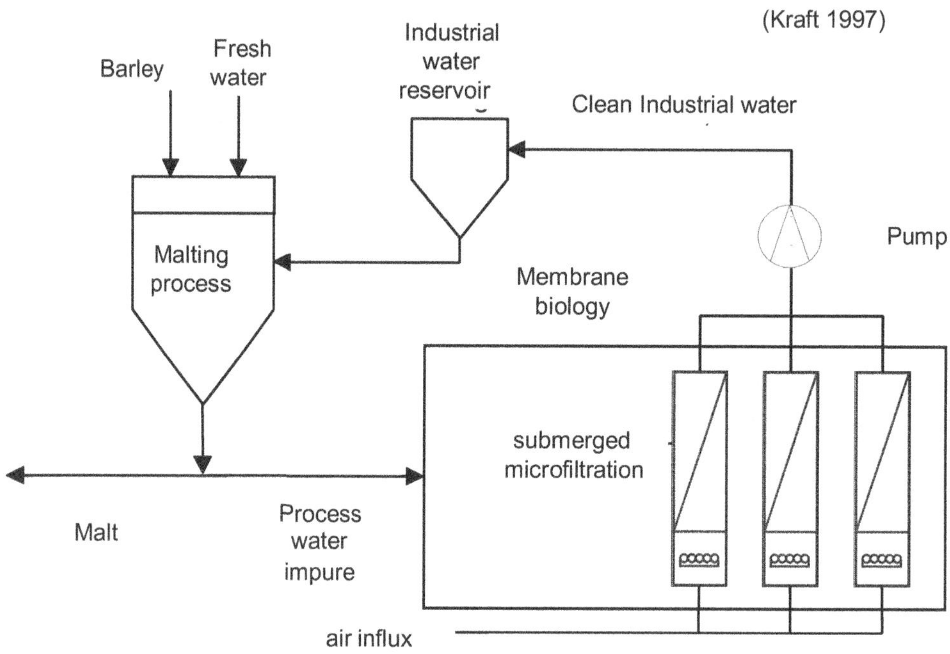

Figure 25.3. Malting industry: integrated process water treatment (Kraft 1997).

25.5.3 Beverage industry (ATV Handbuch Industrieabwasser 2000)

Table 25.4 shows some of the part streams of the beverage industry which may be recycled. In some instances these part streams are used in full-scale; in others, like the effluent of a treatment plant for boiler water, it is only subjected to a scientific study.

Table 25.4. Water reutilisation in the beverage industry (ATV Handbuch Industrieabwasser, 2000)

Part stream	Recycling as
Bottle cleaning lye	Lye cleaning and recycle
Bottle cleaning (last flushing step)	Recycling
Ion exchanger regenerate	Freshwater
Effluent treatment plant	Boiler water
Vacuum pumps	Recycling

The most important water demand in the beverage industry is bottle washing machines – the aim with these machines being on the one hand to recycle the water after filtration or nano-filtration, and on the other hand also to recycle the regenerated tenside solution in order to reuse chemicals and reduce costs. A study shows that the best technology for the recycling of tensides is micro-filtration. In other instances, tensides and COD go into the concentrate (Table 25.5).

Table 25.5. Water re-utilisation in the brewery (according to Schildbach 2000).

Lye / lime treatment with Micro, Ultra-, Nanofiltration			
	COD retention	Lye retention	Tenside retention
Microfiltration	24 %	0 %	50 %
Ultrafiltration	34 %	0 %	80 %
Nanofiltatrion	61 %	0 %	85 %
advantages: clean lye	disadvantages: loss of tensides		

A study was carried out in Hanover regarding the further treatment of pre-treated waste water for use as boiler water in a brewery that shows that this technology is possible with the system of reverse osmosis, but that it is too expensive for use in Germany (the price for water produced by reverse osmosis is higher than fresh water).

Table 25.6. Re-use of treated wastewater as boiler water (ATV Handbuch Industrieabwasser, 2000).

Treatment of waste water effluent and usage as boiler water	
Effluent treatment plant	COD = 22 mg/L Suspended solids = 2.7 mg/L Cl = 607 mg/L
System	Reverse osmosis
Results	good permeability good salt retention good Ca retention (Ca<10 mg/L)

25.5.4 Agrar Bio Recycling GmbH in Wietzendorf (Theilen and Richter 2005)

Description of the treatment plant in 2002

The wastewater treatment and biogas plant of Agrar Bio Recycling ABR GmbH was constructed during 2001 and 2002 to convert, as far as possible, high load wastewater from the Wietzendorf potato starch works of Emsland-Stärke GmbH. As an independent concern, there is a requirement for economical operation as well as the tasks of disposal and conversion. The plant began regular operation in August 2002, on commencement of the 2002/2003 season/campaign.

The plant is divided into five operational units that comprise various individual components and modules. The original 2002 concept looked as follows:

- **Operational unit 1:** Protein separation for fruit and process water through acid and heat coagulation
  - heat exchanger
  - acid dosage
  - sedimentation
  - decanter system
  - drying plant.

- **Operational units 2 and 3:** Biogas plant (2-step process)
  - 2 hydrolysis reactors of 750 m³
  - 4 biogas fermenters of 2600 m³
  - 1 secondary fermentation reactor of 2600 m³
  - 2 thickening decanters
  - 1 dehydration decanter.

- **Operational unit 4:** Gas utilisation
  - gas accumulator 6,000 m^3
  - CHP plant with 4 CHP modules (3+1) of 2.1 MWel
  - 4 steam generators (3+1).

- **Operational unit 5:** additional water treatment and fertiliser production involving:
  - softening
  - stripping/acid wash
  - neutralisation
  - ultra-filtration
  - reverse osmosis plant
  - utilisation as service water.

The following material streams within the plant may be treated as input (design specifications):

- potato wash water: approx. 10 m^3/h
- pulp: approx. 20 m^3/h
- fruit water: approx. 50 m^3/h
- process water: approx. 60 m^3/h.

The following valuable substances are produced from wastewater within the plant:

- plant based protein;
- biogas, is converted into electric energy, (max. 8 MWelect/10 MWtherm);
- fermentation residue (fertiliser);
- fertilising lime;
- ammonia sulphate (liquid fertiliser);
- service water of various qualities;
- concentrate of reverse osmosis (liquid potassic fertiliser).

A partial stream is channelled towards a membrane activation plant with an upstream de-nitrification step and downstream nitrification. The runoff of this membrane unit 1 is passed into receiving water.

Figure 25.4 shows the 2002 process diagram, as the plant was initially commissioned.

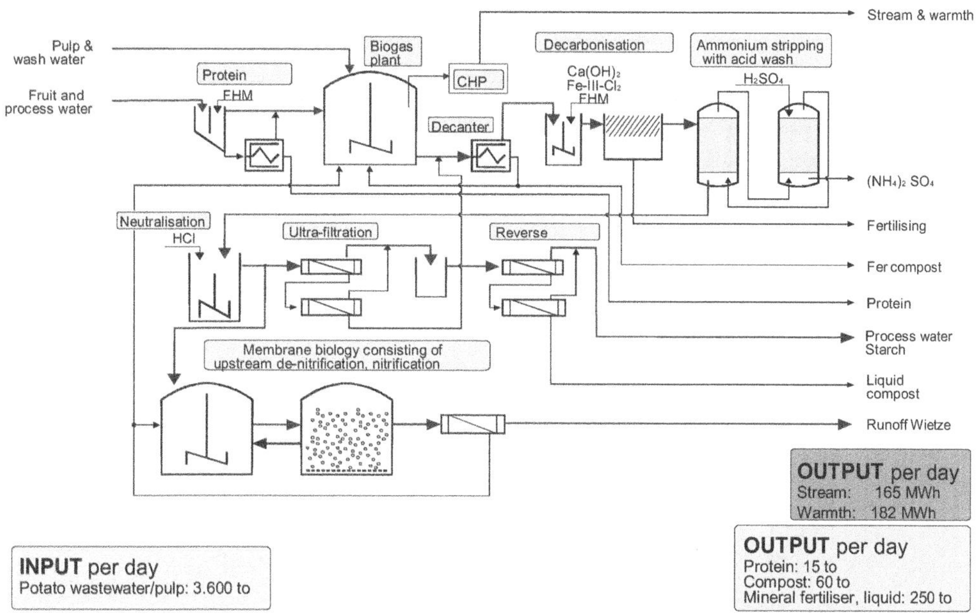

Figure 25.4. Overall flowchart of the wastewater treatment plant of ABR GmbH Wietzendorf at the commencement of operation in 2002 (target values).

Consequences of changes during the 2002/2003 campaign

Based on operational problems during the 2002/2003 campaign, extensive conversion and supplemental installations have already been carried out.

1) Expansion of the sedimentation in the area of operational unit 1 (protein separation) by a dual route flotation for improved protein separation, separately for potato fruit water and process water (after previous semi-commercial trials).
2) Installation of a new heat exchanger unit in order to raise the temperature level for improved protein coagulation.
3) Expansion of decarbonisation/stripping by a second route.
4) Decommissioning of pipe module ultra-filtration.
5) New construction of a second membrane activation plant with submerged micro-filtration membranes as a replacement for the pipe module ultra-filtration plant after earlier semi-commercial trials.
6) Conversion and expansion of the existing reverse osmosis plant.

These conversion measures resulted in a new flowchart from the beginning of the 2003/2004 campaign, as shown in Figure 25.5.

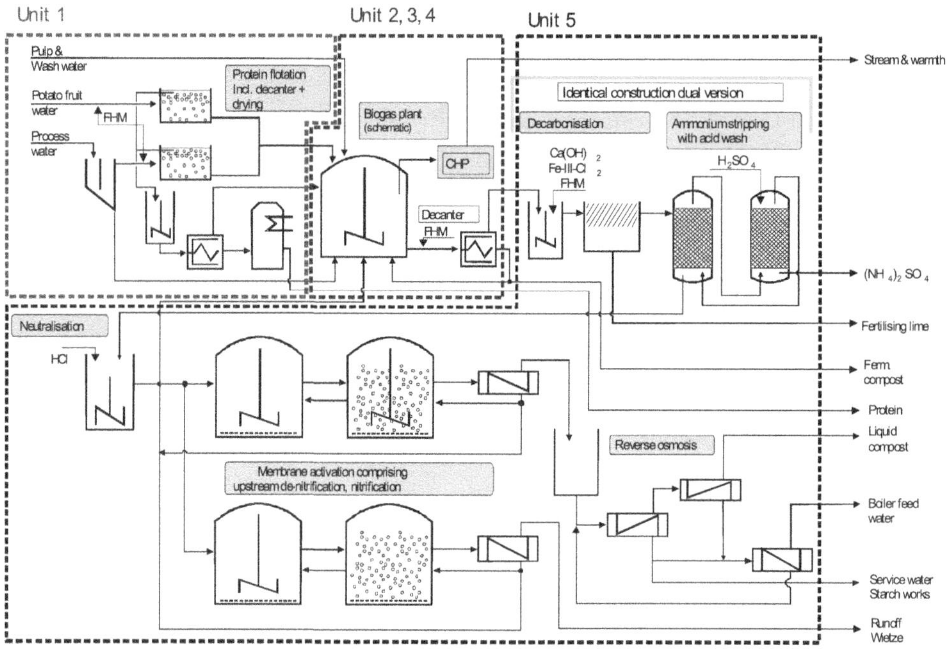

Figure 25.5. ABR flowchart (Wietzendorf plant) as at 2004.

Conclusions and results for the business and industry

The ABR GmbH plant in Wietzendorf needed to undergo extensive conversion and expansion after an initial operation campaign in 2002/2003. Having carried out these measures, plant operation may now be deemed to be stable.

Biogas production, and therefore the generation of energy, is increased outside the campaign by other substrates from agriculture and the food industry. In this regard, it will be possible to achieve a stabilisation of operation based on long-term contracts.

In the event that wastewater from the starch industry – and also various other branches of the food industry – is to be treated to such an extent that recycling within an operation is possible, as a rule, a variety of process technologies need to be combined. These are:

- **biological process technologies**
  - anaerobic for the extraction of biogas, generation of energy;
  - aerobic for the decomposition of organic ingredients as a form of pre-treatment prior to additional physical processes, such as reverse osmosis, in order to reduce the fouling potential.
- **chemical-physical processes**
  - to extract valuable substances; and
  - as a form of pre-treatment prior to additional processes like membrane processes.
- **physical processes**
  - for the separation of valuable substances; and
  - for the production of service water.

These processes then need to be combined into a more or less complex process chain and analysed from the perspective of cost efficiency.

At any rate, prior to the commercial implementation of a concept, it should first be tested sufficiently within the scope of technical laboratory and semi-commercial tests. Here, it is particularly important that the planned combination of processes is tested within the chain as a whole. This aspect needs to be taken into account within the context of scheduling and feasibility considerations.

25.5.5 Economy of water re-cycling

One of the most important factors in terms of the economy of water recycling in Germany is the cost of supplied water and wastewater treatment (Figure 25.6). Supplied water costs approximately 1.75 €/ m³ (~ 2.15 USD/m³), and for end-of-pipe treatment the industry has to pay between 1.5 and 5 €/m³ (or ~ 1.85 to 6.15 USD/m³), depending on the concentration and the technology involved. For recycling, the costs also depend on the concentration and the technology involved. For part streams with membrane technology this amounts to approximately 2 to 5 €/m³ (or ~ 4.46 to 6.15 USD/m³) permeate. So, in each instance, whether the necessary quality for recycling can be attained with the given technology and whether the disinfection achieved meets health requirements has to be researched in detail. Therefore, the cost balance must show if water recycling is an economic method of water management.

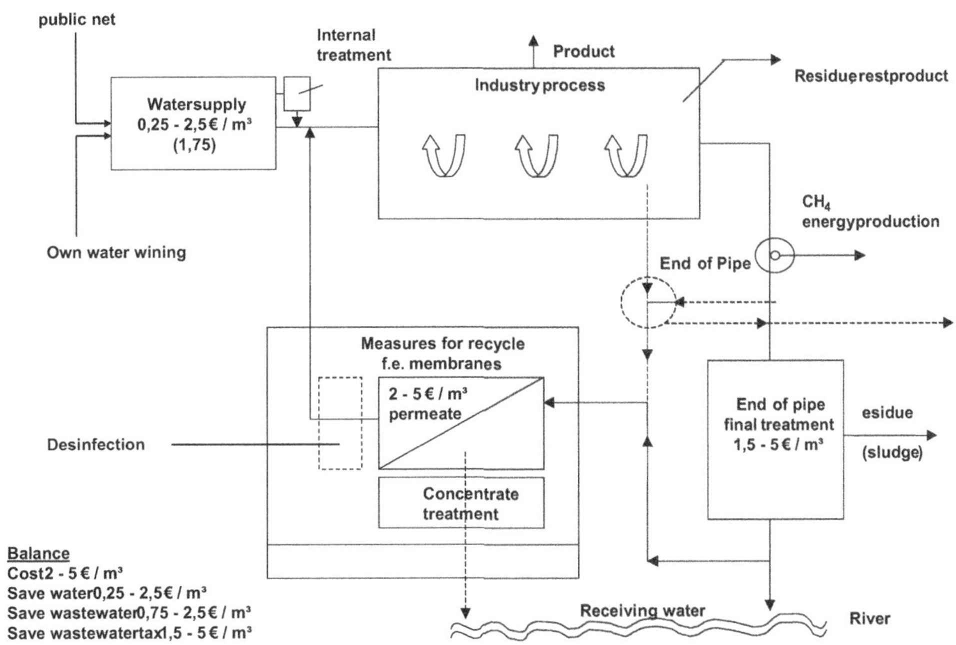

Figure 25.6: The economics of water re-cycling.

25.6 CONCLUSION

In conclusion, it has to be remarked that technical possibilities for water recycling exist in most cases. The decision in favour of water recycling depends on a number of influencing factors and aspects. The most important aspects that have to be considered are health aspects, the regional water supply, the supply of energy and, in some cases, the supply of waste heat. All these aspects have to be considered in terms of an economic and ecological evaluation. We can therefore state that, for a water recycling management system, a decision support system is necessary in order to determine an economical and ecological solution in every instance. In this context, half- and full-scale experiments exist for several example branches of industry. This information could be transferred to many other areas of industry. The main points to ensure the realisation of a water recycling plant are the economic aspects, the quantity of useable water, operating expense and local legislation.

25.7 REFERENCES

Ahrens, A., Abwasserreinigung und Vermeidung in der Brau-, Malz- und Getränkeindustrie. Brauwelt (1998), Heft 15/16, S. 679. [In German]

ATV 2001: Handbuch Industrieabwasser, Dienstleistungs- und Veredelungsindustrie, 4. Auflage, ISBN 3-433-01468, 2000 [In German]

ATV 2000, ATV Handbuch Industrieabwasser, Lebensmittelindustrie, 4. Auflage, ISBN 3-433-01467-1 [In German]

Several Researchs of the Institute of Water Quality and Waste Managment (ISAH) University of Hanover, Welfengarten 1, 30167 Hanover

Kraft 1997, Kraft, A., Mende. U., Das WABAG Submerged Membrane System für Prozeßwasserreinigung und -recyling am Beispiel der Mälzereiprozeßwasser-Aufbereitung. Colloquium produktionsintegrierter Umweltschutz - Abwasserreinigung, Bremen, 1997. [In German]

Rosenwinkel, K.-H., Kreislaufführung und Ressourcenschonung, ATV-DVWK Fachtagung Kassel, März 2002 [In German]

Rüffer, H., Rosenwinkel, K.-H., Taschenbuch der Industrieabwasserreinigung, München 1991, S. 26 [In German]

Statistisches Bundesamt, Wiesbaden, Germany, Fachserie 19, Reihe 2,2 [In German]

Statistisches Jahrbuch 2004, Statistisches Bundesamt, Wiesbaden, Germany) [In German]

Schildbach, S., 2000, S. Kempf, Artikel in der Brauwelt Nr. 33,34 Seite 1333 [In German]

Theilen, U. Richter, J., 2005, Tagung der Deutschen Bundesstiftung Umwelt, Osnabrück, Anwendung von Membranverfahren in der Lebensmittelindustrie [In German]

Water in China, 2003, Water and Environmental Management Series, IWA Publishing, London, UK, ISBN 1843395010.

26

A new paradigm for urban water management and how industry is coping with it

Ivanildo Hespanhol

26.1 THE NEED FOR A NEW PARADIGM

The lack of fresh water resources and increasing water conflicts has prompted the emergence of water reclamation and reuse as formal components of water resources management. The inherent benefits associated with reclaimed treated wastewater for beneficial uses instead of discharge or disposal include preservation of higher quality water resources, environmental protection, economic advantages and social benefits (Asano, 1998).

Water scarcity is not exclusive to arid or semi-arid regions. Many areas receiving significant precipitation but not enough to satisfy excessive local demands will experience water conflicts and restrictions, which may affect economic development and quality of life. Metropolitan São Paulo (MSP) with a population of about 18 million and one of the world's largest industrial complexes, has an average precipitation of about 1,490 mm/year (Prefeitura Municipal de Sao Paulo, 2001). Located at the head of the High Tietê Basin, it is not able to cope with growing water demands, leading to a continuous search for new water resources at neighboring watersheds, bringing as a consequence increasing tariffs and institutional conflicts associated with interbasin conveyance of water. Basin transposition is becoming a very restrictive practice in largely urbanized areas of Brazil due to common lack of water resources and stringent permit conditions, imposed by the recently created Basin Committees.

Low quality waters such as wastewater, waters from agricultural drainage and brackish waters should be, where possible, taken as alternate sources of water for less restrictive uses. The application of adequate technologies for the use of such sources, in association with water conservation and demand management, is the basic strategy to solve the universal problem of water shortage.

Through the hydrological cycle, water is a renewable resource. When recycled through natural systems it becomes a safe resource, which by man's action becomes re-polluted to different extents. However, once polluted, water can be reclaimed and reused for beneficial purposes. Water quality, as well as the specific reasons for reuse, will define the levels of treatment, the security conditions to be adopted and the related capital and O&M costs. The possibilities and types of reuse will depend on the characteristics, conditions and local factors, such as policies, institutional frameworks, technical availability and social, cultural and economic constraints (Hespanhol, 1999).

Water consumption in the RMSP is close to 6 220 800 m^3/d (72 m^3/s), generating about 3 801 600 m^3/d (44 m^3/s) of wastewater (SABESP 2004). The installed capacity of the five existing wastewater treatment plants (Barueri, ABC, Suzano, São Miguel e Parque Novo Mundo) is of 1 555 200 m^3/d (18 m^3/s), but only 924 480 m3/d (10.7 m^3/s) is treated (about 24%) according to SABESP 2004 , since part of the lines and interceptors are still to be built.

Several proposals are being considered to increase the water offer in the region. All of them focus on basin transposition. Consideration is given to the Capivari Monos watershed, relatively close to the RMSP and the Ribeira de Iguape River, in the Ribeira de Iguape Valley, about 100 km south. Due to the political, institutional and environmental problems hampering reversion from the Capivari Monos, the most probable choice is the Juquiá River. The first phase of the project, if approved, will pump 3 456 000 m^3/d (40 m^3/s) from the Ribeira de Iguape, at a cost of US$ 826 million (R$ 1.9 billion). Energy for transportation will cost US$ 76 million per year. No consideration has been given to the additional volumes of wastewater to be produced (estimated as 2 073 600 m^3/d or 24 m^3/s), which will be disposed of, certainly untreated, into the already polluted water bodies of the metropolitan region.

The policy of pumping large volumes of water from distant sources and increasing pollution through generation of wastewater should no longer be accepted both on economic as well as on environmental terms. Since the closest and less polluted sources have already been used, capital costs associated to new projects tend to be more expensive than current ones. A study made by the World Bank in 1992 has shown that the cost of a cubic meter of the "next" project could be two or three times the cost of the existing, as shown in Figure 26.1. On the other hand, this procedure, associated with lack of investment in sanitation, is promoting river pollution in most developing countries, as shown in terms of dissolved oxygen concentration in water bodies (Figure 26.2).

Importing water from distant water bodies to satisfy increasing demand in urban areas is a practice more than two thousand years old. Once the closest sources had been polluted, the Romans used to go further and further to get new sources of water. This practice gave rise to the construction of Roman aqueducts, the remnants of which are still visible in many parts of Europe.

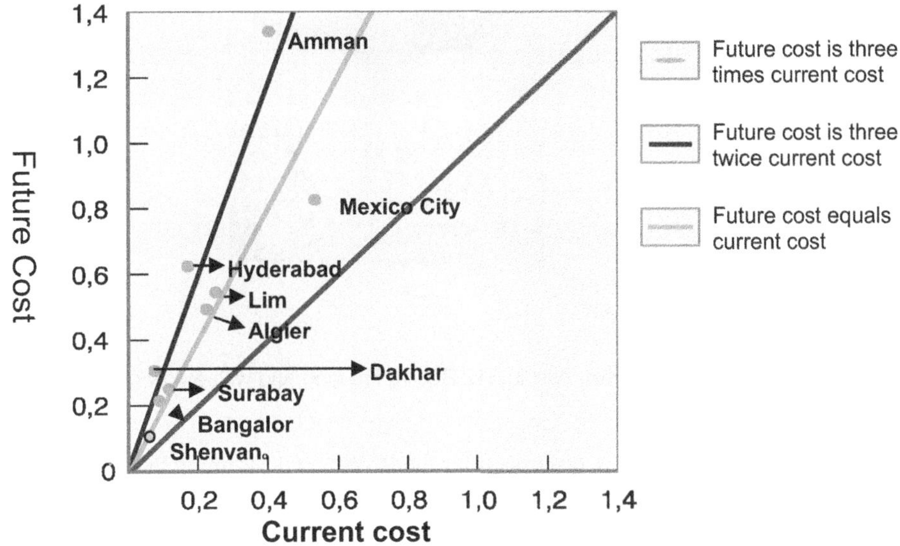

Figure 26.1. Current costs of cubic meters of water versus costs of new projects (World Bank, 1992).

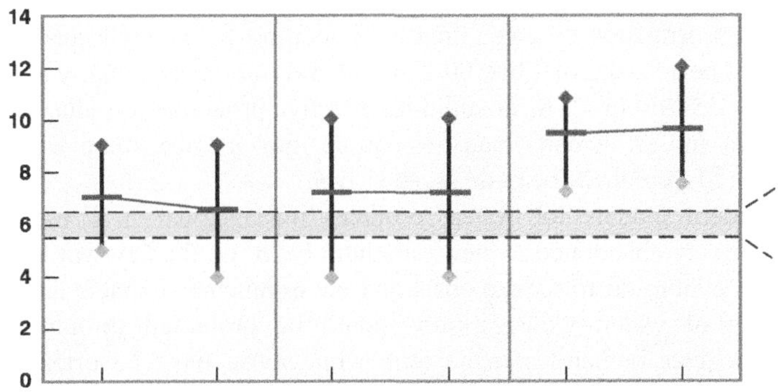

Figure 26.2. Dissolved Oxygen in rivers according to levels and trends across country income groups (World Bank, 1992).

During the Roman Empire, eleven aqueducts were constructed, totalling 502 km in length and with a flow of 1,123 m^3/d (or around 13L/s). The first of them, the Aqua Appia, was built in 312 BC by Appius Claudius Caecus and C. Plautius (Figure 26.3). It was only 16 km long and had a flow of about 73 000 m^3/d. Work on the Aqua Cláudia, shown in Figure 26.4, began in 38 AD under Calígula and was ready in 52 AD under Claudius. It was 69 km long, carrying 184 220 m^3/d. The largest of the Roman aqueducts, at 91 km and with a flow of 187 600 m^3/d, was the Aqua Márcia (Swansea University, 2006; Bowdoin, 2006).

Unfortunately the practice is still the same, precariously solving a water supply problem but increasing pollution in the area receiving the water, to the detriment of the one from where it was drawn. There is, therefore, a need for a new paradigm changing the Roman version of systematically transporting large volumes of water long distances and disposing of wastewater, with little or no treatment at all, into neighboring water bodies.

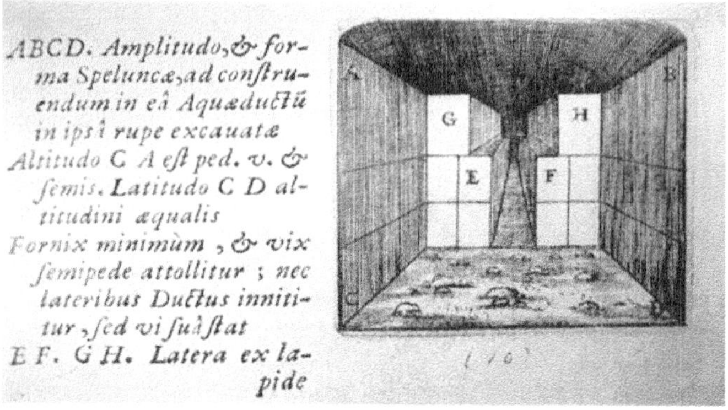

Figure 26.3. Section of the Aqua Appia, built in 312 BC, 16 km long and with a flow of 73,000 m^3/d (from Swansea University, 2006).

Looking again at the proposals in discussion for the RMSP, and assuming that the losses in the distribution system are of about 20% and the service is extended to 100% of the population, the average per capita value will be $0.80 \times 72.10^3 \times 86{,}400/18.10^6 = 276$ litres per inhabitant per day. Assuming conservatively that 140 litres per inhabitant per day is used for potable purposes and the remaining 136 is substituted by reclaimed water and utilized for domestic non-potable purposes (such as flushing and washing of floors), non-potable urban uses (such as washing of vehicles and streets, irrigation of green areas and sporting fields, civil construction) and by industry, there will be a saving of 136x100/276 = 49.3% of potable water, which is equivalent to 28.4 m^3/s or 2 453 760 m^3/d. If, in addition, effective programs to reduce losses from the distribution system and on demand management are implemented, there will be no need to pump 3 456 000 m^3/d from the Ribeira de Iguape River.

Owing to modern models for water resources management it is essential that old, orthodox principles are abandoned. A new paradigm based on the key words "conservation" and "reuse" must evolve, to minimize costs and environmental impacts associated to inter basin conveyance of water. Conservation should be promoted through environmental education and adequate demand management, while reuse must be oriented towards the changing conventional sources of water for alternate sources such as reclaimed water, rain water and groundwater augmented by artificial recharge.

Figure 26.4. The Aqua Claudia, built from 38 to 52 AD and 69 km long, carrying a flow of 184.220 m^3/d (Boudoin, 2002).

26.2 INDUSTRIAL PRODUCTION FEASIBILITY

Today, industry is submitted to two main instruments of pressure: on the one hand, global environmental and public health conditions, imposed by international trade relationships and, on the other, legal constraints, particularly the ones related to charging for the use of water resources (Pio, 2005).

In order to adapt to this new scenario, industry is improving its manufacturing processes, developing environmental management structures to fulfill legal impositions as well as global market specifications, and implementing systems and procedures to reduce water consumption and effluent production.

Depending on water availability, local markets and the adequate use of emerging technologies, the feasibility of industrial production is associated with the analysis of the following options:

i. maintaining the actual procedure, by getting water from public supplies and from surface or groundwater sources;
ii. buying reclaimed water from water authorities; and
iii. reclaiming and reusing, as far as possible, its own effluents after appropriate treatment.

26.2.1 Public supply and natural sources

Industry is supplied by variable proportions of surface or groundwater or by public supplies. Whatever the source, the charges associated to the use of water or disposal of effluents must be taken into account. If a public supply is being used, in addition to those costs, the corresponding tariffs must also be added. In general, surface waters are polluted, leading to high treatment costs, to achieve adequate water quality levels. When available, the tendency is to choose groundwater sources which are, in general less polluted, calling for relatively simple treatment systems.

Intake from surface sources of water by industry is diminishing all over the country. In the state of São Paulo, the reduction was of about 15% from 1990 to 2000 as shown in Table 26.1.

Table 26.1. Decrease of surface water intake by industries in the state of São Paulo from 1990 to 2000 (São Paulo, 1999).

INDUSTRIAL SECTOR	SURFACE WATER INTAKE			
	1990		2000	
	m^3/s	m^3/d	m^3/s	m^3/d
Sugar and Alcohol	46.24	3 995 136	42.3	3 654 720
Chemical and Petrochemical	17.97	1 552 608	15.9	1 373 760
Pulp and Paper	13.20	1 140 480	11.6	1 002 240
Metallurgical	10.64	919 296	7	604 800
Food and Beverages	10.55	911 520	6.7	578 880
Textile	4.19	362 016	4	345 600
TOTAL	102.79	8 881 056	87.5	7 560 000

In the same way, the water demand from public distribution systems is being substantially reduced. The São Paulo Water Authority (SABESP) reports a reduction of about 40%, from 1988 to 2004.

The decrease of industrial water demand may also be attributed to programs implemented by most industrial sectors but the basic reason behind it is the high cost of water practiced by most water authorities in the state of São Paulo. Table 26.2 shows the tariffs practiced by SABESP in the largest part of Metropolitan São Paulo, after the 7% increase imposed after August 31, 2005 (SABESP, 2005).

Table 26.2. Tariffs for Industrial Water Supply and Wastewater Collection in the largest part of the RMSP – SABESP.

m^3/MONTH	WATER TARIFFS (US$)	WASTEWATER TARIFFS (US$)
0 to 10	9.76/month	9.76/month
11 to 20	1.89/ m^3	1.89/ m^3
21 to 50	3.66/ m^3	3.66/ m^3
Over 50	3.80/ m^3	3.80/ m^3

1 US$ = R$ 2.30

26.2.2 Buying reclaimed water from public supplies

Several water authorities are developing projects to make reclaimed secondary effluents available to industry and other non potable consumers.

SABESP, the water authority of São Paulo, has developed a project evaluating the potential of reuse within a radius of 5 km around its five wastewater treatment plants (Barueri, ABC, Suzano, São Miguel and Parque Novo Mundo). The study evaluated the consumer markets, actual and future demands as well as the willingness to pay. The budget for the project is of about US$ 61 million, from which US$ 31 million is from SABESP itself and US$ 30 million from external funding. The feasibility study has assumed a period of project of 25 years at interest rate of 15% per year. The payback period was estimated in 7 years and the internal rate of discount on 18%.

The study has identified a large potential for the industrial sector, requesting reuse water with quality to be utilized as make up water for cooling towers, as shown in Table 26.3.

Table 26.3. Water Quality requested for potential industrial consumers to be used as make up water For Cooling.

Variable	Numerical Values$^{(1)}$	Variable	Numerical Values$^{(1)}$
SiO_2	50	Hardness $^{(2)}$	650
Al	0.1	Alkalinity$^{(2)}$	350
Fe	0.5	pH	6.0 – 9.0
Mn	0.5	MBAS	1
Ca	50	$CCl_3^{(3)}$	1
Mg	0.5	COD	75
NH_3	1	H_2S $^{(4)}$	-
HCO_3	24	O_2 $^{(4)}$	-
SO_4	200	Temperature$^{(4)}$	-
Chloride	500	TSS	100
TDS	500		

(1) Variables in mg/l, except pH. Numerical values must not be surpassed. No sample may have all the maximum numerical values indicated;
(2) as $CaCo_3$;
(3) extract;
(4) accepted as usually received.

The basic prices per cubic meter of reclaimed water with this quality as well as the willingness to pay value (spontaneous value) are presented in Figure 26.5. It can be noted that the average price (0.69 US$/$m^3$) is approximately the same as the spontaneous value (0.68 US$/$m^3$).

The tariff scheme proposed is shown in Figure 26.6. Even the price (0.82 US$) proposed for the smallest range of demand (up to 10,000 m^3/month) is much smaller than the prices established by SABESP, whatever the range of demand.

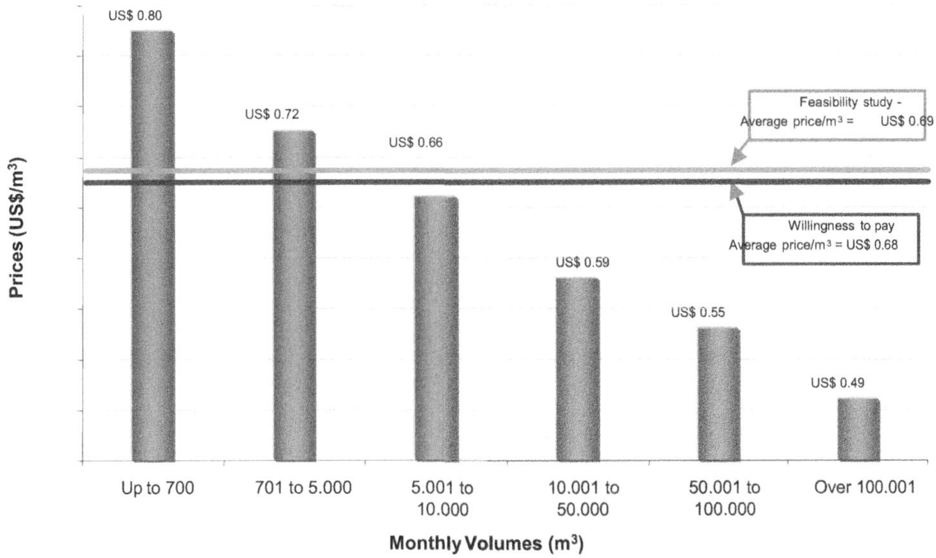

Figure 26.5. Costs of monthly volumes. The feasibility study provided an average value of 0.69 US$ /m^3, and a spontaneous value of us$ 0.68/m^3.

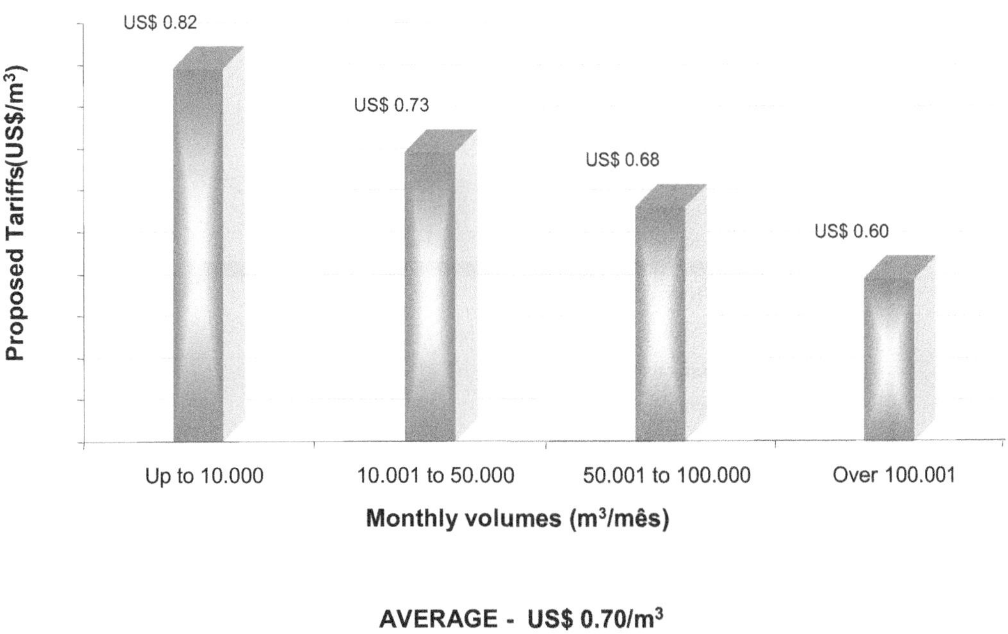

Figure 26.6. Proposed tariffs according to monthly volumes.

By comparing the tariffs shown in Table 26.2 (3.80 US$/$m^3$ for volumes over 50 m^3/month) and the average price estimated for reclaimed water shown in Figure 26.6 (0.70 US$/$m^3$) it can be assumed that there is no interest by SABESP or any other water company, to sell reclaimed water since it may lead to a substantial decrease in income. Even if the new volumes provided by reclaimed water is directed to new consumers there will be a general clamour by consumers paying prices five times higher than others. Therefore, the practice of water reclamation and reuse by Brazilian water authorities can be implemented on a large scale only if new policies and adequate institutional frameworks are established to regulate the use of urban waters for non potable purposes.

26.2.3 Reclaiming and reusing its own effluents

The implementation of the water charging policy, even in basins where the legislation has as yet not been approved led industries to two basic actions: demand management and reduction of effluents produced. In the State of São Paulo, Law no. 12.183 was enacted on 29 December, 2005, to regulate permits and charges for the use of water to all consumers.

Anticipating its enforcement, the Federation of Industries for the State of São Paulo (FIESP) has, with the support from the International Reference Center on Water Reuse (CIRRA/IRCWR) and DCT Engineering, a consulting firm, produced a manual on water conservation and reuse for industry (FIESP/CIESP, 2004). Several other documents, directed to specific industrial sectors such as sugar and alcohol, petrochemical, steel, paper and pulp, pharmaceutical and others will be produced during the coming years. Following the FIESP initiative, the Civil Construction Union for the State of São Paulo (SindusCOn) produced, with the technical support of IRCWR and several other firms, a manual on conservation and reuse for buildings (SindusCon *et al.*, 2005). The methodology to be applied to reduce water demand and generation of effluents is summarised in Figure 26.7.

The basic steps for water management in industry follow below.

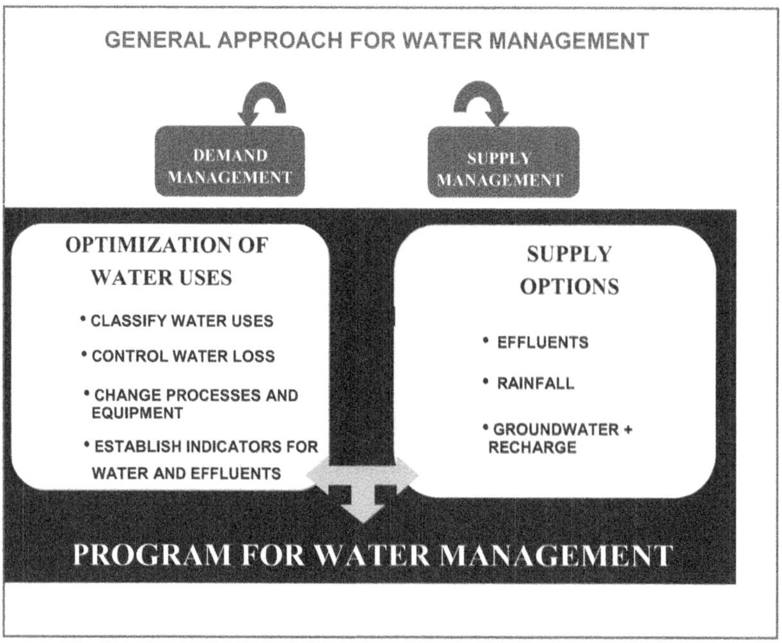

Figure 26.7 Integrated approach for water management in industry.

Demand management

This step is aimed at the reduction of water volumes taken from any source. Since water, not long ago, was assumed to be a free resource, most industries were not prepared to control demand and to measure consumption in each industrial sector. Therefore the first step is to measure and register daily volumes of water utilized in specific areas such as industrial processes, power generation, heating and cooling, washing of parts, reactors and floors, irrigation of green areas, human consumption and so on.

At the same time a program for losses control should be implemented. Main points to consider are physical losses in pipes, connections, valves, reservoirs, machines and equipment. In the same way losses associated to inadequate operational procedures or negligence should be evaluated and corrected through specific training or implementation of additional instructions. In parallel, a program for detection and correction of invisible losses should also be implemented using specialized personnel.

Further actions to reduce demand may imply allocation of significant expenditure since it may lead to the acquisition of new machines, reactors, pumps as well as instrumentation and control systems. The basic objective is the use of modern manufacturing systems demanding less water and generating smaller volumes of effluents.

Consumption indicators (m^3 of water consumed/unit of product) and effluent indicators (m^3 of effluent produced/unit of product) are important tools for control to be taken into account and should be adopted according to benchmarking established in Brazil as well as worldwide. The document "Water Conservation and Reuse in the Industry – Guiding Manual for the Industrial Sector, volume 1" (Mierzwa and Hespanhol, 2005) presents (in Annex I) international indicators, for 1990, for several industrial sectors. Most of those indicators should be adapted to local conditions as well as to improved industrial processes and new advances on effluent treatment systems.

Water supply management

Water supply management aims to exchange traditional sources of water for alternate sources. Many industries use water from public supplies, assuming the quality is adequate for their purposes. In many cases drinking water has quality superior than needed, so it may be possible that local effluents could be made available at lower costs.

The options to eliminate or reduce the water intake from traditional sources are:

- reuse of locally generated effluents with or without treatment. Application of the "pinch" methodology would improve the reuse schemes;
- use of rain water collected from roofs or from paved areas;
- local groundwater, augmented by artificial recharge with treated effluents.

As a general procedure for industrial water supply management, the International Reference Center on Water Reuse (IRCWR) at the University of São Paulo has established the following steps:

1. identification and characterization of effluents in terms of quality and flow distribution; and
2. identification of the reuse potential, including specific needs in terms of flow, water quality and a preliminary lay out of the distribution system.

The information gathered will allow for the design of a reclamation plant to comply with the needs for water quality and quantity. This system is to be designed under a stepwise fashion, starting with simplified units (such as primary treatment) units to produce water for less restrictive uses, and going to more advanced units to comply with the needs of more stringent uses. Once the full system is defined, specific mass balances should be performed throughout the overall layout, including treatment units, reservoirs and the distribution system, in order to identify concentration cycles in the reuse scheme. As a final step, pinch methodology is to be applied to optimize the overall reuse system (Mierzwa and Hespanhol, 2005).

While relatively limited, with the experience gained with the implementation of industrial conservation and reuse projects in the State of São Paulo, it can be estimated that the average costs, if transportation and final disposal of sludge are not needed, varies from 0.35 US\$ /m^3 to 0.52 US\$/m^3.

The feasibility of industrial production in Metropolitan São Paulo, with regard to water supply and effluent disposal can be evaluated considering the costs presented in items 26.2.1 to 26.2.3 above.

Industrial water from public supplies is charged from US\$ 0.97 to US\$ 3.80, depending on monthly volumes (Table 26.2), while reuse water provided by water authorities varies from US\$ 0.60 to US\$ 0.82 as shown in Figure 26.6. As the estimated costs of in-house treatment and reuse varies from US\$ 0.35 to US\$ 0.52, it seems logical that this is the preferred choice of water supply by industrial managers.

However, industrial water treatment and reuse costs should be considered under several local conditions and constraints. First of all, costs are dependent on the conditions prevailing in each industry, and it is mandatory to perform a specific feasibility study for each one by considering capital costs, pay back period, interest rates, internal rates of discount and so on. Secondly, the costs associated with environmental protection (in Brazil to attain the emission standards established by Resolution CONAMA 357) should be taken into account in the costs related to reuse systems; however, only costs associated to additional treatment units, needed to attain levels of quality above the emission standards, should be added to reuse costs.

26.3 INDUSTRIAL WATER REUSE IN BRAZIL AND IN THE STATE OF SÃO PAULO

Industrial reuse, like other forms of reuse, is steadily growing all over the country. Most activities are concentrated in the south and southwest regions, but in the dry northern States of Bahia, Ceará, Paraíba, and Rio Grande do Norte, many isolated systems have been installed, particularly for irrigation of forage crops. In the Paraíba do Sul river basin, where charging for the use of water has been enforced for several years, water reuse is being practiced by about 50% of large industries, 16% of medium size industries and 11% of small industries. From over 32 million m^3 of water withdrawn, 25%, or a little more than 8 million m^3, are reused, particularly within the metallurgical sector (IEPA, 2004).

The industries listed in Table 26.4 have already invested significant resources in conservation and reuse, achieving from 40% to 80% on water savings (DAEE, 2000).

Table 26.4. Large companies with investments in water conservation and reuse (DAEE, 2000).

Industry	Main products	Industry	Main products
3M of Brazil	Chemicals	Ferro Enamel of Brasil	Chemicals
Aegis	Microeletronics	Ford Motor Company	Vehicles
Alpargatas Santista Textil	Fabrics	Freios Varga	Brake systems for vehicles
Alsco Toalheiros Brazil	Industrial Laundry	INA Brazil	Bearings
AmBev	Beverages	Jânssen Farmaceutica	Pharmaceuticals
AMP of Brazil	Electrical connections	Kodak of Brazil	Photographic products
Brastemp	Household appliances	Mahle Metal Leve	Metallurgy
BSH Continental	Ranges	Maxion Components	Metallurgy
BSH Continental	Refrigerators	Natura	Cosmetics
Burigotto	Baby carriages and appurtenances	Pilkinton Brazil	Glasses
Cermatex	Textiles	Polo Petroquímico de São Paulo	Petrochemical Products
Coats Correntes	Textiles	Repla/Revap/Petrobrás	Petrochemical Products
Companhia Brasileira de Bebidas	Beverages	Rhodia	Chemicals
Continental	Household appliances	Rohm and Haas Química	Chemicals
Daimler Chrysler	Vehicles	TRW Automotive	Metallurgy
Dow Chemical	Chemicals	Volkswagen	Vehicles
Elekeiroz	Chemicals	Votorantim	Paper and pulp

26.4 REUSE AS A TOOL TO ATTENUATE INDUSTRIAL WATER CHARGES IN THE STATE OF SÃO PAULO

There are about 130,000 industries in the State of São Paulo. From those about 8,000 are large or medium size being he remaining micro industries. A simulation to evaluate the economic benefits accrued to water reuse was made by considering water withdrawal, consumption and effluent discharge from 2.311 industries from the State, according to the State Law No. 12,183 enacted on 25 December, 2005. Table 26.5 shows the industries according to their hydrographic basins and corresponding volumes of water withdrawal and effluent discharge.

The total withdrawal is of 112.75 m^3/s or 9 741 600 m^3/d and the total of effluent discharged to receiving bodies is of 83. 47 m^3/s or 7 211 808 m^3/d. Therefore, the total consumption by the industries is of 112.75 – 83.47 m^3/s, or 29.28 m^3/s (2 5929 792 m^3/d).

The charges are accrued to water withdrawal, water consumed and to effluent discharge. Besides the flow, the charges related to discharge of effluents take in account the concentrations of Settable Solids (SS), Inorganic Load (IL), Biochemical Oxygen Demand (BOD) and Chemical Oxygen Demand (COD), according to the general equation, where C is the monthly cost.

$$C = \text{withdrawal} + \text{consumption} + \text{effluent}\,(BOD+COD+SS+IL) \qquad \text{(Equation 26.1)}$$

Or, under its operational form:

$$C = (FUP_{WITH} * F_{WITH}) + (FUP_{CONS} * F_{CONS}) + (FUP_{BOD} * F_{BOD} + FUP_{COD} * F_{COD} + FUP_{SS} * F_{SS} + FUP_{IL} * F_{IL})$$

where

F_{WITH} : withdrawal flow;
F_{CONS} : consumption flow;
F_{BOD} : effluent flow related to BOD;
F_{COD} : effluent flow related to COD;
F_{SS} : effluent flow related to SS;
F_{IL} : effluent flow related to IL;
FUP_{WITH} : final unit price related to withdrawal = $BUP_{WITH}(x_1, x_2 ..... x_n)$
FUP_{CONS} : final unit price related to consumption = $BUP_{CONS}(x_1, x_2 ..... x_n)$
FUP_{BOD} : final unit price related to BOD = $BUP_{BOD}(Y_1, Y_2 ..... Y_n)$
FUP_{COD} : final unit price for COD = $BUP_{COD}(Y_1, Y ..... Y_n)$
FUP_{SS} : final unit price for SS = $BUP_{SS}(Y_1, Y_2 ..... Y_n)$
FUP_{IL} : final unit price for IL = $BUP_{IL}(Y_1, Y_2 ..... Y_n)$
X_i : coefficients related to withdrawal: uses of water, classes of receiving bodies, types of sources (surface or groundwater), seasonality, etc.;
Y_i : Coefficients related to discharge of effluents, levels of dilution, assimilation capacity (origin, classes of receiving bodies, seasonality, etc.);
BUP : Unit basic prices related to withdrawal, consumption and to BOD, COD and IL.

Table 26.5 Water demand by industry in the State of São Paulo (DAEE, 2000).

Hidrográfic Basin	Industries	Withdrawal From (m^3/s)				Effluent (m^3/s)
		Surface Water	Ground Water	Public Supply	Total	
Piracicaba	440	13.88	0.33	0.0	14.91	11.17
Tietê/Sorocaba	383	9.60	0.57	0.22	10.39	7.05
Alto Tietê	593	7.67	1.20	1.17	10.04	7.02
Baixo Tietê	37	1.79	0.03	0.02	1.84	1.62
Tietê/Batalha	19	1.39	0.04	0.09	1.52	1.25
Tietê/Jacaré	77	8.49	0.35	0.01	8.85	6.85
Aguapeí	20	0.76	0.03	0.02	0.81	0.60
Peixe/Sto.Anastácio	63	0.57	0.22	0.06	0.85	0.55
Baixo Paranapanema	56	2.72	0.18	0.01	2.91	2.65
Alto Paranapanema	23	3.48	0.01	0.02	3.51	0.49
Ribeira de Iguape	9	1.33	0.00	0.00	1.33	1.32
Baixada Santista	43	12.97	0.05	0.20	13.22	10.76
Litoral Norte	1	0.00	0.00	0.00	0.00	0.00
Paraíba do Sul	164	7.20	1.13	0.19	8.52	4.51
Mantiqueira	3	0.04	0.00	0.00	0.04	0.01
Alto Pardo/Mogí	182	5.38	0.17	0.11	5.66	4.93
Sapucaí/Grande	26	0.02	0.04	0.04	0.10	0.08
Baixo Pardo/Mogí	98	24.33	0.1	0.14	24.98	20.11
Pardo/Grande	22	1,38	0,03	0,00	1,41	0,70
S.José dos Dourados	5	0,23	0,00	0,00	0,23	0,20
Turvo/Grande	46	1,45	0,16	0,02	1,63	1,51
TOTAL	2311	104.68	5.05	3.02	112.75	83.47

The data used for simulation is as follows:

Class of receiving body – Class 2 (CONAMA 357)
Industrial effluent with only BOD and IL;
Seasonal regime;
$F_{WITH} = 112.5\ m^3/s$ (Table 26.5);
$F_{CON} = 112.5\ m^3/s - 83.47\ m^3/s = 29.8\ m^3/s$;
$F_{BOD} = F_{IL} = 83.7\ m^3/s$;
BUP_{WITH} = US\$ 0.013m^3 (average) giving FUP_{WITH} = US\$ 0.016/m^3;
BUP_{CON} = US\$ 0.026/m^3 (average) giving FUP_{CON} = US\$ 0.031/m^3
BUP_{BOD} = US\$ 0.239/Kg (average) giving FUP_{BOD} = US\$0.465/Kg
BUP_{IL} = US\$ 0.435-US\$4.350/kg giving FUP_{IL} = 0.869 US\$ /kg;
$X_1 = 1.2$ (industry); $X_2 = 1.0$ (Class 2); $X_3 = 1.0$ (seasonality)
$Y_1 = 1.3$ (industry);
$Y_2 = 1.5$ (Class 2);

The simulation was made by applying the above prices for the use of water and by considering reuse from 0% up to 60% of withdrawal. The corresponding daily charges for withdrawal and consumption are shown in Table 26.6. Table 26.7 shows the charges for effluent disposal, in terms of BOD and IL as well as total charges, including withdrawal, consumption and disposal.

Table 26.6. Industrial Water Charges for withdrawal and consumption according to increasing percent of water reuse.

Withdrawal (m^3/s)	Water Reuse (% of withdrawal))	Withdrawal (m3/s)	Consumption (m^3/s)	Charges Withdrawal (US\$/d)	Charges Consumption (US\$/d)	Partial Charges (US\$/d)
112,75	0	112,75	29,28	152,477.22	79,193.46	231,670.70
	10	101,48		137,236,26		216,429.74
	20	90,20		121,981.8		201,175.26
	30	78,93		106,740.83		185,934.30
	40	67,65		91,486,35		170,679.83
	50	56,38		76,245.37		155,438.85
	60	45,10		60,990.87		140,184.35

Table 26.7. Total Industrial Water Charges: withdrawal, consumption and effluent disposal according to increasing percent of water reuse.

Withdrawal (m^3/s)	Reuse (% of withdrawal)	Effluent (m^3/s)	BOD (kg/m^3)	Inorganic Load (kg/ m^3)	Charges for Disposal (US\$/d)	Total Charges (US\$/day)
112,75	0	83.47	0,060	0,195	1,424,175.30	1,655,846.00
	10	72.20			1,231,885.19	1,448,314.93
	20	60.92			1.039,424.46	1,240,599.72
	30	49.64			846,963.72	1.032,898.03
	40	38.37			654,673.61	825,353.44
	50	27.09			462,221.88	617,651.73
	60	15.82			269,922.76	410,107.11

The last table indicates that the total charges will reach about US\$1.7 million per day without reuse and only US\$410,000 with 60% of water reuse.

26.5 THE CONTRIBUTION OF THE INTERNATIONAL REFERENCE CENTER ON WATER REUSE (IRCWR/CIRRA) TO SUPPORT THE PRACTICE OF WATER CONSERVATION AND REUSE IN BRAZIL

IRCWR is a nonprofit organisation for research and development, institutionally linked to the University of São Paulo. This link allows for its integration into an advanced scientific network, including professors and qualified professionals with a wide range of experience to cope with the demands of the water resources sector. Aiming at the promotion and enhancement of human and technical resources to motivate environmentally sound practices, the IRCWR was created to promote research and development and to provide advanced training and information on the technical, environmental, institutional, legal, economic and socio-cultural aspects of water conservation and reuse. The Center is contributing to the institutionalisation, regulation and the promotion of water reuse practice in Brazil and worldwide.

On the technical side, IRCWR develops master plans, programs, studies and projects on water conservation and reuse, rain water catchment and on environmental sanitation, for research centers, universities, public and private enterprises, industries and engineering consulting firms. It provides training activities in the form of technical manuals, short courses, both in-company as well as at its headquarters, focusing on water conservation, reuse and related themes. On its internet homepage (www.usp.br/cirra), the IRCWR maintains a database containing broad information on the sector, including papers, documents, legislation and thesis works related to water conservation and reuse as well as on advanced water treatment systems.

The IRCWR laboratory is equipped with a 1 m^3/hour continuous flow pilot plant containing a complete physical chemical set of units (coagulation, flocculation, high rate settling tank and double layer filtration) followed by advanced treatment processes (microfiltration, ultrafiltration, nanofiltration, reverse osmosis and advanced oxidative processes). Effluent disinfection can be achieved by chlorine, chlorine dioxide, ozone and UV. The laboratory is used for research and for treatability studies, particularly for exotic industrial effluents.

26.6 CONCLUSIONS

More than 2,000 years ago, with the building of the Via Appia, the Romans initiated the practice of importing water from more and more distant watersheds. The same criteria are being applied nowadays to solve precariously local supply problems while increasing pollution.

Owing to new approaches to water resources management it is of utmost importance to abandon orthodox concepts. A new paradigm based on the key words "water conservation" and "water reuse" must take its place to minimize environmental impacts and to reduce costs associated with new development projects.

Conservation can be promoted through environmental education programs and demand management, and reuse, on supply management, through the search for alternate sources of supply, including reclaimed water, rain water and groundwater, complemented by artificial recharge of aquifers.

Depending on water availability and relying on new emerging technologies, the feasibility of industrial production remains associated to the following non exclusive options:

(i) maintenance of the traditional condition, that is, getting water from public supplies or from surface or ground water sources;
(ii) buying reclaimed water from water companies; or
(iii) reclaiming and reusing, as far as possible, the in-house effluents.

The Brazilian experience, particularly in the State of São Paulo, allows the conclusion that demand management and in house reuse, leads to a unitary cost of water inferior to the tariffs practiced by water companies, either from drinking water systems or from reclaimed water supplies.

Industrial reuse is growing steadily all over the country. This practice is providing environmental, social and economic benefits and, under the new regulations of permits and charges, an important tool for industrial feasibility.

However, the dissemination of water reuse practice in Brazil is far away from being attained, owing to the associated reduction of income for water companies (who have the largest potential to have the practice implemented on a large scale). Universal water reuse in Brazil could only be achieved through a strong institutional decision, followed by the enactment of comprehensive federal legislation, which could be enforced by the basin committees.

26.7 REFERENCES

Asano, T. (1998) Wastewater Reclamation and Reuse, Water Quality Management Library, vol. 10, Technomic Publishing Inc., Lancaster, Pennsylvania, USA.

Bell, D. (1977) The Coming of Post-Industrial Society, Futures 5: 415-431, New York.

Bowdoin (2006) http//www.academic.bowdoin.edu/classics/research/moyer/html/intro.shtml De Masi, D. (1999) A Sociedade Pós-Industrial,, 443 pp., Ed. Senac, São Paulo, Brasil [In Portuguese]

DAEE (2000) Demanda de Água Industrial no Estado de São Paulo, Departamento de Águas e Esgotos, Secretaria de Energia e Recursos Hídricos do Estado de São Paulo, Brazil. [In Portuguese]

FIESP/CIESP (2004) Conservação e Reúso de água – Manual de Orientações para o Setor Industrial, Vol. 1, Centro Internacional de Referência em Reúso de Água-CIRRA/IRCWR e DCT Engenharia, 92 pp., São Paulo. [In Portuguese]

FIESP (2006) http//www.fiesp.org.br (adaptado). [In Portuguese]

Hespanhol, I. (1999), Água e Saneamento Básico-Uma Visão Realista in: Águas Doces do Brasil, Capital Ecológico, Uso e Conservação: 240-304, Coord. Rebouças, A.C.,Braga, B., Tundisi, J.G, Ed. Escrituras, São Paulo [In Portuguese]

IPEA (2004) " Relatório da Pesquisa de Campo sobre Uso Industrial da Água:Estimação de Funções de Demanda de Água e Ccusto de Controle de Poluição", Rio de Janeiro. [In Portuguese]

Mierzwa, J.C and Hespanhol, I. (2005) Agua na Indústria-Uso Racional e Reúso, 143 pp., Oficina de Textos, São Paulo [In Portuguese].

Pio, A.A.B. (2005) Reflexos da Gestão de Recursos Hídricos para o Setor Industrial Paulista, Master Thesis Escola Politécnica da Universidade de São Paulo, 164 pp., São Paulo. [In Portuguese]

Prefeitura Municipal de São Paulo (2001), Precipitação em Estações Distritais no Município de São Paulo – Médias Mensais, Secretaria das Administrações Regionais, Comissão Municipal de Defesa Civil, COMDEC, São Paulo. [In Portuguese]

SABESP, (2004) Sistemas de Captação e Tratamento da Região Metropolitana de São Paulo http://www.sabesp.com.br/o-que-fazemos/coleta-e-distribuição- de- água/. [In Portuguese]

SABESP (2004) Tratamento de Esgotos da Região Metropolitana de São Paulo http://www.sabesp.com.br/o-que-fazemos/coleta-e-tratamento/ [In Portuguese]

SABESP (2004) Reúso Urbano na Região Metropolitana de São Paulo, Diretoria de Produção e Tecnologia-T, Unidade de Negócios de Tratamento de Esgotos–TE, São Paulo [In Portuguese].

SABESP (2005) Comunicado -03/05-Tarifas e Demais Condições a vigorarem a partir de 31 de agosto de 2005, São Paulo. [In Portuguese]

SABESP (2006) http://www.sabesp.com.br [In Portuguese]

São Paulo (1999) Síntese do Relatório de Situação dos Recursos Hídricos do Estado de São Paulo", Departamento de Águas e Energia Elétrica-DAEE, São Paulo. [In Portuguese]

SindusCon, FIESP and COMASP (2005) Conservação e Reúso de Água na Indústria, diversos autores e colaboradores, 151 pp., São Paulo. [In Portuguese]

Swansea University (2006) http//www.swan.ac.uk/classics/staff/ter/grst/What's%20what%20Things/aqueducts/htm

The World Bank (1992) World Bank Development Report, Development and the Environment, World Bank Development Indicators, Oxford University Press, 308 pp., USA.

27

Israel as a case study

Marcelo Juanicó

27.1 INTRODUCTION

Israel has performed massive irrigation with treated effluents for decades. Today, about 75% of the sewage of the country is reused. This chapter summarizes the developments of this practice in the country, the experience gained and the issues that remain controversial or problematic.

27.1.1 Geography and climate

The country can be divided into two main regions: Northern-Central Israel with a Mediterranean climate (short temperate rainy winters and long warm-sunny dry summers) and Southern Israel with an arid climate (precipitations below 300 mm per year). Droughts are common in the region.

27.1.2 Water resources and demand

Israel uses all of its conventional water resources and water shortage is a chronic problem. Table 27.1 summarizes water resources and water demand in the early 2000s. Freshwater supply to agriculture, which constituted almost 70% of freshwater resources in 1985, diminished to less than 40% in 2000. During drought years, freshwater supply to agriculture is severely cut off, while farmers connected to reclaimed wastewater systems continue to receive a full quota.

Table 27.1. Water resources and water demand in Israel, in the early 2000s, in Mm^3/yr.

Water resources		Water demand	
Source	**Mm^3/y**	**Sector**	**Mm^3/y**
Freshwater	1350	Urban	700
Brackish water	170	Industrial	85
Reclaimed wastewater	350	International agreements (Jordan, Palestine)	85
		Freshwater irrigation	500
		Wastewater and brackish water irrigation	500
TOTAL	1870	TOTAL	1870

Sources: Water Commission and others.

27.1.3 Sewage as a water resource – the strategic decision

The practice of wastewater reuse has emerged as an unavoidable answer to the combination of severe water shortage, a concentrated population with high levels of water consumption and sewage production, and the threat of pollution to diminishing water resources (Shelef, 1991; Friedler, 2001). Israel has performed massive reuse of effluents for agricultural irrigation since the early 1970s and is presently reusing almost 75% of all the sewage produced in the country. The Water Law of 1959 and policy enacted by the administration up until today define sewage as an integral part of the water resources of the country.

27.1.4 The typical farmer in Israel

Most of Israel's farmers are organized into different types of communities and cooperatives. The Jewish Sector is mostly organized into "Kibbutzs" and "Moshavs". The "Kibbutz" is a socialist community of 200-2000 people that cultivates land and markets the produce as a single organization. The "Moshav" is a cooperative of farmers where land is privately owned but marketing is generally performed as a single organization. In the Arab Sector, land is owned privately or by families, but farmers are usually organized into local associations for water management and other purposes. These relatively large organizations are able to hire the professionals (agronomists, water engineers, administrators, *etc.*) that have been essential in leading the wastewater reuse revolution at the farmer level. The Ministry of Agriculture also provides professional guidance through an efficient extension service. Part of the success of the practice of wastewater reuse in Israel is due to the capacity of the well-organized and informed farmers to adapt quickly to the switch from water to wastewater.

27.2 CHRONOLOGY OF DEVELOPMENTS

Wastewater reuse has been practiced in the region since historical times. In 1959, a decade after the creation of the State of Israel, parliament approved "The Water Law" that defines sewage as a "water resource". However, until the 1970s, water reuse in the country was irregular, based on isolated small projects without a clear policy on the issue (Table 27.2). The potential transmission of diseases via water reuse was also somehow overlooked until 1970 when an outbreak of cholera in Jerusalem (due to irrigation of vegetables with untreated sewage) obliged the Ministry of Health to assume control of water reuse practice.

Table 27.2. Chronology of wastewater reuse developments in Israel.

Up to 1970	1970s	1980s	1990s	2000s
Reuse projects				
Isolated initiatives. Wastewater storage capacity in open reservoirs ~ 20 Mm^3	Multiple new local projects. Wastewater storage capacity in open reservoirs increased by another 50 Mm^3 during the decade.	Multiple new small projects are commissioned. Wastewater storage capacity in open reservoirs increased by another 65 Mm^3 during the decade. Kishon Complex Project – Haifa. First large Interregional project. Good quality restricted irrigation.	Some new medium size projects are commissioned. Wastewater storage capacity in open reservoirs increased by another 30 Mm^3 during the decade. Jeezrael Valley Project. Regional project covers 7 towns. Good quality restricted irrigation. Dan Region Project - Tel Aviv. Second stage for unrestricted irrigation implemented ~ 1989-1990. Largest Interregional Project.	Several new medium-sized projects are commissioned. Western Jerusalem. Interregional Project with unrestricted irrigation.
Main events				
Uncontrolled reuse. 1970 Cholera outbreak in Jerusalem due to irrigation of vegetables with untreated sewage; government starts to control wastewater reuse. Government starts to promote wastewater reuse (Hershkovitz *et al.*, 1969).	World Bank funds beginning of National WW Project. WW reuse is proclaimed national policy. Start massive cotton irrigation with low quality wastewater. Start first in-depth surveys on public health effects of wastewater reuse (Fattal *et al.*, 1981; Vasl and Kott, 1981; Fattal *et al.*, 1986)	70% reuse is achieved by the early 1980s. Drip irrigation becomes the dominant irrigation technology. Cotton market starts to decline by the middle '80. Better quality effluents widen the spectrum of crops irrigated with wastewater Ministry of Environment is created in 1989 and addresses environmental effects of wastewater reuse.	Most efforts oriented to improve quality of treated wastewater towards sustainable reuse. Treated wastewater proposed as a source for recovery of dry rivers and landscape, and also proposed as a source of non-potable water for the urban sector. First unrestricted irrigation (Dan Region). Agrotechnic effects of wastewater reuse are addressed (mainly soil salination and clogging capacity on drippers). Salination is recognized as a serious potential problem. A full campaign to reduce addition of salts to wastewater is started.	Treated wastewater is proposed as a source of water for future aquaculture development. Reuse reaches 75% of all sewage in the country in spite of quick population growth. Plans to reach almost 100% are set. First results are available from studies on environmental effects of long-term massive wastewater reuse. Sustainable reuse becomes the main R&D and discussion issue.

Table 27.2 (continued)

Up to 1970	1970s	1980s	1990s	2000s
Legislation, standards, guidelines (sources: Goldman, 1996; and others)				
Drainage and Flood Protection Law (1957) establishes that water projects have priority over other projects. The Water Law (1959) establishes that water resources are public property and controlled by the State. Wastewater is defined as a water resource.	Public Health Law (1973) Minister of Health will define and control treatment for WW irrigation. The Shelef Commission (1977) sets different wastewater quality standards for irrigation of different crops, including unrestricted irrigation.	Public Health Law (1981) restricts wastewater irrigation to list of allowed crops. All reuse projects require a permit.	Public Health Law (1995) All towns with population>10,000 must treat effluents to BOD< 20 mg/L, TSS<30 mg/L The Halperin Commission (1999) derogates Shelef Commission guidelines. Sets public health requirements for wastewater irrigation following California school. Several rules and standards are applied to reduce addition of salts to WW.	Halperin and Aloni (2003) develop public health guidelines for WW reuse in the urban sector, landscape and industry. Inbar Commission (2003) proposes stricter environmental requirements for wastewater reuse including nutrients, salts and other pollutants. Inbar proposal approved in 2005 while this report was being written. Additional rules and standards are applied to reduce addition of salts to wastewater.

27.2.1 The seventies

In the early 1970s water scarcity was already a serious problem and the Water Commission started to promote water reuse by giving incentives to the construction of sewage treatment and storage units and funding R&D (Hershkovitz *et al.*, 1969; Pano, 1975). The development of a textile industry opened a good market for cotton farming and numerous small reuse systems were constructed to provide low-quality effluents for cotton irrigation during the dry summer. It was a multiple win-win situation: the urban sector got rid of sewage at low cost, discharge of sewage to water bodies was drastically reduced, farmers received the water they had been lacking to grow cotton (enriched with valuable fertilizers) and the textile industry expanded on low-cost cotton, produced round the corner. The basic sewage treatment and storage unit was made of two anaerobic ponds in parallel, followed by a wastewater seasonal storage reservoir that accumulated wastewater during the winter and released it for irrigation during summer (Juanicó and Shelef, 1991 and 1994). The effluents provided by these simple systems were of bad quality, especially by the end of the irrigation season when the reservoirs were almost empty and acted as a simple pipe between the anaerobic ponds and the irrigated fields, but irrigation of cotton did not require special quality.

27.2.2 The eighties

In the 1980s, the cotton market started to decline and farmers had to look for alternative crops. New crops required effluents of better quality and the simple anaerobic pond + reservoir units were not able to provide them. The government invested large sums in R&D looking for feasible ways to improve the performance of wastewater storage reservoirs and supplementary systems (Shelef *et al.*, 1987). Drip irrigation was massively introduced in the country during this decade and sub-surface irrigation started to be developed (Oron and DeMalach, 1987). The first large water reuse system (Haifa–Kishon Complex) was commissioned in 1984 (activated sludge followed by two reservoirs in series) and an interdisciplinary multi-year monitoring program was run to learn, control and forecast the system's performance (Rebhun *et al.*, 1987; Juanicó, 1989; Weber and Juanicó, 1990; Azov and Juanicó, 1991).

27.2.3 The nineties

In the 1990s two other large projects were commissioned: the Dan Region project based on activated sludge followed by SAT (Soil Aquifer Treatment) providing effluents for unrestricted irrigation (Azov *et al.*, 1991 and 1992; Icekson-Tal *et al.*, 2003), and the Jeezrael Valley Project based on semi-intensive technologies providing high-quality but restricted-irrigation effluents (Friedler, 1999; Juanicó and Milstein, 2004). The technology to optimize the design and operation of wastewater storage reservoirs was ready (Juanicó and Dor, 1999). The agrotechnic aspects of wastewater irrigation were also addressed (Adin and Elimelech, 1989; Feigin *et al.*, 1991; Teltsch *et al.*, 1991; Friedler and Juanicó, 1996). The growing demand for the limited freshwater resources and the increasing production of high-quality effluents resulted in several proposals to expand the use of reclaimed effluents for river recovery (Gafni and Bar-Or, 1995; Juanicó and Friedler, 1999), landscape development and non-potable urban uses (Lahav, 1995). Several monitoring and R&D programs pointed out salination of soils and aquifers as a potential by-product of irrigation with salty wastewater and the Ministry of the Environment started a full campaign to reduce the addition of salts to water during its industrial and urban use (Weber *et al.*, 1996; Weber and Juanicó, 2004).

27.2.4 Present situation

By 2005, production of sewage in the country was estimated at 500 Mm^3/yr of which about 425 500 Mm^3/yr reached sewage treatment plants and 370 Mm^3/y were reused. Thus, almost 75% of the sewage of the country was reused, mainly in agricultural irrigation. Most treated effluents were of restricted irrigation quality.

Storage of treated wastewater is made in more than 200 open reservoirs (Figures 27.1 and 27.2). The treated wastewater from Metropolitan Tel Aviv is stored in underground reservoirs obtained by isolating areas of the Coastal Aquifer. Both open and underground reservoirs act as storage units and also as equalization and treatment units.

The coexistence of projects of different sizes and characteristics (Table 27.3) has proved to be not only possible but also desirable. Large projects have a large-scale effect on the economy and development but they are difficult to plan, finance and execute. Small projects have only a limited local effect and generally release effluents of restricted quality, but they are much easier to implement and operate, and the sum of numerous small projects has a total effect comparable to that of large ones. The national policy is to promote all sizes.

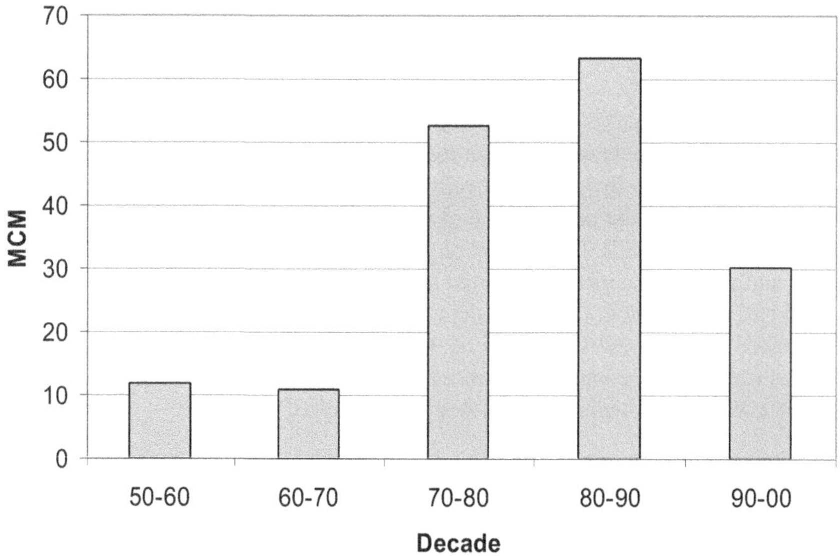

Figure 27.1. Wastewater storage reservoirs constructed in Israel, by decade, as storage volume in million cubic meters, MCM (Mm^3) (data from Dr. Gabi Eitan, personal communication).

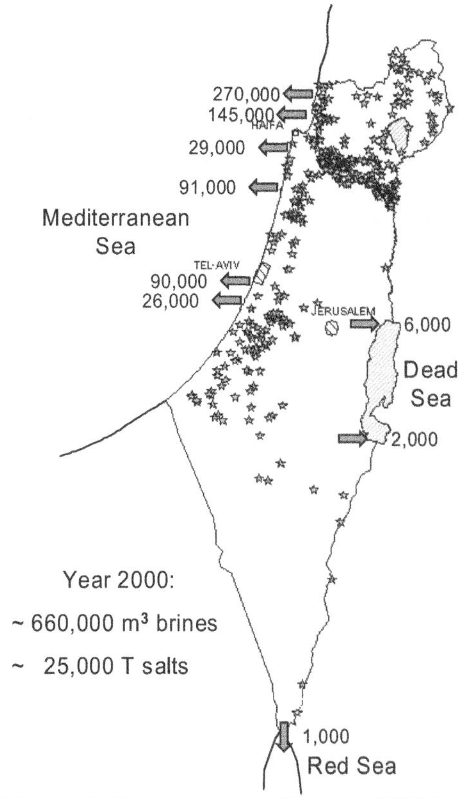

Figure 27.2. Release of brines to the sea during the year 2000 in m^3. Stars represent open wastewater storage reservoirs.

The largest systems, and some of the medium sized ones, release effluents of unrestricted irrigation quality. Smaller projects release effluents of lower quality and are restricted to the irrigation of canned fruits, vegetables for cooking, fruits with non-edible peels, industrial field crops (mainly cotton), fodder crops, forests and pastures (Figure 27.3).

Water availability is the main constraint to maintain and expand an important aquaculture production that exists in the country (Mires, 2000). Treated wastewater is now weighted as a potential source of water for aquaculture and the Water Commission is funding the first R&D efforts to develop a wastewater quality standard for this practice (Feldite *et al.*, submitted).

Table 27.3. Wastewater reuse projects of different sizes.

Project	Capacity [Mm^3/yr]	Scope
Kibbutz Getaot (old)	0.1	Local single-town
Gedera Council	1.5	Local multi-towns
Jeezrael Valley	10.0	Regional
Haifa Metropolitan	25.0	Inter-Regional
Tel Aviv Metropolitan	130.0	Inter-Regional

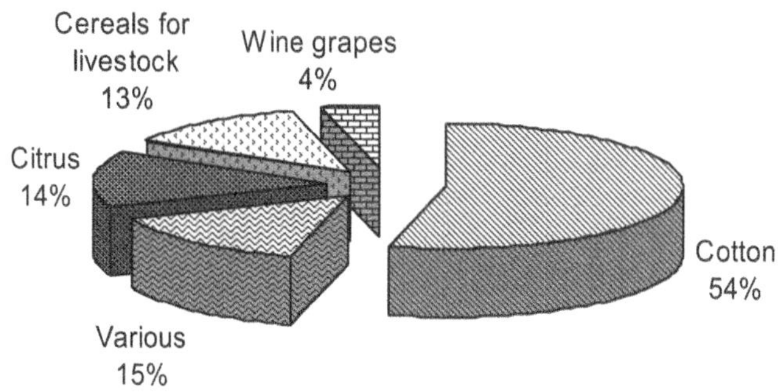

Figure 27.3. Crops irrigated with treated effluents in Central Israel in 1999. (modified from Leshem, 2000).

27.3 CONTROVERSIAL ISSUES

27.3.1 Institutional organization

There are numerous institutions involved in wastewater reuse in Israel. Some of them are:

- Water Commission: in charge of the management of water resources such as planning and control of the development of new water resources including treated effluents, economy of the water sector, substitution of water allocations by wastewater allocations, protection of water resources from any kind of pollution, *etc*.
- Ministry of Health: responsible for public health and develops the guidelines and standards regarding the level of treatment required to irrigate different crops, public parks, *etc*.
- Ministry of the Environment: takes care of all aspects related to the environment *e.g.*, toxic compounds in effluents, salts, excess nutrients, *etc*., as well as the environmental impact of the alternatives to wastewater reuse (discharge of treated effluents to rivers, sea, lakes, *etc*).
- Ministry of Agriculture: addresses the agrotechnic parameters of the treated effluents (e.g. salts, Boron, clogging capacity, nutrients) and the organization of the rural sector regarding this practice, including agronomic guidance to farmers.
- Administration for Water and Sewage (Ministry of Infrastructure): finances and subsidizes sewage treatment works, and supervises project engineering.
- Ministry of Treasure: sets economic policy and provides the funds for the Administration for Water and Sewage. It also evaluates the economic impact of standards and guidelines on wastewater quality.
- Ministry of Interior: approves physical planning and specific projects.

This division of roles sounds nice in theory, but in practice is rather problematic. Israeli law is not clear enough and there is overlapping of responsibilities. For example, it is not unusual for the Ministry of Health to set requirements on the administrative and financial aspects of a project or on the agrotechnic parameters of the treated effluents. There is a conflict between the Ministry of the Environment and the Water Commission on who has the prerogative on soil and/or groundwater protection. There is a foggy overlap between the responsibilities of the central agencies (Ministries) and the local ones (municipalities or local councils). Some inter-ministerial commissions have been created to overcome these problems but these commissions, while effective for the development of policy and legal instruments, have proved to be too heavy and bureaucratic for giving permits or approving specific projects. The Israeli government decided to create a "Water & Sewage Agency" to enter into activity by 2007. The Water & Sewage Agency will concentrate and coordinate all (most of?) the activities regarding the water sector, but meanwhile this is just a decision not yet implemented. The actual prerogatives of this Agency are being discussed and will surely suffer re-definiton.

27.3.2 Is 75% reuse the limit?

The percentage of reclaimed water has been close to 70% for more than a decade (75% in 2005) in spite of multiple efforts to increase it. This is due, in part, to quick population growth due to massive immigration: new treatment plants and reuse schemes are constructed but more sewage is produced and the reuse percentage remains the same. A second more serious problem is that it seems difficult to surpass the limit of 75% reuse. The first commissioned projects were naturally the most promising ones. Now, after more than three decades of treatment and reuse effort, the projects remaining are those much less promising due to high costs, engineering difficulties or lack of demand for reclaimed water in the area. The Israel administration has again taken the decision to reach almost 100% reuse but the decision is controversial. The government has also started massive sea-water desalination to supply the increasing freshwater demand, and the relative costs of sea water desalination versus further sewage reclamation are constantly compared and discussed.

27.3.3 Nutrients in wastewater are not accounted for by farmers

It has long been claimed that the nutrients in the treated wastewater are fertilizers for better crop growing and thus it is not necessary to remove them from wastewater: thus nutrient recycling is a positive by-product of water reuse. This claim has proved to be false. N and P concentration in treated wastewater is in many cases higher than required by crops, leading to problems of vicious crop growth and pollution of soil, aquifers and water bodies. Phosphorus build-up in soil and nitrogen build-up in groundwater have been confirmed. Worse, the characterization and quantification of the nutrients supplied with the effluents is difficult and farmers take the conservative side by only marginally reducing the dosing of fertilizer. Thus, most of the nutrients supplied with the effluents are not recycled and further burden is added to the already serious problem of overfertilization (Juanicó, 1993; Avnimelech, 1997). A recently long-term survey by Tarchitzqui *et al.* (2005) confirms that after more than ten years of efforts to solve this problem, most farmers continue to ignore the nutrients in effluents when dosing fertilizers. This sounds strange in a country where most farmers are well informed, but it seems that sampling and laboratory analyses to determine the variable nutrients content of wastewater is beyond the scope of the farmers' capabilities. The Inbar Commission is now proposing to require nutrient removal at sewage treatment plants in order to cope with the problem, while others insist that it is necessary to better train the farmers to calculate and account for the fertilizers in wastewater.

27.3.4 The contractual relationship between urban and rural sectors

The anaerobic pond + reservoir units of the seventies were a simple and cheap solution for cotton growing. In most cases the reservoirs were constructed by the farmers, avid for water of any available quality, with some financial help from the urban sector for the construction of pipes, pumping stations and the anaerobic ponds. In the eighties, with the decline of the market for cotton, many farmers refused to continue to receive low quality sewage and closed the inlets to the reservoirs; consequently, untreated sewage started to flow into the rivers again. In some cases a contractual obligation existed for farmers to receive all the sewage produced by the urban sector, but farmers were *de facto* unable to fulfil this obligation.

The lack of a clear separation of responsibilities between the urban and the rural sector regarding the treatment and disposal of sewage led to numerous conflicts and problems during the eighties and early nineties (Juanicó, 1993). The regulating agencies found it difficult to apply regulations because urban and rural sectors accused each other of being responsible for the pollution, and government found itself acting not as the controller but as the arbitrator between the two parties. Finally, the Public Health Protocol of 1995 set the whole responsibility for sewage treatment and disposal on the urban sector (the producer of pollution now has sole responsibility for sewage treatment and disposal, and this responsibility can not be transferred to the rural sector or other parties). In cases where the rural sector is in charge of sewage treatment, storage and reuse, it acts as a subcontractor of the urban sector (as any other private firm) while final responsibility in front of the regulatory agencies remains with the urban sector.

Presently, there are numerous different schemes between the urban and the rural sectors. Large reuse schemes are owned and operated by Mekorot (the National Water Company) that sells treated effluents to farmers. Some medium-size treatment plants are owned and operated by municipalities or regional councils who also sell treated wastewater to the rural sector. Other medium-size projects are operated as BOT schemes where the municipality pays a private firm or a farmers' association to construct and operate the sewage treatment and reuse system, or pays per cubic meter of treated sewage. Private firms sell treated wastewater to farmers; water associations also sell it but to their members and at much lower prices. Small rural communities usually have their own sewage treatment and reuse system and treated wastewater is reused in the fields of the community.

27.3.5 The use of wastewater storage reservoirs as treatment units

Wastewater storage reservoirs have proved to be reliable and efficient units for wastewater treatment when operated in series or in batch mode (Juanicó and Dor, 1999; Juanicó and Milstein, 2004). However, only a limited number of sewage treatment and reuse systems are using reservoirs in this way. In most cases treatment is completed at the sewage treatment plant and the reservoirs are operated in continuous-flow mode for three purposes:

- seasonal storage;
- equalization and treatment of potential failures of the sewage treatment plant (a role that has proved to be very important);
- limited treatment provided by the continuous-flow mode (only in small systems).

Regulatory agencies are not interested in addressing these reservoirs as treatment units because most wastewater reservoirs are owned by farmers and sewage treatment falls neither under their legal responsibility nor within the scope of their know-how (see above). In order to use the reservoirs as treatment units, it would be necessary to transfer their operation to the operator of the sewage treatment plant (which belongs to the urban sector) and this change is difficult to implement. Farmers are not interested either, because the operation of the reservoirs in batch mode reduces the amount of water the reservoir can supply during the irrigation season. They prefer to receive fully-treated wastewater from the municipal sewage treatment plant and use their reservoirs for storage only. Thus, the switch of the reservoirs from "storage units" to "treatment units" is limited not by technical problems but by institutional and administrative ones.

27.3.6 Guidelines on wastewater treatment for agricultural irrigation

The Shelef Commission (1977) (Ministry of Health) defined four categories of crops with different quality requirements. Cotton and other industrial crops could be irrigated with effluents with BOD<60 mg/L, TSS<50 mg/L and Dissolved Oxygen > 0.5 mg/L. On the opposite side, unrestricted irrigation required BOD<15 mg/L, chlorination with 2 hours contact time, residual chlorine>0.5 mg/L and Coliforms< 12 MPN/100 mL.

The Halperin Commission (1999) (Ministry of Health) adopted the American Title 22 with few modifications, against the opinion of those academics consulted. Regarding the "unrestricted irrigation" category, the guidelines require "mechanical-biological treatment" and filtration in deep granular media followed by chlorination.

The Inbar Commission (2003) (Ministry of Environment) proposed the adoption of a single high quality for wastewater irrigation (unrestricted irrigation) without quality categories by crops. The quality requirements are not limited to the protection of public health (sufficiently covered by the Halperin Commission Guidelines) but address the protection of the environment with regard to sustainable reuse. The introduction of the sustainability concept is considered a landmark. The Inbar guidelines address numerous parameters (organic matter, nutrients, pathogens, salts, heavy metals, detergents, cyanides and others). These guidelines recognize the existence of regions which are less environmentally sensitive, where less strict requirements can be applied. Thus, the Inbar Commission abandoned the previous approach of "different qualities for different crops" and adopted a single unrestricted irrigation level but with "different qualities for different environmental regions". It did not address the "mechanical-biological treatment" requirement of the Halperin Commission. It did address the requirements for the release of effluents to rivers within the framework of a national plan for river recovery. The proposals of the Inbar Commission were approved as a 10-year goal in May 2005 but some modifications are presently being rediscussed.

The whole issue of quality guidelines is highly controversial. Many professionals consider the requirements of the Halperin and Inbar Commissions as unnecessarily conservative and technically wrong. Others sustain that all the sewage should be treated to an even higher quality including nitrification/denitrification and flocculation/filtration (*e.g.* Rebhun, 2003) or to drinking water standards by membrane technology including desalination (*e.g.* Zaslavski, 2001). The effect that the expected increase in the cost of energy (petroleum, coal) may have on the cost of different levels of wastewater treatment has also started to be discussed.

27.3.7 Salination of soils and aquifers - a threat to sustainability

Sewage is more saline than the supplied freshwater due to the addition of salts during industrial and domestic use (Table 27.4), and the salts are recycled together with the water. There is a clear salination process of soils and aquifers in Israel and a multi-stage approach is used to fight it (Figure 27.4). The main causes of this process are being studied and the issue is controversial. Some hydrologists and soil scientists believe that salination is due mainly to natural processes (primary salination) while others conclude that irrigation with salty wastewater may be the main reason (secondary salination). A review of the whole salination issue is presently underway. It is clear that main salination may be primary in some areas while secondary in others. Whichever, almost everybody agrees that secondary salination due to irrigation with salty wastewater is a threat to sustainability, even in those areas where primary salination dominates.

Table 27.4. Addition of sodium to sewage in two cities of Israel in 1994 (average values, from Mercado and Banin, 1994).

City	Supplied water [Na mg/l]	Sewage [Na mg/l]	Addition [Na mg/l]
Tel Aviv	107	236	129
Haifa	110	256	146

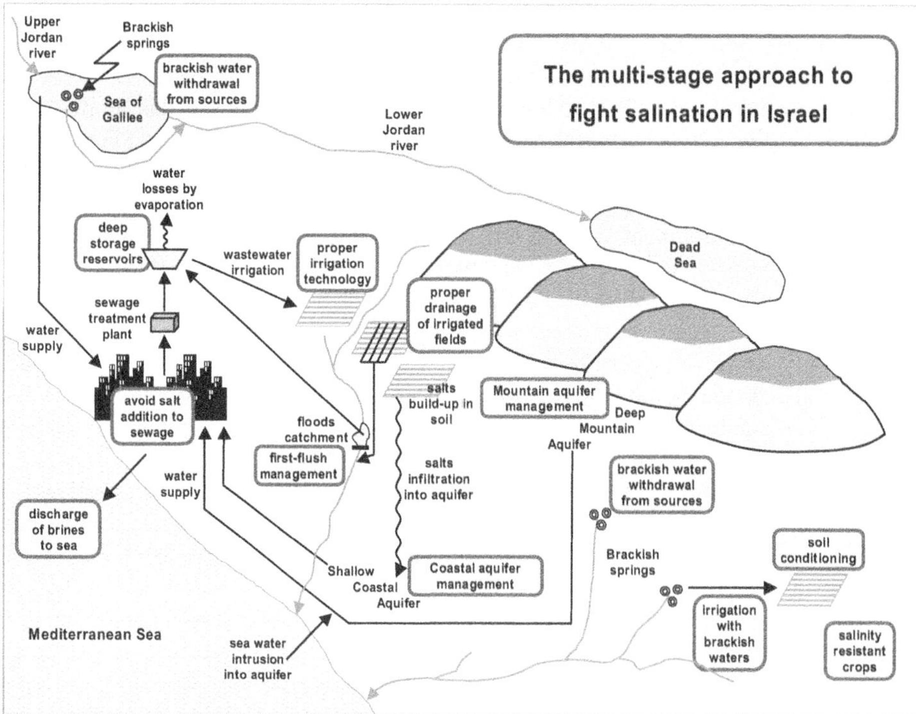

Figure 27.4. The multi-stage approach to fight salination in Israel.

There are no inexpensive ways to remove salts once they enter sewage and the prevention of sewage salt enrichment by controlling sources is the most immediately available solution. The Ministry of the Environment has been engaged in a campaign to reduce the addition of salts to sewage since the early 1990s. The early stages of this campaign were described by Weber *et al.* (1996) while a list of all main regulations and the first long-term results of the campaign are described by Weber and Juanicó (2004). Meanwhile, desalination of reclaimed wastewater is also weighted (Harussi *et al.*, 2001).

Some early surveys indicated detergents as a main source of salts and Boron in sewage in the early nineties (Table 27.5) but not all the detergents had the same salt and Boron content (Figure 27.5). A regulation on the formulation of industrial and laundry detergents was approved in 1999 and it may be extended to dishwasher detergents in the future. Other important sources were water softening for industry, the meat koshering process, and pH neutralization of industrial effluents. The discharge of brines to sewage was first limited and later forbidden, while a system was created to discharge brines to the sea at nine different points off/on the coast (Figure 27.2 above). Some of the discharge points are outfalls from large factories or from industrial areas in which numerous small factories are concentrated.

Other discharge points are located at the outlet of the cooling system of power stations, where cistern trucks can discharge brines. All the discharges are monitored and controlled by the Ministry of the Environment and comply with the requirements of the Barcelona Convention for the Protection of the Mediterranean Sea. The amount of salts discharged to the sea with the brines reached 35,000 Ton in 2003 (Figure 27.6). The range of pH allowed in industrial effluents discharged to sewers was enlarged. The Ministry of the Environment has encouraged the substitution of Na by K or Ca in industrial processes (softening, neutralization). It has also encouraged the substitution of softening by reverse osmosis or Elgressy EST (Electrolysis Scale Treatment), and of water-based air conditioning by air-based technologies that do not lead to water losses by evaporation and the production of brines. Many other initiatives, regulations and activities to reduce addition of salts to sewage have been implemented during the last decade (Table 27.6). Israeli industry has undergone a radical change in recent years. Many factories have adopted K and Ca for softening and neutralization, while others have shifted to reverse osmosis or Elgressy EST - Electrolysis Scale Treatment, and/or from water-based air conditioning to air-based systems. All hospitals substituted the softening technology by 2002, thus reducing the discharge of salts to sewage by 1000 T/yr. These efforts have positive results: data from different treatment plants indicate a steady decrease in salts and Boron in sewage (Figures 27.7, 27.8 and 27.9).

Table 27.5. Addition of sodium to sewage in Tel Aviv, by use, in 1994 (average values, from Mercado and Banin, 1994).

Use	Na [Tons/yr X 1000]	Na [mg/L]
detergents	18	53
household	9	26
water softening	8	22
other industrial uses	6	17
physiological use	4	11
Total	45	129

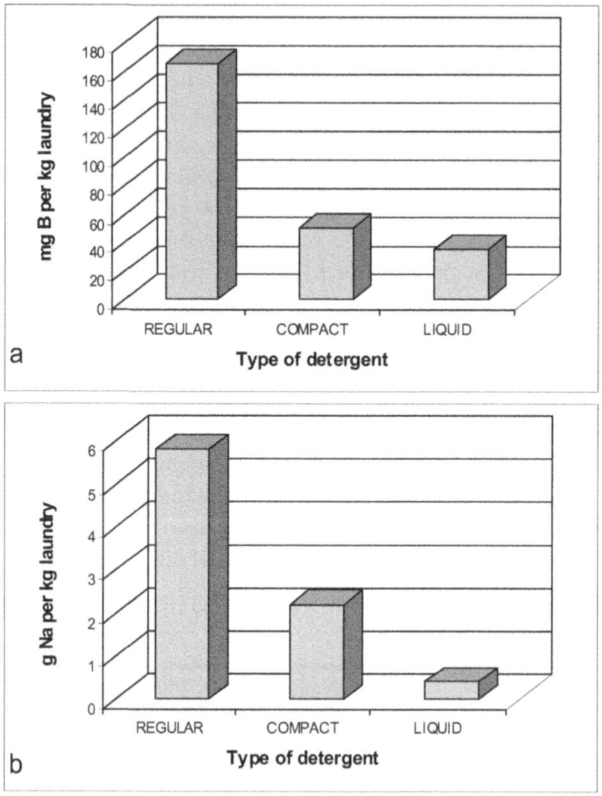

Figure 27.5 Average Boron (a) and Sodium (b) contribution to sewage by different type of laundry detergents used in Israel in the early 1990s.

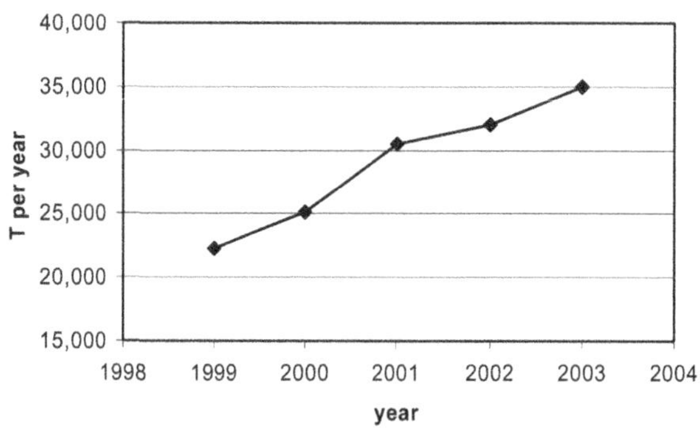

Figure 27.6. Amount of salts discharged to the sea as brines, in tons per year.

Table 27.6 Development of main regulations and activities against sewage salination in Israel.

Year	Main regulations and activities
1991	Industries which consume above a certain amount of salt for the regeneration of ion exchangers were required to use Potassium salts (mainly Potassium chloride).
1993	Some factories are requested to discharge softening brines to the sea.
1994	Regulations on the use of salts in the regeneration of ion exchangers.
1995	Guidelines for controlling salt discharges from slaughterhouses.
1996	construction of network of sites to discharge brines to the sea is begun. Discharge of brines to sewers is limited (business permit law).
1997	Standards on the proper construction and operation of evaporation ponds.
1998	Prohibition of brine discharge to sewers (water law).
1999	A new standard on the formulation of domestic and industrial detergents: reduced Boron, Sodium and Chloride contents.
2000	Recommendations to switch disinfection of swimming pools from hypochlorite, trichloride and chlorine gas to salt electrolysis.
2002	Within the frame of business permit law: • Protocol on ion exchangers. • Limitations on the use of water based air conditioning and refrigeration. • EC limitations to effluents from pickle factories.
2003	Regulations limiting the concentration of salts in all industrial effluents: • Chloride: no more than 200 mg/L above supply water • Sodium: no more than 130 mg/L above supply water • Fluoride: 6 mg/L • Boron: 1.5 mg/L
Proposed new regulations and activities	Prohibition on the use of domestic ion-exchangers. Further restrictions on the formulation of dishwasher detergents. Public education: • Use of salts in dish-washers • Use of detergents

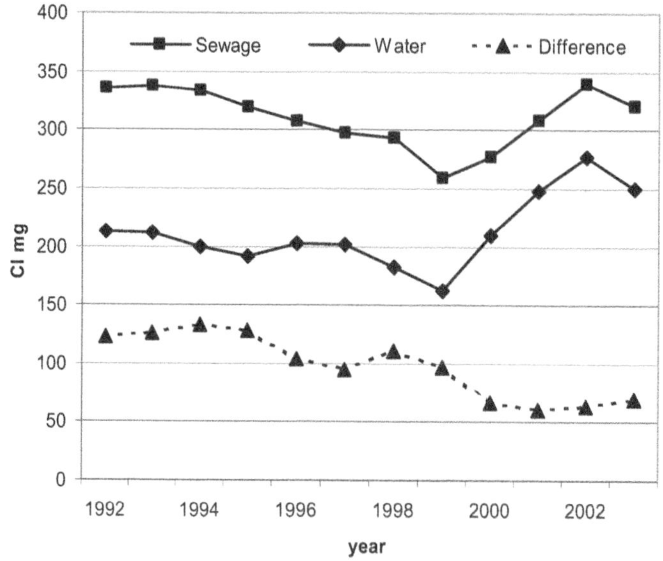

Figure 27.7. Cl (mg/L) measured in the water supplied to Tel Aviv Metropolitan and the sewage reaching the treatment plant.

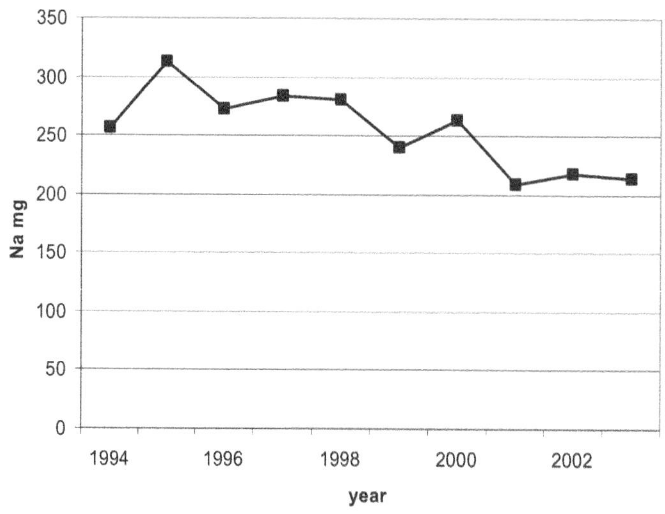

Figure 27.8 Na (mg/L) measured in the sewage of Haifa Metropolitan.

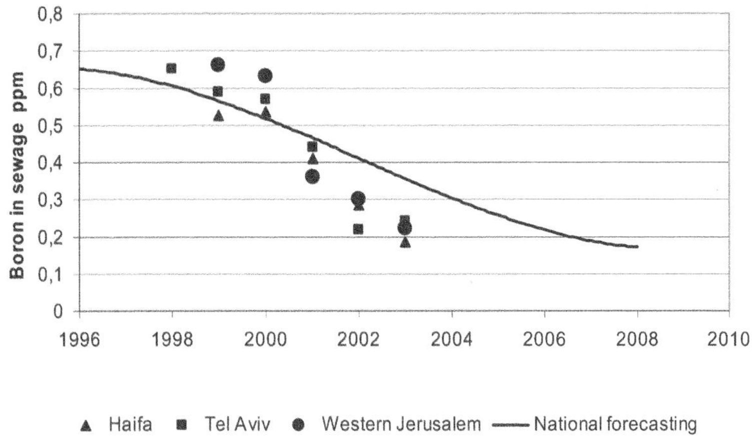

Figure 27.9 Boron measured in the sewage of three main cities and a forecasted national average.

27.4 SUMMARY: WHAT CAN BE LEARNED FROM THE ISRAELI EXPERIENCE

Sewage can be considered an integral part of the water resources of a region.

The ownership of sewage and the responsibility for sewage treatment and disposal must be clearly stated by law.

The proper integration of sewage to the water resources of the region requires efforts at multiple levels: institutional, financial, engineering, agronomic, legislation, R&D, etc.

Not all the problems must necessarily be addressed and solved before starting reuse practice. It is possible to evolve with time. Starting with small local projects for restricted irrigation is a potential approach. But developments must be monitored, discussed and coordinated in order to constantly adapt the switch from water to wastewater irrigation to changing conditions.

In a country that has practiced massive wastewater reuse for decades and is presently reusing 75% of its sewage, most treated wastewater is still dedicated to restricted irrigation. Restricted irrigation liberates freshwater resources for unrestricted irrigation. The controversial issue of "unrestricted irrigation with wastewater" is of secondary importance in many cases.

The development of reuse schemes for irrigation with effluents of very low quality (for cotton or similar crops) may lead to unstable situations when the market for these limited number of crops disappears. Effluents of higher quality allow the irrigation of a wider spectrum of crops adding stability to agricultural development.

Professional advice to farmers has been essential in the successful switch from water to wastewater irrigation in Israel. Israel's farmers obtain this advice mainly through farmers organizations that can hire professional advice, and through the Ministry of Agriculture.

Open wastewater reservoirs can be excellent treatment units if operated as required. But, if the reservoirs belong to the farmers, they operate the reservoirs following irrigation needs and not treatment needs. The reservoirs should be under the control of whoever is responsible for sewage treatment in order to be operated as treatment units.

The coexistence of reuse schemes of different sizes and characteristics is not only possible but also desirable. Small systems are less spectacular than large ones, but their effect on the rural sector is conspicuous.

Proper development and execution of the wastewater reuse policy requires the involvement of several institutions in order to cover the numerous aspects of the practice. But, too many institutions and/or a foggy division of roles, may lead to administrative conflicts and execution delays.

First commissioned projects are naturally the most promising ones while the remaining ones are the most difficult. Israel has been stacked around 70-75% reuse for about two decades in spite of many efforts to reach almost 100% reuse.

Farmers find it difficult to account for the nutrient content of effluents when dosing fertilizers. Most farmers just ignore the nutrients, adding to the problem of over-fertilization.

There are many different schemes between the urban and rural sectors in Israel. The urban sector may treat sewage and sell it to farmers, the farmers may organize to treat the sewage of the urban sector and "sell" it to themselves, third party private firms may act as BOT contractors of the urban sector and sell the wastewater, etc. All the schemes seem to work properly when responsibilities are clearly set.

The first concern when starting reuse practice is the potential transmission of diseases and protection of public health. Agronomic parameters may be also need to be addressed. Later on, sustainable water reuse requires the addressing of environmental issues that were neglected at the beginning. There is a consensus regarding some sustainability issues such as salination of soil and aquifers. Other issues are controversial and Israel is still discussing them.

Reduction of the addition of salts and Boron to sewage during industrial and domestic use is a feasible practice.

27.5 REFERENCES

Adin, A. and Elimelech, M. (1989) Particle filtration for wastewater irrigation. *J. of Irrigation and Drainage Engineering* **115**(3), 474-487.

Avnimelech, Y. (1997) Wastewater recycling in Israel: Past, Present and Future. *Intern. Water Eng.* **17**(9), 46-50.

Azov, Y. and Juanicó, M. (1991) Changes in the Chemical Structure of the Effluents of the Kishon Complex (salts). In The Kishon Reuse Complex Monitoring Program, Technion, Haifa, Annual Report 7, 71-82 [in Hebrew]).

Azov, Y., Juanicó, M., Shelef, G., Kanarek, A. and Priel, M. (1991) Monitoring the quality of secondary effluents reused for unrestricted irrigation after underground storage. *Wat. Sci. Technol.* **24**(9), 267-276.

Azov, Y., Juanicó, M. and Shelef, G. (1992) Monitoring large scale wastewater reclamation systems - policy and experience. *Wat. Sci. Technol.* **26**(7-8), 1545-1553.

Fattal, S., Shuval, H.I., Wax, Y. and Davies, A.M. (1981) Study of enteric disease transmission associated with wastewater utilization in agricultural communities in Israel. *Proc. Water Reuse Symposium II*, Vol. 3, AWWA, Denver, 2200-2215

Fattal, B., Wax, Y., Agursky, T. and Shuval, H. (1986) Comparison of three studies performed in Israel on health risk associated with wastewater irrigation. In Environmental Quality and Ecosystem Stability, Z. Dubinsky and Y. Steinberger, (Eds.), Vol 3(A), pp. 783-794, Bar-Ilan Univ. Press, Ramat Gan, Israel.

Feigin, A., Ravina, I. And Shalhevet, J. (1991) Irrigation with Treated Sewage Effluent. Adv. Ser. Agr. Sci. 17, Springer-Verlag, Berlin.

Feldlite, M., Juanicó, M., Karplus, I. and Milstein, A. Towards a safe standard for heavy metals in reclaimed water used for fish aquaculture. (submitted for publication to J. of Water and Health).

Friedler, E. (1999) The Jeezrael Valley project for wastewater reclamation and reuse, Israel. *Wat. Sci. Technol.* **40**(4-5), 347-354.

Friedler, E. (2001) Water reuse - an integral part of water resources management: Israel as a case study. *Water Policy* **3**, 29-39.

Friedler, E. And Juanicó, M. (1996) Treatment and storage of wastewater for agricultural irrigation. *Water Irrig. Review* **16**(4), 26-30.

Gafni, A. and Bar-Or, Y. (1995) Proposal for the solution of the effluents excess problem by using them for the revival of the main rivers in Israel. *Water and Irrigation* **345:** 45-48 (in Hebrew).

Goldman, D. (1996) Management of Water Systems – The legal frame in Israel. *Report of the Florsheimer Institute for Policy Research*, Jerusalem, Israel [in Hebrew].

Halperin, R. (Committee Chairman) (1999) Principles for the approval of permits for irrigation with wastewater. Ministry of Health - Israel, Division of Environmental Health, Jerusalem, Israel (in Hebrew).

Halperin, R. and Aloni, U. (2003) Rules for wastewater reuse in the city, landscape and industry. Report of the Ministry of Health-Israel, Division of Environmental Health, Jerusalem, Israel (in Hebrew).

Harussi, Y., Rom, D., Galil, N. and Semiat, R. (2001) Evaluation of membrane processes to reduce the salinity of reclaimed wastewater. *Desalination* **137**, 71-89.

Hershkovitz, S.Z., Mor, A., Noi, Y., Feinmesser, A., Fleisher, M. and Kishoni, S. (1969) Utilization of sewage for crop irrigation. Agricultural Publications Division, Water Commission, Ministry of Agriculture, No. 85, Israel (in Hebrew).

Icekson-Tal, N., Avraham, O., Sack, J. and Cikurel, H. (2003) Water reuse in Israel – the Dan Region Project: evaluation of water quality and reliability of plant's operation. *Wat. Sci. Technol.: Water Supply* **3**(4), 231-237.

Juanicó, M. (1989) A Database for a Multi-Institutional Environment Monitoring Program. *Environm. Monitor. Assess.* **12,** 181-190.

Juanicó, M. (1993) Alternative schemes for municipal sewage treatment and disposal in industrialized countries: Israel as a case study. *Ecol. Engineering* **2,** 101-118.

Juanicó, M. and Dor, I. (Eds.) (1999) Reservoirs for Wastewater Storage and Reuse: Ecology, Performance and Engineering Design. Springer-Verlag, Environ. Science Series, 394 pp. Berlin.

Juanicó, M. and Friedler, E. (1999) Wastewater Reuse for River Recovery in Semi-Arid Israel. *Wat. Sci. Technol.* **40**(4-5), 43-50.

Juanicó, M. and Milstein, A. (2004) Semi-intensive treatment plants for wastewater reuse in irrigation. *Wat. Sci. Technol.* **50**(2), 55–60.

Juanicó, M. and Shelef, G. (1991) The performance of Stabilization Reservoirs as a function of the design and operation parameters. *Wat. Sci. Technol.* **23**(7-9), 1509-1516.

Juanico, M. and Shelef, G. (1994) Design, Operation and Performance of Stabilization Reservoirs for Wastewater Irrigation in Israel. *Wat. Res.* **28**(1), 175-186.

Lahav, O. (1995) Wastewater reuse in the urban sector. M.Sc. Thesis, Fac. of Civ. Eng., Technion. (in Hebrew).

Leshem, E. (2000) A solution for the disposal of effluents in the Sharon Region (Central Israel). *Field Health* **8**, 50-53 [in Hebrew].

Mercado, A. and Banin, A. (1994) Addition of dissolved solids to the sewage. *Report to the Ministry of the Environment, Israel (in Hebrew).*

Mires, D. (2000) Development of inland aquaculture in arid climates: water utilization strategies applied in Israel. *Fish. Manag. Ecol.* **7**, 189-195.

Oron, G. and DeMalach, J. (1987) Reuse of domestic wastewater for irrigation in arid zones: a case study. *Water Resources Bulletin* **23**(5), 777-783.

Pano, A. (1975) Storage of wastewater and floodwater in Sarid and Mizra reservoirs. Report by Tahal Consulting Engineers Ltd., Tel Aviv, Israel [in Hebrew].

Rebhun, M. (2003) – Sustainable Wastewater Reuse – Treatment level and wastewater quality required for Israel. *Water & Water Eng.* **57***:18-22* [in Hebrew]).

Rebhun, M., Ronen, D., and Eren, J. (1987) Monitoring and study program of an inter-regional wastewater reclamation system for agriculture. *Journal Wat. Poll. Control Fed.* **59**(5), 242-248.

Shelef, G. (Committee Chairman) (1977) Final report on wastewater quality standards for agricultural irrigation. Advisory Committee for determining wastewater quality standards for agricultural irrigation, Minsitry of Health – Israel [in Hebrew].

Shelef, G. (1991) The Role of Wastewater Reuse in Water Resources Management in Israel. *Wat. Sci. Technol.* **23**(10-12):2081-2090.

Shelef, G., Juanicó, M., and Vikinsky, M. (1987) Reuse of stabilization pond effluent for agricultural irrigation in Israel. *Wat. Sci. Tech.* **19**(12), 299-305.

Tarchitzqui, J., Bar-Hay, M., Levingert, A., Puzin, Y., Sokolobsky, E., Peres, M.., Silverman, A., Einskoot, E., Menashe, Y., Gal, Y., Kanig, E. and Eizenstedet, Y. (2005) National Effluents Survey 1998-2003. Water Irrigat. 459, 8-24 [in Hebrew].

Teltsch, B., Juanicó, M.., Azov, Y., Ben-Harim, I.. and Shelef, G. (1991) The clogging capacity of reclaimed wastewater: a new quality criterion for drip irrigation. *Wat. Sci. Technol.* **24**(9), 123-132.

Vasl, R. and Kott, Y. (1981) Fate of enteroviruses at Haifa's Municipal Wastewater Treatment Plant. In Developments Arid Zone Ecology and Environmental Quality, H. Shuval (Ed.), pp. 233-238, Balaban International Science Services, Rehovot.

Weber, B. and Juanicó, M.. (1990) Variability of effluent quality in a multi-step complex for wastewater treatment and storage. *Water Res.* **24**(6), 765-771.

Weber, B., Juanico, M. and Avnimelech, Y. (1996) Salt enrichment of municipal sewage - New approaches to prevent it in Israel. *Environ. Manag.* **20**(4), 487-495.

Weber, B. and Juanicó, M. (2004) Salt reduction in municipal sewage allocated for reuse: the outcome of a new policy in Israel. *Wat. Sci. Technol.* **50**(2), 17–22.

Zaslavski, D. (2001) The technological, legal and administrative aspects of the shift from potable water to reclaimed effluent. *Water Fluids and Irrigation Engineering* **11**, 18-20 [in Hebrew].

28

Economic analysis of wastewater reuse projects: A methodology for private reuse and public reuse cases

Luis Seguí, Luis Cabrera and Oscar Alfranca

28.1 INTRODUCTION

The economic aspects of wastewater reclamation and reuse appear to be less discussed, because only one component is considered: "private costs", whereas other components, such as external effects (negative and positive) or the cost of water opportunity are relegated to a series of pronouncements about the benefits of wastewater reclamation and reuse. As a result, in many cases only a cost-efficiency analysis is performed. The possible reason is that experts on wastewater reclamation and reuse consider it difficult to evaluate external effects, negative or positive, and to measure them in monetary units. Hence, more interaction with specialists of other areas of knowledge is required. Even though technical, environmental and social factors are considered in the planning of projects, technical and monetary factors usually prevail over others such as environmental, social and cultural ones when deciding on the implementation of these projects.

Technicians have defined the economic analysis necessary as a tool for an assessment of Wastewater Reclamation Reuse Systems (WWRRS) implementation in monetary terms (Asano 2001, Katz *et al.* 2003). A project is considered justified if total benefits exceed total costs.

Another aspect of the economic analysis to mention is that only flows of future investments or derived from the project are considered. Past investments are considered irrelevant for future investment decisions. In this way, interest generated from previous investments was not included in an economic analysis. A typical mistake is the confusion between the price of water and the cost of water.

It would be wrong, in any economic analysis, to suppose initially that reclaimed water reuse involves lower costs than new water supply. This assumption is generally correct but only in cases where wastewater reclamation treatment facilities are situated close to a potential user, as is the case of big agricultural areas or industrial facilities, since water transport is then negligible. Reclaimed wastewater transport and distribution can represent high costs in a water reuse project, and consequently they limit the economic viability of Wastewater Reclamation and Reuse Systems (WWRRS).

In the same way, wastewater sanitation has often been an issue of political interest originated basically from a social request and, as a result, sanitation programs were implemented without considering an economic evaluation; infact, both economic and social aspects need to be considered in the evaluation and analysis of a wastewater sanitation project.

The decision-making process involved in implementing regeneration systems has consisted basically in developing several alternatives and then selecting the one involving minimum costs, instead of considering an economic viability criteria - that is, an evaluation showing the benefits of the actions taken that justify the investment, operation and maintenance of the systems. In this context, it should be mentioned that planned water reuse is another option in integral water management and it can be an economically viable alternative in some cases.

28.2 METHODOLOGY

Traditionally, any economic and financial analysis of WWRRS has focused its attention on the private costs and benefits of the system. The methodology developed here, besides considering these private impacts, incorporates the external factors of the project through the analysis of positive as well as negative external impacts.

The methodology presented is suited to the peculiarities of WWRRS so that it turns into a "custom made" tool, allowing the decision makers to assess the implementation of this type of system. The methodology evaluates WWRRS from a multidisciplinary and interdisciplinary perspective. The main objective is to determine the maximization of the benefits of the project.

The methodology (Seguí 2004) is formed of seven steps that should be executed fully. The steps (see Figure 28.1) are:

1. definition of objectives;
2. definition of study scope;
3. impacts of the project;
4. identification of involved agents;
5. study of financial necessities and possibilities;
6. aggregation of costs and revenues; and
7. sensitivity analysis.

One of the most important contributions to this methodology is referred to as identification of the impacts of the project, since this describes positive as well as negative impacts related to the WWRRS in detail. The six impact groups described are:

1. hydraulic infrastructure;
2. conditioning and reuse of contaminants;
3. use of the resource;
4. public health;
5. environment; and
6. education.

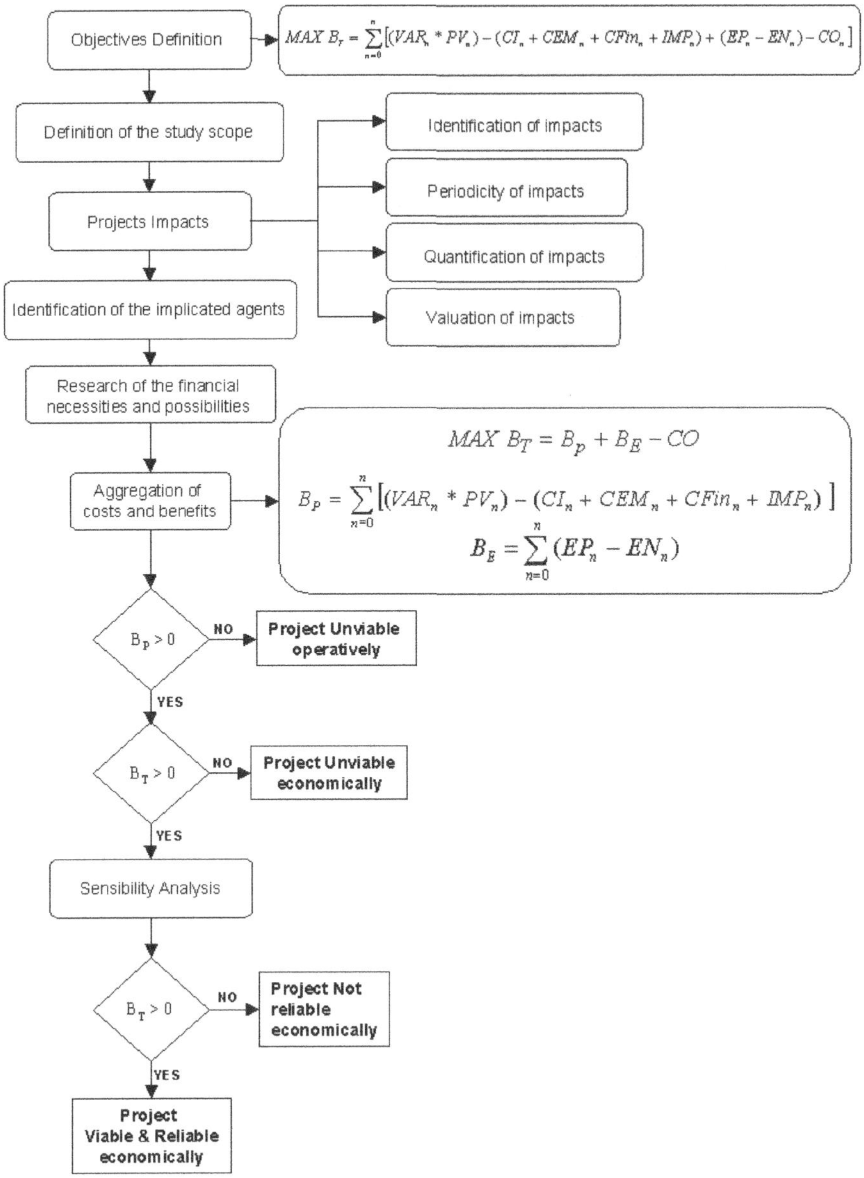

Figure 28.1. Stages for the economic analysis of reclamation and reuse wastewater projects. (Seguí 2004).

This information is based on bibliographic review, expert consultation and professional experience.

Another important contribution concerns the aggregation of costs and revenues, since at this point the maximization of benefits is given by the sum of private and external benefits. This allows two separate situations to be imagined:

1. The running of the WWRRS being economically and financially feasible, which is defined by private benefit determination (normally of interest to technicians and politicians); and
2. The WWRRS being economically, financially and environmentally feasible (which is the interest of economists and society).

The methodology described will be applied in the case studies presented below. It is important to mention that due to issues regarding access to information and confidentiality, adjustments were made so the data could be applied to the methodology.

28.3 PUBLIC AND PRIVATE REUSE OF RECLAIMED WASTEWATER – A CASE STUDY

28.3.1 Wastewater reuse in Spain

A study by Olcina (2002) focused on water resources management in Spain, particularly non-conventional resources. The study presented a deep and detailed analysis of wastewater reclamation and reuse in Spain.

The author considered that effective water consumption in Spain was $57{\cdot}10^6$ m^3/d and $40{\cdot}10^6$ m^3/d of it returned to the river basin. On the other hand, $7{\cdot}10^6$ m^3/d was treated and $0.6{\cdot}10^6$ m^3/d of it reused in a planned way.

"Centro de Estudios y Experimentación de Obras Públicas" (CEDEX), established four zones in Spain where reuse is implemented (Figure 28.2). These are: a) Spain Levante, b) Canary and Balearic Islands, c) areas where varied reuse is planned and d) areas with incidental reuse (Iglesias 2005).

Today, over 255 direct water reuse activities are identified. Agricultural reuse is the most widespread (82% of total volume), 6% for recreational use and golf courses, 7% for municipal use, 4% for environmental requirements and 1% for industrial use.

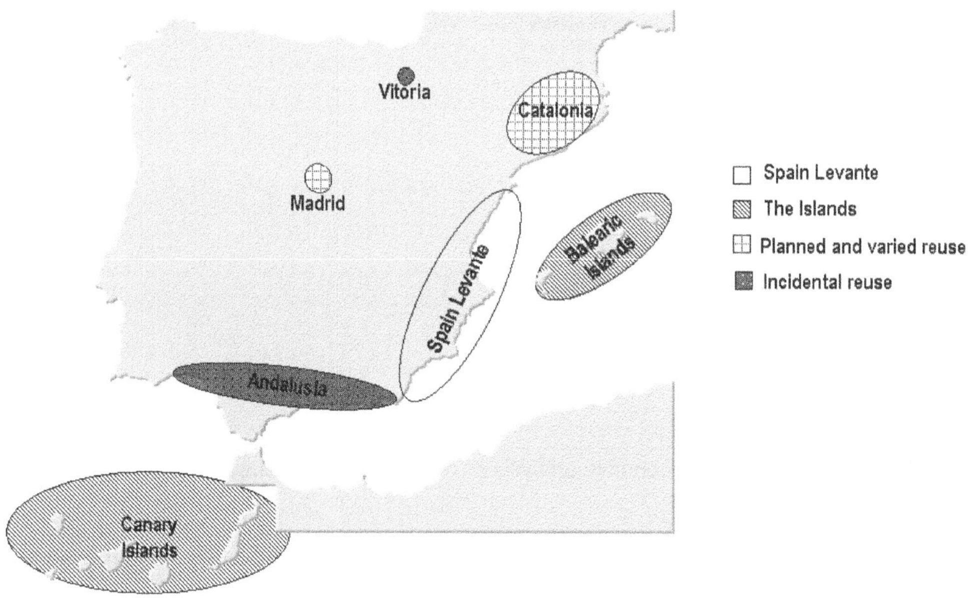

Figure 28.2. Zones in Spain where reuse is implemented (Iglesias, 2005).

28.3.2 Objective

The main objective is to compare the technical and economic viability of two systems of wastewater reclamation and reuse from two different approaches: a) wastewater reuse for private use and b) wastewater reuse for public use.

28.3.3 Location

Figure 28.3 shows the location of the WWRRS case studies, situated in the Mediterranean region of Spain.

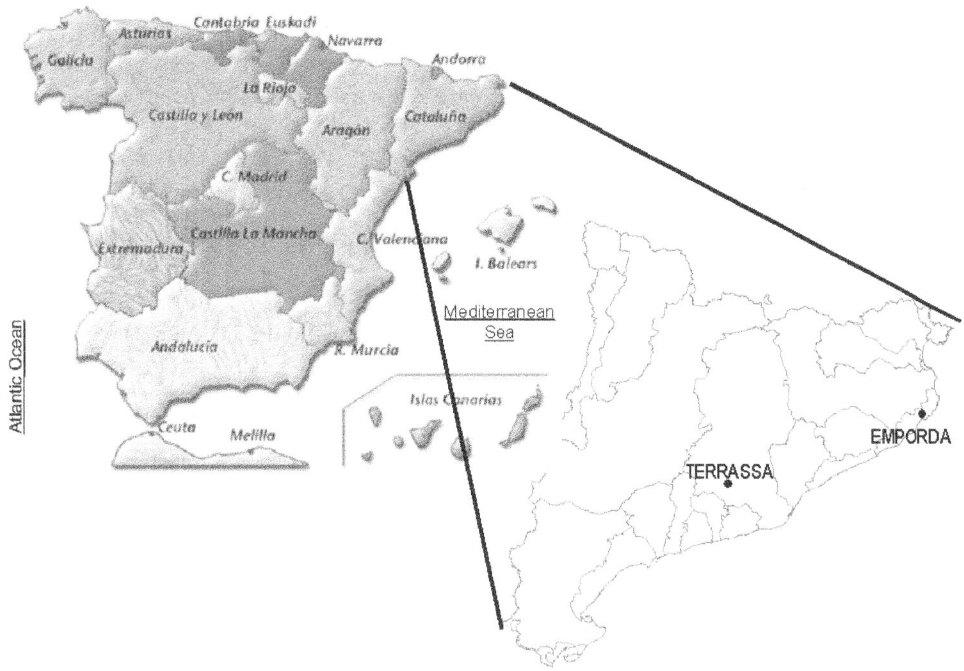

Figure 28.3. Location of the WWRRS case studies.

Terrassa

This city is situated in Valles Occidental county, in the interior of the Barcelona province, 286 meters above sea level. Terrassa has 200,000 inhabitants and is one of the most important industrial cities in the region.

From the total water supply, 60% corresponds to domestic use and 40% to industrial and commercial sectors. The main important industrial sectors situated in Terrassa are construction, textiles, furriery and the chemical industry.

Wastewater discharges are collected and carried to the Terrassa-Les Font wastewater treatment plant (WWTP), with a capacity of 7 200 m^3/d. This WWTP constitutes pre-treatment, primary decantation, activated sludge and secondary sedimentation. Ten per cent of the secondary effluent is used to produce reclaimed water by a tertiary treatment followed by desalination.

There are several treatment processes to eliminate salinity from wastewater, osmosis and electrodialysis being the most important ones. In Spain, several electrodialysis systems have been implemented, and this is the case of the Terrassa WWTP which is the first to receive an inlet with salinity higher than 2 000 mg/L, receiving the fourth highest flow in Spain.

Reclaimed water is reused in golf course irrigation, approximately 8 and 4 km away. Water is pumped to storage ponds for this purpose.

Empuriabrava

Aiguamolls de l'Empordà Natural Park is one of the best known and emblematic areas in Catalonia since its creation in 1983. The park is located in Alt Empordà county and covers a surface of 4 866 ha. Today Aiguamolls de l'Empordà Natural Park is the second Wetland area in Catalonia. It comprises several lakes and meadows in the confluence of the Muga and Fluvia rivers, forming a privileged habitat for fauna, especially for aquatic birds (PNAE 2003).

The WWRRS situated in Empuriabrava municipality supplies reclaimed water to Cortalet Lagoon, particularly during the summer period. This lagoon is the first point of contact for visitors with the park, and tends to dry up due to water consumption upstream used for agricultural irrigation.

28.3.4 Technical description

Table 28.1 describes the main technical aspects related to the two WWRRS analyzed, types of WWRRS established, reused flow, final use and users of reclaimed water as well as the water quality criteria met by the system.

Table 28.1. Characteristics of the WWRRS analyzed.

Location	Treatment systems	Reclamation systems	Flow (m^3/d)	Reuse	Users	Quality criteria
Terrassa[a]	Pre-treatment Primary decantation Activated sludge Secondary sedimentation	Coagulation-flocculation, filtration, reversible electrodialysis and chlorination	7197.1	Ornamental: Irrigation of golf courses	Royal Golf Club of "El Prat" and Golf Course "Sant Joan"	2 NTU 5 mg TSS/L 1200 μS/cm
Empuriabrava[b]	Pre-treatment, biological reactor, sedimentation basin	Stabilization ponds, constructed wetlands	1270.1	Environmental: Ecosystem Maintenance	Aiguamolls de l'Empordà natural park	California Health Laws 2001

References: a) TECNOMA (Seguí *et a.*, 2005) b) Consorci de la Costa Brava (CCB 2003).

28.3.5 Materials and methods

28.3.5.1 Materials

Data corresponding to the Terrassa case study have not been published, and are obtained from empirical work by researchers. In the case of Empuriabrava, information was obtained from:

- CCB (Consorci Costa Brava) offered all the data related to establishment and operation of the Colera and Empuriabrava WWRRS, as it is the organization responsible for these systems; similarly, way technical information was also given (CCB 2003).
- PNAE (Parc Natural dels Aiguamolls de l'Empordà) facilitated information regarding statistics on the number of visitors, as well as information related to transport used to arrive at the park, duration of visits, places where people stay, interests and motives for their visit (PNAE 2002).

28.3.5.2 Methods

Determining the cost and the price of reclaimed water is important in the design and operation of a WWRRS. A detailed WWRRS cost approach can be made but not when we deal with the price of reclaimed water – basically because of the absence of a reclaimed water market that would allow us to determine a price. Therefore, the only point of reference is the price of water from conventional sources. Nevertheless, this price does not reflect all the impacts that a project of reclamation and reuse of wastewaters involves.

The method of our approach is specifically about identification, periodicity, quantification and evaluation of the impacts of the project, considering a particular situation in relation to a particular agent. These impacts are incorporated into an evaluation that combines cost-efficiency analysis (CEA), cost-benefit analysis (CBA) and the Net Actual Value (NAV) in order to assess economic viability of the technical alternatives proposed, with the purpose of determining the alternative that provides the maximum total benefit.

The aim is to determine whether the technical alternative proposed to regenerate and reuse water maximizes the difference between income and costs related to the production of regenerated water and its following reuse. The analysis is performed from the point of view of the final user of regenerated water.

This optimization approach has been selected due to its intuitive interpretation as well as its applicability.

The objective function to optimize is:

$$MAX\ B_T = \sum_{n=0}^{n}[(VAR_n * PV_n) - (CI_n + CEM_n + CFin_n + IMP_n) + (EP_n - EN_n) - CO_n] \quad \text{(eq. 28.1)}$$

Where, B_T= Total Benefit; VAR = Annual Volume of Reclamation Water; PV = Selling Price of Reclamation Water; CI = Investment Costs ; CEM = Maintenance and Operational Costs; $CFin$ = Financial Costs; IMP = Taxes; EP = Positive Externalities of the impacts; EN = Negative externalities of the Impacts ; CO = Opportunity Cost; n = Year.

28.3.6 Project Impacts

An impact is defined as any outcome result of a WWRRS implementation, either wished or not, promoted or accidental, generally susceptible to measurement, in a specific area of hydraulic management. In this project a distinction was made between internal and external impacts. The internal or private impacts are those directly related to the production process of the reclamation water and its later reuse. They correspond basically to the income (positive private impacts) for selling reclaimed water or some of their sub-products. They also correspond to the investment, operation and maintenance costs of the WWRRS (negative private impacts).

External impacts (for example pollution control, rise in water availability or supply guarantee), although sometimes difficult to determine, are quite important since an impact of these characteristics can be the reason for stopping a project or preventing its economic viability. The analysis of the impacts considered in this investigation are described and classified in five groups (for detailed information see Seguí, 2004). Tables 28.2 and 28.3 show the analysis of impacts corresponding to each particular case.

Table 28.2. Impacts produced by construction and operation of the WWRRS Terrassa-Les Fonts.

Set of impacts	Impact description	Identification		Periodical distribution		Quantification	
		Negative (Costs)	Positive (Benefits)	Negative (Costs)	Positive (Benefits)	Negative (Costs)	Positive (Benefits)
Hydraulic infrastructures	Reclamation and reuse wastewater [(a)]	Production and distribution of regenerated water		Initial investment and constant during lifetime of the project		2646 m^3/d	
Preparation and reuse of pollutants	Fertilizers[(b)] (nitrogen)		Nutrient contribution to agriculture		Constant during lifetime		19.6 ton N/yr 2.6 ton P/yr 21.1 Ton K/yr
Resource use	Water quantity[(c)]	Availability of water for a more profitable activity	Increase in resource availability	Constant during lifetime	Constant during lifetime	2646 m^3/d	965 712 m^3/year
	Supply guarantee[(d)]		Reliability in water supply		Constant during lifetime		100% guarantee in supply
	Water quality[(e)]		Quality according to use needs		Constant during lifetime		90% reliability in water produced
Environment[(f)]	Pollution of water bodies		Improvement of underground water quality		Constant during lifetime		75 mg NO_3/L reduction
			Improvement of superficial water quality		constant during lifetime		19.6 Ton N/yr removal 2.6 Ton P/yr 21.1 Ton K/yr
	Protection of the environment		Availability of water for forest fires		constant during lifetime		43 ha of forest
Education[(g)]	Technique		Improvement in secondary treatment operation		constant during lifetime		90% reliability in water produced

References: (a) Renzetti 2003; (b) Asano 2001 and Randall 2003; (c) Faux and Perry 1999; (d) Ward and Michelsen 2002; (e) Lazarova *et al.* 2001 and Kamizoulis *et al.* 2003; (f) Devaux *et al.* 2001, and Chu *et al.* 2003; (g) CCME 2002.

Table 28.3 Impacts produced by construction and operation of the WWRRS of Empuriabrava.

Set of impacts	Impact description	Identification		Periodical distribution		Quantification	
		Negative (Costs)	Positive (Benefits)	Negative (Costs)	Positive (Benefits)	Negative (Costs)	Positive (Benefits)
Hydraulic infrastructures	Reclamation and reuse wastewater[a]	Production and distribution of regenerated water		Initial investmen t and constant during lifetime		1507 m^3/d of reclaimed water	
Preparation and reuse of pollutants	Not considered						
Resource use	Water quantity[b]	Availability of water for a more profitable activity	Increase in resource availability	constant during lifetime	constant during lifetime	1507 m^3/d of reclaimed water	1507 m^3/d of reclaimed water
Environment[c]	Pollution of water bodies		Improvemen t of bath marine water quality		constant during lifetime		Reduction of bacterial pollution in 3 logarithmic units
	Wetlands and rivers habitat		Preservation and recovery of wetlands		constant during lifetime		Major number of species and improvement of landscape
Education[d]	Environ-mental education of society		Divulgation of water reuse benefit		constant during lifetime		10,480 Students attended

References: (a) Renzetti (2003); (b) Faux and Perry (1999); (c) Devaux *et al.* (2001) and Chu *et al.* (2003); (d) CCME (2002).

28.3.7 Financial requirements

The Terrassa-Les Font WWRRS was totally funded by private money from the two main users (Royal Golf Club of "El Prat" and Golf Course "Sant Joan"). Given this 100% private investment, the users (owners) expect a recovery of the investment over the next five years (in the case of the desalination system) and over 20 years (for the rest of the system).

The Empuriabrava WWRRS received an investment of 80% for the construction of the reclamation system from the European Union's Cohesion Fund (Serra *et al.* 2002). All of the initial capital is considered to be a non-returnable subsidy.

28.3.8 Income

Among the identified impacts that provide income (Tables 28.2 and 28.3), only the following were evaluated economically in this study: a) nutrient contribution in water reclamation, b) increase in water availability, c) improvement in superficial water quality. In the case of the Empuriabrava WWRRS the only economically evaluated impact is related to the existence and preservation of the Aiguamolls de l'Empordà Natural Park.

Terrassa

Nutrient contribution in water reclamation – The technique of "cost-substitute" was used to evaluate this impact (Edwards-Jones *et al.*, 2000), which consists of assessing a good or an environmental service by considering the cost of a substitute that: 1) must provide the same function as the substitute service; 2) cost less; and 3) customers are willing to pay for the substitution. Considering this the saving in fertilization would be 0.065 €/m^3.

Increase in water availability – We also used the cost-substitute technique. A commercial strategy is to offer reclaimed water at the same price as water from conventional sources, to be capable to compete in the market. In Catalonia (ACA, 2005) costs of water production from conventional sources are fixed by a water law. The respective price of water supply from conventional sources in this area is 1.6100 €/m^3.

Improvement in superficial water quality – This impact is evaluated through the restoration cost technique (Edwards-Jones *et al*,. 2000). This method computes present or potential environmental cleaning costs. In this study, the law of wastewater discharge is established according to these criteria, and will be the reference to calculate this income. In 2005, the estimated price for removed pollutants is 1.4099 €/m^3.

Empuriabrava

The existence and preservation of the park – This impact is fixed by the Use Value (*UV*) of the park. With the purpose of determining UV the travel cost technique was used, which is a method widely used in the evaluation of natural areas.

The aim of this technique of economic evaluation focuses on determining the costs families or persons are willing to incur to get to a recreational facility (Turner *et al.* 2003). These costs are used as a measure of the paying intention to do a recreational activity (Park and Leeworthy 2002). It is important to emphasize that this technique tries to quantify the value of an environmental use for a group of persons but it is not able to inform about the different values that non-use would have for a higher number of people.

Taking into consideration all the information provided by the Aiguamolls de l´Emporda Natural Park public information center, the number of trips over 2002 year were calculated. In this study only the Catalonian visitors were considered, which corresponds to 70% of the total visits. The remaining 30% corresponds to visitors from other parts of Spain (5%), European countries (24.5%) and other continents (0.5%). Unfortunately these visitors cannot be included in this study due to lack of more detailed information required by this technique (Seguí, 2004).

Using the travel cost technique (Azqueta, 1994), the total accumulated benefit is 660 801 €/year. On the other hand, total capacity provided by Empuriabrava is 550 000 m^3/year. Considering this, it is estimated that Aiguamolls de l´Empordà Natural Park has a use value of 1.2015 €/m^3.

28.3.9 Costs

The impacts causing cost in the project are a) infrastructure and b) use of water resources.

a) Infrastructure (private costs)

Private costs determine the cost per cubic meter, which is considered to be equal to the Minimum Selling Price (MSP). The MSP is defined as the minimum price in which the agent should sell regenerated water to guarantee the cost retrieval and expected benefit, in a manner that the investment is made profitable under the criterion of Net Present Value (NPV). In a process of optimization and when the market presents perfect competence conditions, this cost can be considered per cubic meter as an approximation to Marginal Cost (*MgC*). To obtain the MSP we have decided to turn to the economic technique of (NPV). According to Tietenberg (1992) efficiency is achieved when marginal revenue equals marginal costs, which is the same as saying that net benefit equals zero. Under this criterion, the cost per cubic meter in the Terrassa-Les Fonts WWRRS is 0.6900 €/m^3, whereas in Empuriabrava it is 0.3596 €/m^3.

b) Use of water resources

The costs considered so far correspond to private costs but this methodology also includes the opportunity cost (OC) of the project. Opportunity Cost of water (OC) can be defined as the difference between the benefits obtained with the best use of water and the benefits with the current use. In the two case studies, the most profitable alternative activity is water supply for domestic use in surrounding municipalities, and the cost in this case, according to Ley de la Ordenación, Gestión y Tributación del Agua (ACA, 2005), is 0.3167 €/m^3. It should be explained that the reclaimed water is not supposed to be used for domestic purposes but only the exchange of the water flows in concession.

28.3.10 Results and discussion

Once identified, quantified and valuated the project impacts are added to get the total benefit. Impacts are expressed in euros per cubic meter. Table 28.4 shows the economic valuation of the impacts, negative (costs) and positive (benefits), for each of the analyzed systems as well as the benefit obtained for each of the WWRRS.

The sensitivity analysis allows evaluating sensitivity of the model to changes in some of the main variables affecting reclaimed water production. The selected variables for the sensitivity analysis are: a) discount rate, b) fiscal depreciation, c) opportunity cost, d) project duration, and e) reclaimed water price. The study shows that both WWRSS are economically robust (Seguí, 2004).

Table 28.4. Summary of the impacts assessment for the analyzed WWRRS.

Set of impacts	Terrassa-Les Fonts		Empuriabrava	
	Negative (Costs)	Positive Benefits)	Negative (Costs)	Positive (Benefits)
Hydraulic Infrastructures	0.6900		0.3596	
Preparation and reuse of pollutants		0.0650	Not considered	
Resource use	0.3167	1.6100	0.3167	N.V.
Environment		1.4099		1.2015
Education		N.V.		N.V.
Total	1.0067	3.0849	0.6763	1.2015
Benefit	**2.0782**		**0.5252**	

N.V.- Not Valuated; All quantities are expressed in €/m^3.

28.4 CONCLUSIONS

Traditionally, an economic-financial analysis of WWRRS focuses exclusively on the costs and private benefits. The methodology presented here not only considers the private impacts but also includes the external impacts, both negative and positive, through an analysis of the latter.

The Terrassa-Les Fonts WWRRS generates a maximum benefit of 2.0782 €/m^3 whereas for Empuriabrava the benefit is about 0.5252 €/m^3. Even though not all the impacts have been economically evaluated, this does not affect the results of the decision of this investigation because the impacts that were not evaluated, will increase the profitability of the system. The sensitivity analysis reveals that in case of a fluctuation in the main variables, the total benefit of the WWRRS remains positive.

In order to regain the costs of the WWRRS and consequently guarantee economic and technical viability, the lowest selling price of regenerated water must be 0.6900 €/m^3 for the Terrassa-Les Fonts WWRRS and 0.3596 €/m^3 for Empuriabrava.

28.4.1 Economic policy proposals

In both cases, the local water authorities now use a technical-economic tool to calculate and determine their negotiation margin for selling reclaimed water. Additionally, if water shortages remain, it is probable that in the short term, the price of water from conventional source will increase. This situation would support the actions of reclamation and reuse of wastewater.

In the case of the Empuriabrava WWRRS, only the economic and environmental impacts related to the existence of the Aiguamolls de l'Empordà Natural Park were evaluated. Some studies state the possibility of increasing the water resources of the area. If the responsible water management authorities in this area (Agencia Catalana del Agua) encourage an exchange between conventional water resources and regenerated water, volumes of water from natural resources will be released and used in other activities with higher economic benefit, without involving a change in the user rights of water resources.

28.5 ACKNOWLEDGEMENTS

The authors wish to thank Lluís Sala and Manel Serra for their help in the accomplishment of the case study of Empuriabrava, as well as those responsible at the Royal Golf Club of "El Prat" and Golf Course "Sant Joan" for all the help in the case study of Terrassa-Les Fonts.

28.6 REFERENCES

ACA (2005). Regulatory guidelines for water cannon. Law 6/1999, 12 July. http://www.gencat.net/aca/ [in Spanish]

Asano, T. (2001). "Water from (waste)water – The dependable water resource the dependable water resource (The 2001 Stockholm Water Prize Laureate Lecture)". *Water Science & Technology* Vol 45 No 8 pp. 23–33

Azqueta, D. (1994). Economical Evaluation of Environmental Quality. Mc Graw Hill, España. [in Spanish]

California Health Laws (2001). *The Purple Book*. California Health Laws Related to Recycled Water June 2001. http://www.dhs.ca.gov/

CCB (2003). Consorci de la Costa Brava. Plaça Josep Pla, 4, 3er. 17001 Girona. Tel. 972 – 201467, Fax: 972 – 222726. E-mail: ccb@ddgi.es Persona de contacto: Lluís Sala.

CCME (2002). *Linking water science to policy workshop series water reuse and recycling*. Canadian Council of Ministers of the Environment. http://www.ccme.ca/

Chu W., Wang, J. and Kao, C. (2003) "A simplified risk-based approach for process screening in municipal wastewater reclamation and reuse". *Water Science & Technology* **47**(1): 257–262

Devaux I., Gerbaud L., Planchon C., Bontoux J. y Glanddier Ph.Y. (2001). "Infectious risk associated with wastewater reuse: an epidemiological approach applied to the case of Clermont-Ferrand, France". *Water Science & Technology* Vol 43 No 12 pp. 53–60

Edwards-Jones G, Davies G y Hussain S. (2000). *Ecological Economics*, Blackwell Science, Londres, Inglaterra.

Faux, J. y Perry G. M. (1999). "Estimating Irrigation Water Value Using Hedonic Price Analysis: A Case Study in Malheur County, Oregon". *Land Economics* vol. 75, no.3, pp. 440-452

Iglesias R (2005). Actual situation and proposals for the advance in water reclamation and reuse in Spain. *Integration of reclaimed water in the management of resources Technical conference*, Lloret de Mar, Costa Brava, Girona, Octubre 2005 [in Spanish].

Kamizoulis G., Bahri A., Brissaud F., y Angelakis A.N. (2003). Wastewater Recycling and Reuse Practices in Mediterranean Region: Recommended Guidelines. MED-REUNET Mediterranean Network on Wastewater Reclamation and Reuse. Case Studies http://www.med-reunet.com/home.asp

Katz R. (2003). Water Recycling 2030. Recycled Water Task Force. California Departament of Water Resources. http://www.owue.water.ca.gov/recycle/docs/FinalReport.pdf

Lazarova V., Levine B., Sack J., Cirelli G., Jeffrey P., Muntau H., Brissaud F., y Salgot M. (2001). "Role of water reuse for enhancing integrated water management in Europe and Mediterranean countries". *Water Science & Technology* Vol 43 No 10 pp. 25–33

Olcina C. J. (2002). Hydrological plan and non conventional water resources in Spain. Edition A. Gil Olcina and A. Morales Gil. Digital edition based on the edition Alicante, Caja de Ahorros del Mediterráneo, Instituto Universitario de Geografía, 2002. http://cervantesvirtual.com/ [in Spanish].

Park, B. y Leeworthy (2002). "Valuing snorkeling visits to the Florida Keys with stated and revealed preference models". *Journal of Environmental Management.* Vol. 65, no. 3, pp. 301-312.

PNAE (2002). Parc Natural dels Aiguamolls de l'Empordà. Information center: "El Cortalet" cra. de Sant Pere Pescador km 13,6. Tel.: 972 45 42 22, Fax: 972 454 474, 17486 Castelló d'Empúries. Contact person: Rosa Llinàs Viola.

PNAE (2003). Parc Natural dels Aiguamolls de l'Emporda. http://www.parcsdecatalunya.net

Randall C.W. (2003). "Potential societal and economic impacts of wastewater nutrient removal and recycling". *Water Science & Technology* Vol 48 No 1 pp. 11–17

Renzetti S. (2003). "Full Cost Accounting for Water Supply and Sewage Treatment: A Case Study of the Niagara Region, Canada". *World Bank's Water Resources Management Group on Economic Instruments Conference*. http://worldbank.org/

Seguí L. and Cabrera L. (2005). "Integral management of a reclamation water Project. The case of WWTP Terrassa-Les Fonts tertiary treatment." II Jornades Tècniques de Gestió d'Estacions Depuradores d'Aigües Residuals. 18 - 20 de enero de 2005. Barcelona, España. http://mediambient.gencat.net/aca/es/inici.jsp [in Spanish]

Seguí, L. (2004). Wastewater reclamation and reuse systems. Methodology for technical-economical analysis and case studies. PhD thesis. Universidad Politécnica de Cataluña. http://tdx.cesca.es [in Spanish]

Serra, M., Sala, L., and Mujeriego, R. (2002) Actual situation and recently advances of planned water reuse in Costa Brava. Third Iberian Conference on Water, Seville, Spain. November 2002 [in Spanish].

Tietenberg, T. (1992). *Environmental and Natural Resource Economics*, 3rd. ed., Harper Collins Publishers. EE.UU.

Turner, R., Paavola, J., Cooper, P., Faber, S., Jessamy, V., y Georgiou, S. (2003). "Valuing nature: lessons learned and future research directions". *Ecological Economics*. Vol. 46, pp. 493 – 510.

Ward A. y Michelsen A. (2002). "The economic value of water in agriculture: concepts and policy applications". *Water Policy*, Vol. 4, no. 5, pp. 423-446.

29

Wastewater reclamation and reuse in Spain

Lluís Sala, Raquel Iglesias and Enrique Ortega

29.1 INTRODUCTION

From the mid 1980s until today, Spain has been an increasingly active country in the field of wastewater reuse. Areas where reuse is practiced today include the Mediterranean coast, the Madrid metropolitan area, the city of Vitoria in the Basque Country, Andalucía, and the Balearic and Canary Islands. Dry summer conditions and eventual lack of water during drought events have fostered the use of alternative resources such as wastewater, which in some cases was used by farmers with little treatment, especially during dramatically dry years. The construction of literally thousands of wastewater treatment plants throughout the country during the last 20 years has allowed a steady improvement in the quality of the wastewater to be reused, whereas the implementation of tertiary treatments in many of these plants located in the water reuse 'hot' areas has given a much greater degree of public health protection, and has also favoured the development of planned water reuse projects.

According to data compiled by the Spanish research center, CEDEX, in 2002 (Catalinas and Ortega, 2002), an estimate of the annual volume of wastewater reused, including all kinds of uses and all kinds of qualities, accounted for 0.99 Mm^3/d (346 Mm^3/yr), of which more than 82% corresponded to agricultural irrigation (Table 29.1). Since this has been a fast advancing activity, estimates for 2005 had already increased to 400 Mm^3/yr (Ortega *et al.*, 2005), especially because of the drought affecting the Iberian peninsula since mid 2004. At this moment and by request of the Spanish Environment Ministery, CEDEX is working on a National Water Reuse Database to be completed and publicized in June 2008 in Zaragoza, which will be used as the source of basic information for the definition and implementation of a subsequent National Water Reuse Plan.

According to the estimates in a more recent CEDEX report (Iglesias, 2005), the basins with greater amounts of reused wastewater are the Jucar basin (0.35 Mm^3/d or 128 Mm^3/yr) and the Segura basin (0.29 Mm^3/d or 106 Mm^3/yr), both on the eastern coast of Spain. Right after them, there are the Canary and Balearic Islands (0.03 Mm^3/d and 0.07 Mm^3/d or 40 Mm^3/yr and 25 Mm^3/yr respectively) and lesser flows in the other parts of Spain mentioned above. Even though these are the best available figures, it is sometimes hard to track what the actual reuse figures can be in these highly water-demanding regions, since in many cases effluents are discharged where no dilution flows exist and are captured somewhere downstream for irrigation. Because of this, over the last few years the regional governments of Valencia and Murcia have included tertiary treatments from the beginning at their newly built wastewater treatment plants, as a way to provide the desired level of public health protection since the water will be reused for irrigation anyway, either directly or indirectly.

Since December 8th, 2007, Spain has had a set of national regulations about the production and use of reclaimed water for urban and recreational uses, agricultural and landscape irrigation, industrial uses, environmental enhancement and aquifer recharge. After several progressive amendments, the long-time circulating draft has finally been passed and it is expected to increase the production and supply of reclaimed water, now that the regulatory framework has been established.

Table 29.1. Breakdown of uses of reclaimed water in Spain (Catalinas and Ortega, 2002).

Uses	Volume reused Mm^3/d (Mm^3/yr)	Percentage %
Agricultural irrigation	0.78 (284.9)	82.3
Non-potable urban uses	0.07 (24.0)	7.0
Golf course and landscape irrigation	0.57 (20.6)	6.0
Industrial uses	0.007 (2.5)	0.7
Aquifer recharge and ecological uses	0.04 (14.0)	4.0
Total	0.099 (346.0)	100.0

29.2 CASE STUDIES

Documenting every single reuse project in Spain would prove an arduous task, well beyond the scope of this chapter. Case studies chosen below are those that are possibly the most well-known and documented ones, and they certainly set the tone of wastewater reclamation and reuse in Spain.

29.2.1 Canary Islands

The Canary Islands are a region that has pioneered the use of reclaimed water for irrigation in Spain from the mid 1980s. On the island of Tenerife, the main water reuse project consists of a pipeline crossing the island from north to south following the eastern shore which delivers reclaimed water, produced in Santa Cruz de Tenerife, for irrigation of the banana plantations. Other islands actively pursuing wastewater reclamation and reuse are Gran Canaria and Lanzarote. According to Ibrahim (2005), by 2012 it is expected that reclaimed water could reach approximately 25% (around 0.22 Mm^3/d or 80 Mm^3/yr) of all water resources used in the Canary Islands (0.82 Mm^3/d or 300 Mm^3/yr). Given its sheer limitation of water resources, desalination is a well-known technique in the Canary Islands, and it is also commonly applied to reclaimed waters.

29.2.2 Catalonia

Though it is documented that wastewater reuse was already taking place in 1932 in the village of Reus (Cuadras *et al.*, 2002), in Catalonia the serious efforts for planned water reuse started in 1989, when the first project for reclaimed water supply was implemented in Castell-Platja d'Aro (Costa Brava). A very important project soon to follow (in the early 1990s) was the construction of the tertiary treatment plant in Vilaseca-Salou for the supply of irrigation water to the amusement park of Port Aventura. However, until the early 2000s, when other important water reuse projects were about to be completed, the hot spot for wastewater reclamation and reuse in Catalonia was the area under management of the Consorci de la Costa Brava (Costa Brava Water Agency, a supramunicipal level of administration), where reclaimed water is being reused for several kinds of uses, ranging from golf course and landscape irrigation, to environmental enhancement, agricultural irrigation, urban non-potable uses, and aquifer recharge, accounting for 0.02 Mm^3/d (5.5 Mm^3/yr) of reclaimed water produced out of 30 Mm^3/yr of treated wastewater in 2004 (Sala, 2005). Some of these projects in Costa Brava were funded by EU Cohesion Funds in the mid 1990s, whereas the most recent ones have been funded by the Agència Catalana de l'Aigua (ACA, Catalan Water Agency), a public organization newly created from the merging of the former Water Resources Authority and the Wastewater Authority. The ACA is currently elaborating a Wastewater Reclamation and Reuse Plan in order to identify where water reuse is feasible in Catalonia and what benefits the region can expect, specially if evaluated together with the results of the Water Supply Plan. Efforts to turn treated wastewater into a new resource are being made at private level (golf course irrigation in Llavaneres and Calafell), city level (Sabadell) or even regional level (Baix Llobregat). The latter project, with funding from EU Cohesion Funds, Spanish Environment Ministery and ACA, is aimed at producing 0.14 Mm^3/d (50 Mm^3/yr) of highly treated tertiary effluent to be reused for agricultural irrigation, environmental enhancement and as a barrier against the intrusion of salt water in the lower Llobregat aquifer (Cazurra, 2005).

29.2.3 Costa del Sol (Málaga)

Costa del Sol is internationally known for its beaches, mild-climate, excellent food and, for the last few years, for being a top golf destination. Though the mountainous geography of the region has allowed the construction of several reservoirs which cope with demand in rainy years, the area is prone to regular droughts. Because of this, Acosol SA, the publicly-owned company in charge of drinking water supply and wastewater treatment in the area, has been promoting wastewater reclamation and reuse in the area, mainly for golf course irrigation. Secondary effluents are tertiary treated through filtration and disinfection and supplied to some of the golf courses in the area. Given the high number of golf facilities and the surplus of treated effluent, Acosol is actively looking to increase the supply of reclaimed water and make a more rational management of precious water resources.

29.2.4 Madrid

The city of Madrid has an ambitious plan for the construction of a network of pipelines to deliver high quality tertiary effluent for the irrigation of the main parks and recreational areas. The plan started in 2002 and even though it is expected to be completed in 2008, after a total investment of about 90 million euros it is partially in operation at this moment (December 2007). The projected saving of drinking water is estimated at 0.06 Mm^3/d (22 Mm^3/yr) (De Santiago and Lirola, 2005). Canal de Isabel II (a company owned by the government of the autonomous region of Madrid that deals with water and wastewater management for the whole province) has also launched an ambitious water reuse program for the period 2005-2010 aimed at saving 40 Mm^3/yr of surface waters, until now used for irrigation, by exchanging them for high-quality reclaimed water (Canal de Isabel II, 2005).

29.2.5 Valencia and Murcia

As explained in the introduction, the Jucar and Segura basins, located in the regions of Valencia and Murcia, are areas of intense use of water resources for irrigation. The huge demand for water has historically prompted farmers to use any kind of water available, including wastewater with different degrees of treatment that was being discharged by cities or wastewater treatment plants. Important efforts have been made by the Spanish administration to improve the situation, and Title-22 tertiary treatments are already being installed in existing wastewater treatment plants and included from the beginning in the construction of new ones.

The estimate of the amount of wastewater reused in both areas at the present day is 0.06 Mm^3/d (234 Mm^3/yr), whereas Quesada (2005) estimates the amount at 282 Mm^3/yr. Since in many cases water is reused after being discharged (and thus, there is no direct pipeline connection between the reclamation plant and the point of use) it is difficult to make accurate estimates of how much water is actually reused in the area. However, this is the leading area in terms of percentages and volumes of water reused in Spain.

29.2.6 Vitoria

Since 1995, some 35,000 m^3/day of the secondary effluent of the city of Vitoria undergoes coagulation, flocculation, sedimentation, filtration and disinfection with sodium hypochlorite, and it is used with enormous success for agricultural irrigation of vegetables of all kinds, even those to be eaten raw. The results gathered until the present day show that the microbiological quality of the reclaimed water is equivalent or close to that of potable water. The use of reclaimed water has allowed the production of high-value crops. Investments for a better conservation and supply of vegetables to the markets have been made, turning regional agriculture from a decaying economic activity to a vigorous one. Until now, reclaimed water production and use was limited only to the summer months, but in the near future use of the newly constructed 7 Mm^3 reservoir will allow use of reclaimed water all year round. This reservoir will be used to store reclaimed water produced in winter, so that the irrigated area can be more than doubled. Though the construction of the reclamation plant and the irrigation scheme were initially funded by the Diputación de Álava, the wealth created by reclaimed water among farmers in Vitoria has clearly overtaken the value of the money invested.

29.3 CONCLUSIONS

Spain is an increasingly active country in the field of wastewater reclamation and reuse. Typical mediterranean climatic conditions with frequent droughts, the increase of overall water demand and the generalization of tertiary treatments in the water scarce areas (mostly, eastern and southern Spain, plus the Balearic and the Canary Islands) have allowed an important development of this activity in recent years. The passing of the long-awaited national regulations on December 8th 2007 is also expected to further boost the production of this alternative resource, now that the regulatory framework has been established.

29.3 REFERENCES

Canal de Isabel II (2005) Madrid Dpura. Plan de Depuración y Reutilización del agua en la Comunidad de Madrid, 2005-2010. Information available at the website http://www.madriddpura.com/index.html

Catalinas, P. and Ortega, E. (2002) La reutilización de los efluentes depurados en el marco de una gestión sostenible del agua. *Ingeniería Civil* nº 128. [In Spanish]

Cazurra, T. (2005) La reutilización de aguas residuales en el Baix Llobregat. Workshop "Actualización de los criterios de reutilización de agua residual tratada", Madrid, 20-23 June 2005.

Cuadra, N., Comas, J. and Cañameras, N. (2002) Reutilización de aguas residuales. EUETAB, Barcelona. Internet source: http://mie.esab.upc.es/arr/T4.htm[In Spanish]

De Santiago, F., and Lirola, J.A. (2005) Plan de reutilización de agua de Madrid. Workshop "Actualización de los criterios de reutilización de agua residual tratada", Madrid, 20-23 June 2005. [In Spanish]

Ibrahim, J.C. (2005) La experiencia en Canarias en proyectos de reutilización. Workshop "Actualización de los criterios de reutilización de agua residual tratada", Madrid, 20-23 June 2005. [In Spanish]

Iglesias, R. (2005). Escenarios existentes y propuestas para el avance de la regeneración y reutilización de aguas en España. Technical Workshop "The integration of reclaimed water in water resource management: the fostering role of the territorial region", Lloret de Mar (Costa Brava, Girona, Spain), 19-20 October 2005. Internet source: http://www.ccbgi.org/jornades2005/ponencies/10_iglesias.pdf

Ortega, E., Batanero, G. and Iglesias, R. (2005) La situación de la reutilización de las aguas residuales en España. Problemas actuales y perspectivas futuras. Workshop "Actualización de los criterios de reutilización de agua residual tratada", Madrid, 20-23 June 2005. [In Spanish]

Quesada, J.C. (2005) Reutilización de agua depurada en la Comunidad Valenciana. Workshop "Actualización de los criterios de reutilización de agua residual tratada", Madrid, 20-23 June 2005. [In Spanish]

Sala, L. (2005) Criterios de gestión de la reutilización en la Costa Brava y propuestas de futuro. Workshop "Actualización de los criterios de reutilización de agua residual tratada", Madrid, 20-23 June 2005. [In Spanish].

30

Trying to set a common framework to rule water reuse in the Mediterranean region

François Brissaud and Akiça Bahri

30.1 INTRODUCTION

The total population of the Mediterranean region is around 430 million inhabitants, with 145 million living near the sea and an additional 180 million tourists each year. By 2025, the population is expected to increase by 17–19% and the tourist population by 40%. The impact of the demographic evolution, characterized by a dramatic growth in Eastern and Southern countries and a stabilization in Northern ones, is aggravated by a very intensive urbanization often along the coastal areas.

The average annual *per capita* renewable water has been rapidly decreasing since 1950. All the Mediterranean countries of the EU are expected to maintain themselves at or above 3000 m^3/*capita*.yr when, in the major part of the other Mediterranean countries, the projected water availability is below the level of "chronic water scarcity" (< 1000 m^3/*capita*.yr), due mainly to very high population growth. Some countries, such as the Palestinian Territories of Gaza and the West Bank, Jordan, Tunisia, and Malta, experience "absolute water stress" with per capita water availability of less than 500 m^3/*capita*.yr. In Malta, domestic water consumption exceeds 50% of the available water resources. In such places, conventional water resources will be insufficient to even meet the domestic water demand.

The Mediterranean basin is today depending for its economic and social development on agriculture (largest water use share reaching 61% on average) and tourism and, to a lesser

©2008 IWA Publishing. *Water Reuse - An International Survey of current practice, issues and needs* by B. Jimenez *et al.* ISBN: 9781843390893. Published by IWA Publishing, London, UK.

extent, on industry and other economic activities. Irrigated agriculture, in competition with other sectors, will face increasing problems of water quantity and quality. Competition with other sectors often makes water the main factor that limits agricultural development. Policy makers have, then, been compelled to develop additional water resources as well as to preserve the existing ones. Reclaiming and recycling water is or is becoming, among various measures, an important component of several national resources policies.

Agricultural wastewater reuse, planned or unplanned, is common practice in most countries of the Mediterranean region. However, reused water quality is highly variable. In several countries, wastewater treatment plants equip a limited number of towns, and existing plants are often overloaded; wastewater is then reused raw or insufficiently treated to prevent health risks and prevalent water-related diseases are reported. In other countries, wastewater is submitted to adequate reclamation systems and treated wastewater is being reused without any proved risk for human health. In these cases, conditions for safe reuse are met and reclaimed water is becoming a significant alternative resource for food production and sustainable water supply.

Though the situation may change in the near future, up to now only a few Mediterranean countries (Cyprus, Israel, Jordan, Tunisia) have included water reuse in their water resources planning; few have official policies calling for water reuse.

The main reuse projects in the region are dedicated to agricultural irrigation, landscape irrigation and groundwater recharge. Industrial reuse is very seldom practiced. Wastewater content is often rich in beneficial organic matter and nutrients; conversely, mineral and organic trace substances and pathogens represent a risk for human health and the environment. Pathogen content represents a crucial concern for most water reuse applications, either urban or agricultural, since humans may be in contact, directly or indirectly, with wastewater. Organic and inorganic trace elements may be transferred to animals or humans through different pathways and cause human health effects depending on their concentration. As a consequence, wastewater treatment constitutes a prerequisite for reuse; its level should be adjusted in order not to exceed risks that are acceptable for users, workers, consumers (human and/or animal), and for the environment.

Most Mediterranean countries (Albania, Algeria, Bosnia and Herzegovina, Croatia, Egypt, Greece, Lebanon, Libya, Malta, Monaco, Morocco, Slovenia, Syria, and Turkey) have neither water reuse regulations nor guidelines. Countries where reuse is developing within an organised institutional setting have elaborated and implemented their own regulations or guidelines. However the standards significantly differ between countries and even within a given country such as in Spain. Some countries (France, Tunisia) and regions (Andalusia and the Balearic Islands in Spain, and Sicily in Italy) have adopted a set of water quality criteria based on the WHO guidelines (WHO, 1989) while others (Cyprus, Italy, Israel) have elaborated regulations or guidelines close to the more conservative California Water Recycling Criteria (State of California, 1978). Some countries have taken the approach of minimising any risk, whereas others have adopted a protective approach of reasonably-anticipated adverse effects. This has led to substantial discrepancies in the standards adopted by Mediterranean countries. Therefore, to date, there are no regulations of water reuse at a Mediterranean level. Nevertheless, there is a need for sharing common regulations or, at least, a common rationale for developing water reuse standards on both sides of the Mediterranean. This need is based on the consideration that (1) an agricultural Mediterranean market is developing with large amounts of agricultural products (vegetables, fruits, etc.) imported and exported among Europe and other Mediterranean countries; (2) cross-border tourism is an essential part of the economic activity of the region, and its development might be jeopardized in the long term by disease outbreaks linked to wastewater mismanagement; and (3) there is growing consumer concern about food quality and health hazards.

Contemplating setting up Mediterranean guidelines raises several questions:

1. which methodology will be used for developing reuse guidelines and standards?
2. on which rationale will this development be based?
3. how can the discrepancies between regulations and guidelines that already exist in the Mediterranean region be overcome?
4. should the economic development of the region be considered when deriving health guidelines for water reuse and, if the answer is positive, how should it be taken into account?

To be useful and efficiently contribute to the improvement of human health and the alleviation of water resource shortages, guidelines must take local conditions into account. Whilst, taken as a whole, countries of the North bank are developed, industrialised, and more and more equipped with wastewater treatment plants complying with the EU Directive on wastewater disposal (EU, 1991), the economy of the South bank (some countries excepted) lags far behind, with poor wastewater management policy, slow wastewater treatment development and endemic diseases linked to weaknesses of public hygiene. Farmers of both the North and South banks are not always used to paying for water; when they do, the fees cover only a small part of the costs and farmers are more than reluctant to pay more. Too stringent regulations will not be enforced and will eventually be ignored.

Overcoming the discrepancies between the water reuse standards existing in the region does not mean unifying the national regulations, which would be far too ambitious. Indeed every country is inclined to keep its own approach, particularly those having already adopted very stringent regulations. However, understanding the reasons for the discrepancies between the existing guidelines would clear the way for setting minimum accepted requirements that would guarantee safe water reuse in both the north and south.

Setting up health guidelines should be supported by a clear rationale based, whenever possible, on scientifically established data. In this respect, two approaches are essential: epidemiologic investigations and health risk assessment. Water reuse is assumed to be safe when the risk encountered does not exceed the risk considered as tolerable; therefore, setting health-related targets requires a consensus on what is a tolerable risk. For the sake of integrated water management and to gain public understanding and acceptance, water reuse guidelines should be part of a set of consistent water regulations applying to drinking water, bathing water, irrigation and discharge in the environment. Guidelines on water reuse and other water exposures (*e.g.* drinking water and recreational water contact) should assure equivalent protections. The *WHO guidelines for the safe use of wastewater in agriculture* have been revised according to the same principles (WHO, 2006).

30.2 BENCHMARK STANDARDS

In Mediterranean countries, water reuse regulations and guidelines, when they exist, have been derived from either the California water reuse standards (1978) or the World Health Organization Guidelines (1989). These two benchmark standards result from different historical processes and do not have the same objectives.

30.2.1 California water reuse standards

The first water reuse regulations were established in 1918 by the State of California, USA. At that time, the only application considered was irrigation. In 1933, the first microbial effluent standards for the "irrigation of garden truck produce eaten raw" were set up by the

California State Board of Health at a coliform concentration of ≤ 2.2 MPN/100 mL (Ongerth and Jopling, 1977). The coliform concentration was equivalent to that required for drinking water and based on the concept of "zero risk". Since then, standards have been continuously revised to address new reclaimed water applications and to take into account advances in wastewater treatment technology and updated knowledge in public health protection (Crook, 1998). However, they have been neither substantially changed nor liberalized.

In 1978, the California Wastewater Reclamation Criteria were issued by the California Department of Health Services (DHS). They have been revised recently (State of California Title 22 Water Recycling Criteria, 2000). These standards, applying to wastewater reclamation, include water quality standards, treatment process requirements, operational and treatment reliability requirements. Water quality criteria and treatment requirements are shown in Table 30.1.

No bacterial limit is set and a secondary treatment is required for agricultural reuse restricted to the irrigation of fodder, fiber, seed crops, orchards* and vineyards*, processed food crops, non food-bearing trees, ornamental nursery stock*, and sod farms* and for the flushing of sanitary sewers. For the irrigation of pasture for milking animals, landscape areas (with restricted or controlled access), ornamental nursery stock and sod farms where public access is not restricted, landscape impoundments, for industrial or commercial cooling water where no mist is created, for non structural fire-fighting, industrial boiler feed, soil compaction, dust control, for cleaning roads, sidewalks and outdoor areas, the total coliforms content limit is 23/100 mL; the required reclamation is a secondary treatment plus disinfection.

For other applications (irrigation of food crops and open access landscape areas, supply of impoundments, fish hatcheries, toilet and urinal flushing, industrial process water, decorative fountains, commercial laundries and car washes, snow-making, structural fire-fighting, industrial or commercial cooling where mist is created), the bacterial quality criteria is 2.2 TC/100 mL. The treatment required depends on the applications and turbidity values; it ranges from secondary treatment followed by disinfection to secondary treatment followed by coagulation, clarification, filtration and disinfection.

The standards are based on a microbiological quality, which is expected to guarantee that infectious risks are reduced to an acceptable level, and on treatment requirements which determine the reliability of the process in the achievement of this microbiological quality. As a result, different treatment requirements may correspond to the same microbiological quality, depending on the evaluation of the risks related to the respective types of use: the higher the risks, the higher the required reliability.

When direct or indirect human contact is likely to happen, reclaimed water should essentially be free of pathogens, particularly enteroviruses, because very low virus contents can initiate human infection. The corresponding required microbiological quality is characterised by a total coliform 7-day median value of less than 2.2/100 mL, no more than 23 TC/100 mL in more than one sample in any 30-day period and a maximum value of 240 TC/100 mL in any sample. Furthermore, the turbidity should not exceed 2 NTU after filtration on a continuous monitoring base. This criterion has been shown to determine the virus removal capability of the treatment process. It is assumed that treatment processes controlling viruses provide a water free of parasites, particularly *Giardia lamblia* and *Cryptosporidium parvum*.

For all types of uses resulting in a possible direct or indirect human contact, California is promoting stringent water quality standards. However, as coliforms have proved to be inadequate indicators for those pathogens that are more resistant to disinfection and less readily removed by physical treatments, more reliance is placed on the pre-determined ability

---

* with the restrictions mentioned in Table 30.1.

of a process to reduce all types of pathogens (Cooper and Olivieri, 1998). Confident that advanced technologies provide a safe water quality (*i.e.* free of enteric viruses), technologically-based requirements to reduce the presence of pathogens by treatment or a combination of treatment and use restrictions were established.

Table 30.1 California water recycling criteria: treatment and quality requirements for nonpotable uses of reclaimed water (State of California Title 22 Water Recycling Criteria, 2000).

Type of use	Total coliform limits[a]	Treatment required
Irrigation of fodder, fiber, & seed crops, orchards[b] and vineyards[b], processed food crops, nonfood-bearing trees, ornamental nursery stock[c], and sod farms[c]; flushing sanitary sewers	• None required	• Secondary
Irrigation of pasture for milking animals, landscape areas[d], ornamental nursery stock and sod farms where public access is not restricted; landscape impoundments; industrial or commercial cooling water where no mist is created; nonstructural fire fighting; industrial boiler feed; soil compaction; dust control; cleaning roads, sidewalks, and outdoor areas	• ≤23/100 mL • ≤240/100 mL in more than one sample in any 30-day period	• Secondary • Disinfection
Irrigation of food crops[b]; restricted recreational impoundments; fish hatcheries	• ≤2.2/100 mL • ≤23/100 mL in more than one sample in any 30-day period	• Secondary • Disinfection
Irrigation of food crops[e] and open access landscape areas[f]; toilet and urinal flushing; industrial process water; decorative fountains; commercial laundries and car washes; snow-making; structural fire fighting; industrial or commercial cooling where mist is created	• ≤2.2/100 mL • ≤23/100 mL in more than one sample in any 30-day period • 240/100 mL (maximum)	• Secondary • Coagulation[g] • Filtration[h] • Disinfection
Non restricted recreational impoundments	• ≤2.2/100 mL • ≤23/100 mL in more than one sample in any 30-day period • 240/100 mL (maximum)	• Secondary • Coagulation • Clarification[i] • Filtration[h] • Disinfection

[a] Based on running 7-day median.

[b] No contact between reclaimed water and edible portion of crop.

[c] No irrigation for at least 14 days prior to harvesting, sale, or allowing public access.

[d] Cemeteries, freeway landscaping, restricted access golf courses, and other controlled access areas.

[e] Contact between reclaimed water and edible portion of crop; includes edible root crops.

[f] Parks, playgrounds, schoolyards, residential landscaping, unrestricted access golf courses, and other uncontrolled access irrigation areas.

[g] Not required if the turbidity of the influent to the filters is continuously measured, does not exceed 5 NTU for more than 15 minutes and never exceeds 10 NTU, and there is capability to automatically activate chemical addition or divert the wastewater if the filter influent turbidity exceeds 5 NTU for more than 15 minutes.

[h] The turbidity after filtration through filter media cannot exceed 2 nephelometric turbidity units (NTU) within any 24-hour period, 5 NTU more than 5% of the time within a 24-hour period, and 10 NTU at any time. The turbidity after filtration through a membrane process cannot exceed 0.2 NTU more than 5% of the time within any 24-hour period and 0.5 NTU at any time.

[i] Not required if reclaimed water is monitored for enteric viruses, *Giardia*, and *Cryptosporidium*.

California Health Services have chosen not to use epidemiological studies as a basis for determining water quality standards, mainly because, at low exposure, epidemiological studies are not sensitive.

To evaluate the safety of wastewater reclamation and reuse applications, the USEPA's Surface Water Treatment Rule (SWTR) for domestic water supply was used as a point of reference. Acceptable risks for this evaluation were defined as meeting the 10^{-4} infection risk criterion (probability of one infection per 10,000 population in one year) at least 90 and 95% of the time, (Asano *et al.*, 1992; Tanaka *et al.*, 1998). The calculation used data from the monitoring of the virus content of unchlorinated secondary effluents from 4 wastewater treatment plants in California. Treatments were characterized by their respective enterovirus removal.

As can be seen in Table 30.2, all applications of unchlorinated effluents lead to unacceptable risks. Direct chlorination of secondary effluents means annual risks lower or much lower than acceptable, with the exception of recreational impoundments where the risks to swimmers are higher than acceptable. A full treatment as required by California standards reduces the risks far below those that are acceptable, except for recreational impoundments where bathing is allowed and the risks are about acceptable. The results of Tanaka *et al.* (1998), though some assumptions of the work might be questioned, provide some grounds for the criticism addressed to the Californian water reuse criteria, that they are too conservative for most water reuse applications.

Table 30.2. Expectation of the annual risk of enteric virus infection for the upper 95% confidence limit using Monte Carlo simulations.

Tertiary treatment	Application	Annual risk of infection
Full treatment (5.2 log removal of enterovirus)	Golf course irrigation	$1.1\ 10^{-7} - 6\ 10^{-6}$
	Food crop irrigation	$3.7\ 10^{-10} - 2.1\ 10^{-8}$
	Recreational impoundment	$1.3\ 10^{-5} - 6.8\ 10^{-4}$
Direct chlorination of secondary effluents (3.9 log removal of enterovirus)	Golf course irrigation	$2.2\ 10^{-6} - 1.2\ 10^{-4}$
	Food crop irrigation	$7.5\ 10^{-9} - 4.1\ 10^{-7}$
	Recreational impoundment	$2.6\ 10^{-4} - 1.3\ 10^{-2}$
Unchlorinated effluents (0 log removal of enterovirus)	Golf course irrigation	$1.7\ 10^{-2} - 5.3\ 10^{-1}$
	Food crop irrigation	$5.9\ 10^{-5} - 3.3\ 10^{-3}$
	Recreational impoundment	$6.2\ 10^{-1} - 1\ 10^{0}$

Adapted from Tanaka *et al.*, 1998.

30.2.2 WHO water reuse guidelines

In 1973, the World Health Organization (WHO) proposed health criteria and treatment processes for reuse applications ranging all the way from irrigation of crops not intended for human consumption up to potable reuse. It was recognized that:

1. applying drinking water-type standards (2.2 coliforms/100 mL) for water reuse was unrealistic and lacked an epidemiological basis;
2. few if any rivers worldwide used for unrestricted irrigation of vegetables eaten uncooked carry water of such quality; and
3. few if any developing countries could meet such standards for water reuse.

Nevertheless, relatively stringent guidelines were recommended for the quality of the effluent to irrigate crops to be consumed raw: a guideline value of 100 coliforms/100 mL was set for unrestricted irrigation. These recommendations were based on the concept of "zero risk".

In 1985, the guidelines were reviewed and the nature of health risks associated with agriculture and aquaculture were revised. The epidemiological evidence on pathogens was reconsidered, updated and the approach of microbiological risk assessment was confirmed (Feachem *et al.*, 1983; Shuval *et al.*, 1986; Strauss and Blumenthal, 1989).

The epidemiological approach for health risks assessment allowed an evolution of the water reuse guidelines, the goal remaining no actual risk of infection to the exposed population that can be attributed to water reuse. In 1989, WHO issued a new set of microbiological quality guidelines for water use in agriculture and aquaculture (Table 30.3). Other non potable uses were not considered.

Table 30.3. WHO guidelines for the safe use of wastewater in agriculture (WHO, 1989).

Category	Reuse condition	Exposed group	Intestinal nematodes[b] (arithmetic mean no. of eggs per liter[c])	Faecal coliforms (geometric mean no. per 100 mL[c])	Wastewater treatment expected to achieve the required microbiological quality
A	Irrigation of crops likely to be eaten uncooked, sports fields, public parks[d]	Workers, consumers, public	≤ 1	≤ 1000[d]	A series of stabilization ponds designed to achieve the microbiological quality indicated, or equivalent treatment
B	Irrigation of cereal crops, industrial crops, fodder crops, pasture and trees[e]	Workers	≤ 1	No standard recommended	Retention in stabilization ponds for 8–10 days or equivalent helminth and faecal coliform removal
C	Localized irrigation of crops in category B if exposure of workers and the public does not occur	None	Not applicable	Not applicable	Pretreatment as required by the irrigation technology, but not less than primary sedimentation

[a] In specific cases, local epidemiological, socio-cultural and environmental factors should be taken into account, and the guidelines modified accordingly.
[b] *Ascaris* and *Trichuris* species and hookworms.
[c] During the irrigation period.
[d] A more stringent guideline (<200 faecal coliforms per 100 mL) is appropriate for public lawns, such as hotel lawns, with which the public may come into direct contact.
[e] In the case of fruit trees, irrigation should cease two weeks before fruit is picked, and no fruit should be picked off the ground. Sprinkler irrigation should not be used.

The WHO recommendations (1989) took into account the wastewater treatment process, the irrigation system, the exposed group of population, and the crops to be irrigated. They covered the various options for health protection such as treatment of wastewater, crop restrictions, application controls, and control of human exposures. A multi-barrier approach throughout the water cycle combining different measures was considered an important element. Other precautions such as wearing protective clothing and increased levels of hygiene, cooking, provision of adequate washing facilities, human exposure control, promotion of hygiene, etc., were also recommended.

The WHO guidelines (1989) have allowed a real development of water reuse and enhanced the acceptance of water reuse among decision-makers, engineers, health authorities, and the public in several countries. They are, however, a matter of controversy, particularly on the faecal coliforms geometric mean ≤ 1000 /100mL criterion for unrestricted irrigation; they are questioned on the ability of the faecal coliforms limit to protect against viruses and on the nematode eggs limit to protect against protozoan parasites since viruses and protozoa are not easily removed by conventional treatment processes and disinfection (Blumenthal *et al.*, 2000).

Even though there is no documented scientific evidence that they failed to protect public health, WHO guidelines (1989) are currently under revision.

30.2.3 Comments

Major differences between the WHO and Californian approaches are illustrated by some examples:

- Concerning wastewater treatment, the WHO guidelines recommend a series of stabilization ponds or an equivalent treatment to meet the microbiological water quality requirements for irrigation of food crops eaten raw, when the California criteria stipulate secondary conventional biological wastewater treatment followed by coagulation, filtration and chlorine disinfection.
- The WHO guidelines require the monitoring of intestinal nematodes and faecal coliforms, whereas the California criteria rely on treatment systems and the monitoring of the total coliforms density for assessment of microbiological quality. The WHO requirements are less stringent than the Californian ones: for unrestricted irrigation, the WHO guidelines is ≤ 1000 FC/100 mL compared to the Californian criteria of ≤ 2.2 TC/100 mL, and for irrigation of pasture and commercially processed and fodder crops only a guideline limit on the presence of nematode eggs is set by WHO, whereas the Californian guideline requires ≤ 23 TC/100 mL.
- Californian water reuse criteria are neither based on epidemiological investigations nor on mathematical risk assessment modelling data (Crook *et al.*, 2001), whilst the WHO guidelines are based on the epidemiological and technological evidence available concerning health risks associated with wastewater irrigation.

30.3 PROPOSAL FOR MEDITERRANEAN WATER REUSE GUIDELINES

Meetings in Barcelona, Spain (27–28 January, 2000) and Rabat, Morocco (8–10 October, 2001) have shown that establishing Mediterranean guidelines for urban wastewater reuse is not an easy task. Unfortunately, in this small part of the world, existing guidelines and regulations that should have been the basis of a well-shared consensus, are inspired by two very different benchmark standards, leading to large discrepancies between regulations and guidelines in the region.

30.3.1 Principles

The rationale behind the existing guidelines has a limited scientific basis; water reuse criteria are not based on comprehensive epidemiological studies. Literature data on epidemiological studies are rather limited and mainly evoked for microbiological contamination of bathing or drinking waters. Few epidemiological studies have been conducted on non-potable reuse: the Jerusalem and Mexico studies are exceptional. The existing guidelines were not based on risk assessment modelling data. Therefore it was not possible to propose the adoption of an already existing benchmark standard to set up Mediterranean guidelines. A different approach, derived from as clear a rationale as possible given the actual available international knowledge, had to be searched for.

The first step was to look for references in order to evaluate the safety of wastewater reclamation and reuse applications and set the health-related targets: international potable water and bathing standards were considered. Not all non potable uses, if any, require drinking water quality, nor all bathing water quality. However, infection risks considered as acceptable when related to bathing or drinking potable water can serve as references for reuse guidelines.

The proposed guidelines should provide an efficient protection to the populations of both banks, including the tourists of one bank travelling to the other bank and the consumers of exported vegetables and fruits, but without requiring unaffordable treatments. Thus, the proposed water reuse criteria should not aim at preventing risks lower than those considered acceptable internationally when drinking potable water.

Though the methodology followed to elaborate the WHO (1989) guidelines has been questioned over the years, their microbiological criteria may, after the protection they offer has been re-assessed, be re-invested in Mediterranean guidelines.

30.3.2 Foremost health risks

Potential health risks offered by wastewater reuse are related to the microbial and chemical composition of the reclaimed water.

Possible transmission of infectious disease by waterborne pathogens is the most common concern associated with non potable reuse. The majority of documented disease outbreaks have been related to contamination by bacteria and parasites (Crook, 1998).

Chemical constituents are a major health concern for indirect potable reuse but not for urban uses of reclaimed water; they may also affect agricultural and industrial applications. Toxic chemicals may present adverse health effects when adsorbed or inhaled or where there is contact with water. Heavy metals accumulate in the environment and may be toxic to plants; cadmium, copper, molybdenum, nickel and zinc can affect animals and humans. Heavy metal contents in secondary treated effluents are generally within acceptable levels for most non potable reuses. Attention must be paid to the effect of organics and heavy metals in reclaimed water used for crop irrigation when industrial effluents are a significant fraction of wastewater. The main impact of nutrients is the potential build up of nitrates in aquifers.

Non potable applications are and will long remain the large majority of reclaimed water reuse projects in the Mediterranean. Therefore, the proposed health guidelines focus on the microbiological criteria. Providing guidelines on toxic elements is beyond the scope of this chapter. Guideline values for the chemical quality of irrigation water are available in Ayers and Westcot (1985) and Chang *et al.* (1998). The impact of eventual stable toxic organic constituents on crops should be considered on a case by case basis when harmful discharge of industrial effluents in a public sewer is suspected.

30.3.3 Criteria

Three water quality criteria are proposed: nematode eggs, faecal coliforms or *Escherichia coli*, and suspended solids (TSS). Treatments likely to achieve the required water quality are also suggested (Table 30.4).

Due to uneven levels of development, health risks related to waterborne parasites remain an important concern in several Mediterranean countries. Therefore, nematode egg limits still have to be included in Mediterranean guidelines.

Faecal coliforms or *E. coli* are, together with total coliforms, the indicator bacteria most commonly used in water quality standards, particularly for water reuse, bathing in fresh water and drinking water. Faecal coliforms is a widely used indicator of treatment performances. Faecal coliforms contents are currently monitored in many Mediterranean countries.

Turbidity is a criterion used in several reuse standards, particularly in conservative regulations; for instance, California standards state that where direct or indirect contact with reclaimed water is likely to occur, the turbidity shall not exceed a daily average of 2 NTU (Nephelometric Turbidity Units). Turbidity standards are linked to the virus removal capability of tertiary treatment processes. Though turbidity offers the advantage of allowing a continuous monitoring and is being controlled more and more commonly, TSS has been preferred for this proposal of Mediterranean guidelines for it is a widespread measurement in the region. It also allows establishing links between reuse guidelines and the EU Directive on wastewater disposal (EU, 1991) meaning that secondary treatment is required.

30.3.4 Categories of reuse applications and water quality

Four categories of reclaimed water uses are considered, in order to facilitate the implementation of the guidelines. Water reuse cost-effectiveness was also taken into account in the sense that a reclaimed water supply network must serve as many reuse applications as possible in the same area. The categories are shown in Table 30.4.

The quality of reclaimed water has often been reported to evolve during conveyance and storage; it may be degraded or improved, depending on the initial water quality, the residence time, the conveyance and storage systems characteristics and the environment. This evolution raises the question of where the reclaimed water quality should be monitored. As no specific measures, such as chlorination, are recommended to preserve the degradation of the microbial quality along the supply and storage systems, the microbial quality should be monitored at the outlet of the treatment system (treatment plant + storage, if this storage is included in the treatment process) and also at the point of use in order to evaluate eventual improvements or degradation of the water quality.

Wastewater treatments expected to meet the criteria were defined for each water category. In the case of agricultural reuse, the capacity of reclaimed water application systems or methods, such as surface trickle or subsurface irrigation, to provide a high degree of protection against contamination (and to use water more efficiently) was taken into account.

Table 30.4 Categories of reuse applications and water quality.

Category I: urban and residential reuses, landscape and recreational impoundments
• Residential reuse: private garden watering, toilet flushing, vehicle washing. • Urban reuse: irrigation of areas with free admittance (greenbelts, parks, golf courses, sport fields), street cleaning, fire-fighting, fountains, and other recreational places. • Landscape and recreational impoundments: ponds, water bodies and streams for recreational purposes, where incidental contact is allowed (bathing is excluded).
Category II: unrestricted irrigation, landscape impoundments (contact with water not allowed), and industrial reuses
• Irrigation of vegetables (surface or spray irrigated), green fodder and pasture for direct grazing, sprinkler-irrigated fruit trees. • Landscape impoundments: ponds, water bodies and ornamental streams, where public contact with water is not allowed. • Industrial reuse (except for food industry).
Category III: restricted agricultural irrigation
• Irrigation of cereals and oleaginous seeds, fibre, and seed crops, dry fodder, green fodder without direct grazing, crops for canning industry, industrial crops, fruit trees (except sprinkler-irrigated), plant nurseries, ornamental nurseries. • Landscape irrigation: wooden areas, green areas with no access to the public.
Category IV: irrigation with reclaimed water application systems or methods providing a high degree of protection against contamination
• Irrigation of vegetables (except tuber, roots, etc.) with surface and subsurface trickle systems (except micro-sprinklers) and/or using practices (such as plastic mulching, support, etc.) guaranteeing absence of contact between reclaimed water and edible part of vegetables. • Irrigation of crops in category III with trickle irrigation systems (such as drip, bubbler, micro-sprinkler or subsurface). • Irrigation with surface trickle systems of greenbelts and green areas with no access to the public. • Irrigation of parks, golf courses, sport fields with subsurface irrigation systems.

30.3.5 Guidelines

Water Category I

A unique water category is proposed for residential, urban and landscape reuses and for recreational impoundments and stream augmentation. The reasons are comparable levels of risk and, also, to facilitate the implementation of dual water supply systems in Mediterranean towns. Developing urban and landscape reuses, a major challenge for the next few years, should allow the rate of wastewater that is reused to dramatically increase. To be cost-effective, a reclaimed water supply network must serve as many non potable applications as possible in the same area. The non potable reclaimed water applications which entail the highest health risks are listed in this section. Therefore, water category I should comply with the most stringent criteria.

Bacterial criterion

In the WHO guidelines (WHO, 1989), irrigation of sports fields and public parks has been classified in category A irrigation, together with irrigation of vegetables to be eaten raw; the recommended bacterial guideline being a geometric mean of 1000 FC/100 mL (WHO, 1989).

A more stringent guideline (≤ 200 FC/100 mL) was recommended for public lawns with which the public may come in direct contact. Actually, the Scientific Group which established these guidelines considered that where the public has direct access to lawns and parks, the potential risk may be higher than that associated with irrigation of vegetables eaten raw. Therefore, a guideline of 200 FC/100 mL was adopted, as recommended by Durand *et al.* (1986), who performed an epidemiological investigation on health effects of landscape irrigation with reclaimed water at Colorado Springs.

Because bathing is associated with the inevitable absorption of an important amount of water, bathing water criteria deserve a special consideration. The mean oral absorbed bathing water is hypothesised to be 100 mL per day (Tanaka *et al.*, 1998; López-Pila and Szewzyk, 2000), which represents the highest daily water ingestion after drinking consumption (1 to 2 L/d) and an exposure higher than in any non potable water reuse. The European Directive 76/160/EEC on bathing water quality sets guideline maximum contents of 500 TC/100 mL, 100 FC/100 mL and 100 faecal streptococci (FS)/100 mL, the mandatory limit values being 10,000 TC/100 mL and 2,000 FC/100 mL (EU, 1976). The European Commission has proposed a revision of this Directive, reflecting and adapting the new *WHO guidelines for safe recreational water environments* (WHO, 2003) prepared on the basis of a review of the available scientific literature and the epidemiological investigations carried out by Kay *et al.* (1994) and Fleisher *et al.* (1996). While the WHO recommends an A quality water of ≤ 40 intestinal enterococci per 100 mL in 95% of the samples, which corresponds to less than one case of gastroenteritis illness per 100 exposures, the new guideline maximum contents proposed by the Commission are 100 intestinal enterococci/100 mL and 250 *E. coli*/100 mL with a mandatory limit of 200 intestinal enterococci/100 mL and 500 *E. coli*/100 mL based upon 95 percentile evaluations (COM, 2002). A 200 FC/100 mL limit value (as a geometric mean) was recommended by USEPA (1976) for recreational (bathing) waters. A USEPA investigation calculated that for a content of 200 FC/100 mL (a limit value currently enforced in many states of the USA) the added seasonal gastro-intestinal illness rate is 8 per 1000 swimmers in freshwater and 19 per 1000 swimmers in seawater (USEPA, 1998). Similar figures can also be derived from López-Pila and Szewzyk (2000). In 1986, USEPA recommendations were revised:

- in freshwater, bacterial density should not exceed one or the other of the following: 126 *E. coli* per 100 mL or 33 Enterococci per 100 mL.
- in seawater, the geometric mean of Enterococci content should not exceed 35 per 100 mL (USEPA, 1986).

In California, standards vary between 20 and 2000 FC/100 mL; limits were also set for Enterococci, in the range of 12 to 50 CFU/100 mL. New Zealand has adopted a limit value of 126 *E. coli* /100 ML in freshwater and a seasonal median of 35 Enterococci/100 mL in seawater.

Though a general trend at strengthening bathing water guidelines can be noticed, contents of 100–200 FC or *E. coli* per 100 mL seem to be a widely accepted standard for freshwaters. Recreational water regulations are associated to health risks very significantly higher than those accepted for potable water. That these risks seem to be accepted world-wide is explained by the costs that higher standards would require and the high annual risk of gastro-intestinal disease in the global population.

Accepted risks associated to drinking water constitute another key reference. Several authors have assessed the safety of reuse applications using the USEPA Surface Water Treatment Rule for domestic water supply, according to which an annual risk equal or less than 10^{-4} per person from enteric virus infection was considered as acceptable (Tanaka *et al.*, 1998; Blumenthal *et al.*, 2000). In *Guidelines for drinking-water quality* (WHO, 2004), the acceptable risks set at 10^{-6} DALYs (Disability Adjusted Life Years) per person per year is equivalent to annual risks of infection of respectively $9x10^{-4}$, $8x10^{-4}$ and $3x10^{-3}$ per person for *Cryptosporidium*, *Campylobacter* and rotavirus in high-income regions.

The applications itemized in Category I entail limited or occasional contacts with the reclaimed water (bathing in recreational impoundments has been excluded) and possible ingestion of no more than 1 mL of reclaimed water in one exposure. Following the hypothesis of López-Pila and Szewzyk (2000) and assuming a geometric mean content of 200 FC or *E. coli* /100 mL and no reduction of the pathogens in the environment, a rough risk assessment results in an annual rotavirus infection risk of 3.2 10^{-3} and 3.2 10^{-4} for 100 and 10 exposure frequencies respectively. The figures related to 100 exposures are one order of magnitude higher or of the order of magnitude of the acceptable annual risk for potable water drinking, depending on the reference considered, USEPA or WHO. They are one order of magnitude less than the risk considered as acceptable when bathing. Despite the actual large uncertainties and approximations attached to epidemiological investigations and QMRA, it is concluded that the 200 FC or *E. coli* /100 mL guideline limit offers an acceptable protection to the users and the public.

It is proposed to include toilet flushing in this category of applications, while some countries have adopted much more conservative standards or guidelines for this application. It should be noticed that the criterion which is enforced in Japan, the country where this application is the most developed, is TC content ≤ 1000 CFU/100 mL (Ogoshi *et al.*, 2001), which is considered close to *E. coli.* ≤ 200/100 mL.

Nematode egg criterion

Most of reclaimed water uses listed in Category I imply incidental direct contact and possible absorption of reclaimed water. Therefore, the strictest Nematode egg criterion is recommended. Investigations reviewed by Blumenthal *et al.* (2000) showed that a guideline limit of 1 and even 0.1 egg/L is required for an efficient protection of the most exposed populations, namely children. This requirement should address the risk encountered by children playing on lawns irrigated with reclaimed water.

TSS criterion

Whatever the process, a low TSS content is required for disinfection effectiveness. Therefore, a guideline of 10 mg/L is proposed, which will imply filtration prior to disinfection in most reuse projects.

Minimum treatment recommended

Though the 200 FC/100 mL guideline does not greatly differ from the 1000 FC/100 mL standard applying to Category II in terms of bacterial water quality, it implies an important difference in the treatments required. Extensive treatments, such as lagoons, stabilisation reservoirs and infiltration percolation, aimed at polishing secondary effluents, often cannot guarantee that the 200 FC/100 mL guideline is reliably met. This limitation is linked to the processes, their implementation and also to recontamination in open storage and the environment. Therefore, more conventional disinfection treatments and more secured storages have to be used, which means more costly water reuse projects (Brissaud *et al.*, 2003).

Water category II

The applications listed in Category II do not imply direct contact of humans with reclaimed water. However, the irrigation of vegetables, including those likely to be eaten raw, pasture for direct grazing, aspersion of fruit trees are applications that require high quality water. It is the unrestricted irrigation category, which has been the subject of so intense controversies among the experts. Vegetables to be eaten cooked, such as potatoes, leeks, beans, etc. and not exclusively grown for the canning industry, are included in the same category as vegetables to be eaten raw, for they are often grown in the same fields, irrigated with the same water. The same quality criteria are proposed for landscape impoundments and water bodies where public contact with water is not allowed because, though forbidden, direct contact of people, mainly young children, with water may occur.

The water quality required for industrial purposes is site-specific and depends on the particular use. As the use of reclaimed water for cooling is frequent and presents potential hazards from aerosols and windblown spray, it is proposed to apply unrestricted irrigation criteria as a minimum quality requirement to industrial reuses.

Bacterial criterion

Investigations that may contribute to assess the validity of the WHO guideline limit of 1000 FC/100 mL have been reviewed by Blumenthal *et al.* (2000). The results of epidemiological studies performed in Mexico suggest that the risk of enteric infection due to the consumption of vegetables is significant but low when the guideline limit is exceeded by a factor 10 in the irrigation water. Tests performed in Portugal showed that:

- lettuces sprinkler irrigated with low quality secondary effluents were initially highly contaminated but fell within the quality recommended by the International Commission on Microbiological Specifications for Food (ICMSF) 5 days after the irrigation ceased (Vaz da Costa-Vargas *et al.*, 1991);
- the microbiological quality of crops irrigated with water just exceeding the WHO guideline complied with the ICMSF standard.

In dry weather, the microbiological quality of radish and lettuce drip and furrow irrigated with water slightly exceeding the WHO guideline was well below the ICMSF standard and of better quality than that of locally sold lettuce; however, rainy weather deteriorated the microbiological quality of lettuce. (Bastos and Mara, 1995). Shuval *et al.* (1997) found that the annual risk to be contaminated from eating, on alternate days, lettuce irrigated with reclaimed water meeting the WHO guideline ranged from 10^{-5} to 10^{-7} for hepatitis A, from 10^{-5} to 10^{-6} for rotavirus and was about 10^{-6} for cholera. Asano *et al.* (1992) calculated an annual risk associated with the consumption of spray irrigated food crops between 10^{-6} and 10^{-9} for a 1 virus unit/100 L and between 10^{-4} and 10^{-7} for a 100 times higher virus content; however, the assumption that the irrigation is stopped two weeks before harvesting is not realistic. Extending the work of Shuval *et al.* (1997), Mara *et al.* (2005) estimated the median rotavirus annual infection risk from the consumption of 100g lettuce on alternate days at 10^{3}-10^{-4} and from the weekly consumption of 100g crude onions for 5 months at $4.5x10^{-2}$-$5.6x10^{-3}$ when the *E. coli* content of the irrigation water is 1000/100 mL and assuming a rather low rotavirus die-off (0.1–1 log unit) between harvest and consumption. Despite some concerns related to the consumption of crude vegetables harvested during rainy periods, the *E. coli*/FC 1000/100 mL limit is considered to offer a protection allowing unrestricted irrigation.

Assuming an amount of ingested water of 1 mL and an exposure frequency of 10 days, the annual risk of rotavirus infection resulting of contacts with 1000 *E. coli*/100 mL reclaimed water of landscape impoundments where public contact with wastewater is not allowed is estimated to be $< 2.10^{-4}$. The daily risk when falling into such reclaimed water and swallowing 100 mL is estimated around 5.10^{-3}.

Table 30.5 Recommended guidelines for water reuse in the Mediterranean Region.

	Quality criteria			
	Intestinal nematode[a] (eggs/L)	**FC or *E. coli*[b] (cfu/100 mL)**	**TSS [c] (mg/L)**	**Wastewater treatment expected to meet the criteria**
Category I				
a) Residential reuse: private garden watering, toilet flushing, vehicle washing. b) Urban reuse: irrigation of areas with free admittance (greenbelts, parks, golf courses, sport fields), street cleaning, fire-fighting, fountains, and other recreational places. c) Landscape and recreational impoundments: ponds, water bodies and streams for recreational purposes, where incidental contact is allowed (except for bathing purposes).	≤ 0.1[h]	≤ 200 [d]	≤ 10	Secondary treatment + filtration + disinfection
Category II				
a) Irrigation of vegetables (surface or sprinkler irrigated), green fodder and pasture for direct grazing, sprinkler-irrigated fruit trees b) Landscape impoundments: ponds, water bodies and ornamental streams, where public contact with water is not allowed.	≤ 0.1[h]	≤ 1000 [d]	≤ 20 ≤ 150 [f]	Secondary treatment or equivalent [g] + either filtration + disinfection or storage or well-designed series of maturation ponds or infiltration percolation
c) Industrial reuse (except for food industry).	-			
Category III				
Irrigation of cereals and oleaginous seeds, fiber, & seed crops, dry fodder, green fodder without direct grazing, crops for canning industry, industrial crops, fruit trees (except sprinkler-irrigated)[e], plant nurseries, ornamental nurseries, wooden areas, green areas with no access to the public.	≤ 1	None required	≤ 35 ≤ 150 [f]	Secondary treatment or equivalent [g] + a few days storage or Oxidation pond systems

Table 30.5 Recommended guidelines (continued).

	Quality criteria			
	Intestinal nematode[a] **(eggs/L)**	**FC or *E. coli***[b] **(cfu/100 mL)**	**TSS** [c] **(mg/L)**	**Wastewater treatment expected to meet the criteria**
Category IV				
a) Irrigation of vegetables (except tuber, roots, etc.) with surface and subsurface trickle systems (except micro-sprinklers) using practices (such as plastic mulching, support, etc.) guaranteeing absence of contact between reclaimed water and edible part of vegetables. b) Irrigation of crops in category III with trickle irrigation systems (such as drip, bubbler, micro-sprinkler and subsurface). c) Irrigation with surface trickle irrigation systems of greenbelts and green areas with no access to the public. d) Irrigation of parks, golf courses, sport fields with sub-surface irrigation systems.	None required	None required		Pretreatment as required by the irrigation technology, but not less than primary sedimentation

[a] *Ascaris and Trichuris* species and hookworms; the guideline limit is also intended to protect against risks from parasitic protozoa.

[b] FC or *E. coli* (cfu/100mL): faecal coliforms or *Escherichia coli* (CFU: colony forming unit/100 mL).

[c] TSS: Suspended Solids

[d] Values must be conformed at the 80% of the samples per month, minimum number of samples 5.

[e] In the case of fruit trees, irrigation should stop two weeks before fruit is picked, and no fruit should be picked off the ground. Sprinkler irrigation should not be used.

[f] Stabilization ponds

[g] such as advanced primary treatment (APT) (Jimenez *et al.*, 1999 and 2001)

[h] As very few investigations, if any, have been carried out on how to reach < 0.1 nematode egg /L, this criterion is considered a medium term objective and will be provisionally replaced by <1 nematode egg /L.

Nematode egg criterion

It was inferred from experimental studies in north-east Brazil and Leeds (Ayres *et al.*, 1992) and epidemiological studies in central Mexico (Cifuentes, 1998; Peasey, 2000; Blumenthal *et al.*, 2001), that it would be convenient to adopt a guideline limit of ≤0.1 egg/L. This guideline will allow protecting vegetable consumers (particularly when climatic conditions favour the survival of helminth eggs), farm workers and their children.

TSS criterion

A TSS guideline limit of 20 mg /L (150 mg/L for stabilisation ponds) is proposed in order to ensure that the microbial limits are reliably met when conventional treatments are used.

Minimum treatment recommended

Secondary treatment can be performed in conventional or extensive low technology facilities. Advanced primary treatment may achieve sufficient performances. The limit of 1000 FC/100 mL allows using extensive technologies, such as maturation ponds, reservoirs and infiltration percolation, to polish wastewater in order to reach the microbial guidelines. The possibility of using low technology and O&M cost techniques is regarded as a key factor of the development of planned water reuse in many Mediterranean countries. As mentioned in the recommendations of Blumenthal *et al.* (2000), special attention should be paid to the design and operation of reservoirs, stabilisation and maturation ponds, as their performance is highly dependent on the climate and the feeding/withdrawal schedule. Conventional disinfection techniques might be preferred in densely populated areas.

Water category III

The applications listed in category III (restricted irrigation) exclude direct contact of humans and animals with reclaimed water (with the exception of incidental contacts with workers). Crops cultivated for the canning industry will be disinfected in the canning process. No direct grazing of green fodder will be allowed. Other crops, cereals, fibre, industrials, seeds, dry fodder, etc. are harvested a long time after irrigation has ceased. Therefore health related risks are considerably reduced.

Bacterial criterion

It is proposed not to include any bacterial limit for restricted irrigation. Essential eventual health risks are not related to crop consumption but to workers and neighbours contamination in case of sprinkler irrigation. Therefore, in case of aspersion, setback distances between irrigation sites and residential areas, roadways, sports fields, etc must be established. This measure is more protective than setting a bacterial limit of 10^5 FC/100 mL as suggested by Blumenthal *et al.* (2000), though such a quality would be reached after the secondary or advanced primary treatments required to meet the TSS guideline limit. Mara *et al.* (2005) attempted to assess the risks resulting from soil ingestion by workers or children playing in irrigated fields. However, risks are not related to the *E. coli* irrigation water content but to the *E. coli* soil content. The relationship between *E. coli* soil and water contents being rather complicated, this approach would require more field investigation. Where frequent contact of children or workers with wastewater is observed and cannot be avoided, a limit of 10^3 FC/100 mL should be set and enforced; thus, setback distances would not be relevant, as shown by the review of Blumenthal *et al.* (2000).

Nursery plants should not be irrigated with water category III for at least two weeks before being sold.

Nematode egg criterion

As for the bacterial criterion, no sufficient reason is found to shift from the limit of 1 nematode egg /L to a more stringent guideline, whatever the irrigation technique (sprinkler, furrow and flood irrigation). However, in countries where frequent contact of children and workers with wastewater is observed and can not be avoided, the limit of 0.1 egg/L would be recommended when technologies allowing this limit to be met are available.

TSS criterion

The guideline limit of 35 mg TSS/L (150 mg/L for stabilisation ponds) is proposed so that the wastewater has to go through a secondary, or an advanced primary treatment, before being reused.

Minimum treatment recommended

As mentioned above, a secondary treatment, or an advanced primary treatment is required, followed by a few days storage, in order to meet the nematode egg limit. Oxidation ponds or a secondary treatment followed by maturation ponds may also be used to reach the same figures.

Category IV

A water category IV is proposed when the combination of the irrigation technique and the agricultural practice results in very low microbiological health risks. The applications include:

- irrigation and cultivation practices which guarantee the absence of contact of vegetables and fruits with wastewater and the absence of aerosols and run off;
- irrigation with techniques able to prevent aerosols and run off of crops listed in category III and of green areas where the public has no access;
- irrigation of green areas open to the public and sports fields which guarantee the absence of contact of the public with the irrigation water.

Microbial criteria

Given the very low level of health risks related to the applications listed in category IV, no microbial guideline is set.

TSS criterion and treatment recommended

No TSS guideline is proposed. However, trickle irrigation techniques (drip, bubbler, micro-sprinkler and subsurface) require a treatment of wastewater in order to avoid the clogging of the distribution network and the emitters. A primary sedimentation is highly recommended as a minimum; primary effluents should go through a filtration process appropriate to the irrigation technology which has been selected.

30.4 TOWARDS MORE SCIENTIFICALLY-BASED REGULATIONS?

This proposal of Mediterranean guidelines has been elaborated in 2002 within the activities of the Strategic Action Programme to Address Pollution from Land-based Activities of the MED POL programme for the Assessment and Control of Marine Pollution in the Mediterranean. It involved attempts at: (a) linking health-related criteria to risk levels that are considered acceptable, referring to internationally accepted risks related to drinking and bathing, and (b) assessing the risks using epidemiological data and QMRA results.

Since then, the process of revision of the WHO guidelines for the safe use of wastewater in agriculture (WH0, 2006) has been completed. It is characterized by important steps towards:

- a harmonization of health-based guidelines related to different water uses;
- a deterministic approach for setting health-related guidelines.

A draft of Australian National Guidelines for Water Recycling has been elaborated on the same lines (NRMMC – EPHC, 2005).

30.4.1 Harmonisation of health-based guidelines

The revision of the WHO guidelines has been undertaken within the Stockholm Framework presented in *Water Quality - Guidelines, Standards and Health: Assessment of Risk and Risk Management for Water-related Infectious Disease* (Fewtrell and Bartram, 2001) which involves a harmonized approach to risk assessment and management, whatever the route of water-borne disease transmission. The WHO water and sanitation related guidelines were recently developed in accordance with this framework (WHO, 2003; WHO, 2004; WHO, 2006).

Setting up guidelines requires prior definition of the level of risk that is considered as acceptable or should not be exceeded. There seems to be no reason why this risk level would differ according to the nature and conditions of the water use or reuse. Comparing disease outcomes from different exposure pathways and vectors of disease requires the use of a common metric. To this end, the burden of disease related to a specific disease and pathway is expressed through the Disability Adjusted Life Years (DALYs) lost for a population over a time period. DALYs includes the years of life lost due to premature death and the years lived with a disability caused by the disease of concern. Evaluating risks in DALYs allows comparison of different health outcomes and the working out of balanced risk management policies. Therefore, the DALY metric has been used to propose what could be considered as the acceptable health-related risk when reusing water. In the third edition of the *Guidelines for Drinking-water Quality*, WHO has adopted a tolerable burden of waterborne disease from drinking fully treated water of $\leq 10^{-6}$ DALY loss per person per year (pppy). The same level of protection is recommended for water reuse in agriculture (WHO, 2006). For the sake of the harmonization of health-related water uses guidelines, it should also be recommended for any other water reuse application.

30.4.2 Deterministic approach for setting health-based targets

An example of a simplified deterministic approach for setting health-based targets is presented here (other examples can be found in NRMMC – EPHC, 2005). Microbial health-based targets are derived from the acceptable risk level of $\leq 10^{-6}$ DALY loss pppy through QMRA procedures. The tolerable annual risk of disease due to a pathogen of concern can be derived from the tolerable DALY loss through the following expression:

Tolerable disease risk pppy = Tolerable DALY loss pppy / (DALY loss per case of disease) x (Susceptibility fraction)

The tolerable annual infection risk per person is obtained by dividing the tolerable annual disease risk by a disease/infection ratio.

Assuming that the tolerable annual infection risk per person, $P_A(d)$, results of n exposures in a year to the same dose of pathogens d, then d is considered the maximum dose of pathogen that can be absorbed or ingested by a person in one exposure without exceeding an acceptable risk. The annual risk and the risk per exposure event, $P_I(d)$, are related as follows:

$$P_A(d) = 1 - [1 - P_I(d)]^n$$

The dose *d* is derived from the risk per exposure event, $P_I(d)$, through a dose-response model; the most commonly used being the exponential or β-Poisson models. The volume of water absorbed or inhaled or the quantity of product consumed during one exposure event being estimated, the maximum pathogen concentration that does not lead to an unacceptable risk can be derived.

Then, the numbers of pathogens present in the untreated water being known or estimated, the degree of pathogen reduction to be achieved is determined. This reduction can be achieved either through wastewater treatment alone or by combination of wastewater treatment, health-protection control measures and natural pathogen abatement in the environment.

The health protection control measures related to irrigation water reuse involve:

- crop restriction;
- wastewater application technique;
- time elapsed between last irrigation and consumption;
- food preparation measures; and
- human exposure control.

Similar protection measures can be associated with other water reuses. Wastewater treatment can also be considered in this list.

An evaluation of the pathogen reduction resulting from different measures has been derived from the literature (NRMMC – EPHC, 2005). For instance, localized irrigation of high growing crops, the harvested parts of which are not in contact with the soil (tomatoes, eggplants, etc) provides a protection equivalent to a pathogen reduction of 4 log units; peeling fruits or root vegetables is equivalent to 2 log units reduction, etc.

Verification monitoring

The reduction of pathogens to be performed by the wastewater treatment must be consistently monitored. For viral and bacterial infections, a monitoring in terms of *E. coli* or faecal coliforms is proposed while helminth eggs would be used for helminthic infections. Viral and bacterial pathogens are assumed to be removed in the same proportion to that of *E. coli*.

30.4.3 Limits of the deterministic approach

This approach is a major breakthrough in the methods implemented for the elaboration of reuse guidelines. Its mathematic formulation should allow a more efficient dialogue between the countries and significant progress in the adoption of common policies. At the same time, it makes pointing out the uncertainties attached to the relationships between the microbial quality of the reused water and heath risks easier and, by the way, stressing the need for research to reinforce the scientific basis of reuse guidelines. The methodology itself is not devoid of limitations.

The values of several parameters, such as DALY losses per case of disease, susceptibility fractions, dose/response model parameters, volumes absorbed per exposure, numbers of exposure, numbers of pathogens in raw water etc. are uncertain and variable. Though figures are available in the literature, more experimental data are needed for a better evaluation of the risks related to water reuse. The variability of some parameters has already driven statistical approaches (Tanaka *et al.*, 1998, NRMMC – EPHC, 2005; Mara *et al.*, 2005). Pathogen reductions that can be ascribed to health protection control measures require more in-depth investigations. The validity of dose/response models at low exposure remains a matter of concern. Moreover, the verification monitoring of viral and bacterial reduction is based on the questionable assumption that pathogens and faecal indicators (*E. coli*) are removed in the same proportion through wastewater treatments.

Despite the above mentioned drawbacks and uncertainties, this new approach may pave the way to reach a regional consensus on the setting up of Mediterranean guidelines.

30.5 CONCLUSIONS

Explanations of the dramatic discrepancies in the existing Mediterranean water reuse regulations and guidelines can be found in the methodology that has been used for their elaboration and particularly (a) the reference to either the State of California Title 22 criteria or the former WHO guidelines and (b) the absence of reference to a clear rationale supported by scientifically established data as a solid basis for setting up or adapting water reuse guidelines.

Most actual guidelines and regulations reflect more a perception of a risk hierarchy – which differs from one country to another – than scientifically based risk assessments.

The guidelines proposal presented here is an attempt to keep health risks related to water reuse within the range of risks that are internationally considered acceptable for other water uses. Water reuse risks were assessed using available epidemiologic data and QMRA results.

This proposal may be considered in a possible concerted effort to adapt the new WHO guidelines to the Mediterranean context.

30.6 REFERENCES

Asano, T., Leong, L.Y.C., Rigby, M.G. and Sakaji, R.H. (1992). Evolution of the California wastewater reclamation criteria using enteric virus monitoring data. *Wat. Sci. Tech.*, **26** (7-8), 1513-1524.

Ayers R. S. and Westcot D. W. (1985). *Water quality for agriculture*, Food and Agriculture Organization of the United Nations, FAO Irrigation and Drainage, Paper 29, Rome. Italy.

Ayres, R.M., Stott, R., Lee, D.L., Mara, D.D. and Silva, S.A. (1992). Contamination of lettuces with nematode eggs by spray irrigation with treated and untreated wastewater. *Wat. Sci. Tech.* **26**(7-8), 1615-1623.

Bastos, R.K.X. and Mara, D.D. (1995). The bacteriological quality of salad crops drip and furrow irrigated with waste stabilization pond effluent: an evaluation of the WHO guidelines. *Wat. Sci. Tech.* **31**(12), 425-430.

Blumenthal, U.J., Mara, D.D., Peasey, A., Ruiz-Palacios, G. and Scott, R. (2000). Guidelines for the microbiological quality of treated wastewater used in agriculture: recommendations for revising WHO guidelines. *Bulletin of the WHO* **78** (9), 1104-1116.

Blumenthal, U., Cifuentes, E., Bennett, S., Quigley, M. and Ruiz-Palacios, G. (2001). The risks of enteric infections associated with wastewater reuse: the effect of season and degree of storage of wastewater. *Transactions of the Royal Society of Tropical Medicine and Hygiene* **95**, 1-7.

Brissaud, F., Xu, P. and M. Auset (2003). Extensive reclamation technologies for irrigation reuse. *Wat. Sci. Tech / Wat. Supply* **3**(4), 209-216.

Chang, C., Page, A.L. and Asano, T. (1998). Evaluating methods of establishing human health-related chemical guidelines or cropland application of municipal wastewater. In Asano T. ed. *Wastewater Reclamation and Reuse*. Lancaster, PA, Technomic Publishing, 581-626.

Cifuentes, E. (1998). The epidemiology of enteric infections in agricultural communities exposed to wastewater irrigation: perspectives and risk control. *International Journal of Environmental Health Research* **8**, 203-213.

COM (2002). Proposal for a directive of the european parliament and of the council concerning the quality of bathing waters. http://www.europa.eu.int/eur-lex/en/com/pdf/2002/com2002_0581en01.pdf

Cooper, R.C. and Olivieri, A.W. (1998). Infection disease concerns in wastewater reuse. In Asano T. ed. *Wastewater Reclamation and Reuse*. Lancaster, PA, Technomic Publishing, 489-520.

Crook, J. (1998). Water reclamation and reuse criteria. In Asano T. ed. *Wastewater Reclamation and Reuse*. Lancaster, PA, Technomic Publishing, 627-703.

Crook, J., Johnson, L.J. and Thompson, K. (2001). California's new water recycling criteria and their effect on operating agencies. In *Proceedings of the American Water Work Association 2001 Annual Conference, June 17-21, Washington D.C.*

Durand, R.E. *et al.*, (1986). Epidemiological investigation of community health effects of landscape irrigation with reclaimed water: the Colorado Springs study. Colorado Springs, Wastewater Division of the City of Colorado Springs.

European Union (1976). Bathing water quality directive 76/160/EEC. http://www.europa.eu.int/water/water-bathing/directive.html

European Union (1991). *Council Directive Concerning Urban Wastewater Treatment.* 91/271 EEC of May 21, 1991, OJ NO L135/40 of May 30, 1991.

Feachem, R.G., Bradley, D.H., Garelick, H. and Mara, D.D. (1983). *Sanitation and Disease: Health Aspects of Excreta and Wastewater Management.* John Wiley and Sons, New York.

Fewtrell, L. and Bartram, J. (eds), (2001). *Water quality guidelines, standards and health: assessment of risk and risk management for water-related infectious disease.* IWA Publishing, 413 p.

Fleisher, J.M., Kay, D., Wyer, M. and Merrett, H. (1996). The enterovirus test in the assessment of recreational water-associated gastroenteritis. *Wat. Res.* **30**, 2341-2346.

Jiménez B., Chávez A., and Hernández C. (1999). Alternative wastewater treatment intended for agricultural use in Mexico. *Wat. Sci. Tech.,* 40, No. 4-5, pp. 355-362.

Jiménez B., Chávez A., Maya, C. and Jardines, L., (2001). Removal of microorganisms in different stages of wastewater treatment for Mexico City. *Wat. Sci. Tech.,* 43, No. 10, pp. 155-162.

Kay, D., Fleisher, J.M., Salmon, R.L., Jones, F., Wyer, M.D., Godfree, A.F., Zelenauch-Jacquotte, Z. and Shore, R. (1994). Predicting likelihood of gastroenteritis from sea bathing: results from randomised exposure. *Lancet* **344**, 905-909.

López-Pila, J. and Szewzyk, R. (2000). Estimating the infection risk in recreational waters from the faecal indicator concentration and from the ratio between pathogens and indicators. *Wat. Res.* **34** (17), 4195-4200.

Mara, D.D., Sleigh, P.A., Blumenthal, U.J. and Carr, R. (2005). Health risks in wastewater irrigation: comparing estimates from quantitative microbial risk analyses and epidemiological studies. *J. Water and Health* (submitted).

NRMMC-EPHC (2005) *National guidelines for water recycling. Managing health and environmental risks.* Natural Resource Management Ministerial Council and Environment Protection and Heritage Council – Australia. ISBN Print 0 642 323968.

Ongerth, H. and Jopling, W.F. (1977). Water reuse in California. In "Water Renovation and Reuse" edt. by Shuval H.I., Academic Press, New York, 219-256.

Ogoshi, M., Suzuki, Y. and Asano, T. (2001). Water reuse in Japan. *Wat. Sci. Tech.* 43(10), 17-23.

Peasey, A., (2000). *Human exposure to ascaris infection through wastewater reuse in irrigation and its public health significance* (PhD thesis). University of London, London, England.

Shuval, H., Adin, A., Fattal, B., Rawitz, E. and Yekutiel P. (1986). Wastewater irrigation in developing countries: Health effects and technical solutions. World Bank Technical paper 51, The World Bank, Washington, D.C.

Shuval, H., Lampert, Y. and Fattal, B. (1997). Development of a risk assessment approach for evaluating wastewater reuse standards for agriculture. *Wat. Sci. Tech.* **35**(11-12), 15-20.

State of California. (1978). Wastewater Reclamation Criteria, An Excerpt from the California Code of Regulations, Title 22, Division 4, Environmental Health, Dept. of Health Services, Sacramento, California.

State of California (2000). Title 22, Code of regulations, 24 p. November 2000.

Strauss, M. and Blumenthal, U.J. (1989). Health Aspects of Human Waste Use in Agriculture and Aquaculture –utilisation Practices and Health Perspectives. International Reference Center for Waste Disposal, Dubendorf (Report N° 08/88).

Tanaka, H., Asano, T., Schroeder, E. D. and Tchobanoglous, G. (1998). Estimating the safety of wastewater reclamation and reuse using enteric virus monitoring data. *Wat. Environ. Res.* **70** (1), 39-50.

U.S. Environmental Protection Agency (1976). *Quality Criteria for Water. EPA-R3-73-033.* A report of the National Academy of Science – National Academy of Engineering Committee on Water Quality Criteria. US Environmental Protection Agency, Washington, DC.

U.S. Environmental Protection Agency (1986). Ambient water quality for bacteria – 1986. Bacteriological ambient water quality criteria for marine and fresh recreational waters. EPA440/5-84-002. U. S. Environmental Protection Agency, Washington D.C.

U.S. Environmental Protection Agency (1998). *Bacterial Water Quality Standards for Recreational Waters (Freshwater and Marine Waters).* 4305 EPA-823-R-98-003. Office for Water. U S Environmental Protection Agency.

Vaz da Costa-Vargas, S.M., Mara, D.D. and Vargas-Lopez, C.E. (1991). Residual faecal contamination on effluent-irrigated lettuces. *Wat. Sci. Tech.* **24**(9), 89-94.

World Health Organization (1989). Health guidelines for the use of wastewater in agriculture and aquaculture, Tech. Bull. Ser. 77, WHO, Geneva, Switzerland.

World Health Organization (2003). Guidelines for safe recreational water environments; vol.1. Coastal and freshwaters. WHO, Geneva, Switzerland.

World Health Organization (2004). Guidelines for drinking-water quality; 3ed ed. WHO, Geneva, Switzerland

World Health Organization (2006). Guidelines for the safe use of the wastewater, excreta and grey water; vol. 2: Wastewater use in agriculture. WHO, Geneva, Switzerland.

31

Wastewater use in high rainfall riverine cities: comparisons from Cameroon, Nepal and Vietnam

Liqa Raschid-Sally, Ives Magloire Kengne, Nguyen Viet Anh and Dominique Endamana

31.1 INTRODUCTION

Wastewater use or reuse for agricultural purposes is practiced across the globe, both in the developed world and in developing countries. In the developed world, its use is termed recycling and is associated with treatment and other controls imposed for safety purposes. In the developing world planned wastewater recycling is uncommon because it is unaffordable. There is usually no treatment or at best it is rudimentary because investments for treatment are high and thus beyond the actual priorities of urban development. What tends to happen is *ad hoc* (and therefore unplanned) use of the wastewater which is generated by cities and disposed of in bulk.

In water-scarce areas or under localized water penury in and around cities, farmers extract the water from the source (a canal, ditch or even a broken sewer) and use it for cultivating vegetables and other cash crops which are in high demand. For these farmers this water is the only source available to them. In Pakistan, farmers buy this wastewater from municipalities who see it as a profitable way of disposing of wastewater (Van der Hoek *et al.* 2002).

©2008 IWA Publishing. *Water Reuse - An International Survey of current practice, issues and needs* by B. Jimenez *et al.* ISBN: 9781843390893. Published by IWA Publishing, London, UK.

In this context, its value as a source of income and livelihoods to the farmers involved can be argued, especially where irrigation water is in short supply and industrial contamination insignificant. The patterns of generation and disposal are specific to the urban context of the city, the topography and climate which in turn drives the manner in which it is used. Farming practices under these conditions are described in a number of references notably by Scott *et al.* 2004. National wealth or a country's GDP which translates into better infrastructure investments, is clearly a key factor which influences the way wastewater is used.

What is noteworthy, however, but has never been specifically discussed is that two distinct contexts of use exist even within the developing world of unplanned wastewater use. The case of farmers in **humid** countries with high rainfall contrasts with that of **dry** countries with lower rainfall. In the former, water is available so in principle farmers do not need to resort to wastewater, as is the case with their counterparts in dry regions. In spite of this, they do and it is often argued that wastewater is being used for its nutrient value. This chapter aims to contribute to a better understanding of the drivers of the practice under humid conditions.

The chapter focuses on three cities, one from Africa, namely Yaounde, Cameroon, and two from Asia, namely Kathmandu, Nepal and Hanoi, Vietnam (Figure 31.1). All these cities are in high rainfall regions located in large river valleys where the use of wastewater for agriculture in urban and peri-urban areas is a common occurrence. The main source of the data for the three cities was a survey of 50 cities under the Comprehensive Assessment of Water for Agriculture program of the International Water Management Institute, to understand the drivers of wastewater irrigation. The survey used rapid assessment techniques. Information relevant to this chapter was extracted from the survey and is discussed here. This information was supplemented with data from two more detailed case studies from Yaounde, (Endamana *et al.* 2003, and Raschid *et al.* 2004), and Kathmandu valley (Rutowski 2004). For Hanoi, Vietnam, the case of the district of Thanh Tri where wastewater is used extensively for agriculture and aquaculture (Phuong Anh *et al.* 2004) is discussed. Whilst there are differences between the African and Asian cases, and between countries (and regions), it is interesting that there are startling similarities as well in the way wastewater is perceived and therefore used in these climates. These examples will provide an insight into practices and how issues might be addressed under these conditions.

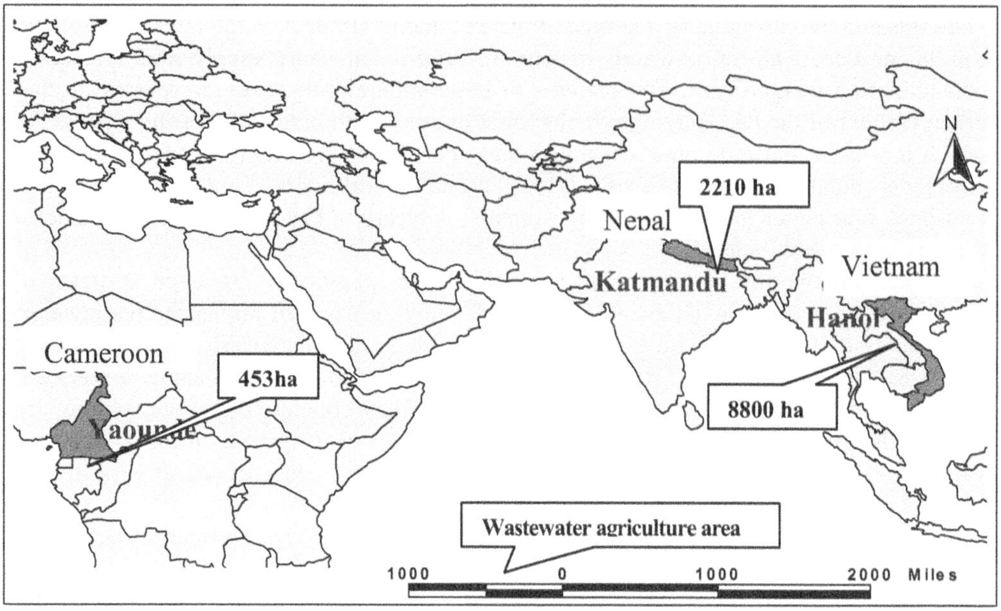

Figure 31.1. Study locations.

31.2 THE AFRICAN CASE

31.2.1 Yaounde, Cameroon

The city of Yaounde is an inland city at an altitude of 700-850m, situated in the 1500-2000mm annual rainfall range. It receives rainfall during two seasons September to mid-November and mid-March to June. It has a dense hydrographical network drained by the river Mfoundi with about 15 watersheds each with a lowland watercourse being used for irrigation livestock, aquaculture and household activities. The city has a total land area[1] of 256 km^2, a population of 1.5 million inhabitants, and a population density of 4441 inhabitants per km^2. The city poverty index is 19.9% which is only half of the national index showing good urban economic growth. Estimates of annual population growth rate for the city are 7% which is more than twice the national figure. The GDP was US$30.6 billion in 2004.

31.3 THE ASIAN CASES

31.3.1 Kathmandu Valley, Nepal

The Kathmandu valley lies at an average altitude of 1250m above sea level, and receives an annual rainfall of 1868mm, essentially 90% of which falls in the summer months of June to September. It comprises five municipal towns, Kathmandu, Lalithpur, Bhaktapur, Kirtipur and Madhyapur Thimi drained by the Bagmati river together with its major tributaries the Manahara, Dhobi, Bisnumati, and Nakhu Rivers which are the sources of irrigation and water supply to the valley. The total area of the combined municipalities is 97 km^2 with the maximum population density of 13,600 inhabitants per km^2 being in the Kathmandu

1 The wide variation in city area as seen from the figures for the three cities may be explained by the way urban administrative boundaries are defined in different countries.

municipality. The city poverty index is 25 to 30% which is about 10% lower than the national index. Annual growth rate of the municipalities averages around 4.4%. The GDP was US$39.53 billion in 2004.

31.3.2 Hanoi, Vietnam

Hanoi is situated in the Red River delta at an altitude of 1-30 meters above sea level and receives an average annual rainfall of 1689mm with over 80% of it occurring between May and October. The dry season extends from November to April. The largest river draining Hanoi is the Red River with its two main basins the To Lich and the Nhue River. Drainage in Hanoi occurs through four drainage rivers: To Lich, Set, Lu and the Kim Nguu rivers. These rivers are the sources of irrigation and water supply to the city and the agricultural areas. Additionally, 27 lakes and ponds characterize Hanoi. Urban and peri-urban Hanoi has a total area of 921 km^2, with a population density of 3352 inhabitants per km^2 but in urban Hanoi the density is much higher at 17,865 persons. The annual growth rate of the city is 1.1%. The GDP is US$227.2 billion in 2004.

31.4 COMPARING AND CONTRASTING

31.4.1 The urban environment: impacts of poor sanitation and wastewater management

For the three cities, environmental sanitation conditions were reported as poor to adequate (Table 31.1). Children's health, related to poor sanitation and wastewater disposal (e.g. hookworm) and ingestion of contaminated food (e.g. ascaris), was the worst for Hanoi which has been confirmed by a general health review by Van der Hoek *et al.* (2003). Per capita water consumption is relatively low in Yaounde and Kathmandu which is not related to physical water scarcity (see *per capita* freshwater resource for the three countries). In Vietnam there is a much higher urban water use per capita, but the low per capita freshwater is a reflection of the very high population in Vietnam.

All three urban centers have high sanitation coverage, through a combination of onsite systems and sewerage but wastewater management and disposal practices are not adequate. Where sewers exist, they collect blackwater, greywater, storm water and industrial wastewater. In all the cases, only some of the wastewater that is collected (varying from 0% to 30% in the best case) is transferred to treatment plants. In all of the cities with treatment plants, none were fully functional. Yaounde has very many small activated sludge plants whereas Kathmandu has lagoons. Hanoi has almost no treatment and it is only recently that two pilot treatment plants have been constructed. In all three cities in spite of high population densities, septic tanks are still widely used. Their efficient functioning requires adequate soakage space and this condition is not satisfied in urban areas. Wastewater accumulates leading to overflow and poor waste stabilization. Overflows are directed to the drainage canals and networks of the city. This is particularly so in Hanoi because of high building density. Tanks have to be emptied frequently, which is undertaken by a service provider (public and private) mechanically or manually as in Yaounde. The high rainfall aggravates the situation and householders empty their septic tanks through the evacuation siphon during these periods. Faecal sludge is disposed of under unsanitary conditions, in the river and other water bodies in the city (Kathmandu, Yaounde). In Hanoi the tanks are emptied infrequently and the sludge is dumped in isolated open spaces and landfill sites which may not always be regulated. High rainfall events in these cities wash the sludge that is disposed of onto land, into the waterways.

Table 31.1. Profile of cities.

Characteristics	Yaounde, Cameroon	Kathmandu Valley, Nepal	Hanoi, Vietnam
Country GDP/PPP[1] (2004) US$	30.6 billion	39.53	227.2
Country GDP/PPP *per capita* (2004) US$	1900	1500	2700
Country Real economic growth% (2004)	4.9	3.0	7.7
Environmental Sanitation Conditions [2]	Adequate	Poor to adequate	Adequate
Health Status	15% positive for ascaris 8.4% positive for hookworm in a sample of 408 children (under 5 years)[3]	10.4% positive for ascaris 8.7% positive for hookworm in a sample of 460 children (all age groups)[4]	60-65% positive for ascaris 0-2% positive for hookworm in two studies of children 6-10 years old. Total sample in the two studies was 827[5]
Per capita freshwater resource m^3/capita	16969	8029	4513
Per Capita water consumption lL/*capita*/d	43	60	120
Water supply coverage (% population)	60	70	70
Sanitation coverage (%population)	90% with onsite sanitation 9% to sewerage (closed)	25% with onsite sanitation 75% to sewerage (open and closed)	60% with onsite sanitation 40% to sewerage (open)
% treated	2-5% (partial treatment). Sludge from septic tanks emptied into water or on land	15-30% (partial treatment) Sludge disposed of in river	No treatment Sludge disposed of in landfill site
Type of treatment	Mainly activated sludge	Mainly stabilization ponds	-
Industrial pollution	12% of total	10% of total	20-25% of total
Area under wastewater agriculture (ha)	450	2200	8400

[1] Gross Domestic Product/Purchasing Power Parity.
[2] Adequate to poor: at best only 60% of the population had sanitation coverage, less than half the city had a covered drainage system and at best only half the city had some form of solid waste collection and disposal.
[3] Ongoing study on "*Maîtrise de l'assainissement dans un écosystème urbain à Yaoundé au Cameroun et impacts sur la santé des enfants âgés de moins de cinq ans.*"
[4] DOHS, 2003, Teku Hospital outpatient records.
[5] Do Thi Hoa et al. 2000.

31.4.4 Impacts of industrial development

All three cities were characterized as residential with a few industries. Yaounde has a low level of industrial development, most of it being in the industrial zone located a little outside the city. In Kathmandu, the 1340 industrial units are mostly scattered around the city but there is one industrial zone for light industry. Hanoi has five industrial zones, 10-20 km outside the city boundaries and nine areas concentrating various small scale industries within the city limits. None of the three cities can be classified as highly industrialized. Even the industrial development within the industrial zones is small to medium scale for the more polluting industries, and textile and garment are the only large scale activities.

Industrial waste is not treated before discharge, and in all three cities the drainage systems collect both domestic and industrial wastewater. This system also collects storm water during the rainy season. In Yaounde the small closed sewer network is separate from the storm water drain which is open. These open drains however also receive overflows from septic tanks and grey water from households. In terms of industrial contaminants, in Yaounde the major sources of industrial waste are food processing units, so the waste is not of chemical origin. In Hanoi, the preferred industrial development is in the electronic, food-processing and the garment sector with a few chemical industries generating under 10% of industrial wastewater. The overall industrial wastewater contribution is under 25% (Table 31.1). In Kathmandu the carpet dyeing industry is the most polluting (Green, 2003) generating chemical contaminants, particularly colored chemical dyes, which are difficult to treat.

In all the cities, hospital waste, which is usually classified as hazardous, is also discharged into the common sewer system.

31.4.5 Wastewater disposal point, and quality of water bodies

All three cities have perennial streams and rivers running through them. The presence of extensive natural waterways within these cities results in many wastewater drains and storm-water canals being linked to these water networks. City wastewater in all the cases was disposed of in the river or its tributaries at multiple locations. Even in the dry season, though there is less dilution, the level of pollution in the streams is much less than in the case of arid cities where there is no dilution. The example of the Musi River in Hyderabad is typical of a seasonal river in arid regions where in the dry season wastewater is the main source of water for the river (Buechler *et al.* 2002).

Table 31.3 Typical water quality parameters from city water bodies receiving wastewater.

Parameters	Yaounde[1]	River Bagmati, Kathmandu[2]	Hanoi[3]
pH	6.71-7.24	6.5-7	6.56 – 7.77
Turbidity (NTU)	-	75-180	18.9 – 239.2
TSS (mg/L)	23-956	92-304	19 – 245
BOD (mg/L)	8.5-271	50-240	67.5 – 139.3
COD (mg/L)	12-739	110-317	85 – 186.4
TDS (mg/L)	-	120-360	170 – 450
DO (mg/L)	1.1-3.4	0.4-1.9	0 – 1.17
NO_3-N (mg/L)	0.7-3.7	0.6-10 and above	0.28 – 2.62
NH_4-N (mg/L)	1.3-59	4-20	1.95 – 37.4
PO_4-P (mg/L)	1.7-50	1-1.7	1.3 – 4.4
Fecal Coliform (counts.per 100 mL)	$1.5x10^3$-$824x10^4$	$1.8x10^4$-$230x10^4$	-
Conductivity(μS/cm)	127-500	-	340 – 910
Total Coliform (counts.per 100 ml.)	-	-	$44x10^2$ - $74x10^4$

[1] Measured between 2001 and 2002, Lami, 2002

[2] Measured between May-Dec 2002, ENPHO 2003 .

[3] To Lich and Kim Nguu rivers, Hanoi city. Samples taken from 4 points (2 per each river), once in every 2 months, between March 2002 - August 2003. Source: data from the monitoring campaigns conducted by the Center for Environmental Engineering of Towns and Industrial Areas (CEETIA), Hanoi University of Civil Engineering, under the National Environmental Monitoring program, Vietnam Environmental Protection Agency (VEPA).

31.4.6 Profile of Wastewater agriculture

The three cities have extremely different extents of agriculture using polluted/waste water varying from 450 ha to 8400 ha (Table 31.1). Usually agriculture using polluted water is nested in the broader frame of urban and peri-urban agriculture. If a city does not accommodate agriculture due to densification and value of urban land, or for any other reason, it will usually not have wastewater agriculture. Urban agriculture usually sources wastewater only in the absence of other water. Therefore, during the rainy season whatever the average rainfall, urban and peri-urban agriculture is essentially rainfed (except in the case of backyard cultivation when municipal water is sometimes used). In the dry season however, cities in river valleys still have access to perennial streams and, with a few exceptions where direct extraction of wastewater from sewers does occur, most farmers utilize water from the streams and river, and shallow wells which are ubiquitous due to the high water table. The quality of the water varies but in most cases in these climates, different degrees of dilution have occurred to the wastewater before use.

31.4.7 Location of sites, sources and field application of wastewater

In Yaounde, the main sites where farmers are in contact with wastewater are in the lowland area of the inland valleys where the streams receive the wastewater from the city drains. Some sites are found within the city as well as in densely populated shanty areas of the town. The riverine areas and floodplains are technically state-owned, according to Cameroon land legislation, but in some areas of the south of Cameroon customary tenure rights are recognised (Diaw, 1997).

The quality of the irrigation water depends on the topography of the site. Water from water courses running through the lowland valley, which almost all the farmers use at some time, is the most polluted, receiving the sewage flows from the city. About 20% of the farmers dig shallow wells in their plots or close to them to avoid carrying water over long distances. Well water quality (1500 *E. Coli*/100 mL) was closer to the WHO microbiological standards for wastewater irrigation. A survey showed that around 40% of farmers used inland valley water because of its abundance and proximity, 14% used it instead of treated pipe-borne water because it was free, while 20% of the farmers perceived it as a source of nutrients for the crops (Raschid-Sally *et al.* 2004).

Hand watering of vegetables is the typical practice, using two 10 to 15 litre watering cans carried in either hand. The water is then sprayed on the surface of the vegetables increasing the risk of contamination. The use of mechanical means (motorized or manual irrigation pumps) for lifting the water was not reported.

In Nepal, the municipalities in the valley which is drained by the Bagmati River are surrounded by agricultural lands. About 60% of the valley comprises agricultural land, of which 80% is under informal irrigation. A detailed study of two municipalities in the Kathmandu valley showed that 60% of farmers used the polluted rivers and streams in the valley whereas only 30% used wastewater directly from sewers. This pattern of wastewater usage is likely typical for the entire lowland area of the Kathmandu Valley (Rutkowski *et al.*, 2007). All lowland rivers including the Dhobi, Bisnumati, and Bagmati Rivers, are polluted with municipal wastewater. Although large sections of these rivers' corridors are urban, large agricultural areas adjacent to these polluted rivers still exist. Significant numbers of these farmers are likely to be irrigating with diluted wastewater, especially in the dry season when river water quality is at its worst.

Flooding of land closest to the river is a significant hazard for farmers. In the study in Kirtipur, three out of every four farmers interviewed stated that their land is regularly flooded during the monsoon season. Furthermore, 38% of the interviewed farmers stated that their crops had been damaged due to flooding. Frequent flooding and exposure to polluted water can physically damage crops, as well as lead to over-fertilization. Flooding occurs in the monsoon season when the crop grown is rice. Another consequence is the accumulation in the fields of mounds of garbage, animal bones, and glass when flood waters recede. Some farmers spend considerable effort raising their plots to avoid chronic flooding and others are obliged to re-plant their crop as many as four times because of flood damage.

The hydrology of the valley is such that in many sites gravity irrigation was possible. Pumps were used when this was not possible. Manual watering was much less common (only 15% of the farmers in the study used this method).

In Hanoi, agricultural land occupies an important place in urban planning, constituting 46% of the total geographic area of the city (district)[2]. Urban agricultural land alone occupies 17% but most agricultural land is in the peri-urban area. Nearly half this agricultural land is in water bodies, used extensively for aquaculture. Only about 42% of agricultural land is under a formal irrigation system, with 60% of the land using water from reservoir ponds and lakes, many of which are polluted by wastewater as explained earlier. An extensive drainage system exists (because of the low-lying nature of the land) and the usual method in Vietnam is that drainage water is also recycled for irrigation in other parts of the system during the dry months of the year. In the wet months, the drainage water has to be pumped out to the rivers to limit flooding of land. The system comprises numerous irrigation and drainage pumps to regulate it. Four of the waterways serving irrigation show high levels of pollution (CEETIA, 1996). The entry of wastewater from the city drainage systems into the irrigation drainage system is therefore not uncommon and much of the urban agriculture both in the formal and informal system use polluted water.

In the formal irrigation system, water is pumped into the irrigation network whence the water flows by gravity to the plots. In the areas without formal infrastructure, the farmers utilize the numerous ponds and lakes that dot the landscape directing this water mainly under gravity to their plots. Some of the canals and waterways are also used for harvesting aquatic plants. For aquaculture, every household has a pond where fish are bred. These ponds also use polluted water in many cases.

31.4.8 Crop production systems and land tenure

In Yaounde, vegetable cultivation and horticulture were the main production systems. The most frequently grown vegetables were the indigenous leafy vegetables. Other vegetables like salad, leeks, ladies fingers (okra), and basil were grown by a few farmers (less than 20% of the sample) as these have a more limited market and are generally subject to parasite attacks. In the rainy season however as much of the lowland areas are flooded, there is a shift from leafy vegetable production in inland valleys to rain-fed mixed crop systems, and salads become commercially more viable.

Of the farmers, 96% were producing for commercial purposes (both vegetables and flowers) and only 4% for exclusive family consumption. Even the commercial farmers consumed a portion of their own produce. Family consumption amounted to 18% of the total production from these systems.

The average area cultivated was reported as 0.13 ha per farmer (Endamana et al., 2003). The most frequently cited modes of land tenure rights in Urban and Peri-urban Agriculture (UPA) were squatting (30-40%) on public land, and borrowing or renting of land (around 40%) from customary landowners. Around 20% of farmers had inheritance rights to their land. Less than 10% had other forms of agreement like sharecropping (the tenure distribution for UPA was extrapolated to the case of wastewater agriculture as well, as it probably exhibits a similar pattern).

In Kathmandu valley, the production system varied from the wet season to the dry season. In the wet season all the land is put under cultivation with rice, maize and beans and some

[2] Note that the boundaries for a city were defined locally and therefore varied between countries.

vegetables. In the dry season only 75% of the land is cultivated mostly with leafy vegetables, beans and potatoes and some wheat.

Ownership of land is different in Kathmandu, with 75-85% of farmers owning their plots and the rest renting them. The farmers using wastewater are originally from the valley, not migrants, and have strong property and tenancy rights. In the case of sharecropping systems, the tenants get half the harvest. They have property rights as well in case the landlord wishes to sell the land which now has high value due to urbanization. They have to approve of the sale, and receive monetary compensation or sometimes half of the property. All farmers are smallholders. The average size of a landholding is below 0.2 ha.

In Hanoi, the land is owned by the commune that distributes it to households following the principles of equitable land distribution after land reformation changes in 1986. Households were given land based on "good with bad" and "high with low". Hydrologically the land is differentiated as "high", "moderate" and "low" elevations, which is relative to its particular location within the communal land areas designated for agriculture; and this designation is not directly linked to a particular quality of land. Given the low capacity of water control facilities, irrigation and drainage conditions correlate to land elevation. The high elevation land has good drainage. This land, depending on soil type, might permit the farmers to grow (in addition to rice) subsistence crops such as vegetable, potatoes and maize. However, it suffers from insufficient hydraulic head, thereby forcing farmers to irrigate manually at certain times. The moderate elevation land has enough water, thus it is good for rice but, due to its limited drainage capacity, it is not suitable for subsistence crops. Low elevation land suffers from inundation due to poor drainage; this sometimes causes crop damage (Doan and Satoh, 1998), particularly if the quality of water is poor.

Each household could own anything between 3-11 plots and sometimes up to 15. In terms of farm area, this would be equivalent to a total of 0.2-0.3 ha per household. Rice is the main crop in the two seasons but more rice is cultivated during the summer season due to rainfall and availability of water. Additionally in spring, maize and soybean are grown. Depending on quality and location of the plot, vegetables like beans, tomato, cucumbers, chillies, squash, potatoes and maize may also be grown. Rice was cultivated under flood irrigation conditions but manual watering with cans and sometimes drip irrigation systems were used for cultivating vegetables and other field crops.

Though unrelated to climatic conditions, land tenure is discussed here because it is an important factor influencing wastewater crop production systems and therefore for effecting change. Three distinct systems of tenure exist in the three cases. Where farmers were not the owners of land (as the majority in Yaounde) but had some form of temporary tenancy rights only, one cannot expect farmers to invest in improvements that require infrastructure changes. In the case of Nepal, land is clearly owned by the farmers and even in the case of tenants, their rights are very strong. Thus, farmers would feel responsible for their land and make the necessary investments for improving the quality of their land and the water they use for irrigation. In Hanoi, the political context plays a strong role, in that the commune is the ultimate "owner" of the land. As such, the commune can play a strong role in influencing change.

One more factor affecting crop production in these systems is pesticide and fertilizer use. In dry climates, even the use of raw sewage is more for its water value than for the nutrients it contains, and thus influences pesticide and fertilizer application patterns; in humid areas the water being used is simply polluted water (or diluted wastewater) with low levels of nutrients, and fertilizer and pesticide application and use is nearly the same as with clean irrigation water[3]. Use of raw sewage where nutrient concentrations are higher and therefore requiring lower inputs of chemical fertilizer (Van der Hoek *et al.,* 2002), encourages leaf and seed growth consequently leading to a higher need for pesticide application. This is not the case with wastewater use in humid climates.

31.4.9 Gender and social diversity of users and related impacts

In all three cities urban farming involves women and men, and therefore by extension so does wastewater farming. In Yaounde, and in the Kathmandu valley, both men and women are involved to nearly equal degrees whereas in Hanoi there were more women involved in farm tasks (men were likely employed in commerce and industry as well given the high level of such activity in Hanoi). A rapid global assessment across 53 cities showed the common pattern that 57% men and 43% women were involved, with women either helping on their husbands' or family plot, or being farmers in their own right. The farm tasks were the same (land preparation, planting, weeding, watering, harvesting) and incomes gained by women were either for their own use or for family use depending on their marital status. Hired labor was also used and if women were farm owners, heavy work like watering with cans, was undertaken by men who might be hired labor. When helping in farm activities, tasks like weeding which are labor-intensive were given to women. They were also the main persons involved in sorting, storing, and washing and selling of farm produce. Additionally in Hanoi, women were involved in carrying blackwater/nightsoil to the fields, as this is a common feature of agriculture, though less applied now in urban areas possibly due to closed sanitation systems.

In Yaounde, wastewater users were from three specific ethnic groups and most of the farmers were from the urban poor. These groups were immigrants from other geographic areas of Cameroon who had come to Yaounde in search of employment. Some of these farmers have other activities to complement their revenue but 92% of the farmers have earnings below the poverty line (ECAMI, 2002).

In Katmandu, the indigenous valley population belongs to one ethnic group, who mostly owns the land and farms it. Urban and peri-urban land value is high and because of the strict tenancy laws which favor the tenants, landowners do not easily rent the land to migrants. Land which is being cultivated now with wastewater previously used the same sources of water which were then clean, so in that sense this land is in no way marginal. It is difficult therefore to categorize the farmers as being from a particular economic class as these were the original valley inhabitants and there is a mix of all income groups amongst them. Some of the tenants who farm small plots of land of less than 0.5 ha are considered poor by the valley standards.

In Hanoi, men and women are involved in agriculture and though women-headed households do exist, usually women work in the fields with their men folk. They are involved in the same activities as the men in general though weeding and harvesting are considered female tasks while land preparation is considered male. The social system in Hanoi and the attribution of land does not allow for large inequities - the land mainly belongs to the household and there is very little renting of land. Use of wastewater is therefore not an activity of marginalized population groups in urban areas in these climates but part of the normal agricultural fabric. Land fed by wastewater is usually not popular among the farmers but ownership is not related to a particular social group.

---

[3] Raschid-Sally L, Doan Tuan D and Jayakody P. Preliminary findings from ongoing field research in Nam Dinh, Vietnam.

Farming using wastewater under these climatic conditions particularly in the lowland areas with higher probability of inundation, does increase women's exposure to wastewater compared to a drier climate. The presence of insect vectors that transmit malaria and filariasis is higher in these lowland areas where water collects, and about 59% of farmers complained of malaria, though this could not be attributed to the presence of wastewater. Nevertheless pooling of wastewater as is common in low-lying areas is an additional risk of exposure to waterborne and vector-borne disease.

Women are also key players in the contamination pathway as they may infect the larger family through their involvement in food preparation activities, if there is poor personal hygiene. Children in particular are in greater contact with women, since they are in the care of women. If women and children are infected with worms, there is a higher likelihood of contamination at household level through contact with fecal matter coupled with poor hygiene.

Like land tenure, social diversity is unrelated to climatic conditions but is a factor which has to be considered in dealing with wastewater use in the developing world. Clearly, in Hanoi and Kathmandu, the social class represented is not a migrant class as is portrayed in Yaounde. In dealing with farmers who have owned or farmed the land for generations, offering alternatives may not be very well received and this must be factored into policy recommendations.

31.4.10 Preferences and perceptions of users

Users of wastewater in the three cities have to face the problem of inundation of fields during the rainy season and consequent losses/reduction of crop yields. In Yaounde, the land being used for cultivation is the low-lying land in the inland valleys where all the wastewater drains. In the Katmandu valley, some of the lands closest to the river are subject to flooding. Although this is not a new problem it is compounded by the poor quality of the water – farmers are unhappy to have no alternative but to irrigate with "dirty water".

The positive view that wastewater is a reliable source of water, additionally containing nutrients, is dependent on the ability of farmers to control the application of this source of water. Under the climatic conditions described here, this is therefore not the common perception. In areas where the original water sources now resemble wastewater, this water is not always appreciated by the farmers, particularly in the rainy season (Rutkowski, 2004).

Vietnam has a long history of applying human excreta to farm-plots and emptying night-soil into fishponds as fish feed. Nevertheless in Nam Dinh, a small city south of Hanoi, farmers who had plots of land close to the wastewater drain complained about being forced to use it when irrigation authorities directed wastewater into the irrigation system that served their plots, at times of low flow in the canal system (Raschid-Sally *et al.*, 2006). This negative response may be due to the perception amongst farmers that "wastewater" often contains industrial chemicals and is therefore dangerous.

Certain agricultural areas in Hanoi receive the city's wastewater through the network of drainage canals and rivers; the geographic location of the city in the delta and the climatic condition of high rainfall are factors which contribute to extensive use of wastewater in agriculture.

In Yaounde about 70% of farmers were aware of the health risk associated with wastewater use but only about 40-60% took any precautions to protect themselves from exposure by wearing footwear (boots). This was also not popular in hot weather when open shoes were used more as protection against stones and rough surfaces rather than against the water. However, all of them wash their hands and feet for personal hygiene which affords some protection. In some urban farming locations, about 44% suffered from skin rash and itching (possibly linked to the constituents of the wastewater but this was not verified) which the farmers clearly associated with wastewater - but ill health from consuming the produce was not attributed to the use of wastewater. They consume their own produce as well showing that they have no reservations about the quality of the produce. Farmers wash produce before sale but more for purposes of cleaning the soil and dirt and keeping it fresh than for washing off potential contamination. The same sources of water are used ,particularly when they do not see it as contaminated, as is the case with many streams and rivers in these climates which receive wastewater but have a "clear/clean" appearance. The sense of sight and smell are the parameters used at the field level to judge the quality of water.

The pattern is replicated in Katmandu and Hanoi. In Katmandu data showed that 90% of farmers were aware of health risks but took no protection measures except to wash exposed parts of bodies with soap and water and use creams and mineral oils as added protection. It was estimated that only 5% used some form of protection. At least 50% of the farmers interviewed in the study complained of skin infections. An ongoing study in Nam Dinh, Vietnam (Raschid-Sally *et al.*, 2006) found similar results, which are likely in the case of situations such as those in Hanoi. In Hanoi in particular, exposure of farmers (men and women) to polluted water may be higher due to their immersion when cultivating and harvesting aquatic vegetables.

The same sources of water used for cultivation were used to clean the produce for the reasons cited above. Consumers and vendors did not usually distinguish between the source of products, one reason being that they were unaware that wastewater was used for cultivation of the produce they were consuming. In recent times with the publicity given to health food, consumers make a distinction between produce from "organic" and regular farming. It is ironic that produce cultivated with a good dose of domestic wastewater would technically fall into the category of organically grown food as no (or less) chemical fertilizers would have been used! The current preoccupation of consumers is the level of pesticides found in produce.

Interestingly, in Hanoi, where the use of polluted water sources for vegetable production is widespread, farmers did not complain about industrial contamination and the market did not differentiate between clean and wastewater produce. A separate study (Phuong Anh *et al.*, 2004) showed that people's concern about quality in Hanoi centered on appearance and other sensory aspects (color, shape, size, freshness, insect attack) and taste (succulence, firmness). The quality of produce from a biological and chemical contamination viewpoint is not usually questioned and only the awareness created in recent years on the effects of pesticide residues has raised this concern in the eyes of the public.

31.5 CONCLUDING REMARKS

Urban expansion and increased use of water has generated large volumes of wastewater which need to be disposed of. In humid climates with their perennial waterways, authorities have always in the past taken advantage of these natural drainage pathways to evacuate urban storm runoff and city wastewater. Presently, this practice persists compounded by the poor sanitation conditions and insufficient resources for wastewater management. Thus the term "wastewater irrigation" as associated with urban and peri-urban agriculture under these climatic conditions refers to the use of water from perennial streams (which have always served multiple purposes including irrigation), which are now polluted through a mixing with city wastewater. In these cases it seldom describes the deliberate use of raw sewage. In high rainfall areas therefore, farmers do not in general choose to use wastewater; rather its use is imposed upon them due to the nature of pollution of surface waters used as irrigation sources.

Also in these areas, water sources dilute the pollution but at the same time hide it. The water does not appear contaminated and therefore people do not avoid it in the same way they would undiluted wastewater. All too often children play in it. Around the world, many water sources which were originally clean and used for irrigation are now polluted and the consequences for irrigated agriculture can be significant.

The impacts of such "dispersed" pollution on agriculture are more difficult to control, unlike in dry climates where wastewater rarely mixes with fresh water sources. Legislation and norms on water quality for agriculture for these countries have to be reviewed as many of these perennial sources (which are considered clean) have higher levels of pollution than those recommended by international guidelines. Similarly, contamination from industrial pollution being dispersed through such natural drainage systems can be critical for agriculture.

In arid regions, the situation is very different where undiluted wastewater and sewage is used because farmers seek the wastewater for its water value, although the higher concentration of nutrients in wastewater can reduce the application of additional chemical fertilizer significantly. In humid climates there is no difference in the use of chemical fertilizer between wastewater and clean-water farmers.

The type of crops grown in these production systems is influenced by the presence of perennial water sources and the climate. For instance, in dry areas, vegetables are the main crop grown using wastewater, with some cereal and fodder crops, whereas in humid high rainfall areas, in addition to vegetables (which are grown for their high value) rice is more common. In such a case, the manner in which chemical contaminants mobilized in water affect the crop and the transmission pathways become critical. An example is the rice production system where cadmium contamination of rice grains may be an issue if cadmium pollution of water and soil occurs.

Many of the areas using wastewater are prone to flooding in wet weather, and farmers have little control over water application to crops. This results in lower potential benefit from the "positive" side aspects of pollution such as nutrients in wastewater.

The quality of agricultural lands prone to flooding is further affected when the soils and crops are subject to constant immersion in polluted water. Economic losses due to loss of harvest and ill health of farmers from exposure during farming are other impacts to contend with. Farmlands in such river valleys are not of recent origin and have been cultivated for generations. These farmlands have been cultivated precisely because of the quality of the soils. The loss of productive land due to wastewater immersion is an issue in these climates.

Whilst both women and men are affected under these climates, the exposure of women due to the type of tasks they undertake, and the impacts this may have for the family at large requires additional consideration. In these humid areas, it is surmised that risks from other types of waterborne disease may also be higher though no comparative work has been done on this aspect. Exposure through immersion is more likely where flooding occurs and this is of particular concern in Hanoi where aquatic vegetables are cultivated.

In the lobby to recycle wastewater in order to release fresh water for beneficial purposes, the arguments that will hold sway are different to those used for arid climates. Under climatic conditions where sufficient water is available, the value of wastewater as a water source and its reliability year round are not the reasons that would push these countries into recycling their wastewater in agriculture. From this perspective, internalizing the costs involved in treating wastewater because it has value for the farmer as a source of water is false. Rather it would be the reduction of environmental pollution and health risks that should be considered in any evaluation.

3.6 ACKNOWLEDGEMENTS

This report was funded in part by a grant from the Government of the Netherlands to the Comprehensive Assessment of Water Management in Agriculture.

3.7 REFERENCES

Buechler, S., Gayathri Devi, M. and Raschid, L. (2002) Livelihoods and Wastewater Irrigated Agriculture along the Musi river in Hyderabad City, India. *Urban Agriculture Magazine* No 8 December 2002.

CEETIA (Center for Environmental Engineering of Towns and Industrial Areas) (1996) Report on survey and evaluation of environmental pollution caused by industries in Hanoi city,Vietnam.

Diaw, M.C. (1997) Si, Nda Bot et Ayong : culture itinérante, occupation des sols et droits fonciers au Sud-Cameroun. Forest Policy and Environment Group, ODI.

Department of Health Services (DOHS) 2003, Teku Hospital Outpatient Records, Ministry of Health, Katmandu, Nepal

ECAMI (2002) Enquête Camerounaise auprès des ménages.

Green, H.M. 2003. The Effects of Carpet Dye on the Bagmati River. Masters of Engineering in Civil and Environmental Engineering Thesis. Massachusetts Institute of Technology, Cambridge, MA, USA.

Lami, 2002. Evaluation de la pollution due au rejet des déchets liquids et solides à Yaoundé et mise au point d'une carte de pollution. Mémoire de fin d'étude, ENSP, 65 p

Phuong Anh, M.T., Ali, M., Lan Anh, H. and Ha T.T.T. (2004) Urban and pri-urban agriculture in Hanoi – Opportunities and constraints for safe and sustainable food production. The World Vegetable Center (AVRDC) Technical Bulletin no 32

Van der Hoek, W., Mehmood ul Hassan., Ensink, J.H.J., Feenstra, S., Raschid-Sally, L., Safraz Munir, S., Aslam, R., Hussain, R. and Matsuno, Y. (2002) Urban wastewater: a valuable resource for agriculture, a case study from Haroonabad, Pakistan. Research Report 63, IWMI, Colombo, Sri Lanka.

Do Thi Hoa, Le Thi Phuong, Do Manh Cuong, andTran Hoang Tung. (2000) Nutrition and helminth egg contamination status of the children in 2 primary schools in Hanoi suburbs. *Journal of Practical Medicine*, 6-2000. Hanoi. pp. 43-48 (in Vietnamese).

Doan Doan Tuan and Satoh, M. (1998) An analysis of land redistribution and its impact on agricultural practices in the rural communities in the Red River Delta, Nothern Vietnam. *Trans. of JSIDRE*, No 196, pp. 125-139 (1998.8)

Endamana, D., Kengne, I..M., Gockowski, J., Nya, J., Wandji, D., Nyemeck, J., Soua, N.N. and Bakwowi, J.N. (2003) Wastewater reuse for urban agriculture in Yaounde Cameroon: opportunities and constraints. In Butterworth, J.A., Moriarty, P.B., & Van Koppen, B. (2003). *Water, poverty, and productive uses of water at the household level: practical experience, new research, and policy implications from innovative approaches to the provision and use of household water supplies.* Proceedings of an international symposium held in Pretoria, South Africa, 21 – 23 January 2003. Delft, the Netherlands, IRC. [on line] Available at www.irc.nl/prodwat

Environment and Public Health Office (ENPHO). (2003) *South Asian Trans-boundary Water Quality* Monitoring (SATWQM) Data. Kathmandu, Nepal. www.enpho.org

Raschid-Sally, L., Bradford, A.M. and Endamana, D. (2004) Productive use of wastewater by poor urban and peri-urban farmers: Asian and African case studies in the context of the Hyderabad Declaration on Wastewater Use. In Butterworth, J.A., Moriarty, P.B. and Van Koppen, B. (eds.) *Beyond Domestic: Case studies on poverty and productive uses of water at the household level.* Technical Paper Series 41, IRC International Water and Sanitation Centre, Delft, The Netherlands, 2004 95-110.

L.Raschid-Sally, Tuan, DD and Jayakody P. (2006). Impact of wastewater use on farm household in Nam Dinh Vietnam. In 'Wastewater reuse in agriculture in Vietnam: water management, environment and human health aspects. Workshop Proceedings', May 2006, CPIM, VIWRR, Hanoi Vietnam.

Rutkowski, T. (2004) "Wastewater Irrigation in the Kathmandu Valley", MSc Thesis submitted to the University of Colorado at Boulder, in partial fulfilment of Master's Degree in Environmental Engineering.

Rutowski, T., Raschid-Sally, L. Buechler, S., 2007. *Wastewater Irrigation in developing countries – two case studies form the Kathmandu Valley in Nepal.* Agriculture Water Management 88 (2007) pp.83-91

Scott, C.A., Faruqui, N.I. and Raschid-Sally, L. (eds.) 2004. *Wastewater Use in Irrigated Agriculture: Confronting the Livelihood and Environmental Realities.* CABI Publishing, Wallingford, U.K.

Van der Hoek, W., Nguyen, Van D., Konradsen, F., Cam, P.D., Thi Viet Hoa, N., Toan, N.D. and Cong, L.D. (2003) Current status of soil transmitted helminths in Vietnam. *Southeast Asian Journal of Tropical Medicine and Public Health* Vol 34 1:1-11.

32

Case Studies in Middle Eastern and North African countries

Akiça Bahri

The diversity of physical, technical and environmental conditions, institutional setting, implementation approaches, and regulatory aspects in the Middle East and North Africa are addressed in the following 17 case studies, by territory.

32.1 ALGERIA

Algeria suffers a significant lack of fresh water. Rainfall varies from over 2000 mm/yr on the heights near the Mediterranean coast to less than 100 mm/yr in the Northern Sahara, with 50% of the runoff flowing into the sea or the chotts (saline lakes). The country is confronted by several water issues including increasing demand, pollution and salinity problems, drought, damaged distribution and sewerage networks, dam siltation and wastewater treatment facilities out of order. Algeria is presently looking at improving water availability (361 m^3/*capita*.yr) by adopting a new water policy and strategy and new alternatives such as desalination and water reuse that will enable the crisis to be eased. Treated wastewater represents a promising alternative that is not only constantly available but also increasingly available, with the development of cities, tourism and industry.

The volume of wastewater disposed of in Algeria in 2004 was approximately 600-700 Mm^3 and is expected to reach 1500 Mm^3 by 2010. Biological treatment is provided to meet environmental discharge requirements. Of the 45 existing wastewater treatment plants, only 15 are operating, and these treat a total of 182,000 m^3/d. They provide 6.3% of the national sanitation needs and 30% of the actual treatment capacity. The majority of the wastewater treatment plants are out of use because of poor O&M conditions. Currently, around 600,000 m^3

©2008 IWA Publishing. *Water Reuse - An International Survey of current practice, issues and needs* by B. Jimenez *et al.* ISBN: 9781843390893. Published by IWA Publishing, London, UK.

of raw wastewater is dumped daily directly to the sea or to wadis. Reuse is considered but there is not yet a planned water reuse policy.

A National Sanitation Agency was set up in 2000 to operate wastewater treatment plants. An ambitious sanitation program has been initiated by the Algerian authorities in the last few years (1) to rehabilitate the existing infrastructure (28 WWTP), (2) to build new WWTP for large (32) cities of more than 100,000 inhabitants (5 plants are under construction) and (3) to construct wastewater stabilization ponds (8) and stabilization ponds or sedimentation basins (435) for smaller cities. The Water Law in Algeria prohibits the use of raw and treated wastewater for the irrigation of raw-eaten vegetable crops, but allows it for the production of fodder crops, pasture and trees. Authorization is required for the reuse of non-conventional waters (Angelakis *et al.*, 1999).

32.2 BAHRAIN

Bahrain's natural resources are under serious threat due to rapid economic development and urbanization, high population growth rate and restricted space. Total annual surface runoff is negligible at only 4 Mm^3 and groundwater (recharged at about 110 Mm^3/yr by underflow) has been declining in quality and quantity due to overabstraction. A high percentage of the water supply comes from the desalination of seawater (three desalination plants with a total capacity of 160,000 m^3/d). Effort is being made to control and conserve available water resources. Annual water consumption is estimated at 300 Mm^3, of which 57% is used in the agricultural sector, 40% for domestic purposes, and 3% in the industrial sector.

Around 87% of the population is connected to the sewerage network and it is anticipated that this will reach 95% by the year 2020. The estimated annual volume of wastewater in Bahrain ranges between 84 and 103 Mm^3. The country has a wastewater treatment capacity of more than 200,000 m^3/d (Al-Zubari, 1997). There are 11 WWTP in Bahrain, most of them small packaged types, which use trickling filters or activated sludge processes.

Out of the total treated wastewater (154,000 m^3/d), only 23% receive tertiary treatment and is reused for fodder crops, gardens and highway landscaping, whilst the rest is discharged into the sea. The newest technologies such as ozonation are being employed to ensure reliable availability of the intended production of 70 Mm^3/yr (192,000 m^3/d) of treated effluent. Reuse of treated effluent has been tried and successfully practiced for about 20 years. Groundwater is being substituted with treated effluent for irrigating farms in the western agricultural areas. Aquifer recharge of treated wastewater is under study.

Strict regulations regarding agricultural reuse ($<$ 2.2 coliforms/100 mL, $\leq$ 1 helminth/L) are applied and periodical quality control tests are conducted. Comparison of costs of seawater desalination (US\$ 0.67/m^3 using multistage flash technology), brackish water desalination (US\$ 0.61 to 0.88/m^3 using reverse osmosis technology) and wastewater treatment in Bahrain (US\$ 0.24 to 0.84/m^3 without and with RO unit) shows that wastewater reclamation can be cost-effective (UNEP, 2001).

32.3 EGYPT

With a population estimated at 73 million in 2004, Egypt is the most populated country in the region. With an average growth rate of 1.8% per year, it is expected to reach 119 million by the year 2025. The major part of the country is desert (about 97%) and rainfall is negligible except along the Mediterranean coast with less than 200 mm/yr. The main water resource is the River Nile (with an annual inflow of 55,500 Mm^3), which contributes about 82% of the available water from different resources. Fossil groundwater is exploited in a small area in the western desert. Egypt is a unique country that depends mainly on irrigation and 84% of the Nile River is used by the agricultural sector. Approximately 45% of its population is urban, of which 80% have access to safe water and 20% to sanitation facilities. Half of the rural population has access to safe water, but few have access to sanitation facilities. Seventy-five per cent of the population is connected to the sewerage system but only 19% of the collected wastewater is secondary treated. Twenty-five per cent of the remainder is partially treated and 60% is discharged raw via open canals to the Nile River or to the Mediterranean Sea.

The water strategy in Egypt includes wastewater reclamation and reuse. Domestic wastewater receives either primary or secondary treatment. At present, the volume of untreated wastewater is estimated at 5.4 Mm^3/d (1,955 Mm^3/yr). The number of treatment plants is increasing from 22 treatment plants in 1992 treating about 1.8 Mm^3/d (650 Mm^3/yr) to a potential of 123 plants in 2008. Approximately, the volume of treated effluent was about 1.9 Mm^3/d (700 Mm^3/yr) in 1995-96 and 2.6 Mm^3/d (940 Mm^3/yr) in the year 2000. Cairo has six treatment plants (three secondary treatment plants and three primary plants) operating, or under construction, with a total design capacity of 2.85 Mm^3/d. The total discharge from Cairo is 1,000 Mm^3/yr. Two major treatment plants (a waste stabilization pond and an activated sludge treatment plant) have recently been completed and are operational for Cairo West and Cairo East. Alexandria, the second largest city, has a new activated sludge treatment plant with a capacity of 1.3 Mm^3/d. Most of the major cities will have wastewater treatment plants with priority given to coastal and tourist cities. The amount of treated effluent should reach 13.7 Mm^3/d (5,000 Mm^3/yr) in the year 2020, which can irrigate about 210,000 ha. Irrigation with treated effluent will take place in the Sinai, the Northwest Coast, and on both sides of the Nile Valley, from Aswan to Cairo (Arar, 2003). Sewage sludge is also used in agriculture as organic fertilizer and soil conditioner.

Wastewater reuse has long been practiced in Egypt. Since 1900, wastewater has been used to cultivate orchards in a sandy soil area around Cairo in a farm covering 1515 ha at El-Gabal El-Asfar planted with citrus (630 ha), palm (30 ha), pecan (34 ha), and forest trees (260 ha), fava beans and maize (294 ha), a nursery (1.7 ha) and 264 ha of bare land. The area has gradually increased to about 4,500 ha. Wastewater has also been used directly in several other areas in the country such as Abu Rawash (Giza) and El Salaam city, and indirectly through discharges to the Nile, to irrigation canals in the delta area, or to agricultural drains. A pilot project was launched in 1980 in the Abu Rawash area (Giza) using primary treated effluent to irrigate 36 ha of citrus and other trees such as Eucalyptus and Casuarina. Near Sharm El-Sheikh, citrus trees have been irrigated with effluents from stabilization ponds. Similar projects are being considered for the major cities and some villages in the Sinai area. A total of 42,000 hectares are presently irrigated with treated wastewater or blended water. The present policy is to reuse effluents for irrigating and producing timber trees and green belts planted in the desert to fix sand dunes, but not for any other crops like fruits, vegetables, and field crops. Farmers either flood or spate irrigate their fields directly from canals. Drip or sprinkler irrigation systems are used in Ismaïlia (Figure 32.1) to irrigate commercial trees and mulberry trees for silkworm raising and silk production with stabilization ponds' effluents (80 000 m^3/d).

Several problems are affecting water reuse and negatively impacting both health and environment in Egypt (Shaalan, 2003) such as (1) lack of infrastructure to treat the amounts of wastewater produced, (2) low number of sewerage connections of the urban and rural populations, (3) direct discharge of a significant volume of untreated municipal and industrial wastewaters into water bodies, and (4) overloaded and/or not properly maintained existing wastewater treatment facilities.

Figure 32.1. View of Ismaïlia's aerated ponds (80,000 m3/d) and of the reuse scheme with spray- and drip-irrigated forest trees.

Seven ministries and several institutions are involved in wastewater reclamation and reuse with unclear definition of responsibilities and limited coordination among them. Ministries with responsibilities for water management and protection are the Ministry of Water Resources and Irrigation (MWRI) and the Ministry of Health and Population (MHP). The MWRI authorizes surface water extraction from the Nile, canals, and drains and also controls the discharges into the Nile and irrigation canals. It is also in charge of policies and procedures for improved urban wastewater discharge and reuse (2001). The MHP is responsible for quality control of the drinking water supply and for establishing standards for drinking water quality and wastewater discharges. The Ministry of Agriculture and Land Reclamation controls crop selection and defines potential sites for reuse.

There are not yet specific regulations dealing with water reuse but comprehensive guidelines are under development for managing reclaimed water, focusing on restricted irrigation of crops such as wood trees, palm trees, citrus, pomegranates, castor beans, olives, and field crops, such as lupines and beans. Water pollution from municipal and industrial effluents is mainly regulated by three laws: Law No. 48/1982 for the protection of the Nile River, Law No.12/1984 on irrigation and drainage, and Law No. 4/1994 on environmental protection. Law No. 48/1982 requires secondary treatment of effluent prior to discharge into the Nile River and irrigation canals and specifies limits for maximum allowable fecal coliform of 5,000/100 mL and BOD of 60 mg/L. The 1984 martial law regulation prohibits the use of treated effluent for irrigating vegetables and the use of effluent for crops unless treated to the required standards for agricultural drainage water. However, the absence of a clear policy and action plan for wastewater management as well as standards that are difficult to enforce worsen the situation and favour the spreading of diseases (Shaalan, 2003).

32.4 IRAN

Covering a surface area of 165 million ha, Iran has a population of almost 70 million which has tripled from 1956 to 1996 (Mahmoodian, 2003). The annual average rainfall is around 251 mm and varies across the country from 1400 mm in the north to less than 50 mm in the Central Desert. The potential renewable water resources are about 137,500 Mm^3/yr (currently 88,500 Mm^3 are mobilized and around 91% allocated to agriculture). More than 60% of the population is urban and the number of cities with more than one million inhabitants is expected to increase to 11 and 19 in the years 2010 and 2020 respectively.

At present, around 19,000 Mm^3 of agricultural, industrial and urban effluents are discharged to surface resources and around 10,000 Mm^3 to groundwater resources whereas the total of treated effluents and returned waters remains negligible compared to the potential. The volume of municipal wastewater generated by the urban and rural areas and susceptible to be collected and treated is equivalent to 5.09 Mm^3/d (1,859 Mm^3/yr) and is expected to reach 16.2 Mm^3/d (5,900 Mm^3) in the year 2020 (USEPA 2004).

Although by the end of 2000, there were many wastewater projects either planned or implemented in several towns, in most urban areas (with the exception of Isfahan and some others), raw wastewater is still discharged through cesspits. Some cities have wastewater collection systems. At present, 39 wastewater treatment plants with a total capacity of 712,000 m^3/d (i.e. about 260 Mm^3/yr) are operating throughout the country. The city of Isfahan generates 64% of the total outflow. This city of 1,374,000 people is the second largest in the country and the first to have a full wastewater collection system nearly 30 years ago. Treated wastewater from the city of Isfahan and from other urban areas is discharged into the river before indirect reuse in agriculture or groundwater recharge. Seventy-nine wastewater treatment plants with a total capacity of 1.9 Mm^3/d are under construction and 112 treatment plants with a total capacity of 1.6 Mm^3/d are under study for completion by the year 2010. Accordingly, by 2025, the volume of treated urban wastewater will reach 10.5 Mm^3/d (3,800 Mm^3/yr).

In some cities without WWTP, raw wastewater is actually either discharged directly to water courses, or used for irrigation, resulting in some health related problems (Mahmoodian, 2003). Groundwater recharge is also taking place, leading to signs of pollution such as high levels of nitrates in the underground waters.

Only about 5% of the municipal wastewater is reused via planned projects (USEPA, 2004). Farmers are considering treated effluents as an irrigation water resource and primary effluent is used for irrigation in the suburban farms, mainly for fodder crops such as corn, millet and alfalfa.

The government recognizes the value of reclaimed water and provides financial incentives in an attempt to encourage water reuse (Shanehsaz *et al.*, 2001). Research studies examining the use of treated wastewater for agricultural irrigation have been carried out. They have shown increased yields and reduced fertilizer consumption due to the nutrient content of the effluents. However, no water reuse policy has yet been set up in Iran and responsibility and authority for reuse is fragmented over several institutions with no or limited coordination. The following ministries and organizations are involved in water and wastewater management: The Ministry of Energy and the Regional Water Boards, the Ministry of Agriculture and Jihad, the Ministry of Health, Medicine and Medical Education, the Ministry of Industries and Mines, and the Department of the Environment and the Environment Protection Organization. Actions need to be taken to restrain industrial pollution and prevent raw wastewater application. The legal, regulatory and institutional frameworks have to be established. There is no specific national standard for the reuse of treated wastewater. The "Effluent Discharge Standard" developed by the Department of the Environment in 1994 is the only existing wastewater code that sets effluent discharge limits to surface waters, soak pits, and agricultural irrigation. This standard does not define any criteria to use treated effluent for industrial, fisheries and recreational activities. International guidelines developed by the WHO and by the USEPA are currently used to regulate water reuse.

32.5 IRAQ

Iraq's population is estimated at 26 million and should reach 49 million by the year 2025 (average growth rate of 3.1% per year). Before the conflict, agricultural production in Iraq was limited by the water flow of the rivers Euphrates and Tigris. Therefore, considerable efforts have been given to use other resources of water such as drainage and groundwater for the production of cereal and vegetable crops in the desert and in regions where no fresh water is available. Due to expanding agriculture and a growing population in Iraq, wastewater has been taken into consideration, as an additional resource, for expansion of irrigated agriculture and the saving of fresh water for other social and industrial purposes. It was envisaged that treated wastewater could be used as an integral part of water management strategy in Iraq and programs for water reuse in irrigation were to be initiated.

Of the total population, approximately 71% live in urban areas. The wastewater collection and treatment system mainly serves Baghdad and reaches approximately 80% of the population. Only 9% of the urban population outside Baghdad is connected to a sewerage system. The rural population and the North of Iraq do not have piped sewerage systems. Iraq has 13 major wastewater treatment plants. Most of the plants consist of a conventional activated sludge process followed by chlorination. They were providing partial treatment before the conflict due to power supply shortages, lack of spare parts and consumables. None of them is presently operational as the recent conflict has damaged several facilities or led to degraded water quality. Raw effluents are being discharged into rivers and water bodies creating a serious environmental and health concern. Deteriorated sewers are contaminating potable water networks and underground water. Diseases related to unsafe water and poor sanitation are reaching alarming rates. Baghdad's three WWTP, which correspond to three-quarters of the country's wastewater treatment capacity, are inoperable. The waste from 3.8 million people is directly discharged without treatment to the Tigris River. The biggest two wastewater treatment plants built in Baghdad County were "Al-Rustumia" designed to handle an average flow of 204 Mm^3/yr, and "Al-Karkh" with an average flow of 150 Mm^3/yr (Salih, 2003). At present, neither guidelines nor a code of practice have been adopted for the use of wastewater in irrigation.

32.6 JORDAN

Jordan covers a land area of 88,945 km^2, with only 9% receiving more than 200 mm of rain annually. Total annual surface water resources are about 400 Mm^3/yr, in addition to 500 Mm^3 of groundwater (of which 143 Mm^3 fossil groundwater). Twenty-one per cent of total resources are transboundary waters. Irrigation uses about 75% of the total demand. Renewable water resources are seriously limited and are far below under the water poverty line of 1000 m^3/*capita*.yr. Jordan is one of the most water stressed countries in the Middle East and is one of the 7 poorest countries in the world in terms of water resources, with only 180 m^3/*capita*.yr. To meet future water demands of the growing population (5.6 million at present, 9.9 million expected in 2020, growth rate of 3.4%/yr), non-conventional water resources, including treated effluents, have to be considered as a major source of irrigation water (FAO, 2000).

In Jordan, the government's policy in the 1980s was to achieve the provision of improved wastewater collection, conveyance, treatment, disposal, and reuse systems. Therefore, urban sanitation coverage, including piped sewerage, was developed at a fast rate, with 80% of urban areas covered, i.e. about 63% of the total population. Most of the cities of Jordan were then equipped with wastewater treatment plants (the largest WWTPs, greater than 4,000 m^3/d, are As Samra, Baqa's, Wadi Arab, Irbid, and Madaba) and it was decided to treat wastewater up to the secondary level and meet WHO guidelines for the use of treated wastewater for irrigation (Tuffaha, 1996). The different treatment processes are stabilization ponds (6 WWTPs producing 88% of the treated effluents), activated sludge, or conventional or extended aeration, and a few trickling filters and rotating biological contactors. As Samra (Figure 32.2.) is one of the largest waste stabilization pond systems (181 ha) in the world located outside Amman with a design capacity of 68,000 m^3/d and inflow estimated at around 190,000 m^3/d. The actual trend in Jordan is to introduce more mechanical treatment plants. The replacement of the existing As Samra wastewater treatment plant to be constructed, operated and maintained according to a 25-year Build-Operate-Transfer (BOT) agreement was awarded in 2000 to a consortium led by Ondeo Degremont.

In Jordan, the total amount of wastewater generated is estimated at 493,000 m^3/d (180 Mm^3/yr). Approximately 112 Mm^3/yr are treated at the 22 existing wastewater treatment plants. Pre-treatment of industrial wastes is also implemented and it has reduced discharges of organic and mineral elements since regulations were issued in 1991. By the year 2020, when the population is projected to be about 9.9 million, the expected available volume of treated wastewater is estimated to be 726,000 m^3/d (265 Mm^3/yr), which is about 25% of the total water available for irrigation (McCornick *et al.*, 2002). Treated wastewater has long been recognized as a valuable resource for use in irrigation (UNDP *et al.*, 1992) and is considered an important water resource in the Jordan Water Strategy (Ministry of Water and Irrigation, 1997). Treated wastewater from Amman has been used for irrigation in the Jordan River Valley (JRV) since the completion of its first activated sludge treatment plant (Ain Ghezal) in 1968. The major proportion of reclaimed water is discharged to wadis and reaches reservoirs used for irrigation. Therefore almost 100% of the treated effluent is utilized in irrigation either directly at the outlet of the treatment plants or after being discharged into watercourses, most reuse being unplanned and indirect; it represents about 13% of the total water used for irrigation. Treated effluent provides a major resource potential for the country to meet the future needs of irrigated agriculture. Planned direct use of treated effluent is, actually, limited (2%) but on the increase since 1985. In order to encourage reuse, the government made it mandatory for all new wastewater treatment projects to include a fully designed and feasible reuse component. Investigations on the use of treated effluent have been carried out in several locations by the Ministry of Agriculture since 1997 to determine the effects of varying levels of treated wastewater on different soil types, different plant types (green fodder (ray-grass), oil seed crops (kanola), barley and wheat), agricultural workers, groundwater, etc. and have shown positive results. Reclaimed water from Kufrinjeh and Aqaba wastewater treatment plants are used for recharge of a salty fossil aquifer near the site in Kufrinjeh and a shallow, partly saline, aquifer in Aqaba.

Despite the widespread public support for water reuse, water reuse is giving ground to different debates about the best agricultural uses for the reclaimed water, whether treated effluent should be used near the point of generation in the highlands, or whether it should continue to support large agricultural users in the Jordan Valley. Another issue is the type of crop (wheat or fodder) to be produced based on its water consumption.

Figure 32.2. Overview of As Samra waste stabilization ponds and alfalfa irrigated with reclaimed water.

Treated effluent is, then, mainly reused for agricultural production. The irrigated area is around 13,300 ha of which 13,000 ha is located in the Jordan Valley, 150 ha in Aqaba, 50 ha in Ramtha, 40 ha in Madaba, etc. Except in the Jordan Valley, the treated effluent is used to irrigate fodder crops, palm trees, forest and fruit trees. Irrigation of golf course is planned.

As the springs feeding the Zarqa River have been used to supply the water needs of Amman, the Amman wastewater has partially compensated for the reduction in base flow of that river. The flow of effluent from the stabilization ponds of the As Samra treatment plant used to be blended with water from the King Talal Reservoir (KTR) and used for unrestricted irrigation downstream in the JRV. The treated effluent provides 15 to 80% of the annual inflow to KTR. Reuse of the effluent from the KTR permits winter flow in the JRV to grow one winter crop, and contributes to the irrigation of perennial crops. The area irrigated partially or completely by KTR water is, at present, approximately 13,000 ha (18% of the actual irrigated area) of which 75% produces vegetables, 15% cereals and 10% trees. The remaining area is used for growing fruit trees or field crops (wheat, barley, berseem, etc.). In the future, there could potentially be about 43,100 ha irrigated with treated effluent if all the water was to be used for irrigation purposes. In the central Jordan Valley, 56% of farms use drip irrigation (70% efficiency on average); the remainder use surface irrigation methods (76% efficiency on average).

The cost of wastewater treatment is 20-35 Fils/m^3 (US$ 0.03-0.05/m^3) for oxidation ponds, and 120-240 Fils/m^3 (US$ 0.17-0.34/m^3) for secondary treatment with mechanical methods. The tariff paid to the Jordan Valley Authority for distributing the water does not cover O&M costs of reclaimed water. Farmers located in the vicinity of WWTP pay US$ 143-286/ha.yr to the Ministry of Water and Irrigation and those using treated effluent mixed with fresh water into rivers, reservoirs, and canals pay up to 45 Fils/m^3 (US$ 0.07/m^3), the same price paid for freshwater. The JVA envisages increasing tariffs for conventional irrigation water in the Northern Jordan Valley, while tariffs for blended water in the Southern Valley would remain at their current level in order to compensate farmers for the higher salinity of the reclaimed water (World Bank, 2001). The Jordan Valley Authority provides free irrigation water to farmers during winter for salt leaching. Farmers cannot sell or lease their water rights.

Several institutions are concerned by wastewater treatment and reuse such as the Ministry of Water and Irrigation (Water Authority of Jordan and Jordan Valley Authority), the Royal Scientific Society, the Ministry of Agriculture (National Centre for Agricultural Research and Technology Transfer), the Ministry of Health, the Ministry of Environment, the Ministry of Planning, Municipalities, and the potential users. Involvement of the private sector through the BOT system for wastewater treatment and reuse is under consideration in Jordan.

The standards for wastewater treatment and reuse were introduced in 1982 by a Martial Law. In 1989, a more liberal version of the Martial Law was enacted. In 1995, the Jordanian standard 893/1995 listing 47 specific constituents and specifying the quality standards required for various wastewater uses such as irrigation of cooked vegetables, fruit and forestry trees, fodder crops, lawns and parks, groundwater recharge, and aquaculture or discharge to streams, wadis, and reservoirs was issued and the general criteria for these standards were also established. It prohibited irrigation of vegetables eaten raw. However, because of the potential impact of regulations on the export market of fresh fruit and vegetables, and the possibility of restrictions placed by importing countries (Saudi Arabia, EU) and in order to further promote the commercial viability, the reliability, safety, environmental sustainability, and social acceptability of direct water reuse, the government of Jordan, with support from USAID, revised the existing water reuse standards (Sheikh, 2001). The new standards have been enacted in 2003 under the title JS893/2003 (Government of Jordan, 2003) (see Table 32.1). They are more stringent. The irrigation of vegetables eaten uncooked (raw) with reclaimed water remains prohibited. Recharging groundwater aquifers used for drinking purposes is not allowed.

The Ministry of Water and Irrigation is conducting a detailed master planning project to maximize the water resources of the basin, including groundwater and reclaimed water and has adopted a new overall water strategy and new policy statements including wastewater management. High priority is given for rehabilitating and upgrading the existing treatment plants, for complying with the effluent water quality standards, examining opportunities for smaller communities, and exploring options for direct water reuse to replace freshwater sources in agriculture, industry and landscaping, with emphasis on replacement of groundwater used for irrigation with reclaimed water from future water reclamation plants (McCornick *et al.*, 2004). In the case of Aqaba, with the fast developing free trade zone and the new facility call for zero emissions of effluent into the Gulf of Aqaba, other reuse options such as irrigation of other crops in addition to the date palm trees, golf course irrigation, and urban landscaping are investigated. On the other hand, greywater reuse is presently being tested in Jordan.

Table 32.1. Revised standards of water reuse in Jordan (Government of Jordan, 2003).

		Allowable limits per end use			
Parameter	Unit	Cooked vegetables, parks, playgrounds and sides of roads within city limits	Fruit trees, sides of roads outside city limits, and landscape	Field crops, industrial crops and forest trees	Groundwater recharge
Biological oxygen demand	mg/L	30	200	300	15
Chemical oxygen demand	mg/L	100	500	500	100
Dissolved oxygen	mg/L	>2	-	-	>2
Total suspended solids	mg/L	50	150	150	50
pH	Unit	6-9	6-9	6-9	6-9
Cl_2 residual		0.5-1.0	-	-	0.5-1.0
Turbidity	NTU	10	-	-	2
Nitrate (NO3)	mg/L	45	70	70	45
Ammonia (NH4)	mg/L	10	-	-	5.0
Total nitrogen (T-N)	mg/L	45	45	45	30
Escherichia coli	Most probable number or colony forming unit/100mL	<100	<1000	-	<2.2
Intestinal helminth eggs	Egg/L	≤1	≤1	≤1	≤1

32.7 KUWAIT

Kuwait has a population of about 2.6 million, most of it living in five urban areas. It has experienced a rapid growth over the last four decades (<0.8 million in the 1960s), which in turn has increased the water demand for various activities. The country is arid with negligible rainfall (average annual rainfall less than 125 mm) and no surface sources. Ninety-five per cent of potable water comes from desalination (multi-stage flash distillation plants) (Al-Attar *et al.*, 1997), the other major water source being brackish groundwater. Fifty-two per cent of the water resources are allocated to agriculture and landscape irrigation and 46% is withdrawn for domestic use. Sixty per cent of irrigation water is supplied from groundwater and more than 30% from reclaimed water.

Figure 32.3. Photos of the effluent discharged by Um Al Haimam WWTP after chlorination and UV radiation and of a closed reservoir where reclaimed water is stored before reuse.

About 90% of the urban population is connected to a sewerage system. Wastewater is treated in four main municipal treatment plants (Ardiya, Riqqa, Jahra, and Um Al Haimam) by secondary and tertiary processes with a high disinfection level (extended aeration activated sludge, filtration, and chlorine plus UV disinfection in the case of Um Al Haimam) (Figure 32.3) and an annual treatment capacity of over 255 Mm^3. The annual quantity of wastewater produced ranges between 564 and 696 10^3 m^3/d (206 and 254 Mm^3). Out of the total wastewater treated, approximately 62% is reused for crops, highways and coastal zones and the rest is discharged into the sea (Al-Zubari, 1997).

Water reclamation and reuse has been practiced in the country for many years as a means of extending its limited natural water supply. Secondary effluent from the Giwan treatment plant was used to irrigate plantations on an experimental farm from 1956 (Agriculture Affairs and Fisheries Resources Authority, Kuwait 1988). Following extensive studies and experiments, the government of Kuwait decided to proceed with a wastewater reclamation and reuse program. Initially, the treated secondary effluent from the Ardiyah plant was distributed to the experimental farm of the Department of Agriculture at Omariyah and trials were undertaken. In 1977, a Master Plan for effective use of all treated effluent in Kuwait, covering the period up to the year 2010 was launched by the Ministry of Public Works (Cobham and Johnson, 1988). Reuse commenced in 1976, with later stages in 1984 including major projects for irrigated agriculture, afforestation and landscaping in greenbelts. Pilot and demonstrational applied research in the field of soil aquifer treatment, urban greenery sites, and use of treated industrial wastewater have been carried out. The national environmental policy of resource conservation is to reuse all wastewater.

The institutions involved in sanitation and reuse are the Ministry of Public Works, the Public Authority for Agricultural Affairs and Fisheries Resources, a Council for Environmental Policy Setting and Environmental Protection, the Ministry of Health, the Ministry of Water and Electricity, the Ministry of Industry, the Kuwait Institute for Scientific Research, and the Municipalities of Kuwait City, Hawalli, Ahmadi, Al-Jahra and Farwaniya.

Although landscape irrigation applications of reclaimed water are increasing, agricultural irrigation is by far the main use of the water. A small amount of reclaimed water is used for groundwater recharge via surface percolation basins. A share of the industrial wastewaters is also recycled after treatment. Use of reclaimed water for landscape irrigation is growing in urban areas. Currently, reuse of tertiary treated effluent is taking place for a variety of trees for greening the highways and to reduce mobile sand movement. Considerable reductions in dust storms have been achieved by utilizing wastewater for reforestation and green areas development.

Figure 32.4. Fodder and ornamental plants irrigated with tertiary treated effluent using sideroll, centre pivot, and drip irrigation systems.

Agricultural irrigation with reclaimed water (4,470 ha in 1997) (Figure 32.4) represents 25% of the total irrigated area. Crop selection is strongly enforced and reclaimed water is only allowed for the irrigation of vegetables eaten cooked (potatoes and cauliflower), industrial crops, and forage crops (alfalfa, barley, and winter forage crops). Spray-irrigation is avoided for vegetables and fruit trees. Irrigation of salad crops and strawberries with wastewater is not permitted. Effluent quality standards for water reuse have been established by the Ministry of Public Works. They are rather restrictive; irrigation of food crops eaten raw requires tertiary treatment with water quality limits of 10 mg/L for both BOD and suspended solids and 100 total coliforms /100 mL (Table 32.2).

Table 32.2. Reclaimed water standards in Kuwait.

Parameter	Irrigation of fodder & food crops not eaten raw, forest land	Irrigation of food crops eaten raw*
Level of treatment	Advanced	Advanced
SS (mg/L)	10	10
BOD (mg/L)	10	10
COD (mg/L)	40	40
Chlorine residual (mg/L) After 12 h at 20°C	1	1
Coliform bacteria (count/100 mL)	10 000	100

* Not including salad crops or strawberries.

Tertiary treated effluent is used to irrigate around 1680 ha in Sulaibiya. The project consists of an elaborated system of storage and pumping of chlorinated effluent to the irrigated areas. Reclaimed water from Ardiya treatment plant together with some of the treated effluent from Riqqa is directed to the Data Monitoring Center storage reservoirs (two 170 000 m^3 closed reservoirs) before transfer (36.5 Mm^3/yr), for agricultural and landscape irrigation. Irrigation methods used consist of sprinklers (both centre pivot and side-roll), drip and furrow systems. The farms produce a variety of agricultural products together with animal fodder which utilizes 75% of the total area, horticulture, and vegetables eaten cooked e.g. spinach, potatoes, onions and eggplants. Relatively small quantities of lemons and dates are also grown. Treated effluent supplied to private farms is charged to farmers at US\$ 0.07/m^3. Proposals for expanding irrigation include 3300 ha.

An ambitious program to upgrade the entire wastewater system has been launched in response to irrigation concerns such as excessive pumping of groundwater supplies and the resulting increased salinity of irrigation water. A new plant is constructed in Sulaibiyah to replace Ardiya and Jahra and allow unrestricted non-potable uses. The Kuwait government signed a 30-year concession contract in May 2002 to a consortium to design, build, own, operate, and maintain the facility. The plant will treat in a first stage 375,000 m^3/d of wastewater from 2005 and be extendable to a capacity of 600,000 m^3/d. The wastewater treatment plant should achieve, through the application of advanced treatment processes, biological nutrient removal, sludge treatment, and removal of residual pollutants, dissolved solids and pathogens through ultrafiltration and reverse osmosis (RO). RO will be used to lower the salinity of the municipal effluents which ranges from 1200 up to 3000 mg/L.

The product water from this plant will be very high quality (TDS < 100 mg/L, TSS < 1 mg/L, BOD < 1 mg/L, NH_3-N < 1 mg/L, PO_4-P < 2 mg/L, TOC < 2 mg/L, TC < 2.2/100 mL, enteric viruses < 5) and will allow several choices for end use. Reuse options include mixing with brackish water and supplying the brackish water system, unrestricted irrigation, and groundwater recharge.

When completed, the plant will deliver water of potable water quality at approximately 60 US cents/m^3, with 40 US cents/m^3 for conventional wastewater treatment and pipeline costs and 20 US cents/m^3 for producing water from treated effluent. Another plant is planned in Shuaiba to replace the Rikka plant.

32.8 LIBYA

The Libyan Arab Jamahiriya has a total area of about 1.76 million km^2. About 93% of the country is desert. Annual rainfall is extremely low, with about 93% of the land surface receiving less than 100 mm/yr. Total population was about 5.8 million in 2005. Annual population growth for the period 1995-2000 is estimated at 2.3%. Seventy-five per cent of the population is concentrated over 1.5% of the total area of the country, primarily along the coast. More than half the population is urban, mostly concentrated in the two largest cities, Tripoli and Benghazi. In 2002, 72% of the population had access to improved drinking water sources and about 96% to sanitation services.

The renewable freshwater resources are around 106 m^3/*capita*.yr. The renewable surface water resources are estimated at 100 Mm^3/yr and the renewable groundwater resources at 500 Mm^3/yr (Pallas, 1980). The total groundwater abstraction of 4 200 Mm^3/yr during the period 1975-2000 is about 8 times the annual renewable groundwater resources and therefore Libya depends heavily on fossil groundwater (Alghariani, 1993). This situation has accelerated the process of water transfer through the implementation of "The Great Manmade River Project" (GMRP), which is designed to transport eventually 2 300 Mm^3 of fossil water from large aquifers under the Sahara in the arid south to coastal cities in the North (Salem, 1992). The GMRP is designed mainly to serve irrigated agriculture, but to date the water is used almost exclusively for domestic and industrial use in the major cities of the country. The first phase was inaugurated in 1996. The GMRP is Libya's main water supply scheme. To meet future demand, there have been several attempts during the last 25 years to introduce and expand sea water desalination plants and wastewater treatment facilities. It is estimated that 18 Mm^3/yr is annually desalinated.

Of the total water withdrawal of 4 268 Mm^3, about 83% is used for agricultural purposes, 14% for domestic use and 3% for industrial use. More than 30% of the present domestic water demand is supplied by the GMRP (FAO, 2000). It is estimated that at least 80% of the agricultural production depends on irrigated agriculture. The total water managed area is approximately 470 000 ha, of which 316 000 ha was actually irrigated in 2000. About 99% is irrigated using groundwater, while the remaining 1% is irrigated by treated wastewater and surface water. Agriculture contributes to about 9% of GDP.

A number of sewage treatment plants are already in operation or planned for the near future. The present level of wastewater treatment is estimated at about 110 x 10^3 m^3/d (40 Mm^3/yr). When all the plants become operational, their total output will average 285 000 m^3/d. All treated water is currently used for irrigation.

The use of wastewater started in 1971 at Hadba El Khadra, 5 km from Tripoli, on sandy soils. Wastewater is treated in a conventional treatment plant followed by sand filtration and chlorination (12 mg/L). The reclaimed water is then pumped and stored in tanks with a 3-day storage capacity. Reuse was first conducted over 1,000 ha to irrigate forage crops and windbreaks. An additional area covering 1970 ha (1,160 ha forage, 290 ha vegetables, e.g. potatoes, onions, lettuce, etc. and 230 ha for windbreaks and sand dune stabilization) was also irrigated with recycled wastewater; 110,000 m^3/d was applied using sprinkler irrigation (pivots). Reuse is also taking place in Al Marj (north-east of Bengazi: 50,000 inhabitants) after biological treatment, sand filtration, chlorination and storage (Angelakis *et al.*, 1999).

32.9 MOROCCO

Morocco covers an area of 446,500 km^2 and has a population of approximately 31 million, growing at an annual rate of 1.7%. Despite its relatively abundant precipitation ranging from 750 mm in the Mediterranean area to 50-200 mm in the northern Sahara desert, Morocco is a semi-arid to arid country. The available water resources potential that can be mobilized in Morocco is estimated at 29,000 Mm^3/yr, of which 22,000 Mm^3 of surface water and 10,000 Mm^3 groundwater, the common part estimated at 3000 Mm^3. Approximately 12,600 Mm^3 of water are used annually, including 3,500 Mm^3 from groundwater. Nearly 87% of this amount is used to irrigate 1.2 million hectares. The per capital natural water resources is 934 m^3 and per capita withdrawal of 406 m^3.

In urban areas, about 50% are connected to a sewer network; the other half of the urban population relies on on-site disposal systems, mainly septic tanks. Sanitation coverage data in rural areas is sparse. Estimates for 1985 showed that about 16% of the rural population had "adequate" sanitation.

Most Moroccan towns have sewer systems that also collect industrial effluent but they are currently under-equipped with treatment facilities. Therefore, about 58% of the currently generated wastewater is discharged to the Mediterranean and Atlantic coasts and the other 42% is discharged to rivers and other water bodies not without adverse health impacts, and a high incidence of waterborne diseases in Morocco (Bouhoum *et al.*, 1995; Bazza, 2002). The volumes of wastewater collected are estimated at 1507 Mm^3/d (550 Mm^3) in the year 2000 and 2465 m^3/d (900 Mm^3) in 2020. Due to rapid urbanization and increase of the water supply connections, the volume of wastewater available for reuse will increase with the improvement of sewerage networks.

Out of the 60 largest towns only 7 have a WWTP, with both design and operation considered insufficient. As a consequence, most of the wastewater produced by the inland towns (about 70 Mm^3/yr, 13% of collected wastewater) goes to spontaneous reuse. More than 8,000 ha of all types of crops are irrigated with no precautionary measures although irrigation of vegetables eaten raw is prohibited (Bazza, 2002).

Since 1950, wastewater treatment plants have been constructed in Morocco in the medium cities. Less than 8% of the total urban population of the country is serviced by these plants (Jemmali and Abdel-Majid, 2003). In 1995, out of the existing 69 plants, 31 were out of service, 6 were not connected to sewage networks and 33 were functioning normally. Effluents from Khouribga, Nador, Beni Mellal, Al Houceima, Benslimane and Boujâad are reused for agriculture irrigation or landscape irrigation. In the cities of Nador and Khouribga, activated sludge treated effluent is used for irrigating the municipal nursery. In Khouribga, amenity trees of the city are irrigated with the treated effluent supplied by means of cistern trucks. In the city of Boujâad, waste stabilization pond effluent is used to irrigate market gardening crops. In Ben Slimane, close to Rabat, a landscape irrigation project was implemented in 1997, where 5,600 m^3/d of stabilization pond effluent that meets the WHO guidelines is used for golf course irrigation. After treatment in anaerobic, facultative, and maturation ponds, the effluent (absence of helminth eggs, less than 20 CF/100 ml) is used for golf course irrigation. In Drarga, the involvement of farmers and users' associations in the planning phase of the scheme was a key element of success of the project. The treatment process combines anaerobic ponds, denitrifying basins, infiltration-percolation basins and reed beds and delivers 1,000 m^3/d of effluent that meets the WHO guidelines and that is used for unrestricted irrigation (cereals, fodder, and vegetables).

In Morocco, there was a willingness to develop low-cost wastewater treatment systems. Therefore, pilot water reuse irrigation projects have been carried out in Ouarzazate, Rabat Ben Sergao, Marrakech, and more recently in Draga. In Ouarzazate (Figure 32.5), a UNDP-WHO-FAO pilot project covering 5 ha and combining wastewater treatment system and crop irrigation with effluent, consists of field experiments to determine the effect on soils, crops, and groundwater of the application of raw and treated wastewater (waste stabilization ponds alone or combined with high-rate algal pond) using a variety of irrigation methods (surface, sprinkler, and drip). In Rabat, a pilot plant with high rate algal ponds and irrigation of vegetable crops has been tested. In Ben Sergao in the Agadir area, a version of the soil-aquifer treatment (infiltration-percolation) to treat 750 m^3/d of wastewater was tested (Bennani *et al.*, 1992). Effluent is applied to 5 infiltration basins of 1500 m^2 each after screening, pre-treatment in an anaerobic pond and an oxidation pond. Percolation and groundwater recharge take place before pumping for irrigation of crops, grass, alfalfa, wheat, and corn; 40,000 m^3/d could be treated using such a system. In Marrakech, a series of reuse experiments were started in early 1990, using effluents from a variety of treatment systems such as over-irrigation and drainage, anaerobic ponds, a soil-aquifer system, and overload flow. Stabilization ponds followed by irrigation system have been tested. The amount of wastewater that could be treated and reused in this area was about 192 000 m^3/d (70 Mm^3/yr).

The institutions involved in some aspect of wastewater reclamation and reuse in Morocco are the Higher Council of Water and Climate, the Ministry of Public Works, the Ministry of the Interior, the Local Communities, the Ministry of Public Health, and the National Agency of Drinking Water, the Ministry of Agriculture and Rural Development and the Ministry of the Environment.

The country does not yet have any specific water reuse regulations and usually refers to the WHO recommendations. A joint decision of the Ministry of Equipment and the Ministry of Urban and Regional Planning, the Habitat and the Environment No 1276-01 of 17 October 2002 has established the quality standards of water that is to be used for irrigation.

Figure 32.5. Photo of the Ouarzazate's waste stabilization ponds and of the experiments testing different irrigation systems for crop and forest tree production.

32.10 OMAN

The Sultanate of Oman is an arid country with low rainfall, less than 100 mm/yr, and an annual evaporation rate of 3000 mm in the coastal areas. It has an estimated population of 3.2 million, of which approximately 65% live along the coast. The average population growth rate is high (3.9% per annum since 1965). It is projected to reach 6 million by the year 2025. Approximately 81% of its population is urban and has access to safe water and sanitation facilities. Among the rural population, 56% has access to safe water and 72% has access to sanitation facilities.

Internal renewable water resources are estimated at 1000 Mm^3/yr (388 m^3/*capita*.yr). Oman has scarce surface water resources and relies heavily on its groundwater resources (96% of water supply) and on non-conventional resources (desalinated seawater (43 plants providing 3% of water supply) and treated wastewater (1% of water supply)) to meet its water needs estimated to be 1735 Mm^3. Groundwater resources are overdrawn (withdrawals exceed annual recharge by 20%) resulting in decline in water tables and severe seawater intrusion. The annual withdrawal from its renewable freshwater resources is 124%, of which approximately 94% is used for agricultural purposes (UNEP, 2001).

The annual volume of untreated wastewater is estimated to range around 296 10^3 m^3/d. At present, the amount of treated municipal wastewater is estimated to be 100,000 m^3/d (36.5 Mm^3/yr), which is expected to increase by the year 2010 to 445,000 m^3/d (162.5 Mm^3/yr). Wastewater treatment plants use activated sludge extended-aeration processes, followed by sedimentation, filtration and disinfection. Stabilization ponds with tertiary treatment have been installed in rural areas.

There are about 262 wastewater treatment plants in Oman of which over 50% are located around the capital, Muscat, with overall capacity of 52,000 m^3/d, and 20% are in Dhofar and Al Batinat (USEPA 2004). Most are package-type tertiary treatment plants and are site-specific with a capacity from 8 m^3/d to 15,000 m^3/d (UNEP, 2001). They are producing reclaimed water for landscape and agricultural irrigation. Twelve wastewater treatment plants are scattered along the coast of Oman and daily discharge 8 to 271 m^3 of treated effluents to the sea. The main existing wastewater treatment plants are Darsait in Muscat, Al Ansab, and Shatti al Qurm, where the daily amount of treated effluent is 11,500 m^3/d, 5400 m^3/d, and 750 m^3/d, respectively (USEPA, 2004). Most final effluent from these plants (26 Mm^3/yr) is reused for landscape (drip irrigation of roadside ornamental trees, bushes and grass) and agricultural irrigation; nearly 25% (7.2 Mm^3/yr) is used for aquifer recharge and 0.1 Mm^3 annually is reused for industrial purposes, while the remainder is discharged into the sea. It is planned to expand water reuse for irrigation purposes over 5,600 ha.

In the second largest city of Oman, Salalah, a groundwater recharge project for seawater intrusion barrier and replenishment of groundwater used for agricultural irrigation is presently under implementation (USEPA, 2004). Tertiary treated effluent from the 20,000 m^3/d capacity plant will be discharged to a series of recharge wells to form a barrier against seawater intrusion along the coast of the Arabian Sea.

The government of Oman has recognized the importance of water reuse. It has issued water reuse regulations similar to the WHO guidelines, although with more restrictive fecal coliform requirement of 200/100 mL for the irrigation of food crops eaten raw (Table 32.3).

Table 32.3. Water reuse irrigation standards in Oman.

Parameters	Units	Category A Irrigation of vegetables and fruit to be eaten raw, landscape areas with public access, controlled aquifer recharge, and spray irrigation	Category B Irrigation of cooked vegetables, fodder, cereals, and areas with no public access
pH	-	6-9	6-9
BOD_5	mg O_2/L	15	20
COD	mg O_2/L	150	200
TSS	mg/L	15	30
Nitrates (NO_3-N)	mg/L	50	50
Total P	mg/L	30	30
Chlorides (Cl)	mg/L	650	650
Oil and grease	mg/L	0.5	0.5
Aluminium	mg/L	5	5
Arsenic	mg/L	0.1	0.1
Beryllium	mg/L	0.1	0.3
Boron	mg/L	0.5	1
Cadmium	mg/L	0.01	0.01
Chromium	mg/L	0.05	0.05
Cobalt	mg/L	0.05	0.05
Copper	mg/L	0.5	1
Cyanide	mg/L	0.005	0.1
Fluorides	mg/L	1	2
Iron	mg/L	1	5
Lead	mg/L	0.1	0.2
Lithium	mg/L	0.07	0.07
Manganese	mg/L	0.1	0.5
Mercury	mg/L	0.001	0.001
Molybdenum	mg/L	0.01	0.05
Nickel	mg/L	0.1	0.1
Vanadium	mg/L	0.1	0.1
Zinc	mg/L	5	5
Fecal coliform	FC/100mL	<200	<1000
Helminth eggs	Number/L	<1	<1

FC: Fecal coliform, TSS: Total Suspended Solids.

The Sultan Qaboos University is carrying out a research study on greywater reuse of ablution water and of households after simple aeration, filtration and chlorination treatments. As new urban sewage systems were put in place in the Muscat area replacing septic tanks at houses, the question arose whether providing a subsidy to Omanis to treat and reuse greywater *in situ* would not be more financially beneficial than investing in larger sewage systems to convey and treat wastewater produced at households (Prathapar *et al.*, 2005).

The following institutions are involved in wastewater reclamation and reuse projects: Ministry of Regional Municipalities and the Environment; Directorate General of Environmental Affairs; Directorate General of Nature Reserves; Directorate General of Public Relations and Guidance; Marine Science and Fisheries Center; Ministry of Agriculture and Fisheries, Muscat; Sultan Qaboos University, Department of Fisheries Science and Technology, and Department of Biology (UNEP, 2001).

32.11 PALESTINE

Palestine consists of the West Bank (WB) and the Gaza Strip (GS) with a land area respectively estimated at 5700 km^2, and 367 km^2. The total population of 3.2 million lives in the two geographical areas, with 64% in the West Bank including Jerusalem and 36% in the Gaza Strip. By the year 2025, the population of Palestine is projected to reach 6 million, assuming an average growth rate of 4.8-3.5% for the years 2000-2010 (UNEP, 2001).

Palestine has limited water resources and mainly relies on its groundwater resources (679 Mm^3/yr for the WB (of which 15 to 20% are accessible to Palestinians) and 60 Mm^3/yr for the GS). High concentrations of chloride, sodium, potassium and nitrate have been recorded in the Gaza Strip as well as in the West Bank because of wastewater seepage into aquifers. In 2000, the total annual water use in WB and GS was estimated at 293 Mm^3 (140 Mm^3 in Gaza and 153 Mm^3 in West Bank), of which 60% is accounted for by the agricultural sector. The overall demand in Palestine is projected to more than double in the 20-year period from 2000 (293 Mm^3/yr) to 2020 (700 Mm^3/yr).

In the West Bank and Gaza Strip, 48% of the population live in urban and semi-urban areas, 31% in rural areas and the rest in refugee camps. Few communities have sewage systems and wastewater treatment facilities. About 30-35% of the houses are connected to a sewage network, 65% in Gaza Strip and about 28% in the West Bank; the remaining rely on cesspits, or are without sanitary facilities. Effluent is also often disposed in open channels before being discharged in open pools, nearby wadis or on the sea shore. A part of this effluent infiltrates the ground. From the Gaza coastline, over 40,000 m^3/d of untreated wastewater is discharged into the sea causing marine pollution and considerably exceeding in some places the international standards for bathing water quality (UNEP, 2001). In the major cities, the sewerage system is often incomplete; sewage is often mixed with solid wastes (sand) which renders collection and treatment difficult. Only 5% of the collected municipal wastewater is partially treated. In 1990, a survey conducted in the Gaza Strip showed that 50% of children under 10-years old were infected by Ascaris (Medaware, 2004).

The amount of wastewater generated in WBGS was about 49,300 m^3/d (18 Mm^3/yr) in 1994 (Gearheart *et al.*, 1994). About 50 Mm^3 of wastewater is currently generated in WBGS, 79,450 m^3/d (29 Mm^3) and 57,530 (21 Mm^3) in WB and GS respectively. In the WB, the annual volume of collected wastewater by sewage networks is 10.7 Mm^3, whereas in GS it is 15 Mm^3. The Water Sector Strategic Planning Study prepared by the Palestinian Water Authority (2000) also sets out that, by 2020 in the West Bank and Gaza Strip, 60% of the available treated effluent will be reused for agricultural purposes (107,000 m^3/d (39 Mm^3) in GS and 140,000 m^3/d (51 Mm^3) in WB) and 15% will be recharged to the aquifer (27,400

m^3/d (10 Mm^3) in GS and 35,600 m^3/d (13 Mm^3) in WB). Taking into account just the Gaza Strip, the annual volume of collected wastewater (41,000 m^3/d, 15 Mm^3) is about 20% of the annual recharge to the groundwater reservoir.

There are eight main wastewater treatment plants in the Palestinian Territories; three are located in the Gaza Strip, while five are in the West Bank. Most of the treatment plants are aerated lagoons with polishing ponds. At present, effluents from most treatment plants do not meet discharge and reuse standards. There are also small wastewater treatment facilities in Palestine and different technologies are being tested such as integrated anaerobic pre-treatment followed by duckweed-based ponds (Mahmoud *et al.*, 2004). A project on greywater treatment and reuse for peri-urban horticulture was also initiated in the West Bank in 1999. This pilot project aims at optimizing the design of small-scale trickling filters for individual or small collective systems, for the treatment of greywater for reuse for irrigating products in home gardens, including raw vegetables in hilly, low-density peri-urban areas of the West Bank (Abdo, 2001).

The three treatment plants found in the Gaza Strip are Beit-Lahia, Gaza City and Rafah. Two schemes using reclaimed water for irrigation purposes or groundwater recharge were set up in the Gaza Strip in the mid 1980s but they have never been operated for different reasons (inadequate water quality, water price). In Gaza City in 1986, a storage reservoir (5000 m^3) and three groundwater recharge basins (about 0.5 ha) were built 1000 m south-east of the treatment plant; 150 ha were to be irrigated with the effluent. These facilities have never functioned because of the poor effluent quality. The "partially" treated effluent used to be directly pumped to the Wadi Gaza and to flow down to the sea (Gearheart *et al.*, 1994). The Gaza wastewater treatment plant has recently been rehabilitated by the US Agency for International Development to cope with the expected population of the Gaza Governorate in 2005, with a total expected wastewater inflow of 35,000 m^3/d (UNEP, 2001). The quality of the effluent has been greatly improved and the BOD is now 25 mg/L. At present, about 25% of the total wastewater produced daily from the Gaza wastewater treatment plant is used for groundwater recharge using infiltration basins (10,000 m^3 of 40,000 m^3 of treated wastewater produced). Some informal and uncontrolled recharge via sand dunes filtration is also practiced in Jebalia, Beit Lahia and Rafah.

Several factors such as political unrest, and weak financing, inadequate planning and management have affected the wastewater sector in Palestine (Medaware, 2004). In the West Bank, over eight main cities, five have a wastewater treatment plant. Jenin, Tulkarem, Hebron, and Ramallah constructed at the beginning of the 1970s use the lagoon technology. They are presently overloaded or not functional and hardly achieve any treatment above primary level. Al Bireh, an extended aeration system built in 2000 which serves around 30,000 inhabitants, is the only plant working properly. At Bir Zeit University, reclaimed water coming out of an activated sludge plant is substituted to potable water in a dual distribution system for toilet flushing. Reclaimed water is also used to irrigate landscaped areas within the university. So, the existing wastewater treatment plants in the West Bank are mostly inadequate to serve the whole volume of wastewater being discharged and to deliver an effluent of adequate quality for reuse.

In order to increase the irrigated areas or to save the fresh water resources for other purposes, reclaimed water may be reused in several locations in the West Bank and Gaza Strip. Water reuse may also provide a very good alternative to groundwater abstraction especially for the agricultural sector. The present potential irrigated area with reclaimed water is about 2200 ha (11% of the actual irrigated area) and may rise to 5500 ha. There are also opportunities for groundwater recharge operations in the Gaza Strip to protect underground freshwater against sea water intrusion.

Public health, municipalities and agricultural officials generally emphasize a positive attitude and strong support to wastewater reclamation and reuse which is considered as an important component of water resources management. A wastewater management strategy has been set up which aims to eliminate raw wastewater discharge to the natural environment and the National Water Plan established in 2000 integrates the use of wastewater into its strategic planning targets, with the aim of using treated effluents for agriculture and groundwater recharge. Cultural and social acceptance of water reuse is not a problem as long as the reclaimed water price is acceptable and that a wide range of crops can be irrigated (Abdo, 2001).

Effluent quality criteria have been adopted by the Palestinian Water Authority for agriculture, groundwater recharge and discharge to wadis. Those adopted for agriculture are similar to those recommended by the WHO (1989).

At present, the following ministries and organizations are involved in wastewater management: the Palestinian Water Authority, the Environmental Quality Authority, the Ministry of Agriculture, the Ministry of Health, the Ministry of Planning, the Ministry of Local Government, and the municipalities and the village councils.

32.12 QATAR

Qatar has a total area estimated at 11 437 km^2 and a population of 0.62 million growing at an average rate of 2.6% per annum. Average annual rainfall is around 75 mm and the climate is of the arid desert type. Ninety two per cent of its population lives in urban areas (Doha City mainly) and almost the entire population has access to safe water and sanitation facilities.

The agricultural sector accounts for 1% of the GDP and uses 74% of water resources. Some of the major industries are petroleum and chemical products. Qatar has 31 seawater desalination plants with a total capacity of 850,000 m^3/d. Industrial wastewater from large industrial complexes in Doha City are separately treated.

Qatar has few renewable surface freshwater (1.35 Mm^3/yr) and groundwater resources (45 Mm^3/yr): the natural water resources per capita are around 86 m^3 and the desalinated water withdrawal per capita is 204 m^3. The agricultural sector is the main water consumer. To meet its water consumption needs, the country relies mainly on desalinated seawater. The annual consumption of water in Qatar is estimated at 290 Mm^3 and approximately 110 Mm^3 of wastewater are generated (UNEP, 2001).

Until the mid 1980s, raw sewage was discharged directly into the sea, but since then a sewerage system has been constructed. Currently, there are two large municipal wastewater treatment plants with a total capacity of 80,000 m^3/d, and nine small wastewater treatment plants. The treated effluent is mainly used for irrigation. The treated municipal wastewater has the potential to meet close to 24% of the water needs of the agricultural sector (Al-Qasimi, 1997).

Advanced treatment works (biological treatment followed by a tertiary treatment step) serve the majority of the population of Doha City (>300,000 people at Naijah (Doha South) and Sailiyah (Doha West)). The treatment capacity is approximately 110,000 m^3/d. On average, over the year, approximately 96% of the treated effluent is reused for irrigation purposes. Of the total reused treated wastewater, 25% is supplied to the municipality of Doha, to be used in landscape irrigation on the roads and areas planted with trees, and the remaining 75% is conveyed via pipelines to ponds, which is then pumped to west Doha to be used for irrigation. The remaining effluent is discharged to a lagoon area in the desert.

For future wastewater treatment plants, environmental impact assessments are a prerequisite under the direction of the Environmental Committee (currently under the supervision of the Ministry of Municipalities and Agriculture), which is in charge of all environmental affairs (UNEP, 2001).

Costs of seawater desalination and municipal wastewater treatment in Qatar are US$1.45--1.64 and US$0.24, respectively (ESCWA, 1985). The average price of water is around US$0.43/m^3. The government subsidy is about 61% of the cost of supplying water. Wastewater reclamation and reuse in the agricultural sector and other sectors may substantially reduce the need for additional capacity for seawater desalination (UNEP, 2001).

Qatar has a strong legal framework and is focusing on expanding water reuse using advanced treatment processes. A strategy for municipal water reuse is under consideration. A feasibility study on groundwater-recharge has been carried out with positive findings and a study on industrial effluent reuse is being updated.

32.13 SAUDI ARABIA

Saudi Arabia has scarce renewable water resources. The mean annual precipitation over the whole country range from 50-85 mm and the mean evaporation is high. Total water demand has tripled in the past two decades, reaching 46 Mm^3/d (17,320 Mm^3/yr) by 2003. The agricultural sector's water requirements consume the largest amount of water in the kingdom (89%). Fossil-groundwater reserves account for at least 80% of total water supply. In addition, the country relies on non-conventional water resources (desalinated seawater and treated wastewater). Twenty seven desalination plants currently provide 2.6 Mm^3/d of freshwater, 70% of the kingdom's water supply and 30% of the world's desalination water production. The average water use per citizen is 430 L/d. Annual *per capita* water consumption in Saudi Arabia is 1040 m^3.

In 2003, Saudi Arabia had an estimated population of 24.9 million. It is expected to reach 54 million by the year 2025 (average growth population rate of 4.4% per annum). Eighty-five per cent of the total population live in urban areas and have access to safe water and sanitation facilities. Approximately half of the rural population have access to safe water and sanitation facilities.

There are 30 major sewage treatment facilities in the country with a total design capacity of 1,426,000 m^3/d (UNEP, 1999), of which there are 11 activated sludge plants, 11 trickling filters, 4 aerated lagoons, 3 tertiary or advanced plants, and one RBC. The wastewater is treated at the secondary or tertiary level (aerobic biological treatment technologies). Pretreatment of industrial wastewater is carried out by several small units prior to discharge to the central wastewater treatment plant as well as closed water cycles to minimize wastewater disposal (Abderrahman, 1997).

Saudi Arabia has adopted a national policy strictly enforcing the reuse of treated effluent particularly for agriculture. Reclaimed water can help meet increasing demands for agriculture and landscape irrigation, industrial reuse and groundwater recharge. The level of wastewater treatment required is dependent upon the intended use of the effluent.

Regulations, measures, and *fatwa* (legal ruling on an issue of religious importance) have been introduced. According to Abderrahman (2001), the Council of Leading Islamic Scholars (CLIS 1978) of Saudi Arabia, after a thorough consultation with both technical and health experts, stated in a special *fatwa* in 1978 "that impure waste water can be considered as pure water and similar to the original pure water, if its treatment using advanced technical procedures is capable of removing its impurities with regard to taste, colour and smell, as witnessed by honest, specialized and knowledgeable experts. Then it can be used to remove body impurities and for purifying, even for drinking. If there are negative impacts from its direct use on the human health, then it is better to avoid its use, not because it is impure but to avoid harming the human beings." (Abderrahman, 2001).

The 1978 *fatwa* allowed water reuse expansion in Saudi Arabia for different purposes depending on its degree of treatment, such as restricted and non-restricted irrigation, toilet flushing, ablution, etc., and in a variety of locations (Abderrahman, 2001). It has resulted in the reuse of 217 Mm^3/yr of treated effluents (representing about 32% of the treated effluent or 674 Mm^3), the majority being reused in large cities like Riyadh and Jeddah. About 36% is used for agricultural irrigation, the remaining is discharged to land (34%), disposed to sea (18%), and about 12% is reused for industrial purposes, groundwater recharge and landscaping (Al-Morgin, 2003). Riyadh has at least 6 wastewater treatment plants with a total daily capacity of 464,000 m^3 and 9,600 km of pipelines that serve 275,000 house connections. About 9000 hectares of date palms, forage crops, such as alfalfa, wheat, citrus, and deciduous trees for afforestation projects are irrigated near Riyadh, at Dirab and Dariyah. Central pivot, sprinklers, drip, bubbler, and surface irrigation systems are used. Wastewater is also reused for irrigating landscape plants, trees, and grass in municipal parks in several cities, such as Dhahran, Jeddah, Jubail, Riyadh, and Taif. Recycling of treated industrial wastewater has been encouraged, and has been implemented by various industrial plants at the plant level for industrial and landscape purposes and in closed water cycles. In addition, ablution water is recycled for toilet flushing at the two holy mosques in Mecca and Medina, thus conserving expensive desalinized seawater (Abderrahman, 2001). The operation and maintenance costs of brackish water desalination in Saudi Arabia are US\$0.26/m^3 (Al-Mudaiheem *et al.*, 1998) and of seawater desalination US\$0.6/m^3 with multi-stage flash system (MSF) and US\$0.8/m^3 for reverse osmosis (RO) (Al Sahlawi, 1999).

Several plants have been upgraded in Taif, Jeddah, and Jubail. Future plans include dual piping for residences in Taif city by the year 2020 (Ukayli and Hussain, 1988 in FAO/RNE and WHO/EMRO, 2003). A water reuse master plan for the capital city Riyadh has been established for 2021. It takes into account water supply, wastewater collection and treatment, and water reuse together in an integrated planning process of the infrastructure plan, and recommends the construction of 12 satellite treatment plans to meet the non-potable water demand (Sheikh *et al.*, 2000). Reclaimed water will be used for agriculture irrigation and irrigation of landscape areas, residential gardens, parks, and for industrial purposes.

Field experiments have been carried out testing different crops and showed that no adverse effects were observed. Higher yields were recorded when using reclaimed water. High total dissolved salts (TDS) could be of some concern and ways to mitigate its negative impacts have been considered such as selection of salt-tolerant plants, or dilution of effluents using groundwater irrigation. It was also found that sea disposal of secondary effluents could have some impact on marine life.

Royal Order No. M/34 issued in 1980 forms the legislative framework for the protection of water resources including wastewater. National wastewater quality standards have been issued drawing up limits for BOD, total coliforms, suspended solids and nitrate concentrations for secondary and tertiary treatment. They request advanced wastewater treatment for unrestricted irrigation and maximum concentrations for total coliforms of 2.2 count/100 mL, BOD and TSS of 10 mg/L and turbidity of 1 NTU (Table 32.4). Irrigation of vegetables of any kind is forbidden.

The main institutions involved in the field of wastewater reclamation and reuse are: The Ministry of Agriculture and Water, the Ministry of Municipalities and Rural Affairs, the Ministry of Health, the Ministry of Industry and Electricity, the Meteorology and Environment Protection Agency, the National Directorates of Water and Sewerage, the Research Institute of the King Fahd, and the University of Petroleum and Minerals.

Table 32.4. Reclaimed water standards for unrestricted irrigation in Saudi Arabia.

Parameter[(a)]	Maximum contaminant level
BOD	10
TSS	10
pH	6-8.4
Total coliform (count/100 ml)	2.2
Turbidity (NTU)	1
Aluminum	5
Arsenic	0.1
Beryllium	0.1
Boron	0.5
Cadmium	0.01
Chloride	280
Chromium	0.1
Cobalt	0.05
Copper	0.4
Cyanide	0.05
Fluoride	2
Iron	5
Lead	0.1
Lithium	0.07
Manganese	0.2
Mercury	0.001
Molybdenum	0.01
Nickel	0.02
Nitrate	10
Selenium	0.02
Zinc	4
Oil & grease	Absent
Phenol	0.002

Note: (a) in mg/L unless otherwise specified

32.14 SYRIA

Syria, with a total area of 185,180 km², is relatively well endowed with water resources. Different factors are however driving water shortages such as a high annual growth rate in population, an increase in average water consumption, economic activities upstream of the Euphrates and Tigris basin, water flow restriction from Turkey, expansion of the irrigated areas, and water use efficiency problems in the agriculture sector.

The Syrian population is estimated at 18.4 million in 2005, which during the last three decades has grown at an average rate of 3.3%. More than half of its urban population has access to safe water and about 45% of the whole population are connected to the sewer system. By the year 2025, Syria's population is expected to reach 32.2 million.

The total renewable freshwater resources, both surface and groundwater, are 26,260 Mm^3/yr. The annual volume of water withdrawal is 19,950 Mm^3, and the availability *per capita* is 1095 m^3; 95% are used by the agricultural sector and 3% by the domestic sector (UNEP, 2001). In Syria, irrigated agriculture plays an important role in enhancing food security. The total irrigated area was about 1.2 million ha in 2000, with 61% irrigated from groundwater and the rest from surface water. To meet its water demand needs, the country also relies on desalinated water and wastewater. The volume of wastewater in Syria is expected to reach 1,642 Mm^3/yr in 2025.

Construction of several treatment plants such as in Damascus (Adra), Aleppo, Homs, Salamyeh, Ras El Ein, and Haramil Awamid has been started during the last decade. Several other wastewater treatment plants are under planning or construction (such as in Tartus, As Sweida Idleb, Al Raqqua, Al Nabik and Deraa). Except Salamieh, which uses the oxidation ponds process, all these new facilities are activated sludge treatment plants with provision for reuse. The total capacity of the seven treatment plants which have been completed for major cities including Damascus is 1,182,000 m^3/d (431 Mm^3/yr).

Except in part of Damascus, the collected untreated wastewater, if not disposed of into the sea, is discharged into water bodies or on agricultural lands and used for unrestricted irrigation (Table 32.5) (FAO, 2001). This practice led to the degradation of surface water quality, of groundwater quality and of soil, and to high concentrations of pollutants, especially in the Barada River and Aleppo southern plains, resulting in the spread of diseases such as typhoid, hepatitis, and dysentery in southern Aleppo and Damascus Ghouta populations (Zulita, 2003). Animals fed with fodder irrigated with untreated wastewater were also affected by diseases (Bazza, 2002). The total area irrigated with wastewater is estimated at around 40,000 ha, with 20,000 ha in Aleppo (Zulita, 2003).

In cities like Damascus, which are facing major challenges due to water shortages, groundwater degradation and pollution, treated effluent from the Adra plant (average 458,000 m^3/d, 110 Mm^3/yr) is used to irrigate the eastern part of Ghouta, with a total area of 9,000 ha. Water is distributed to farmers by open channels, plots are flood irrigated (the main crops are fruits and vegetables).

In Syria, organizations involved in wastewater reclamation and reuse include: the Directorate of Water Supply and Wastewater at the Ministry of Housing and Public Utilities responsible for policy implementation and regulation, the municipalities in charge of construction and operation of wastewater collection and treatment, the Ministry of Agriculture and Agrarian Reform and the Ministry of Irrigation (Directorate of Irrigation and Water Uses) in charge of effluent reuse.

Table 32.5. Farming areas irrigated with wastewater in Syria (after Zulita, 2003).

Province	**Area (ha)**	**Crops**
Rural Damascus	3654	Fruit trees, fodder crops, vegetables
Quneitra	-	No wastewater except for Ba'ath city, Khan Arnabeh and Qahtanieh where wastewater is delivered to Al-Raqad valley
Deraa	-	Wastewater is discharged into the following valleys (Zidi, Al-Ghar, Al-A'aram, Abo Al-Labn, Ghazaleh)
Sweida	-	Wastewater is discharged into the open
Homs	1960	Winter and summer crops, fruit trees (wastewater is discharged within irrigation canals branching from the Orontes)
Hama	159	Wheat, cotton, maize, sunflower, ground peanuts, fruit trees
Al-Ghab	224	Wastewater is discharged via covered canals to pour into the Orontes river.
Idleb	2000	Cotton, wheat, sunflower
Aleppo	19400	Wheat, cotton, maize, fruit trees, vegetables. Wastewater is discharged into Qweiq, Sajjour and Affrin rivers.
Raqqa	2155	Winter and summer crops, vegetables
Deir Ezzor	-	-
Hassakeh	1800	Cereals, cotton, vegetables
Tartous	-	Wastewater is discharged into the sea
Lattakia	500	Citrus, summer vegetables. Wastewater is discharged into the sea.
Total	**31852**	

There are currently no guidelines nor regulations for water reuse in Syria. To prevent risks associated with raw wastewater, the Ministry of Agriculture issued a resolution (No. 2823, dated 29/8/1990) to exclude vegetable irrigation with polluted water sources. Consequently, the use of wastewater is restricted to fodder, industrial crops and fruit trees. There are specifications for water reuse in agriculture, discharge into public water bodies and when wastewater is used in industry for cooling and other purposes (Zulita, 2003).

32.15 TUNISIA

Tunisia has a population of 10 million. Rainfall and evaporation is unevenly distributed across the country. In the north, center and south of Tunisia, the average annual rainfall is 594 mm, 289 mm, and 156 mm, respectively. Evaporation ranges from 1,300 mm in the north to 2,500 mm in the south. Only half of the country's 4,700 x 10^6 m^3/yr of available water has a salt content that is low enough for use without restrictions. In 2000, water availability was 440 m^3/*capita*/yr with withdrawals accounting for 78% of the renewable resources. These water deficits are projected to increase with population growth, an increase in living standards, and accelerated urbanization. According to recent forecasts, increased domestic and industrial water consumption by the year 2020 may cause a decrease in the volume of fresh water available for Tunisian agriculture.

Most residents of large urban centers have access to various, adequate sanitation systems and wastewater treatment facilities (78%, whilst the figure is 61% for all of the population and 40% in rural areas). Concerning industry, compliance with the Tunisian standards (INNORPI 1989a) to discharge wastewater into the sewerage system is required. So, preliminary treatment plants to fulfill the discharge requirements stated in the regulations must be supplied. Subsidies are given to equip industrial units with pre-treatment processes.

In 2003, the volume of wastewater collected was 240x10^6 m^3/yr, 187x10^6 m^3/yr of which was treated in 70 treatment plants utilizing activated sludge, oxidation ditches, and stabilization ponds. Several of them are located along the coast to protect coastal resorts and prevent sea pollution. The annual volume of reclaimed water is expected to reach 290 Mm^3 in the year 2020. The expected amount of reclaimed water will then be approximately equal to 18% of the available groundwater resources, and it could be used to replace groundwater currently being used for irrigation in areas where excessive groundwater mining is causing salt water intrusion in coastal aquifers.

Municipal wastewater is mainly domestic (about 82% domestic, 12% from industries and 6% from tourism) and goes through secondary biological treatment. The treatment processes vary depending on wastewater origin and local conditions. The treatment plants located in the Tunis area produce about 60 Mm^3/yr of treated wastewater. They account for 54% of the country's reclaimed effluent, estimated at 187 Mm^3 in 2003.

In Tunisia, 66% of sanitation costs, including full O&M costs and part of the capital costs for sanitation from municipal and industrial users are recovered. The cost of wastewater treatment to the secondary treatment level is estimated to average US\$ 0.15/m^3, with a marginal treatment cost of US\$ 0.34/m^3. Investment cost is about 80% of the total marginal cost. Low-cost technologies are sought to reduce the cost of treatment.

Tunisia has a national water reuse policy launched at the beginning of the 1980s with treatment and reuse coordinated from the planning stage. Water reuse has been made an integral part of overall environmental pollution control and water management strategy. Water reuse is now a part of Tunisia's overall water resources balance. It is actually considered not just as an additional water resource and as a potential source of fertilizing elements (UNDP,

1987; Bahri, 1998) but also as a complementary treatment stage, and consequently as a way of protecting coastal areas, water resources, and sensitive receiving bodies.

Although some pilot projects have been launched or are under study for groundwater recharge, irrigation of forests and highways, and wetlands development, the water reuse policy, launched in the early 1980s favors planned water reuse for agricultural and landscape irrigation (Bahri, 2000). In 2003, $43x10^6$ m^3 of treated effluent were used for agricultural and landscape irrigation, almost a four-fold increase since 1990.

A gradual approach to expanding reuse since the mid 1960s has been adopted (UNDP *et al.*, 1992). The strategy has consisted of 1) extending wastewater treatment to all urban areas; 2) conducting pilot- and demonstration-scale irrigation operations on agricultural and green areas; 3) establishing large scale irrigation schemes; and 4) implementing a policy calling for an increase in the percentage of treated effluent that is to be reused. Different development phases can then be distinguished.

Reclaimed water from La Cherguia-Tunis wastewater treatment plant has been used since the early 1960s for agricultural irrigation to reduce the impact of saltwater intrusion due to excessive pumping of groundwater. La Cherguia (Figure 32.6) was constructed in 1958 and provides secondary wastewater treatment. The 600 ha area of La Soukra is located to the north-east of Tunis and reclaimed water has been used since then to irrigate citrus and olive trees. Irrigation of raw-eaten vegetables was prohibited from the beginning. The area currently irrigated with reclaimed water is about 8,000 ha, 62% of which is located around Tunis and a few other locations near Hammamet, Sousse, Monastir, Sfax, and Kairouan. Farmers are urged to adopt water reuse through subsidized reclaimed water prices.

The nine existing Tunisian golf courses (Tunis, Hammamet, Sousse, Monastir, Tabarka, Tozeur and Jerba) (Figure 32.7) covering 760 ha, some hotel gardens in Jerba and Zarzis, and other green spaces (340 ha), are also irrigated with recycled wastewater (Bahri, 2003). Secondary treated wastewater is stored in a series of ponds on the golf courses. Irrigation is conducted at night with low range sprinklers to prevent public exposure.

Seasonal recharge of the shallow and sandy aquifer of Nabeul has been performed since 1985. Activated sludge effluents that were not used for irrigation during the winter season were infiltrated and stored in the aquifer, thus increasing the volume available that farmers can pump during the summer season to irrigate the mainly citrus orchards. Groundwater recharge efficiency was proven not only by the increase of the water level in the wells, but also by the improvement of the production of the surrounding wells. This experiment allowed an underground storage and an additional treatment step (UNDP, 1987). Performing coastal aquifer recharge, where the hydrogeological context is favourable, would make water reuse better accepted by farmers. This subject is still under study and other sites are screened for further studies including the comparison of different treatment processes preceding recharge.

Figure 32.6. La Cherguia activated sludge treatment plant and Côtière Nord waste stabilization ponds whose effluents are used to irrigate citrus trees in La Soukra area.

Figure 32.7. Aerated ponds used for reclaimed water storage on the golf course of El Kantaoui.

The National Sewerage and Sanitation Agency is responsible for the construction and operation of all sewage and treatment infrastructure in the larger cities of Tunisia. Prior to wastewater reclamation and reuse projects, environmental impact assessment studies are carried out. When effluent is to be used for agricultural irrigation, the Ministry of Agriculture and Water Resources is responsible for execution of the projects, which include the construction and operation of all facilities for pumping, storing, and distributing the reclaimed water. Various ministries (Environment and Land Use Planning, Public Health, Tourism and Handicrafts, etc.) and departments of the Ministry of Agriculture and Water Resources are responsible for several functions, while regional agricultural departments supervise the water reuse decree and collection of charges. Farmers pay for the reclaimed water they use to irrigate their fields. Reclaimed water is directly used by farmers through distribution systems managed by the local agencies of the Ministry of Agriculture. In order to promote water reuse, the tariff of reclaimed water was set at US$ 0.015/m^3, below the tariff for conventional surface water used in other irrigation systems (US$ 0.022-0.077/m^3). However, farmers still prefer conventional water, if they have a choice.

Water reuse in agriculture is regulated by the 1975 Water Law and by the JORT Decree No. 891047 (1989). The reclaimed water quality criteria for agricultural reuse were developed using the guidelines of FAO (1985) and WHO (1989) for restricted irrigation (less than 1 nematode egg/l), and other Tunisian standards related to irrigation or water supply (Table 32.6). The Water Law prohibits both the use of raw wastewater in agriculture and the irrigation with reclaimed water of any vegetable to be eaten raw. The 1989 decree specifically regulates reuse of wastewater in agriculture and allows the use of secondary treated effluent for growing all types of crops except vegetables, whether eaten raw or cooked. Therefore, recycled water is used to irrigate vineyards, citrus and other fruit trees (olives, peaches, pears, apples, pomegranates, etc.), fodder crops (alfalfa, sorghum, and berseem), sugarbeet, and cereals.

The most common irrigation methods are sprinklers (57% of the equipped area) and surface irrigation (43%). Specifications regarding the terms and general conditions of reclaimed water reuse (and the precautions that must be taken in order to prevent any contamination to workers, residential areas, and consumers) have also been established.

There are several on-going studies to screen reuse options, such as groundwater recharge, municipal, industrial, and environmental uses. A study aimed at developing a strategy to promote water reuse was also carried out (Bechtel and Scet, 1998; ONAS-SERAH, 2001). It showed that the strategy should be oriented towards the substitution of conventional water by reclaimed water for the high-rated water activities, or the creation of a new demand based on strategic projects. A new wastewater treatment plant is planned for the city of Tunis, "The Tunis-West project," with a designed capacity, in the year 2016, of 105,000 m^3/d (41 Mm^3/yr) and an irrigation reuse component. The treatment plant will be partly operated as a BOT and will include disinfection of part of the effluent and seasonal storage.

Table 32.6. Tunisian standards for reclaimed water reused in agriculture (1989).

Parameters [(a)]	Maximum allowed concentration
pH	6.5 - 8.5
Electrical conductivity (EC) ($\mu S\ cm^{-1}$)	7000
Chemical oxygen demand (COD)	90 [(b),(c)]
Biochemical oxygen demand (BOD_5)	30 [(b),(c)]
Suspended solids (SS)	30 [(c)]
Chloride (Cl)	2000
Fluoride (F)	3
Halogenated hydrocarbons	0.001
Arsenic (As)	0.1
Boron (B)	3
Cadmium (Cd)	0.01
Cobalt (Co)	0.1
Chromium (Cr)	0.1
Copper (Cu)	0.5
Iron (Fe)	5
Manganese (Mn)	0.5
Mercury (Hg)	0.001
Nickel (Ni)	0.2
Lead (Pb)	1
Selenium (Se)	0.05
Zinc (Zn)	5
Intestinal nematodes (arithmetic mean no. of eggs per liter)	< 1

[a] all units in mg/L unless otherwise specified;
[b] 24-hr composite sample;
[c] except special authorization for stabilization ponds.

32.16 UNITED ARAB EMIRATES

The United Arab Emirates is a federation of seven emirates: Abu Dhabi, Dubai, Sharjah, Ras Al Khaimah, Fujairah, Umm ul Quwain, and Ajman. The population is approximately 3 million, mostly urban (83%). The average annual rainfall is 25 mm (range in between 0.2 and 85.7 mm). The United Arab Emirates have limited renewable water resources, 150 Mm^3 or 50 m^3/*capita*.yr in 2000.

The annual water demand of 754 m^3/*capita*/yr is met by over-drafting non-renewable aquifers and desalinization (USEPA, 2004). The annual volume of renewable freshwater is used mainly in the agricultural sector (68%). The volume of wastewater produced in urban areas in the Emirates is 500 Mm^3/yr. Four large sewage treatment plants are equipped with tertiary treatment capabilities. Their capacity is 295,000 m^3/d (UNEP, 1999) and the amount of treated water is 280,000 m^3/d. About 61% of the water discharged by these is mainly used for irrigation of parks, golf courses, highways and urban water features. It is estimated that by the year 2025 the annual volume of water consumed in the United Arab Emirates can reach 3185 Mm^3 and the volume of municipal wastewater produced is expected to range between 745 and 917 Mm^3. This has the potential to meet approximately 40% of the needs of the country's agricultural sector (UNEP, 2001).

The most extensive and notable reclaimed water usage occurs in *Abu Dhabi*, the "Garden City", where nonpotable reuse has been practiced since 1976. Abu Dhabi has two population centers, the town of Abu Dhabi and the oasis city of Al-Ain. Its population has grown from 1500 in the mid 1950s to almost a million in the late 1990s. With the development of the city, satisfaction of the population's water requirements became more difficult. It was therefore decreed that water should be use wisely.

The Sewerage Project Committee, established in 1975, developed a policy of using treated effluent, from Mafraq's wastewater treatment plant (commissioned in 1982), for irrigation in order to conserve the potable water supplies. The implemented infrastructure provided over 90% of the irrigation water requirements (around 200,000 m^3/d of tertiary treated effluent) of 15,000 ha of forest land, landscaped areas, public parks and gardens, ornamental plants, a golf course and an animal fodder farm at Al Wathba (23,000 m^3/d supplied to 250 ha). The green area covers 1,300 ha with around 1,200,000 trees. The irrigation infrastructure capacity has since been increased to 350,000 m^3/d. The greening of the city with planted trees and shrubs and the development of other green areas has enhanced the environment, offset pollution, and counteracted desertification. An afforestation program was implemented and desert land was put into production with farms producing fodder crops, dates, milk, cereals and vegetables as well as being involved in poultry farming, and cattle raising.

Treated effluent is also provided along highways for irrigation of roadside planting, local supply to residential areas, and to the international airport. A phased implementation of the sewerage system expansion and reclamation and reuse infrastructure has been planned to take into account the development of the city. Reuse of treated effluent for irrigation has been incorporated to the different sewerage projects.

Mafraq WWTP is an activated sludge process followed by filtration and chlorine disinfection (Figure 32.8). The effluent water quality discharged by the treatment plant is presented in Table 32.7. Each treatment plant includes tertiary treatment and chlorine disinfection. Storage is provided before reuse. Storage infrastructure includes 55 reservoirs with a total storage capacity of 98,400 m^3 at different locations in the city to facilitate the planting and irrigation system installation. The digested sludge is dried on beds and further composted with household solid wastes and used as a soil conditioner and fertilizer on the planted areas.

In Al-Ain (population of about 250,000), wastewater treatment includes dual media filtration and chlorination for disinfection before restricted irrigation reuse 12 km outside the city in designated areas. In Sharjah (2600 km^2), the treatment capacity (over 100,000 m^3/d) has been recently increased (37,400 m^3/d) to meet a projected increase in population. The activated sludge effluent will be disinfected with UV and pumped to storage prior to distribution for landscape (500 ha and 150 ha) and horticulture irrigation. The wastewater recycling program enables valuable groundwater supplies to be conserved and expands green spaces. Regulations for the safe use of recycled wastewater for irrigation have been established (TC < 23 MPN/100 mL and 1mg/L Cl_2 after CT 30 mn).

Figure 32.8. Overview of Mafraq's WWTP (156 000 m3/d) and of the urban landscaping in Abu Dhabi (Photo courtesy of Faray Al Mazroui).

Table 32.7. Effluent quality (in mg/L) at Mafraq WWTP.

Parameter	Influent	Effluent	Effluent standards
EC (dS/m)	2.9	3	-
pH	6.9	6.9	-
BOD_5	228	0.9	<10
COD	515	17	-
Total alkalinity	223	37	-
SS	179	2.4	<10
NH_4-N	30	0.5	-
NO_3-N	-	6.6	-
TC MPN/100 mL	-	3	<100 (in 80% of the samples)
Cl_2 residual	-	1.6	-

Major institutions in the United Arab Emirates involved in wastewater reclamation and reuse are the Federal Environment Authority, the Marine Environment Science Section, the and Desert and Marine Environment Research Center, and the United Arab Emirates University, Al-Ain. As in Saudi Arabia, the religious scholars issued a *fatwa* in the United Arab Emirates stating that wastewater could be safely reused for different applications when it returns after treatment in a pure water source free from impurities.

Costs for seawater desalination and wastewater treatment are US\$1.58/m^3 (using MSF) and US\$0.52/m^3, respectively (Hamoda, 1995). If RO treatment is used, the cost of a cubic meter of tertiary-treated water ranges between US\$0.95 and 1.14, depending on the size of the plant and the type of membrane, which makes water reuse a viable option (UNEP, 2001).

32.17 YEMEN

Yemen is located in the south-western part of the Arabian Peninsula. Its population is around 20.7 million and mainly rural (66%); it has been growing at a rate of 3.7% and is expected to reach 25 million by the year 2015. The climate is semi-arid to arid with an annual precipitation averaging 500-800 mm in the Western Highlands, and less than 50 mm along the coasts of the Red Sea and Gulf of Aden. The total renewable water resources are estimated at 4,100 Mm^3/yr, with surface water about 2,575 Mm^3/yr and renewable groundwater around 1,525 Mm^3/yr. The total demand is estimated at 6,630 Mm^3/yr. Over 90% of the water resources are allocated to the agricultural sector. Water resources are under pollution threat, and more particularly the groundwater resources in coastal areas are under threat due to seawater intrusions.

The urban sewerage coverage is relatively low in the north of Yemen (Sana'a, Taiz, and Hodeida) while in the south, most of the high density areas of Aden are sewered. Currently, nine wastewater treatment plants (2 activated sludge (AS), 1 trickling filter, and 6 waste stabilization ponds (WSP)) discharge annually about 33.5 Mm3 (92,000 m3/d, i.e. 69% of their capacity) of reclaimed water (Qahtan, 2003). The quality of the treated effluent discharged by four treatment plants in Yemen is given in Table 32.8. Additional WWTPs of the stabilization pond type are under construction or planned. With progressive coverage of water supply and sewerage services, some 74 Mm^3/yr of effluent should be potentially available for reuse and 15,000 ha may be irrigated with treated effluent (Angelakis, 1997). However, most wastewater treatment plants in Yemen are suffering from lack of operation and maintenance (Sorour, 1999). The treatment costs vary from 0.03 US\$/m^3 for WSP up to 0.25 for US\$/m^3 for AS (Table 32.9).

Water reuse is common, although unregulated and without health controls, and both treated and untreated wastewater are widely used for irrigation throughout the country.

Controlled water reuse for irrigation is practiced in government projects to build green belts, mainly in the coastal plain cities of Aden and Hudeidah, as well as for sand dune fixation and desertification control in the affected areas. Unplanned wastewater irrigation is widely practiced by farmers to grow corn and fodder crops in some areas such as Taiz and Aden, whereas in other areas such as Sana'a, they grow crops such as vegetables and fruit trees.

Table 32.8. Quality of the treated effluent discharged by four treatment plants in Yemen (after Qahtan, 2003).

Parameter	Sana'a	Aden	Dhamar	Al Hudiedah	Draft Yemeni Standards
BOD (mg/L)	24	NA	102	106	150
COD (mg/L)		NA	189	348	500
TDS (mg/L)		1 695	700	3 110	450 – 3000
SS (mg/L)	28	NA	580	128	50
EC (μS/m)		2 840	920	5 186	700 – 4000
FC (/100 mL)		80 000	110 000	1 366	<1000

Table 32.9. Treatment process, cost and irrigated crops (after Qahtan, 2003).

WWTP	Capacity (m^3/d)	Type	Actual flowrate (m^3/d)	Treatment cost US$/$m^3$	Reuse or disposal	Irrigated crops
Sana'a	50 000	AS	20 000	0.25	Irrigation	Corn, Fruit, Fodder
Ta'aiz	17 000	WSP	17 000	0.03	Irrigation	Corn, Fodder
Al Hudeidah	18 000	WSP	18 000	0.03	Sea + Irr.	Green area
Aden	15 000	WSP	15 000	0.03	Sea + Irr.	Forest+Cotton
Ibb	7 000	AS	7 000	0.25	Irrigation	Corn, Fodder
Dhamar	10 000	WSP	6 000	0.03	Irrigation	Corn, Fodder, Qat
Hajja	5 000	TF	1 150	-	Irrigation	Corn, Fodder
Mukalla	8 000	WSP	6 000	0.025	Sea + Irr.	
Rada'a	2 800	WSP	1 500	0.025	Irrigation	Corn, Fodder
Total	**132 800**		**91 650**			

Experiments related to water reuse have been carried out by the Ministry of Agriculture and Irrigation. Primary treated effluent has been successfully used to irrigate an experimental tree farm (29 species of trees: Eucalyptus, Acacia, and indigenous tree species). Effluents from Sana'a, Taiz, and Hodeida's WWTPs have been tested in a pilot water reuse project carried out for the irrigation of fodder, industrial crops, and fruit trees. Experiments on groundwater recharge with treated effluent in coastal plain areas to prevent seawater intrusion are planned. These pilot studies served for drafting the national water reuse strategy (2001).

Water related strategies, policies, and laws in Yemen call for considering treated effluent as a water resource that should be properly and safely used. The National Water Supply and Sewerage Authority established in November 1973, under the Ministry of Electricity and Water, undertook the development of water supply and sewerage for the major cities of Yemen and has been advocating water reuse. Currently, several institutions deal with water and wastewater within their respective mandates and responsibilities; these are the National Water Resources Authority, the Ministry of Agriculture & Irrigation, the Ministry of Electricity & Water, the Ministry of Local Administration, the Ministry of Tourism & Environment, and the Ministry of Public Works & Urban Planning. However, there is no formal national policy on water reuse as yet, although officials in the Ministry of Agriculture and Irrigation support the practice (Bazza, 2002). A draft national water reuse strategy was elaborated in 2001.

32.18 REFERENCES

Abderrahman, W.A. (1997) The use of closed water cycle in industrial plants in Saudi Arabia, In: Proceedings of the Conference on Development and Environmental Impact, 21–23 September, Ministry of Municipal and Rural Affairs, Riyadh.

Abderrahman, W.A. (2001) Water demand management in Saudi Arabia, In: Faruqui, N.I., A.K. Biswas, and M.J. Bino, eds., Water Management in Islam. Tokyo: UN University Press.

Abdo, K.M. (2001) Water Reuse in Palestine. Ministry of Agriculture, General Directorate of Soil & Irrigation, Ramallah, Palestine, 22 p., Paper presented at the Regional Workshop on Water Reuse in the Middle East and North Africa, organized by the World Bank's MENA Regional Water Initiative in Cairo, Egypt, July 2-5, 2001.

Al-Attar, M.H., Al-Awadhi, N., Al-Sulaimi, J. and Leitner, G.F. (1997) Water reuse concepts and programs in Kuwait leading to optimum use of all water resources. *Desalination & Water Reuse*, Feb.-March 1997, (6/4), 46-53.

Alghariani, S.A. (1993) Satisfying future water demands of northern Libya.

Al-Morgin, S. (2003) Saudi Arabia country paper. Proceedings of the expert consultation for launching the regional network on wastewater re-use in the Near East. TC/D/Y5186E/12.03/100. Food and Agriculture Organization of the United Nations. Regional Office for the Near East, World Health Organization. Regional Office for the Eastern Mediterranean. Cairo, 2003, 94-106.

Al-Mudaiheem, R.I.S., Al-Yousef, S.O.A., Sharif, T. and Islam, A.K.M.A. (1998) Performance evaluation of ten years operation experience of brackish water RO desalination in Manfouha Plant, Riyadh. *Desal.* **120**(1-2), 115-120.

Al-Qasimi, H.A. (1997) Management of wastewater in Qatar. Paper presented at the Regional Workshop on the Technologies on Wastewater Treatment and Reuse, Bahrain, 2-4 June.

Al-Sahlawi, M.A. (1999) Seawater desalination in Saudi Arabia: Economic review and demand projections. *Desal.* **123,** 143-147.

Al-Zubari, W. (1997) Towards the establishment of a total water cycle management and re-use program in the GCC countries. Paper presented at the Regional Workshop on the Technologies on Wastewater Treatment and Reuse, Bahrain, 2-4 June.

Angelakis, A.N. (1997) Wastewater reclamation and reuse practices for urban areas of Yemen. FAO Report. UTF/YEM/024. Rome, Italy.

Angelakis, A.N., Marecos do Monte, M.H., Bontoux, L. and Asano, T. (1999) The status of wastewater reuse practice in the Mediterranean Basin. *Wat. Res.,* **33**(10), 2201-2217.

Arar, A. (2003) Institutional aspects of wastewater reuse. Proceedings of the expert consultation for launching the regional network on wastewater re-use in the Near East. TC/D/Y5186E/12.03/100. Food and Agriculture Organization of the United Nations. Regional Office for the Near East World Health Organization. Regional Office for the Eastern Mediterranean. Cairo, 2003. 35-46.

Bahri, A. (1998) Wastewater reclamation and reuse in Tunisia. Wastewater Reclamation and Reuse, T. Asano (Ed.). Water Quality Management Library. Vol. 10, Technomic Publishing Company, Inc.

Bahri, A. (2000) The experience and challenges of reuse of wastewater and sludge in Tunisia, 15p., Water Week 2000, 3-4 April 2000, World Bank, Washington D.C., USA.

Bahri, A. (2003) Tunisia country paper. Proceedings of the expert consultation for launching the regional network on wastewater re-use in the Near East. TC/D/Y5186E/12.03/100. Food and Agriculture Organization of the United Nations. Regional Office for the Near East, World Health Organization. Regional Office for the Eastern Mediterranean. Cairo, 2003, 116-128.

Bazza, M. (2002) Wastewater reuse in the Near East Region: experience and issues. *Wat. Sci. Tech.* **3**(4), 33-50.

Bechtel International Inc. and Scet-Tunisia. (1998) Development of a strategy to promote reclaimed water reuse in the agricultural sector or others sectors, Water Sector Study, Final Report, December 1998, 112 p. + Appendices.

Bennani, A.C., Lary, J., Nrhira, A., Razouki, L., Bize, J. and Nivault, N. (1992) Wastewater treatment of greater Agadir (Morocco): an original solution for protecting the bay of Agadir by using the dune sands, In: Waste water management in coastal areas, March 31-April 2, 1992, Montpellier, CFRP-AGHTM, Paris, France.

Bouhoum, K., Habbari, K., and Schwartzbrod, J. (1995) Epidemiological study of helminthic infections in Marrakech wastewater spreading zone, Proc. IAWQ Second International Symposium on Wastewater Reclamation and Reuse; Iraklio, Greece, Oct. 1995, 2, 679, 1995.

Council of Leading Islamic Scholars (CLIS) (1978) Judgement regarding purifying wastewater: Judgement no. 64 on 25 Shawwal, 1398 ah, Thirteenth meeting of the Council of Leading Islamic Scholars during the second half of the Arabic month of Shawwal, 1398 ah (1998)," Taif: Journal of Islamic Research 17, pp. 40–41.

Cobham R.O. and Johnson P.R. (1988) The use of treated sewage effluent for irrigation: case study from Kuwait, Treatment and use of sewage effluent for irrigation. M.B. Pescod and A. Arar (eds). Butterworths, London.

ESCWA. (1985) Wastewater reuse and its applications in western Asia. Economic and Social Commission for West Asia, E/ESCWA/NR/84/2/Rev.1.

FAO (1985) Water Quality for Agriculture. FAO Irrigation and Drainage Paper 29 (Rev. 1), R.S. Ayers and D.W. Westcot (Eds), Food and Agriculture Organization of the United Nations, Rome, Italy.

FAO (2000) Aquastat-FAO's information system on water and agriculture, http://www.fao.org/ag/agl/aglw/aquastat/dbase/index.stm

FAO (2001) Strengthening capacity for the reuse of treated wastewater in irrigation. Draft project document for Syria. FAO Regional Office for the Near East, Cairo, Egypt.

Jemmali, A. and Abdel-Majid, K. (2002) Wastewater reuse: the Case of Morocco. First WDM Forum on Wastewater Reuse, IDRC, Rabat, Morocco, 26–27 March.

FAO/RNE and WHO/EMRO (2003) Proceedings of the expert consultation for launching the regional network on wastewater re-use in the Near East. TC/D/Y5186E/12.03/100. Food and Agriculture Organization of the United Nations. Regional Office for the Near East World Health Organization. Regional Office for the Eastern Mediterranean. 138 p. + appendixes. Cairo, 2003.

Gearheart, R.M., Bahri, A. and Al-Hamaidi M. (1994) Wastewater treatment and reuse strategy for Gaza and West Bank: Water and wastewater sector. Palestinian Economic Council for Development and Reconstruction (PECDAR), West Bank, Palestine.

Government of Jordan (1995) Jordanian Standards for Water Reuse. JS893/1995. Standard Publishers, Amman, Jordan.

Government of Jordan. (2003) Technical regulation for reclaimed domestic wastewater. JS893/2002, Jordan Institution for Standards and Meterology, Amman, Jordan.

Hamdallah, H. (2000) Reuse of treated wastewater for agriculture, Paper presented at the Aqua 2000 Conference, April 28 to May 2, 2000, Abu Dhabi, UAE.

Hamoda, M.F. (1995) Evaluation of wastewater treatment and reverse osmosis applications for water reuse in UAE. Presented at the International Desalination Association World Congress on Desalination and Water Sciences, Abu Dhabi, UAE, 18-24 November.

Institut National de la Normalisation et de la Propriété Industrielle (1989) Environment Protection – Effluent discharge in the water bodies – Specifications relative to discharges in the marine environment, hydraulic environment and in the sewers (in French), Tunisian standards, INNORPI, NT 106.002.

Institut National de la Normalisation et de la Propriété Industrielle (1989) Environment Protection – Use of reclaimed water for agricultural purposes – Physical, chemical and biological specifications (in French), Tunisian standards, INNORPI, NT 106.03.

Jiménez, B. and Asano, T (2004) Acknowledge all approaches: the global outlook on reuse. Water 21, December 2004, 32-37.

Mahmoud N., Zimmo, O., Zeeman G., Lettinga G. and Gijzen H. (2004) Perspectives for integrated sewage management in Palestine and the Middle East. Water 21, April 2004. 24-29.

Mahmoudian, S. (2003) Water from water: A review of wastewater reuse in Iran. Proceedings of the expert consultation for launching the regional network on wastewater re-use in the Near East. TC/D/Y5186E/12.03/100. Food and Agriculture Organization of the United Nations. Regional Office for the Near East, World Health Organization. Regional Office for the Eastern Mediterranean. Cairo, 2003, 64-72.

McCornick, P.G., Taha, S. and El Nasser, H. (2002) Planning for Reclaimed Water in the Amman-Zarqa Basin and Jordan Valley. American Society of Civil Engineering – Environmental and Water Resources Institute, Conference and Symposium on Droughts and Floods, Roanoke, Virginia, 10 p.

McCornick, P.G., Hijazi. A. and Sheikh B. (2004) From wastewater reuse to water reclamation: progression of water reuse standards in Jordan. In: Wastewater Use in Irrigated Agriculture – Confronting the Livelihood and Environmental Realities, Edited by Scott C., Faruqui N.I. and Raschid L., CABI/IWMI/CRDI 2004 ISBN 1-55250-112-4 240 p.

Medaware (2004) Development of tools and guidelines for the promotion of the sustainable urban wastewater treatment and reuse in the agricultural production in the Mediterranean countries, (Medaware), European Commission, Euro-Mediterranean Partnership, Task 1: Determination of the Countries Profile, PART E: water related problems associated with health and environment, Task 2: Evaluation of the existing situation related to the operation of urban wastewater treatment plants and the effluent disposal practices with emphasis on the reuse in the agricultural production, March 2004.

Ministry of Water and Irrigation (1997) Water resources strategy. Ministry of Water and Irrigation, The Hashemite Kingdom of Jordan, Amman, Jordan.

Ministry of Water Resources and Irrigation (2001) Application of policies and procedures for improved urban wastewater discharge and reuse. Report No. 34 Agricultural Policy Reform Program. Cairo, Egypt.

Official Journal of the Tunisian Republic (1989) Decree No. 89-1047 issued on July 28, 1989 regulating the use of reclaimed water for agricultural purposes (in French).

Office National de l'Assainissement (Ministère de l'Environnement et de l'Aménagement du Territoire) /SERAH (2001) Etude de la stratégie nationale de valorisation des eaux usées traitées, Phase I: Diagnostic de la situation actuelle; Phase II: Orientations stratégiques pour l'amélioration de la réutilisation des eaux épurées, Dec 2001.

Pallas, P. (1980) Water resources of the Socialist People's Libyan Arab Jamahiriya. In: The Geology of Libya. Proceedings of the Second Symposium on the Geology of Libya. Academic Press, London.

Prathapar, S.A., Ahmed, M. and Jamrah, A. (2005) Is greywater reuse a realistic option in Oman. Arab Water World, April 2005. 12-13.

Qahtan Y.A. (2003) Yemen country paper. Proceedings of the expert consultation for launching the regional network on wastewater re-use in the Near East. TC/D/Y5186E/12.03/100. Food and Agriculture Organization of the United Nations. Regional Office for the Near East World Health Organization. Regional Office for the Eastern Mediterranean. Cairo, 2003. 133-138.

Salem, O.M. (1992) The Great Manmade River Project. In: Water Resources Development, 8(4).

Salih, H. M. (2003) Iraq country paper. Proceedings of the expert consultation for launching the regional network on wastewater re-use in the Near East. TC/D/Y5186E/12.03/100. Food and Agriculture Organization of the United Nations. Regional Office for the Near East World Health Organization. Regional Office for the Eastern Mediterranean. Cairo, 2003. 73-75.

Shaalan, N.S. (2003). Egypt country paper. Proceedings of the expert consultation for launching the regional network on wastewater re-use in the Near East. TC/D/Y5186E/12.03/100. Food and Agriculture Organization of the United Nations. Regional Office for the Near East World Health Organization. Regional Office for the Eastern Mediterranean. Cairo, 2003. 59-63.

Shanehsaz, S.M.J., Hosseinifar, A.R. and Sabetraftar, A. (2001) Water reuse in Iran. Proceedings, Regional Workshop on Water Reuse in the Middle East and North Africa, Sponsored by the World Bank and the Swiss Agency for Development and Cooperation, July 2-5, 2001, Cairo, Egypt, The World Bank, Washington, D.C.

Sheikh, B., Aldukair A., Walker C., and Zahid W.M. (2000) Riyadh, Kingdom of Saudi Arabia - A vision of 2021: Introducing water reuse to a world capital, Proceedings, Water Reuse 2000 Conference, American Water Works Association, January 30-February 2, 2000, San Antonio, Texas.

Sheikh, B. (2001) Standards, regulations & legislation for water reuse in Jordan. Water reuse component, Water Policy Support Project, Ministry of Water and Irrigation, Amman, Jordan 42 pp.

Sorour, M.T. (1999) Mission report. GCP/YEM/026/NET. Watershed management and wastewater reuse in peri-urban areas of Yemen. Cairo, Egypt.

Tuffaha, A.Sh. (1996) Technologies, practices, and performances of wastewater treatment plants in Jordan. Workshop on wastewater treatment and reuse in Jordan, Water Authority of Jordan, Aqaba, 3-5 December 1996.

United Nations Development Program (1987) Reclaimed water reuse in agriculture, Technical report (in French), Vol. 1, 3rd Part, 69 p.+ appendixes, Project RAB/80/011, Water Resources in the North African Countries, Rural Engineering Research Center (Tunisia) – United Nations Development Program, May 1987.

United Nations Development Program/Food and Agriculture Organization/The World Bank/World Health Organization (1992) Wastewater treatment and reuse in the Middle East and North Africa Region – Unlocking the potential, Review Draft, Report of a joint mission to Cyprus, Egypt, Jordan, Kuwait, Morocco, Saudi Arabia, Syria, Tunisia, Turkey, and Yemen, Project RAB/88/009, July 1992.

United Nations Environment Program (1999) Overview on land-based sources and activities affecting the marine environment in the ROPME Sea Area. United Nations Environment Program, Regional Seas Reports and Studies No. 168.

United Nations Environment Program (2001) Overview of the socio-economic aspects related to the management of municipal wastewater in West Asia (including all countries bordering the Red Sea and Gulf of Aden). UNEP/PERSGA/ROPME Workshop on Municipal Wastewater Management in West Asia Bahrain, 10-12 November 2001. UNEP/ROWA-GPA SEWAGE.RW.1/5 2, October 2001. 59 pp.

U.S. Environmental Protection Agency (2004) Guidelines for water reuse. EPA/625/R-04/108, U.S. Environmental Protection Agency, Washington D.C. 286 pp. + Appendixes.

World Bank (2001) Regional Water Initiative: Water reuse in the Middle East and North Africa, Proceedings, Regional Workshop on Water Reuse in the Middle East and North Africa, Sponsored by the World Bank and the Swiss Agency for Development and Cooperation, July 2-5, 2001, Cairo, Egypt, The World Bank, Washington, D. C.

World Health Organization (1989) Health guidelines for the use of wastewater in agriculture and aquaculture, Report of a WHO Scientific Group. WHO Technical report series 778. World Health Organization, Geneva, Switzerland.

Zulita, A. (2003) Syria country paper, Proceedings of the expert consultation for launching the regional network on wastewater re-use in the Near East. TC/D/Y5186E/12.03/100. Food and Agriculture Organization of the United Nations. Regional Office for the Near East, World Health Organization. Regional Office for the Eastern Mediterranean. Cairo, 2003, 107-115.

Annex 1

Water Availability and Water Intensity Use index for different Countries

Data base: Water Availability and Water Intensity Use Indices per region, with information from Earth Trends, 2007.

Region/Countries	Water Availability Index In 2006 m^3/capita·yr	Water Intensity Use Index In 2000 %
Middle East and North Africa	**1,383**	**62.8**
Afghanistan	2,091	42.3
Algeria	350	54.0
Bahrain	157	7,500.0
Cyprus	923	30.8
Egypt	773	3,794.4
Gaza Strip	15	ND
Iran, Islamic Rep	1,955	56.7
Iraq	2,552	121.3
Israel	244	273.3
Jordan	151	148.5
Kuwait	7	ND
Lebanon	1,219	28.8
Libyan Arab Jamahiriya	101	711.3
Morocco	908	43.4
Oman	377	138.1

Region/Countries	Water Availability Index In 2006 m^3/capita·yr	Water Intensity Use Index In 2000 %
Qatar	63	568.6
Saudi Arabia	95	721.7
Syrian Arab Rep	1,346	285.0
Tunisia	450	62.9
Turkey	2,879	16.5
United Arab Emirates	32	1,533.3
Yemen	190	161.7
Asia (excluding Middle East)	**3,990**	**19.3**
Armenia	3,502	32.5
Azerbaijan	3,574	212.6
Bangladesh	8,382	75.6
Bhutan	42,967	0.4
Brunei Darussalam	22,251	1.1{8}
Cambodia	33,176	3.4
China	2,137{1}	22.4
Georgia	14,283	6.2
India	1,694	51.2
Indonesia	12,587	2.9
Japan	3,354	20.6
Kazakhstan	7,400	46.4
Korea, Dem People's Rep	3,416	13.5
Korea, Rep	1,453	28.7
Kyrgyzstan	3,865	21.7
Lao People's Dem Rep	55,059	1.6
Malaysia	22,484	1.6
Maldives	89	11.2{9}
Mongolia	12,990	1.3
Myanmar	20,498	3.8
Nepal	7,595	5.1
Pakistan	1,381	323.3
Philippines	5,670	6.0
Singapore	137	31.7{10}
Sri Lanka	2,391	25.2
Tajikistan	2,425	18.0
Thailand	6,330	41.5
Turkmenistan	5,046	1,812.5
Uzbekistan	1,868	357.0
Viet Nam	10,443	19.5

Region/Countries	Water Availability Index In 2006 m^3/capita·yr	Water Intensity Use Index In 2000 %
Mexico, Central America and Caribbean	**6,740**	**8.5**
Antigua and Barbuda	634	9.6{11}
Bahamas	61	ND
Barbados	296	112.5
Belize	67,473	0.9
Costa Rica	25,551	2.4
Cuba	3,375	21.5
Dominican Rep	2,327	16.1
El Salvador	3,605	7.2
Guatemala	8,618	1.8
Haiti	1,621	7.6
Honduras	13,030	0.9
Jamaica	3,533	4.4
Mexico	4,221	19.1
Nicaragua	35,123	0.7
Panama	45,006	0.6
Puerto Rico	1,785	ND
Saint Kitts and Nevis	558	ND
Trinidad and Tobago	2,934	8.1
United States and Canada	**19,649**	**9.3**
Canada	89,111	1.6
Greenland	10,578,947	ND
United States	10,135	17.1
Europe	**10,680**	**6.4**
Albania	13,251	6.4
Andorra	0	ND
Austria	9,470	3.8
Belarus	5,979	7.5
Belgium	1,753	75.3{12}
Bosnia and Herzegovina	9,586	ND
Bulgaria	2,777	50.0
Croatia	23,156	ND
Czech Rep	1,288	19.6
Denmark	1,102	21.2
Estonia	9,666	1.2
Finland	20,905{2}	2.3
France	3,355	22.4
Germany	1,862	44.0
Greece	6,665	13.4
Hungary	10,327	127.3

Region/Countries	Water Availability Index In 2006 m^3/capita·yr	Water Intensity Use Index In 2000 %
Iceland	572,391	0.1
Ireland	12,352	2.3
Italy	3,290	24.3
Latvia	15,446	1.8
Lithuania	7,287	1.7
Luxembourg	6,582	4.0{9}
Macedonia, FYR	3,142	ND
Malta	125	100.0
Moldova, Rep	2,777	231.0
Netherlands	5,560	72.2
Norway	82,274{3}	0.6
Poland	1,600	30.2
Portugal	6,515	29.6
Romania	9,798	54.8
Russian Federation	31,622	1.8
Serbia and Montenegro	19,863	ND
Slovakia	9,276	ND
Slovenia	16,211	ND
Spain	2,570	32.0
Sweden	19,184	1.7
Switzerland	7,365	6.4
Ukraine	3,035	70.7
United Kingdom	2,456	6.6
Sub-Saharan Africa	**7,209**	**3.1**
Angola	9,024	0.2
Benin	3,033	1.3
Botswana	6,955	8.1
Burkina Faso	917	6.4
Burundi	1,600	2.9
Cameroon	17,198	0.4
Cape Verde	578	7.3
CÃ´te d'Ivoire	4,397	1.2
Central African Rep	35,280	0.0
Chad	4,286	1.5
Comoros	1,465{4}	0.8
Congo	221,035	0.0
Congo, Dem Rep	21,629	0.0
Djibouti	372	6.3
Equatorial Guinea	50,485	0.4
Eritrea	1,382	20.8{13}
Ethiopia	1,539	4.6
Gabon	116,643	0.1

Region/Countries	Water Availability Index In 2006 m^3/capita·yr	Water Intensity Use Index In 2000 %
Gambia	5,141	1.0
Ghana	2,359	3.2
Guinea	23,534	0.7
Guinea-Bissau	18,972	1.1
Kenya	875	7.6
Lesotho	1,687	1.0
Liberia	69,130	0.1
Madagascar	17,639	4.4
Malawi	1,313	6.3
Mali	7,185	10.9
Mauritania	3,610	425.0
Mauritius	2,190{5}	26.4{13}
Mozambique	10,770	0.6
Namibia	8,633	4.9
Niger	2,333	62.3
Nigeria	2,130	3.6
Réunion	6,281	ND
Rwanda	1,029	1.6
Sao Tome & Principe	13,625	0.3{8}
Senegal	3,251	8.6
Sierra Leone	28,174	0.2
Somalia	1,730	55.0{13}
South Africa	1,051	27.9
Sudan	1,744	124.4
Swaziland	4,383	39.5
Tanzania	2,467	6.2
Togo	2,331	1.5
Uganda	2,211	0.8
Zambia	8,869	2.2
Zimbabwe	1,529	34.3
Oceania	**53,290**	**1.6**
Australia	24,158{6}	4.9
Fiji	33,431	0.2
New Zealand	80,482	0.6
Papua New Guinea	133,478	0.0
Solomon Islands	91,225	ND
South America	**45,400**	**1.3**
Argentina	20,800	10.6
Bolivia	66,552	0.5
Brazil	43,588	1.1
Chile	55,998	1.4

Region/Countries	Water Availability Index In 2006 m^3/capita·yr	Water Intensity Use Index In 2000 %
Colombia	46,068	0.5
Ecuador	31,627	3.9
French Guiana	701,571	ND
Guyana	320,479	0.7
Paraguay	53,325	0.5
Peru	67,407	1.2
Suriname	269,912	0.8
Uruguay	39,862	5.3
Venezuela	45,311	1.2
DEVELOPED COUNTRIES	**11,392**	**9.0**
DEVELOPING COUNTRIES	**7,693**	**8.9**
HIGH INCOME COUNTRIES	**10,554**	**10.1**
MIDDLE INCOME COUNTRIES	**10,171**	**6.9**
LOW INCOME COUNTRIES	**5,894**	**12.1**
WORLD	**8,462**{7}	**8.9**

ND: No data

{1}Population data do not include Hong Kong and Macao, Special Administrative Regions (SAR) of China.

{2}Population data includes Ãland Islands

{3}Population data includes Svalbard and Jan Mayen Island.

{4}Population data includes the island of Mayotte.

{5}Population data includes Agalega, Rodrigues, and Saint Brandon.

{6}Population data includes Christmas Island, Cocos (Keeling) Islands, and Norfolk Island.

{7}Value calculated by WRI

{8}1995 value

{9}1985 value

{10}1975 value

{11}1990 value

{12}1980 value

{13}2005 value

Annex 2:

Agricultural irrigation: surface irrigated and volume used

Different values appear when several references were used

Country	Ha with WW	Ha with treated WW	Treated ww used for irrigation, m^3/d	Reference
Argentina	23,700 24,000	23,930(1) 20,000 3,460(1)	129,600	Bartone and Arlosoroff, 1987; EPA, 2004 and WRIS
Australia		20,000		EPA, 2004
Bahrain		1,250	27,000	WRIS
Belgium		800		WRIS
Bolivia	1,560+	1,560(1)	43,200	WRIS
Brazil	+		23,328	WRIS
Chile	40,000* 16,000	130,000	380,000	EPA, 2004 and WRIS
China	1,300,000*	0+	1,238,860	WRIS
Colombia	26,000			WRIS
Costa Rica	1,200 +			WRIS
Cyprus	+	38,200	68,493	Fatta and Skoula, 2004 and Jimenez, 2006
Dominican Rep.	250			WRIS

Country	Ha with WW	Ha with treated WW	Treated ww used for irrigation, m^3/d	Reference
Ecuador		80	4,752	WRIS
Egypt	4,500[(1)] 1,000[(1)]	42,000*	1,917,808	WRIS; AAST, 2004; EPA, 2004 and Jimenez, 2006
France		4,200 1,250[(1)]	19,178	WRIS and EPA, 2004
Germany		2,800		Alagadwai in Raschid-Sally and Abayarwardana, 2003
Ghana	11,800			Raschid-Sally and Abayarwardana, 2003
Greece	+		20,030	WRIS
Guatemala		5[(1)]	1,642	WRIS
India	73,000 *			WRIS; AAST, 2004 and EPA, 2004
Iraq		+		WRIS
Iran	+	+	421,918	WRIS; AAST, 2004; EPA, 2004 and Jimenez, 2006
Israel		65,000[(1)] 40,000 28,000[(1)]	767,123	EPA, 2004
Italy	+	28,285* 4,000* 450[(1)]	741,262	Lazarova and Bahri, 2005; EPA, 2004 and WRIS
Japan				
Jordan	+	13,300 10,000	224,658	EPA, 2004; Lazarova and Bahri, 2005; WRIS and Vallentin, 2006
Kenya	+			Tole in Jimenez and Asano, 2004
Kuwait	4,470+	9,000 6,150 4,470[(1)]	431,520	EPA, 2004; WRIS Alagadwai in Raschid-Sally and Abayawardana, 2003
Kyrgyzstan			384	Jimenez, 2006
Lebanon	+		5,479	AAST, 2004 and Jimenez, 2006
Libya		2,970[(1)]	110,000	WRIS
Malta		600[(1)]	26,000	WRIS and Jimenez, 2006
Mexico	190,000	70,000	4,492,800	Jimenez, 2006 and WRIS
Morocco	8,000	+	6,600	WRIS
Nepal	2,010[(1)] +			Scott *et al.*, 2004
Oman		5,600[(1)]	67,000	WRIS and EPA, 2004
Pakistan	32,500			WRIS and EPA, 2004
Palestine	+			Medaware, 2004 and AAST, 2004
Paraguay	+			WRIS

Country	Ha with WW	Ha with treated WW	Treated ww used for irrigation, m^3/d	Reference
Peru	9,346 5,000 1,341[(1)]	1,350 610		EPA, 2004 and WRIS
Poland		0[(1)]		AAST, 2004
Portugal	1,000+	1,000	0	WRIS and Jimenez, 2006
Qatar			80,000	WRIS
Saudi Arabia		2,850	594,521	WRIS and Alagadwai in Raschid-Sally and Abayawardana, 2003
Senegal	+			Scott *et al.*, 2004
South Africa	22,000*	1,800		EPA, 2004
Spain	+	+	931,507	WRIS and Lazarova and Bahri, 2005
Sudan	2,900			Raschid-Sally and Abayawardana, 2003
Sweden		+		EPA, 2004
Syria	40,000 31,852	9,000+	1,182,000	WRIS; EPA, 2004 and AAST, 2004
Thailand	+	+		AAST, 2004 and WRIS
Tunisia	2,900	8,570	117,802	WRIS
Turkey	9,165	9,165 7,000	136,986	WRIS, EPA, 2004 and Fatta-Skoula, 2004
United Arab Emirates		16,950[(1)] 15,000	200,000	WRIS and EPA, 2004
United States		15,000 14,675	911,000	EPA, 2004 and WRIS
Vietnam	9,500			WRIS
Yaounde	453[(1)]			WRIS
Yemen	+			WRIS
Zimbabwe	+	0*		EPA, 2004

*Data are confusing

+No data are available, although the practice is reported

[(1)]Area might be greater

WRIS Data comes from this survey

REFERENCES

AAST, Australian Academy of Technological Sciences and Engineering (2004) Water recycling in Australia. A review undertaken by the Australian Academy of Technological Sciences and Engineering Water Recycling in Australia.

Bartone, C.R. and Arlosoroff, S. (1987) Reuse of pond effluent in developing countries. *Wat. Sci. Tech.* **19**(12), 289-297.

EPA, Environmental Protection Agency (2004) Guidelines for Water Reuse EPA/625/R-04/108 Washington, D.C.

Fatta, D. and Skoula, I. (2004) Development of Tools and Guidelines for the Promotion of the Sustainable Urban Wastewater Treatment and Reuse in the Agricultural Production in the Mediterranean Countries. Task 2 Medaware.

Jimenez, B (2006) Irrigation in developing countries using wastewater. *International Review for Environmental Strategies (IRES)* **6**(2), 229-250.

Jimenez, B. and Asano, T. (2004) Acknowledge all approaches: The Global outlook on Reuse. *Wat. 21*, December, 32-27 pp.

Lazarova, V. and Bahri, A. (eds) (2005) Water Reuse for Irrigation: Agriculture, Landscapes, and Turf Grass. Catalog no. 1649, CRC PRESS, 456 p. ISBN: I-56670-649-I.

Medaware (2004) Development of Tools and Guidelines for the Promotion of the Sustainable Urban Wastewater Treatment and Reuse in the Agricultural Production in the Mediterranean Countries.

Raschid-Sally, L. and Abayawardana, S. (2003) Wastewater as a non traditional source of livelihoods, food security and water for agriculture. IWA Symposium on Water and Sustainable Development, 14-17 September 2003, Capetown, South Africa.

Scott, C.A., Faruqui, N.I. and Raschid-Sally, L. (eds) (2004) Wastewater use if irrigated agriculture confronting the livelihood and the environmental realities. CABI Publishing, Wallingford.

Vallentin, A. (2006) Agricultural use of reclaimed water: Experiences in Jordan. *Wat. Practice Tech.* **1**(2), 1-5.

Index

IWA Publishing's authorised EU representative for General Product Safety Regulations is Diane D'Arras, 15 rue Duret, 75116 Paris, France, e-mail: safety@iwap.co.uk.

Printed and bound by CPI Group (UK) Ltd, Croydon, CR0 4YY
16/06/2026
02140400-0003